Group Representations
and
Special Functions

Mathematics and Its Applications (*East European Series*)

Antoni Wawrzyńczyk

Department of Mathematical Methods in Physics,
Faculty of Physics, University of Warsaw

Group Representations and Special Functions

Examples and Problems prepared by
Aleksander Strasburger

Department of Mathematical Methods in Physics,
Faculty of Physics, University of Warsaw

D.Reidel Publishing Company
A MEMBER OF THE KLUWER ACADEMIC PUBLISHERS GROUP
Dordrecht/Boston/Lancaster

PWN – Polish Scientific Publishers
Warszawa

Library of Congress Cataloging in Publication Data

Wawrzyńczyk, Antoni.
Group representations and special functions.
(Mathematics and its applications. East European series)
Translation of: Współczesna teoria funkcji specjalnych.
English edition published by Polish Scientific Publishers, Warszawa, Poland, in co-publication with D. Reidel Publishing Company ... " — T.p. verso.
Bibliography: p.
Includes index.
1. Representations of groups. 2. Functions, Special. I. Strasburger, Aleksander. II. Title. III. Series: Mathematics and its applications (D. Reidel Publishing Company) East European series.
QA171. W3513 1984 512'.22 83-9687
ISBN 90-277-1269-7

Translated by Bogdan Ziemian.

This translation has been made from *Współczesna teoria funkcji specjalnych*, published by Państwowe Wydawnictwo Naukowe, Warszawa, 1978.

English Edition published by Polish Scientific Publishers, Warszawa, Poland, in co-publication with D. Reidel Publishing Company, P. O. Box 17, 3300 AA Dordrecht, Holland.

Distributors for Albania, Bulgaria, Cuba, Czechoslovakia, German Democratic Republic, Hungary, Korean People's Democratic Republic, Mongolia, People's Republic of China, Poland, Rumania, the U.S.S.R., Vietnam, and Yugoslavia
ARS POLONA
Krakowskie Przedmieście 7, 00-068 Warszawa 1, Poland

Distributors for the U.S.A. and Canada
Kluwer Boston, Inc.,
190 Old Derby Street, Hingham, MA 02043, U.S.A.
Distributors for all remaining countries
Kluwer Academic Publishers Group,
P. O. Box 322, 3300 AH Dordrecht, Holland

Printed in Poland

Editor's Preface

Growing specialization and diversification have brought a hor'st of monographs and textbooks on increasingly specialized topics. However, the "tree" of knowledge of mathematics and related fields does not grow only by putting forth new branches. It also happens, quite often in fact, that branches which were thought to be completely disparate are suddenly seen to be related.

Further, the kind and level of sophistication of mathematics applied in various sciences has changed drastically in recent years: measure theory is used (non-trivially) in regional and theoretical economics; algebraic geometry interacts with physics; the Minkowsky lemma, coding theory and the structure of water meet one another in packing and covering theory; quantum fields, crystal defects and mathematical programming profit from homotopy theory; Lie algebras are relevant to filtering; and prediction and electrical engineering can use Stein spaces. And in addition to this there are such new emerging subdisciplines as "completely integrable systems", "chaos, synergetics and large-scale order", which are almost impossible to fit into the existing classification schemes. They draw upon widely different sections of mathematics.

This programme, Mathematics and Its Applications, is devoted to such (new) interrelations as exempli gratia:

— a central concept which plays an important role in several different mathematical and/or scientific specialized areas;

— new applications of the results and ideas from one area of scientific endeavor into another;

— influences which the results, problems and concepts of one field of enquiry have and have had on the development of another.

The Mathematics and Its Applications programme tries to make available a careful selection of books which fit the philosophy outlined above. With such books, which are stimulating rather than definitive, intriguing rather than encyclopaedic, we hope to contribute something towards better communication among the practitioners in diversified fields.

Because of the wealth of scholarly research being undertaken in the Soviet Union, Eastern Europe, and Japan, it was decided to devote special attention to the work emanating from these particular regions. Thus it was decided to start three regional series under the umbrella of the main MIA programme.

Special functions as an area of research in mathematics have had their ups and downs. There used to be a time when every (applied) mathematician and physicist knew at least a large chunk of special function theory and used this knowledge regularly. On the other hand, Dick Askey once told me that when he started to work in the field some twenty years ago, this was generally regarded as quixotic to a degree; the field being considered practically dead.

As has happened before, it turned out far otherwise—as illustrated, e.g., by a very recent rather multidisciplinary and hugely stimulating conference in Oberwohlfach on special functions which seem to have reassumed their habit of popping up in the most varied parts of mathematics.

Just what a special function is, and what properties it must have to be called special, is perhaps better left open. But since Vilenkin's 1968 book on the topic it has become clear that very many of them turn up as matrix coefficients of the representation of a Lie group and that their relations with harmonic analysis on Lie groups are very intimate. This certainly accounts for at least some of the importance of special functions in the general scheme of things mathematical.

This book in the MIA (East European) Series is precisely about special functions in their relation to analysis on homogeneous spaces, and is designed for students and researchers in one of the many fields where special functions are useful (such as, e.g., representation theory, mathematical physics, harmonic analysis).

The unreasonable effectiveness of mathematics in science ...

Eugene Wigner

Well, if you knows of a better 'ole, go to it.

Bruce Bairnsfather

What is now proved was once only imagined.

William Blake

As long as algebra and geometry proceeded along separate paths, their advance was slow and their applications limited.
But when these sciences joined company they drew from each other fresh vitality and thenceforward marched on at a rapid pace towards perfection.

Joseph Louis Lagrange

Amsterdam, April 1983

Michiel Hazewinkel

Table of Contents

TO PROFESSOR KRZYSZTOF MAURIN

Preface

This book contains the basic elements of group representation theory and of the theory of special functions; it mainly deals, however, with the relations between those two branches of mathematics.

Strictly speaking, we study only those special functions which are generally known as the functions of mathematical physics. Although we restrict ourselves to this class of special functions, we still have to contend with an extremely wide domain of study, formulated by mathematicians of the highest rank, whose subject interest was by no means marginal. In fact, achievements in the fields of Euler's integrals, of the hypergeometric series of Gauss, of the Fourier series and of the Bessel, Legendre, Jacobi and Hermite functions constitute a significant part of the contribution of the great investigators after whom the above mentioned classes of special functions have been named. They derived inspiration to study "their" functions directly from definite problems of physics, geodesy, and astronomy, and there is no doubt that it was their sensitivity and mathematical intuition which made them discern certain deeper features of the function in question. Owing to those features, special functions appear again and again in new roles in the various fields of mathematics. Applications in the description of a vibrating string; distribution of temperatures in a solid body; motion of planets; attraction of ellipsoids; have played a significant role in the history of those functions. Their vitality, however, derives from the associations they provided in the fields of geometry, topology, calculus of probability, number theory and harmonic analysis.

Every mathematical theory builds up its own store of special functions which serve to describe models and to solve detailed problems. The properties of those functions, when studied in connection with the initial theory, show up clearly and logically, which is not always the case when they are treated in a summary way.

One of the most important concepts uniting several fields of mathematics since the middle of the nineteenth century is the notion of a transformation group. Groups appeared in mathematics in 1830 in Galois's theory as the groups of permutations of roots of a polynomial and have been used to resolve the classical problems of solving algebraic equations by radicals. A complete understanding of their role was possible thanks to F. Klein and S. Lie. In his famous Erlangen program in 1972 F. Klein defined geometry as a theory of the invariants of a certain fixed transformation group. In this way, to a classification of groups corresponds a classification of geometry.

The investigations of S. Lie, aimed at the use of groups in differential equations by studying the structure of the invariant group of a suitable differential operator. It was due to him that the theory of continuous groups, known as Lie groups, was created. The work of Lie showed also that the group theory is a basis of classification in geometry, mechanics and in ordinary and partial differential equations.

Lie's investigations were continued by E. Cartan, who obtained fundamental theorems concerning the structure of semi-simple Lie groups and their homogeneous spaces.

The next stage of the application of group (representation) theory began with the rise of quantum mechanics. The precursors in this field where H. Weyl and E. P. Wigner. It is through their research that in the 1920's the direction and rate of development of representation theory was determined by physics. Physicists aroused interest in the representations of non-compact groups, and their results concerning "physical" transformation groups (E. P. Wigner, 1939, V. Bargmann, 1947) suggested ideas of general solution, which were developed in papers by J. M. Gelfand, G. Mackey, Harish-Chandra and others.

As soon as the group concepts and methods had penetrated into physics, it became natural and desirable to present in the same way special functions employed in the description of physical models.

It was E. P. Wigner who used representation theory as a uniform method of studying special functions. However, his lectures delivered at Princeton in 1955 were not published. The first monographic presentation of the theory of special functions by the method of representation theory was given by N. J. Vilenkin (1965), who made use of the outstanding achievements of the Gelfand school, including his own in the field of spherical functions.

In 1968 J. D. Talman published a book based on Wigner's lectures. In the same year there appeared a monograph by W. Miller, who treated by uniform methods a particulary wide class of special functions.

In the theory of special functions, groups play the same role as in a number of other branches of mathematics and physics. To begin with, they provide a method of classifying these functions. A given class of functions is associated with a pair consisting of a matrix group and its subgroup, or in other words, with a fixed homogeneous space. In the simplest cases, described in the present book, these homogeneous spaces are a plane, a unit sphere in $\boldsymbol{R}^n$, a hyperboloid in $\boldsymbol{R}^3$ and $\boldsymbol{R}^4$, and the corresponding pairs of groups consist of the groups of isometries of these spaces and the isotropy subgroups at a point. A given class of special functions appears as the set of matrix elements of irreducible representations of a given group. The algebraic properties of the group and its Lie algebra are then reflected in the functional and differential equations satisfied by a given family of special functions. The geometry of the homogeneous space, on the other hand influences the nature of the integral transformation associated with that family.

In the spirit of the Erlangen program every class of special functions shows itself as a geometric object whose structure depends on the group which determines that class.

It is to the above problems that Part II of the present book is devoted differing from Vilenkin's monograph in a more extensive treatment of the geometric aspects of the theory.

Part III contains an introduction to the geometry of general symmetric space and to harmonic analysis on those spaces. It seems to me to constitute a natural extension of the results presented in the previous section. The general approach given in Part III permits a more profound understanding of the nature of the phenomena described earlier in a number of examples. A presentation of Fourier transformation theory on a symmetric space together with proofs would have necessitated a considerable increase in the size of this book. It seemed to me that in order to preserve its textbook character it was more important to include the basic information on the geometry and the structure of semi-simple groups and algebras. I leave it to the reader to decide whether the choice of proofs was right.

In the preparation of Part III of the book I mostly used S. Helgason's monograph and his publications presenting Harish-Chandra results and his own. He has been sending me off-prints of his current papers for many years, and hereby I wish to express my deep gratitude to him. A very important component within this work are the examples and problems prepared by Dr. A. Strasburger, who was also the first reader and critic of the book, exerting considerable influence on its final form.

Also, the reviewer of the Polish version, Dr. W. Wojtyński, has contributed a great deal to the improvement and construction of the book and has drawn my attention to a number of inaccuracies.

The work of the translator, just as that of the author, can only be assessed by the reader. On my part I am full of appreciation for the work of Mr B. Ziemian, and hereby thank him for having undertaken the translation of the present book, in which task he has shown both thoroughness and zest.

I would like to wish all authors as good cooperation with their publishers as mine has been thanks to the editors, Mrs Z. Osek. and K. Regulska.

The credit for the fact that the present book has appeared at all goes to Prof. Dr. K. Maurin. He introduced me to the world of mathematics and then encouraged me to take up representation theory. It was also his inspiration which made me write this book and his advice which assisted me in the course of its preparation. I dedicate this book to him.

ANTONI WAWRZYŃCZYK

January 1, 1983

PART I

Chapter 1

Groups and Homogeneous Spaces

1.1. GROUPS

DEFINITION. A non-empty set G is called a *group* if to every pair $(g, g_1) \in G \times G$ an element $gg_1 \in G$ is assigned, called the *product* of g and g_1, so that, for arbitrary $g_1, g_2, g_3, \in G$

$$g_1(g_2g_3) = (g_1g_2)g_3 \text{ (the law of associativity)}, \tag{1.1.1}$$

there exists an $e \in G$, called the *identity* of the group, such that for

$$\text{arbitrary } g \in G \;\; ge = eg = g, \tag{1.1.2}$$

for each $g \in G$ there exists an element g^{-1} called the *inverse* of g,

$$\text{satisfying the relation } gg^{-1} = g^{-1}g = e. \tag{1.1.3}$$

The identity of a group is also called the *neutral element*. In a group there can be only one identity. For if $e' \in G$ also satisfies the condition (1.1.2), then $e' = e'e = e$.

Also the inverse of g is determined uniquely since every $(g^{-1})'$ satisfying the equation of the inverse element can be represented in the following way:

$$(g^{-1})' = (g^{-1})'gg^{-1} = ((g^{-1})'g)g^{-1} = eg^{-1} = g^{-1}.$$

The inverse element of ab is $b^{-1}a^{-1}$ and a is the inverse of a^{-1}.

A group is called *commutative* or *abelian* if for arbitrary $g, h \in G$ the relation $gh = hg$ is satisfied.

The group structure often appears as part of a certain richer algebraic structure in a given set (e.g. linear space structure, field structure, etc.). Many examples arise from such richer structures by "forgetting" certain conditions which define them.

Example 1

(a) Let $\boldsymbol{K}$ be one of the following fields: either the field $\boldsymbol{R}$ of real numbers, or the field $\boldsymbol{C}$ of complex numbers. The set $\boldsymbol{K}$ together with the operation of addition forms a commutative group called the additive group of the field K. The set $K_* = K - \{0\}$, in turn, is a *commutative group* with respect to multiplication—the multiplicative group of the field K.

(b) A vector space V, considered only with the operation of the addition of vectors is a commutative group.

(c) Let X denote a non-empty set (finite or not). The symbol $S(X)$ will denote the set of all bijections of X onto itself. Choosing the composition of mappings as the operation, we can easily prove that $S(X)$ is a group.

The identity element here is the identity mapping. The inverse element of a bijection f is the inverse bijection f^{-1}.

In the case where X contains n elements we write S_n for $S(X)$.

Let $Y \subset X$ be a non-empty subset. The set $S_Y(X)$, containing all the bijections $f \in S(X)$ for which $f(Y) = Y$, is a group with respect to composition (a subgroup of $S(X)$—see below).

Similarly the set $PS_Y(X)$ of all bijections f such that $f(y) \doteq y$ for all $y \in Y$ is a group with respect to composition.

(d) Denote by $\mathrm{GL}(n, K)$ the set of all invertible matrices of degree n with terms in K, i.e. $\mathrm{GL}(n, K) = \{A \in M_n(K)\colon \det A \neq 0\}$. This set together with the operation of matrix multiplication forms a group—called the *general linear group of degree n.*

(e) Let $G = \{(a, b) \in R^2\colon a \neq 0\}$. We define the multiplication of elements of G as follows:

$$(a, b)\,(c, d) := (ac, ad+b).$$

We easily check that the operation defined in this way is associative and non-commutative, that $(1, 0)$ is the neutral element and $\left(\frac{1}{a}, -\frac{b}{a}\right)$ is the inverse of (a, b).

DEFINITION. A *homomorphism* of a group G into G_1 is a mapping $\varphi\colon G \to G_1$ such that $\varphi(gh) = \varphi(g)\varphi(h)$.

We can say that a homomorphism preserves the group operation. A homomorphism of a group onto itself is called an *automorphism.* The group of automorphisms of a group G is denoted by $\mathrm{Aut}\, G$.

LEMMA 1.1.1. *If φ is a homomorphism of G into G_1, then $\varphi(e) = e_1$ (the identity of the group G_1) and $\varphi(g^{-1}) = \varphi(g)^{-1}$.*

Proof. We find $\varphi(e) = \varphi(ee) = \varphi(e)\varphi(e)$. Applying to both sides the element $\varphi(e)^{-1}$, we get $\varphi(e) = e_1$. We also have $e_1 = \varphi(gg^{-1}) = \varphi(g)\varphi(g^{-1})$, which proves the second part of the lemma. □

Example 2

(a) The mapping $S_n \ni \sigma \to \operatorname{sgn}(\sigma) \in \boldsymbol{R}_*$ is a homomorphism—here $\operatorname{sgn}(\sigma)$ denotes the sign of a permutation σ.

(b) The function log: $\boldsymbol{R}_+ \to \boldsymbol{R}$ is an isomorphism of the multiplicative group of positive numbers onto the additive group of real numbers.

(c) The mapping $\mathrm{GL}(n, K) \ni A \to \det A \in \boldsymbol{K}_*$ is a homomorphism, which follows from the well-known formula $\det AB = \det A \det B$.

(d) The function $\boldsymbol{R} \ni t \to \mathrm{e}^{2\pi i t} \in \boldsymbol{C}_*$ is a homomorphism.

We have already become acquainted with examples of groups whose elements are transformations of a certain set into itself. Such group classification is of special interest for us.

DEFINITION. A *transformation group* is a triple $(G, X, \cdot)$, where G is a group, X is a space, and "$\cdot$" is an action of G on X, i.e. a mapping $G \times X \ni (g, x) \to g \cdot x \in X$ satisfying

$$(g_1 g_2) \cdot x = g_1 \cdot (g_2 \cdot x) \quad \text{and} \quad e \cdot x = x \quad \text{for } x \in X.$$

Thus a transformation group of a set X is defined by producing a homomorphism of G into $S(X)$.

Example 3

(a) The mapping $\mathrm{GL}(n, K) \times \boldsymbol{K}^n \ni (g, x) \to gx \in \boldsymbol{K}^n$ defines the transformation group $(\mathrm{GL}(n, K), \boldsymbol{K}^n)$. Here $x \in \boldsymbol{K}^n$ is regarded as a column vector, i.e. a matrix with n rows and one column.

More generally, if V is an n-dimensional vector space over $\boldsymbol{K}$ and $(b_i)_1^n$ is a fixed basis in V, then an action of $\mathrm{GL}(n, \boldsymbol{K})$ on V is defined in the following way: if $v = \sum_{i=1}^{n} v^i b_i$, $g = [g_j^i]$, then $g \cdot v := w$ where $w = \sum_{i=1}^{n} w^i b_i$ with

$$w^i = \sum_{j=1}^{n} g_j^i v^j, \qquad i = 1, \ldots, n.$$

(b) Let G be the group from Example 1(e). For $g = (a, b) \in G$ and $x \in \boldsymbol{R}$ we denote $g \cdot x := ax+b$. Thus the transformation group $(G, R, \cdot)$ is the group of affine transformations of the real line. Quite often it is called the group "$ax+b$".

(c) The same group G as in (b) leads to the transformation group $(G, \boldsymbol{R}^2, \cdot)$ with the action of G on $\boldsymbol{R}^2$ defined by the homomorphism $\varphi: G \to \mathrm{GL}(2, \boldsymbol{R})$, where $\varphi((a, b)) = \begin{bmatrix} a & b \\ 0 & 1 \end{bmatrix}$.

Every abstract group G can be treated in a number of ways as a transformation group of the set G.

DEFINITION. The *left translation by an element* $g \in G$ is the mapping $l_g: G \to G$ given by the formula

$$l_g x = gx.$$

Analogously the *right translation* r_g is defined by the equation

$$r_g x = xg^{-1}.$$

The *inner automorphism* α_g is the transformation $\alpha_g: G \to G$ defined as follows:

$$\alpha_g(x) = gxg^{-1}.$$

Note that α is in fact an automorphism of the group G. The reader can easily check

LEMMA 1.1.2. (G, G, l), (G, G, r), (G, G, α) *are transformation groups.*

DEFINITION. We say that a subset $H \subset G$ is a *subgroup* if for arbitrary $g, h \in H$ the element gh^{-1} belongs to H.

A subgroup is a group with the operation obtained by restricting the group operation of G to the set $H \times H$.

Example 4

We shall prove that if $\varphi: G \to G_1$, is a homomorphism, and $H \subset G$ and $H_1 \subset G_1$ are subgroups, then $\varphi(H) \subset G_1$ and $\varphi^{-1}(H_1) \subset G$ are subgroups in G and G_1, respectively.

Indeed, if $y_1 = \varphi(x_1)$, $y_2 = \varphi(x_2)$, where $x_1, x_2 \in H$, then $y_1 y_2^{-1} = \varphi(x_1)\varphi(x_2^{-1}) = \varphi(x_1 x_2^{-1}) \in H$. Analogously, if $x_1, x_2 \in \varphi^{-1}(H_1)$, then $\varphi(x_1 x_2^{-1}) = \varphi(x_1)\varphi(x_2)^{-1} \in H_1$.

The subgroup $\varphi^{-1}(\{e\}) \subset G$ is called the *kernel* of the homomorphism φ and denoted by $\ker\varphi$.

Example 5

We now list a number of subgroups of $\mathrm{GL}(n, \boldsymbol{K})$ leaving it to the reader to check that they are indeed subgroups:

(a) $\mathrm{SL}(n, \boldsymbol{K}) = \{A \in \mathrm{GL}(n, \boldsymbol{K})\colon \det A = 1\}$,

(b) $\mathrm{GL}_+(n, \boldsymbol{R}) = \{A \in \mathrm{GL}(n, \boldsymbol{R})\colon \det A > 0\}$,

(c) $\Delta(n, \boldsymbol{K}) = \{A \in \mathrm{GL}(n, \boldsymbol{K})\colon A$—diagonal matrix$\}$,

(d) $T(n, \boldsymbol{K}) = \{A \in \mathrm{GL}(n, \boldsymbol{K})\colon T$—upper triangular matrix$\}$.

Recall that a matrix $A = [a_j^i]$ is called *upper triangular* if $a_j^i = 0$ for $i > j$.

Example 6

Let $\beta\colon \boldsymbol{K}^n \times \boldsymbol{K}^n \to \boldsymbol{K}$ be a non-degenerate bilinear form given by a matrix $B \in M_n(\boldsymbol{K})$ by the formula $\beta(x, y) = y^t Bx$. The superscript t denotes the *transposition* of a matrix. We say that $A \in M_n(\boldsymbol{K})$ preserves the form β if $\beta(Ax, Ay) = \beta(x, y)$, $x, y \in \boldsymbol{K}^n$, which holds if and only if $A^t BA = B$.

It can easily be checked that the set $G(\beta)$ of all matrices preserving the form β is a subgroup of $\mathrm{GL}(n, \boldsymbol{K})$, called the *invariant subgroup* of the form β. Here it is the non-degeneracy of the form that ensures invertibility of matrices preserving it. Among the special cases we consider:

(a) the symmetric form $\beta(x, y) = \sum_{i=1}^{n} x^i y^i$, where x^i, y^i are the coordinates of x and y, respectively. In this case the group $G(\beta)$ is called the *n-dimensional orthogonal group* and denoted by $O(n, \boldsymbol{K})$;

(b) the antisymmetric form $\eta(x, y) = \sum_{i=1}^{n} (x^i y^{n+i} - y^i x^{n+i})$ in the space $\boldsymbol{K}^2$ (this form is non-degenerate only in the spaces of even dimension). The group determined by η is called the *2n-dimensional symplectic group* and denoted by $\mathrm{Sp}(2n, \boldsymbol{K})$.

One can also associate to a hermitian form the group of transformations preserving it. The construction and examples are given in Problem 10.

Example 7

Denote by $(e_i)_1^n$ the canonical basis in $\boldsymbol{K}^n$, i.e. $e_i = (\delta_i^j)$ for $i = 1, \ldots, n$, where δ_i^j is the "Kronecker delta" ($\delta_i^j = 1$ for $i = j$ and $\delta_i^j = 0$ for $i \neq j$). The smallest subgroup in $\boldsymbol{K}^n$ containing the set $\{e_i\}_1^n$ (called the *subgroup generated* by $\{e_i\}_1^n$) is the subgroup Z^n consisting of all vectors in $\boldsymbol{K}^n$ with integer coordinates.

Let $(G, X, \cdot)$ be a transformation group. Fix a set $A \subset X$. The set G_A of all $g \in G$ for which $g \cdot x = x$, $x \in A$ is called the *isotropy group* of A or *the stability group* of A. The set $G \cdot A := \{x \in X\colon \bigvee_{y \in A} \bigvee_{g \in G} x = g \cdot y\}$, in turn, is called the *orbit* of A under the action of G. Choose for instance, as a transformation group, a group G acting on the set G by inner automorphisms. The orbit of $x \in G$ is $C_x = \{y \in G\colon \bigvee_{g \in G} y = gxg^{-1}\}$. In this case the orbit of a point x is called the *conjugacy class* of x.

Now, a subgroup $H \subset G$ is said to be *normal* if for arbitrary $g \in G$ $gHg^{-1} \subset H$.

Example 8

We shall prove for instance that if $\varphi\colon G \to G_1$ is a homomorphism of a group G into a group G_1 and $H_1 \subset G_1$ is a normal subgroup, then $H = \varphi^{-1}(H_1)$ is also a normal subgroup (of G). Indeed, for $g \in G$ and $h \in H$, $\varphi(ghg^{-1}) = \varphi(g)\varphi(h)\varphi(g)^{-1} \in H_1$; hence $ghg^{-1} \in \varphi^{-1}(H_1)$. In particular—for an arbitrary homomorphism $\varphi\colon G \to G_1$—$\ker\varphi$ is a normal subgroup of G.

Let $H \subset G$ be a subgroup. The orbits of the transformation group (H, G, r) are called the *left cosets*, while these of the group (H, G, l)—the *right cosets*. The orbit of a point $x \in G$ is denoted by $[x]$ or xH. The group G acts in a natural way on the space of the left (right) cosets. Namely, we define $g \cdot [x] := [gx]$ ($g \cdot [x] = [xg^{-1}]$). The transformation group obtained is denoted by $(G, G/H)$ $((G, H\backslash G))$.

Definition. A transformation group (G, X) is called *transitive* if the orbit of every point coincides with X.

Obviously, it is enough to check that for one point $x_0 \in X$ the equality $X = G \cdot x_0$ is satisfied. Note that the transformation groups $(G, G/H)$

and $(G, H\backslash G)$ are transitive. We shall prove that these examples are typical. But first the notion of a homomorphism (of an isomorphism) of transformation group will be introduced.

DEFINITION. A transformation $\varphi\colon X \to X_1$ is said to be a *homomorphism* of (G, X) into (G, X_1) if for all $x \in X$ and $g \in G$ we have $\varphi(g \cdot x) = g \cdot \varphi(x)$.

If in addition φ is a bijection, we call it an *isomorphism of transformation groups.*

THEOREM 1.1.3. *Let (G, X) be a transitive transformation group of X. Let G_{x_0} be the isotropy group at a point $x_0 \in X$. Then (G, X) is isomorphic to $(G, G/G_{x_0})$.*

Proof. First we map the group G into X in the following way:

$$\tilde{\varphi}\colon G \ni g \to g \cdot x_0 \in X.$$

Owing to the transitivity of (G, X) $\tilde{\varphi}$ is surjective. We now define

$$\varphi(gG_{x_0}) = \tilde{\varphi}(g).$$

The definition is correct since $G_{x_0} \cdot x_0 = x_0$. Also, we have

$$g \cdot \varphi(hG_{x_0}) = gh \cdot x_0 = \varphi(ghG_{x_0}).$$

It only remains to check that φ is invertible. Let us suppose that $\varphi(hG_{x_0}) = \varphi(gG_{x_0})$; thus $h \cdot x_0 = g \cdot x_0$, and so $h^{-1}g \in G_{x_0}$ and finally $hG_{x_0} = gG_{x_0}$. □

Example 9

Denote by S^n the n-dimensional sphere in R^{n+1}, i.e. $S^n = \{x \in R^{n+1}: ||x|| = 1\}$, where $||x|| := \left(\sum_{i=1}^{n+1} (x^i)^2\right)^{\frac{1}{2}}$ is the Euclidean norm in R^{n+1}. The group $G = O(n+1, R)$ acts in a natural way on the sphere S^n according to the formula

$$O(n+1, R) \times S^n \ni (g, x) \to gx \in S^n.$$

The action is transitive since an arbitrary point in the sphere can be obtained by acting with an orthogonal matrix on the "north pole of the sphere S^n", i.e. on the point $p = e_{n+1} = (0, \dots, 0, 1)$. The isotropy group at the point p

$$G_p = \left[\begin{array}{ccc|c} & A & & \begin{matrix}0\\ \vdots\\ 0\end{matrix} \\ \hline 0 & \dots & 0 & 1 \end{array}\right], \qquad A \in O(n, \mathbf{R}),$$

is isomorphic to $O(n, \mathbf{R})$. Thus we conclude that the homogeneous spaces $(O(n+1, \mathbf{R}), S^n)$ and $(O(n+1, \mathbf{R}), O(n+1, \mathbf{R})/O(n, \mathbf{R}))$ are isomorphic.

THEOREM 1.1.4. *Let $H \subset G$ be a normal subgroup. Then the quotient space G/H has a group structure with the group operation*

$$(gH)(g_1H) := gg_1H.$$

Proof. The correctness of the definition follows from the calculations $gHg_1H = gg_1g_1^{-1}Hg_1H = gg_1H$. Note that H is the identity element, and the inverse of gH is $g^{-1}H$. □

Example 10

The subgroup $\mathrm{SL}(n, \mathbf{R})$ is normal in $\mathrm{GL}(n, \mathbf{R})$. The quotient group $\mathrm{GL}(n, \mathbf{R})/\mathrm{SL}(n, \mathbf{R})$ is isomorphic to the multiplicative group $\mathbf{R}_*$. The isomorphism is given by the function $\det(\cdot)$, which is constant on the cosets $\mathrm{GL}(n, \mathbf{R})/\mathrm{SL}(n, \mathbf{R})$.

To end this section, the basic constructions, i.e. those of the direct and the semidirect products of groups, will be given.

DEFINITION. Let groups G_1 and G_2 be given. The *direct product* $G_1 \times G_2$ is the group with an underlying set consisting of the Cartesian product of the sets G_1 and G_2 and the operation

$$(g_1, g_2)(h_1, h_2) := (g_1h_1, g_2h_2).$$

Clearly the pair (e_1, e_2) is the identity element, and the inverse of (g_1, g_2) is the pair (g_1^{-1}, g_2^{-1}).

Suppose in addition that a homomorphism $\tau\colon G_1 \to \mathrm{Aut}\, G_2$ is given. The *semidirect product* $G_1 \times_\tau G_2$ is the group with an underlying set consisting of the Cartesian product of the sets G_1 and G_2 and the group operation defined as follows:

$$(g_1, g_2)(h_1, h_2) := (g_1h_1, g_2\tau(g_1)h_2).$$

Example 11

Suppose that a group G contains two subgroups N and K such that $G = NK$ (here NK denotes the set $\{nk: n \in N, k \in K\}$), $N \cap K = \{e\}$ and N is normal. Denote by τ the homomorphism $\tau: K \to \mathrm{Aut}N$, where $\tau(k)n = knk^{-1}$. It can easily be proved that the transformation $K \times N \ni (k, n) \to nk \in G$ is an isomorphism of the semidirect product $K \times_\tau N$ onto G.

The direct product is a special case of the semidirect product, namely it corresponds to the trivial homomorphism $\tau(g) = e_2$. The "component" groups G_1 and G_2 are isomorphic to the following subgroups of $G_1 \times G_2$:

$$G_1 \to (G_1, e_2) =: G_1',$$

$$G_2 \to (e_1, G_2) =: G_2'.$$

The subgroup G_2' is normal in $G_1 \times_\tau G_2$ and the quotient group $G_1 \times_\tau G_2/G_2'$ is isomorphic to G_1.

So far only the purely algebraic structure of an abstract group has been studied. The real role of group theory will come to light only when the connections of the algebraic structure with other structures, defined on the group space or on the space on which it acts, are investigated.

A large part of this book deals with Lie groups, which will be studied in subsequent sections. Many facts and theorems formulated there, are also valid for wider classes of groups—for locally compact or even arbitrary topological groups. In order to make our exposition of the theory of representations more complete these notions are defined below.

Definition. A group G equipped with a topology τ is called a *topological group* provided the mapping $G \times G \ni (g, g_1) \to gg_1^{-1} \in G$ is continuous.

A continuous homomorphism of topological groups is called a *topological homomorphism*. An isomorphism (automorphism) which is a topological homomorphism is said to be a *topological isomorphism* (*automorphism*). The inner automorphisms of a topological group are obviously topological automorphisms.

The topology of a topological group G is totally determined by a local base of neighbourhoods at an arbitrary point in G. Namely, it follows directly from the definition that if $\{O_i(x)\}_{i\in I}$ is a local base at a point $x \in G$ then the sets $gx^{-1}O_i(x) := \{y \in G: \bigvee_{h\in O_i(x)} y = gx^{-1}h\}$ form a local base at a point $g \in G$.

We say that G is a *locally compact group* if the topology of G is locally compact.

Example 12

(a) The additive group of real numbers is a topological group with respect to the natural topology in $\boldsymbol{R}^1$; so are the additive group of complex numbers and the multiplicative groups of these fields.

(b) Every group equipped with discrete topology is a topological group.

(c) If G is a topological group and H its subgroup (in the algebraic sense), then H equipped with the subspace topology is also a topological group, since continuity of the transformation $H \times H \ni (x, y) \to xy^{-1} \in H$ follows immediately in view of the properties of restriction of continuous mappings.

(d) $\mathrm{GL}(n, \boldsymbol{K})$ can be regarded as an open subset of the space $\boldsymbol{K}^{n^2}$ defined by the condition of non-vanishing of the determinant. With the induced subset topology, the group $\mathrm{GL}(n, \boldsymbol{K})$ is a topological group since the entries of the matrix AB^{-1} are continuous functions of the entries of matrices A and B.

(e) On account of point (c) every group of matrices, as a subgroup of the respective $\mathrm{GL}(n, \boldsymbol{K})$, is a topological group with respect to the topology induced from $\mathrm{GL}(n, \boldsymbol{K})$. Since $\mathrm{GL}(n, \boldsymbol{K})$ is an open subset of the space $\boldsymbol{K}^{n^2}$, the topology coincides with that obtained by embedding the initial group of matrices in the respective space $\boldsymbol{K}^{n^2}$. In particular, the continuity of a function on a group of matrices is equivalent to the continuity with respect to the entries of matrices.

A transformation group (G, X) is called a *topological* (or *continuous*) *transformation group* if G is a topological group, X is a topological space and the mapping $G \times H \ni (g, x) \to g \cdot x \in X$ is continuous.

A transformation group which is both continuous and transitive is called a *homogeneous topological space* provided the following condition is satisfied:

$$\text{for each } x \in X \text{ the mapping } G \ni g \to g \cdot x \in X \text{ is open.}$$

If H is a closed subgroup of a locally compact topological group G, then the quotient topology in the homogeneous space G/H is also locally compact. Recall that $O \subset G/H$ is open in the quotient topology if and only if $\pi^{-1}(O)$ is open in G (π denotes the canonical projection $G \to G/H$). The action of G on G/H satisfies the condition of continuity; hence the pair $(G, G/H)$ is a homogeneous topological space.

THEOREM 1.1.5. *Let (G, X) be a transitive topological transformation group and let the topologies in G and X be locally compact and second countable. Then for an arbitrary $x \in X$ the isotropy subgroup G_x is closed in G and the mapping $G/H \ni gG_x \to g \cdot x \in X$ is a topological homeomorphism. In particular (G, X) is a homogeneous space.*

Example 13

To convince oneself that the isomorphism (in the algebraic sense) of homogeneous spaces established in Example 9 is a homoemorphism, it suffices, on account of Theorem 1.1.5, to check if $(O(n+1, \boldsymbol{R}), S^n)$ is a homogeneous topological space, that is, if the mapping $O(n+1, \boldsymbol{R}) \ni g \to g \cdot x \in S^n$ is open for each $x \in S^n$. Since the translations $g \to gg_0$ in a topological group are homeomorphisms, it is enough to see that the mapping $O(n+1, \boldsymbol{R}) \ni g \to gp \in S^n$ carries every neighbourhood in a fixed local base at the identity $I \in O(n+1, \boldsymbol{R})$ onto a neighbourhood of the point $Ip = p \in S^n$. This condition becomes obvious if we observe that a local base at the identity in $O(n+1, \boldsymbol{R})$ can be chosen so that the last columns of the matrices belonging to a chosen neighbourhood form a neighbourhood of the point $p \in \boldsymbol{R}^{n+1}$.

It follows from the definition of the quotient topology that a function f on a homogeneous space is continuous if, and only if, the function $f \circ \pi$ is continuous on G. This function satisfies the condition $f \circ \pi(gh) = f(ghH) = f(gH) = f \circ \pi(g)$ for $h \in H$. Thus we say that $f \circ \pi$ is *right-invariant with respect to the subgroup H*. Conversely, given a function φ right-invariant with respect to H, we can "project" it to G/H by putting $f(gH) := \varphi(g)$. The function obtained is continuous on G/H provided that φ is a continuous function. It is often convenient to identify the function f on G/H with the function φ defined on the group G. Making use of this identification, we shall often denote both functions by the same symbol f.

PROBLEMS

1. Prove

(a) If $\varphi: G \to G_1$ is a group isomorphism, then $\varphi^{-1}: G_1 \to G$ is also an isomorphism.

(b) If $\varphi: G \to G_1$, $\psi: G_1 \to G_2$ are group homomorphisms, then $\varphi \circ \psi: G \to G_2$ is also a homomorphism.

2. Let G be a topological group.

(a) Prove that the transformation $G \ni x \to x^{-1} \in G$ is a homeomorphism and derive hence that the identity $e \in G$ has a local base consisting of symmetric sets (a subset $A \subset G$ is called *symmetric* if the subset $A^{-1} := \{x \in G: x^{-1} \in A\}$ equals A).

(b) Prove that an arbirtary neighbourhood $U \subset G$ of the identity element contains a neighbourhood $V \subset G$ of the identity such that $V \cdot V := \{xy: x, y \in V\} \subset U$.

(c) Prove that if $H \subset G$ is a subgroup (abelian or normal), then its closure $\bar{H}$ is also a subgroup (abelian or normal, respectively).

(d) Prove that G_0, the connected component of the identity, is a closed normal subgroup of G, and that the quotient group G/G_0 is totally disconnected (i.e. the only connected sets are points).

3. Let G be a topological group, $H \subset G$ a subgroup (in the algebraic sense) and $\pi: G \to G/H$ the canonical projection.

(a) Prove that the quotient topology in G/H satisfies the Hausdorff axiom if, and only if, H is a closed subset of G.

(b) Prove that the transformation π is continuous and open.

(c) Prove that if the subgroup H is normal, then G/H is a topological group with the quotient topology.

4. Let G_1, G_2 be topological groups.

(a) Prove that $G_1 \times G_2$ equipped with the Cartesian product topology is also a topological group.

(b) Suppose that a homomorphism $\tau: G_1 \to \mathrm{Aut}(G_2)$ into the group $\mathrm{Aut}(G_2)$ of topological automorphisms of G_2 is given, such that the mapping $G_1 \times G_2 \ni (g_1, g_2) \to \tau(g_1)g_2 \in G_2$ is continuous. Prove that $G_1 \times_\tau G_2$ equipped with the Cartesian product topology is a topological group (a topological semidirect product).

5. Prove that no group operation can be defined on the interval $[0, 1]$ to make it a topological group with the natural topology.

6. Let H be a Hilbert space and G the group of all unitary transformations from H into H. Prove that G is a topological group with respect to the weak operator topology, i.e. the weakest topology for which all the transformations $G \ni U \to (Ux|y) \in C$ are continuous.

7. (a) Let N be a normal subgroup in G and let π: $G \to G/N$ be the natural projection, i.e. $\pi(g) := gN$. Show that π is a homomorphism and $\ker\pi = N$.

(b) Let φ: $G \to H$ be a homomorphism and let i: $\varphi(G) \to H$ denote the natural inclusion mapping. Show that there exists a unique isomorphism $\bar{\varphi}$: $G/\ker\varphi \to \varphi(G)$ such that $i \circ \bar{\varphi} \circ \pi = \varphi$.

(c) Give an example showing that $\bar{\varphi}$ need not a be topological isomorphism.

8. (a) Let S be an arbitrary non-empty subset of a group G. Prove that there exists a smallest subgroup H in G containing S. (H is called the *subgroup generated by* S and S—the set of generators of H.)

(b) Prove that the subgroup $D(G)$ generated by the set $\{y^{-1}x^{-1}yx\colon x, y \in G\}$ is the smallest normal subgroup in G such that the quotient subgroup it defines is commutative.

9. Let A be an arbitrary non-empty subset of a (topological) group G. Show the following:

(a) The *centralizer* $Z(A)$ of A, i.e. the set $Z(A) = \{g \in G\colon ga = ag$ for all $a \in A\}$, is a (closed) subgroup of the group G. If $A = G$ then $Z(G) = Z$ is a commutative subgroup called the *centre* of the group G.

(b) The *normalizer* $N(A)$ of A, i.e. the set $N(A) = \{g \in G\colon gAg^{-1} = A\}$, is a (closed) subgroup of G.

(c) $Z(A)$ is a normal subgroup of $N(A)$.

(d) If A is a subgroup of G and H a subgroup of $N(A)$, then HA is a subgroup of G and A is a normal subgroup of HA.

10. Let β: $C^n \times C^n \to C$ be a non-degenerate, hermitian form and $B \in M_n(C)$ the matrix representing the form β, i.e.

$$\beta(x, y) = y^*Bx,$$

where "$*$" denotes the hermitian conjugation of matrices. Let $G(\beta) := \{A \in M_n(C)\colon A^*BA = B\}$. Show that

(a) $G(\beta)$ is a closed subgroup of $\mathrm{GL}(n, C)$.

(b) Linear transformations of C^n into itself preserve the form if and only if their matrices (with respect to the canonical basis) belong to $G(\beta)$.

(c) If β is the standard scalar product in $\boldsymbol{C}^n$, i.e. $\beta(x, y) = \sum_{i=1}^{n} \bar{y}_i x_i$, then $G(\beta)$ is the group of unitary matrices, denoted by $U(n)$.

(d) If η is a hermitian form equivalent to β, i.e. if the canonical form of η and β is the same, then $G(\beta)$ is conjugate to $G(\eta)$ in $\mathrm{GL}(n, \boldsymbol{C})$. (Subgroups H, H_1 are said to be conjugate if $H_1 = gHg^{-1}$ for some $g \in G$.)

11. Let $\mathrm{SO}(n, \boldsymbol{K}) := \mathrm{SL}(n, \boldsymbol{K}) \cap O(n, \boldsymbol{K})$ and similarly $\mathrm{SU}(n) = \mathrm{SL}(n, \boldsymbol{C}) \cap U(n)$.

(a) Prove that $\mathrm{SO}(n, \boldsymbol{K})$ and $\mathrm{SU}(n)$ are closed normal subgroups of $O(n, \boldsymbol{K})$ and $U(n)$, respectively.

(b) Prove that the following groups are isomorphic: $O(n, \boldsymbol{K})/\mathrm{SO}(n, \boldsymbol{K}) \simeq Z_2$, and $U(n)/\mathrm{SU}(n) \simeq S^1$, where S^1 is the multiplicative group of complex numbers of absolute value one and $Z_2 = \{1, -1\}$ with the operation of multiplication.

(c) Prove that the matrices of the form

$$\left[\begin{array}{ccc|c} & & & 0 \\ & S & & \vdots \\ & & & 0 \\ \hline 0 & \dots & 0 & \det S \end{array}\right], \qquad S \in O(n, \boldsymbol{R})$$

form a closed subgroup of $\mathrm{SO}(n+1, \boldsymbol{R})$ isomorphic to $O(n, \boldsymbol{R})$. Construct a surjective homomorphism of the homogeneous space $(\mathrm{SO}(n+1, \boldsymbol{R}), S^n)$ onto the homogeneous space $(\mathrm{SO}(n+1, \boldsymbol{R}), \mathrm{SO}(n+1, \boldsymbol{R})/O(n, \boldsymbol{R}))$.

REMARK. The last of the above homogeneous spaces is called an *n-dimensional real projective space*. This space will be dealt with in more detail in the next section.

12. Show that the following topological groups are isomorphic:

(a) $\boldsymbol{C}_* \simeq \boldsymbol{R}_+ \times S^1, \qquad \boldsymbol{R}_* = \boldsymbol{R}_+ \times Z_2$,

(b) $\mathrm{GL}(n, \boldsymbol{K}) \simeq \mathrm{SL}(n, \boldsymbol{K}) \times_\tau \boldsymbol{K}_*, \qquad U(n) \simeq \mathrm{SU}(n) \times_\tau S^1$,

(c) $\boldsymbol{R}^n/Z^n \simeq S^1 \times \dots \times S^1$.

Hint for (b): Immerse $\boldsymbol{K}_*$ (resp. S^1) in $\mathrm{GL}(n, \boldsymbol{K})$ by assigning to a number λ a diagonal matrix with all diagonal terms equal to 1, except one term equal to λ.

13. (a) Show that $T(n, \boldsymbol{K})$ is the topological semidirect product of $\Delta(n, \boldsymbol{K})$ and $N(n, \boldsymbol{K}) = \{A \in T(n, \boldsymbol{K}) \colon a_{ii} = 1,\ i = 1, \ldots, n\}$.

(b) Show that the affine isomorphisms of the space $\boldsymbol{K}^n$, i.e. transformations of the form $\boldsymbol{K}^n \ni x \to Ax + b \in \boldsymbol{K}^n$, where $A \in \mathrm{GL}(n, \boldsymbol{K})$, $b \in \boldsymbol{K}^n$, form a group which is the topological semidirect product of $\mathrm{GL}(n, \boldsymbol{K})$ with $\boldsymbol{K}^n$.

14. Describe the orbits of the transformation group of Example 3 (c).

15. Prove that any subgroup of the group S^1 of complex numbers of absolute value one with the natural topology is either finite or dense in S^1.

1.2. DIFFERENTIABLE MANIFOLDS

DEFINITION. Let M be a topological space. An n-dimensional chart on M is a homeomorphism $\varkappa$ of an open subset $U \subset M$, called the *domain of the chart*, onto an open subset of $\boldsymbol{R}^n$.

The existence of a chart on a domain U means that the part of the space M does not differ topologically from an open subset of the space $\boldsymbol{R}^n$. Thus, the existence of a rich family of continuous functions of the form $U \ni m \to f \circ \varkappa^{-1}(m) \in \boldsymbol{R}$ is ensured, where f is a continuous function on the image of the chart. Among the continuous functions on U a class of functions originating from functions f differentiable in $\boldsymbol{R}^n$ can be distinguished. This family of functions has to play the role of differentiable functions. If there are two different charts $(\varkappa, U)$ and $(\varkappa', U')$ defined on M, then the family of differentiable functions on the intersection of the domains $U \cap U'$ can be defined in two different ways by applying either of the charts $\varkappa$ and $\varkappa'$. One can easily impose a condition to ensure that both charts define the same class of differentiable functions.

DEFINITION. Charts $(\varkappa, U)$ and $(\varkappa', U')$ are said to be *C^∞-compatible* if the mapping $\varkappa' \circ \varkappa^{-1}$ is a diffeomorphism on its domain $\varkappa(U \cap U')$.

Obviously, either the dimension of compatible charts are equal or $U \cap U' = \emptyset$.

DEFINITION. A set $\mathscr{A} = (\varkappa_i, U_i)_{i \in I}$ of C^∞-compatible, n-dimensional charts whose domains cover M is called an *n-dimensional atlas* for the space M.

By a *differentiable manifold* we mean a topological space M equipped with an atlas $\mathscr{A}$ which is maximal in the sense that every chart C^∞-compatible with all charts $\varkappa_i$, $i \in I$, belongs to $\mathscr{A}$. The common dimension of charts in $\mathscr{A}$ is called the *dimension of the manifold M.*

A function f on a manifold $(M, \mathscr{A})$ is called *differentiable* if for any chart $(\varkappa, U)$ in $\mathscr{A}$ the composite $f \circ \varkappa^{-1}$ is a differentiable function on its domain $\varkappa(U) \subset \boldsymbol{R}^n$. The space of differentiable functions on M will be denoted by $\mathscr{E}(M)$. It is an algebra with respect to the operations of addition and pointwise multiplication of functions.

Given an open subset $O \subset M$, we can consider differentiable functions defined only on O. We denote this space by $\mathscr{E}(O)$ in accordance with the notation $\mathscr{E}(M)$, since an open subset of a differentiable manifold possesses the natural manifold structure with the atlas $\mathscr{A}_0$ obtained by restriction of charts $(\varkappa, U)$ to $U \cap O$.

Let $m \in M$. By $\mathscr{E}(m)$ we denote, the set of functions such that each one belongs to some $\mathscr{E}(O(m))$, where $O(m)$ is a neighbourhood of m. Only multiplication by a number is defined naturally in that space. We define the addition and the multiplication of elements of $\mathscr{E}(m)$ as follows: if $f \in \mathscr{E}(O(m))$ and $f' \in \mathscr{E}(O'(m))$, then $f+f'$ and $f \cdot f'$ are functions with the domain $O(m) \cap O'(m)$ given by the formula

$$(f+f')(p) = f(p)+f'(p), \qquad (1.2.1)$$
$$ff'(p) = f(p)f'(p).$$

One can also define analytic structures in a way analogous to that in which differentiable structures have been defined, thus making it possible to distinguish the class of analytic functions. All we have to do is to define analytically compatible charts.

DEFINITION. Charts $(\varkappa, U)$ and $(\varkappa', U')$ on M are said to be *analytically compatible* if $\varkappa' \circ \varkappa^{-1}$ is an analytic diffeomorphism.

By an *analytic manifold* we mean a system $(M, \mathscr{A})$, where $\mathscr{A}$ is a maximal atlas of analytically compatible charts.

A function f on M is analytic if $f \circ \varkappa^{-1}$ is analytic on the image of the chart.

A continuous mapping Φ: $M \to M'$ between two (analytic) manifolds $(M, \mathscr{A})$, $(M', \mathscr{A}')$ is called *differentiable* or *smooth* (*analytic*) if for arbitrary charts $(\varkappa, U)$ and $(\varkappa', U')$ on M and M' the mapping $\varkappa' \circ \Phi \circ \varkappa^{-1}$ is differentiable (analytic) on its domain.

By a *diffeomorphism* Φ between two (analytic) manifolds we mean a differentiable (analytic) homeomorphism such that Φ^{-1} is also a smooth (analytic) mapping.

The condition of compatibility for an atlas guarantees that any chart on a manifold is a diffeomorphism of its domain onto the image.

Example 1

The space $\boldsymbol{R}^n$ possesses a natural structure of a differentiable manifold provided by the identity transformation of $\boldsymbol{R}^n$ onto itself. The space $\boldsymbol{C}^n$, in turn, is equipped with the manifold structure given by the natural $\boldsymbol{R}$-linear isomorphism of $\boldsymbol{C}^n$ onto $\boldsymbol{R}^{2n}$.

Example 2

Let S^n be the n-dimensional unit sphere in $\boldsymbol{R}^{n+1}$ (cf. Example 9 in §1.1.) with the topology induced by the Euclidean metric in $\boldsymbol{R}^{n+1}$. Let $p = e_{n+1} = (0, \ldots, 1)$, $q = -e_{n+1}$ (the "south" and the "north" poles of S^n) and write $U_p = S^n - \{p\}$, $U_q = S^n - \{q\}$. Let $\varkappa_p\colon U_p \to \boldsymbol{R}^n$ be the stereographic projection from the pole p, i.e. for a point $x \in S^n$, let the point $(\varkappa_p(x), 0)$ be the point at which the line passing through x and p interesects the hyperplane $x^{n+1} = 0$. For $x = (x^1, \ldots, x^{n+1}) \in U_p$ we easily find

$$\varkappa_p(x) = \frac{1}{1-x^{n+1}}(x^1, \ldots, x^n).$$

For the analogous stereographic projection from the pole q we get

$$\varkappa_q(x) = \frac{1}{1+x^{n+1}}(x^1, \ldots, x^n)$$

and for $y = (y^1, \ldots, y^n) \in \varkappa_p(U_p) \cap \varkappa_q(U_q) = \boldsymbol{R}^n - \{0\}$ we obtain

$$\varkappa_p \circ \varkappa_q^{-1}(y) = \frac{1}{||y||^2} y,$$

where $||y||^2 = \sum_{i=1}^{n} (y^i)^2$ is the euclidean norm of y.

Thus the charts $\varkappa_p$ and $\varkappa_q$ are compatible and because their domains cover S^n, the set $\{\varkappa_p, \varkappa_q\}$ is an atlas for the sphere S^n. Therefore, a differentiable (or even analytic) manifold structure is defined on S^n (cf. Problem 1). Moreover, the natural inclusion mapping $i\colon S^n \to \boldsymbol{R}^{n+1}$ is smooth with respect to that structure. In Problem 4 another atlas defining the same manifold structure on S^n is described (also cf. Problem 5).

Example 3

Let $O \subset \boldsymbol{R}^m$ be an open subset, $f\colon O \to \boldsymbol{R}^k$, $k < m$, a smooth mapping and let a be a point in $\boldsymbol{R}^n$ such that the level set $M_a := \{x \in O\colon f(x) = a\}$ is non-empty. Assume that the Jacobian matrix of f, i.e. the matrix $\left[\dfrac{\partial f^i}{\partial x^j}\right]$, is of constant rank k for all $x \in M_a$, where the set M_a is equipped with the topology induced from $\boldsymbol{R}^n$, a structure of $(m-k)$-dimensional differentiable manifold can now be defined satisfying the conditions:

(1) the natural inclusion mapping $M_a \to \boldsymbol{R}^n$ is smooth,

(2) for each point $x \in M_a$ indices $i_1, i_2, \ldots, i_{m-k}$ can be chosen so that the mapping $x \to (x^{i_1}, \ldots, x^{i_{m-k}}) \in \boldsymbol{R}^{m-k}$ is a chart for M_a defined on some open neighbourhood of the point x. (Suggestions for the proof will be found in Problem 5.)

To exemplify this method of defining a differentiable manifold structure, consider $O = \mathrm{GL}(2, \boldsymbol{R}) \subset \boldsymbol{R}^4$ and the function det: $\mathrm{GL}(2, \boldsymbol{R}) \to \boldsymbol{R}^1$. Then the group $\mathrm{SL}(2, \boldsymbol{R}) = \{x \in \mathrm{GL}(2, \boldsymbol{R})\colon \det x = 1\}$ is a level set of the function det and $\nabla \det x \neq 0$ for all $x \in \mathrm{SL}(2, \boldsymbol{R})$. In this case we can choose for charts the pairs $(\varkappa_i, U_i)$, $i = 1, 2, 3, 4$, where

$$U_i = \left\{\begin{bmatrix} x^1 & x^2 \\ x^3 & x^4 \end{bmatrix} \in \mathrm{SL}(2, R)\colon \frac{\partial \det}{\partial x^i}(x) \neq 0\right\} \quad \text{and} \quad \varkappa_i \begin{bmatrix} x^1 & x^2 \\ x^3 & x^4 \end{bmatrix}$$

$= (x^1, \overset{i}{\vee}, x^4)$, where the sign $\overset{i}{\vee}$ denotes the omission of the i-th coordinate.

Example 4 (The n-dimensional projective space $P^n\boldsymbol{R}$).

Let the multiplicative group $\boldsymbol{R}_* = \boldsymbol{R} - \{0\}$ act in a natural way on the set $\boldsymbol{R}^{n+1} - \{0\}$ by the formula $\lambda \cdot x := \lambda x$. Then the orbit of a point x is the line passing through x and 0, with the point 0 excluded. Write $P^n\boldsymbol{R}$ for the set of all punctured lines (orbits) and let $\pi\colon \boldsymbol{R}^{n+1} - \{0\} \to P^n\boldsymbol{R}$ be the natural projection assigning to a point its orbit. Equip $P^n\boldsymbol{R}$ with the strongest topology in which π is continuous. For k

$= 1, 2, \ldots, n+1$ define $U_k := \pi(\{x \in \mathbf{R}^{n+1}\colon\ x^k \neq 0\})$ and set $\varkappa_k$: $U_k \ni \pi(x) \to \frac{1}{x^k}(x^1, \ldots, x^{k-1}, x^{k+1}, \ldots, x^{n+1}) \in \mathbf{R}^n$. The proof that the above charts $\varkappa_k$ define in fact an atlas on $P^n\mathbf{R}$ and that π is smooth is left to the reader. $P^n\mathbf{R}$ endowed with this differentiable structure is called the *n-dimensional real projective space*. The complex projective spaces are defined analogously.

Example 5

A continuous mapping γ of an open interval $]a, b[$ into M is smooth if, and only if, for any chart $(\varkappa, U)$ such that $U \cap \gamma(]a, b[) \neq \emptyset$, the real functions $\varkappa^i \circ \gamma$ are C^∞ on the interval $]a, b[$. Such a mapping γ is called a *smooth curve* on the manifold M.

Tangent Vectors

Fix a point $m \in M$ and assume that the domain U of a chart $\varkappa$ contains m. One can define differentiation of $f \in \mathscr{E}(U)$ by means of the coordinates in $\mathbf{R}^n$, denoted by $(x^1, \ldots, x^n)$. Let us set

$$X_i(m)f := \frac{\partial}{\partial x^i} f \circ \varkappa^{-1}(\varkappa(m)), \tag{1.2.2}$$

and for $v = (t^1, \ldots, t^n)$

$$X_v(m)f := \sum_{i=1}^{n} t^i X_i(m). \tag{1.2.3}$$

Each of the vectors $X_v(m)$ defines a functional on the space $\mathscr{E}(m)$ by the formula $f \to X_v(m)f$. Obviously, for this functional the following identities hold:

$$X_v(m)(f+f') = X_v(m)f + X_v(m)f',$$
$$X_v(m)(ff') - (X_v(m)f)f' + f(X_v(m)f').$$

Definition. A *tangent vector to M at a point m* is a mapping $\mathscr{E}(m) \ni f \to X(m)f$ satisfying the conditions:

$$X(m)(f+\alpha f') = X(m)f + \alpha X(m)f', \qquad \alpha \in \mathbf{R}, \tag{1.2.4}$$
$$X(m)(ff') = (X(m)f)f' + f(X(m)f').$$

We already know that there exist non-zero tangent vectors on manifolds of dimension > 0. It follows directly from the definition that the set of tangent vectors at a point m constitutes a vector space; it is denoted by $T_m(M)$. Actually, we are already fully acquainted with the space of tangent vectors at a point. This fact is stated by

PROPOSITION 1.2.1. *For any $X(m) \in T_m(M)$ and any chart $(\varkappa, U(m))$ there exists a $v \in R^n$ such that $X(m) = X_v(m)$.*

The reader can construct a simple proof of this proposition using the hints to Problem 9.

Note that to the same vector $v \in \boldsymbol{R}^n$ correspond different tangent vectors at the point m depending on the chart used for the construction of X_v. We now investigate the relations between the systems $X_i(m)$ arising from different coordinate systems.

Let X_i' denote the tangent vector corresponding to the differentiation $\frac{\partial}{\partial x^i}$ for a chart $\varkappa'$. Then

$$X_i' f = \frac{\partial}{\partial x^i} f \circ (\varkappa')^{-1}(\varkappa'(m)) = \frac{\partial}{\partial x^i} (f \circ \varkappa^{-1})(\varkappa \circ (\varkappa')^{-1})(\varkappa'(m)).$$

Writing $\varkappa \circ (\varkappa')^{-1}(x^1, \ldots, x^n) := (\tau^1(x^1, \ldots, x^n), \ldots, \tau^n(x^1, \ldots, x^n))$ we arrive at

$$X_i' = \sum_{j=1}^{n} \frac{\partial \tau^j}{\partial x^i} (\varkappa'(m)) X_j.$$

The Jacobi matrix $g_i^j := \dfrac{\partial \tau^j}{\partial x^i}$ is of course non-singular since the mapping $\varkappa \circ (\varkappa')^{-1}$ is a diffeomorphism. Furthermore, the vectors X_i are linearly independent. The easiest way to verify this is to act with vectors X_i on functions $M \ni p \to \varkappa(p)^i \in R$. As a result we get $X_i \varkappa^j = \delta_i^j$, which is just enough to prove linear independence of the system. We have thus proved

PROPOSITION 1.2.2. *The system $\{X_i\}$ of tangent vectors given by a chart $(\varkappa, U)$ by formula* (1.2.2) *forms a basis for the tangent space $T_m(M)$. The bases $\{X_i\}$ and $\{X_i'\}$ corresponding to charts $\varkappa$ and $\varkappa'$ are related by a right action of the Jacobi matrix*

$$g_i^j = \frac{\partial(\varkappa \circ (\varkappa')^{-1})^j}{\partial x^i} (\varkappa'(m)), \tag{1.2.5}$$

the action being defined by the formula

$$X_i' = \sum_{j=1}^{n} g_i^j X_j. \tag{1.2.6}$$

Frame Bundle

We shall now carry out the construction of a space canonically connected with differentiable manifolds—the so called *frame bundle*. This bundle is useful in introducing the concept of vector fields, tensor fields and differential forms on a manifold.

The frame bundle denoted in the sequel by $\boldsymbol{B}(M)$, is a collection of tuples $b = (m;\ X_1, X_2, \ldots, X_n)$, where $m \in M$ and $\{X_i\}_1^n$ is a basis for the space $T_m(M)$. In this space we define a projection $\pi\colon \boldsymbol{B}(M) \to M$ by the formula

$$\pi(b) := m,$$

together with a right action of the full matrix group:

$$b \cdot g := (m; g \cdot X_1, \ldots, g \cdot X_n), \tag{1.2.7}$$

where $g = [g_i^j]$ and $g \cdot X_i = \sum_{j=1}^{n} g_i^j X_j$.

A differentiable manifold structure can be defined on $\boldsymbol{B}(M)$ to make π a smooth function and the pair $(G, \boldsymbol{B}(M))$—a transformation group.

To prove this, fix a chart $(\varkappa, U)$ on a manifold M together with the natural coordinates g_j^i in $\mathrm{GL}(n, \boldsymbol{R})$. Let $\varkappa^i$ be a function on U assigning to a point $m \in M$ the value of the i-th coordinate of the point $\varkappa(m)$. First we define a mapping $\pi^{-1}(U) \to \varkappa(U) \times \mathrm{GL}(n, \boldsymbol{R})$: $\pi^{-1}(U) \ni b = (m;\ X_1, \ldots, X_n) \to \big(x^1(m), \ldots, x^n(m);\ [X_j(x^i)]\big) \in \varkappa(U) \times \mathrm{GL}(n, \boldsymbol{R})$. A system of charts determining the differentiable structure on the frame bundle is defined by composing the above mapping with the charts for the manifold.

In the sequel, by a frame bundle we shall always mean $\boldsymbol{B}(M)$ endowed with the differentiable structure, together with the transformation π and the action of the group $\mathrm{GL}(n, \boldsymbol{R})$.

Example 6

Let $O \subset \boldsymbol{R}^n$ be open. Define an action of the group $\mathrm{GL}(n, \boldsymbol{R})$ on $O \times \mathrm{GL}(n, \boldsymbol{R})$ as the right action on the second factor, i.e. $(x, g) \cdot h$

$:= (x, gh)$, $x \in O$, $g, h \in \mathrm{GL}(n, \boldsymbol{R})$. Let $\pi_1 \colon O \times \mathrm{GL}(n, \boldsymbol{R}) \to O$ denote the canonical projection of the product on the first factor. We show how to define a diffeomorphism ψ of the frame bundle $B(O)$ onto $O \times \mathrm{GL}(n, \boldsymbol{R})$ satisfying

$$\pi_1 \circ \psi = \pi, \qquad \psi(b \cdot g) = \psi(b) \cdot g,$$
$$\text{for all} \quad b \in B(O), \quad g \in \mathrm{GL}(n, R). \tag{$*$}$$

For $x \in O$, set $D_i(x)$ to be the tangent vector corresponding to differentiation in the direction of the i-th axis, i.e. $D_i(x)f := \dfrac{\partial f}{\partial x^i}(x)$. The vectors $D_1(x), \ldots, D_n(x)$ form a basis for the space $T_x(O)$; hence an arbitrary basis $(X_i(x))_1^n$ for this space gives rise to a unique matrix $g(X_1, \ldots, X_n) = [g_i^j] \in \mathrm{GL}(n, \boldsymbol{R})$ such that $X_i(x) = \sum_{j=1}^{n} g_i^j D_j(x)$, $i = 1, \ldots$ $\ldots, n$. It is easy to check that the mapping: $B(O) \ni (x;\ X_1(x), \ldots, X_n(x)) \to (x;\ g(X_1, \ldots, X_n)) \in O \times \mathrm{GL}(n, \boldsymbol{R})$ is smooth and satisfies the required conditions (also cf. Problem 11). In the case where the frame bundle $B(M)$ of an n-dimensional manifold M is diffeomorphic to the product $M \times \mathrm{GL}(n, \boldsymbol{R})$ in such a way that identities ($*$) are satisfied, the bundle $B(M)$ is called *trivial*.

Suppose a representation of the group $\mathrm{GL}(n, \boldsymbol{R})$ acts on a linear space V. We define an action of this group on the Cartesian product $B(M) \times V$ as follows:

$$(b, v) \cdot g = (b \cdot g, g^{-1} \cdot v).$$

By a bundle over M with (a typical) fibre V, denoted by $\boldsymbol{V}$, we mean the set of orbits of the group $G = \mathrm{GL}(n, \boldsymbol{R})$ in the space $B(M) \times V$. The projection in the bundle $\boldsymbol{V}$ is defined by the formula $p((b, v)G) := \pi(b) \in M$.

The set $p^{-1}(m)$ is called the *fibre over a point* $m \in M$. Any fibre can be bijectively mapped onto the space V. Let us check that the mapping $V \ni v \to (b, v)G \in \boldsymbol{V}$ is a bijection onto $p^{-1}(m)$ for an arbitrary fixed b satisfying $\pi(b) = m$. If $(b', v') \in p^{-1}(m)$, then there exists a $g \in \mathrm{GL}(n, \boldsymbol{R})$ such that $b' = b \cdot g$, and in that case $(b', v') = (b, g^{-1}v')g$, so $(b', v')G = (b, g^{-1}v')G$. Thus it has turned out that the mapping in question is surjective, and its injectiveness is obvious. Now we equip every fibre with the linear space structure carried over from V. Next, we describe a differentiable structure on the space V. Fix a chart $\varkappa$ on $U \subset M$. We have already come across a mapping which sends the set $\pi^{-1}(U)$ onto $\mathrm{GL}(n, \boldsymbol{R})$:

$$\chi\colon (m; X_1, \ldots, X_n) \to [X_i(x^j)] \in \mathrm{GL}(n, \boldsymbol{R}).$$

By means of χ we define a mapping from $p^{-1}(U)$ onto $U \times V$: $p^{-1}(U) \ni (b, v)G \to \big(\pi(b), \chi(b)v\big) \in U \times \mathrm{GL}(n, \boldsymbol{R})$. The reader will easily check that this mapping is 1-1 and onto. Composing this mapping with the charts for the manifold $U \times \mathrm{GL}(n, \boldsymbol{R})$ we define a differentiable structure on V.

Let us repeat the above construction in the case where $V = \boldsymbol{R}^n$ and the action of $g \in \mathrm{GL}(n, \boldsymbol{R})$ is the natural action of a matrix

$$\boldsymbol{R}^n \ni (x^1, \ldots, x^n) \to \Big(\sum_{i=1}^{n} g_i^1 x^i, \ldots, \sum_{i=1}^{n} g_i^j x^i, \sum_{i=1}^{n} g_i^n x^i\Big).$$

We can easily verify that the fibre of V over $m \in M$ is isomorphic to the space tangent to M at the point m. The isomorphism is given by $p^{-1}(m) \ni \big((m; X_1, \ldots, X_n), (t^1, \ldots, t^n)\big)G \to \sum_{i=1}^{n} t^i X_i \in T_m(M)$. This mapping is well defined since the action of $g \in \mathrm{GL}(n, \boldsymbol{R})$ on $\boldsymbol{R}^n$ is transposed to the action on the space of bases. Surjectiveness is obvious, and as the dimensions of the fibre and of the image are equal, injectiveness follows.

Thus we have returned to a description of tangent vectors, applying for this purpose the frame bundle. But now, in addition, a differentiable structure is defined on the set of all tangent vectors at all points in a manifold. This space is called the *tangent bundle* over a manifold M and denoted by $T(M)$. The general construction provides further examples of vector bundles of the greatest importance.

Example 7

The group $\mathrm{GL}(n, \boldsymbol{R})$ acts on $\boldsymbol{R}^{n*}$—the dual space of $\boldsymbol{R}^n$—by a representation contragredient to the natural representation in R^n, i.e. according to the formula

$$\langle g \cdot w, v\rangle := \langle w, g^{-1} \cdot v\rangle, \qquad w \in \boldsymbol{R}^{n*},\ v \in \boldsymbol{R}^n.$$

Identifying $\boldsymbol{R}^{n*}$ with $\boldsymbol{R}^n$ by the natural duality form

$$\boldsymbol{R}^n \times \boldsymbol{R}^n \ni (w, v) \to \langle w, v \rangle = \sum_{i=1}^{n} w_i v^i \in \boldsymbol{R},$$

we can express this action by the formula $g \cdot w = (g^{-1})^t w$, where the right-hand side is the multiplication of a matrix by a vector. The bundle over M with a typical fibre $\boldsymbol{R}^n$ given by the representation of $\mathrm{GL}(n, \boldsymbol{R})$ contragredient to the natural one is called a *bundle of* 1-*forms* (*cotangent bundle*) and denoted by $T^*(M)$. Below we construct an isomorphism of the fibre $T_m^*(M) := p^{-1}(m)$ of this bundle onto the space $T_m(M)^*$—the dual space of $T_m(M)$.

Example 8

Consider a representation of the group $\mathrm{GL}(n, \boldsymbol{R})$ in the space $T_s^r := \underbrace{\boldsymbol{R}^n \otimes \ldots \otimes \boldsymbol{R}^n}_{r} \otimes \underbrace{\boldsymbol{R}^{n*} \times \ldots \times \boldsymbol{R}^{n*}}_{s}$ of tensors of type (r, s) given on the simple tensors by the formula

$$\begin{aligned} & g \cdot (v_1 \otimes \ldots \otimes v_r \otimes w_1 \otimes \ldots \otimes w_s) \\ & := g v_1 \otimes \ldots \otimes g v_r \otimes g w_1 \otimes \ldots \otimes g w_s \end{aligned}$$

where g acts on vectors in $\boldsymbol{R}^n$ by the natural representation while the action on forms is given by the representation contragredient to the natural one. We extend this transformation to the whole space by linearity. The bundle over M with typical fibre T_s^r is called the *bundle of tensors of type* (r, s) and denoted by $T_s^r(M)$. Also, as in the previous case, an isomorphism of the fibre $(T_s^r)_m(M)$ of this bundle onto the space $T_m(M) \otimes \ldots \otimes T_m(M) \otimes T_m^*(M) \otimes \ldots \otimes T_m^*(M)$ can be contsructed. The details are left to the reader.

Example 9

Recall that if a group G acts on a vector space V by $v \to gv$, then this action extends in a natural way, preserving exterior products and the homogeneous spaces $\bigwedge^k V$, to an action on the Grassmann algebra $\bigwedge V$. Thus on the simple vectors of order k we have $g(v_1 \wedge \ldots \wedge v_k) := g v_1 \wedge \ldots \wedge g v_k$.

Applying the above, in the case $V = \boldsymbol{R}^{n*}$, to the action of $\mathrm{GL}(n, \boldsymbol{R})$ on V contragredient to the natural representation in $\boldsymbol{R}^n$, we get representations of the group $\mathrm{GL}(n, \boldsymbol{R})$ in the spaces $\bigwedge^k \boldsymbol{R}^{n*}$ and $\bigwedge \boldsymbol{R}^{n*}$. The bundle with typical fibre $\bigwedge \boldsymbol{R}^{n*}$, $\bigwedge^k \boldsymbol{R}^{n*}$ constructed according to the general procedure is called the *bundle of forms* (*k-forms*) and denoted by $\bigwedge T^*(M)$ ($\bigwedge^k T^*(M)$ resp.). The fibres of this bundle are isomorphic to $\bigwedge T^*_m(M)$ ($\bigwedge^k T^*_m(M)$ resp.). For $V = \boldsymbol{R}^n$ and the representation of $\mathrm{GL}(n, \boldsymbol{R})$ in $\bigwedge^k \boldsymbol{R}^n$ defined as above, the bundle obtained by the general procedure is the so-called *bundle of k-vectors* over M, denoted by $\bigwedge^k T(M)$.

Extending this definition, we can define the bundle of forms with values in a vector space W. This would be the bundle with typical fibre $\bigwedge^k \boldsymbol{R}^{n*} \otimes W$ and the action on a fibre defined in a natural way: $g \cdot (v_1 \wedge \ldots \ldots \wedge v_k \otimes w) := g v_1 \wedge \ldots \wedge g v_k \otimes w$.

Later we show how to describe, in another way, k-forms on manifold M by making use of the naturalduality between $T(M)$ and $T^*(M)$.

Duality between $T(M)$ and $T^*(M)$.

Assume $X \in T_m(M)$, $\omega \in T^*_m(M)$. We can write $X = (b, t)G$, $\omega = (b, v)G$, where b is a basis for $T_m(M)$, $t \in \boldsymbol{R}^n$, $v \in \boldsymbol{R}^{n*}$. Define a function on $T_m(M) \times T^*_m M$ by setting

$$\langle \omega, X \rangle := \langle v, t \rangle,$$

where the right-hand side bracket denotes the natural duality form in the case of $\boldsymbol{R}^n$: $\langle v, t \rangle = \sum_{i=1}^{n} v_i t^i$. Independence of this definition of representatives of the classes X and ω once again follows from the transformational properties of t and v. The fibre $T^*_m(M)$ can thus be identified in a natural way with the dual space of $T_m(M)$.

Definition. By a *section of a bundle* $V = (W, p, M, V)$ over the base M, with the projection p and typical fibre V we mean a mapping s: $M \to W$ such that $ps(m) = m$.

Thus a *section of a bundle* is a mapping from M into W with the value at a point m belonging to the fibre $V_m = p^{-1}(m)$ over that point.

Sections of the bundle $T(M)$ are called *vector fields*, and the sections of $T^*(M)$ are called *1-forms on M*. Both the smooth vector fields and the forms, form linear spaces, denoted by $\mathscr{T}(M)$ and $\mathscr{T}^*(M)$, respectively. On both spaces an action of the algebra $\mathscr{E}(M)$ can be defined uniformly by the formula

$$fs(m) := f(m)s(m), \qquad f \in \mathscr{E}(M),$$

s being a section of the respective bundle.

Returning to the initial definition of tangent vectors, a vector field X can be viewed as a linear mapping of $\mathscr{E}(M)$ into itself:

$$\mathscr{E}(M) \ni f \to Xf \in \mathscr{E}(M),$$

where $Xf(m) := X(m)f$. This mapping satisfies the identity

$$X(ff_1) = (Xf)f_1 + f(Xf_1). \tag{1.2.8}$$

Now, forms acquire an interpretation as $\mathscr{E}(M)$-linear mappings of $\mathscr{T}(M)$ into $\mathscr{E}(M)$; namely for $\omega \in \mathscr{T}^*(M)$ we set

$$\langle \omega, X \rangle (m) := \langle \omega(m), X(m) \rangle \qquad \text{for } X \in \mathscr{T}(M).$$

Then the following formula holds:

$$\langle \omega, f(X+X_1) \rangle = f(\langle \omega, X \rangle + \omega \langle \omega, X_1 \rangle)$$
$$\text{for arbitrary} \quad X, X_1 \in \mathscr{T}(M) \quad \text{and} \quad f \in \mathscr{E}(M).$$

Next, the duality so defined can be extended to sections of tensor bundles, which is the case described in the following examples.

Example 10

A chart $(\varkappa, U)$ on M, as we have seen before, defines for each $m \in U$ a basis for the space $T_m(M)$ given by formula (1.2.2), i.e. $X_i(m)f := \dfrac{\partial f \circ \varkappa^{-1}}{\partial x^i}(\varkappa(m))$. The assignement $U \ni m \to X_i(m) \in T_m(M)$ is thus a smooth section of the bundle $T(M)$, however, defined not on the whole of M, but only on the domain U of the chart. Sections (resp. smooth sections) of the bundle $T(M)$, defined on an open subset U are called (*smooth*) *vector fields* on U. Every vector field on U can be uniquely represented as a combination of the fields X_i, namely for $X \in \mathscr{T}(U)$ we have $X = \sum_{i=1}^{n} f^i X_i$, where $f^i = X\varkappa^i \in \mathscr{E}(U)$. The vector fields X_i will be called *base vector fields determined by the chart* $(\varkappa, U)$. Define 1-forms $\omega^1, \ldots, \omega^n$ on U (i.e. sections of the bundle $T^*(M)$ over U) which form the dual basis to the base X_i. Thus the forms ω^k are linear transformations $\mathcal{T}(U) \to \xi(U)$ given for $X = \sum_{i=1}^{n} f^i X_i \in \mathcal{T}(U)$ by the formula $\omega^k X = f^k$, $k = 1, \dots, n$. Just as in the case of vector fields X_i, which form a basis for $\mathcal{T}(U)$, 1-forms ω^i are a basis for $\mathcal{T}^*(U)$, and each form $\omega \in \mathcal{T}^*(U)$ can be written uniquely as $\omega = \sum_{i=1}^{n} h_i \omega^i$, $h_i \in \xi(U)$ with $h_i = \omega(X_i) := \langle w, X_i \rangle$.

Example 11

(a) The bundles $T^r_s(M)$ and $T^s_r(M)$ are dual to each other. In order to construct the duality take

$$(T^r_s)_m(M) \ni t = X_1 \otimes \dots \otimes X_r \otimes W_1 \otimes \dots \otimes W_s,$$
$$X_i \in T_m(M), \qquad W_i \in T^*_m(M),$$
$$(T^s_r)_m(M) \ni u = Y_1 \otimes \dots \otimes Y_s \otimes Z_1 \otimes \dots \otimes Z_r,$$
$$Y_i \in T_m(M), \qquad Z_i \in T^*_m(M),$$

and set $\langle t, u \rangle := \langle Z_1, X_1 \rangle \cdot \dots \cdot \langle Z_r, X_r \rangle \cdot \langle W_1, Y_1 \rangle \cdot \dots \cdot \langle W_s, Y_s \rangle$, where the symbol $\langle \cdot, \cdot \rangle$ on the right-hand side of this formula stands for the duality, defined above, between $T_m(M)$ and $T^*_m(M)$. For any sections $\tau\colon M \to T^r_s(M)$, $\sigma\colon M \to T^s_r(M)$ of these bundles we define $\langle \tau, \sigma \rangle$ to be a function in $\mathscr{E}(M)$ given by the formula

$$\langle \tau, \sigma \rangle (m) := \langle \tau(m), \sigma(m) \rangle .$$

For $f \in \mathscr{E}(M)$ we easily check that $\langle f\tau, \sigma \rangle = f\langle \tau, \sigma \rangle = \langle \tau, f\sigma \rangle$.

(b) Differential k-forms can be regarded as mappings of $\mathcal{T}(M) \times \dots \times \mathcal{T}(M)$ into $\mathscr{E}(M)$, $\mathscr{E}(M)$-linear in each argument and alternating, i.e. changing sign after a transposition of any pair of arguments. Hints for the proof that this characterization of k-forms is in fact a "nice" one, will be found in Problem 15.

The Lie Algebra of Vector Fields

With vector fields regarded as mappings of the space $\mathscr{E}(M)$ into itself it makes sense to consider the composition of vector fields. Such a mapping, however, is not a vector field since identity (1.2.8) is no longer satisfied. Still it is satisfied by a mapping called the *commutator* or the *Lie bracket* given by

$$[X, Y] := XY - YX. \tag{1.2.9}$$

We check that $[X, Y]$ is indeed a vector field:

$$\begin{aligned} XY(fh)-YX(fh) &= X\big((Yf)h+f(Yh)\big)-Y\big((Xf)h+f(Xh)\big) \\ &= (XYf)h+(Yf)(Xh)+(Xf)(Yh)+(XYh)f-(YXf)h \\ &\quad -(Xf)(Yh)-(Yf)(Xh)-(YXh)f = ([X,Y]f)h+([X,Y]h)f, \end{aligned}$$

which was to be proved.

The following properties of the commutator follow immediately from the definition:

$$[X, Y] = -[Y, X], \tag{1.2.10}$$

$$[\alpha X+\beta Y, Z] = \alpha[X, Z]+\beta[Y, Z], \qquad \alpha, \beta \in \boldsymbol{R}, \tag{1.2.11}$$

$$[X, [Y, Z]]+[Z, [X, Y]]+[Y, [Z, X]] = 0. \tag{1.2.12}$$

The last identity is known as the *Jacobi identity*. Here we come across a special case of a new algebraic structure.

DEFINITION. A vector space A over $\boldsymbol{R}$ or $\boldsymbol{C}$ in which an operation $[\cdot, \cdot]: A\times A \to A$ is defined satisfying conditions (1.2.10)–(1.2.12) is called a *Lie algebra*.

Example 12

The vector space $M_n(\boldsymbol{K})$ of square matrices of degree n over $\boldsymbol{K}$ provides an important example of a Lie algebra with the commutator defined as $[A, B] := AB-BA$, where AB denotes the product of matrices. The reader can easily check that conditions (1.2.10)–(1.2.12) hold for the commutator defined above. $M_n(\boldsymbol{K})$ together with the Lie algebra structure will be denoted by $\mathfrak{gl}(n, \boldsymbol{K})$ owing to a close connection with the group $\mathrm{GL}(n, \boldsymbol{K})$ (cf. § 1.3, Example 2).

More generally, suppose V is a vector space over the field $\boldsymbol{K}$. We equip the space $L(V)$ of linear mappings of V into itself with a Lie algebra structure by defining, for arbitrary $X, Y \in L(V)$, their commutator as $[X, Y] := X\circ Y-Y\circ X$, where $\circ$ on the right denote composition. For this Lie algebra we use the notation $\mathfrak{gl}(V)$—by choosing a base we can map it isomorphically onto the Lie algebra $\mathfrak{gl}(n, \boldsymbol{K})$ with $n = \dim V$.

By a homomorphism of a Lie algebra A into A_1 we mean a linear mapping: $A \to A_1$ satisfying $A[X, Y] = [AX, AY]$.

A subalgebra of A is a linear subspace $A_1 \subset A$ closed under the operation $[\cdot, \cdot]$. A subspace $A_1 \subset A$ such that $[X, Y] \in A_1$ for arbitrary $X \in A$, $Y \in A_1$ is called an *ideal*.

Example 13

Suppose $\mathfrak{a}$ is a Lie algebra and fix $X \in \mathfrak{a}$. Then the mapping $\mathfrak{a} \ni Y \to [X, Y] \in \mathfrak{a}$ is a linear mapping of the vector space $\mathfrak{a}$ into itself. Denoting this mapping by the symbol $\mathrm{ad}(X)$, we can write the above definition in the form $\mathrm{ad}(X)Y = [X, Y]$. The reader can easily check that the mapping $\mathfrak{a} \ni X \to \mathrm{ad}(X) \in L(\mathfrak{a})$ is linear and that the Jacobi identity implies

$$\mathrm{ad}([X, Y]) = \mathrm{ad}(X) \circ \mathrm{ad}(Y) - \mathrm{ad}(Y) \circ \mathrm{ad}(X),$$

which means that ad is a homomorphism of the Lie algebra $\mathfrak{a}$ into the Lie algebra $\mathfrak{gl}(\mathfrak{a})$.

$\mathrm{Ker\,ad} = \{x \in \mathfrak{a}\colon \mathrm{ad}(X) = 0\}$ is the set of elements commuting with all elements of the algebra $\mathfrak{a}$. It is called the *centre* of the algebra $\mathfrak{a}$ and is an ideal in $\mathfrak{a}$.

A derivation of a Lie algebra $\mathfrak{g}$ is an endomorphism $D\colon \mathfrak{g} \to \mathfrak{g}$ such that $D[X, Y] = [DX, Y] + [X, DY]$ for all $X, Y \in \mathfrak{g}$. Owing to the Jacobi identity (1.2.12) the endomorphisms $\mathrm{ad}(X)$, for $X \in \mathfrak{g}$ are derivations of the Lie algebra $\mathfrak{g}$—they are called *inner derivations*.

The Tangent Mapping

Let $\Phi\colon M \to N$ be a smooth mapping of a manifold M into a manifold N. Such a mapping induces a smooth mapping $\mathrm{d}\Phi$ of the tangent bundle $T(M)$ into $T(N)$ such that a fibre $T_m(M)$ is mapped into $T_{\Phi(m)}(N)$. Namely, for $X \in T_m(M)$ and $f \in \mathscr{E}(\Phi(m))$ we define

$$\mathrm{d}\Phi_m(X)f = X(f \circ \Phi). \tag{1.2.13}$$

By direct calculations we check that composition of differentiable mappings implies composition of the respective tangent mappings. Thus if $\psi\colon N \to P$ then

$$\mathrm{d}(\psi \circ \Phi)_m = \mathrm{d}\psi_{\Phi(m)} \circ \mathrm{d}\Phi_m. \tag{1.2.14}$$

In the case where Φ is a diffeomorphism we can assign to the tangent mapping $\mathrm{d}\Phi$ a linear operator of $\mathscr{T}(M)$ into $\mathscr{T}(N)$. Given a vector field $X \in \mathscr{T}(M)$, we define a field $\mathrm{d}\Phi(X) \in \mathscr{T}(N)$ by setting

$$\mathrm{d}\Phi(X)\big(\Phi(m)\big) = \mathrm{d}\Phi_m\big(X(m)\big). \tag{1.2.15}$$

On the other hand, for an arbitrary smooth mapping we can distinguish the pairs of vector fields on M and N, respectively, related by the tangent transformation. Vector fields $X \in \mathscr{T}(M)$ and $Y \in \mathscr{T}(N)$ are said to be *Φ-related* if for each $m \in M$

$$d\Phi_m(X(m)) = Y(\Phi(m)).$$

If this is the case, we also write $d\Phi(X) = Y$. Now let the pairs X, Y and X_1, Y_1 of vector fields be Φ-related. Then

$$d\Phi[X, X_1] = [Y, Y_1] = [d\Phi(X), d\Phi(Y)]. \qquad (1.2.16)$$

This property follows immediately from the definition of the differential and that of the commutator.

Now we assume Φ to be surjective. Formula (1.2.16) means that the set of vector fields on M possessing Φ-related counterparts on N forms a Lie subalgebra of the algebra $\mathscr{T}(M)$ and the mapping $d\Phi$ is a homomorphism of this subalgebra into $\mathscr{T}(N)$. If, on the other hand, Φ is a diffeomorphism of M onto itself, then, as is shown by formula (1.2.16), $d\Phi^{-1}$ is the inverse of $d\Phi$, which means that $d\Phi$ is an automorphism of the Lie algebra $\mathscr{T}(M)$.

Let us return to the general case of a smooth mapping $\Phi\colon M \to N$. This mapping also induces a linear operator on the space of differential forms on the manifold N.

To $\omega \in \vartheta(N)$ we assign a form $\Phi^*\omega \in \vartheta(M)$ as follows. At a point $m \in M$, its action on $X(m) \in T_m(M)$ is defined by:

$$\langle \Phi^*\omega(m), X(m)\rangle = \langle \omega(\Phi(m)), d\Phi_m(X(m))\rangle.$$

Then for $\Phi\colon M \to N$ and $f\colon N \to P$ we have

$$(\psi \circ \Phi)^* = \Phi^* \circ \psi^*. \qquad (1.2.17)$$

Example 14

Let $f \in \mathscr{E}(M)$ be an arbitrary point in a manifold M and $X \in T_m(M)$. Regarding f as a smooth mapping of M into $\boldsymbol{R}^1$ we have for an arbitrary function $g \in \mathscr{E}(f(m))$ the identity

$$df_m(X)g = X(g \circ f).$$

Every vector Y in the space $T_{f(m)}(\boldsymbol{R}^1)$ can be represented in the form $Y = y\left.\frac{d}{dt}\right|_{t=f(m)}$, where $y \in \boldsymbol{R}^1$ and $\frac{d}{dt}$ denotes differentiation with respect to the natural coordinate in $\boldsymbol{R}^1$. Since $y = Y(\mathrm{id})$, where id $:\boldsymbol{R}^1 \to \boldsymbol{R}^1$ is defined by $\mathrm{id}(t) = t$ we obtain $\mathrm{d}f_m(X) = X(f)\left.\frac{\mathrm{d}}{\mathrm{d}t}\right|_{t=f(m)}$.

On the other hand, if $\omega \in T^*_{f(m)}(\boldsymbol{R}^1)$ satisfies $\left\langle \omega, \left.\frac{\mathrm{d}}{\mathrm{d}t}\right|_{t=f(m)} \right\rangle = 1$, then $\langle f^*\omega, X\rangle = \langle \omega, \mathrm{d}f_m(X)\rangle = X(f)$; so $f^*\omega$ is an element of $T^*_m M$ assigning to a vector X the number $X(f)$. Since, by tradition, the tangent space to $\boldsymbol{R}^1$ is identified with $\boldsymbol{R}^1$ itself via the mapping $Y \to \omega(Y)$, the form $f^*\omega$ is denoted by the symbol $\mathrm{d}f(m)$ and called the *differential* of f. The reader will easily note that the forms ω^i defined in Example 10 are nothing but the forms $\mathrm{d}\varkappa^i$—traditionally denoted by $\mathrm{d}x^i$.

The operators $\mathrm{d}\Phi$ and Φ^* can also be defined, respectively, on tensor fields $\mathscr{T}^r(M)$ and $\mathscr{T}_s(M)$ of higher orders. Namely set

$$\mathrm{d}\Phi_m(X_1 \otimes \ldots \otimes X_r) = \mathrm{d}\Phi_m X_1 \otimes \ldots \otimes \mathrm{d}\Phi_m X_r \tag{1.2.18}$$

for $X_1, \ldots, X_r \in T_m(M)$. The formula defines an action $\mathrm{d}\Phi_m\colon \underset{r}{\otimes}\, T_m(M) \to \underset{r}{\otimes}\, T_{\Phi(m)}(M)$. Whenever $\mathrm{d}\Phi$ is well defined on the space $\overset{r}{\mathscr{T}}(M)$, we can set

$$\mathrm{d}\Phi(\tau)\big(\Phi(m)\big) = \mathrm{d}\Phi_m\big(\tau(m)\big) \qquad \text{for} \qquad \tau \in \mathscr{T}^r(M). \tag{1.2.19}$$

Similarly

$$\langle \Phi^*\omega, \tau\rangle(m) := \langle \omega(m), \mathrm{d}\Phi_m\big(\tau(m)\big)\rangle$$
$$\text{for} \qquad \omega \in \mathscr{T}_s(M), \ \tau \in \mathscr{T}^s(M).$$

DEFINITION. A vector field $\tau \in \mathscr{T}^r(M)$ ($\omega \in \mathscr{T}_s(M)$, resp.) is called *invariant* with respect to a diffeomorphism Φ of a manifold M provided $\mathrm{d}\Phi(\tau) = \tau$ ($\Phi^*\omega = \omega$, resp.).

Example 15

Let $\boldsymbol{T}^n = \boldsymbol{R}^n/\boldsymbol{Z}^n$ be the n-dimensional torus and let $p\colon \boldsymbol{R}^n \to \boldsymbol{T}^n$ be the natural projection. For $a = (a_1, \ldots, a_n) \in \boldsymbol{Z}^n$ define a translation $t_a\colon \boldsymbol{R}^n \ni x \to (x+a) \in \boldsymbol{R}^n$ and observe that t_a is a diffeomorphism of $\boldsymbol{R}^n$. We shall prove that if X is a vector field on $\boldsymbol{R}^n$, then there exists a vector field Y on $\boldsymbol{T}^n$ p-related to X if, and only if, the field X is invariant under all t_a for all $a \in \boldsymbol{Z}^n$.

Here, only the sufficiency of this condition will be proved while the proof of its necessity is left to the reader as an exercise (Exercise 16).

Define a vector field Y by the formula $Y(s) = \mathrm{d}p_x(X(x))$, where $s \in \boldsymbol{T}^n, x \in \boldsymbol{R}^n$ and $p(x) = s$. Since $p(y) = p(x)$ if, and only if, $y = t_a(x)$ for some $a \in \boldsymbol{Z}^n$, the assumption of invariance of X ensures that Y is well defined. To show that Y is smooth note that for each point $s \in \boldsymbol{T}^n$ and any $x \in \boldsymbol{R}^n$ such that $p(x) = s$ there is an open neighbourhood $U \ni x$ such that $p|_U$ is a diffeomorphism of U onto $p(U)$; in fact, we can choose $(p|_U)^{-1}$ as a chart with the domain $p(U)$. Hence, it follows that the restriction of the field Y to the open neighbourhood $p(U)$ of the point s is smooth, thus in view of the arbitrariness of s the field Y itself is smooth.

Submanifolds

DEFINITION. A subset N of a manifold M is said to be a *submanifold* provided N is equipped with a manifold structure and the differential of the natural inclusion mapping $i: N \to M$ is injective at every point $n \in N$.

One-Parameter Transformation Groups

A curve $\gamma:]a, b[\to M$ defines a tangent vector at any point in the manifold M through which the curve passes. Write

$$\gamma'(t)f = \frac{\mathrm{d}}{\mathrm{d}t}(f \circ \gamma)(t). \tag{1.2.20}$$

The functional $\gamma'(t)$ is a tangent vector to M at the point $\gamma(t)$. Let $X \in \mathscr{T}(M)$. A smooth curve γ is said to be an *integral curve of the field* X if $\gamma'(t) = X(\gamma(t))$.

By a one-parameter transformation group of a manifold M we mean an action $\boldsymbol{R} \times M \ni (t, m) \to \varphi(t, m) \in M$, smooth in both variables making $(\boldsymbol{R}, \varphi)$ a transformation group of the set M.

In the sequel we shall use the notation

$$\varphi_t: M \ni m \to \varphi(t, m) \in M, \qquad \varphi_m: t \to \varphi(t, m) \in M.$$

For a fixed point $m \in M$ the mapping $\varphi_m: \boldsymbol{R} \to M$ is a smooth curve. Thus a transformation group G induces a vector field on M according to the formula $X_\varphi(m) = \varphi'_m(0)$. This field is called *the infinitesimal generator of the group G*. Owing to the group property of the function φ, the vector $X_\varphi(\varphi(t, m))$ is tangent to the curve φ_m at the point $\varphi(t, m)$. This will become clear by differentiating at the point $s = 0$ the identity

$$\varphi_{t+s}(m) = \varphi_s(\varphi(t, m)).$$

In other words, the curve φ_n is the integral curve of the field X. Thus we have seen that there is a close relation between the following two problems:

(1) Does there exist for every vector field $X \in \mathcal{T}(M)$ a one-parameter subgroup φ such that $X = X_\varphi$?

(2) Does there exist for every vector field $X \in \mathcal{T}(M)$ and a point $m \in M$ an integral curve of X passing through the point m?

The theorem quoted below gives an answer to both questions, fully explaining at the same time the relation between them.

THEOREM 1.2.3 (cf. Lang, 1972). *Let X be a vector field on M. There exists an open subset $D \subset \boldsymbol{R} \times M$ and a smooth mapping $\varphi: D \to M$ satisfying*

(1) $\{0\} \times M \subset D$,

(2) *if* $(t, m) \in D$ *and* $(s, \varphi_t(m)) \in D$ *then also* $(t+s, m) \in D$ *and* $\varphi_{t+s}(m) = \varphi_s(\varphi_t(m))$,

(3) *for each t the set $D_t = \{m \in M\colon (t, m) \in D\}$ is open in M and $\varphi_t\colon D_t \to M$ is a diffeomorphism of D_t onto D_{-t} with inverse φ_{-t},*

(4) *for every $m \in M$ the mapping φ_m is the maximal curve of the field X, such that $\varphi_m(0) = m$.*

This theorem implies a positive answer to question (2). Thus every point in the manifold lies on some integral curve of the field X. However, in the general case, the integral curves are defined only on some open interval and cannot be extended to the whole axis $\boldsymbol{R}$. Therefore the mapping φ described in the theorem does not form a one-parameter transformation group of the manifold. Because of the group properties stated in (2) and (3), mappings of this kind are called *local one-parameter groups*. Thus, while every one-parameter group induces a vector field, some vector fields define only local groups. Vector fields having integral curves defined on the whole axis (i.e. vector fields such that $D = \boldsymbol{R} \times M$) are called *complete vector fields*.

Example 16

Examples of non-complete vector fields will be found in Problem 18. We shall now establish a sufficient condition for the completeness of a vector field. Namely, we shall prove that every smooth vector field on a compact manifold is complete, in fact somewhat more generally, that every field with compact support is complete. The support of a field is the closure of the set of points at which the field is non-zero.

First, note that if a field vanishes at a point on an integral curve, then it vanishes on the whole curve; thus the integral curves passing through a point at which the field is non-zero lie entirely in the support of the field.

One the other hand, by Theorem 1.2.3 every point $m \in M$ has an open neighbourhood V such that for a certain $\delta > 0$ we have $]-\delta, \delta[\times V \subset D$, i.e. the integral curve passing through an arbitrary point $n \in V$ is defined at least on the interval $]-\delta, \delta[$. Thus it follows that for any compact set $K \subset M$ there exists an $\varepsilon > 0$ such that every integral curve passing through a point in K is defined on the interval $]-\varepsilon, \varepsilon[$ at least. Now we easily conclude that $\boldsymbol{R}^1$ is the domain of any integral curve contained in the support of the field. The case of an integral curve with the zero vector field is left to the reader as an exercise (Problem 20).

PROBLEMS

1. (a) Prove that every atlas for a topological space M is contained in a unique maximal atlas, and thus it defines fully a differentiable manifold structure on M.

(b) Let $\varkappa_1$ be the identity chart on $\boldsymbol{R}^1$ (i.e. $\varkappa_1(x) = x$) and let $\varkappa_2$: $\boldsymbol{R}^1 \ni x \to \varkappa_2(x) := x^3 \in \boldsymbol{R}^1$. Let $\mathscr{A}_1, \mathscr{A}_2$ be the maximal atlases containing $\varkappa_1$ and $\varkappa_2$, respectively. Prove that the manifolds $(\boldsymbol{R}^1, \mathscr{A}_1)$, $(\boldsymbol{R}^1, \mathscr{A}_2)$ are not identical and yet they are diffeomorphic to each other.

(c) Prove that an arbitrary maximal analytic atlas for M is contained in exactly one maximal differentiable atlas for M.

2. (a) Let M, N be differentiable manifolds. Show that there is a unique differentiable structure on $M \times N$ for which the natural projections p_1: $M \times N \to M$, p_2: $M \times N \to N$ are smooth mappings. Prove that the mappings ν_m: $N \ni n \to (m, n) \in M \times N$, ν_n: $M \ni m \to (m, n)$

$\in M \times N$ for $m \in M$, $n \in N$ are smooth. Also prove that the tangent space $T_{(m,n)}(M \times N)$ is isomorphic in a natural way to $T_m(M) \oplus T_n(N)$.

Hint. Consider the images of $T_m(M)$ and $T_n(N)$ under $d\nu_n$ and $d\nu_m$ respectively.

(b) Let M, N, P be differentiable manifolds. Suppose the following conditions are satisfied:

(i) there exist differentiable mappings $p\colon P \to M$, $q\colon P \to N$ such that, for any $m \in M$, q is a bijection of the set $p^{-1}(\{m\})$ onto N and, for any $n \in N$, p is a bijection of the set $q^{-1}(\{n\})$ onto M.

Denote $\nu_m := (q|_{p^{-1}(\{m\})})^{-1}$, $\nu_n := (p|_{q^{-1}(\{n\})})^{-1}$,

(ii) for every $m \in M$ and $n \in N$ the mappings $\nu_m\colon N \to P$ and $\nu_n\colon M \to P$ are smooth and the tangent mappings $d\nu_m$, $d\nu_n$ are injective.

Then prove that the mapping $P \ni x \to (p(x), q(x)) \in M \times N$ is a diffeomorphism.

3. Suppose $\tilde{M}$ is a manifold, M—a topological space and let $p\colon \tilde{M} \to M$ be a local homeomorphism. Show that there exists at most one manifold structure on M such that p is a local diffeomorphism (i.e. for each $\tilde{x} \in \tilde{M}$ and $x = p(\tilde{x})$ there are open neighbourhoods $\tilde{U}$ and U such that $p|_{\tilde{U}}$ is a diffeomorphism of $\tilde{U}$ onto U). In the case where $\tilde{M} = \boldsymbol{R}^n$, $M = \boldsymbol{R}^n/\boldsymbol{Z}^n$ with $p\colon \boldsymbol{R}^n \to \boldsymbol{R}^n/\boldsymbol{Z}^n$ being the natural projection, show that such a differentiable structure on M exists and is identical with that given by the identification of $\boldsymbol{R}^n/\boldsymbol{Z}^n$ with the Cartesian product of n copies of the unit circle S^1.

4. For $k = 1, 2, \ldots, n+1$, write $U_{k,+} = \{x = (x^1, \ldots, x^{n+1}) \in S^n\colon x^k > 0\}$ and $U_{k,-} = \{x = (x^1, \ldots, x^{n+1}) \in S^n\colon x^k < 0\}$ and let $\varkappa_{k,\pm}\colon U_{k,\pm} \ni x \to \varkappa_{k,\pm}(x) := (x^1, \ldots, x^{k-1}, x^{k+1}, \ldots, x^{n+1}) \in \boldsymbol{R}^n$. Show that $\varkappa_{k,\pm}$ are compatible charts defining a manifold structure identical with that described in Example 2.

5. (a) Let $O \subset \boldsymbol{R}^m$ be an open subset and let $f\colon O \to \boldsymbol{R}^k$, $k < m$, be a smooth mapping. Suppose that $f'(x)\colon \boldsymbol{R}^m \to \boldsymbol{R}^k$ is surjective for all $x \in M_c$ whenever the level set $M_c = \{x \in O\colon f(x) = c\}$ is non-empty. Then there exists an $(m-k)$-dimensional differentiable manifold structure on M_c, such that the natural inclusion mapping $i\colon M_c \to O$ is both smooth and constitutes a homeomorphism onto the image (in the induced topology).

(b) Generalize this proposition to the case of a smooth mapping

$f\colon M \to N$ where $\dim N < \dim M$, with the surjective tangent mapping df_m for all m.

Hint. Equip M_c with the topology induced by the Euclidean metric in $\boldsymbol{R}^m$. Assuming that in the Jacobi matrix $\left[\frac{\partial f^j}{\partial x^i}(x_0)\right]_{\substack{i=1,\dots,m\\ j=1,\dots,k}}$ the columns with the indices $m-k+1, \dots, m$ are linearly independent (which can always be obtained by changing the numbering of the coordinates), show that the mapping $\tilde{f}\colon O \to \boldsymbol{R}^{m-k} \times \boldsymbol{R}^k\colon \tilde{f}(x) = (x^1, \dots$ $\dots, x^{m-k}, f(x))$, satisfies the assumptions of the inverse function theorem. Then there exists a neighbourhood U of x_0 such that $f|_{U \cap M_c}$ is a homeomorphism onto a certain open subset $W \subset \boldsymbol{R}^{m-k}$. Prove that every pair of charts defined in this way is compatible.

6. Construct differentiable structures on $\mathrm{SL}(n, \boldsymbol{R})$ and $\mathrm{SO}(n, \boldsymbol{R})$ by using the method from Problem 5.

Hint. Consider, respectively, the function det and the mapping assigning to a matrix the scalar products of the pairs of columns.

7. Let $\pi\colon \boldsymbol{R}^{n+1} \setminus \{0\} \to P^n\boldsymbol{R}$ be the natural projection and denote by p the restriction $\pi|_{S^n}$. Show that p is a smooth surjection and that for every point $m \in P^n\boldsymbol{R}$ there exists an open connected neighbourhood U such that $p^{-1}(U)$ is the sum of two open, disjointed subsets diffeomorphic to U (i.e. S^n is a two-fold covering of $P^n\boldsymbol{R}$).

8. Let M be a differentiable manifold and $m \in M$. Show that for an arbitrary function $f \in \mathscr{E}(m)$ there exists a function $f_1 \in \mathscr{E}(M)$ such that $f|_W = f_1|_W$ for a certain open subset $W \ni m$.

Hint. By taking local charts the problem can be reduced to one concerning an open subset $U \subset \boldsymbol{R}^n$. Apply the function h defined by

$$h(t) := \begin{cases} e^{-\frac{1}{t}}, & t > 0, \\ 0, & t \leqslant 0, \end{cases}$$

to construct a smooth function g_ε on $\boldsymbol{R}^n$ such that

$$g_\varepsilon(x) = \begin{cases} 1, & \|x\| < \varepsilon, \\ 0, & \|x\| > 2\varepsilon. \end{cases}$$

9. Prove Proposition 1.2.1 in the following steps:

(a) For an arbitrary chart $(\varkappa, U)$ in a neighbourhood of $m \in M$ and arbitrary $f \in \mathscr{E}(m)$

$$f(m') = f(m) + \sum_{i=1}^{n} (\varkappa^i(m') - \varkappa^i(m)) X_i(m) f$$

$$+ \sum_{j,i=1}^{n} (\varkappa^i(m') - \varkappa^i(m)) (\varkappa^j(m') - \varkappa^j(m)) f_{ij}(m')$$

for all points m' in a certain neighbourhood of m and for certain smooth functions f_{ij} in that neighbourhood;

(b) If $g \in \mathscr{E}(m)$ is constant on an open neighbourhood of m then for an arbitrary $X(m) \in T_m(M)$, $X(m)g = 0$;

(c) From (a) and (b) derive

$$X(m)f = \sum_{i=1}^{n} (X(m)\varkappa^i) X_i(m) f.$$

Hint. ad (a) Use the Taylor expansion of the function $f \circ \varkappa^{-1}$.

ad (b) Using the properties of differentiation compute $X(m)g$ for a function g equal to one in a neighbourhood of the point m.

10. Let $M = f^{-1}(\{c\})$ stand for the manifold described in Problem 5 and let $(\varkappa, U)$ be a chart in a neighbourhood of a point $m \in M$. Suppose that $\varkappa(m) = 0$. Prove that the image of $\boldsymbol{R}^{m-k}$ under the linear transformation $(\varkappa^{-1})'(0)\colon \boldsymbol{R}^{m-k} \to \boldsymbol{R}^n$ is independent of the choice of the chart and its translation to the point m coincides with the tangent space to the level M as defined in a course of analysis.

11. Denote by r_g the right translation in the frame bundle $B(M)$ for a manifold M ($\dim M = n$), i.e. $r_g(b) := b \cdot g$ for $g \in \mathrm{GL}(n, \boldsymbol{R})$. Let $\pi\colon B(M) \to M$ be the canonical projection. Prove the following statements:

(a) $\pi^{-1}(\{m\})$ is a submanifold of $B(M)$ of dimension n^2;

(b) $r_g\colon B(M) \to B(M)$ is smooth for each g;

(c) Let $(\varkappa, U)$ be a chart on M and $(X_i(m))$ the basis of $T_m(M)$ determined by this chart. Define $s_U\colon U \to B(M)$ by the formula $s_U(m) := (m, X_1(m), \ldots, X_n(m))$. Then s_U is a smooth section of the bundle $B(M)$ over U and the mapping

$$\varphi\colon U \times \mathrm{GL}(n, \boldsymbol{R}) \ni (m, g) \to \varphi(m, g) := r_g s_U(m) \in \pi^{-1}(U)$$

is a diffeomorphism;

(d) Let U, V, W be the domains of some charts on M. Then $s_{UV}: U \cap V \to \mathrm{GL}(n, \boldsymbol{R})$ given by $s_U(m) s_{UV}(m) = s_V(m)$ is a smooth mapping satisfying the identity $s_{UW}(m) = s_{UV}(m) s_{VW}(m)$;

(e) Every mapping f_b: $\mathrm{GL}(n, \boldsymbol{R}) \to B(M)$ defined by the formula $f_b(g) = b \cdot g$ is a diffeomorphism of G onto the fibre containing b. Moreover, for all $b, c \in B(M)$ such that $\pi(b) = \pi(c)$, $f_b^{-1} \circ f_c$ is a left translation in G.

12. (a) Prove that the differentiable structure on $T(M)$ is given by the charts $(\theta_U, p^{-1}(U))$, where for a chart $(\varkappa, U)$ on M θ_U is defined by the formula

$$\theta_U(m, X(m)) := (\varkappa^1(m), \ldots, \varkappa^n(m), X(m)\varkappa^1, \ldots, X(m)\varkappa^n) \in \boldsymbol{R}^{2n}$$

and p is the projection of $T(M)$ onto M.

(b) Prove that a vector field X on M is smooth, if and only if, for every chart $(\varkappa, U)$ on M the functions $U \ni m \to X(m)\varkappa^i$, $i = 1, \ldots, n$ are smooth.

(c) Prove that a vector field X on M is smooth if, and only if, for every function $f \in \mathscr{E}(M)$ the function $M \ni m \to X(m)f \in \boldsymbol{R}$ belongs to $\mathscr{E}(M)$.

13. Prove that the differentiable structure on $T^*(M)$ is given by the charts $(\Xi_U, p^{-1}(U))$, where $(\varkappa, U)$ is a chart on M, and $\Xi_U(m, \omega(m)) := (\varkappa^1(m), \ldots, \varkappa_n(m), \langle\omega(m), X_1(m)\rangle, \ldots, \langle\omega(m), X_n(m)\rangle) \in \boldsymbol{R}^{2n}$, where $(X_i(m))$ is a basis for $T_m(M)$ corresponding to the chart $\varkappa$. Formulate and prove the smoothness conditions for 1-forms on M analogous to those for vector fields given by (b) and (c) in Problem 12.

14. Prove that if an n-dimensional manifold M admits smooth vector fields $X_1, \ldots, X_n$ such that $X_1(m), \ldots, X_n(m)$ is a basis of $T_m(M)$ for every $m \in M$, then the bundle $B(M)$ is diffeomorphic to $M \times \mathrm{GL}(n, \boldsymbol{R})$. Moreover, the diffeomorphism mentioned above can be chosen as an isomorphism of the transformation groups $(B(M), \mathrm{GL}(n, \boldsymbol{R}))$ and $(M \times \mathrm{GL}(n, \boldsymbol{R}), \mathrm{GL}(n, \boldsymbol{R}))$ with the action of $\mathrm{GL}(n, \boldsymbol{R})$ on $M \times \mathrm{GL}(n, \boldsymbol{R})$ given by the right multiplication of the second factor. (In this case the bundle $B(M)$ is said to be *trivializable* and the manifold M—*parallelizable*.)

Also prove that in this case the bundle $T(M)$ is diffeomorphic to the Cartesian product $M \times \boldsymbol{R}^n$ and the space $\mathscr{T}(M)$ is isomorphic to the space $\mathscr{E}(M, \boldsymbol{R}^n)$ of $\boldsymbol{R}^n$-valued smooth mappings on M.

15. (a) Prove that every differentiation of the algebra $\mathscr{E}(M)$ (over $\boldsymbol{R}$), i.e. every linear transformation $X\colon \mathscr{E}(M) \to \mathscr{E}(M)$ satisfying $X(f_1 f_2) = f_1 X(f_2) + X(f_1) f_2$ is a vector field;

(b) Prove that every $\mathscr{E}(M)$-linear transformation of $\mathscr{T}(M)$ into $\mathscr{E}(M)$ is a 1-form;

(c) Let $A_k(\boldsymbol{R}^n)$ be the space of all k-linear antisymmetric forms on the space $\boldsymbol{R}^n$ and define an action of $\mathrm{GL}(n, \boldsymbol{R})$ on this space by means of the formula $g \cdot f(v_1, \ldots, v_k) := f(g^{-1}v_1, \ldots, g^{-1}v_k)$. Prove that the bundle over M with a typical fibre $A_k(\boldsymbol{R}^n)$ is isomorphic to the bundle $\bigwedge^k T^*(M)$ (i.e. there exists a diffeomorphism of one bundle onto the other sending each fibre onto the fibre over the same point of the other bundle and being linear on fibres). Prove that sections of this bundle are mappings on $\mathscr{T}(M) \times \ldots \times \mathscr{T}(M)$ which are simultaneously antisymmetric and $\mathscr{E}(M)$-linear with respect to each argument.

Hint. ad (a) Define the value of a differentiation X at a point $m \in M$ by the formula $X(m)f := Xf(m)$ and prove that $X(m)$ extends to a tangent vector at m by using the method of the extension of functions in $\mathscr{E}(m)$ to the whole of M described in Problem 5.

ad (b) Prove that for an arbitrary $X(m) \in T_m(M)$ there exists a field $\tilde{X} \in \mathscr{T}(M)$ such that $X(m) = \tilde{X}(m)$ and then proceeds as in (a).

ad (c) Show an (algebraic) isomorphism of $\bigwedge^k \boldsymbol{R}^{n*}$ onto $A_k(\boldsymbol{R}^n)$ by making use of the canonical isomorphism $A_k(\boldsymbol{R}^n) \simeq (\bigwedge^k \boldsymbol{R}^n)^*$ and by defining a duality form between $\bigwedge^k \boldsymbol{R}^{n*}$ and $\bigwedge^k \boldsymbol{R}^n$ as follows:

$$\langle w_1 \wedge \ldots \wedge w_k, v_1 \wedge \ldots \wedge v_k \rangle := \det \langle w_i, v_j \rangle,$$

where $\langle w_i, w_j \rangle$ is the natural duality form between $\boldsymbol{R}^{n*}$ and $\boldsymbol{R}^n$. Apply this to prove the second part of (c) in the same way as in (a) and (b).

16. (a) Give the proof of necessity left out in Example 15;

(b) Construct a global basis of vector fields on $\boldsymbol{T}^n$, i.e. find smooth vector fields $X_1, \ldots, X_n$ such that, for each $s \in \boldsymbol{T}^n$, $X_1(s), \ldots, X_n(s)$ is a basis for $T_s(\boldsymbol{T}^n)$.

17. Prove that if $\varphi\colon M \to N$ is smooth, $(\varkappa, U)$ is a chart on M around a point m and (τ, V) a chart on N around $n = \varphi(m)$ and if $(X_i(m))$, $(Y_j(n))$ are bases for $T_m(M)$ and $T_n(N)$ determined by these charts respectively, then the matrix of the mapping $d\varphi_m\colon T_m(M) \to T_n(N)$, with respect to these bases, is the Jacobi matrix of the mapping

$\tau \circ \varphi \circ \varkappa^{-1}$ at the point $\varkappa(m)$. Also find the matrix of the mapping $\varphi^*\colon T_n^*(N) \to T_m^*(M)$ with respect to the bases dual to $(X_i(m))$ and $(Y_i(n))$.

18. (a) Suppose $X \in \mathscr{T}(M)$ is represented in the domain of a chart $(\varkappa, U)$ in the form $X = \sum_{i=1}^{n} f^i X_i$, where $f^i \in \mathscr{E}(U)$, and X_i are the basis fields determined by the chart $(\varkappa, U)$. Prove that a curve, $\gamma\colon]a, b[\to M$, lying in U, is an integral curve of the field X, if, any only if, the functions $\varkappa^i \circ \gamma$ are solutions of the system of differential equations $\frac{\mathrm{d}x^i}{\mathrm{d}t} = f^i$, $i = 1, \ldots, n$;

(b) Give an example of a smooth non-complete vector field on $\boldsymbol{R}^1$;

(c) Find the integral curves of the field $\frac{\partial}{\partial x} + \exp(-y)\frac{\partial}{\partial y}$ on $\boldsymbol{R}^2$. Check whether it is complete or not;

(d) Discuss the behaviour of the integral curves of the field $X = f^1 \frac{\partial}{\partial x} + f^2 \frac{\partial}{\partial y}$ on $\boldsymbol{R}^2$, where f_1, f_2 are linear (homogeneous) functions in x and y.

Hint. (d) Reduce the problem to the study of a linear homogeneous system of differential equations with constant coefficients. Represent the matrix of the system in the simplest (i.e. canonical) form.

19. Let M, N be manifolds and let M be connected. Further, let f, g be smooth mappings of M into N. Prove that $\mathrm{d}f_p \underset{p}{\equiv} 0$ if, and only if, f is a constant mapping. Moreover, if $\mathrm{d}f_p \underset{p}{\equiv} \mathrm{d}g_p$ and $f(q) = g(q)$ for a certain $q \in M$, then $f = g$.

20. Let X be a smooth vector field on M and φ the local one-parameter group defined by X.

(a) Prove that X is invariant with respect to φ, i.e. $(\mathrm{d}\varphi_t)_m X(m) = X(\varphi_t(m))$ whenever $(t, m) \in D$;

(b) If m is a point in M such that $X(m) = 0$, then φ_m maps the whole of $\boldsymbol{R}^1$ onto m;

(c) Complete the missing details in Example 16.

1.3. LIE GROUPS AND LIE ALGEBRAS

DEFINITION. Let a group G be an analytic manifold. We shall call it a *Lie group* provided the mapping

$$G \times G \ni (x, y) \to xy^{-1} \in G \tag{1.3.1}$$

is analytic.

Example 1

(a) The (abelian) group of the vector space $\boldsymbol{R}^n$ or $\boldsymbol{C}^n$ equipped with a natural analytic manifold structure is a Lie group. Similarly the group $\boldsymbol{T}^n$ as well as the direct product of each of the above groups with an arbitrary finite group is a Lie group.

(b) The multiplicative group K_* of the field K (recall that K denotes either of the fields $\boldsymbol{R}$ and $\boldsymbol{C}$) together with the natural structure of an open submanifold of K is a Lie group. The reader will easily check that the mapping $K_* \times K_* \ni (x, y) \to xy^{-1} \in K_*$ is indeed analytic.

(c) The group of all affine orientation preserving transformations, of the real line $\boldsymbol{R}^1$ (the group "$ax+b$"—cf. in § 1.1. Example 3 (b)) is a connected Lie group. In fact, the underlying space of the group is the product $\boldsymbol{R}_+ \times \boldsymbol{R}$ with the operation defined by the formula $(a, b)(a_1, b_1) = (aa_1, ab_1 + b)$, which implies that the mapping (1.3.1) is analytic provided $\boldsymbol{R}_+ \times \boldsymbol{R}$ is equipped with a natural manifold structure.

Example 2

(a) The group $\mathrm{GL}(n, \boldsymbol{R})$ being an open subset of $M_n(\boldsymbol{R}) \simeq \boldsymbol{R}^{n^2}$, possesses a natural analytic manifold structure given by a global chart assigning to a matrix all the n^2 of its entries. Since the entries of the matrix AB^{-1} are rational functions of the entires of matrices A and B with non-vanishing denominators, we conclude that the mapping (1.3.1) is indeed analytic.

(b) Applying the atlas constructed in Example 3, § 1.2, for the group $\mathrm{SL}(2, \boldsymbol{R})$ one can easily check that the entries of the matrix AB^{-1} for $A, B \in \mathrm{SL}(2, \boldsymbol{R})$ are analytic functions with respect to the charts in this atlas. This shows that the group $\mathrm{SL}(2, \boldsymbol{R})$ equipped with the manifold structure determined by this atlas is a Lie group. Later on we shall see that this is a particular case of a more general fact.

We shall now prove that a chart $(\varkappa, U)$ defined in a neighbourhood of the neutral element of the group allows us to construct a whole atlas.

For all $g \in G$ we set $U(g) := gU = \{x \in G\colon x = gy, y \in U\}$ with $\varkappa_g\colon U(g) \to \boldsymbol{R}^n$ given by the formula $\varkappa_g(gx) = \varkappa(x)$. It is condition (1.3.1) which ensures that each pair $(\varkappa_g, U(g))$ is a chart. Obviously we have $\bigcup_{g \in G} U(g) = G$. Compatibility of these charts results also from condition (1.3.1).

It is clear that by equipping a group with a manifold structure a certain class of subgroups is distinguished, namely those admitting a manifold structure compatible with the structure on the group.

It is necessary, in view of the theory of Lie groups, to introduce the following definition.

DEFINITION. Let G be a Lie group. A subgroup $H \subset G$ will be called a *Lie subgroup* provided it is a Lie group and the natural inclusion mapping $j\colon H \to G$ is smooth with an injective differential $\mathrm{d}j_e\colon T_e(H) \to T_e(G)$.

Example 3

(a) We define a Lie group structure on the group $T(n, \boldsymbol{R})$ of upper-triangular matrices of order n by applying a global chart $\bar{\varkappa}\colon T(n, \boldsymbol{R}) \ni g \to (\varkappa_i^j(g)) \in \boldsymbol{R}^k$ with $i \leqslant j$ and $k = \frac{1}{2}n(n+1)$ which assigns to a matrix all the entires lying on and above the diagonal. Now, for the natural inclusion mapping $j\colon T(n, \boldsymbol{R}) \to \mathrm{GL}(n, \boldsymbol{R})$ and a chart $\varkappa\colon \mathrm{GL}(n, \boldsymbol{R}) \ni g \to \varkappa_i^j((g)) \in \boldsymbol{R}^{n^2}$, we have

$$\varkappa_i^j \circ j = \begin{cases} \bar{\varkappa}_i^j, & i \leqslant j, \\ 0, & i > j \end{cases}$$

which proves that $T(n, \boldsymbol{R})$ is a Lie subgroup of the group $\mathrm{GL}(n, \boldsymbol{R})$.

(b) The group $\mathrm{GL}(n, \boldsymbol{C})$ as an open subset of $\boldsymbol{C}^{n^2}$ possesses a Lie group structure given by the global chart $\varkappa\colon \mathrm{GL}(n, \boldsymbol{C}) \ni g \to (\varkappa_i^j(g)) \in \boldsymbol{C}^{n^2}$ where $\varkappa_i^j(g)$ is the (i, j)-th term of a matrix g. On the other hand, $\mathrm{GL}(n, \boldsymbol{C})$ can be treated as a transformation group of the real space $\boldsymbol{R}^{2n}$ obtained by identifying $\boldsymbol{C}^n$ with $\boldsymbol{R}^{2n}$, which in turn defines an imbedding j: GL(n, $\boldsymbol{C}$) → GL($2n$, $\boldsymbol{R}$). If $g \in$ GL(n, $\boldsymbol{C}$) is represented as $g = g_1 + ig_2$, where $g_i \in$ GL(n, $\boldsymbol{R}$), then

$$j(g) = \begin{bmatrix} g_1 & g_2 \\ -g_2 & g_1 \end{bmatrix};$$

hence $j(g)$ is analytic and dj_e is injective. Thus GL(n, $\boldsymbol{C}$) can be regarded as a Lie subgroup of the group GL($2n$, $\boldsymbol{R}$).

Regarding G as a transformation group of the manifold G plays an essential role in the study of Lie groups.

DEFINITION. By a *one-parameter subgroup of a Lie group* G we mean an analytic homomorphism of the group $\boldsymbol{R}$ into G.

A one-parameter subgroup $a(t)$ defines, like any other curve on a manifold, a vector tangent to the group at every point which lies on $a(t)$: $X_a(t) = a'(t) \in T_{a(t)}(G)$.

The vector $X_a(0) \in T_e(G)$ determines the value $X_a(t)$ for an arbitrary $t \in \boldsymbol{R}$. Namely we find

$$X_a(t)f = (f \circ a)'(t) = \frac{\mathrm{d}}{\mathrm{d}s} f(a(t+s))|_{s=0} = \frac{\mathrm{d}}{\mathrm{d}s} f(a(t)a(s))|_{s=0}$$

$$= \frac{\mathrm{d}}{\mathrm{d}s} f \circ l_{a(t)}(a(s)) = X_a(0)\,(f \circ l_{a(t)}) = (\mathrm{d}l_{a(t)})_e(X_a(0))\,f.$$

Thus we arrive at the relation

$$X_a(t) = (\mathrm{d}l_{a(t)})_e X_a(0). \tag{1.3.2}$$

In particular, it follows that the tangent mapping da is injective whenever the one-parameter subgroup is non-trivial (i.e. whenever $a(t) \neq e$ for some t).

A one-parameter subgroup $a(\cdot)$ of a group G defines a one-parameter transformation group of G treated as a manifold. Namely we set:

$$\varphi\colon \boldsymbol{R} \times G \ni (t, g) \to r(t)g := ga(t) \in G.$$

Let us see what the vector field generating that subgroup looks like. We have $X_\varphi(g)f = X_a(0)f \circ l_g = (dl_g)_e X_a(0)$. This field, denoted by X_a, is also completely determined by the value of the tangent vector to $a(t)$ at the identity of the group.

DEFINITION. A vector field X on G is called *left* (*right*)*-invariant* if for an arbitrary $g \in G$ we have

$$X(gh) = (dl_g)_h(X(h)), \quad \text{i.e.} \quad dl_g X = X$$
$$(X(hg) = (dr_{g^{-1}})_h X(h), \quad \text{i.e.} \quad dr_g X = X \text{ respectively})).$$

Any left- or right-invariant vector field is fully determined by its value at an arbitrary point. On the other hand, given a vector $X_0 \in T_e(G)$, we can define a left-invariant vector field with the value X_0 at e. Namely we set:

$$X(g) = (dl_g)_e X_0.$$

This formula defines a bijection of $T_e(G)$ onto the set of left-invariant vector fields. This set will be denoted in the sequel by $\mathscr{T}_l(G)$.

Obviously, the vector field X examined above belongs to $\mathscr{T}_l(G)$.

LEMMA. *$\mathscr{T}_l(G)$ is a Lie subalgebra of the algebra $\mathscr{T}(G)$.*

Proof. Suppose $X, Y \in \mathscr{T}_l(G)$. On account of formula (1.2.16) we obtain $dl_g[X, Y] = [dl_g X, dl_g Y] = [X, Y]$. □

We shall make use of the bijection between $T_e(G)$ and $\mathscr{T}_l(G)$ in order to carry over the Lie algebra structure onto $T_e(G)$.

DEFINITION. Let $X_0, Y_0 \in T_e(G)$ and let X and Y denote the elements of $\mathscr{T}_l(G)$ assuming at the point e the value X_0 and Y_0, respectively. Define

$$[X_0, Y_0] = [X, Y](e).$$

The Lie algebra $(T_e(G), [\cdot, \cdot])$ thus obtained is called the *Lie algebra of the group* G and denoted by $\mathfrak{g}$.

We have already noted that every one-parameter subgroup $a(\cdot)$ defines both an element of the algebra $\mathfrak{g}$, namely the value of $a'(0)$, and a left-invariant vector field with $a(\cdot)$ constituting its integral curve. We shall now prove that the above correspondence is 1–1 and onto.

THEOREM 1.3.1. *For every $X \in \mathfrak{g}$ there exists a unique one-parameter subgroup $R \ni t \to a_X(t) \in G$ such that $a'_X(0) = X$.*

Proof. Let $\tilde{X}$ be the left-invariant vector field on G assuming the value X at e. Denote by φ the local transformation group related to X (Th.1.2.3).

The curve $t \to a(t) = \varphi_e(t)$ is the integral curve of the field X passing through $e \in G$. We want to show that this is the very one-parameter subgroup that we are looking for. Let us compute the tangent vector to the curve $\gamma\colon s \to a(t)a(s)$: $\gamma'(s) = dl_{a(t)}(a'(s)) = dl_{a(t)}(dl_{a(s)}(X)) = dl_{a(t)a(s)}(X) = \tilde{X}(a(t)a(s))$. Thus the curve γ is the integral curve of the field $\tilde{X}$ satisfying the condition $\gamma(0) = a(t)$. Hence in virtue of Th. 1.2.3, p. (2) we get $\gamma(s) = \varphi_s(a(t)) = \varphi_s(\varphi_t(e)) = \varphi_{s+t}(e)$. All those s and t for which φ_{s+t} is well-defined satisfy the condition $a(t)a(s) = a(t+s)$. Now it is enough to note that the above equality allows us to extend the domain of the curve $a(\cdot)$ onto the whole of the real axis. □

DEFINITION. The assignment $\mathfrak{g} \ni X \to \exp X := a_X(1) \in G$ is called the *exponential mapping*.

Below we list a number of properties of the mapping exp. We start with the following simple remark.

The one-parameter subgroups $a_s\colon t \to a_X(s \cdot t)$ and $b_s\colon t \to a_{sX}(t)$ are both integral curves of the field sX. Hence both subgroups are identical, which means that

$$\exp tX = a_X(t). \tag{1.3.3}$$

The following theorem plays a fundamental role in the Lie theory.

THEOREM 1.3.2. *There exists a neighbourhood U of the zero in $\mathfrak{g}$ and a neighbourhood O of the neutral element in G such that* $\exp\colon U \to O$ *is an analytic diffeomorphism.*

Example 4

As an illustration of the above considerations we shall find the Lie algebra of the group $\mathrm{GL}(n, \boldsymbol{R})$ as well as the form of its one-parameter subgroups and of the left-invariant vector field. Let $\varkappa = [\varkappa^j_i]$ be the natural global chart for $\mathrm{GL}(n, \boldsymbol{R})$. With the help of this chart we define an isomorphism of the space $T_g(\mathrm{GL}(n, \boldsymbol{R}))$ for an arbitrary $g \in \mathrm{GL}(n, \boldsymbol{R})$ onto the space of matrices $M_n(\boldsymbol{R})$. Namely we put

$$T_g(\mathrm{GL}(n, \boldsymbol{R})) \ni X(g) \to [X(g)\varkappa^j_i] \in M_n(\boldsymbol{R}). \tag{$*$}$$

Also we define an isomorphism of the space of vector fields $\mathscr{T}(\mathrm{GL}(n, \boldsymbol{R}))$ onto the space of $M_n(\boldsymbol{R})$-valued functions on $\mathrm{GL}(n, \boldsymbol{R})$ by assigning to a vector field X the function $g \to [X(g)\varkappa^j_i]$. (Note that for fixed i, j

the functions $g \to X(g)\varkappa_i^j$ are the coordinates of the field X with respect to the basis of vector fields determined by the chart $\varkappa$—cf. § 1.2, Ex. 10.) Assuming that $X_0 \in T_e(\mathrm{GL}(n, \boldsymbol{R}))$, we have

$$[dl_g X_0 \varkappa_i^j] = [X_0(\varkappa_j^i \circ l_g)] = \Big[\sum_{i=1}^{n} \varkappa_i^k(g) X_0 \varkappa_k^j\Big] = g[X_0 \varkappa_i^j].$$

Thus, under this isomorphism, to left-invariant vector fields correspond functions of the form $g \to gh$, where $h = [X(e)\varkappa_i^j] \in M_n(\boldsymbol{R})$ is a fixed matrix.

If Y is another left-invariant vector field with the corresponding function $g \to gk$, then the function $g \to g[h, k]$ corresponds to the commutator $[X, Y]$. Indeed we get

$$\begin{aligned}[X, Y](e)\varkappa_i^j &= X(e)Y\varkappa_i^j - Y(e)X\varkappa_i^j \\ &= X(e)\Big(\sum_{k=1}^{n} \varkappa_i^k(g)Y(e)\varkappa_k^j\Big) - Y(e)\Big(\sum_{k=1}^{n} \varkappa_i^k(g)Y(e)\varkappa_k^j\Big) \\ &= \sum_{k=1}^{n} X(e)\varkappa_i^k Y(e)\varkappa_k^j - \sum_{k=1}^{n} Y(e)\varkappa_i^k X(e)\varkappa_k^j.\end{aligned}$$

It follows that the Lie algebra of the group $\mathrm{GL}(n, \boldsymbol{R})$ is isomorphic in a natural way to the Lie algebra $\mathfrak{gl}(n, \boldsymbol{R})$—the isomorphism is given by formula (*) with $g = e$.

Let us now recall some facts concerning the exponential function of the matrix argument. It is defined by the formula

$$\mathrm{e}^A := \sum_{n=0}^{\infty} \frac{1}{n!} A^n, \qquad A \in M_n(\boldsymbol{R}).$$

It can be proved that the series is convergent almost uniformly with respect to the norm in the space $M_n(\boldsymbol{R})$. Hence the function $M_n(\boldsymbol{R}) \ni A \to \mathrm{e}^A \in M_n(\boldsymbol{R})$ is differentiable and the derivative $D\mathrm{e}^A\colon M_n(\boldsymbol{R}) \to M_n(\boldsymbol{R})$ is given in the case where $[A, T] = 0$ by the formula $D\mathrm{e}^A(T) := \mathrm{e}^A \circ T$. Moreover, we have

$$\mathrm{e}^A \circ \mathrm{e}^B = \mathrm{e}^{A+B}$$

whenever $[A, B] = 0$ and

$$\det \mathrm{e}^A = \mathrm{e}^{\mathrm{tr} A}.$$

The last identity shows that the exponential function indeed maps $M_n(R)$ into $\mathrm{GL}(n, \boldsymbol{R})$ and the previous identity provides that $t \to e^{tA}$ is a one-parameter subgroup of $\mathrm{GL}(n, \boldsymbol{R})$. The tangent vector to the subgroup $t \to e^{tA}$, at the point $t = 0$, is equal to A and the arbitrariness of A implies in view of Th. 1.3.1 that every one-parameter subgroup of $\mathrm{GL}(n, \boldsymbol{R})$ is of the form $t \to e^{tA}$, with some $A \in M_n(\boldsymbol{R}) = \mathfrak{gl}(n, \boldsymbol{R})$. The same result can be obtained by applying the theorem on the existence and uniqueness of solutions of the initial value problem for ordinary differential equations; cf. Problem 10 (f).

Note that in that case Theorem 1.3.2 is an immediate consequence of the classical inverse function theorem since $De^A|_{A=0} = I$, I—the unit matrix.

Let $\varphi\colon G_1 \to G_2$ be a Lie group homomorphism and let X be a left-invariant vector field on G_1. The fact that φ is a homeomorphism can be expressed in the form $l_{\varphi(x)} \circ \varphi = \varphi \circ l_x$. Hence follows the relation $\mathrm{d}\varphi_x(X(x)) = \mathrm{d}l_{\varphi(x)}(\mathrm{d}\varphi)_e X(e)$. This means that the left-invariant vector field on G_2 determined by the element $(\mathrm{d}\varphi)_o X(e) \in \mathfrak{g}_2$ is φ-related to the field X. Moreover if $(\mathrm{d}\varphi)X_i =: Y_i$, then, on account of formula (1.2.16), the commutator $[Y_1, Y_2]$ is φ-related to the vector field $[X_1, X_2]$. This property is stated by the following

PROPOSITION 1.3.3. *If $\varphi\colon G_1 \to G_2$ is a homeomorphism of group G_1 into G_2, then $(d\varphi)_e\colon \mathfrak{g}_1 \to \mathfrak{g}_2$ is a homomorphism of their Lie algebras.*

Example 5

The differential $\mathrm{Ad}g$ of an inner automorphism $\alpha_g\colon x \to gxg^{-1}$ is an automorphism of the algebra $\mathfrak{g}$.

Let us investigate the relationship between the exponential transformations of the groups related by a homomorphism $\varphi\colon G_1 \to G_2$.

The mapping $t \to \varphi(\exp tX) \in G_2$ is a one-parameter subgroup of G_2. The tangent vector at the identity of the group G_2 is equal to $(d\varphi)_e X$ which means that

$$\varphi \circ \exp X = \exp(\mathrm{d}\varphi)_e X. \tag{1.3.4}$$

As a special case we get the formula

$$g \exp X g^{-1} = \exp \mathrm{Ad}(g) X. \tag{1.3.5}$$

From (1.3.4) we derive

Proposition 1.3.4. (1) *If a homomorphism* $\varphi\colon G_1 \to G_2$ *is injective then* $d\varphi\colon \mathfrak{g}_1 \to \mathfrak{g}_2$ *is injective.*

(2) *A homomorphism* φ *is a local diffeomorphism if and only if* $d\varphi$ *is an isomorphism of the Lie algebras.*

(3) *Let* $H \subset G$ *be a connected Lie subgroup of a group* G *and let* $\mathfrak{h}$ *denote the Lie algebra of* H *imbedded in* $\mathfrak{g}$. *Suppose* $X \in \mathfrak{g}$. *Then* $X \in \mathfrak{h}$ *if and only if* $\exp tX \in H$ *for all* $t \in \boldsymbol{R}$.

Example 6

We have already seen that the group $\mathrm{SL}(n, \boldsymbol{R})$ is a (n^2-1)-dimensional submanifold of the group $\mathrm{GL}(n, \boldsymbol{R})$. Moreover, the topology on $\mathrm{SL}(n, \boldsymbol{R})$ given by the submanifold structure agrees with the relative topology induced from $\mathrm{GL}(n, \boldsymbol{R})$. One can easily prove that $\mathrm{SL}(n, \boldsymbol{R})$ equipped with this manifold structure is a Lie subgroup of the group $\mathrm{GL}(n, \boldsymbol{R})$—the proof is left to the reader.

Denote $\mathfrak{sl}(n, \boldsymbol{R}) := \{X \in \mathfrak{gl}(n, \boldsymbol{R})\colon \operatorname{tr} X = 0\}$. The identity $\operatorname{tr} AB = \operatorname{tr} BA$ implies that $\mathfrak{sl}(n, \boldsymbol{R})$ is a Lie subalgebra of $\mathfrak{gl}(n, \boldsymbol{R})$. Restrict to $\mathfrak{sl}(n, \boldsymbol{R})$ the previously defined isomorphism of $\mathfrak{gl}(n, \boldsymbol{R})$ onto the Lie algebra of the group $\mathrm{GL}(n, \boldsymbol{R})$. Then the algebra $\mathfrak{sl}(n, R)$ is mapped onto the Lie algebra of the group $\mathrm{SL}(n, \boldsymbol{R})$ (imbedded of course in the Lie algebra of the group $\mathrm{GL}(n, \boldsymbol{R})$). Namely, it follows from (3) of Proposition 1.3.4, that an $X \in \mathfrak{gl}(n, \boldsymbol{R})$ corresponds to an element of the Lie algebra of the group $\mathrm{SL}(n, \boldsymbol{R})$, if and only if $\det \mathrm{e}^{tX} = 1$ for all t. Owing to the identity $\det \mathrm{e}^{A} = \mathrm{e}^{\operatorname{tr} A}$ the last condition is equivalent to $\operatorname{tr} X = 0$.

We shall now present without proof a list of analytic properties of the exponential mapping from which one can derive more precise information concerning the relationship between group structure on G and the algebraic structure of its Lie algebra.

Theorem 1.3.5. (I) *Let* $x \in G$ *and* $X \in \mathfrak{g}$. *For* $f \in \mathscr{E}(x)$ *write*

$$X^k f(x) := \frac{\mathrm{d}^k}{\mathrm{d}t^k} f(x \exp tX)|_{t=0}.$$

Then any analytic function f *satisfies for sufficiently small* t *the identity*

$$f(x \exp tX) = \sum_{n=0}^{\infty} (X^n f)(x) \frac{t^n}{n!}.$$

(II) *For* $X, Y \in \mathfrak{g}$ *the following formulae hold:*

$$\exp tX \exp tY = \exp\left\{t(X+Y)+\frac{t^2}{2}[X, Y]+O(t^3)\right\}, \tag{1.3.6}$$

$$\exp tX \exp tY \exp(-tX) = \exp\{tY+t^2[X, Y]+O(t^3)\}, \tag{1.3.7}$$

$$\exp tX \exp tY \exp(-tX)\exp(-tY) = \exp\{t^2[X, Y]+O(t^3)\}. \tag{1.3.8}$$

The symbol $O(t^3)$ stands for an analytic, vector-valued function of the form $t^3 v(t)$ with an analytic v. The property (1.3.8) of the exponential mapping can be applied to define the commutator in a Lie algebra. Namely we note that the tangent vector at $s = 0$ to the curve

$$[0, \infty[\ni s \to \exp s^{\frac{1}{2}} X \exp s^{\frac{1}{2}} Y \exp(-s^{\frac{1}{2}}) X \exp(-s^{\frac{1}{2}}) Y$$

is equal to $[X, Y]$. Formula (1.3.7), in turn, allows us to explain the relationship between the adjoint representation of a group G and the Lie algebra of G.

PROPOSITION 1.3.6. $\mathrm{Ad}(\exp X) = e^{\mathrm{ad} X}$.

Proof. Let us examine the following one-parameter subgroups of $GL(\mathfrak{g})$: $a_1(t) = \mathrm{Ad}(\exp tX)$ and $a_2(t) = e^{t\,\mathrm{ad}X}$. It is enough—in order to show that they are identical—to compare the vectors tangent to them at the identity. Obviously $a_2(0) = \mathrm{ad}X$. The formulae (1.3.5) and (1.3.7) imply that $\exp \mathrm{Ad}(\exp tX)tY = \exp tX \cdot \exp tY \exp(-tX) = \exp(tY+t^2[X, Y]+O(t^3))$. Hence for sufficiently small $t \searrow 0$

$$\mathrm{Ad}(\exp tX)Y = Y+t[X, Y]+O(t^3)$$

and by differentiating we obtain $a_1'(0)Y = [X, Y] = \mathrm{ad}X(Y) = a_2'(0)$. □

The theorems presented below without proof are in fact the deepest results of the Lie theory.

THEOREM 1.3.7. 1° *For every Lie algebra* $\mathfrak{g}$ *there exists a Lie group* G *with its Lie algebra isomorphic to* $\mathfrak{g}$.

2° *If* $\mathfrak{g}$ *is the Lie algebra of a group* G *and* $\mathfrak{h} \subset \mathfrak{g}$ *is a subalgebra, then there exists a unique connected Lie subgroup* $H \subset G$ *with the Lie algebra* $\mathfrak{h}'$ *such that the natural inclusion mapping* $i\colon H \to G$ *satisfies* $di\,(\mathfrak{h}') = \mathfrak{h}$.

3° *Every Lie algebra is isomorphic to a subalgebra of the algebra* $\mathfrak{gl}(n)$.

Note that assertion 3° (the Ado theorem) and 2° lead to assertion 1°, called the Lie theorem. The reader will find the proofs together with a full exposition of the Lie theory in numerous monographs devoted to this subject, e.g. Bourbaki, 1960, 1968, Chevalley, 1946, Varadarajan, 1974 and Helgason, 1968.

PROBLEMS

1. Let G be a Lie group. Find the tangent mappings for $m: G \times G \ni (x, y) \to x \cdot y \in G$ at the point (e, e) and for $i: G \ni x \to x^{-1} \in G$ at e.

2. Show that if $\varphi: G \to H$ is a smooth homomorphism of Lie groups then the rank of $d\varphi_g$ is constant for all $g \in G$.

3. (a) Let G_0 be the connected component of the identity of a Lie group G. Prove that G_0 is an open Lie subgroup of G and that the Lie algebras of G_0 and G are identical.

(b) Suppose G is a connected Lie group. Show that the centre of G coincides with the kernel of the adjoint representation Ad of the group G.

(c) Prove that the Lie algebra of an abelian group is also abelian, i.e. that $[X, Y] = 0$ for all $X, Y \in \mathfrak{g}$. Prove that also the converse theorem is true whenever G is connected. Give an example showing that without the assumption of connectedness the converse theorem is false.

(d) If G is a connected Lie group and $\varphi: G \to H$ is a smooth homomorphism with the discrete kernel, then the kernel is contained in the centre of G.

Hint. A connected component of the identity of a Lie group is generated by any connected, symmetric neighbourhood of the identity.

4. Prove that the integral curves of a left-invariant vector field on a Lie group G are precisely all the curves of the form $t \to ga(t)$ where $g \in G$ and $a(\cdot)$ is the one-parameter subgroup defined by this field.

5. Construct a global basis for the vector space over $\boldsymbol{R}$ of left-invariant vector fields on a Lie group G. Also prove that the frame bundle $B(G)$ is trivial, i.e. isomorphic to the product bundle $G \times \mathrm{GL}(n, \boldsymbol{R})$.

6. If G_i, $i = 1, 2$, are Lie groups, then the direct product $G_1 \times G_2$ equipped with the product manifold structure is also a Lie group.

The Lie algebra of $G_1 \times G_2$ is the direct sum $\mathfrak{g}_1 + \mathfrak{g}_2$ with the commutator given by the formula

$$[(X_1, X_2), (Y_1, Y_2)] := ([X_1, Y_1], [X_2, Y_2])$$

for X_i, $Y_i \in \mathfrak{g}^i$, $i = 1, 2$.

7. Suppose $\{X_i\}_1^n$ is a basis for a Lie algebra $\mathfrak{g}$. The numbers $\{c_{ij}^k\}$ defined by

$$[X_i, X_j] = \sum_{k=1}^{n} c_{ij}^k X_k \qquad (*)$$

are called the *structure constants* of $\mathfrak{g}$. Prove the relations

$$c_{ij}^k = -c_{ij}^k, \qquad \sum_{i=1}^{n} (c_{il}^m c_{jk}^l + c_{jl}^m c_{ki}^l + c_{kl}^m c_{ij}^l) = 0. \qquad (**)$$

Also prove that for an arbitrary system of real numbers $\{c_{ij}^k\}$ satisfying the condition $(**)$ there exists a Lie algebra unique up to an isomorphism with $\{c_{ij}^k\}$ as the structure constants.

Hint. In a vector space V with a basis $\{X_i\}_1^n$ define a commutator by the formula $(*)$ for the basis vectors and extend this definition by linearity to the whole of the space V.

8. Find the Lie algebra of the group "$ax+b$" of affine transformations of the real line and show that it is isomorphic to the Lie subalgebra of the algebra $\mathfrak{gl}(2, \boldsymbol{R})$ consisting of all matrices of the form $\begin{bmatrix} a & b \\ 0 & 0 \end{bmatrix}$ with $a, b \in \boldsymbol{R}$.

9. Let $\mathfrak{g}$ be a three-dimensional Lie algebra with a basis $\{p, q, t\}$ satisfying the commutation relations

$$[p, q] = t, \qquad [p, t] = [q, t] = 0.$$

Show that $\mathfrak{g}$ is isomorphic to the Lie algebra of the group (called the *Heisenberg–Weyl group*) of matrices of the form

$$\begin{bmatrix} 1 & x & z \\ 0 & 1 & y \\ 0 & 0 & 1 \end{bmatrix}, \qquad x, y, z \in \boldsymbol{R}.$$

10. Let $\|\cdot\|$ denote the Euclidean norm in the vector space $\mathfrak{gl}(n, \boldsymbol{R}) \simeq \boldsymbol{R}^{n^2}$.

(a) Prove that the series $\sum_{n=0}^{\infty}(-1)^{n+1}\frac{X^{n+1}}{n+1}$ is convergent in the ball $\{X \in \mathfrak{gl}(n, \boldsymbol{R})\colon \|X\| < 1\}$ and its sum defines a smooth mapping of this ball into $\mathfrak{gl}(n, \boldsymbol{R})$.

(b) Let $U_0 = \{X \in \mathfrak{gl}(n, \boldsymbol{R})\colon \|X\| < 1\}$ and $U_I = \{X \in \mathfrak{gl}(n, \boldsymbol{R})\colon \|X - I\| < 1\}$ be neighbourhoods of the zero and the unit matrix I, respectively. Prove that $X \to e^X$ maps U_0 diffeomorphically onto U_I and the inverse mapping is the function $\log(\cdot)$ defined as $\log X = \sum_{n=0}^{\infty}(-1)^{n+1}\frac{(X-I)^{n+1}}{n+1}$.

(c) Show that $\mathfrak{gl}(n, \boldsymbol{R}) \ni X \to e^X \in GL(n, \boldsymbol{R})$ is not surjective.

(d) Show that $X \to e^X$ is a surjection of the set of symmetric matrices onto the set of positive definite symmetric matrices.

(e) Show that the exponential mapping $A \to e^A$ can be defined for $A \in \mathfrak{gl}(n, \boldsymbol{C}) = M_n(\boldsymbol{C})$ by the same series as in the real case and examine the properties of this mapping in the complex case.

(f) Show, on account of the theorem on the uniqueness of solutions of the initial value problem for ordinary differential equations, that $t \to e^{tA}$ is a general form of the one-parameter subgroups of $GL(n, \boldsymbol{R})$.

Hint to (c) and (d). The exponential mapping maps diagonal matrices onto diagonal ones.

11. (a) Prove that the fields $X = \sum_{i=1}^{n} \alpha^i D_i$ where $\alpha^i \in \boldsymbol{R}$, $D_i f = \frac{\partial f}{\partial x^i}$ are the only left-invariant vector fields on $\boldsymbol{R}^n$.

(b) Let $p\colon \boldsymbol{R}^n \to \boldsymbol{T}^n$ be the natural projection and X a left-invariant vector field on $\boldsymbol{R}^n$, and let $\bar{X}$ be the unique vector field on the torus $\boldsymbol{T}^n$ p-related to X. Show that every vector field $\bar{X}$ is left-invariant and every left-invariant vector field on the torus is of that form.

(c) Let $X = \alpha^1 D_1 + \alpha^2 D_2$ be a left-invariant vector field on $\boldsymbol{R}^2$ and let $\bar{X}$ be the vector field p-related to X. Show that the one-parameter subgroups corresponding to the vector field $\bar{X}$ are either closed or dense in T^2 and the first eventuality occurs if and only if α_1/α_2 is a rational number.

Hint. Cf. Example 15 in § 1.2. To prove the density of the one-parameter subgroup corresponding to the vector field $\bar{X}$ for an irrational

α_1/α_2 show that the family of parallel lines in $\boldsymbol{R}^2$ which get projected onto that subgroup intersects the axis $x_2 = 0$ at a dense subset.

12. Suppose $\mathfrak{g}$ is a Lie algebra and let D: $\mathfrak{g} \to \mathfrak{g}$ denote a differentiation of $\mathfrak{g}$. Show that $\boldsymbol{R} \ni t \to e^{tD} \in \mathrm{GL}(\mathfrak{g})$ is a one-parameter group of automorphisms of the Lie algebra $\mathfrak{g}$.

Hint. Prove by induction the inequality

$$D^k[X, Y] = \sum_{i+j=k} \frac{k!}{i!j!} [D^i X, D^j Y] \qquad \text{for} \qquad i, j, k \geqslant 0.$$

Apply it to compute $e^{tD}[X, Y]$.

13. Prove that any connected abelian Lie group is isomorphic to $R^p \times T^q$ for some non-negative integers p and q.

Hint. Prove first that exp is a homomorphism of $\mathfrak{g}$ onto G.

1.4. TRANSFORMATION GROUPS. INVARIANT TENSOR FIELDS

Let G be a Lie group and M an analytic manifold.

DEFINITION. Suppose the system $(G, M, \cdot)$ is a transformation group. This system is said to be a *Lie transformation group* if the mapping $G \times M \ni (g, m) \to g \cdot m \in M$ is analytic.

The basic example of a Lie transformation group is provided by the coset space of a Lie group by its closed subgroup. The following theorem describes more precisely the analytic structure of this space.

THEOREM 1.4.1. *Let $H \subset G$ be a closed Lie subgroup of a Lie group G. Then the homogeneous space G/H possesses a unique analytic manifold structure with the property that G is a Lie transformation group of the manifold G/H. Namely let π denote the natural projection $G \to G/H$. For each $m \in G/H$ there exists a neighbourhood O of this point and an analytic diffeomorphism j of the set $O \times H$ onto $\pi^{-1}(O)$ satisfying*

$$j(p, h'h) = h' \cdot j(p, h).$$

This theorem implies that the group G is locally diffeomorphic to the product $O \times H$. The diffeomorphism j described above maps the set (p, H) onto the coset $p \subset G$ in such a way that the multiplication by the elements of H is preserved. Thus we get the following commutative diagram:

$$\begin{array}{ccc} O\times H & \xrightarrow{j} & \pi^{-1}(O) \\ & {}_{\pi'}\searrow \quad \swarrow_{\pi} & \\ & O & \end{array}$$

The projection π' is the natural projection on the first coordinate in the product $O\times H$. The identity $\pi' = \pi \circ j$ yields $d\pi' = d\pi \circ dj$. Obviously at every point p the tangent mapping $d\pi'$ maps $T_p(G/H)\times\mathfrak{h}$ surjectively onto $T_p(G/H)$ with the kernel $(p, \mathfrak{h})$. The mapping $(dj)(p, h)$ is an isomorphism onto $T_{j(p,h)}G$. Thus, in particular, $(d\pi)_e\colon \mathfrak{g} \to T_{eH}(G/H)$ is a surjection with the kernel $\mathfrak{h}$.

The following theorem states that the above example is a general model of a transitive transformation group.

THEOREM 1.4.2. *Let (G, M) be a transitive second countable Lie transformation group of an analytic manifold M. Denote by M the isotropy group at a point $m_0 \in M$. Then the mapping $G/H \ni gH \to gm_0 \in M$ is a well defined analytic isomorphism of the homogeneous spaces G/H and M. The inverse mapping is also analytic.*

Example 1

(a) $\mathrm{GL}(n, \boldsymbol{R})$ is a Lie transformation group of $\boldsymbol{R}^n$. Define an action of the group $\mathrm{GL}(n, \boldsymbol{R})$ on the space $M_n(\boldsymbol{R})$ of matrices of degree n by the formula $g\cdot A := gAg^{-1}$. Then the pair $(\mathrm{GL}(n, \boldsymbol{R}), M_n(\boldsymbol{R}))$ is also a Lie transformation group.

(b) In § 1.2 an analytic manifold structure has been defined on the n-dimensional sphere S^n. Earlier we proved that $(\mathrm{SO}(n+1), S^n)$ is a topological homogeneous space with respect to the action

$$\mathrm{SO}(n+1)\times S^n \ni (g, x) \to gx \in S^n. \qquad (*)$$

As a matter of fact, $\mathrm{SO}(n+1)$ is a Lie transformation group of the sphere S^n provided both $\mathrm{SO}(n+1)$ and S^n are equipped with the standard differentiable structures. To verify this it is enough to find a local expression for the mapping $(*)$, using e.g. the charts given by the stereographic projection and noting that a function analytically depending upon the entries of a matrix $g\in \mathrm{SO}(n+1)$ is analytic on $\mathrm{SO}(n+1)$. The reader is invited to prove this together with the fact that the differentiable structure on S^n applied here is the only one up to a diffeomorphism, which is compatible with the standard topology on the sphere and such that $(\mathrm{SO}(n+1), S^n)$ is a Lie transformation group.

(c) We shall now define an action of the group $\mathrm{SL}(n, \boldsymbol{R})$ on the projective space $P^{n-1}\boldsymbol{R}$ in order to make $(\mathrm{SL}(n, \boldsymbol{R}), P^{n-1}\boldsymbol{R})$ a Lie transformation group with respect to standard differentiable structures on $\mathrm{SL}(n, \boldsymbol{R})$ and $P^{n-1}\boldsymbol{R}$. The group $\mathrm{SL}(n, \boldsymbol{R})$ acts analytically on $\boldsymbol{R}^n - \{0\}$ by the usual multiplication of a vector by a matrix. Let π: $\boldsymbol{R}^n - \{0\} \to P^{n-1}\boldsymbol{R}$ be the natural projection. Define

$$g\pi(x) := \pi(gx).$$

This formula makes sense since $\pi(x) = \pi(y)$ if and only if $\pi(gx) = \pi(gy)$. It follows from the definition of the differentiable structure on $P^{n-1}\boldsymbol{R}$ that the action so defined is analytic.

Let us examine more closely the projective line $P^1\boldsymbol{R}$. The points of $P^1\boldsymbol{R}$, i.e. the vector lines in $\boldsymbol{R}^2$, will be described by means of the homogeneous coordinates—a pair (x^1, x^2) will be called the *homogeneous coordinates of the line* $l = \{(tx^1, tx^2): t \in \boldsymbol{R}_*\}$. Let l_0 be the line with the homogeneous coordinates $(t, 0)$ (i.e. the line described by the equation $x_2 = 0$). The set $P^1\boldsymbol{R} - \{l_0\}$ is the domain of a chart $\varkappa$ given in the homogeneous coordinates by the formula

$$\varkappa(x^1, x^2) = x^1/x^2,$$

the quotient x^1/x^2 being called the *inhomogeneous coordinate of the line* $l = \boldsymbol{R}_*(x^1, x^2)$.

Consider the one point compactification (the Alexandrov compactification) of the real line $\boldsymbol{R}$ and denote the point at infinity by ∞. Extend the chart $\varkappa$ to a homeomorphism $\varkappa$: $P^1\boldsymbol{R} \to \boldsymbol{R} \cup \{\infty\}$ by putting $\varkappa(l_0) = \infty$, i.e. interpret ∞ as the inhomogeneous coordinate of the line l_0. Now the action of the group $\mathrm{SL}(2, \boldsymbol{R})$ can nicely by expressed in terms of inhomogeneous coordinates by homographic transformations. If $x \in \boldsymbol{R} \cup \{\infty\}$ is the coordinate of a line $l \in P^1\boldsymbol{R}$ and $g = \begin{bmatrix} a & b \\ c & d \end{bmatrix} \in \mathrm{SL}(2, \boldsymbol{R})$ then the line gl has the coordinate $x' = g \cdot x$ given by the formula $g \cdot x := \dfrac{ax+b}{cx+d}$, where the right-hand side fraction is regarded as ∞ whenever $cx+d = 0$ and as a/c if $x = \infty$. Since the isotropy group at the point $x = 0$ is the group $P = \left\{\begin{bmatrix} a & 0 \\ c & a^{-1} \end{bmatrix} : a \neq 0,\ c \in \boldsymbol{R}\right\}$ we get $P^1\boldsymbol{R} \simeq \mathrm{SL}(2, \boldsymbol{R})/P$.

Let us return to the study of general Lie transformation groups of analytic manifolds. Every such group (G, M) induces an algebra of complete vector fields on the manifold M. Namely, to every element $X \in \mathfrak{g}$ there corresponds a one-parameter subgroup $a_X(\cdot)$ of the group G which, in turn, induces a one-parameter group of transformations of M according to the formula:

$$\varphi(t, m) = a_X(t)m.$$

Denote by X^* the infinitesimal generator of this subgroup.

LEMMA 1.4.3. *The set* $\{X^*\colon X \in \mathfrak{g}\}$ *is a Lie subalgebra of* $\mathscr{T}(M)$.

Proof. Let a function $f \in \mathscr{E}(M)$ be given. Denote by f_m, for a fixed point $m \in M$, a smooth function on G given by the formula $f_m(g) = f(gm)$. Compute $X^*f(m)$ for $X \in \mathfrak{g}$.

$$X^*f(m) = \frac{\mathrm{d}}{\mathrm{d}t} f(a_X(t)m) = Xf_m(e). \tag{1.4.1}$$

On the other hand, $[X, Y]^*f(m) = [X, Y]f_m(e) = XYf_m(e) - YXf_m(e) = X^*Y^*f(m) - Y^*X^*f(m) = [X^*, Y^*]f(m)$, which proves that the set under consideration is closed with respect to the commutator.

REMARK. Formula (1.4.1) shows that the mapping $T_e(G) \ni X \to X^*$ is a homomorphism of Lie algebras, provided the commutator in $T_e(G)$ is defined in terms of the right-invariant vector fields on the group G. Moreover, the Lie algebra so defined is isomorphic to the algebra $\mathfrak{g}$, which implies that the mapping $\mathfrak{g} \ni X \to X^* \in \mathscr{T}(M)$ is a homomorphism.

We shall now state the inverse theorem, which, in the local form, has been proved by Sophus Lie. The version presented below is due to Palais, 1957.

THEOREM 1.4.4. *Let* $\mathscr{G}^*$ *be a finite-dimensional Lie algebra of complete vector fields on a manifold* M. *Let* G *be the subgroup of the group of all diffeomorphisms of* M, *generated by the one-parameter subgroups* *induced by the vector fields in $\mathcal{G}^*$. Then there exists a unique Lie group structure on G such that*

(1) *(G, M) is, with the natural action, a Lie transformation group,*

(2) *the Lie algebra $\mathfrak{g}$ of the group G is isomorphic to $\mathcal{G}^*$ and the isomorphism is provided by the mapping $X \to X^*$.*

Example 2

Define one-parameter subgroups $a_i(\cdot)$ of $SL(2, \boldsymbol{R})$, $i = 1, 2, 3$, by the formula

$$a_1(t) := \begin{bmatrix} 1 & t \\ 0 & 1 \end{bmatrix}, \qquad a_2(t) := \begin{bmatrix} 1 & 0 \\ t & 1 \end{bmatrix}, \qquad a_3(t) := \begin{bmatrix} e^t & 0 \\ 0 & e^{-t} \end{bmatrix}.$$

One can easily check that $a_i(t) = e^{tX_i}$, where

$$X_1 = \begin{bmatrix} 0 & 1 \\ 0 & 0 \end{bmatrix}, \qquad X_2 = \begin{bmatrix} 0 & 0 \\ 1 & 0 \end{bmatrix}, \qquad X_3 = \begin{bmatrix} 1 & 0 \\ 0 & -1 \end{bmatrix},$$

and that X_1, X_2, X_3 is a basis of the Lie algebra $\mathfrak{sl}(2, \boldsymbol{R})$. Direct calculation shows that the following commutation relations hold;

$$[X_3, X_1] = 2X_1, \qquad [X_3, X_2] = -2X_2, \qquad [X_1, X_2] = X_3.$$

Let us find the generators of the one-parameter transformation groups of $P^1\boldsymbol{R}$ defined by these subgroups. Let $f \in \mathscr{E}(P^1\boldsymbol{R})$ and let $\varkappa$ be the chart described in Example 1(c). Denoting by the same symbol f both the function f and the function $f \circ \varkappa^{-1}$, we get

$$X_1^* f(x) = \frac{d}{dt} f(x+t)|_{t=0} = \frac{df}{dx}(x),$$

$$X_2^* f(x) = \frac{d}{dt} f\left(\frac{x}{xt+1}\right)\bigg|_{t=0} = x^2 \frac{df}{dx}(x),$$

$$X_3^* f(x) = \frac{d}{dt} f(e^{2t}x)|_{t=0} = 2x \frac{df}{dx}(x).$$

By an easy calculation we check that $[X_3^*, X_1^*] = -2X_1^* = -[X_3, X_1]^*$, etc., in accordance with the remark following Lemma 1.4.3. We leave it to the reader to find the explicit form of the generators of these one-parameter subgroups in chart $\varkappa'$ defined, in the complement of the line with the homogeneous coordinates $(0, t)$, by the formula $\varkappa'(x_1, x_2) = x_2/x_1$.

Example 3

Consider the Lie algebra of vector fields on the plane $\boldsymbol{R}^2$ generated by the vector fields $L_1 := \frac{\partial}{\partial x}$, $L_2 := \frac{\partial}{\partial y}$, $L_3 := x\frac{\partial}{\partial y} - y\frac{\partial}{\partial x}$. The Lie algebra is three-dimensional owing to the commutation relations $[L_1, L_2] = 0$, $[L_3, L_1] = L_2$, $[L_3, L_2] = -L_1$. These fields are complete—the one-parameter transformation groups φ_{L_i} generated by them are, respectively, translations along the axes and rotations about the point $(0, 0)$, i.e.

$$\varphi_{L_1}(t)(x, y) := (x+t, y), \qquad \varphi_{L_2}(t)(x, y) := (x, y+t),$$
$$\varphi_{L_3}(t)(x, y) = (x\cos t - y\sin t, x\sin t + y\cos t).$$

The group of diffeomorphisms of $\boldsymbol{R}^2$ generated by these one-parameter subgroups, is the group of motions of $\boldsymbol{R}^2$—i.e. of rotations together with translations. We shall examine this group more thoroughly in a later chapter.

We now take up the problem of the existence on a homogeneous manifold G/H of invariant tensor fields of the type $\mathcal{T}^r(M)$. The invariance of a tensor field T means that for arbitrary vectors $X^1, \ldots, X^r \in T_m(M)$ and $g \in G$ the following formula holds

$$(l_g^* T)_m(X^1, \ldots, X^r) = T_{gm}(\mathrm{d}l_g X^1, \ldots, \mathrm{d}l_g X^r) = T_m(X^1, \ldots, X^r) \tag{1.4.2}$$

for all $m \in M$.

All the elements of the isotropy subgroups H_0 at a point m_0 satisfy the identity

$$T_{m_0}(\mathrm{d}l_h X^1, \ldots, \mathrm{d}l_h X^r) = T_{m_0}(X^1, \ldots, X^r).$$

Thus an invariant tensor T defines at every point $m_0 \in M$ an r-linear form on the tangent space, invariant with respect to the automorphisms $(\mathrm{d}l_h)_{m_0}$, $h \in H_0$.

Suppose, on the other hand, that an r-linear form T_0 on $T_{m_0}(G/H)$ is given satisfying the condition:

$$T_0\big((\mathrm{d}l_h)_{m_0} X^1, \ldots, (\mathrm{d}l_h)_{m_0} X^r\big) = T(X^1, \ldots, X^r)$$

for $h \in H_0$ and $X^1, \ldots, X^r \in T_{m_0}(G/H)$. We can define an invariant tensor on G/H by the formula:

$$T_{gm_0}\big((\mathrm{d}l_g)_{m_0} X^1, \ldots, (\mathrm{d}l_g)_{m_0} X^r\big) := T_0(X^1, \ldots, X^r). \tag{1.4.3}$$

It is the transitivity that ensures that the tensor field T is defined on the whole of the space, and the correctness of the definition follows from the invariance of T_0. Namely, assuming $(\mathrm{d}l_g)_{m_0} X^i = (\mathrm{d}l_g)_{m_0} Z^i$ we have $X^i = (\mathrm{d}l_{g^{-1}g'})_{m_0} Z^i$, which yields $T_0(Z^1, \ldots, Z^r) = T_0(X^1, \ldots, X^r)$. The invariance of T is seen directly from the definition. Thus we have proved

PROPOSITION 1.4.5. *The space $\mathcal{T}_I^r(G/H)$ of invariant covariant r-tensors on the homogeneous space G/H is isomorphic to the subspace of r-forms on $T_{m_0}(G/H)$ invariant with respect to the representation $(\mathrm{d}l_h)_{m_0}$ of the isotropy group at a point m_0.*

This isomorphism is given by the mapping $T \to T(m_0)$.

We shall now establish the relationship between the representation $(\mathrm{d}l_h)_{m_0}$ and the adjoint representation of G in $\mathfrak{g}$.

Let $\mathfrak{g}$ be the Lie algebra of a group G and $\mathfrak{h}$ the algebra of the subgroup $H \subset G$. The differential of the natural projection $\pi\colon G \to G/H$ maps $T_g(G)$ surjectively onto $T_{gH}(G/H)$ at every point g. In particular, $\mathrm{d}\pi_e\colon \mathfrak{g} \to T_H(G/H)$ being a surjection with the kernel $\mathfrak{h}$, we can define a linear isomorphism $J\colon \mathfrak{g}/\mathfrak{h} \to T_H(G/H)$. We know that the operators of the adjoint representation $\mathrm{Ad}_G h$ for $h \in H$ preserve the subspace $\mathfrak{h}$ (cf. formula (1.3.9)). Thus a representation of the group H in the space $\mathfrak{g}/\mathfrak{h}$ can be defined by means of the natural lifting of the operators $\mathrm{Ad}_G h$:

$$\mathrm{Ad}_{\mathfrak{g}/\mathfrak{h}}(h)\,[X] := [(\mathrm{Ad}_G h)\,(X)]. \tag{1.4.4}$$

LEMMA 1.4.6. *The following formula holds:*

$$J \circ \mathrm{Ad}_{\mathfrak{g}/\mathfrak{h}}(h) = (\mathrm{d}l_h)_H \circ J.$$

Proof. Note that $\pi(hgh^{-1}) = h \cdot \pi(g)$ for $h \in H$, which can be rewritten as $\pi \circ \alpha_h = l_h \circ \pi$. Passing to the tangent mappings, we arrive at the formula $(\mathrm{d}\pi_e) \circ \mathrm{Ad}_G h = (\mathrm{d}l_h)_{eH} \circ (\mathrm{d}\pi)_e$, which proves the lemma. □

The lemma proved above yields a new description of the invariant tensors $\mathcal{T}_I^r(G/H)$.

PROPOSITION 1.4.7. *The space $\mathcal{T}_I^r(G/H)$ is isomorphic to the space of r-linear forms on $\mathfrak{g}/\mathfrak{h}$ invariant under the representation $\mathrm{Ad}_{\mathfrak{g}/\mathfrak{h}}$ of the group H.*

Example 4

Let us consider the problem of the existence of a non-zero invariant differential n-form on an n-dimensional homogeneous manifold $M = G/H$. It is well known that the function det is the only non-zero n-linear form on an n-dimensional space. Thus we set $T_0(X^1, \ldots, X^n) = \det(X^1, \ldots, X^n)$. We know from elementary algebra that $\det(AX^1, \ldots, AX^n) = \det A \det(X^1, \ldots, X^n)$ for a linear operator A and for a system of vectors $X^1, \ldots, X^n$. This formula leads to the relation

$$T_0\big(\mathrm{d}l_h(X^1), \ldots, \mathrm{d}l_h(X^n)\big) = \det \mathrm{d}l_h T(X^1, \ldots, X^n).$$

Thus, in view of Propositions 1.4.5 and 1.4.7, we infer

COROLLARY 1.4.8. *There exists an invariant n-form on the manifold* $G/H \Leftrightarrow \det \mathrm{Ad}_{\mathfrak{g}/\mathfrak{h}}(h) = 1,\ h \in H$.

Example 5

Let (G, H) be a pair such that the image $\mathrm{Ad}_{\mathfrak{g}/\mathfrak{h}}(H) \subset \mathrm{GL}(\mathfrak{g}/\mathfrak{h})$ is a compact subgroup. Then there exists an $\mathrm{Ad}_{\mathfrak{g}/\mathfrak{h}}$—invariant, symmetric, positive definite form on the space $\mathfrak{g}/\mathfrak{h}$ (see Proposition 1.4.7). Denote this form by g_0. Then it follows from Theorem 1.4.6 that there exists an invariant symmetric tensor on $M = G/H$ given by the formula:

$$g_{xH} = l_x^* g_0.$$

Obviously the tensor g is defined at every point of the manifold by a positive definite form.

Note that the assumptions of Examples 1 and 2 are satisfied for an arbitrary Lie group G in the place of a manifold M (the case of $H = e$).

Example 6

The condition $\mathrm{Ad}_{\mathfrak{g}/\mathfrak{h}}(h) = 1$ holds for the sphere $S^2 = \mathrm{SO}(3)/\mathrm{SO}(2)$ since the image of SO(2) under the continuous homomorphism $\det \mathrm{Ad}_{\mathfrak{g}/\mathfrak{h}}$: $\mathrm{SO}(2) \to \boldsymbol{R}_+$ must be a compact subgroup of $\boldsymbol{R}_+$ and there is only one such subgroup namely $\{1\}$. We shall describe an invariant 2-form on the sphere S^2 by a slight modification of the method presented in the proof of Proposition 1.4.5. Let $\varkappa_+, \varkappa_-$ be charts defined on the upper $(x_3 > 0)$ semisphere U_+ and the lower $(x_3 < 0)$ semisphere U_- by the formula

$$\varkappa_\pm(x_1, x_2, x_3) = (x_1, x_2).$$

Denote by ω the invariant 2-form on S^2. Then ω, restricted to U_+, can be written in the form $\omega(x) = c_+(x)\mathrm{d}x^1 \wedge \mathrm{d}x^2$, where $c_+(\cdot)$ is a smooth function on U_+ and $\{\mathrm{d}x^i\}_1^2$ is the basis of 1-forms on U_+ defined by the chart $\varkappa_+$. Moreover, the form ω can be normalized in such a way that $c_+(p) = 1$ at $p = (0, 0, 1)$—the north pole of the sphere.

Let $H \subset \mathrm{SO}(3)$ be the isotropy group at the point p. We know that

$$H = \left\{\left[\begin{array}{c|c} A & 0 \\ \hline 0 & 1 \end{array}\right] : A \in \mathrm{SO}(2)\right\}.$$

One easily verifies that for any $h \in H$ we have

$$l_h^*(c_+(x)\mathrm{d}x^1 \wedge \mathrm{d}x^2) = c_+(hx)\mathrm{d}x^1 \wedge \mathrm{d}x^2,$$

which in view of the invariance implies that $c_+(\cdot)$ is constant on the orbits of H, i.e. on the "parallels" of the sphere S^2. Take $x = (x^1, 0, x^3)$ $\in U_+$, then for $g = \begin{bmatrix} x_3 & 0 & x_1 \\ 0 & 1 & 0 \\ -x_1 & 0 & x_3 \end{bmatrix} \in \mathrm{SO}(3)$ we have $gp = x$. Let $X_1 = \frac{\partial}{\partial x^1}$, $X_2 = \frac{\partial}{\partial x^2}$ be the basis fields on U_+ determined by $\varkappa_+$. Since

$$\mathrm{d}g_p X_1(p) = x_3 X_1(x), \qquad \mathrm{d}g_p X_2(p) = X_2(x),$$

we obtain $c_+(x) = \omega(x)(X_1(x), X_2(x)) = \frac{1}{x^3}\omega(x)(\mathrm{d}g_p X_1(p), \mathrm{d}g_p X_2(p))$

$$= \frac{1}{x^3}\omega(p)(X_1(p), X_2(p)) = \frac{1}{x^3}.$$

Now, if $c_-(x)\mathrm{d}x^1 \wedge \mathrm{d}x^2$ is the restriction of ω to U_-, we get on account of invariance

$$c_-(-p) = -c(p) = -1.$$

By an argument similar to that applied above we get $c_-(x) = -\frac{1}{x^3}$. Consequently, we see that an SO(3)-invariant 2-form on S^2 is given on the set $U_+ \cup U_-$ by the formula

$$x = \frac{1}{|x^3|} dx^1 \wedge dx^3.$$

1.5. ADDITIONAL STRUCTURES ON MANIFOLDS

The frame bundle which we applied to construct tensor bundles is a special case of a more general notion, namely that of a principal bundle.

DEFINITION. By a *principal bundle* we mean a structure composed of a pair of manifolds $(\boldsymbol{B}, M)$, of a smooth projection $\pi: \boldsymbol{B} \to M$ and of a right action of a Lie group G on $\boldsymbol{B}$, provided the following conditions are fulfilled:

(1) $\pi(p \cdot g) = \pi(p)$,

(2) for each $m_0 \in M$ there exists a neighbourhood $O \ni m_0$ and a diffeomorphism $\psi: O \times G \to \pi^{-1}(O)$ such that $\psi(m, gh) = \psi(m, g) \cdot h$.

The manifold $\boldsymbol{B}$ is called the *space of the bundle* with the base M and the structure group G. The last condition ensures that the fibre $\pi^{-1}(m)$ is isomorphic to the group G, and the action of G on the fibre is both transitive and effective. The frame bundle provides an example of a principal bundle with the structure group $\mathrm{GL}(n, \boldsymbol{R})$. Moreover, there is a theorem which says that any homogeneous space of a Lie group is a principal bundle.

We shall now distinguish a certain class of principal bundles over differentiable manifolds. The problem of the existence of such bundles whenever occurring is strongly connected with the structure of the manifold.

DEFINITION. Let G be a subgroup of $\mathrm{GL}(n, \boldsymbol{R})$. A G-structure is a *principal bundle* $(\boldsymbol{B}_G, M, G)$, where $\boldsymbol{B}_G$ is a submanifold of the frame bundle $\boldsymbol{B}(M)$, $\pi(\boldsymbol{B}_G) = M$ and is such that G acts transitively on the set $\pi^{-1}(m) \cap \boldsymbol{B}_G$.

Example 1

The bundle $\boldsymbol{B}(M)$ itself is a $\mathrm{GL}(n, \boldsymbol{R})$-structure.

Example 2

A $\mathrm{GL}(n, \boldsymbol{R})^+$-structure is called an *orientation* of the manifold M. Defining an orientation on a manifold is equivalent to choosing, at every point, a set of n vectors equivalent modulo $\boldsymbol{R}_+$. To every basis $b \in \boldsymbol{B}_G$ there corresponds an element of $\bigwedge^n T_{\pi(b)}(M)$ defined as

$X_1 \wedge \ldots \wedge X_n$. If $\pi(b) = \pi(b')$ we get, since $b = b' \cdot g$ with $\det g > 0$, that $X_1 \wedge \ldots \wedge X_n = X'_1 \wedge \ldots \wedge X'_n$ modulo $\boldsymbol{R}_+$. An orientation on M can be established by defining a nowhere vanishing n-form on M. Then we define $\boldsymbol{B}_G$ to be the set of all bases such that $\omega(X_1(m), \ldots$ $\ldots, X_n(m)) \in \boldsymbol{R}_+$. Recalling Example 4 of the previous section, we conclude that a homogeneous manifold for which $\det \mathrm{Ad}_{\mathfrak{g}/\mathfrak{h}} = 1$ is orientable.

Example 3

Let a k-dimensional linear subspace $V \subset \boldsymbol{R}^n$ be given. Set $G = \{g \in \mathrm{GL}(n, \boldsymbol{R}) \colon gV \subset V\}$. In this case the manifold $\boldsymbol{B}_G$ is called a *k-dimensional distribution on M*.

Suppose $b \in \boldsymbol{B}(M)$ and let i_b denote the isomorphism $\boldsymbol{R}^n \to T_m(M)$ defined by the formula $i_b(t^1, \ldots, t^n) = \sum_{i=1}^{n} t^i X_i$, when $b = (m;\, X_1, \ldots$ $\ldots, X_n)$. The choice of a G-structure defines in every tangent space T_m a subspace $\mathscr{D}_m = i_b(V)$.

Now the choice of b in the fibre $\pi^{-1}(m)$ is insignificant since for each $b' \in \pi^{-1}(m) \cap \boldsymbol{B}_G$ there exists a $g \in G$ such that $b' = bg$. Hence $i_{b'}(V) = i_{bg}(V) = i_b(gV) = i_b(V)$.

Example 4

An $O(n)$-structure on M is called a *Riemannian structure*, and a manifold equipped with an $O(n)$-structure is called a *Riemannian manifold*. The existence of such a structure is equivalent to choosing a 2-covariant symmetric tensor on M, positive definite at every point in M. Namely, for a fixed $B_{O(n)}$ we can define such a tensor by setting

$$g_m(X(m), Y(m)) := \sum_{i=1}^{n} t^i s^i$$

whenever

$$X(m) = \sum_{i=1}^{n} t^i X_i, \qquad Y(m) = \sum_{i=1}^{n} s^i X_i$$

and $b = (m, X_1, \ldots, X_n) \in B_{O(n)}$.

By a similar argument to that used in the previous example, one checks that the value of g_m does not depend upon the choice of a $b \in \pi^{-1}(m)$. Thus, Example 5 of the previous section shows that a homogeneous space G/H admits an invariant Riemannian structure, provided the group $\mathrm{Ad}_{\mathfrak{g}/\mathfrak{h}}(H)$ is compact.

One can define the length of a smooth curve in a Riemannian space as follows. Given a smooth curve in M

$$\gamma\colon [a, b] \ni t \to \gamma(t) \in M,$$

the number

$$L(\gamma) = \int_a^b g(\gamma'(t), \gamma'(t))^{\frac{1}{2}} dt \tag{1.5.1}$$

is called the *length of the curve* γ.

The value of $L(\gamma)$ is independent of the parametrization of the curve, which means that whenever $f\colon [a', b'] \to [a, b]$ is a mapping smooth inside $[a', b']$, having one-sided derivatives at a and b and $f' > 0$, the length of the curve $\gamma' = \gamma \circ f$ equals $L(\gamma)$. This fact is proved by a change of variables in the integral (1.5.1).

Once the length of a curve is defined, we can introduce a metric on a connected Riemannian manifold as follows. Given arbitrary points m and n in M, consider the family $\Gamma(m, n)$ of all curves joining these points. Define a function d on $M \times M$:

$$d(m, n) = \inf_{\Gamma(m, n)} L(\gamma).$$

The function $d(m, n)$ fulfils the axioms of a metric. It is called the *Riemannian metric* on the manifold (M, g). One can prove that the topology on M defined by the Riemannian metric is equivalent to the initial one. The curve γ is called a *geodesic* if for every $t \in]a, b[$ there exists a neighbourhood U of t such that for all $s \in U$ $\gamma | [s, t]$ is the shortest curve joining $\gamma(s)$ with $\gamma(t)$.

Example 5

An $\mathrm{Sp}(n, \boldsymbol{R})$-structure on a manifold is called a *pseudohamiltonian structure*. Such a structure defines a symplectic form on the tangent space at every point of the manifold.

PROBLEMS

1. Show that every manifold M possessing a global basis (over $\mathscr{E}(M)$) for the space $\mathscr{T}(M)$ of smooth vector fields is orientable. In particular it follows that every Lie group is orientable.

2. Let G denote a Lie group and let $N \subset G$ be a closed normal subgroup.

(a) Show that the group G/N equipped with the analytic manifold structure described in Theorem 1.4.1 is a Lie group.

(b) Show that the Lie algebra $\mathfrak{n}$ of the group N is an ideal in the Lie algebra $\mathfrak{g}$ of the group G. Also prove that the Lie algebra of the group G/N is isomorphic to the quotient algebra $\mathfrak{g}/\mathfrak{n}$.

3. Suppose $H \subset G$ is a closed subgroup of a Lie group G and let $\pi\colon G \to G/H$ be the natural projection. Prove, on account of Theorem 1.4.1, that $(G, G/H)$ is a principal bundle with the structure group H and the projection π.

4. A tensor field $T \in \mathcal{T}^r(G)$ on a Lie group G is called right-invariant provided $r_g^* T = T$ for every $g \in G$.

(a) Let T_0 be an r-covariant tensor on the space $T_e(G)$. Show that there exists a unique right-invariant tensor field $T \in \mathcal{T}^r(G)$ assuming the value T_0 at e. Prove that T is, in addition, left-invariant if and only if $\mathrm{Ad}^*(g) T_0 = T_0$ for every $g \in G$.

(b) Apply point (a) to prove that an oriented, compact and connected Lie group G admits a unique bi-invariant, positively oriented n-form ω such that $\int_G \omega = 1$.

(c) Show that on every compact connected Lie group there exists a bi-invariant Riemannian structure (i.e. a bi-invariant symmetric positive definite 2-covariant tensor field).

Hint ad (c). Show that if w is a symmetric positive definite bilinear form on the Lie algebra $\mathfrak{g}$ of the group G, then $w_0 = \int_G \mathrm{Ad}(g)^* w\, dg$ is a well defined, symmetric bilinear form satisfying $\mathrm{Ad}(g)^* w_0 = w_0$. Here, the integral denotes the Hurwitz integral of the group G (see § 1.6).

5. Write $X_{kl} = [a_i^j]$, where $a_i^j = \delta_k^j \delta_i^l - \delta_i^l \delta_i^k$ and δ_r^p denotes the Kronecker delta ($\delta_r^p = 1$ when $p = r$, $\delta_r^p = 0$ otherwise).

(a) Show that X_{kl} for $1 \leqslant k < l \leqslant n+1$ form a basis for the Lie algebra $\mathfrak{o}(n+1, \boldsymbol{R})$ and prove that

$$X_{kl} = -X_{lk}; \qquad [X_{kl}, X_{pr}] = \delta_{lp} X_{kr} + \delta_{kr} X_{lp} + \delta_{kp} X_{rl} + \delta_{lr} X_{pk};$$
$$k \neq l,\; p \neq r.$$

(b) Find for $n = 2$ the form of the vector fields X_{kl}^* on S^2 generating one-parameter transformation groups of S^2 given by the formula $\boldsymbol{R}\times S^2 \ni (t, x) \to e^{tX_{kl}}$, $x \in S^2$.

(c) Show that the invariant 2-form on S^2 is given in the spherical coordinates by the formula $\omega = c\sin\theta\, d\theta \wedge d\varphi$ and that $\int_{S^2} \omega = 4\pi C$. Also prove that the normalization condition $\int_G \omega = 1$ is satisfied for the constant $c = (4\pi)^{-1}$.

(d) Show that the invariant normalized n-form on S^n is given by the formula

$$\omega = \frac{1}{2}\pi^{-\frac{n}{2}}\Gamma\left(\frac{n}{2}\right)\frac{dx^1 \wedge \ldots \wedge dx^n}{|x^{n+1}|}.$$

6. Let G denote either the group $\mathrm{GL}(n, \boldsymbol{R})$ or $O(n, \boldsymbol{R})$ or $\mathrm{SO}(n, \boldsymbol{R})$ equipped with the natural Lie group structure and let $\pi\colon \boldsymbol{R}^n\backslash\{0\} \to P^{n-1}\boldsymbol{R}$ be the natural projection.

(a) Show that the mapping

$$G\times P^{n-1}\boldsymbol{R} \ni (g, \pi(x)) \to g\pi(x) := \pi(gx) \in P^{n-1}\boldsymbol{R}$$

is (well defined and) analytic and that $(G, P^{n-1}\boldsymbol{R})$ is a transitive Lie transformation group of the projective space $P^{n-1}\boldsymbol{R}$.

(b) Suppose that for some $g_1, g_2 \in G$ the identity $g_1\pi(x) = g_2\pi(x)$ is satisfied for all $x \in \boldsymbol{R}^n\backslash\{0\}$. Prove that $g_1 g_2^{-1}$ belongs to the centre of the group G. Show that the centre Z of the group $\mathrm{GL}(n, \boldsymbol{R})$ consists precisely of the matrices λI, $\lambda \in \boldsymbol{R}_*$.

(c) Equipping the group $\mathrm{GL}(n, \boldsymbol{R})/Z$ with a natural Lie group structure, show that $\mathrm{PL}(n-1, \boldsymbol{R}) := \mathrm{GL}(n, \boldsymbol{R})/Z$ is a Lie transformation group (called the *projective group*) of the projective space $P^{n-1}\boldsymbol{R}$.

(d) Prove that $\mathrm{SL}(n, \boldsymbol{R})\cap Z = \mathrm{SO}(n, \boldsymbol{R})\cap Z = \begin{cases}\{I\}, & n\text{—odd},\\ \{I, -I\}, & n\text{—even},\end{cases}$ and $O(n, \boldsymbol{R})\cap Z = \{I, -I\}$.

(e) Show that the isotropy group at the point $\pi(0, \ldots, 0, 1)$ of the group $\mathrm{PL}(n-1, \boldsymbol{R})$ is isomorphic to the group $\mathrm{GL}(n-1, \boldsymbol{R}) \underset{\tau}{\times} \boldsymbol{R}^{n-1}$. Determine the isotropy groups at $\pi(0, \ldots, 0, 1)$ for the groups $O(n, \boldsymbol{R})$ and $\mathrm{SO}(n, \boldsymbol{R})$ acting on $P^{n-1}\boldsymbol{R}$.

7. Let $\mathrm{SL}(n, \boldsymbol{R})$ act in a natural way on $\boldsymbol{R}^n\backslash\{0\}$.

(a) Show that the form $dx^1 \wedge \ldots \wedge dx^n$, where x^i are the natural coordinates in $\boldsymbol{R}^n$, is an invariant n-form on $\boldsymbol{R}^n\backslash\{0\}$.

(b) Prove that there is no $\mathrm{SL}(n, \boldsymbol{R})$-invariant, $(n-1)$-form on $P^{n-1}\boldsymbol{R}$.

(c) Prove that $P^{n-1}\boldsymbol{R}$ is orientable if and only if n is even.

Hint ad (b). Assuming the contrary, consider the restriction of such a form to the set U_n of all real lines with the homogeneous coordinates $x^1, \ldots, x^n$ with $x^n \neq 0$, which is diffeomorphic to $\boldsymbol{R}^{n-1}$. Note that $\mathrm{SL}(n, \boldsymbol{R})$ contains the group of all affine automorphisms of $\boldsymbol{R}^{n-1}$.

ad (c). Consider the projection $p\colon S^{n-1} \to P^{n-1}\boldsymbol{R}$ obtained by restricting to S^{n-1} the natural projection $\pi\colon \boldsymbol{R}^n \backslash \{0\} \to P^{n-1}\boldsymbol{R}$. Apply the antipodal mapping $S^{n-1} \ni x \to -x \in S^{n-1}$ to the form $p^*\omega$, where ω is a non-vanishing $(n-1)$-form on $P^{n-1}\boldsymbol{R}$.

8. Let $U(n) = \{g \in \mathrm{GL}(n, \boldsymbol{C})\colon gg^* = I\}$ denote the n-dimensional unitary group, $\langle x, y\rangle = y^*x$—the canonical scalar product in $\boldsymbol{C}^n$ and $M = \{x \in \boldsymbol{C}^n\colon \langle x, x\rangle = 1\}$—the unit sphere in $\boldsymbol{C}^n$. Both M and $U(n)$, being levels of analytic functions, each possess natural manifold structures (cf. Example 3 in § 1.2).

(a) Prove that M is diffeomorphic to S^{2n-1}.

(b) Show that the mapping $U(n) \times M \ni (g, x) \to gx \in M$ is analytic and that $U(n)$ is a transitive Lie transformation group of M. Also prove that the isotropy group at a point $m \in M$ is isomorphic to $U(n-1)$ and that M is analytically isomorphic to the homogeneous manifold $U(n)/U(n-1)$.

(c) Show that the group $\mathrm{SU}(n) \subset U(n)$ acts analytically and transitively on M and that M is isomorphic to $\mathrm{SU}(n)/\mathrm{SU}(n-1)$.

(d) Define the complex projective space $P^{n-1}\boldsymbol{C}$ and show that $(\mathrm{SU}(n), P^{n-1}\boldsymbol{C})$ is a Lie transformation group of $P^{n-1}\boldsymbol{C}$, prove also that $P^{n-1}\boldsymbol{C}$ is analytically isomorphic to $\mathrm{SU}(n)/U(n-1)$.

9. (a) Prove that $(\boldsymbol{R}^n \backslash \{0\}, P^{n-1}\boldsymbol{R})$, together with the natural projection π, is a principal bundle with the structure group R_*. Prove that $(S^{n-1}, P^{n-1}\boldsymbol{R})$ with the projection $p = \pi|_{S^{n-1}}$, is a principal bundle with the structure group $Z_2 = \{1, -1\}$.

(b) Prove that $(\boldsymbol{C}^n \backslash \{0\}, P^{n-1}\boldsymbol{C})$, together with the projection $\pi\colon \boldsymbol{C}^n \backslash \{0\} \to P^{n-1}\boldsymbol{C}$, is a principal bundle with the structure group $\boldsymbol{C}_*$. Let S^{2n-1} denote the unit sphere in $\boldsymbol{C}^n \simeq \boldsymbol{R}^{2n}$ and write $p = \pi|_{S^{2n-1}}$. Also show that the above action of $\boldsymbol{C}_*$ on $\boldsymbol{C}^n \backslash \{0\}$, when restricted to the subgroup $S^1 \subset \boldsymbol{C}_*$, leaves S^{2n-1} invariant and makes the pair $(S^{2n-1}, P^{n-1}\boldsymbol{C})$ together with the projection p a principal bundle with the structure group S^1. (The projection p is called the *Hopf mapping* for $n = 2$ and P^1C regarded as the unit sphere in R^3 (the Riemann sphere).)

10. Let M be a k-dimensional submanifold of R^n and denote by i: $M \to R^n$ the natural inclusion mapping. We shall identify the tangent space $T_x(R^n)$ with the space R^n by regarding a vector $v \in R^n$ as the differentiation $D_v = \sum_{i=1}^{n} v^i \frac{\partial}{\partial x^i}\Big|_x \in T_x(R^n)$. On account of this identification we shall equip the space $T_x(R^n)$ with the scalar product induced by the natural scalar product in R^n. Thus the space $T_m(M)$ regarded as a subspace of $T_m(R^n)$ via di_m can be identified with a k-dimensional subspace of R^n. As a result there exists a natural scalar product in $T_m(M)$ denoted by $g_m(\cdot, \cdot)$.

(a) Let g be a 2-covariant tensor field on M given by the formula $g(X, Y)(m) := g_m(X(m), Y(m))$. Prove that g is smooth.

(b) Write $O_m(M)$ for the set of all orthogonal bases for $T_m(M)$ and set $O(M) = \bigcup_{m \in M} O_m(M)$. Endow $O(M)$ with the subspace topology induced from $B(M)$. Denote by π' the restriction of the projection π: $B(M) \to M$ to $O(M)$. Let $(\varkappa, U)$ be a chart for M. Denote by $\{V_i(m)\}$ the basis obtained by the Gram–Schmidt orthonormalization process from the basis $\{X_i(m)\}$ determined by $\varkappa$. Let $\{\omega^i\}$ be the basis of 1-forms on U dual to the basis $\{V_i\}$. Define homeomorphisms $\varphi_\varkappa$: $\pi'^{-1}(U) \to U \times O(k)$ as follows

$$\varphi_\varkappa(m, Y_1(m), \ldots, Y_k(m)) = (\pi'(m), [\langle \omega^j(m), Y_i(m) \rangle]).$$

Show that $\varphi_\varkappa$ define a differentiable manifold structure on $O(M)$, such that $\varphi_\varkappa$ are diffeomorphisms ($U \times O(k)$ are equipped with the standard manifold structure). Prove also that $(O(M))$, M with the projection π' defines an $O(k)$-structure on M and the metric tensor induced by this structure (i.e. the scalar product in $T_m(M)$) coincides with that carried over from R^n.

(c) Suppose $k = n-1$. A smooth mapping $\mathcal{N}$: $M \to T(R^n)$ will be called the *normal vector field on M* provided $\mathcal{N}(m)$ is orthogonal to $T_m(M)$ and $\pi \circ \mathcal{N} = \mathrm{id}$, for the projection π: $T(R^n) \to R^n$ of the tangent bundle. Show that M is orientable if and only if M admits a non-vanishing normal vector field.

(d) Generalize (a) and (b) to the case of an oriented Riemannian manifold M and its submanifold N.

Hint ad (c). Define an $(n-1)$-form on M by the formula $\omega(m) \times \times (X_1(m), \ldots, X_{n-1}(m)) = dx^1 \wedge \ldots \wedge dx^n(\mathcal{N}(m), X_1(m), \ldots, X_{n-1}(m))$.

11. Prove the formula $\det \mathrm{Ad}_{\mathfrak{g}/\mathfrak{h}}(h) = \dfrac{\det \mathrm{Ad}_G(h)}{\det \mathrm{Ad}_H(h)}$.

Hint. Note that $\mathrm{Ad}_G(h)|_H = \mathrm{Ad}_H(h)$. Compute $\det \mathrm{Ad}_G(h)$ in a suitable basis for the Lie algebra $\mathfrak{g}$.

1.6. THE HURWITZ MEASURE

A choice of an orientation for M, splits the space of charts into two classes. Namely, given a chart $(\varkappa, U)$ with connected domain U for a manifold M, with the orientation provided by a $\mathrm{GL}(n, \boldsymbol{R})^+$ structure O, we say that $(\varkappa, U)$ is *positively oriented* if, for every $m \in U$, the basis $X_1, \ldots, X_n$ determined by the coordinates of $\varkappa$ belongs to O.

In the case where the orientation O is defined by a differential n-form this means that $\omega_m(X_1(m), \ldots, X_n(m)) > 0$ for $m \in U$. Since the form ω is nowhere zero this implies that the function $m \to \omega_m(X_1(m), \ldots$ $\ldots, X_n(m))$ does not change sign. Thus for a chart $\varkappa'$ which is not positively oriented we have $\omega_m(X_1(m), \ldots, X_n(m)) < 0$ and such a chart will be called *negatively oriented.* Positively oriented charts cover the manifold. Given two positively oriented charts $(\varkappa, U)$ and $(\varkappa', U')$ the Jacobian of the mapping $\varkappa' \circ \varkappa^{-1}$ is everywhere positive, i.e.

$$\det\left(\frac{\partial(\varkappa_i' \circ \varkappa)}{\partial x_j}\right) > 0,$$

where $\varkappa(m) = (\varkappa_1(m), \ldots, \varkappa_n(m))$.

We shall now relate with a non-vanishing n-form ω, called the *volume element* on M, a positive mesure μ_ω on M.

First, the value of $\mu_\omega(f)$ will be defined for the continuous functions f with support in the domain of a chart $(\varkappa, U)$. We assume that the function $f \circ \varkappa^{-1}$ is integrable on $\varkappa(U)$ with respect to the Lebesgue measure and define:

$$\int_M f \mathrm{d}\mu_\omega := \int_{\varkappa(U)} f\omega(X_1, \ldots, X_n) \circ \varkappa^{-1}(t^1, \ldots, t^n) \mathrm{d}t^1 \ldots \mathrm{d}t^n. \quad (1.6.1)$$

Given another positively oriented chart $(\bar{\varkappa}, \bar{U})$, we get for f such that $\operatorname{supp} f \subset U \cap \bar{U}$:

$$\int_{\bar{\varkappa}(\bar{U})} f\omega(\bar{X}_1, \dots, \bar{X}_n) \circ \bar{\varkappa}^{-1}(t^1, \dots, t^n) \mathrm{d}t^1 \dots \mathrm{d}t^n$$

$$= \int_{\varkappa(U)} f\omega(\bar{X}_1, \dots, \bar{X}_n) \circ \varkappa^{-1}(t^1, \dots, t^n) \det \left[\frac{\partial(\bar{\varkappa}^j \circ \varkappa^{-1})}{\partial t^i}\right] \mathrm{d}t^1 \dots \mathrm{d}t^n$$

$$= \int_{\varkappa(U)} f\omega(X_1, \dots, X_n) \circ \varkappa^{-1}(t^1, \dots, t^n) \mathrm{d}t^1 \dots \mathrm{d}t^n.$$

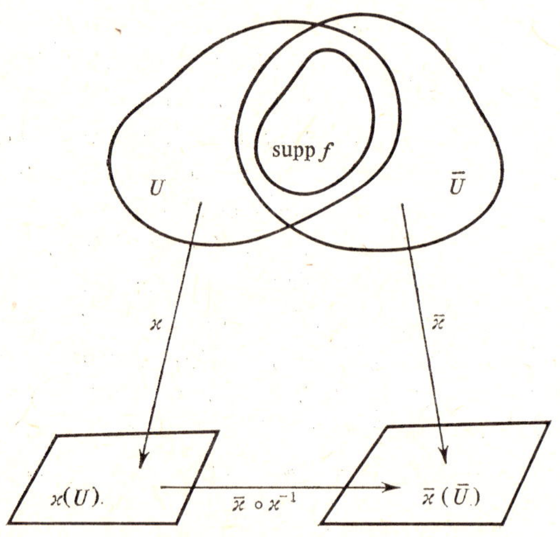

Fig. 1

In the above computations we applied the theorem on the change of variables and the transformation rules for n-forms. The formula obtained shows that the value $\int_M f \mathrm{d}\mu_\omega$ is independent of the choice of a particular chart $(\varkappa, U)$. Write $\mathscr{C}_0(M)$ for the space of compactly supported continuous functions on M.

Now, for $f \in \mathscr{C}_0(M)$ there exists a finite set of charts $\varkappa^1, \dots, \varkappa^k$ whose supports cover $\operatorname{supp} f$. Next, one can define a family of open sets $O_1, \dots, O_k$ satisfying the conditions

(1) $O_i \cap O_j = \emptyset$,
(2) $\operatorname{supp} f \subset \overline{\bigcup_i O_i}$,
(3) $\varkappa^i(U^i \setminus (\bigcup_j O_j))$ is of Lebesgue measure zero in $\boldsymbol{R}^n$.

Define $\int_M f \mathrm{d}\mu_\omega := \sum_j \int_M f \cdot 1_{O_j} \mathrm{d}\mu_\omega$. Obviously, the linear functional on $\mathscr{C}_0(M)$ so defined is positive definite and continuous with respect to convergence on compact sets and therefore it defines a measure on M.

Example 1

Choose as the manifold M, the underlying space of an n-dimensional Lie group G. Let Ω be a G-invariant n-form on G. Denote the corresponding measure $\mathrm{d}\mu_\Omega$ by $\mathrm{d}g$. The invariance of the form Ω implies that

$$\int_G f(xg)\mathrm{d}g = \int_G f(g)\mathrm{d}g. \tag{1.6.2}$$

This condition expresses the so called left-invariance of the measure $\mathrm{d}g$. The construction of the invariant measure on a Lie group was given by Hurwitz, and therefore the measure described above is called the *Hurwitz measure*. The existence of an invariant measure is not a property characteristic of the Lie groups. It is a common feature of all locally compact groups. Namely, we have

THEOREM 1.6.1 (A. Haar, J. von Neumann, A. Weil). *On every locally compact group there exists a left-invariant measure. All left-invariant measures differ from each other by a constant positive factor.*

The left-invariant measure on a group is called the *Haar measure*. Thus the Hurwitz measure is a special case of the Haar measure. The proof of the uniqueness of the Haar measure is given later in this section (Theorem 1.6.5). It is presented in order to provide the reader with complete foundations, at least in the case of Lie groups, of the theory of invariant measures.

Now, let us see how the Hurwitz measure behaves under right translations.

LEMMA 1.6.2. $r_g^* \Omega = \det \mathrm{Ad}(g)\, \Omega$.

Proof. The l_g-invariance of the form Ω implies

$$r_g^* \Omega = l_g^* r_g^* \Omega = \alpha_g^* \Omega.$$

We know, however, that $\alpha_g^*(X_1, \ldots, X_n) = \Omega(\mathrm{Ad}(g)X_1, \ldots, \mathrm{Ad}(g)X_n) = \det \mathrm{Ad}(g)\Omega(X_1, \ldots, X_n)$. □

On account of Lemma 1.6.2 we obtain

COROLLARY 1.6.3.

$$\int_G f(gx)\mathrm{d}g = |\det \mathrm{Ad}(x)| \int_G f(g)\mathrm{d}g. \tag{1.6.4}$$

The appearance of the absolute value of the determinant is caused by the fact that the mapping r_g^* may be orientation reversing.

Example 2

We shall now find the local form of the Hurwitz measure on G. To be precise, if $(\varkappa, U)$ is a chart on G around the identity e, then we shall define such a measure ν on the open subset $\varkappa(U) \subset \boldsymbol{R}^n$ that for every continuous function f with support contained in U

$$\int_G f(g)\mathrm{d}g = \int_{\varkappa(U)} f \circ \varkappa^{-1}(x)\mathrm{d}\nu(x).$$

It will turn out that ν is absolutely continuous with respect to the Lebesgue measure and its density is given by the local expression (via the chart $\varkappa$) of the operation of multiplication $G \times G \ni (g, h) \to gh \in G$.

Let ω be an invariant n-form on G and write $\omega|_U = \varrho\,\mathrm{d}x^1 \wedge \ldots \wedge \mathrm{d}x^n$. If $X_1, \ldots, X_n$ are the basis vector fields on U determined by $\varkappa$, then

$$\begin{aligned}\varrho(g) &= \omega(g)\left(X_1(g), \ldots, X_n(g)\right)\\ &= \omega(e)\left((\mathrm{d}l_{g^{-1}})_g X_1(g), \ldots, (\mathrm{d}l_{g^{-1}})_g X_n(g)\right) = \varrho(e)\det L,\end{aligned}$$

where L is the matrix of the isomorphism $(\mathrm{d}l_{g^{-1}})_g\colon T_g(G) \to T_e(G)$, with respect to the bases $\{X_i(g)\}$, $\{X_i(e)\}$. On the other hand, we have $(\mathrm{d}l_{g^{-1}})_g = ((\mathrm{d}l_g)_e)^{-1}$.

Now let $g, h \in U$ be such that $gh \in U$ and let $x = \varkappa(g)$, $y = \varkappa(h)$ and $\varkappa^i(gh) = \varphi^i(x, y)$. Also suppose that $\varkappa(e) = 0$. Then the matrix of $(\mathrm{d}l_g)_e$ with respect to the bases $\{X_i(e)\}$, $\{X_i(g)\}$ equals $\left[\dfrac{\partial\varphi^i}{\partial y^j}(x, 0)\right]$.

Consequently $\varrho(g) = \varrho(e)\left|\det\left[\dfrac{\partial\varphi^i}{\partial y^j}(x, 0)\right]\right|^{-1}$, $x = \varkappa(g)$. Now, in view of the definition of the Hurwitz measure (neglecting the unimportant constant) we arrive at

$$\mathrm{d}\nu(x) = \left| \det \left[\frac{\partial \varphi^i}{\partial y^j} (x, 0) \right] \right|^{-1} \mathrm{d}x^1 \ldots \mathrm{d}x^n.$$

Making use of this formula we shall describe the Hurwitz measure on some previously encountered groups.

(a) The invariant measure on the multiplicative group R_* has the form $\frac{1}{|x|}$ $\mathrm{d}x$, where $\mathrm{d}x$ is the Lebesgue measure on the real line.

(b) Let G be the affine group of the real line, i.e. the group "$ax+b$". In the natural parametrization an element $g \in G$ is given by $(a, b) \in \boldsymbol{R}_* \times \boldsymbol{R}$ and $(x, y)(a, b) = (xa, xb+y)$. Thus, for the Hurwitz measure we get the expression

$$\int_G f(g)\mathrm{d}g = \int_{R_* \times R} f(a, b)\,|a|^{-2}\mathrm{d}a\,\mathrm{d}b.$$

Let us now apply the parametrization in which $g = (\mathrm{e}^t, b)$ or $g = (-\mathrm{e}^t, b)$ depending on whether g belongs to the connected component of the identity of G or not. We obtain

$$\int_G f(g)\mathrm{d}g = \int_{R^2} f(\mathrm{e}^t, b)\mathrm{e}^{-t}\mathrm{d}t\,\mathrm{d}b + \int_{R^2} f(-\mathrm{e}^t, b)\mathrm{e}^{-t}\mathrm{d}t\,\mathrm{d}b.$$

(c) Splitting every matrix into columns, we shall regard $G = \mathrm{GL}(n, \boldsymbol{R})$ as an open subset of the Cartesian product $\boldsymbol{R}^n \times \ldots \times \boldsymbol{R}^n$. It can easily be seen that a left translation l_g can be represented by a block matrix of degree n^2 with the n blocks equal to g lying on the diagonal and the remaining ones equal to zero. It follows that

$$\int_G f(g)\mathrm{d}g = \int_G f(g)|\det g|^{-n}\mathrm{d}\lambda(g),$$

where $\mathrm{d}\lambda(g)$ denotes the Lebesgue measure on $\boldsymbol{R}^{n^2}$ restricted to G.

(d) Choose a neighbourhood of the identity in the group $\mathrm{SL}(2, \boldsymbol{R})$, of the form

$$U := \left\{ \begin{bmatrix} x & y \\ z & t \end{bmatrix} : x \neq 0 \right\}$$

and let

$$\varkappa\colon U \ni \begin{bmatrix} x & y \\ z & t \end{bmatrix} \to (x, y, z) \in \mathbf{R}^3$$

be a chart for U (cf. Example 3 in § 1.2). A simple calculation shows that

$$\int_G f(g)\mathrm{d}g = \int_{\varkappa(U)} f(\varkappa^{-1}(x, y, z))\frac{1}{|x|}\mathrm{d}x\mathrm{d}y\mathrm{d}z,$$

for any continuous function f with support in U.

We now proceed to the study of the case of an arbitrary locally compact group G. To begin with, we introduce the following notation.

Write $\mathfrak{M}(G)$ for the space of all Radon measures on the group G. We define an action of the group G on $\mathfrak{M}(G)$ by using the right and the left translations. Namely, for $\mu \in \mathfrak{M}(G)$ and $g \in G$ we set

$$\int_G f(x)\mathrm{d}(l_g\mu)(x) := \int_G f(gx)\mathrm{d}\mu(x) \tag{1.6.5}$$

and

$$\int_G f(x)\mathrm{d}(r_g\mu)(x) = \int_G f(xg^{-1})\mathrm{d}\mu(x). \tag{1.6.6}$$

In the sequel we shall also apply the notation $\mathrm{d}(l_g\mu)(x) = \mathrm{d}\mu(g^{-1}x)$ and $\mathrm{d}(r_g\mu)(x) = \mathrm{d}\mu(xg)$.

A left-invariant measure is defined by the condition $l_g\mu = \mu$ whereas the condition $r_g\mu = \mu$ defines a right-invariant one.

The existence of a left-invariant measure on an arbitrary Lie group has already been established. Moreover, we note that the same argument applied to a right-invariant n-form on G yields the existence of a right-invariant measure. On the orher hand, the existence of such a measure evolves from tbe following relation:

Lemma 1.6.4. *Let $\check{\mu}$ denote a measure on G defined by the identity*

$$\int_G f\mathrm{d}\check{\mu} := \int_G f(x^{-1})\mathrm{d}\mu(x). \tag{1.6.7}$$

If the measure μ is left-invariant then $\check{\mu}$ is right-invariant.

Proof.

$$\int_G f\,\mathrm{d}r_g\check{\mu} = \int_G f(xg^{-1})\,\mathrm{d}\check{\mu}(x) = \int_G f(x^{-1}g^{-1})\,\mathrm{d}\mu(x)$$

$$= \int_G f((gx)^{-1})\,\mathrm{d}\mu(x) = \int_G f(x^{-1})\,\mathrm{d}\mu(x) = \int_G f\,\mathrm{d}\check{\mu}. \qquad \square$$

It transpires that we already know all the left- and right-invariant measures on a group. This fact is stated in the following theorem, proved independently by J. von Neuman and A. Weil. (cf. Bourbaki, 1963).

THEOREM 1.6.5. *Let μ and μ' be positive left-invariant measures. Then there exists a $\lambda > 0$ such that $\mu = \lambda\mu'$.*

Proof. Denote by ν the right-invariant measure $\check{\mu}'$. Choose an $f \in \mathscr{C}_0(G)$ such that $\mu(f) \neq 0$. Consider an auxiliary function $D_f(s) := \mu(f)^{-1} \int_G f(t^{-1}s)\,\mathrm{d}\nu(t)$. The continuity of the function D_f follows on account of the Lebesgue theorem. Next by applying Tonelli's theorem, we obtain in view of the properties of invariant measures

$$\mu(f)\nu(\varphi) = \left(\int f\,\mathrm{d}\mu\right)\left(\int \varphi\,\mathrm{d}\nu\right) = \int \mathrm{d}\mu(s) \int f(s)\varphi(ts)\,\mathrm{d}\nu(t)$$

$$= \int \mathrm{d}\nu(t) \int f(s)\varphi(ts)\,\mathrm{d}\mu(s) = \int \mathrm{d}\nu(t) \int f(t^{-1}s)\varphi(s)\,\mathrm{d}\mu(s)$$

$$= \int \varphi(s)\left[\int f(t^{-1}s)\,\mathrm{d}\nu(t)\right]\mathrm{d}\mu(s) = \mu((\varphi \cdot \mu(f)D_f).$$

This implies that for an arbitrary function f' such that $\mu(f') \neq 0$

$$\mu(D_f - D_{f'}) = 0.$$

The support of μ being the whole of G, we infer that $D_f = D_{f'}$ almost everywhere. Since both functions are continuous, they must be identical. This allows to set $D_f(g) := D(g)$. Returning to the definition of D_f we get the formula

$$\mu(f)D(e) = \check{\nu}(f) = \mu'(f). \qquad \square$$

The left translations in $\mathfrak{M}(G)$ commute with the right ones, and this implies that a left translate of the measure $\mathrm{d}g$ is again a left-invariant measure. It follows from the above theorem that there exists a number $\delta_G(x) > 0$ such that $r_x\mathrm{d}g = \delta_G(x)\mathrm{d}g$. In the case of a Lie group this function is given by Corollary 1.6.3 in the form

$$\delta_G(x) = |\det \mathrm{Ad}_G(x^{-1})|. \qquad (1.6.8)$$

It is called the *modular function of the group* G. If $\delta_G = 1$ then the group G is said to be *unimodular*.

Example 3

(a) All abelian groups are (obviously) unimodular, and the same is true of compact groups. The latter follows from the fact that the modular function δ_G is a continuous homomorphism of G into the multiplicative group of positive numbers (see Problem 6 (a) for the proof of the continuity of δ_G in the case of a general topological group).

(b) The group $\mathrm{GL}(n, \boldsymbol{R})$ is unimodular. A direct calculation shows that the Hurwitz measure for $\mathrm{GL}(n, \boldsymbol{R})$ given in Example 1 (c) is bi-invariant.

(c) A group G that coincides with the derived group $DG = \overline{\{xyx^{-1}y^{-1}\colon x, y \in G\}}$ is unimodular. (This property is satisfied, for instance, for connected semisimple Lie groups, see Chapter 17.) Indeed, the modular function is equal to one of the elements of the form $xyx^{-1}y^{-1}$ and the continuity implies that this is true for the whole of G.

(d) The group "$ax+b$" provides an example of a non-unimodular group. By an easy calculation we find that in this case the modular function is of the form $\delta(a, b) = |a|^{-1}$.

(e) The group $\mathrm{SL}(2, \boldsymbol{R})$ is unimodular; in fact this is true for $\mathrm{SL}(n, \boldsymbol{R})$ for all n (cf. Problem 2).

PROPOSITION 1.6.6. *Suppose* μ *is a left-invariant measure. Then*

$$\check{\mu} = \delta_G^{-1}\mu.$$

Proof. First we check that the measure $(\delta_G^{-1}\mu)^\vee$ is left-invariant:

$$\int_G f(gx)\,\mathrm{d}(\delta_G^{-1}\mu)^\vee = \int_G f(gx^{-1})\delta_G^{-1}(x)\,\mathrm{d}\mu(x)$$

$$= \int_G f((xg^{-1})^{-1})\delta_G^{-1}(x)\,\mathrm{d}\mu(x)$$

$$= \int_G f(x^{-1})\,\delta_G^{-1}(xg)\delta_G(g)\,\mathrm{d}\mu(x)$$

$$= \int_G f(x^{-1})\,\delta_G^{-1}(x)\,\mathrm{d}\mu(x) = \int_G f(x)\,\mathrm{d}(\delta_G^{-1}\check{\mu}).$$

Consequently, there is a number $c > 0$ such that $c\mu = (\delta_G^{-1}\mu)^\vee = \delta_G\check{\mu}$. Applying to both sides the operation $^\vee$, we get $c\check{\mu} = \delta_G^{-1}\mu$. Comparing the above identities, we find that $c = 1$, which proves the proposition.□

Proposition 1.6.6 provides a new description of the right-invariant measure on G since according to Lemma 1.6.3 the measure $\mu_r := \check{\mu}$ is right-invariant. The reader is invited to deduce from the above relations the identity

$$l_g\mu_r = \delta_G(g)\mu_r. \tag{1.6.9}$$

Example 4

Combining Proposition 1.6.6 with Examples 2(b) and 3(d) we see that the right-invariant measure on the group "$ax+b$" has the form $d\mu_r(a, b) = \frac{1}{|a|}\,dadb$.

Convolution of Measures and Functions

Let (G, M) be a locally compact transformation group on a locally compact space M. Write $\mathfrak{M}_f(M)$ and $\mathfrak{M}_f(G)$ for the set of finite measures on M and G, respectively.

For any $\nu \in \mathfrak{M}_f(M)$ and $\mu \in \mathfrak{M}_f(G)$ we define a measure on M according to the formula

$$\int_M \varphi(m)\mathrm{d}(\mu*\nu) = \int_G\int_M \varphi(gm)\mathrm{d}\mu(g)\mathrm{d}\nu(x) \tag{1.6.10}$$

for $\varphi \in \mathscr{C}_0(M)$.

The measure $\mathrm{d}(\mu * \nu)$ is well-defined since the function $G \times M \ni (g, m) \to \varphi(gm)$ is both continuous and bounded and therefore integrable with respect to the finite measure $\mu\otimes\nu$ on $G\times M$.

The measure $\mu * \nu$ is called the *convolution* of the measures μ and ν. The associativity of the group operation combined with the associativity of the construction of the product measure implies that for every $\mu_1, \mu_2 \in \mathfrak{M}_f(G)$ and $\nu \in \mathfrak{M}_f(M)$ we have the relation

$$(\mu_1 * \mu_2) * \nu = \mu_1 * (\mu_2 * \nu), \tag{1.6.11}$$

where the convolution $\mu_1 * \mu_2$ is defined in terms of the left-action of the group G into itself.

We note that

$$\delta_g * \nu = l_g \nu, \qquad \delta_g * \delta_h = \delta_{gh}. \tag{1.6.12}$$

It is possible to establish, by fixing a measure on a group or on a homogeneous space, a correspondence between the integrable functions and finite measures

$$L^1(\mu) \ni \varphi \to \varphi \mathrm{d}\mu \in \mathfrak{M}_f. \tag{1.6.13}$$

This correspondence allows us to introduce the operation of convolution of functions. We shall apply it to the Haar measure on the group G. Thus for $f \in L^1(G, \mathrm{d}g)$, $\nu \in \mathfrak{M}_f(M)$ we define

$$\int_M \varphi(m)\mathrm{d}(f * \nu) := \int_G \int_M \varphi(gm) f(g) \mathrm{d}g \mathrm{d}\nu. \tag{1.6.14}$$

The convolution of two functions $\varphi_1, \varphi_2 \in \mathscr{C}_0(G)$ which arises when $M = G$ and $\mathrm{d}\nu = \varphi_2 \, \mathrm{d}g$ can also be written in the form

$$(\varphi_1 * \varphi_2)(g_1) := \int_G \varphi_1(g)\varphi_2(g^{-1}g_1)\mathrm{d}g. \tag{1.6.15}$$

The reader will easily check that

$$(\varphi_1 * \varphi_2)\mathrm{d}g = (\varphi_1 \mathrm{d}g) * (\varphi_2 \mathrm{d}g). \tag{1.6.16}$$

All compactly supported continuous functions form an algebra with respect to the operation of convolution. Whenever $\operatorname{supp}\varphi_i \subset K_i$ we have $\operatorname{supp}(\varphi_1 * \varphi_2) \subset K_1 \cdot K_2$, which yields the compactness of $\operatorname{supp}(\varphi_1 * \varphi_2)$. The continuity follows by the Lebesgue theorem. A detailed proof of the above fact is given in the following chapter.

Compactly supported smooth functions, i.e. the Schwartz test functions on a Lie group G, also form a convolution algebra denoted by $\mathscr{D}(G)$. This in turn follows from the theorem on differentiability of integrals with parameters.

The third important example of a convolution algebra, in fact the most important one from the point of view of group representation theory, is provided by the algebra $L^1(G)$ of functions integrable with respect to the Haar measure on G.

Expression (1.6.15) may not be defined for $\varphi_1, \varphi_2 \in L^1(G)$. Still, the integral under consideration is convergent for $\mathrm{d}g$—almost all g

and the function $\varphi_1 * \varphi_2$ so obtained is in $L^1(G)$. This results from the following estimate provided by the Fubini theorem.

$$||\varphi_1 * \varphi_2||_{L^1(G)} := \int_G \left| \int_G \varphi_1(g)\varphi_2(g^{-1}x)dg \right| dx$$

$$\leqslant \int_G\int_G |\varphi_1(g)||\varphi_2(g^{-1}x)|dgdx = \int_G |\varphi_1(g)| \int_G |\varphi_2(g^{-1}x)|dxdg$$

$$= \int_G |\varphi_1(g)|dg \int_G |\varphi_2(g)|dg = ||\varphi_1||_{L^1(G)}||\varphi_2||_{L^1(G)}. \tag{1.6.17}$$

Therefore the convolution proves to be a continuous operation

$$L^1(G)\times L^1(G) \to L^1(G).$$

Now, on account of Proposition 1.6.6 one can define an involution in $L^1(G)$ to be the isometry

$$f^*(g) := f(g^{-1})\delta_G(g^{-1}). \tag{1.6.18}$$

Directly from the definition of convolution we get the property

$$(\varphi_1 * \varphi_2)^* = \varphi_2^* * \varphi_1^*. \tag{1.6.19}$$

All the three convolution algebras $\mathscr{C}_0(G)$, $\mathscr{D}(G)$, $L^1(G)$ are closed under involution.

On account of formula (1.6.15) one can also define the convolution of functions which are not in $L^1(G)$. Namely, for $\varphi_1 \in L^1(G)$ and $\varphi_2 \in L^\infty(G)$ the integral (1.6.15) exists for almost all $g \in G$ and

$$||\varphi_1 * \varphi_2||_{L^\infty(G)} \leqslant ||\varphi_1||_{L^1(G)}||\varphi_2||_{L^\infty(G)}.$$

Similarly, whenever $\varphi_1 \in L^1(G)$ and $\varphi_2 \in L^2(G)$, the convolution is defined almost everywhere and belongs to $L^2(G)$. Also the following inequality holds:

$$||\varphi_1 * \varphi_2||_{L^2(G)} \leqslant ||\varphi_1||_{L^1(G)}||\varphi_2||_{L^2(G)}. \tag{1.6.20}$$

We compute

$$|\varphi_1 * \varphi_2|^2(y) \leqslant \left[\int_G |\varphi_1|(g)|\varphi_2|(g^{-1}y)dg\right]^2$$

$$= \left[\int_G |\varphi_1|^{\frac{1}{2}}(g)|\varphi_1|^{\frac{1}{2}}(g)|\varphi_2|(g^{-1}y)dg\right]^2$$

$$\leqslant \int_G |\varphi_1|(g)dg \int_G |\varphi_1|(g)|\varphi_2|^2(g^{-1}y)dg.$$

The last integral, being the convolution of integrable functions is finite for almost all y (with respect to the Haar measure dy). Moreover, it follows from the above inequality that

$$\|\varphi_1 * \varphi_2\|^2_{L^2(G)} \leqslant \|\varphi_1\|_{L^1(G)} \| |\varphi_1| * |\varphi_2|^2 \|_{L^1(G)},$$

which in view of (1.6.17) yields

$$\|\varphi_1 * \varphi_2\|_{L^2(G)} \leqslant \|\varphi_1\|_{L^1(G)} \|\varphi_2\|_{L^2(G)}.$$

PROBLEMS

1. (a) Prove that the formula

$$\mu(f) = \frac{1}{2\pi} \int_0^{2\pi} f(e^{i\theta}) d\theta$$

defines an invariant normalized measure on the group $\boldsymbol{T}$ of complex numbers of absolute value 1. Show that this measure can also be defined by the formula

$$\mu(f) = \frac{1}{\pi} \int_{-\infty}^{\infty} f\left(\frac{1+ix}{1-ix}\right) \frac{dx}{1+x^2}.$$

(b) Let $p\colon S^1 \to P^1\boldsymbol{R}$ be the restriction of the natural projection $\pi\colon \boldsymbol{R}^2 \setminus \{0\} \to P^1\boldsymbol{R}$ to $S^1 \subset \boldsymbol{R}^2$. Show that the formula

$$\nu(f) := \mu(f \circ p), \quad f \in C(P^1\boldsymbol{R}),$$

where we naturally identify functions on S^1 with functions on $\boldsymbol{T}$, defines a measure ν on $P^1\boldsymbol{R}$, invariant under the action of the group SO(2) on $P^1\boldsymbol{R}$. Find the local expression of ν in the chart $\varkappa(x^1, x^2) = x^1/x^2$, (x^1, x^2) being the homogeneous coordinates for $P^1\boldsymbol{R}$.

2. On account of the formula $|\det \mathrm{Ad}(x^{-1})| = \delta(x)$, prove in one of the ways indicated below that the group SL(2, $\boldsymbol{R}$) is unimodular. Thus either

(a) compute $\det \mathrm{Ad}(x)$ for the one-parameter subgroups $a_i(t)$ of SL(2, $\boldsymbol{R}$) defined in Example 2 in § 1.4 and show that the elements of the form $x = a_2(t_2)a_3(t_3)a_1(t_1)$ form a dense subset of SL(2, $\boldsymbol{R}$) or

(b) show that the group $\mathrm{Ad}(\mathrm{SL}(2, \boldsymbol{R}))$ is a subgroup of the group $G(\beta)$, β being a bilinear form $\beta(X, Y) := \mathrm{tr}(XY)$ defined on the algebra $\mathfrak{sl}(2, \boldsymbol{R})$.

3. Prove that the left and the right Haar measures on the group $T(n, \boldsymbol{R})$ of triangular matrices are defined, respectively, by

$$\mu_l(f) = \int f(x_{11}, x_{12}, \ldots, x_{nn}) |x_{11}^n x_{22}^{n-1}, \ldots x_{nn}|^{-1} \mathrm{d}x_{11} \mathrm{d}x_{12} \ldots \mathrm{d}x_{nn},$$

$$\mu_r(f) = \int f(x_{11}, x_{12}, \ldots, x_{nn}) |x_{11} x_{22}^2 \ldots x_{nn}^n|^{-1} \mathrm{d}x_{11} \mathrm{d}x_{12} \ldots \mathrm{d}x_{nn},$$

while the modular function δ equals

$$\delta(g) = |x_{11}^{1-n} x_{22}^{3-n} \ldots x_{nn}^{n-1}|,$$

where $g = \begin{bmatrix} x_{11} & x_{12} & \cdots & x_{1n} \\ 0 & x_{22} & \cdots & x_{2n} \\ \vdots & \vdots & \ddots & \vdots \\ 0 & 0 & \cdots & x_{nn} \end{bmatrix}$.

4. Show that every function $f \in \mathscr{C}_0(G)$ is left (right, resp.) uniformly continuous on G, i.e. for every $\varepsilon > 0$ there exists a neighbourhood U of the identity $e \in G$ such, that $|f(xy) - f(x)| < \varepsilon$, for all $x \in G$ and $y \in U$ a neighbourhood $V \ni e$ such that $|f(yx) - f(x)| < \varepsilon$, $x \in G$, $y \in V$ (resp—the neighbourhoods U and V are, as a rule, different.)

Hint. Sets of the forms Ux (xU respectively) constitue a base of neighbourhoods for $x \in G$. Construct a covering of supp$f$ by the sets Ux and use the compactness of supp$f$.

5. Let μ be the Haar measure on G and $f \in \mathscr{C}_0(G)$ a positive non-zero function. Show that $\mu(f) > 0$. Derive from this that every compact set with non-empty interior is of measure greater than zero, which implies that the support of the Haar measure is the whole of G—this fact was used in the proof of Theorem 1.6.5.

Hint. Let f_0 be a positive non-zero function in $\mathscr{C}_0(G)$ such that $\mu(f_0) = 0$. Show that every compact set can be covered with a finite number of translates of the set on which f_0 is different from zero. Then conclude (by the invariance of the integral) that μ vanishes for an arbitrary positive function from $\mathscr{C}_0(G)$.

6. Write μ for the left Haar measure on a locally compact group G. Let δ_G be its modular function defined by the identity $r_x \mu = \delta_G(x) \mu$.

(a) Prove that δ_G is a continuous homomorphism of the group G into the multiplicative group $\boldsymbol{R}_+$.

(b) Given a topological automorphism τ of G, we define for every measure μ a measure $\tau\mu$ by setting

$$\int f(x)\mathrm{d}\tau\mu(x) := \int f \circ \tau^{-1}(x)\mathrm{d}\mu(x).$$

Show that there exists a positive constant $\chi(\tau)$, called the *modulus* of the automorphism τ, satisfying $\tau\mu = \chi(\tau)\mu$. Also prove that $\tau \to \chi(\tau)$ is a homomorphism of the group of topological automorphism of G into $\boldsymbol{R}_+$.

(c) Prove that if α_x is an inner automorphism of G ($\alpha_x(y) = xyx^{-1}$) then $\chi(\alpha_x) = \delta_G(x^{-1})$.

(d) Show that δ_G is invariant under topological automorphisms of the group G, i.e. for any such automorphism τ we have $\delta(\tau(g)) = \delta(g)$ for all $g \in G$.

(e) Suppose we are given a positive measure μ on G satisfying $l_g\mu = \chi_\mu(g)\mu$, $g \in G$ for some (positive) function χ_μ on G. Show that

(i) χ_μ is a continuous homomorphism of G into $\boldsymbol{R}_+$,

(ii) $\nu = \chi_\mu\mu$ is the left Haar measure on G.

Hint. Apply the result of Problem 4 to prove the continuity of δ_G and χ_μ.

7. Prove that the Haar measure on the direct product $G = G_1 \times G_2$ is given by $\mu = \mu_1 \otimes \mu_2$, i.e.

$$\int_G f(g_1, g_2)\mathrm{d}\mu(g_1, g_2) = \int_{G_1}\int_{G_2} f(g_1, g_2)\mathrm{d}\mu_1(g_1)\mathrm{d}\mu_2(g_2),$$

where μ_1, μ_2 are the Haar measures on G_1 and G_2, respectively; moreover, the modular function $\delta_G(g_1, g_2)$ equals $\delta_{G_1}(g_1)\delta_{G_2}(g_2)$.

8. (a) Suppose $G = K \times_\tau H$. Let μ_K, μ_H be the left Haar measures on K and H, respectively, and let δ_K, δ_H be the respective modular functions. Show that if $\chi(k)$ denotes the modulus of an automorphism $\tau(k)$ of the group H then

(i) The left Haar measure on G is of the form $\chi^{-1}(\cdot)\mu_K \otimes \mu_H$, that is

$$f \to \mu_K \otimes \mu_H(\chi^{-1}f).$$

(ii) $\delta_G(k, h) = \delta_K(k)\delta_H^{-1}(h)\chi^{-1}(k)$.

(iii) The right Haar measure on G satisfies $\mu_G^r = \mu_K^r \otimes \mu_H^r$.

(b) On account of (a) find the form of the Haar measure for the group of affine automorphism of $\boldsymbol{R}^n$.

9. Show that the following groups are unimodular:

(a) open subgroups of an unimodular group,

(b) closed, normal subgroups of an unimodular group,

(c) groups possessing a compact neighbourhood of the identity which is invariant under inner automorphisms.

Hint to (a). The Haar measure restricted to such a subgroup is the Haar measure for that subgroup; (b) Apply Theorem 1.7.4; (c) such a neighbourhood is of finite positive measure.

10. (a) Show that for every locally compact group G there exists a base of symmetric, relatively compact neighbourhoods of the identity. (A set is said to be *relatively compact* provided its closure is compact).

(b) Let $\{O_\alpha\}$ be a base of neighbourhoods of the identity as in (a), μ—the Haar measure on G, χ_α—the characteristic function of the set O_α. Then the family $\{\varphi_\alpha\}$ of functions $\varphi_\alpha := \mu(O_\alpha)^{-1}\chi_\alpha$ is a generalized sequence which converges in $\mathfrak{M}_f(G)$ to the point measure δ_e (the Dirac measure) concentrated at the identity e. (The sequence is directed in the way oposite to the relation of inclusion, i.e. $\alpha < \beta$ whenever $O_\alpha \supset O_\beta$.)

(c) Prove that there exists a generalized sequence $\{\psi_\alpha\}$ of continuous functions (of smooth functions in the case of a Lie group) convergent to δ_e.

Hint. (b) Apply the identity

$$\delta_e(f) - \int \varphi_\alpha(x) f(x)\, d\mu(x) = \int \varphi_\alpha(x)[f(e) - f(x)]\, d\mu(x)$$

and make use of the uniform continuity of continuous compactly supported functions on G.

(c) For every compact set K and open $O \supset K$ there is a continuous function f_K equal to 1 on K and vanishing outside O. This follows from the normality of locally compact spaces (the Urysohn lemma). An analogous statement in the case of a Lie group and a smooth function f can be proved by applying Problem 8 from § 1.2.

11. (a) Define mappings L_g: $L^1(G) \to L^1(G)$, R_g: $L^1(G) \to L^1(G)$ according to the formula

$$L_g f(x) := f(g^{-1} x), \qquad R_g f(x) = f(xg)\,\delta_G(g).$$

Show that L_g and R_g are isometries and that $(L_g f)^+ = (R_g f)^+$.

(b) Show that the convolution of functions integrable with respect to the Haar measure on a group G is given by formula (1.6.15), i.e. $\varphi_1 * \varphi_2(g) = \int_G \varphi_1(g)\varphi_2(g^{-1}g_1)dg$ for almost all $g_1 \in G$. Show that for $\varphi_1 \in L^1(G)$ and $\varphi_2 \in L^\infty(G)$ the convolution $\varphi_1 * \varphi_2$ is defined and is a continuous function on G.

(c) In the case where G is compact, prove that the convolution $\varphi_1 * \varphi_2$ is well defined for arbitrary $\varphi_1, \varphi_2 \in L^2(G)$ and is a continuous function. Also prove that the mappings L_g, R_g defined above map $L^2(G)$ into itself and are unitary. Finally, show that the mapping $f \to f^+$ is also unitary.

12. Let G be an unimodular group.

(a) Let $\nu \in \mathfrak{M}(G)$, $f \in \mathscr{C}_0(G)$. Show that $f * \nu$, $\nu * f$ are continuous functions on G and, if in addition ν is a finite measure, then these functions are integrable.

(b) Let $\mu, \nu \in \mathfrak{M}_f(G)$, $f \in \mathscr{C}_0(G)$. Prove that

$$\nu(f * \mu) = \nu * \check{\mu}(f),$$
$$\nu(\mu * f) = \check{\mu} * \nu(f).$$

Hint. (a) Use Problem 4.

13. Let G be a locally compact unimodular group and $H \subset G$ a closed unimodular subgroup. For an arbitrary measure μ on H we define a functional μ^G on $\mathscr{C}_0(G)$ by putting $\mu^G(f) = \mu(f|_H)$ for $f \in \mathscr{C}_0(G)$.

(a) Prove that μ^G is a measure on G and the mapping $\mu \to \mu^G$ is an injection of $\mathfrak{M}(H)$ into $\mathfrak{M}(G)$.

(b) Prove that

(i) $(\check{\mu})^G = (\mu^G)^\vee$,

(ii) $f|_H * \mu = (f * \mu^G)|_H$, $f \in \mathscr{C}_0(G)$,

(iii) $(\mu * \nu)^G = \mu^G * \nu^G$, $\mu, \nu \in \mathfrak{M}/(H)$,

where in (ii) and (iii) the convolution is taken on the left with respect to H and on the right with respect to G.

Hint. (b) To prove property (iii) apply (b) of the preceding problem.

1.7. QUASI-INVARIANT MEASURES

Let (G, M) be a locally compact transformation group. One can define an action of G on the space of all measures on M. Namely, we denote by $g\mu$ the measure satisfying

$$\int_M f(m)\mathrm{d}(g\mu)(m) := \int_M f(g\,m)\mathrm{d}\mu(m). \tag{1.7.1}$$

A measure μ on M will be called *invariant* if $g\mu = \mu$ for all $g \in G$. Such a measure, however, does not necessarily exist. For instance, take as G the group "$ax+b$" acting in a natural way on $\boldsymbol{R}^1$. A G-invariant measure μ on $\boldsymbol{R}^1$ is, in particular, invariant under all translations. Consequently, μ is the Haar measure for the group $\boldsymbol{R}^1$ and therefore of the form $\mu = A\,\mathrm{d}x$, $A \in \boldsymbol{R}_+$. However, the Lebesgue measure $\mathrm{d}x$ is not invariant under transformations $x \to ax$, $a \neq 1$, thus $A = 0$, and hence $\mu = 0$.

Still, in the case of a homogeneous space (G, M), one can distinguish a class of measures which are transformed under the action of an element of G into a measure equivalent to the initial one.

DEFINITION. A measure μ on M is called *quasi-invariant* provided for every $g \in G$ the measure $g\mu$ is equivalent to the measure μ.

Hence, on account of the Radon-Nikodym theorem, we get

PROPOSITION 1.7.1. *For every quasi-invariant measure μ on M there exists a function $G \times M \ni (g, m) \to \dfrac{\mathrm{d}g\mu}{\mathrm{d}\mu}(m) \in \boldsymbol{R}_+$ so that for every $f \in \mathscr{C}_0(M)$*

$$\int_M f(g\,m)\mathrm{d}\mu(m) = \int_M f(m)\frac{\mathrm{d}g\mu}{\mathrm{d}\mu}(m)\mathrm{d}\mu(m). \tag{1.7.2}$$

Denote

$$\mathscr{S}_\mu(g, m) := \frac{\mathrm{d}g^{-1}\mu}{\mathrm{d}\mu}(m).$$

The reader is invited in Problem 5 in § 1.7 to prove the following simple

Proposition 1.7.2. *For arbitrary $g_1, g_2 \in G$ and for almost all $m \in M$ we have the identity*

$$\mathscr{S}_\mu(g_1 g_2, m) = \mathscr{S}_\mu(g_1 g_2, m)\mathscr{S}_\mu(g_2, m). \tag{1.7.3}$$

The following theorem, due to Mackey, deals with the theory of quasi-invariant measures (cf. Mackey, 1952, 1953).

Theorem 1.7.3. *Suppose G is a locally compact, separable group and H a closed subgroup of G, and let* dg, δ_G *and* dh, δ_H *be fixed left-invariant measures and modular functions on G and H, respectively, then*

(1) *for every quasi-invariant measure* dμ *on G/H there exists a measurable positive function ϱ on G verifying*

$$\varrho(gh) = \frac{\delta_H(h)}{\delta_G(h)}\varrho(g), \qquad g \in G, \quad h \in H \tag{1.7.4}$$

and

$$\int_G f(g)\varrho(g)\mathrm{d}g = \int_{G/H} \mathrm{d}\mu(gH)\int_H f(gh)\mathrm{d}h; \tag{1.7.5}$$

the function $\mathscr{S}_\mu$ corresponding to this measure is of the form

$$\mathscr{S}_\mu(g, hH) = \varrho(gh)/\varrho(h), \tag{1.7.6}$$

(2) *for every positive function on G, satisfying* (1.7.4) *there exists a quasi-invariant measure μ on G/H fulfilling* (1.7.5) *and* (1.7.6),

(3) *every Lie group G admits a differentiable function ϱ satisfying* (1.7.4).

From this fundamental theorem we get in particular

Corollary 1.7.4. *For every couple (G, H) of Lie groups with $H \subset G$ there exists a quasi-invariant measure μ such that the function $\mathscr{S}_\mu$ is differentiable on $G \times G/H$.*

Corollary 1.7.5. *There exists an invariant measure on G/H if, and only if,* $\delta_G(h) = \delta_H(h)$ for $h \in H$.

Proof. $\Rightarrow$ The invariance of a measure μ implies that $\mathscr{S}_\mu \equiv 1$. Thus in particular $\mathscr{S}_\mu(g, eH) = \varrho(g)/\varrho(e) = 1$ and the assertion follows from (1.7.4).

$\Leftarrow$ The function ϱ, being identically 1, fulfils the assumptions of Theorem 1.7.3(2), and the corresponding measure defined by (1.7.5), is obviously invariant. □

PROBLEMS

1. Let H be a closed subgroup of G and let μ be the Haar measure on H. Prove that the function F defined for $f \in C_0(G)$ as $F(xH) = \int_H f(xh)\,d\mu(h)$ belongs to $\mathscr{C}_0(G/H)$.

Hint. To prove the continuity of F it sufficies to estimate $\int_H |f(yxh)-f(xh)|\,d\mu(h)$ for y, in a suitable neighbourhood of $e \in G$. Take a compact neighbourhood V of e and let $g \in \mathscr{C}_0(G)$ be a positive function equal to 1 on $V \cdot \operatorname{supp} f$. Apply the inequality $|f(yx)-f(x)| \leqslant |f(yx)-f(x)|g(x)$ and the continuity of f to get desired estimation for the above integral.

2. Suppose μ is a positive measure on G, such, that there exists a function $\lambda\colon G \times G \to R$, continuous and satisfying the identity $\int_G f(gx)\,d\mu(x) = \int_G \lambda(g^{-1}, x) f(x)\,d\mu(x)$ for $f \in \mathscr{C}_0(G)$. Prove that λ is positive and that the identity $\sigma(gx) = \lambda(g, x)\sigma(x)$ is valid for the function $\sigma\colon x \to \lambda(x, e)$. Infer that $\sigma^{-1}\mu$ is the left Haar measure on G.

3. (a) Let λ be a measure on the projective line $P^1\boldsymbol{R}$, defined by the formula

$$\int f\,d\lambda = \int_{-\infty}^{\infty} f(t, 1)\,dt$$

with f on the right-hand side regarded as a function of homogeneous coordinates on $P^1\boldsymbol{R}$. Prove that λ is a quasi-invariant measure with respect to the group $G = \mathrm{SL}(2, \boldsymbol{R})$ and

$$s_\lambda(g, x) = \frac{1}{(cx+d)^2} \quad \text{for } g = \begin{bmatrix} a & b \\ c & d \end{bmatrix} \in G.$$

(b) Suppose $G = \mathrm{SL}(2, \boldsymbol{R})$ acts on the upper half-plane $H = \{z \in \boldsymbol{C}\colon \operatorname{im} z > 0\}$ by homographies:

$$z \to gz := \frac{az+b}{cz+d}, \qquad g = \begin{bmatrix} a & b \\ c & d \end{bmatrix}.$$

Show that the Lebesgue measure λ on H is quasi-invariant with respect to G, and find the form of the function $s_\lambda(g, z)$. Also prove that there exists a G-invariant measure on H. Find this measure.

Hint to (b). Show that the isotropy group G_i equals SO(2) and deduce from Corollary 1.7.4 the existence of an invariant measure. In order to find the explicit form of the measure, note that the function $g \to s_\lambda(g, i)$ is constant on the cosets of the isotropy subgroup and apply a modification of the method of Problem 2.

4. Suppose G contains closed subgroups K and H such that $K \cap H = \{e\}$ and $K \times H \ni (k, h) \to kh \in G$ is a homeomorphism. Set $\varrho(kh) = \delta_H(h)/\delta_G(h)$. Show that ϱ defines a quasi-invariant measure on G/H which corresponds to the Haar measure on K after G/H is identified with K. Moreover, show that the Haar measure μ_G satisfies

$$\int f(x) \mathrm{d}\mu_G(x) = \int\limits_K \int\limits_H f(kh)\, \delta_G(h)/\delta_H(h) \mathrm{d}\mu_H(h) \mathrm{d}\mu_K(k).$$

Use this to compute the Haar measure, the quasi-invariant measure and the Radon–Nikodym derivative of its translate in the following cases

(a) $G = G_1 \times_\tau G_2, \quad K = G_1, \quad H = G_2,$

(b) $G = G_1 \times G_2, \quad K = G_2, \quad H = G_1,$

(c) $G = \mathrm{SL}(2, \boldsymbol{R}), \quad K = \mathrm{SO}(2), \quad H = \left\{ \begin{bmatrix} x & y \\ 0 & x^{-1} \end{bmatrix} : x > 0, y \in \boldsymbol{R} \right\}.$

5. Let μ be a quasi-invariant measure on M. Denote by $\lambda(g, m)$ the Radon–Nikodym derivative of $g\mu$, i.e. such a function that

$$\int\limits_M f(m) \mathrm{d}g\mu(m) = \int\limits_M f(m)\, \lambda(g, m) \mathrm{d}\mu(m), \quad f \in \mathscr{C}_0(M).$$

(a) Prove that λ satisfies the equation

$$\lambda(g_1 g_2, m) = \lambda(g_2, g_1^{-1} m)\, \lambda(g_1, m)$$

for μ-almost all m and deduce hence Proposition 1.7.2.

(b) Prove that

$$\int\limits_M f(m)\, \lambda(g^{-1}, m) \mathrm{d}g\mu(m) = \int\limits_M f(m) \mathrm{d}\mu(m).$$

1.8. ELEMENTS OF THE CLASSIFICATION OF THE LIE GROUPS AND ALGEBRAS

The purpose of the present section is to distinguish the basic types of Lie groups and algebras and to give their general characterization. We shall deal in a parallel fashion with finite-dimensional algebras over the fields $\boldsymbol{R}$ and $\boldsymbol{C}$, since in describing the real case, just as in the classification of endomorphism of a linear space, the way leads through the theory of the spaces over the algebraically closed field $\boldsymbol{C}$.

Consequently let $\mathfrak{g}$ denote a Lie algebra over the field $\boldsymbol{R}$ or $\boldsymbol{C}$. By the adjoint representation of the algebra $\mathfrak{g}$ we understand the mapping

$$\mathfrak{g} \ni X \to \operatorname{ad}X \in L(\mathfrak{g}),$$

where $\operatorname{ad}X(Y) = [X, Y]$.

The mapping ad is a homomorphism of the algebra $\mathfrak{g}$ into $L(\mathfrak{g})$, i.e. a representation on the vector space $\mathfrak{g}$. The properties of this representation constitute a basis for the classification of Lie algebras.

The algebra $\mathfrak{g}$ is said to be *abelian* if $\operatorname{ad}X = 0$, for every $X \in \mathfrak{g}$, i.e. if the commutator is a zero mapping.

Further types of algebras appear as those which differ "as little as possible" from the abelian case.

Let $\mathfrak{a}$ and $\mathfrak{b}$ be subalgebras of G. Denote by $[\mathfrak{a}, \mathfrak{b}]$ the smallest subalgebra of $\mathfrak{g}$ containing the commutators $[X, Y]$, $X \in \mathfrak{a}$, $Y \in b$. If both $\mathfrak{a}$ and $\mathfrak{b}$ are ideals, then $[\mathfrak{a}, \mathfrak{b}]$ is also an ideal of $\mathfrak{g}$ contained in $\mathfrak{a} \cap \mathfrak{b}$. By the derived algebra of the algebra $\mathfrak{g}$ we mean the ideal $[\mathfrak{g}, \mathfrak{g}]$, also denoted by $\mathfrak{D}\mathfrak{g}$. The quotient algebra $\mathfrak{g}/\mathfrak{D}\mathfrak{g}$ is commutative (abelian). The proofs of the above facts are simple and are left to the reader.

We define inductively a descending sequence of ideals of the algebra $\mathfrak{g}$, called the *central sequence of* $\mathfrak{g}$:

$$\mathscr{C}^0\mathfrak{g} = \mathfrak{g}, \quad \mathscr{C}^1\mathfrak{g} = \mathfrak{D}\mathfrak{g}, \quad \mathscr{C}^{n+1}\mathfrak{g} = [\mathfrak{g}, \mathscr{C}^n\mathfrak{g}].$$

We say that the algebra $\mathfrak{g}$ is *nilpotent* if $\mathscr{C}^n\mathfrak{g} = 0$ for a sufficiently large n. As can easily be observed, the last of the non-zero subspaces in the central sequence is contained in the centre of the algebra $\mathfrak{g}$, i.e. all the operators $\operatorname{ad}X$, $X \in \mathfrak{g}$ are zero on it. If the algebra $\mathfrak{g}$ is nilpotent and $\mathscr{C}^n\mathfrak{g} = 0$, then for an arbitrary sequence $X_1, \ldots, X_n \in \mathfrak{g}$

$$\operatorname{ad}X_1 \circ \ldots \circ \operatorname{ad}X_n = 0.$$

In particular, every endomorphism $\operatorname{ad} X$ is a nilpotent operator, i.e. satisfies the equation $A^n = 0$. It follows that the operators of the adjoint representation of a nilpotent algebra have spectrum consisting only of zero.

We now define, inductively, a sequence of ideals called the *derived sequence*:

$$\mathfrak{D}^0\mathfrak{g} = \mathfrak{g}, \quad \mathfrak{D}^1\mathfrak{g} = \mathfrak{D}\mathfrak{g} \ldots \mathfrak{D}^{n+1}\mathfrak{g} = \mathfrak{D}(\mathfrak{D}^n\mathfrak{g}).$$

The algebra $\mathfrak{g}$ is said to be *solvable* if the derived sequence is zero from a certain place on. Thus, solvability means that there exists a sequence of ideals

$$\mathfrak{g} = \mathfrak{g}^1 \supset \ldots \supset \mathfrak{g}^{n-1} \supset \mathfrak{g}^n = 0$$

such that $\mathfrak{g}^{i-1}$ is an ideal in $\mathfrak{g}^i$ and the quotients $\mathfrak{g}^{i+1}/\mathfrak{g}^i$ are abelian algebras. The last of the non-zero components of the derived sequence is a commutative ideal of the algebra $\mathfrak{g}$.

Obviously every nilpotent algebra is solvable.

We now introduce semisimple algebras, which constitute a type of algebras opposite to solvable algebras. Namely, we say that the algebra $\mathfrak{g}$ is *semisimple* if it has no commutative ideal besides the zero one. An algebra is said to be *simple* if it has dimension > 1 and contains no ideals besides the trivial ideals $\mathfrak{g}$ and $\{0\}$.

It follows directly from the definition that a simple algebra is semisimple.

The Killing form of an algebra characterizes the type of the algebra.

DEFINITION. The bilinear function

$$\mathfrak{g} \times \mathfrak{g} \ni (X, Y) \to B_{\mathfrak{g}}(X, Y) = \operatorname{Tr}(\operatorname{ad} X \circ \operatorname{ad} Y)$$

is called the *Killing form of the algebra* $\mathfrak{g}$.

When dealing with a fixed algebra $\mathfrak{g}$, we shall write simply $B := B_{\mathfrak{g}}$.

Here are the elementary properties of the Killing form:

LEMMA 1.8.1.

(1) $B(X, Y) = B(Y, X)$;

(2) *for every automorphism* σ *of the algebra* $\mathfrak{g}$

$$B(\sigma X, \sigma Y) = B(X, Y);$$

(3) *for every differentiation* D *of the algebra* $\mathfrak{g}$ *we have*

$$B(DX, Y) = -B(X, DY).$$

Property (1) follows from the elementary properties of the trace, (2) becomes obvious if the definition of an automorphism of $\mathfrak{g}$ is written in the form $\operatorname{ad}\sigma X = \sigma \circ \operatorname{ad} X \circ \sigma^{-1}$. Assertion (3) follows from the definition of a differentiation as an operator satisfying the equation $\operatorname{ad} DX = [D, \operatorname{ad} X]$, and from an application of the Jacobi identity.

Since $\operatorname{ad} Z$ is a differentiation of the algebra $\mathfrak{g}$, it follows in particular that

$$B(\operatorname{ad} Z(X), Y) = -B(X, \operatorname{ad} Z(Y)), \qquad X, Y, Z \in \mathfrak{g}. \tag{1.8.1}$$

Moreover, we observe

LEMMA 1.8.2. *If* $\mathfrak{a} \subset \mathfrak{g}$ *is an ideal, then*

$$B_{\mathfrak{a}}(\cdot, \cdot) = B_{\mathfrak{g}}(\cdot, \cdot)|_{\mathfrak{a}\times\mathfrak{a}}.$$

Proof. We may select a basis $\{X_i\}$ for $\mathfrak{g}$ in such a way that $\{X_i\}_{i=1,\dots,\dim\mathfrak{a}}$ constitute a basis for $\mathfrak{a}$. Then, in this basis, the matrix $[a_{ij}]$ of the operator $\operatorname{ad} X_i$ for $i \leqslant \dim\mathfrak{a}$ will have the form

$$\left[\begin{array}{c|c} A_i & B_i \\ \hline 0 & 0 \end{array}\right]$$

and the matrix $\operatorname{ad} X_1 \operatorname{ad} X_2$ the form

$$\left[\begin{array}{c|c} A_1 A_2 & A_1 B_2 \\ \hline 0 & 0 \end{array}\right]$$

Then $B_{\mathfrak{g}}(X_1, X_2) = \operatorname{Tr} A_1 A_2 = B_{\mathfrak{a}}(X_1, X_2)$. □

THEOREM 1.8.3 (E. Cartan).

(1) *An algebra* $\mathfrak{g}$ *is semisimple* $\Leftrightarrow$ *the Killing form* $B_{\mathfrak{g}}$ *is non-degenerate.*

(2) *An algebra* $\mathfrak{g}$ *is solvable* $\Leftrightarrow$

$$B_{\mathfrak{g}}(X, [Y, Z]) = 0, \qquad X, Y, Z \in \mathfrak{g}.$$

We shall not quote the proof of this theorem, it can be found in most basic monographs dealing with Lie algebras, e.g. in Chevalley, 1946, Helgason, 1968 and Zhelobenko, 1970.

COROLLARY 1.8.4. *An ideal of a semisimple algebra is itself a semisimple algebra.*

Thanks to Theorem 1.8.3 and Lemma 1.8.2 it remains to show that $B(\cdot, \cdot)|_{\mathfrak{a}\times\mathfrak{a}}$ is a non-degenerate form. So let us examine $\mathfrak{b} = \mathfrak{a} \cap \mathfrak{a}^{\perp}$.

Since by (1.8.1) $\mathfrak{a}^\perp$ is an ideal, then so is $\mathfrak{b}$. Moreover, since for every $Z \in \mathfrak{g}$, $X_1, X_2 \in \mathfrak{b}$

$$B(Z, [X_1, X_2]) = B([X_1, Z], X_2) = 0,$$

it follows that $[X_1, X_2] = 0$ and hence $\mathfrak{b}$ is an abelian ideal.

Let us now compute $B(T, Z)$ for $Z \in \mathfrak{g}$, $T \in \mathfrak{b}$. The operator $\operatorname{ad} T \operatorname{ad} Z$ maps $\mathfrak{b}$ into 0, and every vector which is not in $\mathfrak{b}$ is sent into $\mathfrak{b}$. Thus, the matrix of this operator in a basis containing a basis of $\mathfrak{b}$ has only zero entries in the diagonal; hence

$$B(T, Z) = 0.$$

Finally $T = 0$.

Corollary 1.8.5. *A semisimple algebra is an orthogonal sum, relative to the Killing form, of simple ideals.*

Proof. In the proof of the previous corollary we observed that if $\mathfrak{a} \subset \mathfrak{g}$ is an ideal then $\mathfrak{a} \cap \mathfrak{a}^\perp = 0$. Thus we get a decomposition into the orthogonal direct sum $\mathfrak{g} = \mathfrak{a} + \mathfrak{a}^\perp$. The dimension of the algebra $\mathfrak{g}$ being finite, we may decompose it in finitely many steps into a direct sum of simple ideals. □

Corollary 1.8.6. *The adjoint representation of a semisimple algebra is faithful, i.e.* $\operatorname{ad} X = 0$ *implies* $X = 0$.

Proof. If $\operatorname{ad} X$ were zero for a certain $X \neq 0$, then $B(X, Y) = 0$ for an arbitrary $Y \in \mathfrak{g}$, which contradicts Theorem 1.8.3 (1). □

A connected Lie group is said to be *solvable*, *nilpotent*, *abelian*, *simple* or *semisimple*, provided its Lie algebra has the respective property.

Information concerning the structure of the algebra can be carried over to the group via the exponential mapping. The adjoint representations of the algebra and the group are related according to the following formula, known from § 1.3 (Proposition 1.3.6)

$$e^{\operatorname{ad} X} = \operatorname{Ad}(\exp X).$$

Let $\mathfrak{h} \subset \mathfrak{g}$ be an ideal. Then for an arbitrary $X \in \mathfrak{g}$ and $Y \in \mathfrak{g}$ we have $e^{\operatorname{ad} X} \in \mathfrak{h}$.

It follows from Proposition 1.8.4 that

$$\exp X \exp Y \exp(-X) = \exp X A\mathrm{d}(\exp X)(Y) = \exp e^{\operatorname{ad} X}(Y).$$

Hence $\exp X \exp Y \exp(-X)$ belongs to the connected subgroup $H_{\mathfrak{h}} \subset G$ corresponding to the subalgebra $\mathfrak{h}$.

If the group G is connected then the elements of the form $\exp X$, $X \in \mathfrak{g}$ generate a group. Finally we obtain $ghg^{-1} \in H_{\mathfrak{h}}$ for arbitrary $g \in G$ and $h \in H_{\mathfrak{h}}$. Thus we have proved

COROLLARY 1.8.7. *Let G be a connected group. Then $\mathfrak{h} \subset \mathfrak{g}$ is an ideal $\Leftrightarrow$ the subgroup $H_{\mathfrak{h}}$ is normal.*

The subgroup corresponding to the ideal $\mathfrak{D}\mathfrak{g} := [\mathfrak{g}, \mathfrak{g}]$ is called the *derived group*. We see from Theorem 1.3.5 that this group is generated by elements of the form $ghg^{-1}h^{-1}$, $g, h \in G$.

As in the case of Lie algebra, we denote by $\mathfrak{D}G$ the derived subgroup of a connected group G and inductively $\mathfrak{D}^{n+1}G = \mathfrak{D}(\mathfrak{D}^n G)$.

The definitions given above can be extended to non-connected groups: a group is solvable if and only if $\mathfrak{D}^n G = \{e\}$ for a certain n. Similarly G is simple, if it contains no non-trivial normal subgroups; It is abelian if $\mathfrak{D}G = \{e\}$.

PROBLEMS

1. Prove that if $\mathfrak{a}, \mathfrak{b} \subset \mathfrak{g}$ are ideals of a Lie algebra $\mathfrak{g}$ then $[\mathfrak{a}, \mathfrak{b}]$ and $\mathfrak{a}+\mathfrak{b}$ are ideals of $\mathfrak{g}$. Also prove that if both $\mathfrak{a}$ and $\mathfrak{b}$ are solvable ideals (i.e. they are ideals of $\mathfrak{g}$ and solvable Lie algebras with respect to the Lie algebra structure inherited from that of $\mathfrak{g}$) then $[\mathfrak{a}, \mathfrak{b}]$ and $\mathfrak{a}+\mathfrak{b}$ are solvable. Is the last statement true if the word solvable is replaced by the word nilpotent?

2. Prove that if T is an endomorphism or an derivation of a Lie algebra $\mathfrak{g}$ then $T(D\mathfrak{g}) \subset D\mathfrak{g}$.

3. Verify the following statements ($\boldsymbol{K}$ stands for either $\boldsymbol{R}$ or $\boldsymbol{C}$)

(a) The Lie algebra $\mathfrak{n}(n, \boldsymbol{K})$ consisting of square $n \times n$ matrices with zero on and below the main diagonal is nilpotent,

(b) The Lie algebra $\mathfrak{t}(n, \boldsymbol{K})$ consisting of square $n \times n$ matrices with zeros below the main diagonal is solvable but not nilpotent,

(c) The Lie algebra consisting of all square $n \times n$ matrices of the form $\begin{bmatrix} a & & * \\ & \ddots & \\ 0 & & a \end{bmatrix}$ is nilpotent, though the matrices in the algebra are not necessarily nilpotent.

(In all cases, the bracket is the usual commutator of matrices.)

4. (a) Prove that the Lie algebra $\mathfrak{aff}(1)$ consisting of matrices $\begin{bmatrix} a & b \\ 0 & 0 \end{bmatrix}$ $a, b \in K$ is solvable but not nilpotent. Prove also that every two-dimensional non-abelian Lie algebra is isomorphic to $\mathfrak{aff}(1)$.

(b) Prove that every derivation of $\mathfrak{aff}(1)$ is inner, i.e. if D is a derivation then $D = \mathrm{ad}(X)$ for some $X \in \mathfrak{aff}(1)$,

(c) Prove that if a Lie algebra $\mathfrak{g}$ contains an ideal $\mathfrak{a}$ isomorphic to $\mathfrak{aff}(1)$, then there exists an ideal $\mathfrak{b} \subset \mathfrak{g}$ such that $\mathfrak{g} = \mathfrak{a} \oplus \mathfrak{b}$ (direct sum),

(d) Prove that if the derived ideal of a Lie algebra $\mathfrak{g}$ is one-dimensional and not contained in the centre of $\mathfrak{g}$ then $\mathfrak{g}$ is a direct sum of an abelian subalgebra isomorphic to $\mathfrak{aff}(1)$.

Hint. (b) Use problem 2. (c) Take b equal to the centralizer of $\mathfrak{a}$, i.e. $\mathfrak{b} = \{X \in \mathfrak{g}\colon \mathrm{ad}(X)|_{\mathfrak{a}} = 0\}$ and use (b).

5. (a) Prove that the Lie algebra of the Heisenberg–Weyl group (Problem 9, § 1.3) is the only three-dimensional Lie algebra whose derived algebra is one-dimensional and contained in the centre.

(b) Show that a Lie algebra whose derived subalgebra has dimension one and is contained in the centie, is a direct sum of an abelian subalgebra and a subalgebra isomorphic to the Heisenberg–Weyl algebra.

6. If $\mathfrak{z}$ is the centre of a Lie algebra $\mathfrak{g}$ prove that never dim $\mathfrak{z}$ = dim $\mathfrak{g}-1$.

7. Let $\mathfrak{g}$ be a Lie algebra, $\mathfrak{h} \subset \mathfrak{g}$ a subalgebra, $\mathfrak{k} \subset \mathfrak{g}$ an ideal. Prove the following:

(a) If $\mathfrak{g}$ is nilpoten then $\mathfrak{h}$ and $\mathfrak{g}/\mathfrak{k}$ are nilpotent,

(b) If $\mathfrak{g}$ is solvable then $\mathfrak{h}$ and $\mathfrak{g}/\mathfrak{k}$ are solvable,

(c) If $\mathfrak{k}$ is solvable and $\mathfrak{g}/\mathfrak{k}$ is solvable then $\mathfrak{g}$ is solvable.

By means of an example show that the last statement does not hold when solvable is replaced by nilpotent.

8. Show that solvability of a Lie algebra $\mathfrak{g}$ is equivalent to existence of a sequence of ideals $\mathfrak{g} = \mathfrak{g}^1 \supset \mathfrak{g}^2 \supset \ldots \supset \mathfrak{g}^{n-1} \supset \mathfrak{g}^n = 0$, such that $\mathfrak{g}^{i+1}$ is an ideal in $\mathfrak{g}^i$ and $\mathfrak{g}^i/\mathfrak{g}^{i+1}$ is abelian for $1 \leqslant i \leqslant n$.

9. Let $\mathfrak{g}$ be a Lie algebra and c_{ij}^k its structure constants relative to some basis of $\mathfrak{g}$ (cf. Problem 7, § 1.3). Prove that

(a) $\mathfrak{g}$ is semisimple iff it does not contain any solvable ideal,

(b) $\mathfrak{g}$ is semisimple iff $\det\left(\sum_{i,k} c_{lk}^{i} c_{si}^{k}\right) \neq 0$.

10. Let $\mathfrak{g}$ be a Lie algebra and $\mathfrak{a} \subset \mathfrak{g}$ a simple semisimple ideal. Show that there exists an ideal $\mathfrak{b} \subset \mathfrak{g}$ such that $\mathfrak{g} = \mathfrak{a} \oplus \mathfrak{b}$.

11. Let $\varrho\colon \mathfrak{g} \to \mathfrak{gl}(n, \boldsymbol{K})$ be a homomorhism of a Lie algebra $\mathfrak{g}$. Show that the bilinear form $\tau_\varrho(X, Y) = \operatorname{tr}\varrho(X)\varrho(Y)$ is symmetric and satisfies an identity $\tau_\varrho(\operatorname{ad}(Z)X, Y) + \tau_\varrho(X, \operatorname{ad}(Z)Y) = 0$.

12. (a) Prove that the Lie algebra $\mathfrak{sl}(2, \boldsymbol{K})$ is simple and, more generally, that $\mathfrak{sl}(n, \boldsymbol{K})$ is simple.

(b) Show that $\mathfrak{sl}(n, \boldsymbol{K})$ is the derived subalgebra of $\mathfrak{gl}(n, \boldsymbol{K})$ and that $\mathfrak{gl}(n, \boldsymbol{K})$ is a direct sum of its centre and $\mathfrak{sl}(n, \boldsymbol{K})$.

Hint. Take $X_+ = \begin{bmatrix} 0 & 1 \\ 0 & 0 \end{bmatrix}$, $X_- = \begin{bmatrix} 0 & 0 \\ 1 & 0 \end{bmatrix}$, $H = \begin{bmatrix} 1 & 0 \\ 0 & -1 \end{bmatrix}$ and show that a non-zero ideal would contain either X_+ or X_- and hence all X_+, H_-, H. The case of a general n could be proved similarly by taking subalgebras of triangular (upper and lower) and diagonal matrices or by showing that the Killing form of $\mathfrak{sl}(n, \boldsymbol{K})$ is equal to $4n\operatorname{tr}(XY)$.

13. In the space $\boldsymbol{C}^3$ define the commutator by the formula $[x, y] = x \times y$ (the vector product of x and y). Prove that this turns $\boldsymbol{C}^3$ into a Lie algebra isomorphic with $\mathfrak{sl}(2, \boldsymbol{C})$. Compute the Killing form of this algebra.

Chapter 2

Representations of Locally Compact Groups

The objective of the following chapter is to introduce the notion of a representation and to describe some general examples. We deal here with a general notion of a representation on a locally convex space (since in Part II we shall come across both unitary representations and representations on spaces of smooth functions). Problems of the continuity of representations have been studied rather extensively, in order to avoid the doubts that might otherwise arise, as regards the sense of the integrals of the operator and the vector-valued functions defining the integral transformations, which are investigated later on.

Inducing § 2.2 is a basic construction, which sets the abstract theory of representations in the realities of functional analysis. Since two types of inducing are considered (called inducing in the sense of Mackey and in the sense of Bruhat) this allows us to deal with the elements of both measure theory and distribution theory. The multiplier representations in the range considered here are but a different realization of induced representations.

Another construction of basic importance, for both theoretic and practical reasons, is the Gelfand–Raikov construction, assigning to a positive definite function on a group a cyclic unitary representation of that group. It follows from Theorem 2.5.3 that every cyclic representation can be obtained in this way. The last part of the chapter provides a description of the space of smooth vectors of a representation of a Lie group. Also basic properties of the infinitesimal representation connected with such a representation are described.

There is a number of monographs which develop extensively the theory here outlined: Bourbaki, 1963, Dixmier, 1964, Hewitt, Ross, 1963, Kirillov, 1972, Maurin, 1968, Naimark, 1959, Weil, 1940, and Warner, 1972.

2.1. DEFINITION OF A REPRESENTATION. EXAMPLES

Throughout this chapter G is assumed to be a locally compact group.

DEFINITION. A representation (E, T) of the group G on a linear topological space E over C is said to be *continuous*, provided the mapping, $G \times E \ni (g, e) \to T_g e \in E$ is continuous.

We shall write out three conditions which are satisfied by every continuous representation, and by studying the relationship between them we shall get some convenient criteria of the continuity of a representation:

(1) for every $g \in G$ the operator T_g is continuous,

(2) for every $e \in E$ the mapping $G \ni g \to T_g e \in E$ is continuous,

(3) for every compact set $K \subset G$ the family of operators $\{T_g: g \in K\}$ is equicontinuous.

We immediately observe the implication (3) → (1).

A representation fulfilling (1) and (2) is called *partially continuous*. A partially continuous representation need not be continuous. Nevertheless we have the following:

PROPOSITION 2.1.1. *Any representation* (G, E, T) *on a locally convex space* E, *satisfying* (2) *and* (3) *is continuous.*

Proof. Choose an arbitrary compact neighbourhood O of a point $g \in G$. Condition (3) means that for an arbitrary neighbourhood P_1 of zero in E there exists a neighbourhood P_2 such that $T_g x \in P_1$ for $g \in O$ and $x \in P_2$. Hence we conclude that the continuity of the mapping $G \times E \ni (g, x) \to T_g x \in G$ at the point (g, O) follows directly from condition (3). We shall apply (2) to prove the continuity at an arbitrary point. Every neighbourhood of a point $y \in E$ is of the form $y + P_1$, P_1 being a neighbourhood of zero in E. Therefore we may select neighbourhoods O of g and P_2 of zero in E so that $T_h x \in T_g x + \frac{1}{2} P_1$ and $T_h P_2 \subset \frac{1}{2} P_1$ for $h \in O$. We get $T_h(x + P_2) \subset T_g x + P_1$ for $h \in O$, which proves the continuity of the representation. □

At this point let us defer for a while the exposition of the general theory in order to consider a number of examples.

The Regular Representation

Let $F(G, V)$ denote the space of all functions on G with values in a space V. We define an action of the group G on $F(G, V)$ according to the formula

$$L_g\varphi(h) = \varphi(g^{-1}h), \qquad \varphi \in F(G, V), \quad g, h \in G.$$

Let Φ be a linear subspace of the space $F(G, V)$ such that $L_g\Phi \subset \Phi$ for all $g \in G$. The pair (Φ, L) is called the *left regular representation of the group* G *on* Φ provided Φ is endowed with a topology such that (Φ, L) is a continuous representation.

We now present some special examples.

Let V be a Banach space.

PROPOSITION 2.1.2. *Suppose* Φ *is one of the spaces;*

(a) $\mathscr{C}(G, V)$*—the space of* V*-valued continuous functions on* G *equipped with the topology of almost uniform convergence;*

(b) $\mathscr{C}_0(G, V)$*—the space of compactly supported continuous functions with the topology of compact convergence*

(c) $L^p(G, V)$, $1 \leqslant p < \infty$;

then the pair (Φ, L) *is the left regular representation of* G *on* Φ.

Proof. (a) We first check that condition (3) is satisfied. The group multiplication being continuous, this implies that for arbitrary compact sets $K, C \subset G$ the set

$$K^{-1}C = \{g \in G\colon g = k^{-1}c;\ k \in K,\ c \in C\}$$

is compact. Write $P_{C,\varepsilon}, := \{f \in \mathscr{C}(G, V)\colon \|f(g)\| < \varepsilon \text{ for } g \in C\}$. Whenever $f \in P_{K^{-1}C}$, we have $L_g f \in P_{C,\varepsilon}$ for all $k \in K$, and condition (3) follows since the sets $P_{C,\varepsilon}$ constitute a base of neighbourhoods at zero in the representation space.

It remains to establish the continuity of the mapping $G \ni g \to L_g f$ for a fixed $f \in \mathscr{C}(G, V)$. Let $g \in G$. We have to show that for an arbitrary $\varepsilon > 0$ and a compact $K \subset G$ there is a neighbourhood O of the point g such that

$$\sup_{h \in K} |f(g^{-1}h) - f(g_1^{-1}h)| < \varepsilon$$

for every $g_1 \in O$.

We first assume that g is the identity of the group. All we have to do is to show the uniform continuity of the function f restricted to the compact set K; but this was already proved in Problem 4, § 1.6. The continuity at an arbitrary point $g \in G$ now follows by applying this result to the set $gK \subset G$. Finally, the continuity of the representation follows in view of Proposition 2.1.1.

(b) This is proved by the same method.

(c) In this case the operators of the representation are isometries, hence condition (3) is immediately satisfied.

The proof of (2) is divided into two parts. First, we note that for $f \in \mathscr{C}_0(G, V)$ the continuity of the representation $G \ni g \to L_g f \in L^p(G, V)$ follows from the continuity of the regular representation on $\mathscr{C}_0(G, V)$ and from the continuity of the inclusion mapping $\mathscr{C}_0(G, V) \to L^p(G, V)$. Second, for arbitrary $f, f_1 \in L^p(G, V)$ we apply these estimates

$$\begin{aligned} ||L_g f - L_{g_1} f||_p &= ||L_g(f - L_{g^{-1}g_1} f)||_p \leqslant ||L_g|| \, ||f - L_{g^{-1}g_1} f||_p \\ &\leqslant ||f_1 - f||_p + ||L_{g^{-1}g_1} f - f_1||_p \\ &\leqslant ||f_1 - f||_p + ||L_{g^{-1}g_1} f - L_{g^{-1}g_1} f_1||_p + ||L_{g^{-1}g_1} f_1 - f_1||_p \\ &\leqslant 2||f_1 - f||_p + ||L_{g^{-1}g_1} f_1 - f_1||_p \end{aligned}$$

for every $g, g_1 \in G$. In the above computations we made use of the isometricity of L_g and we also applied Minkowski's inequality. It follows from the last inequality that for every g_1 close enough to g we have $||L_g f - L_{g_1} f||_p$. To this end it suffices to approximate f by an $f_1 \in \mathscr{C}_0(G, V)$ and to apply the continuity of the representation L on $\mathscr{C}_0(G, V)$. □

Proposition 2.1.3. *Suppose G acts on a normed space E and satisfies condition* (3). *Let the mapping $G \ni g \to T_g e \in E$ be continuous for e running over a dense linear subset Ψ of E. Then the representation (E, T) itself is continuous.*

Proof. Let $\varepsilon > 0$ and $e \in E$. Fix a compact neighbourhood O of the identity and let $N = \sup ||T_g||, g \in O$. Choose an $e_1 \in \Psi$ so that $||e_1 - e|| < \varepsilon$ and a neighbourhood $O_1 \subset O$ for which $||T_g e_1 - e_1|| < \varepsilon$.

We have the estimates

$$\begin{aligned} ||T_g e - e|| &\leqslant ||T_g e - T_g e_1|| + ||T_g e_1 - e|| \\ &\leqslant ||T_g|| \, ||e - e_1|| + ||T_g e_1 - e_1|| + ||e_1 - e|| \leqslant (N+2)\varepsilon, \end{aligned}$$

which imply partial continuity at the identity of the group.

Continuity at an arbitrary point $g \in G$ results from the inequalities

$$||T_g e - T_{g_1} e|| \leqslant ||T_{g_1}(T_{g_1^{-1}g} e - e)|| \leqslant ||T_{g_1}|| \, ||T_{g_1^{-1}g} e - e||.$$

Hence in view of Proposition 2.1.1 the assertion follows. □

It may very well be the case that a representation is not continuous although condition (3) is satisfied. Consider, as an example, the left regular representation of the group G on the space $L^\infty(G)$. In the case in question the operators L_g are isometries, and the mapping $G \ni g \to L_g \varphi \in L^\infty(G)$ is continuous if and only if φ is a left-uniformly continuous function on G. Proposition 2.1.3 does not work in this case, for the space of such φ is not dense in $L^\infty(G)$.

In order to throw some more light upon the present subject, certain facts concerning the relationship between the partial continuity and condition (3) will be given below.

There exists a wide class of linear topological spaces for which conditions (1) and (2) imply (3). The partial continuity of a representation is then equivalent to its continuity. That class comprises all locally convex linear spaces for which the assertion of the Banach–Steinhaus theorem remains valid (see e.g. Bourbaki, 1955). In particular, we have:

PROPOSITION 2.1.4. *If E is a linear space, complete in the topology defined by a countable system of semi-norms (i.e. E is a Fréchet space) then every partially continuous representation on E of a locally compact group is continuous.*

Therefore, as long as we deal with Banach spaces (with Hilbert spaces, in particular) there is no need whatsoever to distinguish between continuous and partially continuous representations.

In the special case of a Lie group G ($\dim G = d$) one can define continuous representations on the spaces of smooth functions on G with values in a Banach space.

We start by defining adequate topologies in these spaces.

A sequence of functions $\varphi_n \in \mathscr{D}(X, V)$ is said to be *convergent to zero* if there is a compact set K such that $\operatorname{supp} \varphi_n \subset K$ for all n and the sequence $D(\varphi_n \circ \varkappa^{-1})$ tends to 0 uniformly on $\varkappa(V)$ for every chart $(V, \varkappa)$ and every differential operator D on $\boldsymbol{R}^d$.

PROPOSITION 2.1.5. *The left regular representations of* G *on* $\mathscr{E}(G, V)$ *and on* $\mathscr{D}(G, V)$ *are continuous.*

The proof of this fact is based on the same arguments as that of Proposition 2.1.2. One should only remember that for every compact subset of G there are finitely many charts whose domains provide cover.

DEFINITION. A representation (H, U) on a Hilbert space H is called *unitary* if the operators U_g are unitary for $g \in G$.

The following proposition expresses a fundamental property of unitary representations, which accounts for the special role of such representations in decomposition theory.

PROPOSITION 2.1.6. *Let* (H, U) *be a unitary representation on a Hilbert space* H *and let* $H_1 \subset H$ *be an invariant subspace, i.e.* $U_g H_1 \subset H_1$ *for all* $g \in G$. *Then the subspace* $H_1^{\perp}$ *is also invariant. Moreover, the orthogonal projection* P *on* H_1 *commutes with the representation operators.*

Proof. Let $h_1 \in H_1^{\perp}$. Then for every $h \in H_1$ and $g \in G$ we have $0 = (U_{g^{-1}} h|h_1) = (h|U_g h_1)$, which means that $U_g h_1 \in H_1^{\perp}$. Now, for arbitrary $h, h_1 \in H$, we get the identities:

$$\begin{aligned}(PU_g h|h_1) &= (U_g h|Ph_1) = (h|U_{g^{-1}}Ph_1) = (h|PU_{g^{-1}}Ph_1)\\ &= (U_g Ph|Ph_1|) = (PU_g Ph|h_1) = (U_g Ph|h_1).\end{aligned}$$

In the above manipulations we twice applied the invariance of the subspace H_1. This concludes the proof of the proposition. □

Example 1

The left regular representation on the space $L^2(G)$ of square integrable functions with respect to the Haar measure is unitary. Indeed, the operators L_g are isometric as follows easily from the invariance of the Haar measure

$$||L_g f||^2 = \int_G |f(g^{-1}x)|^2 dx = \int_G |f(x)|^2 dx = ||f||^2.$$

On the other hand, the equalities $L_g \circ L_{g^{-1}} = L_{g^{-1}} \circ L_g = \text{id}$ show that L_g is a unitary operator.

We may define a unitary representation on $L^2(G)$ by means of the right translation. Namely we set

$$(R_g f)(x) := \delta^{\frac{1}{2}}(g) f(xg)$$

(δ being the modular function of G). Then the identities

$$||R_g f||^2 = \delta(g) \int_G |f(xg)|^2 dx = \int_G |f(x)|^2 dx = ||f||^2$$

show that R_g is an isometry. Again we have $R_g \circ R_{g^{-1}} = R_{g^{-1}} \circ R_g = \text{id}$, which allows us to conclude that R is also a unitary representation on $L^2(G)$. It will be established later that both representations are, in a sense, identical (equivalent).

Example 2

Let G denote a locally compact abelian group and $\hat{G}$ the group of characters of G, and let μ be a measure on $\hat{G}$. For every $g \in G$ we define an operator T_g on the space $L^2 = L^2(\hat{G}, \mu)$ by putting $T_g f(\chi) := \chi(g) f(\chi)$, $\chi \in \hat{G}$. It can easily be checked that the assignment $g \to T_g$ is a unitary representation of G. Every measurable subset $S \subset \hat{G}$ with $\mu(S) \neq 0$ defines an invariant subspace of L^2, namely the subspace L^2_S of functions vanishing outside S. Moreover, if $\mu(G-S) \neq 0$ then $L^2_{\hat{G}-S}$ is also a non-zero invariant subspace of L^2, and we get the decomposition

$$L^2 = L^2_S \oplus L^2_{\hat{G}-S}.$$

Example 3

Let (G, M) be a locally compact transformation group. Suppose that there exists a G-invariant measure μ on M, then the formula

$$T_g f(x) := f(g^{-1}x), \qquad g \in G, \quad x \in M$$

defines a unitary representation of the group G on the space $L^2(M, \mu)$. In the following section we present two different (although in a sense equivalent) constructions, namely of an induced representation and of a multiplier representation, being a generalization of the above definition to the case where M does not admit an invariant measure.

2.2. BASIC CONSTRUCTIONS OF REPRESENTATIONS. INDUCED REPRESENTATIONS

Suppose G is a locally compact group and denote by Γ a closed subgroup of G. Every representation (G, E, T) defines a representation of the subgroup Γ on E by restricting T to Γ,

$$T_h^{\downarrow} := T_h.$$

The operation $(G, E, T) \to (\Gamma, E, T)$ is called the *restriction of the representation.*

Inducing is a procedure inverse to restriction. It consists in constructing a representation of the group G by applying a representation of its subgroup.

We shall first deal with the unitary representations induced by a unitary representation of a subgroup, and then proceed to the induced representations on the spaces of smooth functions on the group.

The general construction of inducing with the help of unitary representations was introduced by G. Mackey, who generalized the methods of V. Bergmann, I. M. Gelfand and M. A. Naimark for obtaining families of irreducible representations of classical groups. This construction is also an extension of the theory of induced representations of compact groups, which was developed by A. Weil and which originated from G. Frobenius's theory of finite group representations.

One can see that the above stages of the development of this theory are closely related to the development of the theory of invariant and quasi-invariant measures both on groups and on symmetric spaces. Bruhat's theory of induced smooth representations is, in turn, connected with the appearance of distribution theory.

Let $d\mu$ be a quasi-invariant measure on the homogeneous space G/Γ and let ϱ denote, as before, the function on G related to the measure $d\mu$ by the formula (1.7.5). Let (E, V) be a unitary representation of the subgroup Γ on a separable Hilbert space E.

Denote by H^V the space of E-valued functions on G satisfying the following conditions:

$$\text{for every } a \in E \text{ the function } g \to (f(g)|a)_E \text{ is measurable,} \tag{2.2.1}$$

for every $h \in \Gamma$ and $g \in G$ we have

$$f(gh) = \varrho(h)^{\frac{1}{2}} V(h^{-1}) f(g), \qquad (2.2.2)$$

$$\int_{G/\Gamma} \varrho(g)^{-1} (f(g)|f(g))_E \, d\mu(g\Gamma) < \infty. \qquad (2.2.3)$$

H^V will denote the space of classes of functions in H^V equal dg-almost everywhere. The integral in (2.2.3) should be understood in the following sense:

The function $\psi\colon g \to \varrho(g)^{-1}(f(g)|f(g))_E$, on account of (2.2.2) and property (1.7.4) of the function ϱ, is invariant under all right translations from Γ. Therefore it defines a function on G/Γ by the formula $\tilde{\psi}(g\Gamma) = \psi(g)$. The integral (2.2.3) denotes

$$\int_{G/\Gamma} \tilde{\psi}(g\Gamma) \, d\mu(g\Gamma).$$

We define an inner product on the space H^V by setting

$$(f_1|f_2) := \int_{G/\Gamma} \varrho(g)^{-1} (f_1(g)|f_2(g))_E \, d\mu(g\Gamma). \qquad (2.2.4)$$

To begin with let us verify the correctness of the definition. Choosing an orthonormal basis for E, we can represent the integrand in the form

$$(f_1(g)|f_2(g))_E = \sum_{i=1}^{\infty} (f_1(g)|e_n)_E (e_n|f_2(g))_E.$$

The function $g \to \varrho(g)^{-1}(f_1(g)|f_2(g))_E$, being the pointwise limit of measurable functions, is measurable itself. We shall prove its integrability by applying the polarization formula

$$\begin{aligned} 4(f_1(g)|f_2(g))_E &= \|f_1(g)+f_2(g)\|_E^2 - \|f_1(g)-f_2(g)\|_E^2 \\ &\quad + i\|f_1(g)+if_2(g)\|_E^2 - i\|f_1(g)-if_2(g)\|_E^2. \end{aligned}$$

The same proof as in the case of the classical Riesz–Fischer theorem yields the following result.

PROPOSITION 2.2.1. *The pair* $(H^V, (\cdot|\cdot))$ *is a Hilbert space.*

PROPOSITION 2.2.2. *The space* H^V *is invariant under the left regular representation. The representation* (H^V, L) *is unitary and continuous.*

Proof. Since the left translations commute with the right ones, a left translate of a function in H^V is again a function satisfying (2.2.2). We shall verify at one stroke both condition (2.2.3) and the unitarity of a left translation

$$\begin{aligned}(L_h^{-1}f|L_h^{-1}f) &= \int_{G/\Gamma} \varrho(g)^{-1}\|f(hg)\|_E^2 d\mu(g\Gamma)\\ &= \int_{G/\Gamma} \frac{\varrho(hg)}{\varrho(g)}\varrho(hg)^{-1}\|f(hg)\|_E^2 d\mu(g\Gamma)\\ &= \int_{G/\Gamma} \varrho(g)^{-1}\|f(g)\|_E^2 d\mu(g\Gamma) = (f|f)\end{aligned}$$

since, as we remember (cf. § 1.7), the function $g \to \dfrac{\varrho(hg)}{\varrho(g)}$ is the Radon–Nikodym derivative of the measure $dh\mu$ with respect to $d\mu$. In view of the unitarity of the representation if suffices to check the continuity of the mapping $G \ni g \to L_g f \in H^V$. In the proof we shall apply Proposition 2.1.3. To this end we shall construct a dense subspace of H^V on which the continuity of the representation will be easier to handle.

Define a mapping $\beta_V: \mathscr{C}_0(G, E) \to H^V$ by requiring

$$\beta_V(f)(g) = \int_\Gamma \varrho(\gamma)^{-\frac{1}{2}} V(\gamma) f(g\gamma) d\gamma, \tag{2.2.5}$$

$d\gamma$ denoting the left Haar measure on Γ. The function $\beta_V f$ is not, in general, compactly supported (unless Γ is compact); nevertheless, we see that

$$\operatorname{supp} \beta_V(f) \text{ is compact in } G/\Gamma. \tag{2.2.6}$$

Denote by $\mathscr{C}_0^V$ the space of functions of the form $\varphi = \beta_V(f)$ for a certain $f \in \mathscr{C}_0(G, E)$. Plainly, $\mathscr{C}_0^V \subset H^V$. Moreover, we shall prove that

LEMMA 2.2.3. *The set $\mathscr{C}_0^V$ is dense in H^V.*

Proof. Suppose $f \in H^V$ is orthogonal to all $\varphi = \beta_V(\psi)$;

$$0 = \int_{G/\Gamma} \varrho(g)^{-1} (f(g)|\varphi(g))_E d\mu(g\Gamma)$$

$$= \int_{G/\Gamma} \int_{\Gamma} \varrho(g)^{-1} (f(g) | L_\gamma \psi(g\gamma))_E \varrho(\gamma)^{-\frac{1}{2}} \mathrm{d}\gamma \, \mathrm{d}\mu(g\Gamma)$$

$$= \int_{G/\Gamma} \int_{\Gamma} \varrho(g\gamma)^{-1} (V_\gamma^{-1} f(g) \, \delta_\Gamma^{\frac{1}{2}}(\gamma) | \psi(g))_E \mathrm{d}\gamma \, \mathrm{d}\mu(g\Gamma)$$

$$= \int_{G/\Gamma} \int_{\Gamma} \varrho(g\gamma)^{-1} (f(g\gamma) | \psi(g\gamma))_E \mathrm{d}\gamma \, \mathrm{d}\mu(g\Gamma)$$

$$= \int_{G} (f(g) | \psi(g))_E \mathrm{d}g$$

in virtue of the formula (1.7.5). It means that $f(g) = 0$ for dg-almost all g and consequently $||f|| = 0$, what proves the lemma. □

The continuity of the mapping $g \to L_g f$ can now be derived by applying the Lebesgue theorem. We shall justify its applicability.

For a compact neighbourhood O and an element $g \in G$ the supports of the functions $\varrho^{-1}||L_h f - f||^2$, $h \in O$ are all contained in the compact subset $O \circ \operatorname{supp} f \subset G/\Gamma$. Therefore, the functions may be majorized by a function constant on this set and vanishing outside. Letting $g_n \to g$, we see that owing to the continuity of f the sequence of functions $\varrho^{-1}||L_{g_n} f - f||^2$ is pointwise convergent to zero. Hence the Lebesgue theorem yields the convergence of $L_{g_n} f$ to f in the norm on H^V. □

DEFINITION. The left regular representation of the group G on the space H^V is termed the *representation induced by the unitary representation* (E, V) of the subgroup Γ and denoted by ${}_G U^V$ or U^V.

In order to distinguish between this representation and the induced representations on other spaces, defined below, we term it the *representation induced in the sense of Mackey.*

Let G be a Lie group countable at infinity, i.e. a countable union of compact subsets. Let Γ denote a closed subgroup of G and let (E, V) be a representation of Γ on a Banach space E.

Write $\mathscr{E}^V$ for the subspace of $\mathscr{E}(G, V)$ consisting of all functions satisfying (2.2.2). Recall that the function $\varrho|\Gamma$ is the quotient of the modular functions of the groups G and H, and therefore condition (2.2.2) is independent of the choice of the function ϱ. Denote by $\mathscr{D}^V$ the space of functions in $\mathscr{E}^V$ verifying the condition:

$\operatorname{supp} f$ is a compact subset of G/Γ.

Definition. By the *differentiable representation* induced by the representation (E, V) of the subgroup Γ we shall understand the regular representation of G on $\mathscr{D}^V$.

We say that this representation is induced by (Γ, E, V) in the sense of Bruhat. The operator β_V (see (2.2.5)) maps the space $\mathscr{D}(G, E)$ into $\mathscr{D}^V$. One can even prove surjectivity (Bruhat, 1956, also see Warner, 1972). This operator plays an important role in representation theory. In Chapter 18 of Part III we shall see that, for some special pairs (G, Γ), β_V can be regarded as a generalized Fourier transformation. One can easily see that for $\Gamma = G = \boldsymbol{R}^n$ this operator indeed coincides with the Fourier transformation (Problem 4).

Multiplier Representations

Let (G, M) be a transformation group and E a linear space.

Definition. By a *multiplier* we mean a function χ on $G \times M$ with values in a space of linear automorphisms of E satisfying the conditions

$$\chi(e, m) = 1 \tag{2.2.7}$$

for an arbitrary $m \in M$ (e being the neutral element of G and 1 the identity operator on E),

$$\chi(gh, m) = \chi(h, m)\chi(g, hm) \tag{2.2.8}$$

for $g, h \in G,\ m \in M$.

Multipliers arise in a natural way upon the construction of representations on function spaces on M. To begin with let us note how they are related to the representations of the isotropy groups at points in M. To this end fix a point $m_0 \in M$ and let $G_0 = \{g \in G\colon gm_0 = m_0\}$. Denote

$$V(g) := \chi(g^{-1}, m_0), \tag{2.2.9}$$

Then the identity (2.2.8) takes the form

$$V(gh) = V(g)\,V(h) \qquad \text{for} \quad g, h \in G_0.$$

Hence V is a representation of the group G_0 on the space E.

Example 1

Let G denote the group of all affine transformations of the space $\boldsymbol{R}^n$, i.e. $G = \mathrm{GL}(n, \boldsymbol{R}) \times_\tau \boldsymbol{R}^n$, τ being the natural action of $\mathrm{GL}(n, \boldsymbol{R})$ on $\boldsymbol{R}^n$. Denote by π the natural projection $\pi\colon G \to G/\boldsymbol{R}^n = \mathrm{GL}(n, \boldsymbol{R})$ and let

ϱ: $\mathrm{GL}(n, \boldsymbol{R}) \to \mathrm{GL}(V)$ be a representation of the group $\mathrm{GL}(n, \boldsymbol{R})$ on the space V. Then $\chi(g, x) := \varrho \circ \pi(g^{-1})$ is a $L(V)$-valued multiplier.

It will be useful to note that $\pi(g) = Dg(x)$, $x \in \boldsymbol{R}^n$, where $Dg(x)$ denotes the derivative of the mapping $\boldsymbol{R}^n \ni x \to gx \in \boldsymbol{R}^n$ at the point $x \in \boldsymbol{R}^n$. In particular, the differentiation rule for composite functions shows that $\chi_0(g, x) := Dg^{-1}(x)$ is a multiplier.

Every multiplier determines a representation of the group G on the space of all functions on M. Namely we put

$$(T^{\chi}(g)f)(m) := \chi(g^{-1}, m)f(g^{-1}, m) \tag{2.2.10}$$

for an E-valued function f on M. Condition (2.2.7) ensures that $T^{\chi}(e) = 1$ whereas the relation

$$T^{\chi}_{g_1 g_2} = T^{\chi}_{g_1} T^{\chi}_{g_2} \tag{2.2.11}$$

is a consequence of condition (2.2.8).

Definition. A pair (Φ, T^{χ}) will be called a *multiplier representation* provided $\Phi \in E^M$ is a linear subspace invariant under the action of the representation T^{χ}.

The Radon–Nikodym derivative relative to a quasi-invariant measure provides us with an example of a multiplier. Let (G, M) be a continuous transformation group of a locally compact space M with a quasi-invariant measure $\mathrm{d}\mu$ on M. Let $\chi_0(g, \cdot)$ denote the Radon–Nikodym derivative of the measure $\mathrm{d}g^{-1}\mu$ with respect do $\mathrm{d}\mu$. By definition the function χ_0 satisfies the condition

$$\int_M f(g^{-1}m)\mathrm{d}\mu(m) = \int_M \chi_0(g, m)f(m)\mathrm{d}\mu(m) \qquad \text{for } f \in C_0(M). \tag{2.2.12}$$

The reader is invited to check that, owing to the associativity of the action of G on M, χ_0 is indeed a multiplier.

Let χ_1 be a multiplier assuming values in the space of unitary operators on a Hilbert space H. Suppose also that χ_1 is a measurable function on $G \times M$. Define

$$\chi = \chi_0^{\frac{1}{2}} \chi_1 . \tag{2.2.13}$$

Proposition 2.2.4. *The representation T^{χ} is a unitary representation on the space $L^2(M, H, \mathrm{d}\mu)$.*

Proof. Formula (2.2.12) defines a measurable function on M. We shall establish the square integrability of this function by showing that the operator $T^{\chi}(g)$ is an isometry.

$$||T_g^{\chi} f||^2 = \int_M \chi_0(g^{-1}, m)||\chi_1(g^{-1}, m)f(g^{-1}, m)||_H^2 \mathrm{d}\mu(m)$$

$$= \int_M \chi_0(g^{-1}, m)||f(g^{-1}m)||_H^2 \mathrm{d}\mu = \int_M ||f(m)||_H^2 \mathrm{d}\mu = ||f||^2.$$

The continuity of the representation is now immediately obtained from Proposition 2.1.3 by choosing $\mathscr{C}_0(M, H)$ as the dense subset and then applying the Lebesgue dominated convergence theorem. □

Next, let us consider the case of a multiplier for a Lie transformation group (G, M). If the $L(E)$-valued function χ is differentiable on $G \times M$ (with respect to the strong topology in $L(E)$) then the representation T^{χ} acts inside both spaces, $\mathscr{E}(M, E)$ and $\mathscr{D}(M, E)$. It can be checked that under these assumptions the pairs $(\mathscr{E}(M, E), T)$ and $(\mathscr{D}(M, E), T)$ are continuous representations.

Example 2

The following example may serve as a typical illustration of situations in which multiplier representations frequently appear.

Setting $M = \boldsymbol{R}^2 - \{0\}$, $G = \mathrm{SL}(2, \boldsymbol{R})$, we consider the Lie transformation group (G, M) with the natural action of G on M. We define a representation T of G on the space $\mathscr{E}(M)$ of smooth functions on M by the formula $T(g)f(x) := f(g^{-1}x)$. Explicitly we set $T(g)f(x_1, x_2) := f(ax_1 + bx_2, cx_1 + dx_2)$, where

$$g^{-1} = \begin{bmatrix} a & b \\ c & d \end{bmatrix}.$$

Obviously the spaces $\mathscr{E}_+, \mathscr{E}_- \subset \mathscr{E}(M)$, consisting of even and odd functions respectively, are invariant under T. Next, each of the spaces $\mathscr{E}_+, \mathscr{E}_-$ contains invariant subspaces $\mathscr{E}_{+,s}, \mathscr{E}_{-,s}$ of positively homogeneous functions of order s. (We recall that a function f on $\boldsymbol{R}^n$ is said to be positively homogeneous of order s, $s \in \boldsymbol{C}$ if $f(tx) = t^s f(x)$ for $x \in \boldsymbol{R}_+$.) Denote by $T_{+,s}(T_{-,s})$ the restriction of T to $\mathscr{E}_{+,s}(\mathscr{E}_{-,s})$. The representations $T_{\pm,s}$ can be realized as multiplier representations on the projective space $P^1\boldsymbol{R}$. We shall sketch this construction for the representations $T_{+,s}$, leaving the case of $T_{-,s}$ to the reader as an exercise.

Let us define an isomorphism $A: \mathscr{E}_{+,s} \to \mathscr{E}(P^1\boldsymbol{R})$ by setting

$$Af(\pi(x)) := ||x||^{-s} f(x) \qquad \text{for } f \in \mathscr{E}_{+,s}$$

(π: $\boldsymbol{R}^2 - \{0\} \to P^1\boldsymbol{R}$ being the natural projection). Use this isomorphism to carry over the representation $T_{+,s}$ to the space $\mathscr{E}(P^1\boldsymbol{R})$ by defining $V_s(g) := AT_{+,s}A^{-1}$. It turns out that the representation V_s is a multiplier representation. Since for $f \in \mathscr{E}(P^1\boldsymbol{R})$ we have $A^{-1}f(x) = ||x||^s f(\pi(x))$ it follows that

$$V_s(g)f(\pi(x)) = ||x||^{-s}||g^{-1}x||^s f(g^{-1}\pi(x)).$$

We leave it to the reader to check that the function

$$\chi_{+,s}(g, \pi(x)) := (||gx||/||x||)^s$$

is a $\boldsymbol{C}$-valued multiplier.

Example 3

Suppose $G = \mathrm{SL}(2, \boldsymbol{R})$ acts on the half-plane $H = \{z \in \boldsymbol{C}\colon \operatorname{Im} z > 0\}$ by homographs, i.e. $z \to gz$, where $gz := \dfrac{az+b}{cz+d}$ for $g = \begin{bmatrix} a & b \\ c & d \end{bmatrix}$.

Define a function j: $G \times H \to \boldsymbol{C}$ by setting $j(g, w) := \dfrac{\mathrm{d}(gz)}{\mathrm{d}z} = (cz+d)^{-2}$.

Obviously, we have $j(I, z) = 1$ and by differentiating the identity $(g_1 \circ g_2) \circ z = g_1 \circ (g_2 \circ z)$ we see that j satisfies the multiplier identity (2.2.8)

$$j(g_1 g_2, z) = j(g_1, g_2 \cdot z) j(g_2, z).$$

It is easily verified that the function $\chi(g, z) := j^{-\frac{1}{2}}(g, z) = (cz+d)$ is a multiplier and so is every composite of χ with a homomorphism of the multiplicative group $\boldsymbol{C}_*$.

The reader will easily check that for every integer m the formula $\mathrm{d}\mu_m(z) := (\operatorname{Im} z)^{m-2}\mathrm{d}x\mathrm{d}y$, $z = x+iy \in H$, defines a quasi-invariant measure μ_m on H. In particular, the measure μ_0

$$\mathrm{d}\mu_0(z) := \frac{\mathrm{d}x\mathrm{d}y}{(\operatorname{Im} z)^2}$$

is invariant. Since $\mathrm{Im}(gz) = \dfrac{\mathrm{Im} z}{|cz+d|^2}$, we have

$$\int_H f(g^{-1}\cdot z)\,\mathrm{d}\mu_m(z) = \int_H f(z)|j(g,z)|^m \mathrm{d}\mu_m(z).$$

Hence we infer that the multiplier representation T^m on the space $L^2(H, \mathrm{d}\mu_m)$ with the multiplier $\chi^{-m}(g, z) = (cz+d)^{-m}$ is a unitary representation. On the other hand, the multiplier χ^{-m} being a holomorphic function on H for a fixed $g \in G$, we can define the representation T^m on the space $\mathscr{A}(H)$ of holomorphic functions on H. In turns out that for $m \geqslant 2$ the space $L^2(H, \mathrm{d}\mu_m)$ contains a closed invariant (non-empty) subspace consisting of holomorphic functions. The restriction of T_m to this subspace is called the *discrete series representation* for the group $\mathrm{SL}(2, \boldsymbol{R})$.

2.3. Further Constructions of Representations

Direct Sum of Representations

Let two representations (E, T) and (E', T') of a group G be given. We define a representation on the space $E+E'$ in a natural way by setting

$$(T+T')(g)(e, e') := (T_g' e,\ T_g' e'). \tag{2.3.1}$$

The reader will easily check that $T+T'$ possesses the relevant algebraic properties. Also it is immediately verified that if the representations T and T' of G on locally convex spaces E and E' are continuous then so is $T+T'$, the direct sum of T and T'.

Suppose now that we are given a family of unitary representations (H^i, U^i), $i = 1, 2, \ldots$ We can define a representation $U = \bigoplus_{i=1} U^i$ on the Hilbert space $H = \bigoplus_{i=1} H^i$ in the following manner

$$U(g)(h_1, h_2, \ldots,) := (U_g^1 h_1, U_g^2 h_2, \ldots).$$

The representation (H, U) so defined is both continuous and unitary.

Tensor Product of Representations

Let (E^i, T^i), $i = 1, 2$ be finite-dimensional representations. We define a representation on $E^1 \otimes E^2$ by the formula

$$(T_g^1 \otimes T_g^2)(a_1 \otimes a_2) := (T_g^1 a_1) \otimes (T_g^2 a_2). \tag{2.3.2}$$

There exists a unique linear operator on $E^1 \otimes E^2$ satisfying (2.3.2).

The verification of the algebraic properties of the representation so obtained presents no difficulty.

The representation $(E^1 \otimes E^2, T^1 \otimes T^2)$ is called a *tensor product* of the representations T^1 and T^2.

If T^1 and T^2 are unitary representations then we can define an inner product on $E^1 \otimes E^2$ invariant with respect to $T^1 \otimes T^2$. Namely, let $(\cdot|\cdot)_i$ denote a T^i-invariant inner product on E^i. Then the function defined on simple tensors as

$$(a_1 \otimes a_2 | b_1 \otimes b_2) := (a_1|b_1)_1 (a_2|b_2)_2 \tag{2.3.3}$$

extends to a $T^1 \otimes T^2$-invariant inner product on $E^1 \otimes E^2$. The space $E^1 \otimes E^2$ equipped with the inner product so defined is denoted by $E^1 \underline{\otimes} E^2$ and the representation on this space by $T^1 \underline{\otimes} T^2$.

The Contragredient Representation

Let (U, E) be a representation of a locally compact group G on a locally convex space E. Denote by E the space of continuous linear functional on E endowed with the strong topology. We define an action of the group on E by setting

$$\langle \hat{U}_g x', x \rangle = \langle x', U_{g^{-1}} x \rangle .$$

Problems of the continuity of the representation so defined turn out to be rather complicated.

Example 1

Let G be a locally compact group and write $L^1(G)$ for the space of all functions on G integrable with respect to the Haar measure. We have the standard duality

$$L^1(G) \times L^\infty(G) \ni (f, h) \to \int_G f(g) h(g) \mathrm{d}g,$$

$L^\infty(G)$ denoting the space of measurable, essentially bounded functions on G. Clearly, thanks to the identity

$$\int_G L_x f(g) h(g) \mathrm{d}g = \int_G f(g) L_{x^{-1}} h(g) \mathrm{d}g,$$

the left regular representation on $L^\infty(G)$ coincides with the representation contragredient to the left regular representation on $L^1(G)$. It can easily be checked that the strong topology in $L^\infty(G)$ agrees with that given by the norm $||f|| = \operatorname{ess\,sup}|f(g)|$ hence we derive that the continuity of the mapping $G \ni g \to L_g f \in L^\infty(G)$ for $f \in L^\infty(G) \cap \mathscr{C}(G)$ implies the uniform continuity of f. Consequently the left regular representation of G on L is not continuous. This shows that the natural representation $g \to {}^tU(g^{-1})$ on the space of continuous linear functionals need not be continuous.

Nevertheless the following theorem holds

THEOREM 2.3.1 (Bruhat, 1956). *There exists a closed subspace $\hat{E}$ of E', dense in the weak topology, invariant under $\hat{U}_g$ and such that $(\hat{U}, \hat{E})$ is a continuous representation of G.*

The pair $(\hat{U}, \hat{E})$ is called the *contragredient representation.* In the case of a unitary representation U on a Hilbert space or on a finite-dimensional space E we obviously have $E' = \hat{E}$.

We have already presented various examples concerning the construction of contragredient representations. To give another one, we note that the action of a group on the space of all measures on that group is contragredient to the regular representation on $\mathscr{C}_0(G)$. Also the regular representation on $L^p(G)$ is contragredient to the regular representation on $L^q(G)$ for $\dfrac{1}{p} + \dfrac{1}{q} = 1$. In this case we also have $E = \hat{E}$.

By definition the representations $U, \hat{U}$ are related by a bilinear invariant form $\langle\ ,\ \rangle$. In the sequel the matrix elements of the representations U and $\hat{U}$ formed with respect to this form will be called the *matrix elements of the representation U.*

Example 2

The representation contragredient to a unitary representation.

Let (U, H) denote a unitary representation of a group G on a Hilbert space H. Denote by $\overline{H}$ the Hilbert space having the same underlying set as H, with the same addition of vectors, but multiplication by complex numbers defined in a different way. Namely, the product of a $\lambda \in C$ and a vector $x \in \overline{H}$ is defined as $\bar{\lambda}x$. Consequently, the inner product $(\cdot|\cdot)_{\overline{H}}$ on $\overline{H}$ is defined as $(x|y)_{\overline{H}} := (y|x)$. Since the linear operators on H are still linear relative to $\overline{H}$, we see that the assignment $g \to U_g$

is a unitary representation on H, denoted in the sequel by $\bar{U}$. The bilinear duality form

$$\bar{H}\times H \ni (x, y) \to \langle x, y\rangle := (x|y) \in C$$

allows us to identify $\bar{H}$ with the space H' of continuos linear functionals on H. Under this identification the representation $\bar{U}$ coincides with the representation contragredient to U, which is seen directly from the identities

$$\langle \hat{U}_g x, y\rangle = \langle x, U_{g^{-1}} y\rangle = (x|U_{g^{-1}} y) = (U_g x|y) = \langle \bar{U}_g x, y)\rangle.$$

Example 3

The representation contragredient to the induced representation.

Let (V, E) denote a unitary representation of a closed subgroup $\Gamma \subset G$ and let (U^V, H^V) be the representation induced by V. Write $(\bar{V}, \bar{E})$ for the representation contragredient to (V, E) and, finally, let $(U^{\bar{V}}, H^{\bar{V}})$ be the representation induced by $(\bar{V}, \bar{E})$.

The representation $(U^{\bar{V}}, H^{\bar{V}})$ can be identified in a natural way with the representation contragredient to (U^V, H^V) by means of a G-invariant duality form between H^V and $H^{\bar{V}}$. Namely, for f in H^V and h in $H^{\bar{V}}$ we set

$$\langle h,f\rangle := \int_{G/\Gamma} (h(x)|f(x))\varrho(x)^{-1}d\mu(x\Gamma),$$

ϱ being the function corresponding to the measure $d\mu$ by formula (1.7.5). Observe that the integrand is constant on the cosets of Γ and is integrable with respect to $d\mu$ (the latter property follows by applying twice the Schwarz inequality). The same computations as in the proof of unitarity of the induced representation shows that $\langle\cdot, \cdot\rangle$ is G-invariant, i.e. that $\langle U^{\bar{V}}_g h, U^V_g f\rangle = \langle h,f\rangle$,

$$\begin{aligned}\langle U^{\bar{V}}_g h, U^V_g f\rangle &= \int_{G/\Gamma} (h(g^{-1}x)|f(g^{-1}x))\varrho(x)^{-1}d\mu(x\Gamma)\\ &= \int_{G/\Gamma} (h(g^{-1}x)|f(g^{-1}x))\frac{\varrho(g^{-1}x)}{\varrho(x)}\varrho(g^{-1}x)^{-1}d\mu(x\Gamma)\\ &= \int_{G/\Gamma} (h(x)|f(x))\varrho(x)^{-1}d\mu(x\Gamma).\end{aligned}$$

The penultimate identity is a consequence of the relation between the function ϱ and the Radon–Nikodym derivative of the translate of a measure.

2.4. INTERTWINING OPERATORS. UNITARY EQUIVALENCE OF REPRESENTATIONS

We have already described a considerable number of representations of Lie groups and of locally compact groups. To study their structure we define a class of mappings between the representation spaces, compatible with the structures of these representations.

DEFINITION. Let representations (E, T) and (E', T') of a group G be given. A linear operator $A: E \to E'$ is said to *intertwine* these representations if $AT_g = T'_g A$ for arbitrary $g \in G$.

The space of continuous operators intertwining the representations T and T' is denoted by $L_G(T, T')$ or by $L_G(E, E')$ when there is no danger of a misunderstanding.

The intertwining operators can be used to direct the set of unitary representation (the relation of inclusion) and to define the concept of the equivalence of representations.

DEFINITION. Let (H, U), (H', U') denote unitary representations of a group G. We say that the representation U *contains* $U'(U \succ U')$ provided there is an isometry A: $H' \to H$ intertwining these representations.

Representations U and U' are said to be *equivalent* (denoted by $U \simeq U'$) if there exists an intertwining operator A: $H \to H'$ which is both an isometry and a linear isomorphism.

In order to illustrate these notions we shall apply them to the constructions introduced in the preceding sections.

Example 1

Consider unitary representations (E, T), (E', T') of a group G. The representations $T \oplus T'$ and $T' \oplus T$ are equivalent in a natural way—the equivalence is provided by $(x, x') \to (x', x)$. Similarly the mapping

$x\otimes x' \to x'\otimes x$ extends to an isomorphism of the Hilbert space $E\underline{\otimes}E'$ onto $E'\underline{\otimes}E$, intertwining the representation $T\otimes T'$ and $T'\otimes T$.

The operations of direct sum and of tensor product are both up to unitary equivalence distributive with respect to each other. More precisely, for arbitrary unitary representations T, T_1, T_2 of G we have

$$T\otimes(T_1\oplus T_2) \simeq (T\otimes T_1)\oplus(T\otimes T_2).$$

It is left to the reader to find the corresponding intertwining isometry.

Further examples arise from the theory of induced and multiplier representations.

THEOREM 2.4.1.

(1) $U^{V_1\oplus V_2} \simeq U^{V_1}\oplus U^{V_2}$.

(2) $U^{V_1\otimes V_2} \simeq U^{V_1}\otimes U^{V_2}$.

(3) $V_1 \prec V_2 \Rightarrow U^{V_1} \prec U^{V_2}$.

(4) *Let $G \supset \Gamma \supset \Gamma_1$. Suppose that Γ, Γ_1 are closed subgroups of G and let V_1 be a unitary representation of the subgroup Γ_1. Then the following equivalence takes place*:

$$_GU^{({}_\Gamma U^{V_1})} \simeq {}_GU^{V_1}.$$

The last assertion is known as the *rule of inducing a representation in "stages"*.

Assertions (1), (2) and (3) are a simple consequence of the definition of an induced representation. The proof of (4) is natural but laborious if we employ our definition of an induced representation. However, it is very simple in the formulation of the theory of induced representations presented by Blattner. The proof in question, due to I. M. G. Fell, together with an exposition of the method of Blattner can be found in the monograph of Warner, 1972.

In order to construct the representation space for an induced representation Blattner does not apply an arbitrary quasi-invariant measure $d\mu$ on G/Γ. Our construction, however, being more explicit seems to depend upon the choice of the measure μ determining the inner product on the space H^V. However, from the point of view of representation theory the final effect is independent of this choice.

Let us proceed to the study of the relationship between the two basic constructions described in this chapter—the induced and the multiplier representations. We shall describe an assignement which relates to every multiplier representation, an equivalent induced representation.

Let (G, M) be a homogeneous space of a group G and let T^χ be an $L(E)$-valued multiplier representation. Fix a point m_0 in M and denote by Γ the isotropy subgroup at that point.

The multiplier χ defines a representation V of the subgroup Γ according to formula (2.2.9). Given an E-valued function ψ on M, we define a function on the group G by putting

$$A\psi(g) = \chi(g, m_0)\psi(gm_0).$$

LEMMA 2.4.2. *$A\psi(gh) = V(h^{-1})A\psi(g)$ for $h \in \Gamma$.*

Proof. We check directly from the definition:

$$\begin{aligned} A\psi(gh) &= \chi(gh, m_0)\psi(ghm_0) = \chi(h, m_0)\chi(g, hm_0)\psi(gm_0) \\ &= \chi(h, m_0)\chi(g, m_0)\psi(gm_0) = V(h^{-1})A\psi(g). \end{aligned}$$ □

LEMMA 2.4.3. *For every $g \in G$ we have $L_g A\psi = AT^\chi\psi$.*

Proof. Again, applying only identity (2.2.8) we find

$$\begin{aligned} A\psi(g^{-1}x) &= \chi(g^{-1}x, m_0)\psi(g^{-1}xm_0) \\ &= \chi(x, m_0)\chi(g^{-1}, xm_0)\psi(g^{-1}xm_0) \\ &= AT_g^\chi\psi(x). \end{aligned}$$

PROPOSITION 2.4.4. *Let T^χ be a multiplier representation on the space $L^2(M, H, \mathrm{d}\mu)$ described in Proposition 2.2.4. Then the operator A intertwines T^χ with the representation induced by the representation*

$$V_1(\gamma) := \chi_1(\gamma^{-1}, m_0)$$

of the subgroup $\Gamma = \{\gamma \in G\colon \gamma \cdot m_0 = m_0\}$.

Proof. On account of Lemma 2.4.2 we conclude that A satisfies condition (2.2.2). On the other hand, Lemma 2.4.3 ensures that the operator A intertwines the multiplier representation with the regular one. It remains to check that A is an isometry. We have

$$\begin{aligned} ||A\psi||^2 &= \int_M \varrho(g)^{-1}||\chi(g, m_0)\psi(gm_0)||_H^2 \mathrm{d}\mu(g\Gamma) \\ &= \int_M \varrho(g)^{-1}\chi_0(g, m_0)||\chi_1(g, m_0)\psi(gm_0)||_H^2 \mathrm{d}\mu(g\Gamma). \end{aligned}$$

Recall that $\chi_0(g, xm_0) = \dfrac{\varrho(gx)}{\varrho(x)}$.

Finally, on account of the unitarity of χ_1, we get

$$||A\psi||^2 = \int_M ||\psi(gm_0)||_H^2 d\mu(g\Gamma) = ||\psi||. \qquad \square$$

The theorem proved above shows that a multiplier representation is a different description of an induced representation. One can also prove that every induced representation can be realized as a multiplier representation. The construction of a multiplier for an induced representation is given in Problem 8.

2.5. POSITIVE DEFINITE MEASURES AND CYCLIC REPRESENTATIONS

DEFINITION. A measure μ on a locally compact group G is said to be *positive definite* (p.d.) if for every function $f \in \mathscr{C}_0(G)$, μ satisfies the condition $\mu(f^* * f) \geqslant 0$.

A function ω locally integrable (with respect to Haar measure dg) is called *positive definite* provided the measure ωdg is positive definite.

PROPOSITION 2.5.1 *The form defined by the formula*

$$\mathscr{C}_0(G) \times \mathscr{C}_0(G) \ni (f, h) \to \mu(f^* * h)$$

is both Hermitian and invariant with respect to the left regular representation.

Proof. The linear properties of this form follow directly from the definition. Next, we compute

$$\begin{aligned} 0 &\leqslant \mu\big((f+\lambda g)^* * (f+\lambda g)\big) \\ &= \mu(f^* * f) + \lambda\mu(f^* * g) + \bar{\lambda}\mu(g^* * f) + |\lambda|^2\mu(g^* * g). \end{aligned}$$

This means that $\operatorname{Im}\big(\lambda\mu(f^* * g) + \bar{\lambda}\mu(g^* * f)\big) = 0$ for all $\lambda \in \boldsymbol{C}$. It follows upon specializing the parameter λ to $\lambda = 1$ and $\lambda = i$ that

$$\mu(f^* * h) = \overline{\mu(h^* * f)}.$$

In order to establish the invariance of the form we find

$$(L_g f)^* * (L_g h)(y) = \int_G \delta_G(x)\bar{f}(g^{-1}x^{-1})h(g^{-1}x^{-1}y)\mathrm{d}x$$
$$= \int_G \bar{f}(g^{-1}x)h(g^{-1}xy)\mathrm{d}x$$
$$= \int_G \bar{f}(x)h(xy)\mathrm{d}x = f^* * h(y).$$

Hence the assertion follows. □

Owing to this observation it is possible to construct a unitary representation of the group G. Write: $N = \{f \in \mathscr{C}_0(G)\colon \mu(f^* * f) = 0\}$. The set N is a linear subspace of $\mathscr{C}_0(G)$ (the Schwarz inequality) invariant under the left regular representation. The quotient space $H_0 = \mathscr{C}_0(G)/N$ is a linear space which can be given a pre-Hilbert structure by defining

$$([f]|[h]) = \mu((f^* * h)^{\vee}), \qquad \text{where } [f] := f+N. \tag{2.5.1}$$

We can also define a representation on H_0 according to the formula

$$U_g[f] = [L_g f]. \tag{2.5.2}$$

The properties of the space N, mentioned above, guarantee that definitions (2.5.1) and (2.5.2) are correct. Let us check for instance that (2.5.2) is well-defined. Suppose that $[f] = [f']$, that is, $f-f' \in N$. Since N is G-invariant, we have $U_g[f'] = [L_g f'] = [L_g f + L_g(f'-f)]$, which proves the correctness of (2.5.2). Let ${}^{\mu}H$ denote the completion of the space H_0 with respect to the norm $||x|| = (x|x)^{\frac{1}{2}}$, $x \in H_0$. Then ${}^{\mu}H$ is a Hilbert space. The operator U_g, being a bijective isometry on H_0 with respect to the above norm, extends to a unitary operator on ${}^{\mu}H$. We shall denote this extension by ${}^{\mu}U_g$.

PROPOSITION 2.5.2. *$({}^{\mu}H, {}^{\mu}U)$ is a continuous unitary representation of the group G.*

Proof. The algebraic properties of the representation on H_0 and of its extension are identical. It remains to verify the continuity of the representation in question. On the basis of Proposition 2.1.3 it suffices to prove the continuity of the representation (U, H_0). We have

$$||U_g[f]-[f]||^2 = ||[L_g f - f]||^2 = \mu\big(((U_g f - f^*) * (U_g f - f))^{\vee}\big).$$

The regular representation of G on $\mathscr{C}_0(G)$ is continuous and so are the involution and the convolution on $\mathscr{C}_0(G)$. By definition $\|\cdot\|$ is a continuous functional on $\mathscr{C}_0(G)$. Consequently, the function $G \ni g \to \|U_g[f]-[f]\|$, $f \in H_0$, is continuous as a composition of continuous mappings.

Example 1

The functions $\boldsymbol{R}^n \ni (t_1, \ldots, t_n) = t \to \mathrm{e}^{i(\xi_1 t_1 + \ldots + \xi_n t_n)} \in \boldsymbol{T}$ are positive definite on $\boldsymbol{R}^n$ for arbitrary $\xi = (\xi_1, \ldots, \xi_n) \in (\boldsymbol{R}^n)^* \simeq \boldsymbol{R}^n$. Indeed, if for a function $f \in \mathscr{C}_0(\boldsymbol{R}^n)$ we denote by $\hat{f}$ the Fourier transform, defined as $\hat{f}(\xi) := \int f(t)\mathrm{e}^{i(t|\xi)}\,\mathrm{d}t$ (we write $(t|\,\xi) := \sum_i t_i \xi_i$), then on account of the well-known properties of the Fourier transformation we obtain the identity $\int (f^* * f)(t)\mathrm{e}^{i(t|\xi)}\mathrm{d}t = |\hat{f}(\xi)|^2 \geqslant 0$. This shows that the function $t \to \chi_\xi(t) = \mathrm{e}^{i(t|\xi)}$ is positive definite for arbitrary $\xi \in \boldsymbol{R}^n$. Actually not only the characters χ_ξ are positive definite. The same is also true of linear combinations of χ_ξ with positive coefficients and most generally of integrals of χ_ξ with respect to finite positive Radon measures on $\boldsymbol{R}^n$. In fact if μ is such a measure then its Fourier transform $\hat{f}(t) := \int \mathrm{e}^{i(t|\xi)}\,\mathrm{d}\mu(\xi)$ is a continuous function. The positive definiteness of $\hat{\mu}$ follows from the computations:

$$\begin{aligned}\int f^* * f(t)\hat{\mu}(t)\mathrm{d}t &= \int f^* * f(t)\int \mathrm{e}^{i(t,\xi)}\mathrm{d}\mu(\xi)\mathrm{d}t\\ &= \int\int f^* * f(t)\mathrm{e}^{i(t,\xi)}\mathrm{d}t\,\mathrm{d}\mu(\xi)\\ &= \int |\hat{f}(\xi)|^2\mathrm{d}\mu(\xi) \geqslant 0,\end{aligned}$$

where we have used the Fubini theorem and the properties of the Fourier transformations.

Returning to the case of the function $\chi(t) = \chi_\xi(t) = \mathrm{e}^{i(t,\xi)}$ one can easily see that the construction carried out at (2.5.1) and (2.5.2) leads in this case to a one-dimensional representation of the group $\boldsymbol{R}^n$ on the vector space, ${}^\chi H \simeq \mathscr{C}$, defined as ${}^\chi U(t)\lambda := \mathrm{e}^{i(t,\xi)}\lambda$ for $\lambda \in {}^\chi H$. (For ${}^\chi H$ we may take here $\{\hat{f}(\xi) \colon f \in \mathscr{C}_0(\boldsymbol{R}^n)\}$.) Consequently we have obtained not only cyclic but irreducible unitary representations of the group $\boldsymbol{R}^n$ (cf. § 2.2).

Every unitary representation defines a family of p.d. continuous functions. We easily check that every matrix element $\omega_x(g) := (x|U_g x)$ is a p.d. function. Let us write

$$U(f)x := \int_G f(g)U_g x\,dg \qquad \text{for } f \in \mathscr{C}_0(G). \tag{2.5.3}$$

The integrand being an H-valued, compactly supported, continuous function implies that the integral is finite. Next, for every function f in $C_0(G)$, we have

$$\begin{aligned}0 \leqslant (U(f)x|U(f)x) &= \Big(\int_G f(g)U_g x\,dg \Big| \int_G f(g)U_g x\,dg\Big)\\ &= \int_G\int_G \overline{f(g)}f(y)\omega_x(g^{-1}y)\,dg\,dy\\ &= \int_G\int_G \overline{f(g)}f(gy)\omega_x(y)\,dy\,dg\\ &= \int_G f^* * f(g)\omega_x(g)\,dg.\end{aligned}$$

The positive definiteness of the function ω is therefore established.

Example 2

Let U be the left regular representation of the group G on $L^2(G)$ and suppose that f is in $L^2(G)$. Then

$$\begin{aligned}\overline{(f|L_g f)} = (L_g f|f) &= \int \overline{f(g^{-1}x)}f(x)\,dx\\ &= \int f(x)\tilde{f}(x^{-1}g)\,dx = f * \tilde{f}(g),\end{aligned}$$

where we denoted by $\tilde{f}$ the function $\tilde{f}(x) := \overline{f(x^{-1})}$. Since $\bar{\varphi}$ is positive definite if φ is such, we see that functions of the form $f * \tilde{f}$, $f \in L^2(G)$, are all positive definite.

We observed above that to every unitary representation there corresponds a family $\{\omega_x\colon x \in H\}$ of p.d. continuous functions on G. Moreover, the sets corresponding to equivalent representations are identical. Conversely, given a set of p.d. functions it is natural to ask whether the construction, previously described, of a representation generated by a p.d. measure (function) allows a reconstruction of the class of representations to which this set of p.d. functions corresponds.

DEFINITION. A vector x of a representation (U, E) is said to be *cyclic* if the linear span of the set $\{U(f)x\colon f \in \mathscr{C}_0(G)\}$ is dense in E. A representation is called *cyclic* if it contains a cyclic vector.

THEOREM 2.5.3 (Gelfand, Raikov, 1943).

(1) *Let U be a cyclic unitary representation and let x be a cyclic vector. Denote by ω_x the p.d. function $G \ni g \to (x|U_g x)$ corresponding to the cyclic vector x. Set $\omega = \omega_x \mathrm{d}g$. Then the representation ${}^{\omega}U$ is equivalent to U.*

(2) *Let ω be a continuous p.d. function on G. Then there exists a cyclic vector x in the representation space of ${}^{\omega}U$ such that $\omega = \omega_x$.*

Proof. (1) We construct an isometry T of the representation space of U into the representation space of ${}^{\omega}U$ by defining $TU(f) := [f]$ on the dense subset of vectors of the form $\{U(f)x\colon f \in \mathscr{C}_0(G)\}$. To prove that T is indeed an isometry we compute:

$$(TU(f)x|TU(f)x) = ([f]|[f]) = \omega((f^* * f))$$
$$= (U(f)x|(U(f)x).$$

It can also be seen that T commutes with the representation operators

$$TU_g(f)x = TU(L_g f)x = [L_g f] = {}^{\omega}U_g[f] = {}^{\omega}U_g TU(f)x.$$

The extension of T to the whole of H is also an intertwining operator and its surjectivity follows from the construction of the representation $({}^{\omega}U, {}^{\omega}H)$. This proves (1).

(2) Let us construct a cyclic vector. To this end let $\{O_\alpha\}$ denote a base of compact symmetric neighbourhoods at the identity $e \in G$. Write $\varphi_\alpha := \frac{1}{g(O_\alpha)} 1_{O_\alpha}$, $g(O_\alpha)$ denoting the Haar measure of the set O_α. This generalized sequence (directed by the relation $O_\alpha > O_\beta$ if $O_\alpha \subset O_\beta$) is convergent to δ_e in the space of measures (Problem 10, § 1.6).

Let us estimate the norm of φ_α in ${}^{\omega}H$. We have

$$\|\varphi_\alpha\|^2 = \int_G\int_G \varphi_\alpha(g)\varphi_\alpha(h)\omega(g^{-1}h)\,\mathrm{d}g\,\mathrm{d}h$$

$$\leqslant \sup_{h \in O_\alpha} \int_G \varphi_\alpha(h)|\omega(g^{-1}h)|\,\mathrm{d}h \leqslant \sup_{O_\alpha^{-1}O_\alpha} |\omega| < \infty,$$

because $O_\alpha^{-1}O_\alpha \to \{e\}$. The ball in a Hilbert space being weakly compact this implies that the sequence has a limit point which will be denoted by x.

Now, for every f in $\mathscr{C}_0(G)$ we find

$$\begin{aligned}([f]|^{\omega}U_h[\varphi_\alpha]) &= \int_G\int_G \bar{f}(g)\varphi_\alpha(h^{-1}gy)\omega(y)\mathrm{d}g\,\mathrm{d}y \\ &= \int_G\int_G \bar{f}(g)\varphi_\alpha(y)\omega(g^{-1}hy)\mathrm{d}y\,\mathrm{d}g \to \int_G \bar{f}(g)\omega(g^{-1}h)\mathrm{d}y \\ &= \bar{f}*\omega(h).\end{aligned}$$

On the other hand $([f]|U_h[\varphi_\alpha]) \to ([f]|U_h x)$, hence $([f]|U_h x) = \bar{f}*\omega(h)$. Replacing f by the functions φ_α and passing again to the limit, we get $(x|U_h x) = \lim_\alpha \varphi_\alpha * \omega(h) = \omega(h)$. It remains to show that x is cyclic. Let $[f], f \in \mathscr{C}_0(G)$ be a vector, normal to all vectors of the form $U(\varphi)x$, $\varphi \in \mathscr{C}_0(G)$. Then

$$\begin{aligned}0 \underset{\varphi}{\equiv} ([f]|U(\varphi)x) &= \int_G \bar{f}*\omega(h)\varphi(h)\mathrm{d}h \\ &= \int_G\int_G \bar{f}(g)\varphi(h)\omega(g^{-1}h)\mathrm{d}g\,\mathrm{d}h = ([f]|[\varphi]).\end{aligned}$$

Therefore we conclude that $f = 0$. □

The relation between positive definite functions and representations plays a fundamental role in representation theory. Namely, it provides a method of connecting this theory with a branch of functional analysis called the convex set analysis. There are a number of monographs dealing extensively with this subject.

In the present book Theorem 2.5.3 will mostly be applied to the study of the properties of special functions arising as matrix elements of representations. To end this section, we list some basic properties of p.d. functions:

PROPOSITION 2.5.4. *Letting ω be a p.d. function, we have*

(1) $\sum_{i,j=1}^{n} \bar{\lambda}_j \lambda_i \omega(x_i^{-1}x_j) \geqslant 0$ *for every system of complex numbers* λ_1, $\ldots$, λ_n *and of points* $x_1, \ldots, x_n \in G$,

(2) $\omega(g^{-1}) = \bar{\omega}(g)$,
(3) $\omega(e) \geqslant 0$,
(4) $|\omega(g)| < \omega(e)$,
(5) $|\omega(g)-\omega(h)|^2 \leqslant 2\omega(e)(\mathrm{Re}(\omega(e)-\omega(gh^{-1}))$.

Proof. Properties (2)–(5) can be derived from (1) in the following fashion:

(2) Apply (1) to the system $\lambda_1 = 1$, $\lambda_2 = \lambda$, $g_1 = g$, $g_2 = e$. We get the inequality

$$\omega(e)+\omega(g)\bar{\lambda}+\omega(g^{-1})\bar{\lambda}+\omega(e)|\lambda|^2 \geqslant 0 \text{ for an arbitrary } \lambda \tag{2.5.4}$$

which means that the value of $\omega(g)\bar{\lambda}+\omega(g^{-1})\lambda$ is real for all λ. In particular $\omega(g)+\omega(g^{-1}) \in \boldsymbol{R}$ and $i(\omega(g^{-1})-\omega(g)) \in \boldsymbol{R}$, which is the case only when $\omega(g^{-1}) = \bar{\omega}(g)$.

(3) Set $g_1 = e$, $\lambda_1 = 1$ in (1).

(4) Assuming $\omega(e) = 0$, we infer from (2.3.2) that $\omega(g) = 0$. Consequently (4) is true in this case. If $\omega(e) > 0$, then the required inequality follows from (2.5.4) upon setting $\lambda = \dfrac{-\omega(g)}{\omega(e)}$.

(5) Take $n = 3, g_1 = e, g_2 = g, g_3 = h, \lambda_1 = 1, \lambda_2 = \dfrac{t|\omega(g)-\omega(h)|}{\omega(g)-\omega(h)}$, $\lambda_3 = -\lambda_2$. It follows from (1) that

$$0 \leqslant \omega(e)(1+2t^2)+2t|\omega(g)-\omega(h)|-2t^2\mathrm{Re}\,\omega(gh^{-1}).$$

Consequently the discriminant of this square trinomial must be negative, which is exactly the required inequality (5).

It remains to prove (1). Given a system of $\lambda_1, \ldots, \lambda_n$ and $x_1, \ldots, x_n$, we construct a function f_α: $\sum_i \lambda_i L_{x_i}^{-1}\varphi_\alpha$ with the same functions φ_α as in the proof of Theorem 2.5.3.

Since the measures $f_\alpha dg$ converge to $\sum_i \lambda_i \delta_{x_i}$, property (5) follows by passing to the limit in the inequality $\omega(f_\alpha^* * f_\alpha) \geqslant 0$. □

2.6. MATRIX ELEMENTS OF REPRESENTATIONS

Every group representation can be realized as a regular representation on a function space. The easiest method is to proceed as follows.

Take a continuous representation (G, E, T). For every a in E define a function $O_a \in \mathscr{C}(G, E)$ by setting $O_a(g) = T_g a$. Denote by I the linear injection

$$I\colon E \ni a \to O_a \in \mathscr{C}(G, E).$$

Then $IT_g = R_g I$, which means that I intertwines the representation T with the right regular representation on $\mathscr{C}(G, E)$.

We can also define a representation of the group on a space of scalar-valued functions. Namely, to an element $e \in E$ we assign the scalar function $G \ni g \to t_{w,a}(g) := \langle w, T_g a\rangle$, $w \in E$ being fixed. The function $t_{w,a}$ is called the *matrix element of the representation* T, with respect to the pair (w, a).

The mapping, $A_w\colon E \ni a \to t_{w,a} \in \mathscr{C}(G)$ is continuous and intertwines the representation T with the right regular representation of G on $\mathscr{C}(G)$.

In the sequal we shall deal, as a rule, with a more specific form of this construction.

Suppose we are given two representations (E^1, T^1), (E^2, T^2) of a group G. Also suppose that there exists on $E^1 \times E^2$ a form which is bilinear or $1\frac{1}{2}$-linear (i.e. antilinear in the first argument and linear in the second) and invariant in the following sense:

$$\langle T_g^1 a, T_g^2 b\rangle = \langle a, b\rangle, \qquad a \in E^1,\ b \in E^2,\ g \in G.$$

In accordance with the preceding notation we define

$$t_{a,b}(g) = \langle a, T_g b\rangle. \tag{2.6.1}$$

Then in the case of the bilinear form the operator

$$F_b\colon E^1 \ni a \to t_{a,b} \in \mathscr{C}(G)$$

intertwines T with the left regular representation, whereas in the case of the $1\frac{1}{2}$-linear form the operator

$$\bar{F}_b\colon E^1 \ni a \to \bar{t}_{a,b} \in \mathscr{C}(G)$$

possesses this property.

The operators F_a and $\bar{F}_a$ are both injective whenever the linear functionals on E^2 of the form $\langle\cdot, T_g b\rangle$, $g \in S$, span a dense subspace in the dual space of E^1. In particular, this is the case if b is a cyclic vector for the representation T and the form $\langle\cdot\,,\,\cdot\rangle$ is non-degenerate, so that $\langle a, b\rangle \underset{b}{\equiv} 0$ implies $a = 0$.

We shall find the operator F_a in the case of the unitary representation $({}^{\omega}H, {}^{\omega}U)$ generated by a continuous positive definite function ω. For b we take the cyclic vector x of the representation. As we already know, the matrix element $t_{[f],x}$ has the form

$$t_{[f],x}(g) = ([f]|U_g x) = \bar{f} * \omega(g), \qquad f \in \mathscr{C}_0(G).$$

Hence we get the formula $F_x[f] = f * \bar{\omega}$. In particular,

$$(F_x x)(g) = (x|U_g x) = \bar{\omega}(g).$$

Thus, in the description of the representation ${}^{\omega}U$ as a subrepresentation of the regular representation of G on the space $\mathscr{C}(G)$, it is the positive definite function $\bar{\omega} = \check{\omega}$ that plays the role of a cyclic vector.

Suppose in addition that there exists a Schauder basis for E^2, i.e. a linearly independent subset $\{e_i^2\}_1^\infty$ with the property that every element $x \in E^2$ can be uniquely represented as the sum of a convergent series: $x = \sum_{i=1}^{\infty} \lambda_i e_i^2$. Also assume that there exists a sequence of vectors $\{e_i^1\}_1^\infty$ in E^1, dual to $\{e_i^2\}_1^\infty$, i.e. $\langle e_i^1, e_j^2 \rangle = \delta_{ij}$. Under these assumptions $x = \sum_{i=1}^{\infty} \langle e_i^1, x \rangle e_i^2$.

Furthermore, we may introduce the matrix calculus in the space of linear continuous operators on E^2. For $A \in L(E^2)$ we have $Ae_i^2 = \sum_{j=1}^{\infty} \langle e_j^1, Ae_i^2 \rangle e_j^2 = \sum_{j=1}^{\infty} a_{ji} e_j^2$, a_{ji} denoting the matrix element of A with respect to e_j^1 and e_i^2. The well-known formulas of finite-dimensional algebra are also valid here.

Let $[a_{ij}]$ be the matrix of an operator A, $[b_{ij}]$ the matrix of an operator B and $[c_{ij}]$ the matrix of the composite $A \circ B$. Then

$$c_{ij} = \sum_{k=1}^{\infty} a_{ik} b_{kj}. \tag{2.6.2}$$

We verify this by computing

$$\langle e_i^1, ABe_j^2 \rangle = \left\langle e_i^1, A\left(\sum_{k=1}^{\infty} b_{kj} e_k^2\right)\right\rangle$$

$$= \sum_{k=1}^{\infty} \langle e_i^1, Ae_k^2 \rangle b_{kj} = \sum_{k=1}^{\infty} a_{ik} b_{kj}.$$

PROPOSITION 2.6.1. *Let $\{e_{i_k}^2\}_1^\infty$ be a Schauder basis for E^2, $\{e_j^1\}_1^\infty$ the dual set and (E^2, T) a representation of a group G. Then for all g, h in G we have*

$$t_{ij}(gh) = \sum_{k-1}^{\infty} t_{ik}(g)\, t_{kj}(h).$$

PROPOSITION 2.6.2. *Let $(E, (\cdot, \cdot))$ be a Hilbert space with an orthonormal basis $\{e_i\}$. A representation (G, E, U) is unitary* $\Leftrightarrow \sum_{k=1}^{\infty} \overline{t_{ki}(g)}\, t_{kj}(g) = \delta_{ij}$.

Proof. On the one hand, the unitariness is equivalent to the system of identities $(U_g e_i | U_g e_j) = \delta_{ij}$. On the other hand,

$$(U_g e_i | U_g e_j) = \sum_{k=1}^{\infty} (U_g e_i | e_k)(e_k | U_g e_j) = \sum_{k=1}^{\infty} \overline{t_{ki}(g)}\, t_{kj}(g).$$

Hence the assertion follows. □

These simple relations give rise to a series of formulas called *the addition formulas* for special functions.

Example 1

Let $T\colon G \to \mathrm{GL}(V)$ be a finite-dimensional representation. We shall compute the matrix elements of the representation $\hat{T}$ contragredient to T. Let $\{e_i\}_1^n$ be a basis for V and $\{e_i'\}_1^n$ the dual basis for $\hat{V}$. By definition

$$\hat{t}_{ij}(g) = \langle e_i, T_g e_j' \rangle = \langle T_{g^{-1}} e_i, e_j' \rangle = t_{ji}(g^{-1}).$$

Since $G \ni g \to [t_{kl}(g)] \in \mathrm{GL}(n, \boldsymbol{K})$ is a group homomorphism, we have $[t_{ji}(g^{-1})] = [t_{ji}(g)]^{-1}$ and, denoting by $\hat{T}(g)$ (by $T(g)$, resp.) the matrix $\hat{t}_{ji}(g)$ ($t_{kl}(g)$, resp.), we conclude that $\hat{T}(g) = (T(g)^t)^{-1}$.

Example 2

Matrix elements of the tensor product of representations.

Choose bases $\{e_i^1\}_1^n$, $\{e_j^2\}_1^k$ for E^1 and E^2, respectively, and let $\{f_i^{(1)}\}_1^n$, $\{f_j^{(2)}\}_1^k$ denote the dual bases, so that representations T^1, T^2

of a group G have matrix elements $t^{(1)}_{rs}(g)$ and $t^{(2)}_{pq}(g)$. Then, relative to the dual bases $\{e^{(1)}_i \otimes e^{(2)}_j\}$, $\{f^{(1)}_i \otimes f^{(2)}_j\}$, we obtain

$$\langle f^{(1)}_l \otimes f^{(2)}_m, T^1(g) \otimes T^2(g)(e^{(1)}_i \otimes e^{(2)}_j)\rangle = t^{(1)}_{li}(g)\, t^{(2)}_{mj}(g)$$

for $1 \leqslant l,\ i \leqslant n,\ 1 \leqslant m,\ j \leqslant k$.

2.7. GROUP ALGEBRA REPRESENTATIONS AND GROUP REPRESENTATIONS

The decomposition theory of group representations is based on the results of the group algebra representation theory. In this section we present theorems relating these two theories.

Suppose (E, U) is a continuous representation of a locally compact group G on a complete, locally convex space E. For every compactly supported measure μ on G we can define the Bochner integral

$$U(\mu)a := \int_G U_g a \, \mathrm{d}\mu(g). \tag{2.7.1}$$

Obviously, the mapping $E \ni a \to U(\mu)a \in E$ is linear. In order to show that it is continuous we let $a_n \to 0$. Then the sequence $U_g a_n$ tends to zero uniformly on compact subsets, since the set of operators U_g, g ranging over a compact set, is equicontinuous. Hence $U(\mu)a_n \to 0$. Consequently, the mapping $\mathfrak{M}_0(G) \ni \mu \to U(\mu) \in L(E)$ is linear and satisfies

$$U(\mu * \nu) = U(\mu)U(\nu). \tag{2.7.2}$$

To verify (2.7.2) we compute

$$U(\mu * \nu)a = \int_G\int_G U_{gh} a \, \mathrm{d}\mu(g)\mathrm{d}\nu(h) = \int_G\int_G U_g U_h a \, \mathrm{d}\mu(g)\mathrm{d}\nu(h)$$

$$= \int_G U_g U(\nu) a \, \mathrm{d}\mu(g) = U(\mu)U(\nu)a.$$

The mapping $\mathscr{C}_0(G) \ni f \to f\mathrm{d}g \in \mathfrak{M}_0(G)$imbeds the convolution algebra $\mathscr{C}_0(G)$ into $\mathfrak{M}_0(G)$ preserving the convolution

$$(f_1 * f_2)\mathrm{d}g = (f_1\mathrm{d}g) * (f_2\mathrm{d}g).$$

If we denote $U(f\mathrm{d}g) =: U(f)$, then it follows from (2.7.2) that

$$U(f_1 * f_2) = U(f_1)U(f_2). \tag{2.7.3}$$

The mapping $f \to U(f)$ is continuous. In addition, if U is a unitary representation on a Hilbert space then we have

$$U(f^*) = U(f)^*, \tag{2.7.4}$$

where the * on the right denotes the Hermitian conjugation of an operator on a Hilbert space.

A mapping $\mathscr{C}_0(G) \ni f \to T(f) \in L(E)$ satisfying (2.7.3) is called a *representation of the algebra* $\mathscr{C}_0(G)$. If E is a Hilbert space and T also satisfies (2.7.4), then such a T is called a *unitary representation*.

Example 1

Define a unitary representation of the additive group $\boldsymbol{R}$ on the space $L^2(\boldsymbol{R})$ by setting $(U(t)g)(s) := e^{its}g(s)$ for $g \in L^2(\boldsymbol{R})$. Take an f in $\mathscr{C}_0(\boldsymbol{R})$. To find $U(f)$ we take $g, h \in L^2(\boldsymbol{R})$ and compute

$$\int_{\boldsymbol{R}} f(t)(U(t)g|h)\mathrm{d}t = \int_{\boldsymbol{R}} f(t) \int_{\boldsymbol{R}} e^{its}g(s)\overline{h(s)}\mathrm{d}s\mathrm{d}t$$

$$= \int_{\boldsymbol{R}}\left(\int_{\boldsymbol{R}} f(t)e^{its}\mathrm{d}t\right) g(s)\overline{h(s)}\mathrm{d}s = \int_{\boldsymbol{R}} \hat{f}(s)g(s)\overline{h(s)}\mathrm{d}s.$$

Therefore we conclude that $U(f)$ is the operator of multiplication by $\hat{f}$, i.e. $(U(f)g)(s) = \hat{f}(s)g(s)$.

One can now ask the following question: what are the conditions which ensure that a representation of the algebra $\mathscr{C}_0(G)$ corresponds to a representation of the group G in the way described above? The following theorem gives an answer.

THEOREM 2.7.1. *Let T be a continuous representation of the algebra $\mathscr{C}_0(G)$. There exists a continuous representation (U, E) of the group G, such that $T(f) = U(f)$ if and only if the following conditions are satisfied:*

(1) *The linear span of the set of vectors of the form $T(f)a$, $f \in \mathscr{C}_0(G)$, $a \in E$ is dense in E.*

(2) *The set of operators $\{T(f)\colon \int |f| \mathrm{d}g \leqslant 1\}$ is equicontinuous.*

The proof of this theorem will be found in the monograph by Warner, 1972.

Another version of this theorem can be found in a number of basic monographs dealing with group representations and algebra representations.

The necessity of condition (1) is a particularly important property of a representation of a group, therefore we shall prove it here. Namely, we shall prove that for a sequence $0 \leqslant \varphi_n \in \mathscr{C}_0(G)$ approximating the Dirac measure (this means that $\operatorname{supp}\varphi_n \to e$ and $\int_G \varphi_n dg = 1$) $U(\varphi_n)x$ tends to x in the space E.

Note that for an arbitrary semi-norm on E we have the estimates

$$||U(\varphi_n)x - x|| = ||\int_G \varphi_n(g)(U_g x - x)dg||$$

$$\leqslant \int_G \varphi_n(g)||U_g x - x||dg \leqslant \sup_{\operatorname{supp}\varphi_n} ||U_g x - x||.$$

Now, the continuity of the representation implies that the right-hand side of the above inequality converges to zero. Since this is true for an arbitrary seminorm on E we obtain the desired result.

2.8. THE UNIVERSAL ENVELOPING ALGEBRA OF A LIE GROUP ALGEBRA. THE DIFFERENTIAL OF A REPRESENTATION

Let G be a Lie group with the Lie algebra $\mathfrak{g}$. Here we regard the elements of $\mathfrak{g}$ as vector fields on G invariant under right translations. Let $X \in \mathfrak{g}$ and denote by a_X the one-parameter subgroup related to X. We identify X with a linear operator on $\mathscr{E}(G)$ whose value on a function $f \in \mathscr{E}(G)$ is defined by the formula

$$Xf(g) = \lim_{t \to 0} \frac{f(a_X(t)g) - f(g)}{t}. \tag{2.8.1}$$

By the same formula we define the value of X on functions in $\mathscr{E}(G, E)$.

The operators X and the identity operator generate over C an associative subalgebra of $L(\mathscr{E}(G))$ called the *universal enveloping algebra* of $\mathfrak{g}$ and denoted by $\mathfrak{G}$.

Let A be an arbitrary associative algebra over the field C and denote by $\mathscr{A}$ a Lie algebra with the bracket $[\cdot, \cdot]$ defined as $[a, b] := ab - ba$. We have the following

THEOREM 2.8.1. *For every homomorphism* α *of the Lie algebra* $\mathfrak{g}$ *into* $\mathscr{A}$ *there exists an associative algebra homomorphism* $\tilde{\alpha}\colon \mathfrak{G} \to \mathscr{A}$ *such that* $\alpha(X) = \tilde{\alpha}(X)$ *for* $X \in \mathfrak{g}$.

Suppose now that we are given a continuous representation (E, U) of a Lie group G on a locally convex space E. We say that $e \in E$ is a *smooth vector* of the representation U provided the function $\tilde{e}$

$$G \ni g \to U_g e \in E$$

is infinitely differentiable on G.

The linear space of smooth vectors of a representation U is denoted by E_∞ or $D(U)$.

We can define an action of an operator $X \in \mathfrak{g}$ on E_∞ as follows:

$$\mathrm{d}U(X)x := \lim_{t\to 0} \frac{U(\exp tX)x - x}{t}. \tag{2.8.2}$$

In other words, the value $\mathrm{d}U(X)$ at a point $x \in E$ is the value of the derivative at the neutral element in the direction X of the function $\tilde{x}$: $g \to U_g x$. We apply this observation to check that the mapping $X \to \mathrm{d}U(X)$ is a representation of the Lie algebra on the space of linear endomorphisms of the space E_∞

$$\begin{aligned}&\big(\mathrm{d}U(X)\mathrm{d}U(Y) - \mathrm{d}U(Y)\mathrm{d}U(X)\big)x\\ &= \big((XY - YX)x\big)(e) = [X, Y]x(e) = \mathrm{d}U[X, Y]x.\end{aligned}$$

It follows from Theorem 2.8.1 that the representation $\mathrm{d}U$ extends to a representation of the enveloping algebra $\mathfrak{G}$. This extension is called the *differential of the representation* U, or the *infinitesimal representation*.

If T is a continuous operator from E into a locally convex space F, then for an arbitrary function $f \in \mathscr{E}(G, E)$ the composite $T \circ f$ is a smooth representation assuming values in F. If there is another representation V of the group G on the space F and T intertwines U and V, then we get the relation:

$$(Tx)^{\tilde{}}(g) = V_g(Tx) = TU_g x = (T \circ \tilde{x})(g). \tag{2.8.3}$$

This immediately implies that for every $x \in E_\infty$ the vector Tx is a smooth vector of the representation V. Moreover it follows directly from definition that

$$T\mathrm{d}U(D)x = \mathrm{d}V(D)Tx \qquad \text{for } D \in \mathfrak{G} \text{ and } x \in E_\infty. \tag{2.8.4}$$

Thus we have obtained another simple formula which has wide applications in the theory of matrix elements (special functions) establishing a series of basic relations of differential character. However, in order that the infinitesimal representation be useful for our investigations we must know that the domain of the operator $\mathrm{d}U(X)$, that is the space E_∞, is large enough. This problem is solved by the following theorem, due to Gårding.

THEOREM 2.8.2. *The space E_∞ is dense in E.*

The proof is based on the following simple observation: vectors of the form $U(f)x$, $x \in E$, $f \in \mathscr{D}(G)$ are smooth. The reader will easily prove this by a direct computation, verifying at the same time the formula

$$\mathrm{d}U(X)\,Uf = U(Xf)x \qquad \text{for } X \in G. \tag{2.8.5}$$

We know from the preceding section that the vectors of the form $U(f)x$ constitute a dense subset of E. This proves Gårding's theorem.

Example 1

Let U be the left regular representation of a Lie group G on $L^2(G)$. We shall check directly from the definition that $D(G) \subset L^2_\infty(G) := (L^2(G))_\infty$. Namely, we shall prove that

$$\lim_{t\to 0} \int_G \left| t^{-1}\big(f(\exp(-tX)g)-f(g)\big) - Xf(g)\right|^2 \mathrm{d}g = 0$$

for $f \in D(G)$. To this end let $M := \sup_{g\in G} |Xf(g)|^2$. Then for $0 < t \leqslant 1$

$$\begin{aligned} &\left|t^{-1}\big(f(\exp(-tX)g)-f(g)\big)-Xf(g)\right|^2 \\ &\leqslant 2\left|t^{-1}\big(f(\exp(-tX)g)-f(g)\big)\right|^2+2\left|Xf(g)\right|^2 \leqslant 4M. \end{aligned}$$

Finally, the assertion follows in view of Lebesgue's dominated convergence theorem.

Example 2

Let G be the group of matrices of the form

$$g = \begin{bmatrix} 1 & a & c \\ 0 & 1 & b \\ 0 & 0 & 1 \end{bmatrix}.$$

The Lie algebra of this group consists of vectors X_a, X_b, X_c, which are differentiations in the direction of the respective variables. We have the following commutation relations:

$$[X_a, X_b] = X_c, \qquad [X_a, X_c] = [X_b, X_c] = 0.$$

The representation of G on $L^2(R)$ defined by the formula

$$U(g)f(x) = e^{ic}e^{-ibx}f(x+a)$$

is called the *Schrödinger representation*. The operators of the infinitesimal representation have the form

$$dU(X_a) = -\frac{d}{dx}, \qquad dU(X_b) = -ix, \qquad dU(X_c) = i\,\mathrm{id}.$$

Thus the elements of the space $D(U)$ are differentiable in the usual sense and, moreover, they decrease at infinity faster than any polynomial.

We leave it to the reader to verify that:

PROPOSITION 2.8.3. *The space of smooth vectors of the representation U is identical with the Schwartz space $\mathscr{S}(R)$ (the definition of $\mathscr{S}(R)$ will be found in Chapter* 3).

Finally, we present a characterization of the space of smooth vectors of a representation induced in the sense of Mackey.

THEOREM 2.8.4 (N. S. Poulsen). *Let (H^V, U^V) be the unitary representation of a Lie group G induced in the sense of Mackey by a unitary representation V of a subgroup $H \subset G$. Then*

$$(H^V)_\infty = \{f \in H^V\colon Xf \in H^V \text{ for each } X \in \mathfrak{G}\}.$$

Xf denotes here the natural action of X regarded as a differential operator on the group.

PROBLEMS

1. (a) Prove that a unitary representation U of a group G on a Hilbert space H is continuous iff for every pair of vectors x, y in H the function $G \ni g \to (U_g x|y) \in C$ is continuous.

(b) Show that if (E, T) is a representation of a group G on a Banach space E, such that the mapping $G \ni g \to T_g \in L(E)$ is continuous in the norm on $L(E)$, then (E, T) is a continuous representation.

(c) Show that a unitary representation (H, U) of the group $\boldsymbol{R}$ of real numbers is continuous in the norm on $L(H)$, if and only if its infinitesimal generator is bounded.

The infinitesimal generator of a continuous unitary representation $t \to U_t$ of $\boldsymbol{R}$, is the unique self-adjoint operator A such that $U_t = \exp(itA)$. The existence of such an A is asserted by the Stone theorem, cf. e.g. Maurin, 1976.

2. Let G be a locally compact group with left Haar measure dg. For every g in G define an operator ϱ_g on $L^2(G, \mathrm{d}g)$ as follows:

$$\varrho_g f(x) := f(xg).$$

(a) Show that $g \to \varrho_g$ is a continuous representation of G on $L^2(G, \mathrm{d}g)$.

(b) Show that the unitary representation $g \to R(g)$ on $L^2(G, \mathrm{d}g)$ with $R(g)f(x) = \delta^{\frac{1}{2}}(g)f(xg)$ is unitarily eqivalent to the left regular representation L_g on $L^2(G, \mathrm{d}g)$.

(c) Set $\nu = \delta^{-1}\mathrm{d}g$. Show that ϱ_g is a unitary operator on $L^2(G, \mathrm{d}\nu)$ and the assignment $f \to f^{\vee}$, where $f^{\vee}(x) = f(x^{-1})$, establishes the equivalence of L_g and ϱ_g.

3. Retaining the notation of Lemma 2.2.3, show that the functions in $\mathscr{C}^V$ are uniformly continuous.

Hint. Reduce the problem to proving the inequality

$$||\beta_V f(x) - \beta_V f(y)|| \leqslant \int ||f(x\gamma) - f(y\gamma)|| \mathrm{d}\gamma .$$

Apply Problem 1 of § 1.7 to estimate this integral.

4. Find the expression for Bruhat's mapping in the case of $G = \Gamma = \boldsymbol{R}^n$ and the representation which is the character $\boldsymbol{R}^n \ni x \to e^{i(p|x)} \in \boldsymbol{T}$ of the group G.

5. Suppose that ϱ is a positive continuous function on G satisfying $\varrho(xh) = \dfrac{\delta_H(h)}{\delta_G(h)}\varrho(x)$, $x \in G$, and $h \in H$, where H is a closed subgroup of G.

Show that the function $\lambda\colon G \times G \to \boldsymbol{R}_+$, $\lambda(x, y) = \dfrac{\varrho(xy)}{\varrho(x)}$ defines a function $\sigma\colon G \times G/H \to \boldsymbol{R}_+$ by the formula $\sigma(x, yH) = \lambda(x, y)$ and that the function σ defined in this way is a multiplier satisfying

$$\sigma(h, eH) = \frac{\delta_H(h)}{\delta_G(h)}.$$

6. Let M be an oriented n-dimensional manifold and denote by ω a nowhere vanishing n-form compatible with the orientation. Also suppose that G is an orientation preserving Lie transformation group of M. Define a function $\chi\colon G \times M \to \boldsymbol{R}_*$ so that $(g^*\omega)_m = \chi(g, m)\omega_m$. Show that χ is a smooth $\boldsymbol{R}_+$-valued multiplier. Find a local expression for the multiplier so defined in the case $M = P^1\boldsymbol{R}$, $G = \mathrm{SL}(2, \boldsymbol{R})$ with ω defining the standard orientation of $P^1\boldsymbol{R}$.

7. Let $G = \mathrm{SL}(2, \boldsymbol{R})$ and let $\Gamma \subset G$ be a subgroup consisting of all matrices of the form $\begin{bmatrix} a & b \\ 0 & a^{-1} \end{bmatrix}$ with $a > 0$.

(a) Prove the reducibility of the representation U^t of the group G induced in the sense of Mackey by a character χ_t of the subgroup Γ where

$$\chi_t \begin{bmatrix} a & b \\ 0 & a^{-1} \end{bmatrix} = a^{it}, \qquad t \in \boldsymbol{R}.$$

Hint. Note that for $s = \begin{bmatrix} 0 & 1 \\ -1 & 0 \end{bmatrix} \in G$ the mapping $f \to R_s f$, where $R_s f(g) = f(gs)$, commutes with the operators $U^t(g)$, $g \in G$. Consider the eigenspaces for this mapping.

(b) Show that the multiplier form of the representations $T_{+,\sigma}$, $T_{-,\sigma}$ for $\sigma = -1+it$ constructed in Example 2 in § 2.2 can be extended to unitary representations on the space $L^2(P^1\boldsymbol{R}, \mathrm{d}\lambda)$, $\mathrm{d}\lambda$ being the measure on $P^1\boldsymbol{R}$ defined in Problem 3 in § 1.7.

(c) Show that the representation U^t is equivalent to the direct sum $T_{+,\sigma} \oplus T_{-,\sigma}$ for $\sigma = -1+it$.

8. Let (G, M) be a homogeneous space and $\chi\colon G \times M \to \mathrm{Aut}(E)$ a multiplier.

(a) Show that if $\tilde{\chi}\colon G \times M \to \mathrm{Aut}(E)$ is such a multiplier that there exists a function $\alpha\colon M \to \mathrm{Aut}(E)$ satisfying

$$\tilde{\chi}(g, x) = \alpha(x)\chi(g, x)\alpha(gx)^{-1}$$

(the multipliers χ and $\tilde{\chi}$ are then said to be *cohomological*) then the mapping $f \to \alpha f$ intertwines the representation T^χ with $T^{\tilde{\chi}}$.

(b) Select a point x_0 in M. Define a function φ by the formula

$$\varphi(g) = \chi(g, x_0)$$

and let V be the representation of the isotropy subgroup G_0 at x_0 determined by this multiplier (cf. (2.2.9)). Prove that φ satisfies the identity

$$\varphi(gh) = V(h)^{-1}\varphi(g), \qquad h \in G_0 \tag{$*$}$$

and that the multipler χ is uniquely determined by φ.

Hint. $\chi(g, g_1 x_0) = (\varphi(g_1))^{-1}\varphi(gg_1)$

(c) Prove that if functions φ, ψ satisfy $(*)$ then the multipliers determined by them are cohomological.

(d) Let σ: $M \to G$ be a mapping such that $\sigma(x)x_0 = x$ for every $x \in M$.[1]

Show that for every $g \in G$ there exists a unique decomposition

$$g = \sigma(gx_0)h(g),$$

where $h(g) \in G_0$. If (V, E) is a representation of the group G_0 then

$$\chi_V\colon\ G \times M \ni (g, x) \to V\big(h\big(g\sigma(x)\big)\big)^{-1} \in \mathrm{Aut}(E)$$

is a multiplier. Moreover, the mapping $f \to f \circ \sigma$ intertwines the left regular representation on the space of functions on G satisfying $(*)$ with the multiplier representation T^{χ_V}.

9. Let $G = \mathrm{SL}(2, \boldsymbol{C})$, $\Gamma = \left\{\begin{bmatrix} a & 0 \\ b & a^{-1}\end{bmatrix} : a, b \in \boldsymbol{C},\ a \neq 0\right\}$.

(a) Show that every one-dimensional unitary representation of the group Γ is given for some constants m, s, $m \in \boldsymbol{Z}$, $s \in \boldsymbol{R}$ by the formula

$$\lambda_{m,s}\begin{bmatrix} a & 0 \\ b & a^{-1}\end{bmatrix} := \left(\frac{a}{|a|}\right)^m |a|.$$

(b) Let G act on $P^1\boldsymbol{C}$ in a natural way and write π: $\boldsymbol{C}^2 - \{(0, 0)\} \to P^1\boldsymbol{C}$ for the canonical projection. Choose $p_0 = \pi(0, 1)$. Show that Γ is the isotropy subgroup at the point p_0 and the mapping σ defined as

$$\sigma(p) = \begin{bmatrix} 1 & x_1/x_2 \\ 0 & 1 \end{bmatrix}$$

((x_1, x_2) being the homogeneous coordinates of the point p, i.e. p

[1] Mackey in (1952) and (1953) showed that such a σ exists for all separable locally compact homogeneous spaces (G, M). In most of the cases considered in the sequel it will be possible to construct a σ which is continuous on a set of full measure.

$= \pi(x_1, x_2))$ is a continuous mapping from the set $P^1C - \{\pi(1,0)\}$ into G satisfying $\sigma(p)p_0 = p$, $p \neq \pi(1,0)$.

(c) Define a family of mappings $T_{m,s}$ of the space $L^2(C, d\lambda)$, $d\lambda$ being the Lebesgue measure, by the formula

$$T_{m,s}(g)f(z) = \left(\frac{az+b}{cz+d}\right)^m |cz+d|^{-2+is} f\left(\frac{az+b}{cz+d}\right),$$

where $g^{-1} = \begin{bmatrix} a & b \\ c & d \end{bmatrix} \in G$. Show that $T_{m,s}$ is a unitary representation of the group G, equivalent to the representation $U^{m,s}$ induced in the sense of Mackey by the character $\lambda_{m,s}$.

10. Let G be the Heisenberg–Weyl group consisting of matrices of the form

$$\begin{bmatrix} 1 & x_1 & x_3 \\ 0 & 1 & x_2 \\ 0 & 0 & 1 \end{bmatrix}, \qquad x_i \in R \text{ for } i = 1, 2, 3.$$

(For brevity such a matrix is denoted by (x_1, x_2, x_3).)

(a) Show that the subset $Z = \{(0,0,z)\colon z \in R\} \subset G$ is exactly the centre of the group G and that $G/Z \simeq R^2$. Denoting by $\pi\colon G \to G/Z$ the canonical projection, show that every unitary representation character) χ of the group R^2 lifts to a unitary representation of G according to the formula $\tilde{\chi}(g) := \chi(\pi(g))$.

(b) Show that $N = \{(x_1, x_2, x_3)\colon x_1 = 0\}$ is a normal subgroup of G such that the quotient group G/N is isomorphic to the additive group of real numbers. Also prove that the homogeneous space $(G, G/N)$ is isomorphic to the space (G, R^1) with the action of G on R^1 defined by the formula

$$(x_1, x_2\ x_3)\cdot t := t + x_1, \qquad t \in R^1$$

Hint. G is a semidirect product of the subgroup N and the subgroup A consisting of all elements of the form $(p, 0, 0)$ with $p \in R^1$.

(c) Let $\chi\colon N \to S^1$ denote a unitary character of N given by the formula $\chi(0, x_2, x_3) = \exp i(ux_2 + vx_3)$, $u, v \in R$. Show that the representation U induced by χ in the sense of Mackey is equivalent to the multiplier representation T acting on the space $L^2(R, d\lambda)$ ($d\lambda$ being the Lebesgue measure) with the multiplier

$$\omega(x_1, x_2, x_3, t) = \exp[-i(ux_2 + vx_3)] \exp iux_2(z + x_1).$$

(d) Check that in the case $u = 0$, $v = 1$ we obtain the Schrödinger representation from Example 2 in § 2.7. Set $U(p) := T^{\omega}(p, 0, 0)$, $V(q) := T^{\omega}(0, q, 0)$. Show that the one-parameter unitary groups U and V satisfy the relation

$$U(p)\,V(q)\,U(p)^{-1}\,V(q)^{-1} = \mathrm{e}^{-iqp},$$

called the *canonical commutation relation in the Weyl form.* (This is the "integrated" form of the canonical commutation relations of Heisenberg: $[p, q] = iI$.)

11. Let (H, U) be a unitary representation of a group G.

(a) Show that for f in $L^1(G)$ the formula

$$U(f) := \int_G f(g)\,U_g\,\mathrm{d}g$$

defines a $*$-representation of the convolution algebra $L^1(G)$ on H, fulfilling the estimate

$$\|U(f)\| \leqslant \|f\|.$$

(By a $*$-representation we understand a representation of a convolution algebra such that $U(f^*) = U(f)^*$.)

(b) If U is the left regular representation of the group G on the space $L^2(G)$ then $U(f)h = f * h$, $h \in L^2(G)$.

Hint. One can prove, without using the Bochner integral theory, that $U(f)$ is well defined by the formula in (a) by considering a sesquilinear form $B_f(x, y) = \int_G f(g)(x|U_g y)\,\mathrm{d}g$ on H and by showing that the operator $U(f)$ is uniquely determined by the condition $B_f(x, y) = (x|U(f)y)$.

12. Let G be a finite group consisting of n elements.

(a) Show that the convolution algebra $\mathscr{C}(G)$ is an n-dimensional algebra with a unit over $\mathbf{C}$. Also prove that there exists a basis $\{e_g\}$ for $\mathscr{C}(G)$ indexed by the elements of G and such that

$$e_g * e_h = e_{gh}, \qquad g, h \in G. \tag{$*$}$$

Show that the multiplication in $\mathscr{C}(G)$ is uniquely determined by the condition of bilinearity and by the relations $(*)$.

Hint. Take for e_g the function equal to one at the point g and to zero outside.

(b) Let $H \subset G$ be a subgroup of G. Show that the algebra $\mathscr{C}(H)$ is a subalgebra of the algebra $\mathscr{C}(G)$.

(c) Regarding G as a subset of the algebra $\mathscr{C}(G)$, show that every representation of the algebra $\mathscr{C}(G)$ determines by restriction a representation of the group G and that the operation of restriction defines a bijection between the set of representations of the algebra $\mathscr{C}(G)$ and the set of representations of the group G. Also prove that a representation of the algebra $\mathscr{C}(G)$ is irreducible if, and only if, its restriction to G is such.

13. Let G be a finite group and $H \subset G$ a subgroup, and denote by (E, V) a finite-dimensional representation of the subgroup G.

(a) Show that the space H^V of the representation induced by V contains a subspace E' with the following properties:

(i) E' is isomorphic to E and invariant under U_h^V for h in H.

(ii) The representation of H obtained by restriction of U^V to E' is equivalent to (E, V).

(iii) $H^V = \bigoplus_{[g \in G/H]} U_g^V E'$

(Note that the image $U_g^V E'$ depends upon g only modulo H, and therefore the above sum is well-defined.)

(b) Prove that $H^V \simeq \mathscr{C}(G) \otimes_{C(H)} E$ and the action of G on H^V corresponds to the natural action of G on the first factor. (The symbol on the right denotes a tensor product of $\mathscr{C}(H)$-moduli, cf. e.g. Kirillov, 1976.)

Hint. Take for E' the space of functions vanishing outside H.

14. Let H be a separable Hilbert space with an orthonormal basis $\{e_i\}_{i=1}^{\infty}$. We say that a linear mapping A of H into itself admits a matrix representation if for an arbitrary vector $x \in H$ the coordinates of the vector Ax relative to the basis $\{e_i\}$ defined by

$$Ax = \sum_{i=1}^{\infty} y_i e_i$$

are expressible in terms of the coordinates of the vector $x = \sum_{i=1}^{\infty} x_i e_i$ in the form

$$y_k = \sum_{i=1}^{\infty} (e_i | Ae_k) x_i .$$

Prove that the set of mappings admitting a matrix representation coincides with the set $L(H)$ of all continuous operators on H. Further prove that A is a Hilbert–Schmidt operator on H (see § 4.1 for the pertinent definition) if, and only if, $\sum_{i,k=1}^{\infty} |(Ae_i|e_k)|^2 < \infty$.

15. (a) Let φ be a continuous function on G. Show that φ is positive definite if, and only if, for every positive integer n and arbitrary $x_1, \dots$ $\dots, x_n \in G$, the matrix $[\varphi(x_i^{-1}x_j)]_{i,j}$ is positive definite.

Hint. Since the condition is merely a reformulation of the definition, necessity is proved as Proposition 2.3.4. To prove sufficiency note that the Riemann sums approximating the integral $\int \varphi(x)f^* * f(x)\,dx$ are positive.

(b) Show that a translate of a continuous positive definite function φ is a linear combination of four positive definite functions.

Hint. Represent φ as a matrix element of a unitary representation.

(c) Show that a function $\omega \in L^\infty(G)$ is positive definite if, and only if, it is equal a.e. to a positive definite continuous function.

Hint. The proof of necessity follows once it is noted that similarly as in the course of the proof of Theorem 2.5.3 it can be shown that

$$\int f(g)\omega(g)\,dg = \int f(g)(x_\omega|^\omega U_g x_\omega)\,dg, \qquad f \in \mathscr{C}_0(G).$$

Chapter 3

Decomposition Theory of Unitary Representations

Having introduced different operations on group representations, we proceed to the description of representations in terms of their simple components—the irreducible representations, i.e. the ones which contain no non-trivial subrepresentations. Schur's lemma (Theorem 3.1.1) provides the basic criterion of irreducibility of a representation.

In the following chapter, we shall see that every representation of a compact group on a Hilbert space, can be represented as a direct sum of irreducible representations which, in this case, are finite-dimensional. However, when we pass to the study of representations of non-compact groups, the situation becomes more complicated. Not only infinite-dimensional irreducible representations may appear but also there are representations whose every subrepresentation contains a non-trivial invariant subspace. We have already encountered such situations in the harmonic analysis on $\boldsymbol{R}$. The reduction of such a representation by means of the construction of direct sum proves to be impossible.

To cope with this problem one can use the construction of a direct integral of Hilbert spaces introduced by J. von Neumann in order to study decomposition of algebra representations. An analogous theorem concerning the decomposition of unitary representations of groups, is due to F. Mautner and R. Godement.

The general decomposition theorem will not be applied in the second part of the book. However, we introduce the notions of a direct integral and of decomposition of representations, so that the expansions which appear in our considerations may be precisely formulated and interpreted in a uniform way.

As an illustration of the decomposition theorem, we present an elementary theory of Fourier integrals and of Fourier series, together with elements of the theory of the Fourier transformation on locally compact abelian groups. This gives us an opportunity to underline the special role of the Gelfand–Raikov theorem (Theorem 3.3.3) in harmonic analysis on groups. This theorem establishes links between representation theory and the convex analysis, since the positive definite functions constitute a convex subset in the space of all bounded functions on the group. The importance of positive definite functions in the analysis on $\boldsymbol{R}^n$ was noted in 1933 by S. Bochner, who proved that every positive definite function can be represented as the Fourier transform of a finite measure. Regarding Bochner's theorem as a statement in the theory of representations of commutative groups, A. Weil proved its general version. Then it became evident that Herglotz's theorem, preceding Bochner's theorem, can be regarded as a special case of the Bochner–Weil theorem corresponding to the case of the group T.

In the third part of this book the methods of convex analysis will be applied to the theory of harmonic functions on symmetric spaces.

The present chapter deals with representations of a locally compact group G on a separable Hilbert space.

3.1. IRREDUCIBLE REPRESENTATIONS. SCHUR'S LEMMA

In the preceding chapter (in § 2.4) we introduced the basic notions of decomposition theory, i.e. the concept of the inclusion and the equivalence of representations together with the notion of an intertwining operator.

DEFINITION. A representation (E, T) of a group G on a locally convex space E is said to be *irreducible* if it has no non-trivial subrepresentations. Otherwise the representation is said to be *reducible*.

Thus, if (E, T) is an irreducible representation and $E_1 \subset E$ is an invariant subspace, then either $E_1 = 0$ or $E_1 = E$.

We shall now give a convenient reformulation of the condition of irreducibility of a unitary representation in terms of intertwining operators.

Denote by $L_G(T^1, T^2)$ the space of operators intertwining representations (E^1, T^1) and (E^2, T^2). We also write $L_G(E^1, E^2)$, when there is no danger of a confusion. If $T^1 = T^2 = T$, we write simply $L_G(T)$ or $L_G(E)$.

THEOREM 3.1.1 (Schur's lemma). *A unitary representation* (H, U) *is irreducible* $\Leftrightarrow$ $\dim L_G(U) = 1$.

Proof. $\Leftarrow$ Let $H_1 \subset H$ be such that $U_g H_1 \subset H_1$ for all $g \in G$. It follows from Proposition 2.1.6 that the orthogonal projection P onto H_1 is in $L_G(U)$; hence is proportional to the identity operator constituting a basis for $L_G(U)$. Therefore either $P = 0$ or $P = \mathrm{id}$.

$\Rightarrow$ Let $A \in L_G(U)$, where (H, U) is an irreducible representation. Then the adjoint operator A^* also belongs to $L_G(H)$:

$$(A^* U_g x|y) = (U_g x|Ay) = (x|U_{g^{-1}}Ay) = (x|AU_{g^{-1}}y)$$
$$= (A^*x|U_{g^{-1}}y) = (U_g A^* x|y), \qquad x, y \in H, \qquad g \in G.$$

Consequently, the operators $B := (A+A^*)$ and $C := i(A-A^*)$ are hermitian elements of $L_G(U)$. It follows that their spectral families consist of orthogonal projections which commute with the representation operators. Since the image of a projection which commutes with all U_g is an invariant subspace, we conclude on account of the irreducibility of the representation, that the space, $L_G(U)$, contains only the identity projection. Therefore $B = \mu\mathrm{id}$, $C = \lambda\mathrm{id}$ and consequently $A = \nu\mathrm{id}$. □

COROLLARY 3.1.2. *Let* U^1 *be a unitary irreducible representation and* U^2 *an arbitrary unitary representation. Then every operator intertwining* U^1 *and* U^2 *is proportional to an isometry.*

Proof. Let $T \in L_G(U^1, U^2)$. Since the operator T^*T is an element of $L_G(U^1)$ it must be of the form $\lambda\mathrm{id}$ for a $\lambda \geqslant 0$. Hence $(Tx|Ty) = (T^*Tx|y) = \lambda(x|y)$. □

One can investigate the components of a given representation (E, T) by studying the spaces of operators intertwining that representation with simpler ones. To this end we shall prove an important

PROPOSITION 3.1.3. $L_G(T, T_1+T_2) = L_G(T, T_1)+L_G(T, T_2)$.

Proof. We shall describe the desired isomorphism. Given a decomposition E_1+E_2 of the representation space T_1+T_2, we define projections P_I: $E \to E_i$, $i = 1, 2$, which commute with the action of the group. Further, we may assign to every operator $A \in L_G(T, T_1+T_2)$ a pair of operators A_1, A_2 where $A_i = P_i T$. Obviously $A_i \in L_G(T, T_i)$. The assignment is injective since $A = A_1 + A_2$ and its surjectivity is obvious. □

Denote by nU the direct sum of n copies of a unitary representation U. It follows from the last proposition and Schur's lemma that

COROLLARY 3.1.4. *Every irreducible unitary representation of a commutative group is one-dimensional.*

Proof. The group being commutative, this implies that $U_g \in L_G(U)$, hence $U_g = \lambda(g)\mathrm{id}$. It follows that every subspace of the representation space is invariant. Therefore the assertion follows from the assumption of irreducibility. □

Example 1

The assumption of unitariness in Corollary 3.1.4 cannot be dropped. To show this, consider the identity representation of the group SO(2) of all rotations on the plane, i.e. of matrices

$$\begin{bmatrix} \cos\varphi & -\sin\varphi \\ \sin\varphi & \cos\varphi \end{bmatrix}.$$

DEFINITION. The number $\dim L_G(U^1, U^2)$ is called the *intertwining number* of the representations U^1 and U^2.

Representations U^1, U^2 are said to be *disjoint* if $L_G(U^1, U^2) = 0$.

If U^1 is an irreducible representation then $\dim L_G(U^1, U^2)$ is called the *occurrence number* of U^1 in U^2.

To justify this terminology we have the following

THEOREM 3.1.5. *Let U^1 and U^2 be unitary representations and suppose that U^1 is irreducible. Then the following conditions are equivalent:*

(a) $\dim L_G(U^1, U^2) = n$,

(b) $U^2 = U^0 + nU^1$, *where U^0 is a representation disjoint with U^1.*

Proof. The implication (b) ⇒ (a) is obvious in view of Proposition 3.1.3. Therefore let us assume (a). If $n = 0$ then (b) is true. If $n \neq 0$ then the representation U^2 contains at least one copy of U^1. Denote by $\mathscr{F}$ the family of all subrepresentations of U^2 of the form kU^1. Write $k(U) := k$ if $U = kU^1$. Obviously $k(U) \leqslant n$ for arbitrary representations $U \in \mathscr{F}$. Choose a $(H, U) \in \mathscr{F}$ with the maximal number $k(U)$ and suppose $k(U) < n$. Then there is a representation U^0 on H disjoint with U^1. In virtue of the implication (b) $\Rightarrow$ (a) we obtain $\dim L_G(U^1, U^2) = k(U)$—a contradiction. Consequently $k(U) = n$ and $U^2 = nU^1 + U^0$. $\square$

In order to illustrate the main idea of decomposition theory, we first consider the case of a finite-dimensional unitary representation.

PROPOSITION 3.1.6. *Every finite-dimensional unitary representation* (H, U) *of a group* G *is equivalent to a direct sum of irreducible ones.*

The proof proceeds by induction on $n = \dim H$. A 1-dimensional representation is obviously irreducible. Suppose that the assertion is true of every representation of dimension $\dim H < n$, and let (H^1, U^1) be a representation with $\dim H^1 = n$. If it is irreducible then there is nothing to prove. So we assume that H^1 contains an invariant proper subspace H^2 and consequently we get the decomposition $H^1 = H^2 \oplus (H^2)^{\perp}$, invariant under the action of the representation. Both component spaces have dimension less than n; thus on account of the induction hypothesis, they are decomposable into a direct sum of irreducible components. This yields the desired decomposition.

It can easily be shown that the assumptions of unitariness and of finite dimension are both significant in this proposition.

Example 2

Let $G = \left\{\begin{bmatrix} 1 & z \\ 0 & 1 \end{bmatrix} : z \in \boldsymbol{C}\right\}$ with the matrix multiplication as the group operation.

Let G act in a natural way on the 2-dimensional space. We have an invariant subspace consisting of vectors of the form $\begin{bmatrix} a \\ 0 \end{bmatrix}$. However this subspace does not admit an invariant complementary subspace, as may be easily seen.

Example 3

Let U denote the regular representation of the group $\boldsymbol{R}$ on $L^2(\boldsymbol{R})$. It can easily be verified that this representation does not contain a one-dimensional subrepresentation. Namely, suppose that $f \in L^2(\boldsymbol{R})$ is a function such that $L_g f = \lambda(g) f$. For an arbitrary function $\varphi \in \mathscr{C}_0(\boldsymbol{R})$

the convolution $f * \varphi$ is a continuous element of $L^2(\boldsymbol{R})$ also satisfying this equation. Thus we can assume that f is continuous. If there exists a g_0 such that $f(g_0) \neq 0$ then $f(gg_0) = \lambda^{-1}(g)f(g_0)$. Consequently, since the representation is unitary, we have $|f(g)| = |f(g_0)||\lambda^{-1}(g)| = |f(g_0)|$. Thus we infer $f =$ const, in contradiction to the assumption that $f \in L^2(\boldsymbol{R})$.

We know from Corollary 3.1.4 that every irreducible representation of $\boldsymbol{R}$ is 1-dimensional. Therefore the regular unitary representation on $L(\boldsymbol{R})$ is not decomposable into a direct sum of irreducible representations.

It follows that the operation of direct sum does not suffice to construct all unitary representations (up to an isomorphism) from the irreducible ones. A construction that meets these demands—the direct integral of representations—was introudced by von Neumann for the purpose of studying the decomposition of representations of algebras. Analogous theorems concerning group representations are due to Naimark and Mautner.

We shall first describe the construction of the direct integral of Hilbert spaces.

Direct Integral of Hilbert Spaces

Let X be a locally compact separable space which will be referred to in the sequel as the *base*. For every $x \in X$ let a Hilbert space $(H(x), (\cdot|\cdot)_x)$, be given, called the *fibre* over x. Put $\boldsymbol{B} = \bigcup_{x \in X} H(x)$. We are concerned with $\boldsymbol{B}$-valued functions on X with the value at a point $x \in X$, belonging to the fibre over x. We shall reformulate this condition by applying the mapping p: $\boldsymbol{B} \to X$, called the *projection* and defined as $p(b) = x$ for $b \in H(x)$. Obviously $p^{-1}(x) = H(x)$.

If a function f: $X \to \boldsymbol{B}$ verifies the condition $pf(x) = x$ then such an f is called a *vector field* on X.

In the definition of a vector field the Hilbert space structure was of no importance and therefore, in an analogous way, we can define a vector field on X relative to a family of vector spaces $\{E(x)\}_{x \in X}$.

Further, suppose that there exist a measure μ on X and a countable family of vector fields Γ on X called the *fundamental family* satisfying:

(1) for every pair $f_1, f_2 \in \Gamma$ the scalar function $X \ni x \to (f_1(x)|f_2(x))_x$ is μ-measurable;

(2) for every $x \in X$ the set of vectors $\{f(x)|f \in \Gamma\}$ spans the space $H(x)$.

DEFINITION. A vector field f on X is said to be *μ-measurable* provided for every $f_1 \in \Gamma$ the function $X \ni x \to (f_1(x)|f(x))_x$ is μ-measurable.

The space $\mathcal{M}$ of μ-measurable vector fields is a vector space under the operations

$$\begin{aligned}(f_1+f_2)(x) &= f_1(x)+f_2(x),\\ (\lambda f)(x) &= \lambda(f(x)).\end{aligned} \tag{3.1.1}$$

We shall prove that for every $f_1, f_2 \in \mathcal{M}$ the scalar function $X \ni x \to (f_1(x)|f_2(x)) \in \boldsymbol{C}$ is also measurable.

For this purpose we shall verify first that the non-negative function $X \ni x \to (f_1(x)|f_2(x)) \in \boldsymbol{R}$ is measurable.

Consider the family Γ' of vector fields which are linear combinations of elements of the fundamental family with the coefficients of the form w_1+iw_2, $w_i \in W$. The set Γ' is countable and for every $x \in X$ the set $\{f(x)\colon x \in \Gamma'\}$ is countable and dense in $H(x)$. Further, for a fixed $f \in \Gamma'$ we define the following vector field assuming values in the unit ball in $H(x)$:

$$\psi(x) = \begin{cases} 0, & f(x) = 0,\\ f(x)/\|f(x)\|_x, & f(x) \neq 0.\end{cases}$$

The set of values at a point $x \in X$ of all vector fields of this form is dense in the unit sphere in $H(x)$. Therefore we can write $(f_1(x)|f_1(x))_x = \|f_1\|_x^2 = \sup_{\psi \in \Gamma'} (\psi(x)|f_1(x))^2$. Thus the function $X \ni x \to \|f_1\|_x^2$, being a countable supremum of measurable functions is itself measurable.

The measurability of the function $X \ni x \to (f_1(x)|f_2(x)) \in \boldsymbol{C}$ for arbitrary $f_i \in \mathcal{M}$ now follows easily from the polarization formula (§ 2.2).

Denote by $\boldsymbol{H}$ the system $(\mathbf{B}, p, X, \Gamma)$ of the objects entering into the above construction and write

$$L^2(\mu, \boldsymbol{H}) = \Big\{f \in \mathcal{M}\colon \int_X \|f(x)\|_x^2 d\mu < \infty\Big\}. \tag{3.1.2}$$

Identifying vector fields equal μ-almost everywhere, we obtain a linear space $L^2(\mu, H)$ (also denoted by $\int_X H(x)\,d\mu(x)$) called the *direct integral* of the Hilbert space $\{H(x)\}_{x\in X}$.

PROPOSITION 3.1.7. *The space* $L^2(\mu, \boldsymbol{H})$ *is a Hilbert space with the inner product*

$$(f|g) := \int_X (f(x)|g(x))_x d\mu(x). \tag{3.1.3}$$

By applying the polarization formula (§ 2.2) it is easy to verify the correctness of the definition. The proof of completeness is actually the same as that of the Riesz–Fischer theorem on the completeness of the space $L^2(\mu)$.

Example 4

The direct sum $\bigoplus_{n=1}^{\infty} H_n$ of Hilbert spaces can be viewed upon as a special case of the above construction. Indeed, let $X = N$ (the set of positive integers), $H(x) := H_n$ for $x = n \in N$ and let μ be the "counting" measure on N (i.e. such a measure that $\mu(\{p\}) = 1$ for every $p \in N$). Choose an orthonormal basis for every space H_n and denote by Γ the set of all functions $N \ni n \to x_n \in \bigcup_{i=1}^{\infty} H_n$ such that x_n is an element of that base for every n. Since $\int_N f(n)\,d\mu(n) = \sum_{i=1} f(n)$ for every f, it follows that $L^2(\mu, \boldsymbol{H}) = \left\{(x_n) \in \bigcup_{n=1}^{\infty} H_n : \sum_{i=1}^{\infty} \|x_n\|^2 < \infty\right\}$ (a function $n \to x_n$ is written as a sequence (x_n)), which is precisely the condition defining the direct sum of Hilbert spaces $\bigoplus_{n=1}^{\infty} H_n$.

Example 5

Suppose that (X, μ) is a space with a measure and let $\boldsymbol{H}$ be a Hilbert space. It is possible to define the space of "square integrable" $\boldsymbol{H}$-valued functions by applying the above method. Namely, we set $\boldsymbol{H}(x) = \boldsymbol{H}$ for arbitrary $x \in X$. Then the vector fields on X are in fact the $\boldsymbol{H}$-valued functions on X.

Let S be a countable dense subset of $\boldsymbol{H}$ and write Γ for the set of all constant functions on X with values in S. It follows from the general construction that the measurable functions on X are precisely those functions for which $x \to (f(x)|h)$ is measurable for every $h \in \boldsymbol{H}$. Therefore the direct integral of the Hilbert space $\boldsymbol{H}$ is the space of all functions measurable in the above sense with the property that $\int_X \|f(x)\|^2 d\mu(x)$ is finite. This space is denoted by $L^2(\mu, \boldsymbol{H})$. The reader will easily check that if $\boldsymbol{H} = \bigoplus_{n=1}^{\infty} H_n$ then $L^2(\mu, \boldsymbol{H}) = \bigoplus_{n=1}^{\infty} L^2(\mu, H_n)$.

Operators on the Space of the Direct Integral

We shall now distinguish, as we did in the case of the direct sum of spaces, a class of "simplest" operators on $L^2(\mu, \boldsymbol{H})$.

We consider the space of vector fields corresponding to the family of vector spaces $\{L(H(x))\}_{x\in X}$.

An operator field A is said to be *measurable* if, for every $f \in \mathcal{M}$, the vector field $X \ni x \to A(x)f(x) \in H(x)$ is measurable. A measurable operator field defines an operator on $\mathcal{M}$ by the formula

$$Af(x) = A(x)f(x). \tag{3.1.4}$$

Operators of this type are said to be decomposable. However, in general the operators obtained in this way are not continuous (and not even everywhere defined) on $L^2(\mu, \boldsymbol{H})$.

PROPOSITION 3.1.8. *If A is an operator field and the function a: $X \ni x \to \|A(x)\|_x =: a(x)$ is of class $L^\infty(\mu)$, then the operator defined by* (3.1.3) *is continuous and satisfies $\|A\|_{L^2(\mu, H)} = \|a\|_{L^\infty(\mu)}$. Moreover, A^* is also a decomposable operator corresponding to the operator field $X \ni x \to A(x)^*$.*

A decomposable operator corresponding to the operator field of the form $X \ni x \to \lambda(x)\mathrm{id}_{H(x)}$ for some function λ on X is called a *diagonal operator*.

If both A and B are decomposable operators, then $A \circ B$ is also a decomposable operator corresponding to the operator field

$$X \ni x \to A(x)+B(x).$$

The decomposable operator A corresponding to an operator field $x \to A(x)$ will be denoted by $\int\limits_X A(x)\,\mathrm{d}\mu(x)$.

Example 6

Let us find the form of decomposable operators in the case of the Hilbert direct sum. If (A_n) is a sequence of operators $A_n \in L(H_n)$, $n \in N$, so that $\sup\|A_n\| < \infty$, then the operator $A = \bigoplus\limits_{n=1}^{\infty} A_n$ is defined as $A(x_n) = (A_n x_n)$.

Note that if $P_i\colon \bigoplus\limits_{n=1}^{\infty} H_n \to H_n$, $i = 1, 2, \ldots$ is the orthogonal projection on the i-th component (we identify H_i with the subspace of $\bigoplus\limits_{n=1}^{\infty} H_n$ consisting of all sequences with the terms different from the i-th term equal to zero) then $AP_i = P_i A$. Conversely, one can easily check that if $A \in L(\bigoplus\limits_{n=1}^{\infty} H_n)$ satisfies this condition then A is decomposable. Namely, it is enough to note, in view of $A(H_i) \subset H_i$, $i = 1, 2, \ldots$, that for $A_i := A|_{H_i}$ we have $\bigoplus\limits_{n=1}^{\infty} A_n = A$.

Example 7

The space $L^2(\boldsymbol{R}^n, \mathrm{d}\lambda^n)$ is a special case of the direct integral. Obviously the decomposable operators on this space are precisely the diagonal operators defined by a function f on $\boldsymbol{R}^n$:

$$Ax(\lambda) = f(\lambda)x(\lambda), \qquad x \in L^2(\boldsymbol{R}).$$

Putting $f(\lambda) = \chi_t(\lambda) := \exp i(\lambda|t)$, $t \in \boldsymbol{R}^n$, we obtain a representation of the group $\boldsymbol{R}^n$ on $L^2(\boldsymbol{R}^n)$ defined as follows:

$$\big(U(t)x\big)(\lambda) = \chi_t(\lambda)x(\lambda).$$

Symbolically we write $U = \int\limits_{\boldsymbol{R}^n} \chi_t \,\mathrm{d}\lambda^n(t)$. Thus we obtain a representation defined as a direct integral of representations which are obviously irreducible. The statement that the representation so defined is equivalent to the regular representation on $\boldsymbol{R}^n$ is the contents of the Fourier analysis on $\boldsymbol{R}^n$. This example will be described more precisely in § 3.2.

We are now in a position to state the fundamental theorem of decomposition theory of unitary representations.

THEOREM 3.1.9 (von Neumann, Mautner, Naimark) (see Dixmier, 1964). *Suppose that (U, H) is a unitary representation of a locally compact, separable group G. Then there exist a locally compact, separable space Λ and a measure μ on Λ such that the representation $\mathscr{U}$ is equivalent to the representation $(U, \mathscr{H})$ with*

$$\mathscr{H} = \int_{\Lambda} H(\lambda) \mathrm{d}\mu(\lambda), \tag{3.1.5}$$

$$\mathscr{U}_g = \int_{\Lambda} U_g(\lambda) \mathrm{d}\mu(\lambda) \tag{3.1.6}$$

and $(U(\lambda), H(\lambda))$ are irreducible representations of the group G for $\mathrm{d}\mu$-almost all λ.

An intertwining unitary operator $H \to \mathscr{H}$ establishing that equivalence will be called a *generalized Fourier transformation* on H.

It is worth-while to note that the decomposition provided by this theorem is not unique. Consequently, for a given representation, the term "Fourier transform" is not univocal. We shall illustrate this phenomenon with the following example.

Example 8

Define a representation U of $\boldsymbol{R}$ on $L^2(\boldsymbol{R})$ as $U_x = (\exp ix)\mathrm{id}$. The action of every operator consists in multiplication by a scalar and, therefore, every decomposition of the representation space into a direct integral of 1-dimensional spaces, determines at the same time a decomposition of the representation into a direct integral of irreducible equivalent representations.

In particular, $L^2(\boldsymbol{R}) = \int_{\boldsymbol{R}^1} H(x) \mathrm{d}x$, where $H(x) \cong \boldsymbol{C}$. On the other hand, since $L^2(\boldsymbol{R})$ is a separable Hilbert space, we have $L^2(\boldsymbol{R}) = \bigoplus_{i=1}^{\infty} H_i$, where $H_i = \boldsymbol{C}e_i$ and $\{e_i\}$ is an orthonornal basis for $L^2(\boldsymbol{R})$.

Thus we see that the assertion of Theorem 3.1.9 can be realized through different decompositions whose bases (the spaces Λ) are not isomorphic either topologically or in the measure-theoretic sense.

The fundamental Theorem 3.1.9 ends the stage of fundamental studies in the representation theory. However, being a theorem of an existential character, it only points out the direction of further development of the theory.

Among the main topics which still remain open we have:

—determination for each class of groups the space of equivalence classes of irreducible representations,

—finding effective realizations of these representations,

—identifying, for a given representation, the space Λ and the measure μ, called the *Plancherel measure.*

The order in which these problems are solved for special types of groups is suggested by applications of the decomposition theory in different branches of mathematics and physics. The representation theory of semi-simple groups is of great interest owing to the wide range of its applications. In fact, the fundamental questions of harmonic analysis on semi-simple groups have already been solved mainly as a result of the researches conducted in two great mathematical centres directed by Harish-Chandra and I. M. Gelfand. At the same time we can see an impressive influence of the representation theory on other branches of mathematics, such as the number theory, algebraic geometry and also the theory of special functions. Therefore representation theory is still alive and still continues to develop.

In subsequent chapters, many examples will be given underlying the achievements in representation theory.

We begin with the simplest ones: the Fourier analysis on $\boldsymbol{R}^n$ and on T presented from the point of view of representation theory.

3.2. CLASSICAL FOURIER TRANSFORMATION

Each continuous irreducible representation of the commutative group $\boldsymbol{R}^n$ is one-dimensional. The dual group is isomorphic to $\boldsymbol{R}^n$. Under this isomorphism, to an element $p \in \boldsymbol{R}^n$, there corresponds a character χ_p given by the formula $\chi_p(x) = \exp i(p|x)$, where $(p|x) = \sum_{i=1}^{n} p^i x^i$. The mapping $\boldsymbol{R}^n \ni p \to \chi_p \in (\boldsymbol{R}^n)^\wedge$ is a group homomorphism which can be seen from the relation $\chi_{p+q} = \chi_p \chi_q$.

We know by Example 3 in § 3.1, that the regular representation on the space $L^2(\mathbf{R}^n)$ does not contain irreducible components and thus there are no operators intertwining $L^2(\mathbf{R}^n)$ with an irreducible representation. However such operators exist for the regular representation on the space $L^1(\mathbf{R}^n)$. One can easily find the form of an operator intertwining $L^1(\mathbf{R}^n)$ with the character χ_p. By fixing a vector which spans the one-dimensional representation we reduce the problem to finding a linear functional $F: L^1(\mathbf{R}^n) \to \mathbf{C}$ satisfying the condition $F(L_g f) = \chi_p(g) Ff$. Multiplying both sides by $f_1 \in L^1(\mathbf{R}^n)$ and integrating with respect to measure dx, we obtain $F(f_1 * f) = \chi_p(f_1) F(f)$, where $\chi_p(f_1) = \int_{\mathbf{R}^n} \chi_p(x) f_1(x) dx$. The convolution on a commutative group being commutative, this implies that $\chi_p(f_1) F(f) = F(f_1 * f) = F(f * f_1) = \chi_p(f) F(f_1)$.

For a fixed p we can find a function f_1 such that $\chi_p(f_1) \neq 0$. Hence we get

$$F(f) = \chi_p(f) \frac{F(f_1)}{\chi_p(f_1)}.$$

Thus the functional in question is equal (up to a constant factor) to the Fourier transforms of f at the point $p \in \mathbf{R}^n$:

$$L^1(\mathbf{R}^n) \ni f \to \int_{\mathbf{R}^n} \exp i(p|x) f(x) dx := \hat{f}(p). \tag{3.2.1}$$

Directly from the definition we verify the following properties of the Fourier transformation:

PROPOSITION 3.2.1.

(1) $(\alpha f_1 + \beta f_2)\hat{} = \alpha \hat{f}_1 + \beta \hat{f}_2$,
(2) $(L_g f)\hat{}(p) = \chi_p(x) \hat{f}(p)$,
(3) $(f_1 * f_2)\hat{} = \hat{f}_1 * \hat{f}_2$.

In order to state further properties we define the Schwartz space on $\mathbf{R}^n$, consisting of all functions in $L^1(\mathbf{R}^n) \cap \mathscr{E}(\mathbf{R}^n)$ satisfying, for arbitrary multi-indices $\alpha = (\alpha_1, \ldots, \alpha_n)$ and $\beta = (\beta_1, \ldots, \beta_n)$ the condition

$$\sup_{x \in \mathbf{R}^n} |x|^\beta |X^\alpha f(x)| < \infty,$$

where $X^\alpha = \frac{\partial^{\alpha_1}}{\partial x_1^{\alpha_1}} \cdots \frac{\partial^{\alpha_k}}{\partial x_k^{\alpha_k}}$ and $|x|^\beta = |x_1|^{\beta_1} \ldots |x_m|^{\beta_m}$. The Schwartz space is denoted by $\mathscr{S}(\mathbf{R}^n)$.

Example 1

The function $R^n \ni x \to e^{-\alpha\|x\|^2}$ for $\alpha > 0$ belongs to the Schwartz space $\mathscr{S}(R^n)$. The reader is invited to check that the Fourier transform of $f(x) = e^{-\alpha\|x\|^2}$ equals $\hat{f}(p) = \left(\frac{\pi}{\alpha}\right)^{\frac{n}{2}} e^{-\frac{\|p\|^2}{4\alpha}}$.

We shall prove the following fundamental properties of $\mathscr{S}(R^n)$:

PROPOSITION 3.2.2. *Every function* $f \in \mathscr{S}(R^n)$ *satisfies the identities*

(1) $\dfrac{\partial \hat{f}}{\partial p_l} = i(x_l f)^\wedge$,

(2) $\left(\dfrac{\partial f}{\partial x_l}\right)^\wedge (p) = -ip_l \hat{f}(p)$,

(3) *The Fourier transformation maps* $\mathscr{S}(R^n)$ *into* $\mathscr{S}(R^n)$.

Proof. (1) $$\frac{\partial}{\partial p_l}\int_{R^n} f(x)\exp i(x|p)\,dx = \int_{R^n} f(x)\frac{\partial}{\partial p_l}\exp i(x|p)\,dx$$
$$= i\int_{R^n} x_l f(x)\exp i(x|p)\,dx = i(x_l f)^\wedge.$$

(2) $$\int_{R^n} \frac{\partial}{\partial x_l} f(x)\exp i(x|p)\,dx = -\int_{R^n} f(x)\frac{\partial}{\partial x_l}\exp i(x|p)\,dx$$
$$= -p_l i\int_{R^n} f(x)\exp i(x|p)\,dx = -ip_l \hat{f}(p).$$

(3) It follows from (2) that for an arbitrary polynomial $w(p_1, \dots, p_n)$ there exists a differential operator D with constant coefficients such that $(Df)^\wedge = w\hat{f}$. Since $Df \in \mathscr{S}(R^n)$, it follows that the function $w\hat{f}$ is bounded as the Fourier transform of an integrable function. From (2) we infer that this property is true of an arbitrary derivative of $\hat{f}$, hence $\hat{f} \in \mathscr{S}(R^n)$. □

The following theorem is of great importance for the Fourier transformation theory.

THEOREM 3.2.3 (Plancherel). *For every* $f \in L^2 \cap L^1(R^n)$
$$\int_{R^n} |f|^2(x)\,dx = \frac{1}{(2\pi)^n}\int_{R^n} |\hat{f}|^2(p)\,dp. \tag{3.2.2}$$

Suggestions concerning the proof of this theorem will be found in Problem 9.

By applying the polarization formula for the inner product in L^2 we obtain from (3.2.2)

$$\int_{R^n} f(x)g(x)\mathrm{d}x = \frac{1}{(2\pi)^n} \int_{R^n} \bar{\hat{f}}(p)g(p)\mathrm{d}p. \tag{3.2.3}$$

We shall now indicate how to obtain the inversion formula for the Fourier transformation by using the Plancherel formula and then we proceed to consider the Fourier transformation from the point of view of decomposition theory of representations.

THEOREM 3.2.4. *For an arbitrary f in $\mathscr{S}(\boldsymbol{R}^n)$ we have*

$$f(x) = \frac{1}{(2\pi)^n} \int_{R^n} \hat{f}(p)\exp[-i(p|x)]\mathrm{d}p. \tag{3.2.4}$$

Proof. We shall make use of a sequence $\{f_j\}$ which approximates the identity of the convolution algebra $L^1(\boldsymbol{R}^n)$, i.e. of a sequence consisting of non-negative functions of class $\mathscr{C}_0(\boldsymbol{R}^n)$ with the property $\int_{R^n} f_j = 1$ and $\operatorname{supp} f_j \searrow 0$. Obviously $|\hat{f}_j| \leqslant 1$ and $\hat{f}_j$ is pointwise convergent to 1. Applying the Plancherel formula to a $g \in \mathscr{S}(\boldsymbol{R}^n)$, we get in view of Lebesgue's theorem and Proposition 3.2.3:

$$g(0) = \frac{1}{(2\pi)^n} \lim_{j\to\infty} \int_{R^n} f_j(x)g(x)\mathrm{d}x = \frac{1}{(2\pi)^n} \lim_{j\to\infty} \int_{R^n} \bar{\hat{f}}_j(p)\hat{g}(p)\mathrm{d}p$$

$$= \frac{1}{(2\pi)^n} \int_{R^n} \hat{g}(p)\mathrm{d}p.$$

The theorem follows upon replacing g by $L_{-x}f$. □

COROLLARY 3.2.5. *The Fourier transformation is an isomorphism of $\mathscr{S}(\boldsymbol{R}^n)$ onto $\mathscr{S}(\boldsymbol{R}^n)$.*

Proof. The Plancherel formula implies in particular the injectivity of the Fourier transformation. Further, the inversion formula says that every function in $\mathscr{S}(\boldsymbol{R}^n)$ is the Fourier transform of a function in $\mathscr{S}(\boldsymbol{R}^n)$, namely of the function $p \to -\frac{1}{(2\pi)^n}\hat{f}(p)$. □

The basic properties of the Fourier transformation are contained in the decomposition theorem for the regular representation of the group $\boldsymbol{R}^n$ (the theorem contains in an implicit way the inversion formula (3.2.4)).

THEOREM 3.2.6. *The regular representation of the group* $\boldsymbol{R}^n$ *on* $L^2(\boldsymbol{R}^n)$ *is decomposable into the direct integral* $U = \int_{\boldsymbol{R}^n} \chi_p \frac{\mathrm{d}p}{(2\pi)^n}$. *The decomposition is realized by the classical Fourier transformation in* $\boldsymbol{R}^n$: $\mathscr{F}: \mathscr{S}(\boldsymbol{R}^n) \ni f \to \hat{f} \in \mathscr{S}(\boldsymbol{R}^n)$.

Proof. It follows from the Plancherel theorem and Corollary 3.2.5 that the Fourier transformation maps $\mathscr{S}(\boldsymbol{R}^n)$ isometrically onto $\mathscr{S}(\boldsymbol{R}^n)$. Now, the density of $\mathscr{S}(\boldsymbol{R}^n)$ in $L^2(\boldsymbol{R}^n)$ implies that this operator extends to an isometry of $L^2(\boldsymbol{R}^n)$ onto $L^2(\boldsymbol{R}^n)$. Condition (2) in Proposition 3.2.1 ensures that this operator is intertwining. □

Once again let us consider the problem of distinguishing a class of functions which is suitable for the analysis of the space $L^2(\boldsymbol{R}^n)$. By considering the group structure of $\boldsymbol{R}^n$ we observe that the set of its unitary characters (the exponential functions) serves our purpose best.

The same class of functions can be distinguished also by considering the geometric properties of the space $\boldsymbol{R}^n$.

The reader will easily prove the following statements.

PROPOSITION 3.2.7. *Every differential operator on* $\boldsymbol{R}^n$ *which is invariant under all rotations and translations is a polynomial in the Laplace operator*

$$\Delta = \sum_{i=1}^{n} \frac{\partial^2}{\partial x_i^2}.$$

PROPOSITION 3.2.8. *Every eigenfunction of the operator* Δ *which is constant on the hyperplanes* $(x-x_0|p) = 0$, *is proportional to the function* $f_{p,\lambda}(x) = A \exp \lambda(x|p) + B$, $A, B \in \boldsymbol{C}$.

3.3. THE FOURIER TRANSFORMS OF FUNCTIONS IN $\mathscr{D}(\boldsymbol{R}^n)$

The Paley–Wiener Theorem

We introduce the following notation: for $z, w \in \boldsymbol{C}^n$, $(z|w) := \sum_{i=1}^{n} z_i w_i$.

The Fourier transformation maps the space $\mathscr{D}(\boldsymbol{R}^n)$ injectively into $\mathscr{S}(\boldsymbol{R}^n)$. It turns out that transforms of this type are uniquely characterized by the asymptotic behaviour of their analytic extensions to $\boldsymbol{C}^n$. Obviously for $f \in \mathscr{D}(\boldsymbol{R}^n)$ the function

$$\boldsymbol{C}^n \ni z \to \int_{\boldsymbol{R}^n} f(x) e^{i(z|x)} dx$$

is defined and holomorphic on the whole of $\boldsymbol{C}^n$; consequently the Fourier transform has an extension to an entire function on $\boldsymbol{C}^n$.

Let $\operatorname{supp} f \subset K(a)$. Then

$$|\hat{f}(p+iq)| \leqslant \int_{K(a)} |f(x)| e^{-(q|x)} dx \leqslant \sup |f| e^{b\|q\|} \int_{K(a)} e^{-(q|x)-b\|q\|} dx$$

for $b > a$. Since $e^{-b\|q\|-(q|x)} \leqslant 1$ on $K(a)$ we have the estimate

$$|\hat{f}(p+iq)| \leqslant C e^{b\|q\|} \qquad \text{for every } b > a.$$

Moreover, it turns out that the above conditions are sufficient to ensure that the Fourier transform of a function is in $\mathscr{D}(\boldsymbol{R}^n)$. Namely, we have the following

THEOREM 3.3.1 (Paley–Wiener, see e.g. Maurin, 1976). *For an entire function h on* $\boldsymbol{C}^n$ *to be the Fourier transform of a function of class* $\mathscr{D}(\boldsymbol{R}^n)$ *it is necessary and sufficient that h restricted to* $\boldsymbol{R}^n$ *be of class* $\mathscr{D}(\boldsymbol{R}^n)$ *and there exist constants* $b > 0$ *and* $C > 0$ *such that*

$$|h(x+iy)| \leqslant C e^{b\|y\|}. \tag{3.3.1}$$

COROLLARY 3.3.2. *For every* $f \in \mathscr{D}(\boldsymbol{R}^1)$ *we have the inversion formula*

$$f(x) = \frac{1}{2\pi} \int_{-\infty+ci}^{\infty+ci} f(z) e^{-ixz} dz, \qquad c \in \boldsymbol{R}^1,$$

where $\int_{-\infty+ci}^{\infty+ci} dz$ *denotes the integral over the line in* $\boldsymbol{C}^1$ *described in Figure* 2.

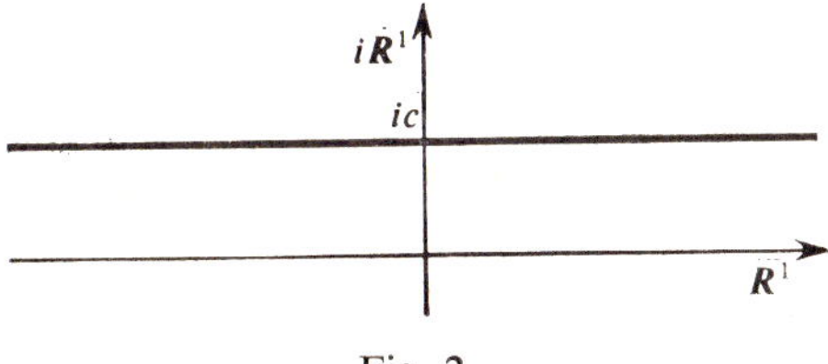

Fig. 2

Proof. Since f is an entire function, we infer that the integral of the function $f(z)e^{-ixz}$ over the contour C_α (Fig. 3) vanishes for arbitrary c and α. We shall prove that the integrals over the perpendicular segments tends to zero as α tends to ∞.

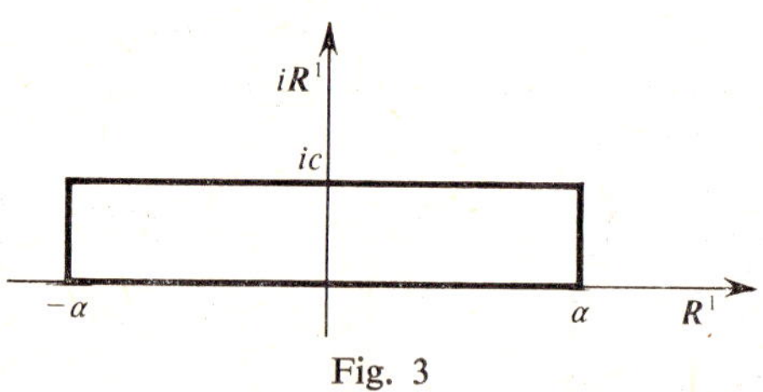

Fig. 3

The function $p \to \hat{f}(p+iy)$ is the Fourier transform of the function $x \to f(x)e^{-yx}$ which is also in $\mathscr{D}(\boldsymbol{R}^1)$. Consequently for $p \to \pm\infty$ we have $|f(p+iy)| \to 0$.

Now, it follows from Lebesgue's theorem that the function $I(\alpha) = \int_0^c f(\alpha+it)e^{-ix(\alpha+it)}dt$ is convergent to zero when α tends to $\pm\infty$, since the integrands are uniformly bounded (condition (3.3.1)) and are pointwise convergent to zero as α tends to $\pm\infty$. Passing to the limit with the integrals $\int_{C_\alpha} f(z)e^{-ixz}dz$ we finally obtain

$$\frac{1}{2\pi}\int_{-\infty+ic}^{\infty+ic} f(z)e^{-ixz}dz = \frac{1}{2\pi}\int_{-\infty}^{\infty} f(t)e^{-ixt}dt = f(z). \qquad \square$$

3.4. ANALYSIS ON THE MULTIPLICATIVE GROUP $\boldsymbol{R}_+$. THE MELLIN TRANSFORMATION

The multiplicative group $\boldsymbol{R}_+$ is isomorphic to $\boldsymbol{R}$. The isomorphism is given by the exponential function $\boldsymbol{R} \ni x \to e^x \in \boldsymbol{R}_+$. We intend to carry over the results of the Fourier analysis on $\boldsymbol{R}$ to $\boldsymbol{R}_+$.

The invariant measure on $\boldsymbol{R}_+$ is given by the formula

$$\int_{\boldsymbol{R}_+} f(t)dm(t) = \int_{\boldsymbol{R}} f(e^x)dx = \int_{\boldsymbol{R}_+} \frac{f(t)}{t}dt. \tag{3.4.1}$$

The characters of the group $\boldsymbol{R}_+$ are of the form

$$\boldsymbol{R}_+ \ni t \to U_\lambda(t) := t^\lambda \in \boldsymbol{C}, \quad \lambda \in \boldsymbol{C}.$$

The unitary characters correspond to the pure imaginary values $\lambda = i \log s$. This correspondence allows us to identify the group dual to $\boldsymbol{R}_+$ with $\boldsymbol{R}_+$ itself. Let $f \in \mathscr{C}_0(\boldsymbol{R}_+)$. Then the Fourier transform of the function $f \circ \exp$ on $\boldsymbol{R}$ has the form

$$R \ni p \to \int_{\boldsymbol{R}} f(\exp t) \exp ipt \, dt = \int_{\boldsymbol{R}_+} f(v) v^{ip-1} dv := \mathscr{M}f(ip). \tag{3.4.2}$$

The mapping $\mathscr{C}_0(\boldsymbol{R}_+) \ni f \to \mathscr{M}f \in \mathscr{E}(\boldsymbol{C})$ is called the *Mellin transformation.*

The Plancherel formula for the Fourier transformation on $\boldsymbol{R}$ yields the inversion formula and the Plancherel formula for the Mellin transformation:

$$f(t) = \frac{1}{2\pi i} \int_{-i\infty}^{i\infty} \mathscr{M}f(x) t^{-x} dx, \qquad f \in \mathscr{C}_0(\boldsymbol{R}_+) \tag{3.4.3}$$

and $\displaystyle\int_0^\infty |f(t)|^2 \frac{dt}{t} = \frac{1}{2\pi} \int_{-\infty}^{+\infty} |\mathscr{M}f(iy)|^2 dy.$

3.5. THE CIRCLE GROUP AND THE FOURIER SERIES

Denote by $\boldsymbol{T}$ the circle group, i.e. the multiplicative group of complex numbers of absolute value 1. This group can also be viewed as the quotient group of $\boldsymbol{R}$ by the subgroup of all integers. The mapping $\boldsymbol{R} \ni t \to e^{2\pi i t} \in \boldsymbol{T}$ is a group homomorphism whose kernel is precisely the subgroup $\boldsymbol{Z}$. This natural parametrization defines a topology and a differentiable structure on $\boldsymbol{T}$. The circle group is both compact and commutative and therefore the harmonic analysis on $\boldsymbol{T}$ is the simplest model of the general theory.

We shall first describe the dual group $\hat{\boldsymbol{T}}$, i.e. the multiplicative group of characters of $\boldsymbol{T}$. Let χ be in $\boldsymbol{T}$. Then the mapping $t \to \chi(e^{2\pi i t})$, being a character of $\boldsymbol{R}$, is of the form $\chi(e^{2\pi i t}) = e^{2\pi i p t}$ for a certain $p \in \boldsymbol{R}$. The condition $\chi(e^{2\pi i n}) = 1$ for $n \in \boldsymbol{Z}$ implies that p is an integer.

Thus every character $\chi \in \hat{T}$ determines an integer $n \in Z$ such that χ can be written as $\chi(z) = z^n$, $z \in T$. Such a character will be denoted in the sequel by χ_n. The assignment $Z \ni n \to \chi_n \in T$ is not only a bijection but also a group homomorphism:

$$\chi_{n+m} = \chi_n \chi_m .$$

Under this isomorphism it is possible to identify $\hat{T} = Z$.

The invariant measure on T is finite and we shall normalize it by requiring

$$\int_T f \mathrm{d}m := \int_0^1 f(\mathrm{e}^{2\pi i\varphi}) \mathrm{d}\varphi \; \frac{1}{2\pi} \int_0^{2\pi} f(e^{it}) \mathrm{d}t .$$

By the Fourier transform of a function $f \in \mathscr{C}(T)$ we shall understand a function on Z defined by the formula

$$\hat{f}(n) = (\chi_{-n}|f)_{L^2(T)} = \int_T f(z) z^n \mathrm{d}m(z) = \int_0^1 f(\mathrm{e}^{2\pi i\varphi}) \mathrm{e}^{2\pi in\varphi} \mathrm{d}\varphi .$$

The functional $F\colon f \to f(n)$ intertwines the regular representation on $\mathscr{C}(T)$ with the irreducible 1-dimensional representation generated by the character χ_n. The compactness of the group T implies that its characters, being continuous functions, are square integrable. Therefore, every irreducible representation of T can be realized as a subrepresentation of the regular representation.

Denote by H_n the 1-dimensional subspace of $L^2(T)$ spanned by the functions χ_{-n}. Then for every $x \in H_n$ we have $L_\alpha x = \alpha^n x$.

A direct computation shows that the system of functions $\chi = \{\chi_n\}_{n \in Z}$ constitues an orthonormal system in $L^2(T)$.

LEMMA 3.5.1. *The system χ constitutes an orthonormal basis for the space $L^2(T)$.*

Proof. It remains to show that a function orthogonal to all elements of the family χ is equal to zero almost everywhere.

The functions of the family χ separate points in the space T since this is already true of the function χ_1. The linear subspace spanned by these functions constitutes an algebra under multiplication of functions. Moreover this algebra is closed with respect to complex conjugation $(\bar{\chi}_n = \chi_{-n})$. Therefore all the assumptions of the Stone–Weierstrass theorem are satisfied and consequently the subspace under consideration is dense in $\mathscr{C}(\boldsymbol{T})$ in the topology of uniform convergence and thereby in the topology in $L^2(\boldsymbol{T})$. Hence a function f orthogonal to χ is orthogonal to $\mathscr{C}(\boldsymbol{T})$ and this implies that $f = 0$ a.e. □

By applying a fundamental formula of the theory of Hilbert spaces, i.e. the Parseval identity, we obtain

$$||f||_{L^2(\boldsymbol{T})} = ||\hat{f}||_{L^2(\boldsymbol{Z})} \quad \text{and} \tag{3.5.1}$$

$$f = \sum_{n \in \boldsymbol{Z}} \hat{f}(-n)\chi_n. \tag{3.5.2}$$

Note that the last formula can also be written as $f = \sum_{n \in \boldsymbol{Z}} f * \chi_n$.

Let us sum up the results that we have obtained so far in the form of a theorem on decomposition of the regular representation of the group $\boldsymbol{T}$.

THEOREM 3.5.2. *The Fourier transformation*

$$L^2(\boldsymbol{T}) \ni f \to \hat{f} \in \bigoplus_{n \in \boldsymbol{Z}} H_n$$

is an isometry which intertwines the regular representation with the representation $\bigoplus_{n \in \boldsymbol{Z}} \chi_n$.

In the course of the proof we have also obtained:

COROLLARY 3.5.3. *Every continuous function on* $\boldsymbol{T}$ *can be uniformly approximated by linear combinations of the characters.*

In the case of an infinitely differentiable function a stronger version of Corollary 2.3.3 can be proved. To this end we introduce the following concept:

DEFINITION. A function φ on $\boldsymbol{Z}$ is said to be *rapidly decreasing at* ∞ if for every $k \in \boldsymbol{Z}$.

$$\lim_{n \to \infty} n^k \varphi(n) = 0. \tag{3.5.3}$$

THEOREM 3.5.4. (a) *Let* $f \in \mathscr{E}(\boldsymbol{T})$. *Then* $\hat{f}$ *is a function rapidly decreasing at infinity and the series* $\sum_{n \in \boldsymbol{Z}} \hat{f}(n)\chi_{-n}$ *is convergent to* f *in the space* $\mathscr{E}(\boldsymbol{T})$.

(b) *For every functon φ on* **Z** *which is rapidly decreasing at infinity, the series* $\sum_{n=-\infty}^{\infty} \varphi(n)\chi_{-n}$ *is convergent in* $\mathscr{E}(\boldsymbol{T})$.

Proof. We begin by showing (b). The proof is based on the theorem on differentiation of integrals with parameters. The series under consideration is convergent uniformly since by (3.5.3) the sequence φ_n is absolutely summable:

$$\sup_{\alpha}\left|\sum_{|n|>N} \varphi(n)\chi_{-n}(\alpha)\right| \leqslant \sum_{|n|>N} |\varphi(n)| < \varepsilon$$

for sufficiently large N. Write $f = \sum_{n\in Z} \varphi(n)\chi_{-n}$. We define a differential operator

$$Xf(e^{it}) = \frac{\mathrm{d}}{\mathrm{d}t} f(\mathrm{e}^{it}).$$

Then $X^k\left(\sum_n \varphi(n)\chi_{-n}\right) = (2\pi)^k \sum_n (-in)^k\varphi(n)\chi_{-n}$.

In a way identical to that followed above, we check that the series on the right is absolutely convergent, and so we can apply the theorem on the differentiation of an integral (sum) with a parameter to obtain

$$X^kf = (2\pi)^k \sum_{n\in Z} (-in)^k\varphi(n)\chi_{-n}.$$

The function $n \to (-in)^k\varphi(n)$ decreases rapidly at infinity; hence the series is uniformly convergent, which proves (b).

We shall now prove that the Fourier transform of a smooth function is rapidly decreasing at infinity. To this end we integrate by parts $k+1$ times the formula

$$\hat{f}(n) = \frac{1}{2\pi}\int_0^{2\pi} f(\mathrm{e}^{it})\mathrm{e}^{int}\mathrm{d}t. \tag{3.5.4}$$

We obtain an expression of $\hat{f}(n)$ in terms of the $(k+1)$-th derivative of the function f:

$$f(n) = \frac{i^{k+1}}{2\pi n^{k+1}}\int_0^{2\pi} \frac{\mathrm{d}^{k+1}}{\mathrm{d}t^{k+1}} f(\mathrm{e}^{it})\mathrm{e}^{int}\mathrm{d}t = \frac{i^{k+1}}{n^{k+1}}(X^kf)\hat{}\,(n).$$

Hence $n^k|\hat{f}(n)| = |(X^kf)\hat{}\,|(n)$.

By assumption the function $X^k f$ is continuous and therefore the sequence on the right is convergent, which was to be shown. The convergence of the series $\sum \hat{f}(n)\chi_{-n}$ to f in the sense of $\mathscr{E}(T)$ now follows from (b) and formula (3.5.4).

As a matter of fact, the theorems proved here for the harmonic analysis on T, comprise the theory of the Fourier transformation on the additive group Z of all integers. Above we have identified this group with the group of characters of T. Now, let χ be a unitary character of Z. Then $\chi(n) = \chi(1)^n$, which means that the mapping $\chi \to \chi(1)$ defines a bijection of the set of characters of Z onto T. Moreover, multiplication of characters corresponds under this bijection to the group multiplication in T. Thus the transformation which played the role of the inverse Fourier transformation in the analysis on T is in fact the Fourier transformation on Z; for a rapidly decreasing function f on Z and $\alpha \in T$ we have

$$\hat{f}(\alpha) := \sum_{n \in Z} f(n)\alpha^n = \sum_{n \in Z} f(n)\chi_n(\alpha). \tag{3.5.5}$$

The Plancherel formula takes the form

$$\sum_{n \in Z} |f(n)|^2 = \frac{1}{2\pi}\int_0^{2\pi} |\hat{f}(e^{it})|^2 dt = \int_T |\hat{f}|^2(\alpha) dm\alpha. \tag{3.5.6}$$

The formula is true of an arbitrary rapidly decreasing function f on Z since by Theorem 3.5.4 (b) every function on Z is the Fourier transform of a smooth function on T. The polarization formula now implies

$$\sum_{-\infty}^{\infty} f(n)\bar{g}(n) = \int_T \hat{f}(\alpha)\bar{\hat{g}}(\alpha) dm(\alpha).$$

Taking $f(n) = \begin{cases} 1, & n = 0, \\ 0, & n \neq 0, \end{cases}$ we arrive at

$$g(0) = \int_T \hat{g}(\alpha) dm(\alpha). \tag{3.5.7}$$

In view of the intertwining property of the transformation (3.5.5) with respect to translations, we get

$$g(n) = \int_T \alpha^{-n}\hat{g}(\alpha) dm(\alpha). \tag{3.5.8}$$

—the inversion formula for the Fourier transformation on Z.

The Poisson Summation Formula

The Lebesgue measure on $\boldsymbol{R}$ can be represented in terms of invariant measures on $\boldsymbol{T}=\boldsymbol{R}/\boldsymbol{Z}$ and on $\boldsymbol{Z}$ according to the general procedure described in § 1.7. Namely, we have the representation

$$\int_{-\infty}^{\infty} f(x)\,\mathrm{d}x = \int_0^1 \sum_{n=-\infty}^{\infty} f(x+n)\,\mathrm{d}x,$$

which is immediate for compactly supported functions.

Given a function $f \in \mathscr{D}(\boldsymbol{R})$, we define an element $\tilde{f} \in \mathscr{E}(\boldsymbol{T})$:

$$\tilde{f}(t) = \sum_{n=-\infty}^{\infty} f(t+n).$$

By applying the inversion formula

$$f(t) = \sum_{-\infty}^{\infty} \hat{f}(n)\mathrm{e}^{-2\pi i m v}$$

($\hat{f}$ denotes the Fourier transform of a function $\tilde{f}$ on $\boldsymbol{T}$) to the function $\tilde{f}$ we obtain

$$\tilde{f}(0) = \sum_{m=-\infty}^{\infty} \int_0^1 \tilde{f}(t)\mathrm{e}^{2\pi i m t}\mathrm{d}t = \sum_m \int_0^1 \sum_n f(t+n)\mathrm{e}^{2\pi i m t}\mathrm{d}t$$

$$= \sum_m \int_{-\infty}^{\infty} f(x)\mathrm{e}^{2\pi i m x}\mathrm{d}x = \sum_{m=-\infty}^{\infty} \hat{f}(2\pi m).$$

Finally

$$\sum_{n=-\infty}^{\infty} f(n) = \sum_{m=-\infty}^{\infty} \hat{f}(2\pi m). \tag{3.5.9}$$

This identity is called the *Poisson summation formula.*

3.6. FOURIER ANALYSIS ON A COMMUTATIVE LOCALLY COMPACT GROUP

Now that we are acquainted with examples of the Fourier analysis on the commutative groups $\boldsymbol{R}^n$, $\boldsymbol{T}$ and $\boldsymbol{Z}$, it is easy to formulate basic ideas of a general theory for locally compact groups. This theory was created by Pontriagin and van Kampen.

In the sequel G will denote a locally compact, commutative separable group.

The set $\hat{G}$ of equivalence classes of (one-dimensional) unitary representations of the group G can be identified with the set of complex-valued functions on G fulfilling

$$\chi(gg_1) = \chi(g)\chi(g_1) \qquad \text{and} \qquad \chi(g^{-1}) = \bar{\chi}(g),$$

i.e. with the set of unitary characters of G.

The set G constitues a group called the *dual* group. This group is also commutative with respect to multiplication of characters:

$$\chi\chi_1(g) = \chi(g)\chi_1(g).$$

We endow $\hat{G}$ with the topology of almost uniform convergence of characters. It can be proved that $\hat{G}$ is locally compact in this topology.

The famous duality of Pontriagin gives

THEOREM 3.6.1. $\hat{\hat{G}} = G$.

The identification is given by the topological group isomorphism: $G \ni g \to \langle g, \cdot \rangle \in \hat{\hat{G}}$ where $\langle g, \chi \rangle = \chi(g)$.

We have already seen examples of this isomorphism corresponding to the case $G = \boldsymbol{R}^n$ and $G = \boldsymbol{T}^1$. The first case is rather special since $G \cong \hat{G}$.

In the sequel, in order to underline the symmetry between G and $\hat{G}$, we shall denote the characters of G by $\hat{g}$ and the value of a character $\hat{g}$ on a g by $\langle g, \hat{g} \rangle$.

The pair $(\boldsymbol{T}, \boldsymbol{Z})$ is a good illustration of a general rule:

THEOREM 3.6.2 (Pontriagin, 1958). (a) *If the group G is compact then $\hat{G}$ is discrete.*

(b) *If G is discrete then $\hat{G}$ is compact.*

In both groups the group operation will be written in the additive way. The set of functions on $\hat{G}$ of the form $\hat{g} \to \langle g, \hat{g} \rangle$ with $g \in G$ coincides, on account of Theorem 3.6.1, with the set of characters of $\hat{G}$. On the other hand, the set of functions $g \to \langle g, \hat{g} \rangle$ for $\hat{g} \in \hat{G}$, agrees with the set of characters of G.

The Fourier transformation on G is an integral operator which to a function $f \in L^1(G)$ assigns a function $\hat{f}$ on $\hat{G}$ defined as

$$\hat{f}(\hat{g}) = \int_G f(g)\langle g, \hat{g}\rangle \mathrm{d}g \tag{3.6.1}$$

($\mathrm{d}g$ denoting the Haar measure on G). We observe that

$$(L_g f)^\wedge(g) = \langle g, \hat{g}\rangle \hat{f}(g).$$

THEOREM 3.6.3 (The Plancherel formula). *Subject to a suitable normalization of the Haar measure* $\mathrm{d}g$ *on* G, *the identity*

$$\|f\|_{L^2(G)} = \|\hat{f}\|_{L^2(\hat{G})} \tag{3.6.2}$$

holds for every function $f \in L^1 \cap L^2(G)$.

In contradistinction to most monographs we shall now proceed along the route indicated in the examples. Thus for an arbitrary locally compact connected, commutative group we define a set of functions, called the *Schwartz functions* and denoted by $\mathscr{S}(G)$, which is dense in $L^2(G)$ and satisfies in addition

(1) $\mathscr{S}(G) \subset L^1(G) \cap L^2(G) \cap \mathscr{C}(G)$,

(2) $\mathscr{S}(G)^\wedge = \mathscr{S}(\hat{G})$ (see Wawrzyńczyk, 1968).

Therefore, for a function of this class we can define the double Fourier transform: $\hat{\hat{f}} := (\hat{f})^\wedge$. Moreover, reasoning as in the case $G = \boldsymbol{R}^n$ (§ 3.2), we deduce from Theorem 3.5.3 that

$$f(-g) = \int_{\hat{G}} \langle g, \hat{g}\rangle \hat{f}(\hat{g}) \mathrm{d}\hat{g}. \tag{3.6.3}$$

THEOREM 3.6.4. *The Fourier transformation* (3.5.1) *extends to an isometry* $L^2(G) \to L^2(G)$, *which realizes the decomposition of the regular representation* U *on* $L^2(G)$ *into the direct integral*

$$U = \int_{\hat{G}} \hat{g} \mathrm{d}\hat{g}.$$

Thus the Plancherel measure for the commutative group G coincides with the Haar measure on the dual group.

The inversion formula (3.6.3) can be regarded as a representation of the Dirac measure in the form

$$\delta_e(f) = \int_{\hat{G}} \hat{g}(f) \mathrm{d}\hat{g}, \qquad \text{where} \quad \hat{g}(f) := \hat{f}(\hat{g}).$$

Symbolically

$$\delta_e(g) = \int_{\hat{G}} \langle g, \hat{g} \rangle d\hat{g}. \tag{3.6.4}$$

This formula represents the (positive definite) measure δ_e as a continuous "linear combination with non-negative coefficients" of the simplest positive definite functions on the group, i.e. characters of the representations. Let us compare this form of the Plancherel formula with the Bochner–Weil theorem:

THEOREM 3.6.5. *For every positive definite function ω on a locally compact, commutative group G there exists a positive, finite measure μ on $\hat{G}$, so that*

$$\omega(g) = \int_{\hat{G}} \langle g, \hat{g} \rangle d\mu(\hat{g}). \tag{3.6.5}$$

Thus every p.d. function on G turns out to be the Fourier transform of a positive measure on G. Formula (3.6.4) represents the Dirac measure as the Fourier transform of the Haar measure on the group $\hat{G}$. It is left to the reader to interpret Theorem 3.6.5 as an expansion into a direct integral of the representation of the group G related to ω through the Gelfand–Raikov construction.

PROBLEMS

1. (a) Let (E, T), (F, V) be finite-dimensional (not necessarily unitary), irreducible representations of a group G. Show that either $L_G(E, F) = \{0\}$ or every element of $L_G(E, F)\setminus\{0\}$ is an isomorphism.

(b) If (E, T) is an irreducible finite-dimensional representation of a group G on a complex vector space E (without the assumption of unitariness), then $L_G(E) = CI$. Prove also that the converse statement is false.

2. Let two unitary representations (H, U) and (K, V) be given and suppose that $T \in L_G(U, V)$.

(a) Show that $\ker T$ and $\overline{\operatorname{Im} T}$ are closed invariant subspaces of H and K, respectively;

(b) Show that the restriction of U to $(\ker T)^{\perp}$ is equivalent to the restriction of V to $\overline{\operatorname{Im} T}$;

(c) Prove that (H, U) and (K, V) are disjoint if, and only if, U contains no subrepresentation which is equivalent to a subrepresentation of V.

Hint. (b) Apply the polar decomposition of bounded operators: $T = (T^*T)^{\frac{1}{2}}R$, R being a partial isometry.

3. Let (H, U) be a unitary representation of a group G.

(a) Prove that $L_G(U)$ is an algebra closed under conjugation and closed in the weak topology in $L(H)$ (i.e., a von Neumann algebra).

Suppose in addition that U decomposes into a direct sum of irreducible representations. Prove that

(b) $L_G(U)$ is commutative if, and only if, every two components of this decomposition are non-equivalent.

(c) The centre of the algebra $L_G(U)$ equals $\{\lambda I: \lambda \in C\}$ if, and only if, all the components of this decomposition are equivalent.

Hint. (b) and (c). Use a construction similar to the decomposition of matrices into blocs, i.e. consider $P_i TP_j$ where P_k denotes the orthogonal projection onto the representation space of an irreducible representation occurring in the given decomposition. Also note that the projection corresponding to a space which is invariant under both U and $L_G(U)$ belongs to the centre of $L_G(U)$.

4. Let G be a commutative locally compact group and $\lambda: G \to T$ a measurable function satisfying the identity $\lambda(x+y) = \lambda(x)\lambda(y)$. Show that λ is continuous.

Hint. Multiply the identity by a function $f \in \mathscr{C}_0(G)$ and integrate. Apply the continuity of the translates $x \to L_x f$ (cf. § 3.2, Lemma 3.2.1).

5. Prove that the Fourier transform of a function in $L^1(\boldsymbol{R}^n)$ is a bounded, uniformly continuous function on $\boldsymbol{R}^n$. Prove also that $\|\hat{f}\|_\infty \leqslant \|f\|_1$ for $f \in L^1(\boldsymbol{R}^n)$. Give an example of a $f \in L^1(\boldsymbol{R}^1)$ such that $\hat{f}$ is not integrable.

6. Prove the Riemann–Lebesgue lemma: for every $f \in L^1(\boldsymbol{R}^n)$ $\hat{f}(x) \to 0$ when $\|x\| \to \infty$ (in other words: $\hat{f}$ vanishes at infinity).

Hint. First prove the theorem for the characteristic functions of n-dimensional cubes, i.e. sets $I_n = \{x: a_i \leqslant x_i \leqslant b_i\}$. Apply the density of linear combinations of such functions in $L^1(\boldsymbol{R}^n)$ and make use of the preceding problem.

7. Let $f, g \in L^1(\boldsymbol{R}^n)$. Prove that $\int \hat{f}(x)g(x)\,\mathrm{d}\lambda^n(x) = \int f(x)\hat{g}(x)\,\mathrm{d}\lambda^n(x)$. *Hint*. Fubini's theorem.

8. For $a > 0$ denote by δ_a an expansion with coefficient a, defined as $(\delta_a f)(x) := f(ax)$. Show that for any $f \in L^1(\boldsymbol{R}^n)$

$$a^n(\delta_a f)\hat{}\,(p) = \hat{f}(a^{-1}p).$$

9. In this problem we sketch a method due to Wiener (1933), of proving the Plancherel theorem.

In the following h_n denotes the Hermite function defined by the formula

$$h_n(x) = (-1)^n \mathrm{e}^{\frac{1}{2}x^2} \frac{\mathrm{d}^n}{\mathrm{d}x^n} \mathrm{e}^{-x^2}, \qquad n = 0, 1, \ldots$$

(a) Prove that h_n is the product of a polynomial of degree n (the Hermite polynomial) and the function $x \to \mathrm{e}^{-\frac{1}{2}x^2}$; show that $h_n \in \mathscr{S}(\boldsymbol{R}^1)$.

(b) Prove the following recurrence formulas for h_n:

$$h_n'(x) - xh_n(x) = -h_{n+1}(x),$$
$$h_n'(x) + xh_n(x) = 2nh_{n-1}(x), \qquad n = 0, 1, \ldots$$

with the convention that $h_{-1} = 0$.

(c) Show that h_n satisfies the differential equation

$$Lh_n := \left(\frac{\mathrm{d}^2}{\mathrm{d}x^2} - x^2\right) h_n(x) = -(2n+1)h_n(x).$$

(d) Show that h_n are eigenfunctions of the Fourier transformation, i.e.

$$\hat{h}_n = \sqrt{2\pi}\, i^n h_n.$$

(e) Show that $(Lh_n|h_m) = (h_n|Lh_m)$, where L is the differential operator defined in (c) and conclude hence that h_n are mutually orthogonal in the Hilbert space $L^2(\boldsymbol{R}^1, \mathrm{d}x)$.

(f) Prove that $\{h_n : n = 0, 1, \ldots\}$ is a basis for $L^2(\boldsymbol{R}^1, \mathrm{d}x)$.

(g) Write $e_n = \|h_n\|^{-1}h_n$ for the normalized Hermite functions and define for an arbitrary $f \in L^2(\boldsymbol{R}^1, \mathrm{d}x)$

$$\hat{f} := \sum_{n=0}^{\infty} (e_n|f)\hat{e}_n = \sum_{n=0}^{\infty} (e_n|f)\sqrt{2\pi}(i)^n e_n.$$

Show that the map $f \to \hat{f}$ is an isometry of $L^2(\boldsymbol{R}^1, dx)$ (up to a scalar factor $\sqrt{2\pi}$) onto itself and that for every $f \in \mathscr{S}(\boldsymbol{R}^1)$ the function $\hat{f}$ defined above coincides with the Fourier transform defined as usual.

Hint. (b) Differentiate the formula which defines h_n.

(c) Apply the recurrence formulas in (b).

(d) Apply the Fourier transform to the identity in (b), note that $\hat{h}_0 = \sqrt{2\pi}\, h_0$ and that the recurrence formulas in (b) determine h_n uniquely for a given h_0.

(f) Make use of the fact that the Fourier transform of the function $x \to f(x)e^{-\frac{1}{2}x^2}$ with $f \in L^2(\boldsymbol{R}^1, dx)$ is equal to zero if, and only if, f is zero. Show that if $(h_n|f) = 0$ for all n, then f is orthonogal to $x^n e^{-\frac{1}{2}x^2}$ for arbitrary n, which in turn implies that the Fourier transform of $f(x)e^{-\frac{1}{2}x^2}$ vanishes identically.

10. *The inversion theorem for the Fourier transformation on* $L^1(\boldsymbol{R})$.[1]

(a) Prove that for $H(p) = \dfrac{1}{2\pi} e^{-|p|}$, $p \in \boldsymbol{R}$, $\sigma > 0$ we have

$$h_\sigma(x) = \int_{\boldsymbol{R}} H(\sigma p) e^{-ipx} d\lambda(p) = \frac{1}{\pi} \frac{\sigma}{\sigma^2 + x^2}$$

and $\int_{\boldsymbol{R}} h_\sigma(x) d\lambda(x) = 1$.

(b) Prove that if $f \in L^1(\boldsymbol{R}^1)$ then

$$f * h_\sigma(x) = \int_{\boldsymbol{R}} H(\sigma p) \hat{f}(p) e^{-ipx} d\lambda(p).$$

(c) Show that $\{h_\sigma(\cdot)\}_{\sigma>0}$ approximates the identity of the convolution algebra in the following sense:

(i) If $f \in L^\infty(\boldsymbol{R}^1)$ and f is continuous at $x \in \boldsymbol{R}^1$ then $\lim_{\sigma \to 0} f * h_\sigma(x) = f(x)$.

(ii) If $f \in L^p(\boldsymbol{R}^1)$ with $1 \leqslant p < \infty$ then $f * h_\sigma \in L^p(\boldsymbol{R}^1)$ and $\lim_{\sigma \to 0} \|f * h_\sigma - f\|_p = 0$.

(d) Prove that if both f and $\hat{f}$ are in L^1 then the function g given as $g(x) = \int_{\boldsymbol{R}} \hat{f}(p) e^{-ipx} d\lambda(p)$ equals f a.e.

---

[1] The method presented in this problem is taken from the book of W. Rudin, 1966. For its generalizations see Stein and Weiss, 1971.

(e) Show that the Fourier transformation on L^1 is injective.

11. Show that the mapping $L^1(\boldsymbol{T}) \ni f \to \hat{f}(k) \in \boldsymbol{C}$ is multiplicative with respect to convolution in $L^1(\boldsymbol{T})$, i.e. $(f * g)^\wedge(k) = \hat{f}(k)\hat{g}(k)$.

Hint. The convolution in $L^1(\boldsymbol{T})$ is defined by the formula

$$f * g(e^{it}) = \frac{1}{2\pi}\int_0^{2\pi} f(e^{ix})g(e^{i(t-x)})\,dx.$$

12. Recall that a non-zero vector x is said to be an eigenvector for an operator A on a Hilbert space $\boldsymbol{H}$, provided there exists an $\lambda \in \boldsymbol{C}$ (called the eigenvalue of x) such that $Ax = x$. An operator A is said to have a *purely point spectrum* if its eigenvectors span the space $\boldsymbol{H}$.

Prove that the convolution operator $f \to h * f$, where $h \in L^2(\boldsymbol{T})$ defined on $L^2(\boldsymbol{T})$ has a purely point spectrum; determine the spectrum. Deduce hence that $f \to h * f$ is a Hilbert–Schmidt type operator (see Chapter 4).

13. Let $\mathscr{C}_0$ denote the Banach space of all functions on $\boldsymbol{Z}$ such that $|\varphi(n)| \to 0$ as $n \to \pm\infty$, with the norm $\|\varphi\| = \sup_n |\varphi(n)|$.

(a) Prove that the Fourier transformation F maps $L^1(\boldsymbol{T})$ into $\mathscr{C}_0$ and that $\|F\| = 1$; (the Riemann–Lebesgue lemma).

(b) Prove that F is injective.

(c) Prove that F is not surjective.

Hint. (a) Observe that the Fourier transform of a trigonometric polynomial belongs to $\mathscr{C}_0$ and then make use of the possibility of approximating functions by these polynomials.

(b) If the Fourier transform of an f is zero then $\int_{\boldsymbol{T}} f(z)\varphi(z)\,dm(z) = 0$ for arbitrary $\varphi \in \mathscr{C}(\boldsymbol{T})$.

(c) Deduce from (a), (b) and Banach's open mapping theorem that the surjectivity of F would imply $\|f\|_1 \leqslant \gamma\|Ff\|$. Consider the trigonometric polynomials $D_n(z) = \sum_{k=-n}^{n} z^k$, $z \in \boldsymbol{T}$.

14. Let $f \in L^1(\boldsymbol{T})$.

(a) Show that the power series $\hat{f}(0) + 2\sum_{k=1}^{\infty} \hat{f}(k) z^k$ is convergent almost uniformly on the unit disc in $\boldsymbol{C}$.

(b) Let $z = re^{i\theta}$. Prove that the series $\sum_{k=-\infty}^{\infty} r^{|k|}\hat{f}(k)e^{ik\theta} =: A(z)$ is convergent in the unit disc $r < 1$ and its sum is a harmonic function.

(c) Prove that A can be represented by the formula (the Poisson integral formula)

$$A(re^{i\theta}) = \frac{1}{2\pi}\int_0^{2\pi} P(r, \theta - t) f(e^{it}) dt,$$

where $P(r, \theta)$ is the Poisson kernel defined as

$$P(r, \theta) = \mathrm{Re}\frac{1+re^{i\theta}}{1-re^{i\theta}} = \frac{1-r^2}{1-2r\cos\theta+r^2}.$$

Hint. (a) Use the boundedness of the Fourier transformation to estimate the sum of the series in question. (b) Compare the series in (a) and (b) in the case where f is a real function. For any arbitrary function consider separately the real and the imaginary part.

15. Let $f \in \mathscr{C}_0^1(\boldsymbol{R}_+)$. Prove that the following formulas hold for the Mellin transformation:

$$\mathscr{M}(f')(p+1) = (1-p)\mathscr{M}f(p),$$
$$\mathscr{M}(vf')(p) = (-p)\mathscr{M}f(p).$$

Chapter 4

Representations of Compact Groups

The exposition of the theory of representations of compact groups presented in this chapter, provides the opportunity to recall the names and dates which determined the directional development of the general theory of representations.

In the first place the case of a compact group constitutes a basic model of this theory. However, the role of a compact group is much greater than this statement suggests. Bur first some history.

During the last two decades of the 19th century, G. Frobenius laid the foundations of the theory of representations of compact groups. In his papers we encounter the notions of an irreducible representation, of the decomposition of a representation, and of the character of a representation, the construction of an induced representation and finally the theorem on duality which describes the decomposition of an induced representation into irreducible components (1897). Frobenius and, later, W. Burnside emphasized in their papers the role of representations in studying structure of finite groups.

An important point in the development of the theory of representations of topological groups was J. Schur's proof of the orthogonality relations for matrix elements of irreducible representations of the rotation group. Herman Weyl was the first to observe that the cause of this phenomenon was the existence of an invariant measure on the group, and it was he who proved the orthogonality relations in the general case. Weyl's observation also gave rise to the Peter–Weyl theory of decomposition of the regular representation of a compact group on the space $L^2(G)$ (1927). A complete extension of the Frobenius theory to arbitrary compact groups was made in A. Weyl's fundamental monograph "L'integration dans les groupes topologiques et ses applications" (1940).

In this chapter the elements of the theory of representations of compacts groups (the Peter–Weyl theory in particular) are deduced from the general theorem on duality. The proof quoted here is taken from M. Rieffel, 1972.

In order to appreciate fully the role of compact groups in the general theory, we need to consider this case together with another fundamental model—the familiar theory of decompositions of representations of a commutative group.

We recall that irreducible unitary representations of a commutative group are all one-dimensional, which allows us to identify them with functions on the group.

Compact groups provide examples of irreducible representation on spaces of dimension greater than one. Therefore we have to consider equivalent representations, which complicates the description of the structure of the dual space G. One of the fundamental theorems of the theory asserts that an irreducible representation of a compact group is, in any case, finite-dimensional. However, from the point of view of decomposition theory, the case of a compact group is much simpler. The construction of a direct sum suffices for constructing from irreducible representations every unitary representation (up to equivalence).

As regards commutative groups (e.g. $\boldsymbol{R}^n$) we have to introduce for this purpose the notion of a direct integral of unitary representations.

We see that each of the fundamental examples brought different problems, concepts and solutions into the theory.

Studying representations of a non-compact, non-commutative group G, we are looking for the possibility of applying the theories already known. In particular, we are concerned with the compact and the commutative subgroups of the group G under consideration. A special role is played by those subgroups which are both compact and commutative. In the case of a Lie group G, the maximal torus $T(G)$ will play a fundamental role in the study of the structure of that group and its representations.

The theory of representations and of spherical functions, which is the subject of the next chapter, is based on a simultaneous application of the results of the theory of representations of compact groups and of the results of commutative analysis.

The theory works when the group G contains a "large enough" compact subgroup $K \subset G$ and the quotient space G/K, which does not have to be a group but possesses a special structure ensuring the commutativity of the convolution algebra $\mathscr{C}_0(K\backslash G/K)$. Then the problem of the decomposition of a representation of the group G on the space $L^2(G/K)$ reduces to problems of commutative analysis.

We shall familiarize ourselves with the manner in which this schedule works, first in a number of examples in Part II and then in the general form in Chapter 18. The schedule shows how the methods of the theory of representations of the commutative and the compact groups intertwine and depend on one another.

All groups considered in this chapter are assumed to be separable.

4.1. OPERATORS OF THE HILBERT–SCHMIDT TYPE

DEFINITION. Let two separable Hilbert spaces H_1 and H_2 be given. A bounded operator T: $H_1 \to H_2$ is said to be of the *Hilbert–Schmidt type* (*H-S type*, for short) if there exists a basis $\{e_i\}_{i=1}^{\infty}$ for the space H_1 such that

$$||T||_{HS}^2 := \sum_{i=1}^{\infty} ||Te_i||_2^2 < \infty .$$

The space of all operators of *H-S* type will be denoted by $HS(H_1, H_2)$. The number $||T||_{HS}$ is called the *H-S norm* of the operator T. By choosing arbitrarily a basis $\{f_j\}_{j=1}^{\infty}$ for the space H_2 we can write

$$\begin{aligned} ||T||_{HS}^2 &= \sum_{i,j} (Te_i|f_j)\,(f_j|Te_i) = \sum_{i,j} (e_i|T^*f_j)(T^*f_j|e_i) \\ &= ||T^*||_{HS}. \end{aligned} \tag{4.1.1}$$

This implies the following statement:

PROPOSITION 4.1.1. (1) $T \in HS(H_1, H_2) \Leftrightarrow T^* \in HS(H_2, H_1)$ *and* $||T_{HS}|| = ||T^*||_{HS}$.

(2) *The number* $||T||_{HS}$ *is independent of the choice of the basis* $\{e_k\}$.

(3) $||T|| \leqslant ||T||_{HS}$.

Proof. (1) follows directly from identity (4.1.1).

(2) The choice of the basis $\{e_i\}$ does not affect the third expression appearing in formula (4.1.1). This proves (2).

$$(3)\quad \|Th\|^2 = \sum_i (Th|f_i)(f_i|Th) = \sum_i (h|T^*f_i)(T^*f_i|h)$$
$$\leqslant \|h\|^2 \sum_i \|T^*f_i\|^2 = \|h\|^2 \|T\|^2_{HS}. \qquad \square$$

The simplest operators of H-S type are the finite-dimensional operators, i.e. the ones with finite-dimensional images. In particular, the operator T_h: $H_1 \ni h_1 \to (h|h_1)h \in H_1$ for $h \in H_1$ belongs to the space $HS(H_1) := HS(H_1, H_1)$ and its norm is equal to $\|h\|^2$.

One can equip the space $HS(H_1, H_2)$ with an inner product by setting

$$(T|S) = \sum_{i=1}^{\infty} (Te_i|Se_i) \qquad \text{for } T, S \in HS(H_1, H_2). \tag{4.1.2}$$

The convergence of the series follows by the Schwarz inequality.

Expression (4.1.2) is independent of the choice of the basis $\{e_i\}$.

THEOREM 4.1.2. *The space $HS(H_1, H_2)$ with the inner product given by* (4.1.2) *is a Hilbert space. The finite-dimensional operators constitute a dense subspace of $HS(H_1, H_2)$.*

Proof. The verification of the algebraic properties of the inner product is immediate. One has to prove the completness of the space. For this purpose let T_n be a Cauchy sequence in the operator norm. Since $L(H_1, H_2)$ is a Banach space, we infer that there exists a $T \in L(H_1, H_2)$ which is the limit of T_n in the operator norm. It remains to prove that $T \in HS(H_1, H_2)$. Fix a basis $\{e_i\}$ for H_1. Every operator T_n defines a sequence $C^n = \{C_i^n\}_{i=1}^{\infty}$ such that $C_i^n = \|T_n e_i\|$, which is square integrable and is a Cauchy sequence in the (complete) space l^2. The limit sequence $c = \{c_i\} \in l^2$ coincides with the pointwise limit of the sequence C^n. Hence $c_i = \lim_n \|T_n e_i\| = \|Te_i\|$ and consequently $\sum_i \|Te_i\|^2 = \sum_i \|c_i\|^2 < \infty$.

Next we shall prove that every element $T \in HS(H_1, H_2)$ can be approximated by the finite-dimensional operators. Let E_n be the subspace of H_1 spanned by the elements $\{e_i\}_{i=1}^n$ of the basis, Set $T_n = T|_{E_n}$. Then $T_n \to T$ in the norm on $HS(H_1, H_2)$. $\square$

We shall now describe the properties of superposition of operators of H-S type with the bounded ones.

PROPOSITION 4.1.3. *Let* $T \in HS(H_1, H_2)$, $A \in L(H_2, H_3)$, $S \in HS(H_2, H_3)$, $B \in L(H_0, H_1)$. *Then*

(1) $||A \circ T||_{HS} \leqslant ||A|| \, ||T||_{HS}$,

(2) $||S \circ T||_{HS} \leqslant ||S||_{HS} ||T||_{HS}$,

(3) $||T \circ B||_{HS} \leqslant ||T||_{HS} ||B||$.

Proof. Property (1) follows directly from the definition of the operator norm and that of the norm $||\cdot||_{HS}$; (2) follows from (1) in view of Proposition 4.1.1 (3).

Property (3) can be proved by applying (1) to the operator $(T \circ B)^* = B^* \circ T^*$ and then by using Proposition 4.1.1 (1). □

Theorem 4.1.3 asserts that the operators of the *H-S* type constitute an ideal in the space $L(H)$ and that $HS(H)$ is an algebra whose multiplication is continuous in the norm $||\cdot||_{HS}$.

The operators of the *H-S* type provide a fundamental example of an H^*-algebra which is defined below.

A Hilbert space A, equipped with an algebra structure and an antilinear operation of conjugation $x \to x^*$, is an H^*-algebra provided the following conditions are fulfilled:

(1) $(x|y) = (y^*|x^*)$,

(2) $(xy|z) = (y|x^*z)$,

(3) $||xy|| \leqslant ||x|| \, ||y||$,

(4) $x^*x = 0 \Rightarrow x = 0$.

The proof that $HS(H)$ is an H^*-algebra is left to the reader as an exercise.

We shall formulate without proof a basic theorem of the theory of the *H-S* type operators.

THEOREM 4.1.4 (the spectral theorem for the *H-S* operators). *Let* $T \in HS(H)$ *be a hermitian Hilbert–Schmidt operator. Then* $T = \sum_{i=1}^{\infty} \lambda_i P_i$, *where* λ_i *is the sequence of non-zero eigenvalues of the operator* T *and* P_i *is the finite-dimensional projector onto the eigenspace corresponding to the value* λ_i. *Moreover*

$$||T||_2^2 = \sum_{i=1}^{\infty} n_i \lambda_i^2, \quad \textit{where } n_i = \dim P_i H.$$

Example 1

An important class of the Hilbert–Schmidt operators is that of the integral operators of the *H-S* type. In fact, it was this particular case which gave rise to the general theory of operators of the *H-S* type. Let M be a separable locally compact space and μ a Radon measure on M. We assign to a function $k \in L^2(M \times M, \mu \otimes \mu)$ the integral operator with kernel k defined by the formula

$$Kf(x) := \int_M k(x, y) f(y) \mathrm{d}\mu(y), \qquad f \in L^2(M, \mu).$$

It follows from the Schwarz inequality and the Fubini theorem that

$$|Kf(x)|^2 \leqslant \int_M |k(x, y)|^2 \mathrm{d}\mu(y) \, ||f||^2$$

which shows that $Kf(x)$ is defined for μ almost all x. Integrating once again we obtain the inequality

$$||Kf||^2 \leqslant \int_M \int_M |k(x, y)|^2 \mathrm{d}\mu(x) \, \mathrm{d}\mu(y|) \, ||f||^2,$$

which implies that K is a continuous operator on $L^2(M)$ with the operator norm not greater than the norm (in $L^2(M \times M, \mu \otimes \mu)$) of the kernel k. Now let $\{\varphi_i\}_1^\infty$ be an orthonormal basis of $L^2(M, \mu)$. Then, as we know, the functions $(x, y) \to \Phi_{ij}(x, y) = \overline{\varphi_i(x)} \varphi_j(y)$, $i, j = 1, \ldots, \infty$ constitute an orthonormal basis for $L^2(M \times M, \mu \otimes \mu)$. Since

$$(\varphi_j | K\varphi_i)_{L^2(M)} = \int_M \int_M k(x, y) \varphi_i(y) \overline{\varphi_j(x)} \mathrm{d}\mu(x) \mathrm{d}\mu(y)$$

$$= (\Phi_{ij} | k)_{L^2(M \times M)}$$

we get $||K||_{HS}^2 = \sum_{i,j} |(\varphi_j | K\varphi_i)|^2 = ||k||_{L^2(M \times M)}^2$.

Thus the rule $k \to K$ is an isometry of $L^2(M \times M, \mu \otimes \mu)$ into the space $HS(L^2(M, \mu))$, actually it is an isometric isomorphism (see Problem 3).

4.2. THE TENSOR PRODUCT OF HILBERT SPACES

Suppose that two Hilbert spaces H_1 and H_2 are given with the corresponding inner products $(\cdot | \cdot)_1$ and $(\cdot | \cdot)_2$, respectively. Let us select orthonormal bases $\{e_i\}$ and $\{f_j\}$ for H_1 and H_2, respectively. We can define an inner product on the space $H_1 \otimes H_2$ by requiring that the basis $v_{ij} = e_i \otimes f_j$ be orthonormal. The inner product so defined is independent of the choice of the bases. The reader will easily check that for arbitrary simple tensors the inner product satisfies the formula

$$(x_1 \otimes x_2 | y_1 \otimes y_2) = (x_1 | y_1)(x_2 | y_2), \tag{4.2.1}$$

which, on the other hand, determines the inner product uniquely.

Definition. By the *Hilbert tensor product* of Hilbert spaces H_1 and H_2 we mean the space obtained by completing the algebraic product $H_1 \otimes H_2$ in the norm determined by the inner product (4.2.1). This space is denoted by $H_1 \underline{\otimes} H_2$.

The well known relationship concerning the spaces of linear operators $A: E_1' \to E_2$ and the tensor product $E_1 \otimes E_2$ in the finite-dimensional case has its counterpart in the theory of operators $HS(H_1, H_2)$ and the product $H_1' \otimes H_2$.

Theorem 4.2.1. *There is a natural isometry* $H_1 \underline{\otimes} H_2 - HS(H_1', H_2)$ *determined by the formula*

$$T_{x \otimes y}(x^*) := \langle x^*, x \rangle y \quad \textit{for } x \in H_1, y \in H_2, x^* \in H_1'. \tag{4.2.2}$$

Proof. We check the isometry of the mapping $x \otimes y \to T_{x \otimes y}$: $\|T_{x\otimes y}\|_{HS}^2 = \sum_i (T_{x\otimes y}(e_i^*) | T_{x \otimes y} e_i^*) = (y|y)_2 \sum_i \overline{\langle e_i^*, x\rangle} \langle e_i^*, x\rangle = (x|x)_1 (y|y)_2$ $= \|x \otimes y\|^2$. An analogous computation shows that the mapping (4.2.2) preserves the orthogonality of simple tensors and therefore is an isometry on the dense subset $H_1 \otimes H_2 \subset H_1 \underline{\otimes} H_2$. The image of this dense subset consists of all finite-dimensional operators which are dense, on account of Theorem 4.1.2, in $HS(H_1', H_2)$. Therefore, the mapping given by (4.2.2) extends to an isometric isomorphism. □

We leave it to the reader to verify the following canonical isomorphisms, which correspond directly to the well known facts in linear algebra.

Proposition 4.2.2. (1) $(H_1 \underline{\otimes} H_2)' \cong H_1' \underline{\otimes} H_2'$.

(2) $(H_1 \underline{\otimes} H_2) \underline{\otimes} H_3 = H_1 \underline{\otimes} (H_2 \underline{\otimes} H_3)$.

(3) $(H_1 + H_2) \underline{\otimes} H_3 = (H_1 \underline{\otimes} H_3) + (H_2 \underline{\otimes} H_3)$.

The concept of the Hilbert tensor product of spaces can be useu to obtain a description of function spaces of type L^2.

THEOREM 4.2.3. *Let (M, μ) be a separable locally compact space with a Radon measure μ. Then for every Hilbert space V the following spaces are isometrically isomorphic*:

$$L^2(M, V) \cong L^2(M) \underline{\otimes} V \cong HS\big(L^2(M), V\big).$$

Proof. The first isomorphism consists in assigning to a tensor $f \otimes v$ the vector function on M:

$$M \ni m \to f(m)v \in V. \tag{4.2.3}$$

It already follows from the construction of the space $L^2(M, V)$ that the image of the tensor product $L^2(M) \otimes V$ is dense in $L^2(M, V)$. Thus the extension of the mapping defined by (4.2.3) is a linear isometric bijection.

To prove the second part we note that by Theorem 4.2.1 there exists an isomorphism $L^2(M) \otimes V \simeq HS\big((L^2(M))', V\big)$. Next we identify $(L^2(M))'$ with $L^2(G)$ via the duality

$$\langle f, \varphi \rangle = \int_M f(g)\varphi(g) \mathrm{d}\mu(g).$$

This concludes the proof of Proposition 4.2.2. □

REMARK. An explicit expression for an isomorphism between $L^2(M, V)$ and $HS\big(L^2(M), V\big)$ can be obtained by composing the above isomorphisms. The reader will easily check that the operator corresponding to a function $\varphi \in L^2(M, V)$ is of the form

$$A_\varphi(f) = \int_M f(m)\varphi(m) \mathrm{d}\mu(m).$$

The Hilbert Product of Representations

We suppose now that (H^1, T^1) and (H^2, T^2) are continuous representations of a group G on Hilbert spaces H^1 and H^2. We define a representations on the space $H^1 \underline{\otimes} H^2$ by setting

$$T^1 \underline{\otimes} T^2(g)(h^1 \otimes h^2) := T^1_g h^1 \otimes T^2_g h^2. \tag{4.2.4}$$

It is worth while to note that if we identify the spaces $H_1 \underline{\otimes} H_2$ and

$HS(H_1', H_2)$ (Theorem 4.2.1) then to the action of the operator $T^1 \underline{\otimes} T^2(g)$ corresponds the action

$$\mathscr{T}^1 \underline{\otimes} \mathscr{T}^2(g)(A) := T_g^2 A \hat{T}_g^1. \tag{4.2.5}$$

on the space of operators. ($\hat{T}^1$ denotes the contragredient representation on $(H_1)'$.)

A new description of the representation (4.2.4) shows that operators defined on the dense subset $H_1 \otimes H_2 \subset H_1 \underline{\otimes} H_2$ have continuous invertible extensions to the whole of the space.

In the case of unitary T^i, $i = 1, 2$ the representation $T^1 \underline{\otimes} T^2$ is also unitary.

The inequality

$$||\mathscr{T}^1 \underline{\otimes} \mathscr{T}^2(g)(A)|| = ||T_g^2 A T_g^1|| \leqslant ||T_g^2|| \, ||T_g^1|| \, ||A||_{HS}$$

shows that the representation operators constitute an equicontinuous family for g ranging over a compact subset $K \subset G$.

Owing to Proposition 2.2.1 we shall prove the continuity of the representation by verifying that the mapping

$$G \ni g \to T_g^1 \underline{\otimes} T_g^2(v) \in H^1 \underline{\otimes} H^2$$

is continuous for v running through the set of simple tensors. But this mapping can be represented as a composite of two continuous mappings: $g \to (T_g^1 h_1, T_g^2 h_2) \to T_g^1 h_1 \otimes T_g^2 h_2$. Thus we have proved

PROPOSITION 4.2.4. *The representation $T^1 \underline{\otimes} T^2$ on $H^1 \underline{\otimes} H^2$ is continuous. It is unitary if both T^i, $i = 1, 2$, are unitary.*

4.3. THE FROBENIUS THEOREM

Now we proceed to study the representations of a compact group G. A basic problem here is the analysis of the regular representation of the group G on $L^2(G)$ ($= L^2(G, dg)$—dg denoting the Haar measure on G).

We shall focus our attention in the left regular representation. The fundamental theorems, among them the Peter–Weyl theorem, can be obtained by studying the space of the operators of the H-S type intertwining the regular representation on $L^2(G)$, with an arbitrary irreducible representation of the group G.

We introduce the following notation: for given representations (H^1, U^1) and (H^2, U^2) of G, we write $HS_G(H^1, H^2) := HS(H^1, H^2) \cap L_G(H^1, H^2)$. Sometimes this space will also be denoted by $HS_G(U^1, U^2)$.

THEOREM 4.3.1. *Let (H, U) be a continuous unitary representation of a compact group G. Then the following spaces are isometrically isomorphic:*

$$HS_G(L^2(G), H) \cong H.$$

The isomorphism is given by the mapping which assigns to an element $h \in H$ the operator T_h defined as

$$T_h(f) = U(f)h = \int_G f(g) U_g h \, dg. \tag{4.3.1}$$

Proof. The intertwining property of T_h has already been established in § 2.7. It remains to prove that the operator is of the H-S type and its norm is equal to $||h||$. The proof is based on Theorem 4.2.3. Owing to the compactness of the group G the continuous function $G \ni g \to U(g)h \in H$ is an element of $L^2(G, V)$.

The isomorphism described in Theorem 4.2.3, assigns to this function an element of $HS(L^2(M), V)$, which in this case coincides with T_h (cf. remark following Theorem 4.2.3). It remains to verify the surjectivity of this mapping, which amounts to describing the inverse mapping.

Let $T \in HS_G(L^2(G), H)$. Applying again Theorem 4.2.3, we describe T with the help of a function $F \in L^2(G, H)$:

$$T(f) = \int_G f(g) F(g) \, dg.$$

The intertwining property $T(L_g f) = U_g T(f)$ means that F satisfies $\int_G f(g) U_x F(g) \, dg = \int_G f(g) F(xg) \, dg$ for arbitrary $x \in G$. By integrating both sides multiplied by a function $\varphi \in \mathscr{C}_0(G)$ over the variable x, we obtain the following identity with respect to f and g:

$$\int_G \int_G f(g) \varphi(x) U_x F(g) \, dg \, dx = \int_G \int_G f(g) \varphi(x) F(xg) \, dg \, dx.$$

It follows that the identity $U_x F(g) = F(xg)$ is valid on $G \times G$ for $dg \otimes dg$

almost all pairs (x, y). Thus there exists a $x_0 \in G$ such that $U_x F(x_0) = F(xx_0)$ for dg almost all x. We define

$$h = U(x_0^{-1})F(x_0).$$

Then

$$T_h(f) = \int_G f(g)U(g)U(x_0^{-1})F(x_0)\mathrm{d}x = \int_G f(g)F(gx_0^{-1}x_0)\mathrm{d}g$$
$$= T(f). \qquad \square$$

The general Frobenius theorem gives a description of the space of operators intertwining an arbitrary induced representation with a unitary one.

Before presenting this theorem we shall first describe the induced representation in a new way by applying the concept of the Hilbert tensor product.

Let G be a compact group with a closed subgroup K. Denote by (V, π) a unitary representation of the subgroup K. In this case the representation space H^π of the induced representation is defined as follows:

$$H^\pi = \{f \in L^2(G, V)\colon f(gk) = \pi(k^{-1})f(g), g \in G, k \in K\}.$$

This space is the orthogonal complement in $L^2(G, V)$ of the linear subspace spanned by the vectors of the form

$$g \to \varphi(gk) - \pi(k^{-1})\varphi(g) \qquad \text{for} \quad k \in K, \ \varphi \in L^2(G, V).$$

Hence it follows that

LEMMA 4.3.2. *The representation (H^π, U^π) is equivalent to a representation of G on the orthogonal complement of the subspace $E \subset L^2(G)\underline{\otimes} V$ spanned by the vectors of the form*

$$R_k f \otimes v - f \otimes \varkappa(k^{-1})v.$$

We assume that G acts on $L^2(G)\underline{\otimes} V$ in terms of the tensor product of the regular representation on $L^2(G)$ and the trivial representation on V.

In view of the isomorphism $L^2(G, V) \cong L^2(G)\underline{\otimes} V$ the assertion follows from the remarks preceding the lemma.

THEOREM 4.3.3 (Frobenius theorem). *Let K be a closed subgroup of a compact group G. Let (H, U) and (V, π) be unitary representations* *of the group G and K, respectively. Then the following spaces are isometrically isomorphic:*

$$HS_G(H^\pi, H) = HS_K(V, H).$$

The isomorphism is defined by assigning to an operator $A \in HS_K(V, H)$ an element $T_A \in HS_G(H^\pi, H)$ which at a tensor $f \otimes v$ assumes the value

$$T_A(f \otimes v) = \int_G f(g) U_g A v \, dg. \tag{4.3.2}$$

Proof. First of all we shall make use of the isomorphisms

$$\begin{aligned} HS(L^2(G) \underline{\otimes} V, H) &= (L^2(G) \underline{\otimes} V)' \underline{\otimes} H = L^2(G)' \underline{\otimes} V' \underline{\otimes} H \\ &= HS(L^2(G), V' \underline{\otimes} H). \end{aligned}$$

We are concerned with the subspace consisting of those operators which satisfy in addition the intertwining condition:

$$T(L_g f \otimes v) = U_g T(f \otimes v).$$

For these operators we have

$$\begin{aligned} HS_G(L^2(G) \underline{\otimes} V, H) &= HS_G(L^2(G), V' \underline{\otimes} H) = V' \otimes H \\ &= HS(V, H). \end{aligned}$$

The isomorphism thus obtained corresponds to the special case of the representation induced by the trivial subgroup $K = \{e\}$. In order to deal with the general case we shall use the description $H^\pi = E^\perp$ given by Lemma 4.3.2. The elements of $HS_G(H^\pi, H)$ are in a one-to-one correspondence with those elements of $HS_G(L^2(G) \underline{\otimes} V, H)$ which vanish on E.

We shall verify that if $A \in HS_K(V, M)$ then formula (4.3.2) indeed defines an operator vanishing on E.

$$\begin{aligned} &T_A(R_k f \otimes v - f \otimes \pi(k^{-1} v)) \\ &= \int_G f(gk) U_g A v \, dg - \int_G f(g) U_g A \pi(k^{-1}) v \, dg \\ &= \int_G f(g) U_g (U_{k^{-1}} A v - A \pi(k^{-1})) v \, dg = 0. \end{aligned}$$

The same computation performed in the reverse order shows that an operator T_A vanishing on E determines an element A satisfying the intertwining condition. □

4.4. THE PETER–WEYL THEORY

We proceed to drawing conclusions from the Frobenius theorem. Theorem 4.3.1 alone allows us to derive the Peter–Weyl theory dealing with the structure of the space $L^2(G)$.

COROLLARY 4.4.1. *Every irreducible representation* (H, V) *of a compact group is finite dimensional. Moreover, the spaces* $L_G(H^\pi, H)$ *and* $L_K(V, H)$ *are linearly isomorphic.*

Proof. It follows from the Frobenius theorem that the space $HS_G(L^2(G), H)$ is non-trivial. Let T be in this space. Then the operator TT^* is of the H-S type and, moreover, intertwines the irreducible representation U with itself. Schur's lemma shows that TT^* is an isometry. But only a finite-dimensional isometry is of the H-S type. Hence the assertion follows. □

COROLLARY 4.4.2. *Every unitary representation of a compact group is completely reducible i.e. decomposable into a direct sum of irreducible components.*

Proof. We shall prove that every unitary representation (H, U) of a compact group G contains a finite-dimensional, non-trivial, invariant subspace.

Theorem 4.3.1 ensures that there exists a non-trivial element T in the space $HS_G(L^2(G), H)$. The hermitian operator $TT^* \in L_G(H)$ is of the H-S type and therefore admits a spectral decomposition as in Theorem 4.1.4. Let $\lambda \neq 0$ be an eigenvalue of the operator TT^* and H_λ the corresponding eigenspace. This subspace is G-invariant since $TT^*U_g h = U_g TT^*h = \lambda U_g h$ for $h \in H_\lambda$, which means that $U_g H_\lambda \subset H_\lambda$. Hence H_λ is the required finite-dimensional subspace.

Now, let us distinguish a family of sub-representations $(\mathscr{H}, U)$ of the representation (H, U) which have the form

$$(\mathscr{H}, U) = (\bigoplus_{\alpha\in\Lambda} H_\alpha, \bigoplus_{\alpha\in\Lambda} U_\alpha), \tag{4.4.1}$$

where H_α are invariant finite-dimensional subspaces of H and $U^\alpha = U|H_\alpha$.

Define a partial order in this family by putting $(\mathscr{H}, U) \prec (\mathscr{H}', U')$ if $\mathscr{H}'$ contains $\mathscr{H}$ and the decomposition of U is a part if the decomposition of U'. A linearly ordered set of representations of this type has an upper bound being the direct sum of the components appearing in the decomposition of the representations belonging to that set. It follows from the Kuratowski–Zorn lemma that there exists a subrepresentation $(H_0\ U_0)$ of form (4.4.1) which is not contained in any other representation of this type. In other words, H_0 does not contain any finite-dimensional subrepresentation. But this means that $H_0^\perp = \{0\}$, which proves the corollary. □

Let us have a closer look at the decomposition of the regular representation of a compact group G on the space $L^2(G)$ and on $L^2(G/K)$. Denote by $\hat{G}$ the space of the equivalence classes of unitary representations of the group G. A representation which belongs to a class $\alpha \in \hat{G}$ is denoted by (H^α, U^α). The dimension of the space H will be denoted by $n(\alpha)$ and called the *degree* of the representation.

Corollary 4.4.2 says that the space $L^2(G)$ is decomposable into a direct sum of finite-dimensional invariant subspaces

$$L^2(G) = \bigoplus_i H_i,$$

so that the representations $U|H_i$ are irreducible. It follows from the elementary decomposition theory that the multiplicity of a representation (H^α, U^α), $\alpha \in \hat{G}$ in the above decomposition is equal to $\dim L_G(L^2(G), H) = \dim HS_G(L^2(G), H)$.

The Frobenius theorem provides us with additional information: the multiplicity of (H^α, U^α) equals $n(\alpha)$. Summimg up the above, we have

THEOREM 4.4.3. *Let G be a compact group and $(L^2(G), L)$ its regular representation. Then*

$$L^2(G) = \bigoplus_{\alpha \in \hat{G}} n(\alpha) H^\alpha,$$

$$L = \bigoplus_{\alpha \in \hat{G}} n(\alpha) U^\alpha.$$

The regular representation of the group G on the space $L^2(G/K)$ can be regarded as the representation induced by the trivial one-dimensional representation of the subgroup K. Thus by applying Theorem 4.3.3 we obtain the following result:

THEOREM 4.4.4. *The regular representation L^K of a compact group G on the space $L^2(G/K)$ decomposes into*

$$L^2(G/K) = \bigoplus_{\alpha\in\hat{G}_K} m(\alpha)H^\alpha,$$

$$L^K = \bigoplus_{\alpha\in\hat{G}_K} m(\alpha)U^\alpha,$$

where $\hat{G}_K \subset \hat{G}$ denotes the set of equivalence classes of these representations which contain K-fixed vectors, and $m(\alpha)$ is the dimension of the space of K-fixed vectors in H^α.

The theorems ensure only the existence of decompositions and do not provide us with their description. We wish to know how to construct the subspaces of $L^2(G)$ and $L^2(G/K)$ corresponding to the irreducible representations of the group. Further, we would like to describe the projections on the spaces $L^2(G)$ and $L^2(G/K)$ which assign to a function its component in the spaces $n(\alpha)H$ and $m(\alpha)H$, respectively.

The constructions carried out during the proof of the Frobenius theorem allow us to expect that these projections will be operators of convolution with certain functions. In the following section we describe the transition from a given irreducible representation (H^α, U^α) to the subspace of $L^2(G)$ $\big(L^2(G/K)\big)$ on which this representation is realized.

Example 1

Let us determine the irreducible representations and the decomposition of the regular representation for the group S_3 of all permutations of three elements. We shall denote by $\sigma_1, \ldots, \sigma_6$ all the different elements of the group S_3, $\sigma_1 = e$ is the identity of the group. The space $L^2(S_3)$ consists of all functions on S_3 and its dimension equals the order of the group S_3, i.e. 6. It follows from Theorem 4.4.3 that

$$6 = \sum_{\alpha\in\hat{S}_3} n(\alpha)^2$$

so that S_3 may have representations of dimensions 1 and 2 only and moreover there is (up to equivalence) at most one two-dimensional representation. We know two one-dimensional representations of the group S_3: the trivial representation (all elements are mapped onto the identity in a one-dimensional space) and the representation

$$S_3 \ni \sigma \to \mathrm{sgn}(\sigma) \in \{1, -1\},$$

where $\mathrm{sgn}(\sigma)$ is the sign (parity) of the permutation σ.

To these representations correspond in $L^2(S_3)$ the subspace of constant functions and the subspace of functions $\sigma \to c\,\mathrm{sgn}(\sigma)$, $c \in \boldsymbol{C}$.

To prove that the group S_3 has only 3 non-equivalent irreducible representations we shall construct a two-dimensional representation of S_3. To this end we observe that S_3 contains a cyclic subgroup C_3 of three elements generated by the cycle (1 2 3). Let us put $\sigma_2 = (1\ 2\ 3)$, $\sigma_3 = \sigma^2 = (1\ 3\ 2)$ and let σ_4 be the transposition (1 2). The subgroup C_3 has two non-trivial one-dimensional representations (which are characters), namely $\varkappa_1$, given by the formula $\varkappa_1(\sigma_2) = e^{\frac{2}{3}\pi i}$, and $\varkappa_2$, defined as $\varkappa_2(\sigma_2) = e^{\frac{4}{3}\pi i}$. The representations induced by these characters of the subgroup C_3 are two-dimensional subrepresentations of L (S_3/C_3 has two elements) and therefore equivalent provided they are irreducible. To verify this we shall write out the matrix elements for $U^{\varkappa_1}(\sigma_2)$ and $U^{\varkappa_1}(\sigma_4)$. The remaining matrix elements are uniquely determined by them since σ_2 and σ_4 generate S_3. As a basis for the representation space of $U^{\varkappa_1}$ we choose the functions f_1 and f_2 satisfying the conditions $f_1(\sigma_1) = 1 = f_2(\sigma_4)$, $f_1(\sigma_4) = 0 = f_2(\sigma_1)$ (these conditions determine f_1, f_2 uniquely because the cosets C_3 and $\sigma_4 C_3$ exhaust S_3). Then we have

$$U^{\varkappa_1}(\sigma_4) = \begin{bmatrix} 0 & 1 \\ 1 & 0 \end{bmatrix}, \qquad U^{\varkappa_1}(\sigma_2) = \begin{bmatrix} e^{\frac{2}{3}\pi i} & e^{\frac{2}{3}\pi i} \\ e^{\frac{2}{3}\pi i} & e^{\frac{2}{3}\pi i} \end{bmatrix},$$

and hence the equality $L_{S_3}(U^{\varkappa_1}) = \boldsymbol{C} \cdot I_2$ follows by an easy computation.

4.5. THE ORTHOGONALITY RELATIONS FOR MATRIX ELEMENTS

In the present section we consider representations of a locally compact group G, dropping for the time being the assumptions of compactness.

DEFINITION. A unitary, continuous representation (H, U) of the group G is said to be *square integrable* (of type L^2) provided for every $x, y \in H$ the function $G \ni g \to w_{x,y}(g) := (x|U_g y)$ is of class $L^2(G)$.

Thus the representations of L^2 type are characterized by square integrable matrix elements.

Every L^2-representation can be intertwined in a natural way with the regular representation. Namely to an arbitrary fixed vector $h \in H$ we assign an operator S_h: $H \to L^2(G)$, defined as

$$S_h x := \overline{w_{x,h}}. \tag{4.5.1}$$

The intertwining properties of this operator are obvious:

$$S_h U_g = L_g S_h.$$

PROPOSITION 4.5.1. *The operator S_h belongs to $L_G(H, L^2(G))$.*

Proof. It is enough to prove the boundedness of this operator. To this end we shall apply the closed graph theorem, which states that every closed linear operator on a Banach space is bounded. So we assume that $H \ni x_n \to 0$ and $S_h x_n \underset{L^2}{\to} w \in L^2(G)$. We wish to prove that $w = 0$. By fixing a $\varphi \in \mathscr{C}_0(G)$ we find

$$\begin{aligned}\int_G \overline{\varphi} w \, dg &= \lim_{n\to\infty} \int_G \overline{\varphi} S_h x_n \, dg = \lim_{n\to 0} \int_G \overline{\varphi}(g) (U_g x_n | h) dg \\ &= \lim_{n\to\infty} (U(\varphi) x_n | h) = 0,\end{aligned}$$

since $U(\varphi)$ is a bounded operator. This identity shows that w is indeed zero. □

The operator S_h is injective whenever h is cyclic. Therefore, the cyclic representations of L^2 type are subrepresentations of the regular representation of G on $L^2(G)$. In particular, the irreducible L^2-representations are direct components of the regular representation. In this case much more information about the operator S_h can be derived from Corollary 3.1.2 which follows Schur's lemma. Namely, we have

LEMMA 4.5.2. *If (H, U) is a irreducible representation of L^2 type then for every h the operator S_h is proportional to an isometry.*

For an arbitrary $h \in H$ there exists a number $c(h) = ||S_h||^2$ such that

$$\int_G \overline{w_{x,h}(g)} w_{x_1,h}(g) dg = c(h) (x_1 | x) \tag{4.5.2}$$

identically for $x, x_1 \in H$. Since $w_{x,h} = \overline{w_{h,x}}$ it follows that

$$c(h)(x\, x) = c(x)(h | h).$$

For $h = 0$ we obviously have $c(h) = 0$ and outside the point zero the function $\frac{c(x)}{(x|x)}$ is non-zero and constant on H. The inverse of this number is called the *formal dimension* of the representation U and denoted by $n(U)$. Thus identity (4.5.2) takes the form

$$(w_{x,h}|w_{x_1,h})_{L^2(G)} = \frac{1}{n(U)}(x_1|x)(h|h). \tag{4.5.3}$$

Both sides being bilinear forms in h this implies on account of the polarization formulas the following

THEOREM 4.5.3 (orthogonality relations). *For every irreducible L^2-representation there exists a positive number $n(U)$ such that*

$$(w_{x,h}|w_{x_1,h_1})_{L^2(G)} = \frac{1}{n(U)}(x_1|x)(h|h_2) \tag{4.5.4}$$

identically in $x, x_1\, h, h_1 \in H$. Further, if (H^1, U^1) and (H^2, U^2) are unequivalent representations then the corresponding matrix elements $w^1_{x,h}$ and $w^2_{x_1,h_1}$ are orthogonal for arbitrary x, x_1, h, h_1.

Proof. It is only the last part of the assertion that remains to be proved. To this end we observe that the operators S^1_h and $S^2_{h_1}$ map H^1 and H^2, respectively, on irreducible subrepresentations of L^2 which are unequivalent and therefore orthogonal. This proves the theorem. □

The properties of matrix elements can now be given in the most general form.

THEOREM 4.5.4. *Retaining the assumptions of Theorem* 4.5.3, *we have*

$$w_{x,h} * w_{x_1,h_1} = \frac{1}{n(U)}(x_1|h)\,w_{x,h_1} \tag{4.5.5}$$

and $w^1_{x,h} * w^2_{x_1,h_1} = 0$.

Proof. We compute

$$\int_G w_{x,h}(g)(x_1|U_{g^{-1}}U_{g_1}h_1)\mathrm{d}g = \int_G w_{x,h}(g)(U_g x_1|U_{g_1}h_1)\mathrm{d}g$$

$$= (w_{U_{g_1}h_1,x_1}|w_{x,h}) = \frac{1}{n(U)}(x|U_{g_1}h_1)(x_1|h).$$

The relation concerning $w^1_{x,h}$ and $w^2_{x_1,h_1}$ is proved in an analogous way. □

Finally, let us have a look at the formal dimension of a representation. For this purpose let us see what the adjoint of S_h looks like: $(S_h^* f|x) = (f|S_h x)_{L^2} = \int_G \overline{f}(g)(U_g h|x)\mathrm{d}g = \big(U(f)h|x\big)$. Thus in the notation of § 4.3, the operator S_h^* assumes the well known form

$$S_h^* f = U(f)h = T_h(f).$$

In the case of a compact group all representations are square integrable. By assuming, in addition, the irreducibility of the representation (H, U) we get

$$||S_h||^2 = c(h) = \frac{1}{n(U)}(h|h). \tag{4.5.6}$$

On the other hand on account of Theorem 4.3.1

$$||S_h||_{HS} = ||S_h^*||_{HS} = ||T_h|| = ||h||. \tag{4.5.7}$$

S_h is an isometry and therefore the norms $||S_h||$ and $||S_h||_{HS}$ are related in the following way:

$$||S_h||^2_{HS} = \dim H ||S_h||^2. \tag{4.5.8}$$

Consequently $\dim H = n(U)$.

Thus the formal dimension of an irreducible finite-dimensional representation has turned out to be equal to the linear dimension of the representation space, which is the reason for the adopted terminology.

We shall encounter a representation of L^2-type of a non-compact group in the chapter devoted to the study of the Laguerre polynomials.

Suppose now that G is a compact group and let (H, T) be its representation on a Hilbert space $(H, [\cdot|\cdot])$.

We define a new inner product on H by putting

$$(x|x_1) := \int_G [T_g x | T_g x_1]\mathrm{d}g. \tag{4.5.9}$$

The invariance of the measure ensures that the product $(\cdot|\cdot)$ is invariant under the action of the group. In other words, the representation $\big((H, (\cdot|\cdot)), T\big)$ is already unitary. The reader will verify that the norms on H induced by the products $[\cdot|\cdot]$ and $(\cdot|\cdot)$ are equivalent. This means that all Hilbert representations of a compact group are "actually" unitary. More precisely all the facts concerning representations of compact group which can be proved for unitary representations by applying only the existence of an invariant inner product regardless of its particular form can be generalized to arbitrary representations on Hilbert spaces. In particular, every Hilbert representation of a compact group is a direct sum of irreducible representations and every irreducible Hilbert representation is finite-dimensional.

4.6. CHARACTERS OF FINITE-DIMENSIONAL REPRESENTATIONS

DEFINITION. By the *character of a finite-dimensional representation* (E, T) of a group G we shall understand the function

$$\chi_T\colon G \ni g \to Tr\, T_g \in \boldsymbol{C}.$$

The following properties of characters are direct consequences of the definition of trace:

$$\chi_T(e) = n(T) = \dim E. \tag{4.6.1}$$

$$\chi_T(ghg^{-1}) = \chi_T(h) \quad (\chi_T \text{ is a central function}). \tag{4.6.2}$$

$$\chi_{T+T_1} = \chi_T + \chi_{T_1}. \tag{4.6.3}$$

$$\chi_{T_1 \otimes T_1} = \chi_T \chi_{T_1}. \tag{4.6.4}$$

If there exists a linear bijection $A\colon E \to E_1$ intertwining T with T_1 then $\chi_T = \chi_{T_1}$. (4.6.5)

If (H, U) is a unitary representation then $\chi^* = \chi$. (4.6.6)

In the case of a compact group property (4.6.6) is true of an arbitrary finite-dimensional representation, since every such representation is "actually" unitary.

Further properties of characters are investigated under the assumption that G is compact.

THEOREM 4.6.1. *Suppose we are given two irreducible representations* (H^α, U^α) *and* (H^β, U^β) *of* G. *Denote by* χ^α *the character of the representation* U^α *and write* $w^\beta_{x,y}$ *for a matrix element of the representation* U^β. *Then*

$$\chi^\alpha * w^\beta_{x,y} = \frac{\delta_{\alpha\beta}}{n(\alpha)} w^\beta_{x,y}, \tag{4.6.7}$$

where $\delta_{\alpha,\beta} = \begin{cases} 0, & \alpha \neq \beta, \\ 1, & \alpha = \beta. \end{cases}$

Proof. We represent the character as $\chi^\alpha(g) = \sum_i u_{ii}(g)$, where $u_{ij}(g) = (e_i|U_g e_j)$, $\{e_i\}$ is a basis for H^α. Next, thanks to formula (4.5.5), we find

$$\chi^\alpha * w^\beta_{x,y} = \sum_i u_{ii} * w^\beta_{x,y} = 0, \qquad \text{whenever } \alpha \neq \beta$$

and

$$\chi^\alpha * w^\beta_{x,y} = \frac{1}{n(\alpha)} \sum_i (x|e_i) w^\beta_{e_i,y} = \frac{1}{n(\alpha)} w^\beta_{x,y}. \qquad \square$$

COROLLARY 4.6.2. *Let (H, U) be a representation of the group G and suppose that a decomposition of the representation into irreducible components has the form $(H, U) = (\oplus H^i, \oplus U^i)$.*

Denote $P^\alpha := n(\alpha) U(\chi^\alpha) = n(\alpha) \int_G \chi^\alpha(g^{-1}) U_g \mathrm{d}g$. Then $P^\alpha|H^i = \mathrm{id}$ if U^i belongs to the class α and $P^\alpha|H^i = 0$ otherwise.

Proof. Assuming $h_1, h_2 \in H^i$, we compute

$$(P^\alpha h_1|h_2) = n(\alpha) \int_G \chi^\alpha(g) (h_1|U^i_{g^{-1}} h_2) \mathrm{d}g = n(\alpha) \chi^\alpha * w^i_{h_1,h_2}$$

$$= \begin{cases} 0, & U^i \notin \alpha, \\ 1, & U^i \in \alpha. \end{cases}$$

Thus, with the help of the character χ^α of the representation, we have constructed a projection operator, from the space H, onto that subspace of H on which the representation is a multiple of the representation U. A representation of this type is called a *factor representation.* Thus we can construct a decomposition of a continuous representation of G on a Hilbert space into factor representations provided we know the irreducible representations and their characters.

THEOREM 4.6.3. *Let (H, U) be a unitary representation of a compact group G and let $\hat{G}$ be the dual group. Then the representation U restricted to $P^\alpha H$ is a multiple of the irreducible representation U^α and*

$$\bigoplus_{\alpha \in \hat{G}} P^\alpha = \mathrm{id}_H. \qquad (4.6.8)$$

Example 1

Let us compute the character of the representation $U := U^{\varkappa_1}$ of the group S_3 constructed in Example 2. S_3 has three different conjugacy classes. (By a conjugacy class we understand an orbit in G of the action of the group G on itself by inner automorphism.) The classes are of the following elements $\{e\}$, $\{\sigma_2, \sigma_3\}$, $\{\sigma_4, \sigma_4\sigma_2, \sigma_4\sigma_3\}$. Characters are constant on these classes, and for $\chi := \chi_U$ we have

$$\chi(e) = 2, \qquad \chi(\sigma_2) = 2\cos\frac{2}{3}\pi = -1, \qquad \chi(\sigma_4) = 0.$$

Example 2

We shall find the irreducible components in the decomposition $U\otimes U$ of the group S_3 (here, as in Example 3, U denotes the irreducible representation of S_3 induced by $\varkappa_1$ (Example 2)). The character of the representation $U\otimes U$ is equal to χ^2 and, on the other hand, for the decomposition $U\otimes U = U^1\oplus U^2\oplus \dots \oplus U^k$ into a direct sum of irreducible representations we have $\chi^2 = \chi_1 + \dots + \chi_k$. Since $\chi^2(e) = 4$, it follows that k may be equal to 2, 3 or 4. The first and the last case are excluded in view of $\chi^2(\sigma_2) = 1$. Now it is easy to see that $U\otimes U$ contains every irreducible representation of the group S_3 with multiplicity 1.

4.7. HARMONIC ANALYSIS ON COMPACT GROUPS AND ON THEIR HOMOGENEOUS SPACES

The decompositions described in Theorems 4.4.3 and 4.4.4 will now obtain the form of the Parseval identities in the Hilbert spaces $L^2(G)$ and $L^2(G/K)$.

For $\alpha \in \hat{G}$, as before, we denote by (H^α, U^α) an irreducible representation belonging to the class α. We fix an orthonormal basis $\{e_k\}$ for H^α. The corresponding matrix elements will be denoted by u^α_{kl}. It follows from the orthogonality relations that the set of functions u^α_{kl}, $\alpha \in \hat{G}$, $k, l \leqslant n(\alpha)$ constitutes an orthogonal system in $L^2(G)$.

THEOREM 4.7.1. *For every $f \in L^2(G)$ we have the decompositions*

$$f = \sum_{\alpha\in\hat{G}} n(\alpha) \sum_{k,l=1}^{n(\alpha)} (u^\alpha_{kl}|f)\, u^\alpha_{kl} = \sum_{\alpha\in\hat{G}} n(\alpha)\chi^\alpha * f$$

$$= \sum_{\alpha\in\hat{G}} n(\alpha) f * \chi^\alpha. \tag{4.7.1}$$

The series are convergent in $L^2(G)$.

Proof. It follows from decomposition (4.4.3) that the space $P^\alpha L^2(G)$ is $n(\alpha)^2$-dimensional. The matrix elements u^α_{kl} constitute a system of n^2-vectors in this space. The orthogonality relations imply that $n(\alpha)^{\frac{1}{2}}u^\alpha_{kl}$ is an orthonormal system. This proves the first decomposition in (4.7.1).

The second decomposition coincides with formula (4.6.8). Finally, it remains to note that $\chi^\alpha * f = f * \chi^\alpha$, since on account of identity (4.6.2)

$$\chi^\alpha * f = \int_G \chi^\alpha(g)f(g^{-1}x)\mathrm{d}g = \int_G \chi^\alpha(xg^{-1})f(g)\mathrm{d}g$$

$$= \int_G \chi^\alpha(g^{-1}x)f(g)\mathrm{d}g = f * \chi^\alpha.$$

Analogous decompositions take place in the space $L^2(G/K)$. We retain the notation of Theorem 4.4.4. In the representation space of an $\alpha \in \hat{G}_K$ we select a basis in such a way that $\{e_i\}$, $i \leqslant m(\alpha)$ is an orthonormal basis for the space of K-fixed vectors in H^α. Then every matrix element u^α_{kl}, $l \leqslant m(\alpha)$ is a right K-invariant function and therefore can be projected onto the space $L^2(G/K)$ by putting

$$u^\alpha_{kl}(gK) := u^\alpha_{kl}(g). \qquad \square$$

THEOREM 4.7.2. *For $f \in L^2(G/K)$ the following decomposition takes place*:

$$f = \sum_{\alpha \in \hat{G}_K} n(\alpha) \sum_{k=1}^{n(\alpha)} \sum_{l=1}^{m(\alpha)} (u^\alpha_{kl}|f)u^\alpha_{kl}. \tag{4.7.2}$$

DEFINITION. A function φ on a group G is said to be *central* if $\varphi(gkg^{-1}) = \varphi(k)$ for $k, g \in G$.

We know that characters are central functions. Moreover, it follows from formula (4.6.7) that

$$\chi^\alpha * \chi^\beta = \frac{\delta_{\alpha\beta}}{n(\alpha)}\chi^\alpha. \tag{4.7.3}$$

Evaluating both sides on the point $e \in G$ we find

$$(\chi^\alpha|\chi^\beta)_{L^2(G)} = \delta_{\alpha\beta}. \tag{4.7.4}$$

THEOREM 4.7.3. *The characters* $\{\chi^\alpha\}_{\alpha\in\hat{G}}$ *form an orthonormal basis for the space of all central elements of* $L^2(G)$.

In the proof we shall apply the following

LEMMA 4.7.4. *An arbitrary matrix element* u^α_{ij} *verifies*

$$\int_G u^\alpha_{ij}(gkg^{-1})\,dg = \delta_{ij}\frac{1}{n(\alpha)}\chi^\alpha(k). \tag{4.7.5}$$

Proof of the lemma. The orthogonality relations imply

$$\begin{aligned}\int_G u^\alpha_{ij}(gkg^{-1})\,dg &= \sum_{k,l}\int_G u^\alpha_{ik}(g)u^\alpha_{kl}(k)u^\alpha_{lj}(g^{-1})\,dg\\ &= \sum_{k,l} u^\alpha_{kl}(k)\int_G u^\alpha_{ik}(g)\,\overline{u^\alpha_{lj}(g)}\,dg\\ &= \delta_{ij}\frac{1}{n(\alpha)}\chi^\alpha(k).\end{aligned}$$

Proof of Theorem 4.7.3. For a central function f we have

$$\chi^\alpha * f = \frac{1}{n(\alpha)}P^\alpha f = \sum_{i,k} u^\alpha_{ik}(u^\alpha_{ik}|f).$$

At the same time identity (4.7.5) implies that

$$\begin{aligned}(u^\alpha_{ik}|f) &= \int_G \overline{u^\alpha_{ik}}(k)\,f(k)\,dk = \int_G\int_G \overline{u^\alpha_{i,k}}(k)f(g^{-1}kg)\,dg\,dk\\ &= \int_G\int_G \overline{u^\alpha_{ik}}(gkg^{-1})\,dg f(k)\,dk = \frac{\delta_{ik}}{n(\alpha)}\int_G \overline{\chi}^\alpha(k)f(k)\,dk\\ &= \frac{\delta_{ik}}{n(\alpha)}(\chi^\alpha|f).\end{aligned}$$

Consequently we obtain the decompositions:

$$\chi^\alpha * f = \frac{1}{n(\alpha)}(\chi^\alpha|f)\chi^\alpha$$

and

$$f = \sum_{\alpha\in\hat{G}} n(\alpha)\chi^\alpha * f = \sum_{\alpha\in\hat{G}}(\chi^\alpha|f)\chi^\alpha. \tag{4.7.6}$$

The decomposition (4.7.6) shows that the orthonormal system of characters is complete in the space of square integrable central functions. □

The decomposition theorems presented so far have referred to the approximation in the sense of the norm on $L^2(G)$. We shall now apply them to obtain the following theorem on uniform approximation of continuous functions.

THEOREM 4.7.5 (Weyl's theorem on approximation). *Every continuous function on a compact group is a uniform limit of linear combinations of matrix elements of irreducible representations of the group.*

Proof. We have to prove that for every function $f \in \mathscr{C}(G)$ and every $\varepsilon > 0$ there exists a finite linear combination of matrix elements such that

$$\sup_{x\in G}\left|f(x) - \sum_{\alpha,i,j} C^{\alpha}_{ij} u^{\alpha}_{ij}(x)\right| < \varepsilon.$$

The function f being continuous on a compact set, is uniformly continuous. Thus for a sufficiently small neighbourhood O of the identity of the group we have

$$\sup_{y\in O}\sup_{x\in G}|f(x) - L_g f(x)| < \frac{1}{2}\varepsilon.$$

Putting $\varphi_\varepsilon = \dfrac{1}{dg(O)} 1_O$, we also have

$$\sup_{x\in G}|f(x) - \varphi_\varepsilon * f(x)| < \frac{1}{2}\varepsilon.$$

It follows from Theorem 4.7.1 that there exist indices $\{\alpha_i\}_{i=1}^n$ and a function $f_n \in \bigoplus_{i=1}^{n} H^{\alpha_i}$ such that

$$\|f - f_n\|_{L^2(G)} \leqslant \frac{1}{2}\varepsilon dg(O)^{\frac{1}{2}}.$$

Consequently

$$\begin{aligned}|f(x) - (\varphi_\varepsilon * f_n)(x)| &\leqslant |f(x) - (\varphi_\varepsilon * f)(x)| + |\varphi_\varepsilon * (f - f_n)(x)| \\ &\leqslant \frac{1}{2}\varepsilon + \|\varphi_\varepsilon\|_{L^2}\|(f - f_n)\|_{L^2} = \varepsilon\end{aligned}$$

in view of the Schwarz inequality in $L^2(G)$.

Since the spaces H^{α_i} are invariant under left translations, we have $\varphi_\varepsilon * f_n \in \bigoplus_{i=1}^{n} H^{\alpha_i}$. Finally we note that the finite-dimensional space $\bigoplus_{i=1}^{n} H^{\alpha_i}$ is spanned by matrix elements of irreducible representations. This concludes the proof of Weyl's theorem. □

Example 1

Decomposition of the regular representation of the group $O(2)$.

Select an element $s \in O(2)$ such that $\det s = -1$ (s represents a reflection of the plane with respect to a line) and let $SO(2) \cong T$ be the rotation group. The choice of the element s defines an isomorphism $O(2) \to Z_2 \times_\tau T$ where $\tau(-1)\alpha = \bar{\alpha}$ for $\alpha \in T$. We shall parametrize the elements of G with the help of the inverse isomorphism by writing $r(\alpha)$ for the element corresponding to $(1, e^{i\alpha})$ and $r(\alpha)s$ for the element corresponding to $(-1, e^{i\alpha})$. Hence we obtain the following expression for the normalized Haar measure on $G = O(2)$:

$$\int_G f(g)\mathrm{d}g = \frac{1}{4\pi}\left(\int_0^{2\pi} f(1, e^{i\varphi})\mathrm{d}\varphi + \int_0^{2\pi} f(-1, e^{i\varphi})\mathrm{d}\varphi\right).$$

The subspaces of $L^2(G)$ corresponding to the two one-dimensional representations, the trivial representation and the representation $g \to \det g$, are composed of the constant functions and of the functions of the form $g \to c \det g$, $c \in C$ respectively. We shall denote these subspaces by E_+, E_-, respectively (we leave it to the reader to find the form of the corresponding projectors).

We introduce the family of representations $\{U^k\}$ (numbered by integers) which act on the space C^2, by defining the matrices of the representation operators with respect to the canonical basis.

$$U^k(r(\alpha)) = \begin{bmatrix} e^{ik\alpha} & 0 \\ 0 & e^{-ik\alpha} \end{bmatrix}, \qquad U^k(r(\alpha)s) = \begin{bmatrix} 0 & e^{ik\alpha} \\ e^{-ik\alpha} & 0 \end{bmatrix}.$$

The representations defined in this way are irreducible—the representations U^l and U^k are unequivalent except for the case $l = -k$.

The matrix elements u^k_{ij} of the representation U^k with $k > 0$ span the orthogonal complement of $E_+ \oplus E_-$ in $L^2(G)$. Indeed, the restrictions of these elements to any of the circles $\{1\}\times T$, $\{-1\}\times T$ give exponential functions $e^{ik\alpha}$ different from the constant function, and as is well known, such functions constitute an orthonormal basis for the complement in $L^2(T)$ of the space of constant functions. U^k and U^{-k} being equivalent we can write decomposition (4.4.3) in the form

$$L = I\oplus\det\oplus\big(\bigoplus_{k=1}^{\infty}(U^k\oplus U^{-k})\big).$$

The reader will surely note that the representations U^k are in fact the representations induced by one-dimensional representations of the subgroup SO(2).

PROBLEMS

1. A linear mapping $T\colon H_1 \to H_2$ (not necessarily continuous) is called a *compact mapping* if the image $T(B)$ of an arbitrary bounded subset $B \subset H_1$ is a relatively compact subset in H_2 (i.e. its closure is compact).

(a) Prove that every compact mapping is continuous (i.e. bounded) and that every Hilbert–Schmidt operator is compact.

(b) Prove that the set of compact operators on a Hilbert space H, is a two-sided ideal in the space $L(H)$ and closed in the operator norm. If H is separable, then the ideal of the finite-dimensional operators on H is dense with respect to the operator norm in the ideal of compact operators.

(c) A unitary operator $U\colon H_1 \to H_2$ is a Hilbert–Schmidt operator (more generally—a compact operator) if, and only if, H_1 and H_2 are finite-dimensional.

(d) Let $T \in HS(H)$ and let $E \subset H$ be the eigenspace of T corresponding to a non-zero eigenvalue. Prove that $\dim E < \infty$.

Hint. To prove the compactness of an operator of the H-S type use the possibility of approximation with respect to the operator norm by finite-dimensional operators and apply (b).

2. Check that for the operators of the H-S type the inner product given by formula (4.1.2) is independent of the choice of the orthonormal basis. Check that for the operators on a finite-dimensional space H this inner product is given by

$$(T|S) = \operatorname{tr}(T^*S), \qquad T, S \in L(H).$$

3. (a) Show that the isometry of the space $L^2(M\times M, \mu\otimes\mu)$ into the space $HS(L^2(M, \mu))$ constructed in Example 1 is surjective and hence is an isomorphism.

(b) Prove that

$$L^2(M, \mu)\otimes L^2(M, \mu) \cong L^2(M\times M, \mu\otimes\mu),$$

where $\cong$ denotes an isometric isomorphism.

(c) Let G be a compact group and μ the normalized Haar measure on G. Show that for an arbitrary $h\in L^2(G, \mu)$ the convolution operator $f\to f*h$ is an operator of the H-S type with the H-S norm equal to norm of the function h.

Hint. Define the kernel corresponding to an operator K by the formula

$$k(x, y) = \sum_{i,j} (\varphi_j|K\varphi_i)\overline{\varphi_i}(x)\varphi_j(y)$$

and compute the matrix elements for the integral operator with this kernel.

4. Let A be an H^*-algebra.

(a) Prove that

$$(xy|z) = (x|zy^*), \qquad x, y, z\in A.$$

(b) Recall that a linear subspace E of the algebra A is called a left-sided ideal of the algebra if it is closed under left multiplication by the elements of A, i.e. if $AE\subset E$ (right-sided and two-sided ideals are defined analogously). Prove that the orthogonal complement of a left-sided (right-sided) ideal in an H^*-algebra is a left-sided (right-sided) ideal.

(c) An element $e\in A$ is called an *idempotent* if $e^2 = ee = e$. Prove that if $e\neq 0$ is an idempotent then $||e||\geqslant 1$ and Ae is a closed left-sided ideal (eA is a closed right-sided ideal).

(d) Let $E\subset A$ be a closed left-sided ideal and let $z\in A$. Denote by L_z^E the operator $x\to zx$ acting on E (in particular, for $E = A$ $L_z^A = L_z$ is an operator of left multiplication in A, i.e. an operator of the left regular representation). Show that

(i) $||L_z^E||\leqslant ||z||$, $z\in A$,

(ii) $(L_z^E)^* = L_{z^*}^E$ ($*$ on the left denotes the hermitian conjugation of an operator).

Hint. (a) Apply $(xy)^* = y^*x^*$ and the identity $(x|y) = (y^*|x^*)$. (c) Observe that the element e is a right-sided unit in Ae. (d) The inner product on E is the restriction of the inner product on A.

5. (a) Prove that if z is a non-zero hermitian element in an H^*-algebra A (z is hermitian, if $z = z^*$) such that $||L_z|| = 1$ then the sequence $\{z^{2n}\}_{n=1}^{\infty}$ is convergent and its limit is a hermitian idempotent.

(b) Prove Ambrose's theorem: Any non-zero left-sided (right-sided) ideal in an H^*-algebra contains a hermitian idempotent (the ideal is not assumed to be closed).

Hint. (a) Prove first that $||L_{z^n}|| = 1$ for all positive integers n by showing that the sequence $||L_{z^n}||$ is non increasing and that $||L_{z^{2^n}}|| = ||L_{z^{2^{n-1}}}(L_{z^{2^{n-1}}})^*|| = 1$. Then observe that the sequence of the numbers

$$\{||z^{2n}||\}_{n-1}^{\infty}$$

is convergent and its limit is not less than 1, and deduce hence that $\{z^{2n}\}_{n=1}^{\infty}$ is a Cauchy sequence.

(b) Construct an element of the ideal, satisfying the conditions given in (a) and prove that $e = \lim z^{2n}$ also belongs to that ideal.

6. Let G be a compact group and T a continuous representation of G on a Hilbert space $(H, [\cdot|\cdot])$. Check that the formula

$$(x|x_1) := \int_G [T_g x | T_g x_1]\, dg$$

defines an inner product on H relative to which T is a unitary representation. Prove that the norms on H defined by $(\cdot|\cdot)$ and $[\cdot|\cdot]$ are equivalent.

Hint. Apply condition (3) of the definition of a representation in § 2.1 to obtain the inequalities

$$[T_g x|T_g x] \leqslant c[x|x], \qquad [x|x] \leqslant c_1[T_{g-1}x|T_{g-1}x].$$

7. Let (H^1, T^1), (H^2, T^2) be two unitary representations of a compact group G.

(a) Prove that the mapping $HS(H^1, H^2) \ni A \to A^0 \in HS(H^1, H^2)$, where

$$A^0 = \int_G T_g^2 A T_{g^{-1}}^1 dg$$

is the orthogonal projection onto the subspace $HS_G(H^1, H^2)$.

(b) Prove that an arbitrary operator belonging to the space $HS_G(L^2(G))$ has the form $f \to f * k$ where $k \in L^2(G)$. The operator is hermitian if and only if $k = k^*$ where $k^*(x) = \overline{k(x^{-1})}$.

8. Let G be a compact group. Define an involution on $L^2(G)$ by putting $f^*(x) := \overline{f(x^{-1})}$.

(a) Prove that $L^2(G)$ together with its Hilbert space structure with convolution as multiplication and with involution * is an H^*-algebra.

(b) Prove that a closed subspace of $L^2(G)$ is invariant under left (right) regular representation if, and only if, it is a left (right)-sided ideal in the convolution algebra $L^2(G)$.

(c) Prove that the H^*-algebra $L^2(G)$ contains a unity if, and only if, G is a finite group.

(d) Denote by $Q(\alpha)$ for $\alpha \in \hat{G}$ a maximal left invariant subspace of $L^2(G)$ containing only the representations belonging to the class α. Prove that $Q(\alpha)$ is a finite-dimensional H^*-subalgebra of the H^*-algebra $L^2(G)$ and that the H^*-algebra Q contains a unity (cf. also Problem 18).

Hint. (b) Observe that the algebra $L^2(G)$ contains an approximate unit (a Dirac sequence). (c) If a function $\varphi \in L^2(G)$ were a unity in the algebra $L^2(G)$, then, for an arbitrary measurable set $S \subset G$,

$$\int_S \varphi(g)\,dg$$

would assume different values depending on whether $e \in S$ or $e \notin S$ (e—the unity of the group G). (d) $Q(\alpha)$ is a two-sided closed ideal invariant under the involution *.

9. Let G be a compact group.

(a) Prove that every unitary, continuous representation of G on a Hilbert space H extends uniquely to a *-representation of the H^*-algebra $L^2(G)$ (i.e. to a representation of the convolution algebra which preserves the involution *).

(b) Prove that if π is a *-representation of an H^*-algebra A on a Hilbert space H then for an arbitrary hermitian idempotent $e \in A$ the operator $\pi(e)$ is an orthogonal projector on H.

Hint. Make use of the inclusion $L^2(G) \subset L^1(G)$ and define the extension by means of a suitable integral as in Chapter 2 (Problem 11).

10. Suppose G is a compact group.

(a) Prove that the mapping $L^2(G) \ni f \to K_f \in HS_G(L^2(G))$, where

$$K_f h(x) = \int_G h(y) f(x^{-1}y) \mathrm{d}y = h * f^{\vee}(x),$$

(here $f^{\vee}(x) = f(x^{-1})$) is an isometric isomorphism of H^*-algebras.

(b) Prove that every finite-dimensional left-sided ideal in $L^2(G)$ is of the form $\{x * e : x \in L^2(G)\}$ for a certain idempotent e.

(c) Show that a finite-dimensional ideal is minimal if, and only if, the corresponding idempotent is indecomposable, i.e. does not admit a decomposition $e = e_1 + e_2$, where e_i are idempotents both different from zero and such that $e_1 * e_2 = e_2 * e_2 = 0$.

(d) Prove that a idempotent e is indecomposable if, and only if, for every $x \in L^2(G)$ $e * x * e = \lambda_x e$, where $\lambda_x \in \mathbf{C}$.

(e) Prove that minimal ideals E_1 and E_2 define equivalent representations of G if, and only if, $\{e_1 * x * e_2 : x \in L^2(G)\} \neq \{0\}$, where e_i is the idempotent corresponding to E_i.

Hint. (b) Apply (a) to the orthogonal projector P which projects onto the ideal under consideration. (c) Make use of the relation between projectors and idempotents established in (b). (d) Consider the mapping $y \to y * e * x * e$ for an indecomposable idempotent e. (e) Relate elements of the form $e_1 * e_2$ to the intertwining mappings of E_1 into E_2.

11. Prove that a continuous function h on G belongs to the centre of the algebra $L^2(G)$ if, and only if, for any $x, y \in G$, $h(xy) = h(yx)$, i.e. h is a central function. Deduce hence that G is abelian if, and only if, $L^2(G)$ is an abelian algebra.

12. Let G be a finite group of r elements (order $G = r$).

(a) If $n(\alpha)$ is the dimension of an irreducible representation of the group G of class $\alpha \in \hat{G}$ then

$$\sum_{G} n(\alpha)^2 = r.$$

(b) Prove that the number of different classes of irreducible representations of G equals the dimension of the centre of $L_G(L^2(G))$ which in turn equals the number of conjugacy classes in G.

(c) Prove that G is abelian if, and only if, all its irreducible representations are one-dimensional.

Hint. Apply Problem 11 and the isomorphism of $L_G(L^2(G))$ with $L^2(G)$.

13. Prove that a finite-dimensional representation of a compact group G is irreducible if, and only if, its character is of norm 1 in $L^2(G)$.

14. Prove that a unitary representation (T, H) of a compact group G is cyclic if, and only if, for any arbitrary irreducible representation (U, K), $\dim L_G(H, K) \leqslant \dim K$.

Hint. Each of these conditions allows us to construct an injective operator intertwining the representation T with the regular representation of G.

15. Let (T^i, H^i), $i = 1, 2, 3$ be unitary representations of a compact group G. Prove that for arbitrary operators $S_1 \in L_G(H^1, H^3)$, $S_2 \in L_G(H^2, H^3)$ such that S_2 is a surjection there exists an operator $R \in L_G(H^1, H_2)$ satisfying $S_2 \circ R = S_1$.

Hint. Apply the complete reducibility of a representation of a compact group.

16. (a) Let G be a compact group (T_1, H_1), and (T_2, H_2) its two finite-dimensional representations. Check that the mapping

$$G \times G \ni (g_1, g_2) \to T_1(g_1) \otimes T_2(g_2) \in L(H_1 \otimes H_2)$$

is a representation of the group $G \times G$ (we call it the *exterior tensor product*).

(b) Prove that a representation (T, H) of the group $G \times G$ is irreducible if, and only if, it is equivalent to an exterior tensor product of two irreducible representations of G.

(c) Prove the Burnside theorem: Let (U, H) be a unitary, irreducible representation of a compact group G and suppose $A \in L(H)$. Then there exist numbers $c_1, \ldots, c_n$ and elements $g_1, \ldots, g_n$ of G such that

$$A = \sum_{i=1}^{n} c_i U(g_i), \qquad \text{i.e. the set } U(G) \text{ spans } L(H).$$

Hint. (b) Prove that the representation $G \ni g \to T(g, e)$ on the space H is of the form kT_1 where (T_1, H_1) is an irreducible representation of G and k is a natural number. Take as (T_2, H_2) the representation on the space $L_G(H_1, H)$ given by the formula $T_2(g)A := T(e, g) \circ A$.

Then the mapping $H_1 \otimes H_2 \ni (v, A) \to Av \in H$ is the required equivalence.

(c) Apply (b).

17. Let G be a locally compact unimodular group and (H, U) a continuous unitary irreducible representation of G. Show that if there exist vectors $x, y \in H$, $x \neq 0 \neq y$ such that $w_{x,y} \in L^2(G)$ then (H, U) is a square integrable representation.

Hint. Prove by fixing in a suitable way the vector x that the set $\{y \in H;\ w_{x,y} \in L^2(G)\}$ is a G-invariant closed subspace in H. Then consider in an analogous way the dependence of the matrix element on the first vector.

18. As in Theorem 4.7.1, let $U^\alpha_{k,l}$ be the matrix elements of an irreducible representation (H^α, U^α) corresponding to an orthonormal basis $\{e_i\}_{i=1}^{n(\alpha)}$ of the space H^α, i.e. let $u^\alpha_{k,l}(g) = (e_k | U^\alpha_g e_l)$.

(a) Denote by $Q(\alpha)$ the subspace of $L^2(G)$ spanned by the functions $\{\overline{u}^\alpha_{k,l}\}_{k,l=1}^{n(\alpha)}$. Prove that $Q(\alpha)$ is a minimal subspace invariant under both the left and the right regular representations of G (a minimal closed two-sided ideal of the algebra $L^2(G)$) and, moreover, that the restriction to $Q(\alpha)$ of the left regular representation contains only representations of class α. The restriction to $Q(\alpha)$ of the right regular representation contains only representations equivalent to the representation conjugate to (H^α, U^α), i.e. equivalent to the representation by the matrices $[\overline{u}^\alpha_{k,l}(\cdot)]_{k,l=1}^{n(\alpha)}$.

(b) Prove that for an arbitrary $1 \leqslant k \leqslant n(\alpha)$ the subspace spanned by $\{\overline{u}^\alpha_{i,k}\}_{i=1}^{n(\alpha)}$ is invariant and irreducible for the left regular representation and the subspace spanned by $\{\overline{u}^\alpha_{k,i}\}_{i=1}^{n(\alpha)}$ is invariant and irreducible for the right regular representation.

(c) Show that the functions $n(\alpha)^{\frac{1}{2}}\overline{u}^\alpha_{i,k}$, $1 \leqslant i, k \leqslant n(\alpha)$ constitute an orthonormal basis for $Q(\alpha)$.

(d) Prove that

$$HS_G(L) = \bigoplus_{\alpha \in \hat{G}} L_G(Q(\alpha), Q(\alpha))$$

and that the algebra $L_G(Q(\alpha), Q(\alpha))$ is isomorphic to the algebra of matrices of degree $n(\alpha)$ over $\boldsymbol{C}$.

19. Let G be a compact group and K its closed subgroup. Prove that for an arbitrary irreducible representation of K there exists an

irreducible representation of G whose restriction to K contains a representation equivalent to the initial one.

Hint. Let χ be the character of an irreducible unitary representation of the group K. Regarding χ as a measure on G, according to Problem 12, § 1.16 observe that the mapping $f \to \chi * f$ is an orthogonal projection of $L^2(G)$ onto a K-invariant subspace containing only the representations of the group K of character χ. Make use of the fact that every irreducible representation of the group G is contained in the left regular representation.

20. Let G be a compact group and g an element of G different from identity. Show that there exists a finite-dimensional irreducible unitary representation U of the group G such that $U(g) \neq U(e)$.

Hint. The Weyl theorem.

21. A representation (U, H) of a group G is called *faithful* if $U: G \to L(H)$ is an injective mapping. Assuming that a compact group G has a finite-dimensional faithful representation, deduce the Weyl theorem from the Stone–Weierstrass theorem.

Chapter 5

Theory of Spherical Functions

Formulas (4.3.6) and (4.3.7), which are fundamental ones in harmonic analysis on homogeneous spaces of compact groups, are obvious analogues of the inversion formulas for the Fourier transformation on R^n and on T. However, compared with the latter, they have essential defects. An example is, in the case of (4.3.6), the arbitrariness of the choice of bases for the spaces H, relative to which the matrix elements are formed, which results in the non-uniqueness of the decomposition. Non-uniqueness does not appear in decomposition (4.3.6) since there is a one-to-one correspondence between the characters and the irreducible representation, but, on the other hand, the characters are not defined on a homogeneous space. Their presence in harmonic analysis on G/K is undesirable for both aesthetic and practical reasons. The components of a decomposition should be functions in the same variables as the function under decomposition.

The search for possibly "elementary" functions on homogeneous spaces, leading to a decomposition of type (4.3.6), gave rise to the theory of spherical functions.

The sphere S^n served as a starting model and the objects sought were defined as solutions of the Laplace equations which are invariant under rotations about one of the axis.

The zonal spherical functions on S^n can be expressed by means of the Gegenbauer polynomials and, in the special case of S^2, by the Jacobi polynomials.

The general theory was initiated by E. Cartan in 1929 and then developed by H. Weyl (1934). Their papers dealt with compact Riemannian manifolds. The next stage—the extension of the theory to non-compact manifolds was started by I. M. Gelfand and his collaborators around 1959. Their paper, as well as those of R. Godement, Harish-Chandra and others, created a new approach to the theory of spherical functions, which regards a spherical function as a solution of an integral equation called the *spherical equation*. This equation, whose basic properties are investigated in § 5.1, resembles the mean value formula for harmonic functions and is indeed related to this formula, as will be seen in Part III.

Choosing the spherical equation as the starting point of the theory, allows us to connect it with the theory of representations of groups and group algebras. At the same time, however, the differential properties of spherical functions are pushed to the background. From the point of view of this book, such an approach has the advantage of allowing easy derivation of numerous integral formulas for the Bessel and the Legendre functions and for the Legendre and the Gegenbauer polynomials.

Our exposition of the elements of the theory of spherical functions is essentially based on the monographs of T. Tamagava, 1960, K. Maurin, 1968, S. Helgason 1968 and G. Warner, 1972.

5.1. THE SPHERICAL INTEGRAL EQUATION

Let G be a locally compact unimodular group and let K be its compact subgroup. Fix a continuous homomorphism $\chi\colon K \to T$.

DEFINITION. A representation (H, U) of the group G on a Hilbert space H is said to be *χ-spherical* provided the space of those vectors v in H which satisfy

$$U_k v = \chi(k) v, \qquad k \in K \tag{5.1.1}$$

is one-dimensional and cyclic.

Fix an inner product on H which is invariant under the action of the subgroup K (cf. (4.5.9)). We retain the notation:

$$u_{x,y}(g) = (x|U_g y), \qquad x, y \in H, \; g \in G.$$

THEOREM 5.1.1. *Any matrix element $u_{x,y}$ satisfies the integral equation*

$$\int_K u_{x,y}(gkh)\bar{\chi}(k)\mathrm{d}k = u_{x,v}(g)u_{v,y}(h), \tag{5.1.2}$$

v being a normalized vector satisfying (5.1.1) *and* dk *the normalized Haar measure on K.*

Proof. The orthogonal projection onto the subspace of vectors satisfying (5.1.1) is the integral operator

$$P_\chi = \int_K \overline{\chi}(k) U_k \mathrm{d}k. \tag{5.1.3}$$

Since, by assumption, P_χ is a one-dimensional projector we also have $P_\chi x = (v|x)v$. Consequently

$$\begin{aligned}\int_K u_{x,y}(gkh)\overline{\chi}(k)\mathrm{d}k &= (x|U_g P U_h y) = (U_{g^{-1}}x|P_\chi U_h y)\\ &= (x|U_g v)(v|U_h y) = u_{x,v}(g)u_{v,y}(h). \qquad \square\end{aligned}$$

During the proof we obtained the following

COROLLARY 5.1.2.

$$P_\chi U_h y = u_{v,y}(h)v.$$

DEFINITION. A function w on G is said to be a *zonal χ-spherical function* if it fulfils the requirement

$$\int_K w(gkh)\overline{\chi}(k)\mathrm{d}k = w(g)w(h). \tag{5.1.4}$$

By a *χ-spherical function associated with* w we mean a function ϕ on G satisfying the equation

$$\int_K \phi(gkh)\overline{\chi}(k)\mathrm{d}k = \phi(g)w(h). \tag{5.1.5}$$

If $\chi \cong 1$ we shall refer to w as the *zonal spherical function* and to ϕ as the *spherical function associated with* w. It follows from Theorem 5.1.1 that $u_{v,v}$ is a zonal χ-spherical function and $u_{x,v}$ is a χ-spherical function associated with $u_{v,v}$.

We proceed to investigate the basic properties of solutions of equation (5.1.4). To begin with we introduce some necessary symbols.

Suppose we are given two homomorphisms χ and $\varkappa$ of the group K into T. If $\mathscr{F}(G)$ is a linear subspace of the space of all functions on the group G (e.g. $\mathscr{F}(G) = \mathscr{C}(G)$ or $L^2(G)$), then we write

$$\begin{aligned}{}^{\chi}\mathscr{F}(G) &:= \{f \in \mathscr{F}(G) \colon f(k^{-1}g) = \chi(k)f(g),\ g \in G,\ k \in K\},\\ \mathscr{F}^{\varkappa}(G) &:= \{f \in \mathscr{F}(G) \colon f(gk) = \varkappa(k)f(g),\ g \in G,\ k \in K\},\\ {}^{\chi}\mathscr{F}^{\varkappa}(G) &:= \{f \in \mathscr{F}(G) \colon\\ &\qquad f(k^{-1}gk_1) = \chi(k)\varkappa(k_1)f(g),\ g \in G;\ k, k_1 \in K\}.\end{aligned}$$

In the case where $\chi \equiv 1$ or/and $\varkappa \equiv 1$ we write ${}^{\chi}\mathscr{F}(G/K) := {}^{\chi}\mathscr{F}^{1}(G)$ and $\mathscr{F}^{\varkappa}(K\backslash G) := {}^{1}\mathscr{F}^{\varkappa}(G)$. Furthermore, $\mathscr{F}(K\backslash G/K) := {}^{1}\mathscr{F}^{1}(G)$. Also, we agree to identify a function φ on the homogeneous space G/K with the following K-invariant function on G:

$$G \ni g \to \varphi(gK) \in \boldsymbol{C}.$$

This ensures that the symbols introduced above are consistent with the previous terms.

In "good" function spaces the elements of ${}^{\chi}\mathscr{F}$ and $\mathscr{F}^{\varkappa}$ are constructed by means of projections:

$$P_{\chi} := \int_K \overline{\chi}(k) L_k \mathrm{d}k, \qquad Q_{\varkappa} := \int_K \overline{\varkappa}(k) R_k \mathrm{d}k.$$

LEMMA 5.1.3. *Let ϕ and w be continuous non-zero solutions of the equation* (5.1.5). *Then*

(a) $P_{\overline{\chi}}\Phi = \Phi(e)w$,

(b) *w is a zonal χ-spherical function,*

(c) $\Phi \in \mathscr{C}^{\chi}(G)$, $w \in {}^{\overline{\chi}}\mathscr{C}^{\chi}(G)$,

(d) $w(e) = 1$.

Proof. Property (a) results upon setting $g = e$ in (5.1.2).

(b) Select a g_1 such that $\Phi(g_1) \neq 0$. On the basis of (5.1.5) we obtain

$$w(h) = \Phi(g_1)^{-1} \int_K \phi(g_1 kh)\overline{\chi}(k)\mathrm{d}k.$$

Hence we conclude that

$$\begin{aligned}\int_K w(gk'h)\overline{\chi}(k')\mathrm{d}k' &= \phi(g_1)^{-1} \int_K\int_K \Phi(g_1 kgk'h)\overline{\chi}(k)\overline{\chi}(k')\mathrm{d}k\mathrm{d}k' \\ &= \phi(g_1)^{-1} \int_K \Phi(g_1 kg)\overline{\chi}(k)\mathrm{d}k\, w(h) = w(g)w(h)\end{aligned}$$

which proves (b).

(c) By fixing a $h \in G$ such that $w(h) \neq 0$ we obtain in this case $\Phi = w(h)^{-1}Q_{\chi}R_h\Phi$, which proves the first part of the assertion. The second part is proved in an analogous manner by using (5.1.4).

(d) Set $g = e$ in equation (5.1.4). It follows that for every $h \in G$, $w(h) = w(e)w(h)$, which implies (d). □

Example 1

Suppose in addition that G contains a closed subgroup S such that the mapping $K \times S \ni (k, s) \to ks \in G$ is a homeomorphism (it need not be a group homomorphism). Then the space G/S is homeomorphic in a natural way to K and the transport of the Haar measure on K is a quasi-invariant measure on G/S (cf. § 1.7, Problem 4).

Let σ: G $G/S \to T$ be a continuous multiplier satisfying the condition $\sigma(k, x) = 1$ for $k \in K$ and $x \in G/S$ (sometimes such multipliers are termed *K-multipliers*). Then the function ω defined as $\omega(g) = \int_{G/S} \sigma(g, x) \mathrm{d}\mu(x)$ ($\mathrm{d}\mu$ being the quasi-invariant measure mentioned above) is a zonal spherical function and the function $g \to \sigma(g, x)$ for a fixed x is a zonal spherical function associated with ω. The later fact is a simple consequence of the former one, which in turn results from the computations

$$\begin{aligned}\int_K \omega(gkh)\mathrm{d}k &= \int_K \int_{G/S} \sigma(gkh, x)\mathrm{d}\mu(x)\mathrm{d}k \\ &= \int_K \int_{G/S} \sigma(g, khx)\sigma(kh, x)\mathrm{d}\mu(x)\mathrm{d}k \\ &= \int_{G/S} \left(\int_K \sigma(g, khx)\mathrm{d}k\right) \sigma(k, x)\mathrm{d}\mu(x) = \omega(g)\omega(h),\end{aligned}$$

where we have applied the bi-invariance of the Haar measure on K to obtain $\int_K f(kx)\mathrm{d}k = \int_{G/S} f(x)\mathrm{d}\mu(x)$ for every $x \in G/S$. Detailed examples of such multipliers and the corresponding spherical functions will be given in subsequent sections (see also Problem 5).

Let us return to the study of the first model which provided us with examples of spherical functions, i.e., to the representation (H, U) with the property that the image of the projection P_χ is one-dimensional and is spanned by a vector $v \in H$.

Let $f \in {}^{\chi}\mathscr{C}_0(G)$. As before, we write $U(f) = \int_G f(g)\, U_g \mathrm{d}g$. Thanks to the identity

$$P_\chi U(f) = U(P_\chi f) = U(f) \tag{5.1.6}$$

the image of the operator $U(f)$ is either zero or equal to $P_\chi H$. Therefore we can define a functional on the space ${}^\chi\mathscr{C}_0(G)$ by setting

$$\lambda(f)v := U(f)v. \tag{5.1.7}$$

The space ${}^\chi\mathscr{C}_0(G)$ constitutes a convolution algebra and the mapping ${}^\chi\mathscr{C}_0(G) \ni f \to \lambda(f) \in G$ is an algebra homomorphism. The above functional is, in fact, defined by a zonal χ-spherical function. Indeed, by Corollary 5.1.2 we find

$$U(f)v = P_\chi U(f)v = P_\chi \int_G f(g) U_g v \mathrm{d}g = \int_G f(g) P U_g v \mathrm{d}g$$
$$= \left(\int_G f(g) u_{vv}(g) \mathrm{d}g\right) v.$$

which, combined with (5.1.7), yields

$$\lambda(f) = \int_G f(g) u_{vv}(g) \mathrm{d}g.$$

This sugg ests the idea of defining, for an arbitrary zonal χ-spherica function, a functional on ${}^\chi\mathscr{C}_0(G)$ by setting

$$w(f) := \int_G w(g) f(g) \mathrm{d}g. \tag{5.1.8}$$

PROPOSITION 5.1.4. *The functional w is linear and multiplicative on the algebra ${}^\chi\mathscr{C}_0(G)$.*

Proof. The linearity is obvious. We shall verify the multiplicativity by a direct calculation based on equation (5.1.4):

$$w(f * \varphi) = \int_G\int_G w(g) f(h) \varphi(h^{-1}g) \mathrm{d}h \mathrm{d}g = \int_G\int_G w(hg) f(h) \varphi(g) \mathrm{d}g \mathrm{d}h$$
$$= \int_G\int_G w(hg) f(h) P_\chi \varphi(g) \mathrm{d}g \mathrm{d}h$$
$$= \int_G\int_G\int_K w(hkg) \overline{\chi}(k) \mathrm{d}k f(h) \varphi(g) \mathrm{d}g \mathrm{d}h$$
$$= \int_G w(h) f(h) \mathrm{d}h \int_G w(g) \varphi(g) \mathrm{d}g = w(f) w(\varphi). \qquad \square$$

The construction of multiplicative functionals on the algebra $\mathscr{C}_0(G)$ in terms of zonal spherical functions is, in fact, universal, as will be seen from the theorems presented below.

THEOREM 5.1.5. *Every multiplicative linear functional continuous in compact convergence topology on* ${}^{\chi}\mathscr{C}_0(G)$ *is uniquely determined by a continuous zonal* χ*-spherical function.*

Proof. Let η be a functional satisfying the assumption of the theorem. Then η defines a Radon measure on the group according to the formula

$$\mu(f) := \eta(P_\chi f).$$

Select an $f_2 \in {}^{\chi}\mathscr{C}_0(G)$ such that $\eta(f_2) \neq 0$. The multiplicativity of η implies that

$$\eta(f_1) = \eta(f_2)^{-1}\eta(f_1 * f_2) = \eta(f_2)^{-1} \int_G \int_G f_1(g) f_2(g^{-1}x)\,\mathrm{d}g\,\mathrm{d}\mu(x)$$

$$= \eta(f_2)^{-1} \int_G f_1(g) \int_G f_2(g^{-1}x)\,\mathrm{d}\mu(x)\,\mathrm{d}g.$$

By the Lebesgue theorem the function w,

$$G \ni g \to w(g) := \eta(f_2)^{-1} \int_G f_2(g^{-1}x)\,\mathrm{d}\mu(x),$$

is continuous on G. It follows directly from the definition that

$$\eta(f_1) = w(f_1).$$

In order to complete the proof it is enough to verify that w is a zonal function. To this end we find that

$$\int_G \int_G \int_K w(gkh)\overline{\chi}(k)\,\mathrm{d}k\, f_1(g) f_2(h)\,\mathrm{d}g\,\mathrm{d}h$$

$$= \int_G \int_K \int_G w(gkh) f_2(h)\overline{\chi}(k)\,\mathrm{d}h\,\mathrm{d}k\, f_1(g)\,\mathrm{d}g$$

$$= \int_G \int_G w(gh) P_\chi f_2(h) f_1(g)\,\mathrm{d}h\,\mathrm{d}g = w(f_1 * f_2) = \eta(f_1 * f_2)$$

$$= \eta(f_1)\eta(f_2) = \int_G w(g) f_1(g)\,\mathrm{d}g \int_G w(h) f_2(h)\,\mathrm{d}h$$

identically with respect to $f_1, f_2 \in {}^{\chi}\mathscr{C}(G)$. Hence, w being continuous, it follows that

$$\int_K w(gkh)\overline{\chi}(h)\,\mathrm{d}k = w(g)w(h).$$

The reader will easily note that owing to the property $w \in {}^{\bar{\chi}}\mathscr{C}^{\chi}(G)$, of a zonal χ-spherical function w, the value of the functional $f \to w(f)$ on an arbitrary function f is completely determined by its values on ${}^{\chi}\mathscr{C}_0^{\bar{\chi}}(G)$.

In the remaining part of our exposition we shall mostly be concerned with the case of the algebra $\mathscr{A} = \mathscr{C}_0(K\backslash G/K)$ of K-biinvariant functions on the group G. This algebra is invariant under the operation $f \to f^*$ where $f^*(g) = \bar{f}^{\vee}(g) = \bar{f}(g^{-1})$.

By a unitary functional on $\mathscr{A}$ we shall mean a multiplicative functional η with the property

$$\eta(f^*) = \overline{\eta(f)} \qquad \text{for } f \in \mathscr{A}.$$

A zonal spherical function w defines a unitary functional whenever

$$w^* = w. \tag{5.1.9}$$

Hence we infer

PROPOSITION 5.1.6. *Multiplicative unitary functionals are in a one-to-one correspondence with the zonal spherical functions satisfying* (5.1.9).

In the examples below we describe the space $\mathscr{C}_0(K\backslash G/K)$ and give explicit expressions of formula (5.1.4) for two simplest groups—SO(3) and SL(2, $\boldsymbol{R}$). A detailed study of these groups is deferred until subsequent sections. A slightly different approach to the case of SO(3), based on the charming book of Dym and McKean, is given in the problems appended to this chapter.

Example 2

Take $G = \text{SO}(3)$ and $K = \text{SO}(2)$. Regard K as the group of rotations about the third axis. If we denote $p = (0, 0, 1) \in \boldsymbol{R}^3$, the mapping $g \to gp$ sets up an isomorphism between the homogeneous space G/K and the unit sphere S^2 in $\boldsymbol{R}^3$. Under this isomorphism to a double coset KgK corresponds the set of points on the sphere of the same "latitude", i.e., the set of points given in the spherical coordinates θ, φ by the condition $\theta = \text{const}$, $0 \leqslant \theta \leqslant \pi$.

It follows that a function f in $C(K\backslash G/K)$ can be regarded as a function in a single variable θ or as a function of $\cos\theta$ given by the formula

$$\tilde{f}(\cos\theta) = f(g(\theta)p),$$

where $g(\theta)$ denotes the positive rotation through an angle θ about the axis x_1.

Note that for every continuous function f on G we have

$$\int_G f(g)\mathrm{d}g = \int_{G/K}\int_K f(gk)\mathrm{d}k\,\mathrm{d}(gK),$$

where $\mathrm{d}k$, $\mathrm{d}(gK)$ denote the normalized invariant measures on K and G/K, respectively. Hence, by applying the above homomorphism between G/K and S^2 and expressing the measure on S^2 in spherical coordinates we get

$$\int_G f(g)\mathrm{d}g = \frac{1}{2}\int_0^{\pi} \tilde{f}(\cos\theta)\sin\theta\,\mathrm{d}\theta, \qquad f \in C(K\backslash G/K).$$

Substituting $g = g(\theta_1)$, $h = g(\theta_2)$ (rotations about x_1 through angles θ_1 and θ_2, respectively) in (5.1.4), we arrive at

$$\tilde{f}(\cos\theta_1)\tilde{f}(\cos\theta_2) = \frac{1}{2\pi}\int_0^{2\pi} \tilde{f}(\sin\theta_1\sin\theta_2\cos\theta + \cos\theta_1\cos\theta_2)\mathrm{d}\theta.$$

Example 3

Let $G = \mathrm{SL}(2, \boldsymbol{R})$, $K = \mathrm{SO}(2)$. The homogeneous space G/K can be realized as the sheet of the hyperboloid $x_1^2 - x_2^2 - x_3^2 = 1$ in $\boldsymbol{R}^3$ distinguished by the condition $x_1 > 0$. For this purpose let us consider the three-dimensional vector space $V = \mathfrak{sl}(2, \boldsymbol{R})$ and let $\{X_i\}_{i=1}^3$ be a basis of V such that

$$X_1 = \begin{bmatrix} 0 & -1 \\ 1 & 0 \end{bmatrix}, \quad X_2 = \begin{bmatrix} 0 & 1 \\ 1 & 0 \end{bmatrix}, \quad X_3 = \begin{bmatrix} 1 & 0 \\ 0 & -1 \end{bmatrix}.$$

Note that the action of the group G on V defined as

$$g \cdot X := gXg^{-1}$$

preserves the surfaces defined by the equation $\det X = \text{const}$.

Define coordinates x_i on V by the condition

$$X = \sum_{i=1}^{3} x_i X_i.$$

The surface $S = \{X \in V\colon \det X = 1\}$ is described in these coordinates by the equation

$$x_1^2 - x_2^2 - x_3^2 = 1.$$

Moreover, the isotropy group at the point X_1 is precisely the group SO(2). It can be proved that G acts transitively on each of two sheets (connected components) of the hyperboloid S (see Problem 11). Therefore the mapping $G \ni g \to g \cdot X_1 \in S$ defines an isomorphism between the homogeneous spaces G/K and S. It is easy to note that the subgroup SO(2) acts on S by rotations about the first axis and consequently the functions in $C(K\backslash G/K)$ can be regarded as functions on S invariant under rotations about the first axis. Introducing the "polar" coordinates on the hyperboloid S,

$$(x_1, x_2, x_3) = (\cosh t,\ \sinh t\ \cos\theta,\ \sinh t\ \sin\theta),$$
$$0 \leqslant t < \infty, \qquad 0 \leqslant \theta \leqslant 2\pi,$$

we assign to a function $f \in C(K\backslash G/K)$ either a function $\tilde{f}$ defined on the half-axis $[1, \infty]$ as

$$\tilde{f}(\cosh t) = f(\cosh t, \sinh t \cos\theta, \sinh t \sin\theta),$$

or a function f^0 defined on the positive half-axis by the formula

$$f^0(t) = \tilde{f}(\cosh t).$$

It can be proved that the invariant measure $\mathrm{d}x$ on S can be expressed in terms of the coordinates (t, θ) in the form

$$\mathrm{d}x = \frac{1}{2\pi} \sinh t\,\mathrm{d}t\,\mathrm{d}\theta$$

(The constant $\frac{1}{2\pi}$ serves the purpose of normalizing the measure so that the measure of K be equal to one.) This being so, it follows that all K-bi-invariant functions satisfy

$$\int_G f(g)\,\mathrm{d}g = \int_0^\infty f^0(t) \sinh t\,\mathrm{d}t.$$

Thus, in the case at hand, formula (5.1.4) leads to the identity

$$\tilde{f}(\cosh t)\tilde{f}(\cosh u) = \frac{1}{2\pi} \int_0^{2\pi} \tilde{f}(\cosh t \cosh u + \sinh t \sinh u \cos\theta)\,\mathrm{d}\theta.$$

In the example which follows we prove that normalized characters of irreducible representations of a compact group K can be treated as zonal spherical functions on the group $K \times K$ and, furthermore, that every zonal spherical function on $K \times K$ arises from the character of a representation of the group K.

Example 4

Let K be a compact group and write $G := K \times K$ and $K_1 := \{(g, g)\colon g \in k\} \subset G$. The mapping $G \ni (g_1, g_2) \to g_1 g_2^{-1} \in K$ induces an isomorphism of the homogeneous spaces G/K_1 onto K, and therefore the functions on G which are constant on the cosets of K_1 can be identified with functions on $K \times K$. Under this identification equation (5.1.4) which defines zonal spherical functions, can be written as

$$\int_K \varphi(gkhk^{-1})\mathrm{d}k = \varphi(g)\varphi(h).$$

To the functions in ${}^1\mathscr{C}^1(G)$ correspond, in turn, the central functions on the group K.

In order to describe the zonal spherical functions on G we note that, if χ is the character of an n-dimensional irreducible representation of the group K such that $\bar{\chi} * \varphi(e) \neq 0$ for a zonal spherical function φ, then we have

$$\begin{aligned}\varphi(s)\bar{\chi} * \varphi(e) &= \varphi(s) \int_K \bar{\chi}(k)\varphi(k^{-1})\mathrm{d}k \\ &= \int_K \int_K \varphi(sk'k^{-1}k'^{-1})\bar{\chi}(k)\mathrm{d}k\mathrm{d}k' \\ &= \int_K \int_K \varphi(sk^{-1})\bar{\chi}(k'^{-1}kk')\mathrm{d}k\mathrm{d}k' \\ &= \int_K \varphi(sk^{-1})\bar{\chi}(k)\mathrm{d}k = \chi * \check{\varphi}(s^{-1}).\end{aligned}$$

Now, utilizing an equation from Chapter 4, we get

$$\chi * \check{\varphi}(s^{-1}) = \frac{1}{n}\chi(s^{-1})(\chi|\hat{\varphi}) = \frac{1}{n}\chi(s^{-1})\bar{\chi} * \varphi(e),$$

and hence $\varphi = \dfrac{1}{n}\chi$. Note that this equation is, in fact, the spherical equation for (G, K_1). Thus we have proved that zonal spherical functions for (G, K_1) are the normalized characters of irreducible representations of the group K.

The theorem presented below is an application of spherical functions to harmonic analysis.

We retain the notation

$$\psi(f) = \int_G \psi(g) f(g) \mathrm{d}g$$

for arbitrary $\psi \in \mathscr{C}(G)$ and $f \in \mathscr{C}_0(G)$.

THEOREM 5.1.7. *Let Φ and w be continuous solution of the spherical equation*

$$\int_K \Phi(gkh) \mathrm{d}k = \Phi(g) w(h).$$

Then

(1) $\Phi * f = \check{\Phi}(f)\, \phi$ *for* $f \in \mathscr{C}_0(G/K)$,

(2) $f * \Phi = \check{\Phi}(f) w$ *for* $f \in \mathscr{C}_0(K \backslash G)$,

(3) $f * w = w * f = \check{w}(f) w$ *for* $f \in \mathscr{C}_0(K \backslash G/K)$.

Proof. By way of example we prove (1):

$$\begin{aligned}\int_G \Phi(g) f(g^{-1}x) \mathrm{d}g &= \int_G \Phi(xg^{-1}) f(g) \mathrm{d}g = \int_G \int_K \Phi(xg^{-1}) f(gk) \mathrm{d}k \mathrm{d}g \\ &= \int_G \int_K \Phi(xkg^{-1}) \mathrm{d}k f(g) \mathrm{d}g \\ &= \Phi(x) \int_G \Phi(g^{-1}) f(g) \mathrm{d}g.\end{aligned}$$

The proof of (2) is analogous and (3) follows from (1) and (2). □

5.2. SPHERICAL FUNCTIONS AND SPHERICAL REPRESENTATIONS

In the present section we are concerned with unitary spherical representations and positive definite zonal spherical functions.

If (H, U) is a unitary spherical representation of a group G then the function $w(g) = (v | U_g v)$ is a positive definite zonal spherical function. On the other hand, to a given continuous p.d. function on G there corresponds, via the Gelfand–Raikov construction, a cyclic unitary representation of the group G. Thus it is natural to ask whether every p.d. zonal spherical function is a matrix element of a spherical unitary representation and whether the Gelfand–Raikov construction establishes a 1-1 correspondence between these objects.

In this section we answer these questions, as well as other, more special ones.

THEOREM 5.2.1. *Let w be a continuous p.d. zonal spherical function and $({}^wH, {}^wU)$ the corresponding representation obtained by means of the Gelfand–Raikov construction. Then $({}^wH, {}^wU)$ is an irreducible spherical representation.*

Proof. It follows from the proof of Theorem 2.5.3 that a cyclic vector x for the representation ${}^wU$ is defined by the condition $([f]| {}^wU_h x) = \bar{f} * w(h)$ for $f \in \mathscr{C}_0(G)$. Hence $([f]| {}^wU_k x) = \bar{f} * w(k) = \bar{f} * w(e) = ([f]|x)$.

The set $\{[f]: f \in \mathscr{C}_0(G)\}$ being dense in ${}^wH$, we obtain

$$ {}^wU_k x = x. \tag{5.2.1}$$

Therefore it remains to prove that every K-fixed vector in H is proportional to x. For this purpose we note that the operator $P_1 = \int_K {}^wU_k dk$ is a projection onto the space of K-fixed vectors in ${}^wH$. We compute

$$ (P_1[f]|{}^wU_h x) = \overline{P_1 f} * w(h) = (\overline{P_1 Q_1 f}) * w(h); $$

we have used the fact that $w \in \mathscr{C}(K\backslash G/K)$, whence $\varphi * w = Q_1 \varphi * w$ for an arbitrary $\varphi \in \mathscr{C}_0(G)$. Consequently, in view of Theorem 5.1.7 (3), we get

$$ (P_1[f]|{}^wU_h x) = \check{w}(\overline{P_1 Q_1 f}) w(h) = \check{w}(P_1 Q_1 f)(x|{}^wU_h x) $$

for an arbitrary $f \in \mathscr{C}_0(G)$. Hence, owing to the cyclicity of x in ${}^wH$, we arrive at the formula

$$ P_1[f] = w(P_1 Q_1 f)x, \tag{5.2.2}$$

which shows that $({}^wH, {}^wU)$ is a spherical representation. Irreducibility results from the following general lemma.

LEMMA 5.2.2. *Every χ-spherical unitary representation (H, U) is irreducible.*

Proof. Suppose that the representation (H, U) decomposes into a nontrivial direct sum as follows:

$$H = H_1 + H_2, \qquad U = U_1 + U_2,$$

where $U_i = U|H_i$, $i = 1, 2$. Then the cyclic vector v satisfying the condition $U_k v = \chi(k)v$ can be decomposed as

$$v = v_1 + v_2, \qquad \text{where } v_i \in H_i.$$

If one of the components of v were zero, this would mean that the corresponding space H_i is zero. However, every v_i satisfies the identity $U_k v_i = \chi(k)v_i$ which contradicts the fact that U is χ-spherical. This contradiction ends the proof of the lemma and of the theorem. □

Denote by $\Omega(G, K)$ the set of zonal spherical functions corresponding to the pair of groups (G, K). By $\hat{G}_K$ we shall understand the set of equivalence classes of unitary spherical representations. Under these notations we have

THEOREM 5.2.3. *The mapping $\hat{G}_K \to \Omega(G, K)$, which assigns to a class $[U] \in \hat{G}_K$ the zonal spherical function of the representation U, is a bijection.*

Proof. First of all, we must verify that equivalent representations define the same spherical function. We note that a unitary transformation $T\colon H_1 \to H_2$ intertwining U^1 with U^2 sends a K-fixed vector onto a vector with the same property. Therefore

$$w^2(g) = (Tx|U_g^2 Tx) = (Tx|TU_g^1 x) = (x|U_g^1 x) = w^1(g).$$

The surjectivity of the mapping in question follows from Theorem 5.2.1 since w is a matrix element of the representation $({}^wH, {}^wU)$. Next, if U^1 and U^2 are irreducible spherical representation defining the same spherical function w, then they are equivalent to ${}^wU$, which completes the proof of Theorem 5.2.3. □

5.3. EXISTENCE OF SPHERICAL FUNCTIONS. GELFAND PAIRS

In the preceding sections spherical functions were described in a threefold manner: as solutions of the spherical equation, as multiplicative functionals on the algebra $\mathscr{C}_0(K\backslash G/K)$, and finally with every spherical function we associated a spherical representation. We now proceed to consider the existence of spherical functions and to investigate their role in harmonic analysis on the homogeneous space G/K.

DEFINITION. A pair (G, K) where G is a locally compact unimodular group and K its compact subgroup is called a *Gelfand pair* if the algebra $L^1(K\backslash G/K)$ is commutative.

We recall that $L^1(K\backslash G/K)$ consists of those integrable functions on G which are bi-invariant under K. This space is closed under both the convolution and the conjugation $f \to f^*$. The space $\mathscr{C}_0(K\backslash G/K)$ constitutes a continuously imbedded, dense subalgebra of $L^1(K\backslash G/K)$.

Every continuous functional on $L^1(K\backslash G/K)$ defines by restriction a continuous functional on $\mathscr{C}_0(K\backslash G/K)$. By Theorem 5.1.5 we may therefore imbed the space of continuous multiplicative functionals on $L^1(K\backslash G/K)$ into the space of zonal spherical functions of the pair (G, K).

Unitary positive definite functionals are given by p.d. spherical function by means of the formula

$$\eta(f) = \int_G f(g)w(g)dg.$$

Conversely, since every p.d. function w is bounded, by using the above formula we can define a functional not only on $\mathscr{C}_0(K\backslash G/K)$ but also on $L^1(K\backslash G/K)$.

If a pair (G, K) is a Gelfand pair, then the space $\Omega(G, K)$ of positive definite zonal spherical functions corresponding to this pair can be identified with the space of unitary homomorphisms of the commutative algebra $L^1(K\backslash G/K)$, i.e. with the spectrum of this algebra.

A description of commutative Banach $*$-algebras is provided by the Gelfand–Naimark theory. We shall refer to it in Chapter 16, where we prove the basic theorems of harmonic analysis on homogeneous spaces. In particular, by applying this theory we shall prove that for arbitrary elements $f_1, f_2 \in L^1(K\backslash G/K)$ the identity

$$w(f_1) = w(f_2)$$

for all $w \in \Omega(G, K)$ implies that $f_1 = f_2$. Therefore spherical functions separate points of the algebra $L^1(K\backslash G/K)$.

Meanwhile we shall prove a theorem which implies in particular that in the case of a Gelfand pair there exists a rich family of spherical functions.

THEOREM 5.3.1. *Suppose that G is a locally compact separable group and let (G, K) be a Gelfand pair. Then every irreducible representation which comprises a K-fixed vector is a spherical representation.*

Proof. We denote by (H, U) an irreducible representation of the group G with the property that the (closed) space

$$H_K = \{h \in H\colon\ U_k h = h,\ k \in K\}$$

is different from zero. We intend to prove that H_K is one-dimensional.

We first note that the space H_K is invariant under the operators

$$U(f) = \int_G f(g)\, U_g \mathrm{d}g, \qquad f \in L^1(K\backslash G/K).$$

Denote by $\mathscr{U}(f)$ the restriction of the operator $U(f)$ to H_K and let $\mathscr{U}$ denote the algebra of all operators on H_K of the form $\mathscr{U}(f)$. $L^1(K\backslash G/K)$ being commutative, this implies that $\mathscr{U}$ is a commutative algebra of bounded operators on a Hilbert space. As in the case of the operators $U(f)$, we also have

$$\mathscr{U}(f)^* = \mathscr{U}(f^*).$$

Hence $\mathscr{U}$ is also closed under hermitian conjugation of operators.

Suppose $\dim H_K > 1$. It follows from the spectral theorem that there exists an orthogonal decomposition $H_K = H_1 + H_2$ into $\mathscr{U}$-invariant non-trivial subspaces H_i, $i = 1, 2$. Select non-zero elements $a_i \in H_i$, $i = 1, 2$. We introduce G-invariant subspaces $E_i \subset H$ by setting

$$E_i = \{U(f)a_i\colon f \in L^1 G\}^-$$

($^-$ denotes the closure in H). Next, by using the K-invariance of a_i we find that

$$\begin{aligned}(a_1|U(f)a_2) &= (P_1 a_1|U(f)P_1 a_1) = (a_1|P_1 U(f) P_1 a_2)\\ &= (a_1|U(P_1 Q_1 f)a_2) = 0,\end{aligned}$$

since $P_1 Q_1 f \in L^1(K\backslash G/K)$ and the space H_2 is $\mathscr{U}$-invariant. Therefore we have proved that $U(f)a_2 \perp a_1$ for an arbitrary $f \in L^1(G)$, i.e. that

$E_2 \perp a_1$. This, however, contradicts the irreducibility of the representation U since E_2 is a non-trivial G-invariant subspace. Consequently $\dim H_K = 1$. □

In order to connect this theorem with the problem of the existence and richness of spherical functions we have to refer to the Frobenius duality. Let us suppose for the moment that the Gelfand pair (G, K) consists of compact groups. Then the regular representation on the space $L^2(G/K)$ decomposes, as we know, into the direct sum of those irreducible representations of the group G which contain K-fixed vectors. Moreover, the occurrence number of a representation (H^α, U^α) equals the dimension of the space of K-fixed vectors in H^α.

In view of the above theorem we get

$$L^2(G/K) = \bigoplus_{\alpha \in \hat{G}_K} H^\alpha. \tag{5.3.1}$$

The following decomposition takes place:

$$f = \sum_{\alpha \in \hat{G}_K} n(\alpha) \sum_{j=1}^{n} (u^\alpha_{j1} | f) u^\alpha_{j1}, \tag{5.3.2}$$

where $u^\alpha_{j1}(g) = (e_j | U^\alpha_g e_1)$, $\{e_j\}$ being an orthonormal basis for H^α and e_1 a K-fixed vector in H^α. Moreover,

$$f = \sum_{\alpha \in \hat{G}_K} n(\alpha) f * w^\alpha, \tag{5.3.3}$$

where $w^\alpha = u^\alpha_{11}$ is the zonal spherical function of the representation. The last formula follows from (4.7.6) upon applying the following important fact.

Proposition 5.3.2. *Let (H^α, U^α) be an irreducible representation of a compact group G. Let K be a subgroup of G so that (G, K) is a Gelfand pair. Then the function*

$$w^\alpha(g) := \int_K \chi^\alpha(gk)\,dk \tag{5.3.4}$$

is equal to the zonal spherical function of the representation U^α if U is a spherical representation and is zero otherwise.

Proof. The right-hand side of formula (5.3.4) can be represented as

$$\sum_j \int_K u_{jj}(gk)\,dk = \int_K (e_j | U_g^\alpha U_k^\alpha e_j)\,dk = \sum_j (e_j | U_g^\alpha P e_j).$$

The operator P is the projection onto the subspace of K-fixed vectors, i.e., $Pe_j = (x_0 | e_j) x_0$ for a fixed K-vector x_0. Hence we obtain

$$w^\alpha(g) = \sum_j (x_0 | e_j)(e_j | U_g^\alpha x_0) = (x_0 | U_g^\alpha x_0). \qquad \square$$

Formula (5.3.3) is simplified for $f \in \mathscr{C}_0(K\backslash G/K)$. Upon applying Theorem 5.1.7 (3) we get

$$f = \sum_{\alpha \in \hat{G}_K} n(\alpha)(w^\alpha | f) w^\alpha. \tag{5.3.5}$$

Consequently, we see that, in the case of a Gelfand pair (G, K) the spherical representations provide a unique decomposition of a representation of the group on the space $L^2(G/K)$. At the same time the zonal spherical functions are direct counterparts of the exponential functions, if formula (5.3.3) is compared with the decomposition into Fourier series or Fourier integrals. In particular, it follows from the formulas presented here, that the spherical functions u_{j1}^α occurring in (5.3.2), separate points of the space G/K and the functions $n(\alpha)^{1/2} w^\alpha$ constitute an orthonormal basis for the space $L^2(K\backslash G/K)$.

The facts proved above for compact Gelfand pairs generalize, subject to obvious modifications, to arbitrary Gelfand pairs. This will be seen in examples presented in Part II; the general proof is deferred to Chapter 16.

To end this section, we present for the time being without proof, a convenient condition which ensures that a pair of Lie groups (G, K) is a Gelfand pair.

DEFINITION. Let G be a locally compact group, and $K \subset G$ a closed subgroup. The pair (G, K) is called a *symmetric pair* if there exists an involutive automorphism $\sigma\colon G \to G$ different from the identity and such that

$$(K_\sigma)_0 \subset K \subset K_\sigma,$$

where $K_\sigma = \{g \in G\colon \sigma(g) = g\}$ and $(K_\sigma)_0$ denotes the connected component of the identity of the group K.

The quotient space G/K corresponding to a symmetric pair is called a *symmetric space*.

THEOREM 5.3.3. *Let G be a connected Lie group with a compact subgroup K. If the pair (G, K) is symmetric, then it is a Gelfand pair.*

The proof will be found in § 15.6.

Example 1

(a) Let G be a locally compact abelian group with the group operation written additively. Put $K = \{e\}$. Then the mapping $G \ni g \to -g \in G$ is an involutive automorphism of G and K is the set of fixed points of this automorphism.

It follows that the symmetric space G/K coincides with G itself. The corresponding spherical functions are characters of the group G and the decomposition of functions on the symmetric space is described by the theory of the Fourier transformation on abelian groups, which is a generalization of the classical Fourier analysis on $\boldsymbol{R}^1$ and $\boldsymbol{T}$.

(b) For a compact group K and a locally compact abelian group A define a semi-direct product $G = K \times_\tau A$ where $\tau\colon K \to \operatorname{Aut} A$ is a homomorphism. The mapping σ defined by $\sigma(k, a) := (k, -a)$ is an involutive automorphism of G, with $K \times \{e\}$ constituting the fixed point subgroup. Indeed, we have

$$\begin{aligned}\sigma\big((k_1, a_1)(k_2, a_2)\big) &= \sigma\big(k_1 k_2, a_1 + \tau(k_1)a_2\big)\\ &= \big(k_1 k_2, -a_1 - \tau(k_1)a_2\big)\\ &= \big(k_1 k_2, -a_1 + \tau(k_1)(-a_2)\big)\\ &= (k_1, -a_1)(k_2, -a_2)\\ &= \sigma(k_1, a_1)\sigma(k_2, a_2),\end{aligned}$$

which shows that σ is an automorphism. The remaining properties of σ are verified in an equally simple way.

In order to illustrate the case under consideration, take $K = \mathrm{SO}(n)$ and $A = \boldsymbol{R}^n$ with τ denoting a natural action of K on A. Then $G = M(n)$ is the motion group of $\boldsymbol{R}^n$ and σ the inversion in $\boldsymbol{R}^n$. We note also that in this case the functions in ${}^1\mathscr{C}_0^1(G)$ depend only upon the length of a vector $x \in \boldsymbol{R}^n$, i.e. $f(kx) = \tilde{f}(\|x\|)$, which allows us to verify by direct calculation the commutativity of convolution in this space.

Numerous examples of symmetric pairs are provided by simple matrix Lie groups. Among those most often encountered we have

(c) $G = \mathrm{SL}(n; \boldsymbol{R})$, $\sigma(A) := A^{t-1}$. Then the subgroup of all fixed points for σ is $K = \mathrm{SO}(n)$. We have seen above that in the case of $n = 2$, G/K can be identified with the upper sheet of a hyperboloid in $\boldsymbol{R}^3$.

(d) $G = \mathrm{SL}(n; \boldsymbol{C})$ and $\sigma(A) := (\bar{A}^t)^{-1}$. Then $K = \mathrm{SU}(n)$.

(e) $G = \mathrm{Sp}(n; \boldsymbol{R})$ and $\sigma(A) := (A^t)^{-1}$. Then $K = \mathrm{Sp}(n; \boldsymbol{R}) \cap O(2n)$ and the symmetric space G/K can be identified with a space called the *upper Siegel half-plane*, i.e., with the set $\mathfrak{s}_n := \{A \in M_{n\times n}(\boldsymbol{C})$: A is symmetric and $\operatorname{Im} A$ is positive definite$\}$. For $n = 1$ we have $\mathrm{Sp}(1; \boldsymbol{R}) \simeq \mathrm{SL}(2; \boldsymbol{R})$, $K = \mathrm{SO}(2)$ and $\mathfrak{s}_n = H$—the upper half-plane in $\boldsymbol{C}^1$.

(f) $G = \mathrm{SO}(n)$, $K = \mathrm{SO}(n-1)$. As usual, we imbed $\mathrm{SO}(n-1)$ in $\mathrm{SO}(n)$ via the mapping

$$\mathrm{SO}(n-1) \ni A \to \left[\begin{array}{ccc|c} & & & 0 \\ & A & & \vdots \\ & & & 0 \\ \hline 0 & \dots & 0 & 1 \end{array}\right] \in \mathrm{SO}(n).$$

The reader will verify that σ defined as $\sigma(B) = SBS^{-1}$ with $S = \begin{bmatrix} -1 & 0 & \dots & 0 \\ 0 & \ddots & & \vdots \\ \vdots & & -1 & 0 \\ 0 & \dots & & 1 \end{bmatrix}$ has the desired properties, and thus the pair $(\mathrm{SO}(n), \mathrm{SO}(n-1))$ for $n \geqslant 2$ is a symmetric pair.

5.4. DIFFERENTIABILITY OF SPHERICAL FUNCTIONS ON LIE GROUPS

Throughout this section G is a Lie group and K a compact subgroup of G.

To begin with we shall prove

LEMMA 5.4.1. *Let $w \neq 0$ be a zonal spherical function on G and ϕ a spherical function associated with w. Then ϕ, $w \in \mathscr{E}(G)$.*

Proof. It follows from the proof of Lemma 5.1.7 that for a zonal spherical function $w \neq 0$ there exists a $\varphi \in \mathscr{C}_0(K\backslash G/K)$ such that $\langle \varphi, w\rangle \neq 0$. Since $\mathscr{D}(G) \subset \mathscr{C}_0(G)$ is a dense subset of $\mathscr{C}_0(G)$, we deduce that there exists an $f \in \mathscr{D}(K\backslash G/K) := \mathscr{D}(G) \cap \mathscr{C}_0(K\backslash G/K)$ such that $\langle f, w\rangle \neq 0$. Consequently,

$$\Phi = \langle f, w\rangle^{-1} \Phi * f.$$

Thus Φ is a convolution of a compactly supported smooth function with a continuous function and hence belongs to $\mathscr{E}(G)$. □

In the sequel Φ and w are regarded as functions on G/K.

THEOREM 5.4.2. *Under the assumptions of Lemma* 5.4.1, Φ *and* w *are solutions of the differential equation.*

$$Df = Dw(o)f \text{ for an arbitrary differential operator } D \text{ on } G/K \text{ commuting with left translations.} \tag{5.4.1}$$

Proof. We write the spherical equation in the form

$$\int_K L_{k^{-1}g^{-1}} \Phi \, \mathrm{d}k = \Phi(g) w. \tag{5.4.2}$$

Next we act on both sides of (5.4.2) with an operator D; since D commutes with left translations, we obtain

$$D\left(\int_K L_{k^{-1}g^{-1}} \Phi \, \mathrm{d}k\right) = \int_K L_{k^{-1}g^{-1}}(D\Phi) \, \mathrm{d}k = \Phi(g) Dw.$$

Finally, upon specializing the argument to $eK = o$, it follows that

$$\int_K D\Phi(gkK) \, \mathrm{d}k = \Phi(gK) Dw(o)$$

that is

$$D\Phi(gK) = \Phi(gK) Dw(o).$$ □

Historically, zonal spherical functions were defined as K-invariant solutions of the system of equations (5.4.1). Here, we shall apply these equations in order to obtain various differential properties of special functions defined by means of relations with matrix elements of representations.

PROBLEMS

In the following the letters G and K will denote an arbitrary unimodular locally compact group and its compact subgroup, respectively, unless otherwise explicitly stated.

1. Let χ be the character of an irreducible unitary representation of K and $\alpha \in \hat{K}$ the class of irreducible representations with the character χ. Let $\xi = n(\alpha)\bar{\chi}$, where $n(\alpha)$ is the dimension of the representations which belong to the class α and let ${}^{\xi}\mathscr{C}_0^{\xi}(G)$ be the subspace of $\mathscr{C}_0(G)$ composed of all functions satisfying the identity

$$\xi * f * \xi(x) := \int_K \int_K \xi(k) f(k^{-1} x k_1) \xi(k_1^{-1}) \mathrm{d}k \mathrm{d}k_1$$
$$= f(x), \qquad x \in G.$$

The $*$ on the left-hand side of this identity denotes convolution with the measure on G determined by the function $\xi \in \mathscr{C}(K)$ (cf. § 1.6, Problem 13).

(a) Show that ${}^{\xi}\mathscr{C}_0^{\xi}(G)$ is a convolution subalgebra of the convolution algebra $\mathscr{C}_0(G)$, closed in the topology of compact convergence, and that the mapping $P_\xi\colon f \to \xi * f * \xi$ is a continuous idempotent operator (projection) $\mathscr{C}_0(G) \to {}^{\xi}\mathscr{C}_0^{\xi}(G)$.

(b) Verify that for $f_1, f_2 \in \mathscr{C}_0(G)$

$$P_\xi\big((P_\xi f_1) * f_2\big) = P_\xi f_1 * P_\xi f_2 .$$

(c) Show that if ξ_1, ξ_2 correspond to different classes of irreducible representations of K then

$$P_{\xi_1} P_{\xi_2} = P_{\xi_2} P_{\xi_1} = 0.$$

(d) Prove that P_ξ extends to an orthogonal projector on $L^2(G)$ and the image of P_ξ is a subspace bi-invariant under K such that, on that subspace, the restriction to K of the left regular representation is equivalent to a multiple of α and the restriction to K of the right regular representation is equivalent to $\hat{\alpha}$ ($\hat{\alpha}$ denoting the class of the representation contragredient to a representation from the class α).

2 (continued). Let (H, U) be a unitary representation of G and let α, ξ have the same meaning as above. Denote by H^α a maximal subspace of H which is $U(K)$-invariant and such that $(U|_K, H^\alpha)$ contains only representations from the class α. Let P^α be the orthogonal projection onto H^α.

(a) Show that H^α is invariant under the operators $U(f)$, where $f \in {}^{\xi}\mathscr{C}_0^{\xi}(G)$ and that $f \to U(f)|H^\alpha$ is a representation of the algebra ${}^{\xi}\mathscr{C}_0^{\xi}(G)$ on H^α.

(b) Suppose $H^\alpha \neq 0$. Show that if (H, U) is irreducible then so is the representation of the convolution algebra ${}^{\xi}\mathscr{C}_0^{\xi}(G)$ on H^α defined in (a). Assuming in addition that H^α is a cyclic subspace for (H, U) (i.e. $U(G)H^\alpha$ spans a subspace dense in H), prove the converse implication.

Hint. (b) If $W \subset H^\alpha$ is ${}^{\xi}\mathscr{C}_0^{\xi}(G)$-invariant and $x \in H^\alpha \cap W^\perp$, then x is orthogonal to the subspace $\{U(f)w: f \in \mathscr{C}_0(G),\ w \in W\}$. To prove the converse implication, show that for every G-invariant subspace $H_0 \subset H$, $P_\xi H_0 \neq 0$.

3. Prove the converse of Theorem 5.1.1, in the following formulation. Let (H, U) be a unitary representation of G such that there exists a cyclic vector $v \in H$ satisfying (5.1.1). If the function $u_{v,v}$ is a zonal $\varkappa$-spherical function, then the space of solutions of equation (5.1.1) is one-dimensional.

Hint. As in the proof of Theorem 5.2.1, show that $P_x U(x)v = u_{v,v}(x)v$.

4. Prove the converse of Theorem 5.1.7, i.e. that

(a) a function $\omega \in \mathscr{C}(K\backslash G/K)$ satisfying the equation

$$f * \omega = \lambda(f)\omega, \qquad f \in \mathscr{C}_0(K\backslash G/K)$$

and the normalization condition $\omega(e) = 1$ is a zonal spherical function.

(b) the function $\varphi \in \mathscr{C}(G/K)$ satisfying the identity

$$f * \varphi = \lambda(f)\omega, \qquad f \in \mathscr{C}_0(K\backslash G/K)$$

for a certain zonal spherical function ω (independent of f) is a spherical function.

Hint. (a) Consider the function $y \to \int_K \omega(xky)\,dk := \omega_x(y)$ for a fixed $x \in G$ and show that, for $f \in \mathscr{C}_0/(K\backslash G/K)$, $f * \omega_x = \omega(x) f * \omega$. Apply this identity to the function $P_1 Q_1 f_n$, where f_n is a Dirac sequence in $\mathscr{C}_0(G)$ and P_1, Q_1 are projections onto $\mathscr{C}(K\backslash G)$ and $\mathscr{C}(G/K)$, respectively.

5. Let $S \subset G$ be a closed subgroup such that $K \times S \ni (k, s) \to ks \in G$ is a homeomorphism. Let $\lambda\colon S \to C_*$ be a continuous homomorphism and $\mu\colon G \to C_*$ its extension obtained by putting $\mu(ks) = \lambda(s)$. Show that the function $\omega(g) = \int_K \mu(gk)\,dk$ is a zonal spherical function.

Hint. Cf. Example 1.

6. Let D be the unit disc in the complex plane and S^1 its boundary.

We denote by G the group SU(1, 1) and by K its subgroup composed of diagonal matrices.

(a) Show that (G, K) is a symmetric pair and the homogeneous space G/K is isomorphic to D, on which G acts by homographies.

(b) For a function f on D define a function $\hat{f}$ on G by the formula $\check{f}(g) = f(g \cdot 0)$.

Prove that a continuous function f on D is harmonic if, and only if, $\hat{f}$ satisfies the equation

$$\hat{f}(g) - \int_K \hat{f}(gkx)\mathrm{d}x, \qquad x, g \in G,$$

i.e., if $\hat{f}$ is a spherical function associated with the function identically equal to 1.

(c) Let $P(z, b)$, $z \in D$, $b \in \boldsymbol{T}$ be the Poisson kernel for D and define $\hat{P}(g, b) = P(g \cdot 0, b)$.

Show that $\hat{P}$ is a K-multiplier on $G \times \boldsymbol{T}$ (G acts on $\boldsymbol{T}$ also by homographies). On the basis of (b) prove also that for an arbitrary function $\tilde{f}$ integrable with respect to the Lebesgue measure on $\boldsymbol{T}$, the function f defined by the Poisson integral

$$f(g \cdot 0) = \int_{S^1} \hat{P}(g, b)\tilde{f}(b)\mathrm{d}b$$

is a harmonic function.

Hint. (a) SU(1, 1) is isomorphic to SL(2, $\boldsymbol{R}$). (b) Harmonic functions on $\boldsymbol{C}$ are characterized by a mean value theorem.

7. Let G be the motion group of the plane $\boldsymbol{R}^2$. Show that by identifying G with the semi-direct product $\boldsymbol{T} \times_\tau \boldsymbol{C}$, where τ is a natural action of $\boldsymbol{T}$ on $\boldsymbol{C}$, every K-multiplier ($K = \boldsymbol{T}$) on $G/\boldsymbol{C}$ with values in the set of positive reals is of the form

$$\sigma_t\big((\zeta, z), \xi\big) = \exp\big(\mathrm{Re}(\bar{\zeta}\xi z t)\big)$$

for a certain $t \in \boldsymbol{C}$.

Hint. Show that if σ is such a multiplier then $\sigma\big((\zeta, z), \xi\big) = \sigma\big((1, \bar{\zeta}\xi z), 1\big)$ and the rule $z \to \sigma\big((1, z), 1\big)$ is a homomorphism of $\boldsymbol{C}$ into $\boldsymbol{R}_+$.

8. Suppose there exists an involutive anti-automorphism τ of the group G such that for certain $k_1, k_2 \in K$ (in general, depending on x) $\tau(x) = k_1 x k_2$.

(a) Show that (G, K) is a Gelfand pair.

(b) Check that this holds for the pairs $(\mathrm{SL}(2, \boldsymbol{R}), \mathrm{SO}(2))$ and $(\mathrm{SO}(3), \mathrm{SO}(2))$ if τ is the operation of taking the inverse element (in the latter case SO(2) is regarded as the subgroup of rotations arounds a fixed, say, the third, axis in $\boldsymbol{R}^3$).

Hint. (a) Observe that for $f \in \mathscr{C}_0(K\backslash G/K)$, $f(\tau(x)) = f(x)$ and utilize this observation to compute the convolution $f * g$ for $f, g \in \mathscr{C}_0(K\backslash G/K)$. (b) In the case of $G = \mathrm{SL}(2, \boldsymbol{R})$ apply the polar decomposition of a matrix and the diagonalization theorem in order to prove the decomposition $G = KDK$ where $D \subset G$ is the subgroup composed of all diagonal matrices with positive entries. As regards the case of $G = \mathrm{SO}(3)$ make use of the fact that every rotation is a rotation around an axis, to derive an analogous decomposition $G = KAK$, where $A \subset G$ is the subgroup of rotations around a fixed axis.

9. [1] Take $G = \mathrm{SO}(3)$ and let $K \simeq \mathrm{SO}(2)$ be the subgroup of all rotations about the third axis. We identify the quotient space G/K with the unit sphere $S^2 \subset \boldsymbol{R}^3$. By p we shall denote the north pole of the sphere, i.e. the point $(0, 0, 1)$. Spherical coordinates (θ, φ) are introduced according to the formula $x_1 = \cos\varphi \sin\theta$, $x_2 = \sin\varphi \sin\theta$, $x_3 = \cos\theta$ where $0 \leqslant \varphi \leqslant 2\pi$, $0 \leqslant \theta \leqslant \pi$. Recall that the invariant measure on S^2 is given in terms of these coordinates by the formula $\frac{1}{4\pi} \sin\theta \, d\theta \, d\varphi$ (Problem 5, § 1.5).

(a) Let $s(\theta, \varphi) \in S^2$ denote the point with spherical coordinates (θ, φ). For a function $F \in \mathscr{C}(K\backslash G/K)$ we define an f by the formula, $f(\cos\theta) = F(s(\theta, \varphi))$. Show that the mapping $F \to f$ sets up an isometric isomorphism of the space $L^1(K\backslash G/K)$ onto the space $L^1([0, \pi], \frac{1}{2} \sin\theta \, d\theta)$ and derive an explicit formula for the convolution of functions in $L^1(K\backslash G/K)$ in the form

$$f_1 * f_2(\cos\theta) = \frac{1}{4\pi} \int_0^{\pi} \sin x \, dx \int_0^{2\pi} f_1(\sin\theta \sin x \cos\varphi + \cos\theta \cos x) f_2(\cos x) \, d\varphi .$$

---

[1] In this problem and the next one we follow the method presented by H. Dym and H. P. Mc Kean in (1972).

(b) Apply the explicit expression for the convolution of functions in $L^1(K\backslash G/K)$ to obtain the identity $f_1 * f_2 = f_2 * f_1$ for $f_1, f_2 \in L^1(K\backslash G/K)$. Prove also that the convolution algebra $L^1(G/K)$ is non-commutative.

(c) Show that the function

$$p_n(\cos\theta) = \frac{1}{2\pi}\int_0^{2\pi} (i\sin\theta\cos\varphi + \cos\theta)^n \mathrm{d}\varphi$$

is a real polynomial in the variable $\cos\theta$ of degree n (precisely) satisfying the equation

$$\left((1-x^2)\frac{\mathrm{d}^2}{\mathrm{d}x^2} - 2x\frac{\mathrm{d}}{\mathrm{d}x} + n(n+1)\right) p_n(x) = 0.$$

The function $P_n \in \mathscr{C}(K\backslash G/K)$ corresponding to p_n is an eigenfunction of the Laplace operator Δ on the sphere, with the eigenvalue $-n(n+1)$. In terms of spherical coordinates the operator Δ has the form

$$\Delta = \frac{1}{\sin\theta}\frac{\partial}{\partial\theta}\left(\sin\theta\frac{\partial}{\partial\theta} + \frac{1}{\sin^2\theta}\frac{\partial^2}{\partial\varphi^2}\right).$$

(d) Show that the functions $\theta \to p_n(\cos\theta)$ constitute an orthogonal basis for the space $L^2[(0,\pi], \frac{1}{2}\sin\theta\,\mathrm{d}\theta)$ satisfying the condition $||p_n||^2 = \dfrac{1}{2n+1}$ and that the corresponding functions P_n on the sphere are eigenfunctions, unique up to a constant of the Laplace operator on the space $L^2(K\backslash G/K)$.

(e) Show that the functions P_n are zonal spherical functions and that every zonal spherical function is equal to one of the functions P_n. Check also that identity (5.1.4), reduces for the functions P_n to the equation given in Example 2.

Hint. (b) To prove non-commutativity of $L^1(G/K)$ note that for the function $x_1(\cdot)$ on the sphere (the first coordinate) and an arbitrary $f \in L^1(G/K)$ we have $f * x_1 = 0$ and compare this with the limit of the sequence $x_1 * f_n$ where f_n is a Dirac sequence in $L^1(G/K)$. (d) The Laplace operator Δ is symmetric. In precise terms this means that for arbitrary $f_1, f_2 \in \mathscr{E}(S^2)$ we have $(\Delta f_1|f_2) = (f_1|\Delta f_2)$

with respect to the inner product on $L^2(S^2)$ corresponding to the G-invariant measure. Then make use of the density of the algebra of polynomials in $L^2([-1, 1])$. (e) Check by using the invariance of Δ with respect to rotations that the function $g \to \int_K P_n(g_1 kg)dk$ is an eigenfunction for Δ and then apply (d) and the invariance property of this function with respect to both g_1 and g.

10 (continued). (a) Let (G, K) be a Gelfand pair where the group G is compact and let ω^α be a zonal spherical function corresponding to a representation (H^α, U^α) of dimension $n(\alpha)$. Prove the identity

$$(L_{g'}\omega^\alpha | L_g\omega^\alpha) = \frac{1}{n(\alpha)}\omega^\alpha(g^{-1}g),$$

where the inner product on the left is the inner product in $L(G/K)$.

(b) Let $\{f_k\}_{k=1}^n$ be an orthonormal basis for a subspace of $L^2(G/K)$ on which the left regular representation is equivalent to (H^α, U^α). Prove the addition formula

$$\sum_{k=1}^{n(\alpha)} \overline{f_k^\alpha(h)} f_k^\alpha(g) = n(\alpha)\omega^\alpha(h^{-1}g), \qquad g, h \in G.$$

In the remaining part of this problem (G, K) will denote the pair $(\mathrm{SO}(3), \mathrm{SO}(2))$ and we also retain the other notations of Problem 9.

(c) Let H^n be a closed subspace of $L^2(S^2)$ spanned by left translates of the function P_n. Show that H^n is the representation space of an irreducible representation of the group G and that $\dim H^n = 2n+1$.

(d) Let $\{Y_n^k\}_{k=1}^{2n+1}$ be an arbitrary orthonormal basis of H^n. Prove that Y_n^k are eigenfunctions of the Laplace operator with the eigenvalue (independent of k) equal to $-n(n+1)$.

(e) Prove that every function in H^n (such functions are called *spherical harmonics of weight n*) is a restriction to S^2 of a homogeneous polynomial of degree n satisfying the equation $\Delta_3 f = 0$, and conversely. Here Δ_3 stands for the Laplace operator:

$$\Delta_3 = \left(\frac{\partial}{\partial x^1}\right)^2 + \left(\frac{\partial}{\partial x^2}\right)^2 + \left(\frac{\partial}{\partial x^3}\right)^2.$$

(f) Define the functions

$$P_n^k(\cos\theta) = \frac{1}{2\pi}\int_0^{2\pi}(i\sin\theta\cos\varphi+\cos\theta)^n e^{-ik\varphi}d\varphi, \qquad k \in Z.$$

Prove that $P_n^k = 0$ for $|k| > n$.

Show that $Y_n^k(\theta,\varphi) := \sqrt{(2n+1)}\, P_n^k(\cos\theta)e^{ik\varphi}$ for $|k| \leqslant n$ constitutes an orthogonal basis for H^n.

11. Let (G,K) be the Gelfand pair $(SL(2,\boldsymbol{R}), SO(2))$ and let $S \subset \mathfrak{sl}(2,\boldsymbol{R})$ be the set of matrices of determinant 1 (cf. Example 3). Consider the action of G on $\mathfrak{sl}(2,\boldsymbol{R})$ by the conjugations

$$X \to gXg^{-1}, \qquad g \in G,\ X \in \mathfrak{sl}(2,\boldsymbol{R}).$$

(a) Prove that G acts on S transitively.

(b) Let $\boldsymbol{C}_+$ denote the cone in $\mathfrak{sl}(2,\boldsymbol{R})$ of matrices with the positive determinant. Prove that the Lebesgue measure on the cone $\boldsymbol{C}_+$ is invariant under the action of G and derive hence the form of the invariant measure on S given in Example 3.

Hint. (a) Every matrix in S can be mapped by means of conjugations onto X_1. (b) Introduce a parametrization of $\boldsymbol{C}_+$ such that one of the coordinates is $\det X$ and those remaining correspond to polar coordinates on S.

PART II

Special functions are introduced and investigated with the help of representation theory, according to the following schedule:

In a separable Hilbert space H, on which acts a continuous representation U of a group G, we fix an orthonormal basis $B = \{e_i\}_{i=0}$. In the general case, choice of the basis is determined by the properties of the group structure of G and of the representation U itself; for instance, it may be a basis consisting of eigenvectors U itself; for instance, it may be a basis consisting of eigenvectors of operators of a certain subgroup of G.

We construct a family $\mathscr{U} = \{u_{ij}\}_{i,j=0}$ of "special functions" on G, where

$$u_{ij}(g) := (e_i | U_g e_j).$$

(1) The formula $U(gh) = U(g)U(h)$ leads to the relation

$$u_{ij}(gh) = \sum_{i=0} u_{ik}(g) u_{kj}(h) \qquad \text{(Chapter 2).}$$

The relationships between the functions of the family $\mathscr{U}$ originating from this relation are called *addition formulas*.

(2) The mapping S_i, which to every vector $x \in H$ assigns a matrix element: the function $G \ni g \to (e_i | U_{g^{-1}} x) = u_{i,x}(g)$, intertwines the representation U with the left regular representation of G on $\mathscr{C}(G)$. Therefore the operator S_i also intertwines the differential of the representation U with the differential of the regular representation (§ 2.6). Hence follow the differential and the recurrence properties of matrix elements. Suppose for instance that the enveloping algebra $\mathfrak{G}$ of the Lie algebra of G contains an element X such that

$$\mathrm{d}U(X) e_j = e_{j+1}.$$

Then $\mathrm{d}L(X)\check{u}_{ij} = \check{u}_{i,j+1}$ is a recurrence differential equation.

(3) If the group G contains a compact subgroup K and forms a Gelfand pair with it, we choose the basis B in such a way that there is a K-fixed vector e_0 in it. Special functions of the form $u_{0j}(g) = (e_0 | U_g e_j)$

Class of functions	Group, subgroups, parametrization	Homogeneous spaces	Representations
1	2	3	4
1. Bessel functions	$G = M(2)$—the proper motion group of $\boldsymbol{R}^2 \simeq \boldsymbol{C}$. Parametrization: $T^1 \times \boldsymbol{C} \ni (\zeta, \xi) \to g(\zeta, \xi) \in G$ $g(\zeta, \xi) \cdot z := \zeta z + \xi, z \in \boldsymbol{C}$, K—the group of rotations about $z = 0$, A—the subgroup (normal) of all translations of the plane $A = \{g(\zeta, \xi) \colon \zeta = 1\}$	$G/K \simeq \boldsymbol{C}$ $G/A \simeq K$	Characters of the subgroup $A\colon \chi^\beta(g(1, z)) := e^{i\mathrm{Re}\bar{\beta}z}$; T^β—the representation of G induced by a character of the group A. In the multiplier form: the representation space $L^2(T^1, \mathrm{d}\varphi)$ $T^\beta_{g(\zeta, z)} f(\varphi) := e^{i\mathrm{Re}\beta\bar{z}\bar{\varphi}} f(\bar{\zeta}\varphi)$, $\varphi \in T$. For $\beta \neq 0$ the representation is irreducible; $T^\beta \cong T^{\beta'} \Leftrightarrow \lvert\beta\rvert = \lvert\beta'\rvert$
2. Jacobi and Legendre polynomials, associated Legendre functions	$G = \mathrm{SU}(2)$ Parametrization: Euler angles $(\varphi, \theta, \psi) \to g(\varphi, \theta, \psi)$ $:= \begin{bmatrix} \alpha & \beta \\ -\bar{\beta} & \bar{\alpha} \end{bmatrix}$, where $\alpha := \cos\frac{1}{2}\theta e^{\frac{1}{2}i(\varphi+\psi)}$, $\beta := i \sin\frac{1}{2}\theta e^{\frac{1}{2}i(\varphi-\psi)}$ and $0 \leqslant \varphi < 2\pi$, $0 < \theta < \pi$, $-2\pi \leqslant \psi < 2\pi$, $K := \{g \in G\colon \alpha = e^{\frac{1}{2}i\varphi}$, $\beta = 0\}$	$G/K = S^2$	(H^l, T^l), $l = 0, \frac{1}{2}, 1, \frac{3}{2}, \ldots$ H^l—the space of homogeneous complex polynomials on $\boldsymbol{C}^2$ of degree $2l$; $T^l_g w(z_1, z_2) := w(\alpha z_1 - \gamma z_2, \beta z_1 - \delta z_2)$. The representations T^l are irreducible and unitary

Special vectors	Basic formula	Definition of the family of special functions
5	6	7
Orthonormal basis for the representation space: $e_n(\alpha) := \alpha^n$, $\alpha \in T^1$. e_n is the eigenvector of the operators T_k, $k \in K$: $T^{\beta}_{g(\zeta,0)}\, e_n = \bar{\zeta}^n e_n$	$t^{\beta}_{mn}(g) := (e_m \vert T_g e_n)$ $= i^{m-n} \times e^{-i(n-m)\theta - in\psi}$ $\times J_{m-n}(\beta r)$ for $g = (e^{i\psi}, re^{i\theta})$	$J_n(z) := \frac{1}{2\pi}\int_0^{2\pi} e^{iz\sin\varphi - in\varphi}\,d\varphi$ $n \in \mathbf{Z}, z \in \mathbf{C}$.
Orthonormal basis for H^l: $e_n(z_1, z_2)$ $:= \frac{z_1^{l+n} z_2^{l-n}}{[(l+n)!\,(l-n)!]^{1/2}}$, n runs over integers or half-integers: $-l \leqslant n \leqslant l$; $\{e^j\}$—the basis dual to $\{e_n\}$. e_n is an eigenvector of the operators T^l_k, $k \in K$: $T^l_k e_n = e^{in\varphi} e_n$ for $k = \begin{bmatrix} e^{\frac{1}{2}i\varphi} & 0 \\ 0 & e^{-\frac{1}{2}i\varphi} \end{bmatrix}$	$t^l_{mn}(g) := \langle e^m, T^l_g e_n \rangle$ $= e^{i(m\varphi + n\psi)} \times P^l_{mn}(\cos\theta)$ for $g = g(\varphi, \theta, \psi)$.	Jacobi polynomials: $P_k^{(\alpha,\beta)}(x)$ $:= \frac{(-1)^k}{2^k k!}(1-x)^{-\alpha}(1+x)^{-\beta}$ $\times \frac{d^k}{dx^k}[(1-x)^{\alpha+k}(1+x)^{\beta+k}]$ $k = 0, 1, 2, \ldots;\ \alpha, \beta \in C$. Associated Legendre functions $P^m_k(x)$ $:= \frac{2^m(k+m)!}{k!}(1-x^2)^{-\frac{m}{2}}$ $\times P^{(-m,-n)}_{k+m}(x)$ Legendre polynomials: $P_k := P_k^{(0,0)}$, $P^l_{ij}(x)$ $:= P^{(i-j,-i-j)}_{l+j}(x)$ $\times \left(\frac{(l+i)!(l-i)!}{(l+j)!(l-j)!}\right)^{\frac{1}{2}}$ $\times (-1)^{i-j}(1-x)^{\frac{1}{2}(i-j)}$ $\times (1+x)^{\frac{1}{2}(-i-j)}$

Class of functions	Group, subgroups, parametrization	Homogeneous spaces	Representations
1	2	3	4
3. Gegenbauer polynomials	$G = \mathrm{SO}(n)$—the group of real orthogonal unimodular $n \times n$ matrices. K—the subgroup of matrices of the form $g = \begin{bmatrix} U & 0 \\ 0 & 1 \end{bmatrix}$, where $U \in \mathrm{SO}(n-1)$.	$G/K = S^{n-1}$	(U^l, H^l), $l = 0, 1, 2, \ldots$ H^l—the space of homogeneous harmonic polynomials ($\Delta w = 0$) on $\boldsymbol{R}^n$. $U^l_g w(x) = w(g^{-1}x)$ for $w \in H^l$, $g \in G$, $x \in \boldsymbol{R}^n$.
Jacobi and Legendre functions	$G = \mathrm{SL}(2; \boldsymbol{R})$—the group of all real 2×2 matrices with determinant 1. K—the subgroup of all matrices of the form: $K(\varphi) = \begin{bmatrix} \cos\frac{1}{2}\varphi & \sin\frac{1}{2}\varphi \\ -\sin\frac{1}{2}\varphi & \cos\frac{1}{2}\varphi \end{bmatrix}$ A—the subgroup of all matrices of the form: $a(t) = \begin{bmatrix} e^{\frac{1}{2}t} & 0 \\ 0 & e^{\frac{1}{2}t} \end{bmatrix}$ N—the subgroup of all matrices of the form: $n(s) = \begin{bmatrix} 1 & s \\ 0 & 1 \end{bmatrix}$, $M := \left\{ \begin{bmatrix} 1 & 0 \\ 0 & 1 \end{bmatrix}, \begin{bmatrix} -1 & 0 \\ 0 & -1 \end{bmatrix} \right\}$ $A_1 := MAN$. Parametrization: $(\varphi, \tau, \psi) \to g(\varphi, \tau, \psi) := k(\varphi + \frac{1}{2}\pi) a(\tau) k \times (\psi - \frac{1}{2}\pi)$, $0 \leqslant \varphi < 2\pi$, $0 < \tau < \infty$, $-2\pi \leqslant \psi < 2\pi$.	$G/K = H_+$ —the upper half-plane $G/MAN \simeq K/M$—compactified real axis in C^1. G acts by homographies.	Parametrization of the characters of the group A_1: $\sigma = (\varepsilon, l)$, $\varepsilon = 0, 1$, $l \in C^1$. $\chi_\sigma\left(\begin{bmatrix} \alpha & s \\ 0 & \alpha \end{bmatrix}\right) := (\operatorname{sgn}\alpha)^\varepsilon \lvert\alpha\rvert^l$ (U^σ, H^σ)—the representation of G induced by the character χ_σ of the subgroup A_1. $H^{(0,l)}$—the space of functions in $L^2(K, d\varphi)$ such that $f(-k) = f(k)$. $H^{(1,l)}$—the space of functions in $L^2(K, d\varphi)$ such that $f(-k) = -f(k)$. $U^\sigma_{a(\tau)} f(k(\theta)) := (\cosh\tau - \sinh\tau\cos\theta)^{l-1} \operatorname{sgn}\varepsilon f(k)$, where $k = \begin{bmatrix} e^{-\frac{1}{2}\tau}\cos\frac{1}{2}\theta & e^{\frac{1}{2}\tau}\sin\frac{1}{2}\theta \\ -e^{\frac{1}{2}\tau}\sin\frac{1}{2}\theta & e^{\frac{1}{2}\tau}\cos\frac{1}{2}\theta \end{bmatrix} \times (\cosh\tau - \sinh\tau\cos\theta)^{-1/2}$; U^σ_k is an operator of the regular representation of the subgroup K.

Special vectors	Basic formula	Definition of the family of special functions
5	6	7
K—fixed vector in H^l: $w_0^l(x)$ $= \frac{l!\Gamma(n-2)}{\Gamma(l+n-2)} C_l^{\frac{n-2}{2}}(x_n)$ for $x = (x_1, \ldots, x_n)$	The spherical function of the representation $U^l: \omega^l(g) = (w_0^l \mid U_g^l w_0^l)$ $= w_0^l(g_{nn})$, g_{nn} denotes the n-th element in the n-th column in the matrix g	$C_l^p = \frac{\Gamma(2p+l)\Gamma(p+\frac{1}{2})}{\Gamma(2p)\Gamma(p+l+\frac{1}{2})}$ $\times P_l^{(p-\frac{1}{2}, p-\frac{1}{2})}$ $l = 0, 1, 2, \ldots$
Basis for H^σ: $e_n^\varepsilon(k(\varphi)) := e^{\frac{1}{2}i(2n+\varepsilon)\varphi}$, $-\infty < n < \infty$, e_n^ε is an eigenvector of each of the operators U_k^σ, $k \subset K$.	$u_{mn}^\sigma(g) := (e_m^\varepsilon \mid U_g^\sigma e_n^\varepsilon)$ $= e^{-i(m+\frac{1}{2}\varepsilon)\varphi - i(n+\frac{1}{2}\varepsilon)\psi}$ $\times i^{n-m}$ $\times \mathfrak{P}^{-\frac{l+1}{2}}_{m+\frac{1}{2}\varepsilon, n+\frac{1}{2}\varepsilon}(\cosh\tau)$; $u_m^{(0,l)}(g)$ $= e^{-i(m+\frac{1}{2}\varepsilon)\varphi} i^{-m}$ $\times \frac{\Gamma(\frac{1}{2}(1-l))}{\Gamma(m+\frac{1}{2}(1-l))}$ $\times \mathfrak{P}^m_{-\frac{1}{2}(l+1)}(\cosh\tau)$; $u_{00}^{(0,l)}(g) =$ $\mathfrak{P}_{-\frac{1}{2}(l+1)}(\cosh\tau)$	Jacobi functions: $\mathfrak{P}_{mn}^l(z)$ $= \frac{\Gamma(l+n+1)}{2^l(l+m+1)(n-m)!}$ $\times F\left(-l-m, n-l;\right.$ $\left. n-m+1; \frac{z-1}{z+1}\right)$ $l \in \mathbf{C}$; m, n—both integers or both half-integers. Associated Legendre functions: $\mathfrak{P}_l^n(z) = \frac{\Gamma(l+n+1)}{\Gamma(l-n\times 1)} n!$ $\times \left(\frac{z-1}{z+1}\right)^{\frac{1}{2}}$ $\times F((l+1, -l; n+1;$ $\frac{1}{2}(1-z))$. Legendre functions: $\mathfrak{P}_l(z) = \mathfrak{P}_l^0(z)$ $= F(l+1, -l; 1; \frac{1}{2}(1-z))$

Class of functions	Group, subgroups, parametrization	Homogeneous spaces	Representations
1	2	3	4
Laguerre polynomials	G—the group of all complex 3×3 matrices of the form: $g(z, x)$ $= \begin{bmatrix} 1 & \bar{z} & \frac{1}{2}\lvert z\rvert^2+ix \\ 0 & 1 & z \\ 0 & 0 & 1 \end{bmatrix}$ $z \in \boldsymbol{C}, x \in \boldsymbol{R}$. $Z = \{g(z, x): z = 0\}$.	$G/Z = C$	$(\mathscr{H}_\lambda, \mathscr{U}_\lambda), 0 < \lambda \in R$ $\mathscr{H}_\lambda$—the Hilbert spaces of functions on C such that: $\int_C \lvert f\rvert^2(z)\exp(-\lambda\lvert z\rvert^2)\mathrm{d}z \wedge \mathrm{d}\bar{z} < \infty$ $(U^\lambda_{g(z,x)}\varphi)(w) := \exp(\lambda(ix - \frac{1}{2}(w\vert w)-(w\vert z))\varphi(z+w)$.
Euler function	$G =$ "$ax+b$" the affine group of the real line $\boldsymbol{R}$ $g(a, b)\cdot x := ax+b$, $a \in \boldsymbol{R}_*, b \in \boldsymbol{R}, x \in \boldsymbol{R}$, $G_1 := \{g(a, b): a = 1\}$.		Parametrization of the space of characters of G_1: $\chi_\lambda(b) := \exp \lambda b, \lambda \in R^1$. Induced representation $U^{\chi\lambda}$ in the multiplier realization: the representation space—$\mathscr{D}(\boldsymbol{R}_*)$ $U^{\chi\lambda}_{g(a,b)}(x) = \exp \lambda \mathrm{b} x\varphi(ax)$ $\hat{U}^\lambda_g := \mathscr{M} U^{\chi\lambda}_g \mathscr{M}^{-1}$, where $\mathscr{M}$ is the Mellin transformation.

are spherical functions of the group G, and therefore they satisfy the spherical equation ((5.1.1))

$$\int_K u_{0j}(gkh)\mathrm{d}k = u_{00}(g)u_{0j}(h).$$

In the examples which we are studying, the orbits of the group K are spheres in the Riemannian spaces G/K and the above formula can be understood as a mean value theorem for a spherical function.

(4) In the case where the matrix elements of the irreducible representation U, which is being examined, are square integrable (in particular if G is a compact group) the orthonormality of the basis B implies the orthogonality of the family of functions $\mathscr{U}$. In particular, systems of

Special vectors	Basic formula	Definition of the family of special function
5	6	7
Basis for $\mathscr{H}_\lambda$: $e_m(w) := \left(\frac{\lambda^m}{m!}\right)^{\frac{1}{2}} w^m$, $m = 0, 1, 2, \ldots$	$u^\lambda_{nm}(g) = (e_n \mid \mathscr{U}^\lambda_g e_m)$ $= \left(\frac{n!}{m!}\right)^{\frac{1}{2}} \exp(i\lambda x$ $- \frac{1}{2}\lambda\lvert z\rvert^2)\lambda^{\frac{1}{2}(m-n)}$ $\times z^{m-n} L_n^{m-n}(\lambda\lvert z\rvert^2)$, if $g = g(z, x)$.	$L_n^\alpha(x) = e^x \frac{x^{-\alpha}}{n!}$ $\times \frac{d^n x}{dx^n} (e^{-x} x^{n+\alpha})$, $\alpha > -1$, $n = 0, 1, 2, \ldots$
	$\hat{U}^\lambda_{g(1,b)} \varphi(p)$ $= \frac{1}{2\pi} \int_{-i\infty}^{i\infty} \varphi(z)\Gamma(p-z)$ $\times (-\lambda b)^{z-p} dz$ $\operatorname{Re} p > 0$, $b > 0$, $\lambda < 0$.	$\Gamma(z) = \int_0^\infty e^{-t} t^{z-1} dt$, $\operatorname{Re} z > 0$.

orthonormal polynomials of Jacobi, Gegenbauer, Laguerre appear in this way.

(5) The theory of decomposition of representations enters into the theory of special functions when we consider families of representations $\mathscr{U}$ which appear in decompositions of representations of the groups $L^2(G)$ or $L^2(G/K)$. The abstract Fourier transformation associated with such decompositions, takes the form of an integral transformation with a kernel, expressable in terms of the special functions of the family $\mathscr{U}$. The Plancherel formula and the inversion formula take the form of expansion of functions with respect to the family of special functions.

The above schedule derives from the papers of E. Cartan, E. Wigner and I. M. Gelfand and was formulated in a uniform way in the monographs of N. J. Vilenkin, 1965, J. D. Taleman, 1968, and W. Miller jr., 1968.

Chapters 7–12 of the present book are mostly based on the N. J. Vilenkin's book and articles. We have only endeavoured to make more natural the transition from the problems of representation theory to harmonic analysis on homogeneous spaces. In the proofs we often refer to the methods applied in the general theory, which is presented in Part III of the book. In solving detailed problems in representation theory, such as the irreducibility of certain representations, we do not try to apply a uniform method; we use this opportunity to present a variety of techniques, at least in examples.

Chapters 6 and 13 do not follow the general schedule outlined above. In Chapter 6 the Γ-function is investigated by means of representation theory but in a way different from that described above.

In Chapter 13 elements of the theory of the hypergeometric function are formulated in a completely traditional way. Consequently, prominence is given to the properties of special functions as solutions of differential equations which in an approach based on the theory of representations are pushed to the background. The tables preceding Chapter 6 represent items of the schedule described in cases of the consecutive classes of special functions. For each of these classes we give a group, used to analyse the family of special functions, a set of representations, systems of vectors with respect to which matrix elements are formed and the final form of the matrix elements expressed in terms of special functions.

Chapter 6

The Euler Γ- and B-functions

6.1. DEFINITION OF THE Γ-FUNCTION

We first define the Γ function on the complex half-plane by the integral with a parameter:

$$\Gamma(z) = \int_0^\infty e^{-t} t^{z-1} dt, \qquad \operatorname{Re} z > 0. \tag{6.1.1}$$

Assuming that $0 < a \leqslant \operatorname{Re} z \leqslant b$, we have the following estimates

$$|t^{z-1}| \leqslant t^{a-1} \qquad \text{on the interval } 0 < t \leqslant 1$$

and

$$|t^{z-1}| \leqslant t^{b-1} \qquad \text{on the half axis } t \geqslant 1.$$

The integral defining the Γ-function is thereby uniformly convergent for $z \in [a, b]$. By applying the theorems on the continuity and differentiability of integrals with a parameter we conclude that the function Γ is holomorphic on the whole of the domain $\operatorname{Re} z > 0$. Clearly

$$\Gamma(1) = 1. \tag{6.1.2}$$

For an arbitrary z in the domain we can perform integration by parts in the integral defining $\Gamma(z+1)$ and obtain

$$\int_0^\infty e^{-t} t^z dt = z \int_0^\infty e^{-t} t^{z-1} dt.$$

Hence we get the relation

$$\Gamma(z+1) = z\Gamma(z). \tag{6.1.3}$$

Taking into account formula (6.1.3), we obtain

$$\Gamma(n) = (n-1)! \qquad \text{for } n = 1, 2, \ldots \tag{6.1.4}$$

Thus the Γ-function is a holomorphic extension of the factorial. Moreover thanks to formula (6.1.3) we can define the value of the Γ-function for $z \in C - \{0, -1, -2, \ldots\}$.

We show by induction that

$$\Gamma(z) = \frac{\Gamma(z+n)}{z(z+1)\dots(z+n-1)}. \tag{6.1.5}$$

Since for a fixed $z \neq 0, -1, -2, \dots$ we may take $n > -\text{Re}\, z$, we can regard this formula as a definition of the value of $\Gamma(z)$ for that z. One can easily examine the behaviour of the function Γ in a neighbourhood of the points at which it is not defined. In particular, formula (6.1.3), implies that inside the punctured disc $0 < |z| < 1$ the function $z\,\Gamma(z)$ is holomorphic and has the limit at zero equal to 1. Therefore, by Riemann's theorem, it can be extended to a holomorphic function on the disc $|z| < 1$. Consequently, the function Γ has a simple pole at zero of order 1 with the residuum equal to one.

Accordingly, by writing formula (6.1.5) in the form

$$\Gamma(z-n) = \frac{\Gamma(z+1)}{z(z-1)\dots(z-n)}$$

we verify in the same manner that the function $z \to z\,\Gamma(z-n)$ has the limit $\dfrac{(-1)^n}{n!}$ at the point $n \in N$, which means that Γ has a simple pole at every point $z = -n$, $n \in N$, with the residuum equal to $\dfrac{(-1)^n}{n!}$.

We shall now examine the bahaviour of the function Γ restricted to the lines of the form $c + i\boldsymbol{R}$, $c > 0$. For this purpose we make a substitution in the integral (6.1.1) by putting $z = c+ip$, and $t = e^x$. We get

$$\Gamma(c+ip) = \int_{-\infty}^{+\infty} e^{-e^x+cx} e^{ixp}\, dx. \tag{6.1.6}$$

This formula represents the function $p \to \Gamma(c+ip)$ as the Fourier transform of the function $x \to e^{-e^x+cx}$, which is of class $\mathscr{S}(\boldsymbol{R})$. Hence, in virtue of Corollary 3.2.5, we have

LEMMA 6.1.1. *The function* $\boldsymbol{R} \ni p \to \Gamma(c+ip)$ *is of class* $\mathscr{S}(\boldsymbol{R})$.

We shall also prove

LEMMA 6.1.2. *If* Re $w > 0$ *and* Re $u > 0$, *then*

$$\int_0^\infty x^{z-1} e^{-xu} dx = \Gamma(z) u^{-z}. \tag{6.1.7}$$

Proof. If $u > 0$, then we obtain the formula by substituting $xu = v$.

For an arbitrary u in the half-plane Re $u > 0$ we write $u = re^{i\varphi}$ with $r > 0$, $-\pi/2 < \varphi < \pi/2$. We again make the substitution $xu = v$: $\int_0^\infty x^{z-1} e^{-xu} dx = u^{-z} \int_L e^{-v} v^{z-1} dv$, where L denotes a line inclined at an angle φ towards the real line. In order to compute the latter integral we make use of the contour W in Fig. 4.

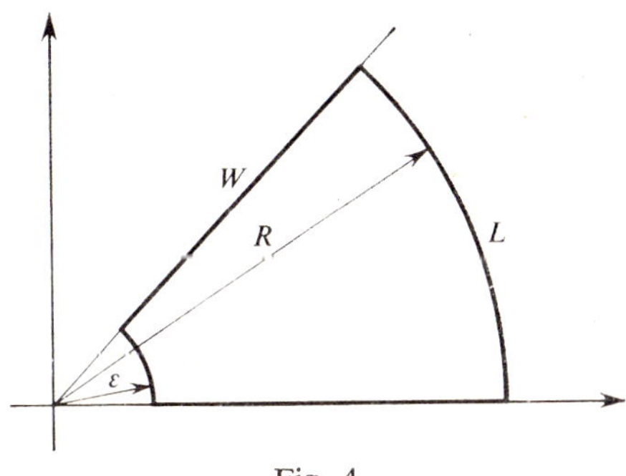

Fig. 4

The integral being non-singular inside the contour W, we have

$$\int_W e^{-v} v^{z-1} dv = 0.$$

Now, the assumption Re $z > 0$ ensures that for $\varepsilon \to 0$ the integral over the small arc tends to zero. On the other hand, the exponential factor occurring in the integrand, guarantees that the integral over the large arc tends to zero if $R \to \infty$. Consequently, we obtain the identity

$$\int_L e^{-v} v^{z-1} dv = \int_0^\infty e^{-x} x^{z-1} dx = \Gamma(z),$$

which immediately yields the desired result. □

6.2. THE FOURIER TRANSFORMATION AND THE MELLIN TRANSFORMATION

To the field $\boldsymbol{R}$ of real numbers there correspond two commutative groups: the additive group $\boldsymbol{R}^1$ and the multiplicative group $\boldsymbol{R}_+$ of positive numbers.

A function f of class $\mathscr{C}_0(\boldsymbol{R})$ with support contained in $\boldsymbol{R}_+$ can be submitted to the Fourier decomposition when regarded as an element of $L^2(\boldsymbol{R})$, and to the Mellin transformation when treated as a vector of the space $L^2\left(\boldsymbol{R}_+, \dfrac{\mathrm{d}t}{t}\right)$.

We shall investigate the connection between the transforms $\mathscr{M}f$ and $\hat{f}$. To this end we apply the inversion formula for the Fourier transformation in the form

$$f(x) = \frac{1}{2\pi} \int\limits_{ci+R} \hat{f}(w)\mathrm{e}^{-ixw}\mathrm{d}w, \tag{6.2.1}$$

where c is a fixed real number. Next we find

$$\mathscr{M}f(z) = \int\limits_0^\infty f(x)x^{z-1}\mathrm{d}x = \frac{1}{2\pi}\int\limits_0^\infty x^{z-1} \int\limits_{ci+R} \hat{f}(w)\mathrm{e}^{-ixw}\mathrm{d}w. \tag{6.2.2}$$

The order of integration can be interchanged provided $\operatorname{Re} z > 0$ and $c < 0$. Consequently

$$\mathscr{M}f(z) = \frac{1}{2\pi} \int\limits_{ci+R} \hat{f}(w)\int\limits_0^\infty x^{z-1}\mathrm{e}^{-ixw}\mathrm{d}x. \tag{6.2.3}$$

Replacing the inner integral by its value given by Lemma 6.1.2, we see that formula (6.2.3) takes the form

$$\begin{aligned} \mathscr{M}f(z) &= \frac{\Gamma(z)}{2\pi} \int\limits_{ci+R} \hat{f}(w)(iw)^{-z}\mathrm{d}w \\ &= \frac{\Gamma(z)}{2\pi}\mathrm{e}^{-i\frac{\pi}{2}z} \int\limits_{ci+R} \hat{f}(w)w^{-z}\mathrm{d}w \end{aligned} \tag{6.2.4}$$

for $c < 0$ and $\operatorname{Re} z > 0$.

Now the Paley–Wiener theorem supplies us with the information that the restriction of the function f, to an arbitrary line $ci+R$ is a function of class $\mathscr{S}(\boldsymbol{R})$, depending uniformly upon the parameter c. This allows us to pass to the limit with $c \to 0$ in the integral (6.2.4). As a result we get the formula

$$\mathscr{M}f(z) = \frac{\Gamma(z)}{2\pi} e^{-i\frac{\pi}{2}z} \int_{-\infty}^{+\infty} f(x)x^{-z}\mathrm{d}x \qquad \text{for } \operatorname{Re} z > 0. \qquad (6.2.5)$$

Hence we observe that the mapping $f \to \mathscr{M}f$ is an integral operator with the kernel

$$G(z, w) = \frac{\Gamma(z)}{2\pi} e^{-i\frac{\pi}{2}z} w^{-z}.$$

Formula (6.2.4) can also be regarded as an expression for the Fourier transform of the distribution

$$\tau_+^{z-1}(t) := \begin{cases} t^{z-1}, & t > 0, \\ 0, & t \leqslant 0. \end{cases}$$

Namely, we have the identity

$$\mathscr{F}\tau_+^{z-1}(x) = \Gamma(z)x^{-z}e^{-i\frac{\pi}{2}z}.$$

Accordingly, for the distribution

$$\tau_-^{z-1}x = \begin{cases} 0, & x \geqslant 0, \\ |x|^{z-1}, & x < 0 \end{cases}$$

we have

$$\mathscr{F}\tau_-^{z-1}(x) = \Gamma(z)x^{-z}e^{i\frac{\pi}{2}z}.$$

6.3. THE REFLEXION FORMULA FOR THE Γ-FUNCTION

By the *reflection formula for the Γ-function* we mean the relation

$$\Gamma(z)\Gamma(1-z) = \frac{\pi}{\sin \pi z}. \qquad (6.3.1)$$

In order to derive this formula we shall represent the Γ-function in a new way, namely as the kernel of an integral operator related to representations of the group "$ax+b$" of affine transformations of the real line.

We keep in mind that "$ax+b$" $= \boldsymbol{R}_+ \times_\nu \boldsymbol{R}$, ν denoting the action of the group $\boldsymbol{R}_+$ on $\boldsymbol{R}$ defined as

$$\nu(a)x = ax, \qquad a \in \boldsymbol{R}_+, \quad x \in \boldsymbol{R}.$$

Just as in the preceding section, the Γ-function describes the relationship between the additive and the multiplicative structures of the field of real numbers.

The space of equivalence classes of irreducible unitary representations of the group, "$ax+b$" is composed of a series, indexed by $\varrho \geqslant 0$ of one-dimensional representations defined as

$$U^{\varrho}_{(a,b)} = a^{\varrho}$$

and of two infinite-dimensional representations $U^+ = U^{\chi_1}$, $U^- = U^{\chi_2}$, which can be realized as representation induced by the characters $\chi_1(b) = \mathrm{e}^{ib}$, $\chi_2(b) = \mathrm{e}^{-ib}$ of the normal subgroup $B = (1, b)$ (isomorphic to $\boldsymbol{R}$).

The partition of $\hat{G}$ into three components corresponds to the fact that there exist, precisely, three orbits of the action of $\boldsymbol{R}_+$ on $\boldsymbol{R}$.

It will be useful to consider a wider class of induced (even nonunitary) representations U^{χ_λ} where $\chi_\lambda(b) = \mathrm{e}^{\lambda b}$.

The homogeneous space G/B is isomorphic to $A = \{(a, 0) \colon a \in \boldsymbol{R}_+\} \cong \boldsymbol{R}_+$. For a quasi-invariant measure on A we take the Haar measure, i.e. the measure $\frac{\mathrm{d}t}{t}$. In the multiplier realization the representation space of the differentiable induced representation U^{χ_λ} is the space $\mathscr{D}(\boldsymbol{R}_+)$ of infinitely differentiable compactly supported functions on $\boldsymbol{R}_+$.

The representation operators have the form

$$U_{(a,b)} f(x) = \mathrm{e}^{\lambda b/x} f(a^{-1}x). \tag{6.3.2}$$

To simplify computation we note that the mapping

$$\mathscr{D}(\boldsymbol{R}_+) \ni f \to \check{f} \in \mathscr{D}(\boldsymbol{R}_+)$$

defined as $\check{f}(x) = f\left(\frac{1}{x}\right)$ intertwines this representation with the representation

$$f \to \mathrm{e}^{\lambda bx} f(ax). \tag{6.3.3}$$

In what follows, by U^{χ_λ} we shall understand the representation defined by (6.3.3).

When restricted to the subgroup B, the representation U acts by diagonal operators: the operators of multiplication by the function $e^{\lambda bx}$. In the case where $\lambda \in i\boldsymbol{R}$, i.e. where U^{χ_λ} is unitary, we may write

$$U^{\chi_\lambda}|_B = \int_{\boldsymbol{R}_+} \chi_{\lambda x} \frac{dx}{x},$$

where $\int_{\boldsymbol{R}_+} \frac{dx}{x}$ is to be understood in the sense of a direct integral.

On the other hand, the representation $U^{\chi_\lambda}|_A$ is equivalent to the regular representation of the group $A = \boldsymbol{R}_+$. As we know, the Mellin transformation plays the role of the generalized Fourier transformation diagonalizing this representation:

$$\mathscr{M} U^{\chi_\lambda}|_A = \int_{\boldsymbol{R}_+} \eta_{it} \frac{dt}{t}.$$

We inted to find a realization of the induced representation having the property that the subgroup A acts by diagonal operators. Thus we are looking for operators $\hat{U}_g^\lambda$ on $L^2(\boldsymbol{R}^1)$ such that

$$\hat{U}_g^\lambda \mathscr{M} := \mathscr{M} U_g^{\chi_\lambda}, \qquad g \in G.$$

For this purpose we compute the right-hand side of this formula

$$\mathscr{M} U_{(a,b)}^{\chi_\lambda} f(p) = \int_{\boldsymbol{R}_+} x^{p-1} e^{\lambda bx} f(ax) dx. \tag{6.3.4}$$

In particular, for the elements of the subgroup B, we have

$$\mathscr{M} U_{(1,b)}^{\chi_\lambda} f(p) = \int_{\boldsymbol{R}_+} x^{p-1} e^{\lambda bx} f(x) dx. \tag{6.3.5}$$

Hence, by applying the inversion formula for the Mellin transformation, we arrive at

$$\mathscr{M} U_{(1,b)}^{\chi_\lambda} f(p) = \frac{1}{2\pi i} \int_{\boldsymbol{R}_+} \int_{c+i\boldsymbol{R}} x^{p-1} e^{\lambda bx} \mathscr{M} f(z) x^{-z} dz\, dx$$

for an arbitrary $c \in \boldsymbol{R}$.

For computational reason we confine ourselves to $b > 0$, $\operatorname{Re} \lambda < 0$,

Re$p > c$. This is enough to ensure that the double integral be absolutely convergent and the order of integration can therefore be reversed

$$\mathscr{M}U^{\chi_\lambda}_{(1,b)}f(p) = \frac{1}{2\pi i}\int\limits_{c+iR} \mathscr{M}f(z)\int\limits_0^\infty e^{\lambda bx}x^{p-z-1}\mathrm{d}x\mathrm{d}z. \tag{6.3.6}$$

Consequently, $\hat{U}_b$ turns out to be an integral operator with the kernel

$$k^\lambda(b,p,z) = \frac{1}{2\pi i}\int\limits_0^\infty e^{\lambda bx}x^{p-z-1}\mathrm{d}x = \frac{1}{2\pi i}(-\lambda b)^{z-p}\Gamma(p-z). \tag{6.3.7}$$

The action of the representation $\hat{U}^\lambda$ on functions which are the Mellin transforms of functions in $\mathscr{D}(\boldsymbol{R}_+)$ is therefore defined as

$$\hat{U}^\lambda_b\varphi(p) = \frac{1}{2\pi i}\int\limits_{c+iR}\varphi(z)(-\lambda b)^{z-p}\Gamma(p-z)\mathrm{d}z \tag{6.3.8}$$

for all values $b > 0$, Re $\lambda < 0$, Re$p > c$.

As a matter of fact, we are mostly concerned with the representations $\hat{U}^\lambda$ for purely imaginary values of λ where $\hat{U}^\lambda$ is a unitary irreducible representation of the group G. An expression for the representations $\hat{U}^\lambda$, for such λ, is obtained by means of an analytic extension of formula (6.3.8). To this end we observe that kernel (6.3.7), as a function in the variable $v = -\lambda b$, is analytic on the whole plane $\boldsymbol{C}^1$, except the negative real half-axis. The same property is also valid for the right-hand side formula (6.3.8). On the other hand, it follows from formula (6.3.4), which represents the same function, that $\hat{U}^\lambda_b\varphi$, as a function in v, is analytic on the whole plane. It follows that expression (6.3.8) is true for all v not lying on the negative half-axis provided Re$p > c$.

The reflection formula arises by expressing the property

$$\hat{U}^\lambda_{b+1} = \hat{U}^\lambda_b\hat{U}^\lambda_1 \tag{6.3.9}$$

in terms of the kernels of the operators $\hat{U}^\lambda_b$. It is enough to consider $\lambda = -1$. In view of formula (6.3.8) we get the identity

$$\int\limits_{c+iR}\varphi(z)(b+1)^{z-p}\Gamma(p-z)\mathrm{d}z$$

$$= \frac{1}{2\pi i}\int\limits_{c_1+iR}\Gamma(p-q)b^{q-p}\int\limits_{c+iR}\varphi(z)\Gamma(q-z)\mathrm{d}z\mathrm{d}q, \tag{6.3.10}$$

which is satisfied identically with respect to φ for Re$p > c_1 > c$.

Lemma 6.1.1 ensures that the double integral over $(c_1+i\boldsymbol{R})\times\times(c+i\boldsymbol{R})$ is absolutely convergent, so it is permissible to reverse the order of integration. Further, we know that the Mellin transforms of functions of class $\mathcal{D}(\boldsymbol{R}_+)$, when restricted to $c+i\boldsymbol{R}$, constitute a dense subset of the space $L^1(c+i\boldsymbol{R})$ defined relative to the measure transported from $\boldsymbol{R}$.

By formula (6.3.10) we thus obtain

$$b^p(b+1)^{z-p}\Gamma(p-z) = \frac{1}{2\pi i}\int_{c_1+i\boldsymbol{R}} \Gamma(p-q)\Gamma(q-z)b^q \mathrm{d}q$$

for $\mathrm{Re}\,p > c_1 > \mathrm{Re}\,z$.

In particular, upon putting $z = 0$ it follows that

$$\left(\frac{b}{b+1}\right)^p\Gamma(p) = \frac{1}{2\pi i}\int_{c_1+i\boldsymbol{R}} \Gamma(p-q)\Gamma(q)b^q \mathrm{d}q$$

for $\mathrm{Re}\,p > c_1 > 0$. The last formula means that the function $b \to \left(\dfrac{b}{b+1}\right)^p$ is the inverse Mellin retransform of the function $q \to \Gamma(p+q)\Gamma(-q)/\Gamma(p)$. Applying the Mellin transformation to both sides of the above formula, we therefore obtain

$$\int_0^\infty b^{p+q-1}(b+1)^{-p}\mathrm{d}b = \frac{\Gamma(p+q)\Gamma(-q)}{\Gamma(p)} \tag{6.3.11}$$

for $\mathrm{Re}(p+q) > 0$ and $\mathrm{Re}\,q < 0$. In particular, for $p = 1$ we get

$$\Gamma(1+q)\Gamma(-q) = \int_0^\infty b^q(1+b)^{-1}\mathrm{d}b$$

for $-1 < \mathrm{Re}\,q < 0$. The integral on the right-hand side can be computed effectively by applying the residue theorem. The final result is $\dfrac{\pi}{\sin \pi q}$. Utilizing this expression, we obtain the reflection formula

$$\Gamma(z)\Gamma(1-z) = \frac{\pi}{\sin \pi z} \tag{6.3.12}$$

for $0 < \mathrm{Re}\,z < 1$. Both sides of (6.3.12) being meromorphic functions on $\boldsymbol{C}$ with singularities at integer points, this implies, in view of the uniqueness of analytic extensions, that identity (6.3.12) extends to $\boldsymbol{C}-\boldsymbol{Z}$.

Formula (6.3.12) leads to the following identity, known as the *duplication formula*:

$$\Gamma(2z) = \frac{2^{2z-1}\Gamma(z)\Gamma\left(z+\frac{1}{2}\right)}{\sqrt{\pi}} \tag{6.3.13}$$

Proof. Formula (6.3.11) can be written as

$$\frac{\Gamma(z)\Gamma(v)}{\Gamma(z+v)} = \int_0^\infty t^{z-1}(1+t)^{-z-v}dt$$

and, after the substitution $s = \frac{t}{1+t}$,

$$\frac{\Gamma(z)\Gamma(v)}{\Gamma(z+v)} = \int_0^1 t^{z-1}(1-t)^{v-1}dt. \tag{6.3.14}$$

For $z = v$ this gives

$$\frac{\Gamma^2(z)}{\Gamma(2z)} = \int_0^1 t^{z-1}(1-t)^{z-1}dt = \int_{-\frac{1}{2}}^{\frac{1}{2}} \left(\frac{1}{4}-u^2\right)^{z-1} du$$

$$= 2\int_0^{\frac{1}{2}} \left(\frac{1}{4}-u^2\right)^{z-1} du.$$

Finally on substituting $4u^2 = t$, we obtain in view of (6.3.14)

$$\frac{\Gamma^2(z)}{\Gamma(2z)} = 2^{-2z+1}\int_0^1 (1-t)^{z-1}t^{-\frac{1}{2}}\,dt = 2^{-2z+1}\frac{\Gamma(z)\Gamma\left(\frac{1}{2}\right)}{\Gamma\left(z+\frac{1}{2}\right)}.$$

The proof of (6.3.13) is therefore complete since $\Gamma(\frac{1}{2}) = \sqrt{\pi}$ by (6.3.11). □

The function $(z, v) \to \dfrac{\Gamma(z)\Gamma(v)}{\Gamma(z+v)}$ is called *Euler's beta function* and denoted by $B(z, v)$. From formula (6.3.14) we see that

$$B(z, v) = \int_0^1 t^{z-1}(1-t)^{v-1}\mathrm{d}t. \tag{6.3.15}$$

6.4. THE RIEMANN ζ-FUNCTION

The ζ-function is defined on the half-plane $\operatorname{Re} z > 1$ by means of the series

$$\zeta(z) = \sum_{n=1}^{\infty} \frac{1}{n^z}. \tag{6.4.1}$$

The series is absolutely and almost uniformly convergent on the half-plane, and so the ζ-function is well defined and holomorphic.

A connection between the ζ-function and the Γ-function can be established by utilizing formula (6.1.7). Since

$$\frac{1}{n^z} = \frac{1}{\Gamma(z)} \int_0^{\infty} x^{z-1} \mathrm{e}^{-xn} \mathrm{d}x,$$

and for $\operatorname{Re} z > 1$ we can interchange the order of summation and integration:

$$\zeta(z) = \frac{1}{\Gamma(z)} \int_0^{\infty} \frac{x^{z-1}}{\mathrm{e}^x - 1} \mathrm{d}x. \tag{6.4.2}$$

Thus the ζ-function is the quotient of the Mellin transform of the function $(\mathrm{e}^x - 1)^{-1}$ by the Γ-function.

Finally, we shall derive the following number-theoretic expression for the ζ-function:

$$\zeta(z) \prod_{p \in \Omega} (1 - p^{-z}) = 1, \qquad \operatorname{Re} z > 1, \tag{6.4.3}$$

where Ω is the set of prime numbers.

We first note that

$$\zeta(z) - \frac{1}{2^z}\zeta(z) = 1 + \frac{1}{3^z} + \frac{1}{5^z} + \ldots$$

Accordingly,

$$\zeta(z)\left(1-\frac{1}{2^z}\right)\left(1-\frac{1}{3^z}\right)=1+\sum\frac{1}{n^z},$$

where the sum is taken over all integers which are not multiples of 2 and 3. Inductively we prove the formula

$$\zeta(z)\prod_{p\in\Omega_n}(1-p^{-z}) = 1+\sum\frac{1}{s^z},$$

where Ω_n is the set of the n initial prime numbers and the sum ranges over the set of all integers, which are not multiples of the elements of Ω_n.

Formula (6.4.3) follows upon passing to the limit with $n \to \infty$ since the sum on the right-hand side tends to zero.

PROBLEMS

1. (a) Prove the identity (for a fixed z such that $\operatorname{Re} z > 0$)

$$\Gamma(z) = \lim_{n\to\infty}\int_0^n \left(1-\frac{t}{n}\right)^n t^{z-1}\mathrm{d}t$$

and deduce hence the Euler limit formula

$$\Gamma(z) = \lim_{n\to\infty}\frac{n!n^z}{z(z+1)(z+2)\ldots(z+n)}.$$

(b) With the help of the recurrence formula (6.1.3) extend the validity of the Euler limit formula to $z \neq 0, -1, \ldots$

(c) Applying the Euler limit formula, give an independent proof of the reflection formula.

Hint. (c) Make use of the representation of the function $\sin\pi z$ in the form of the infinite product

$$\sin\pi z = \pi z\prod_{n=1}^{\infty}\left(1-\frac{z^2}{n^2}\right).$$

2. Prove that the formula

$$\Gamma(z) = \int_1^{\infty} t^{z-1}\mathrm{e}^{-t}\mathrm{d}t+\sum_{n=0}^{\infty}\frac{1}{n!}\,\frac{(-1)^n}{n+z}$$

defines the function Γ in the complement of the set $\{0, -1, -2, \ldots\}$.

3. For $\varepsilon > 0$ denote by C_ε the contour in the complex plane consisting of the interval $[\varepsilon, \infty]$, run over twice in opposite directions, and of a positively oriented circle $|z| = \varepsilon$.

(a) Show

$$\Gamma(z) = \frac{1}{e^{2\pi i z} - 1} \int_{C_\varepsilon} e^{-\zeta} \zeta^{z-1} d\zeta, \qquad z \in \mathbf{C}.$$

(b) Let D_ε be the contour obtained by rotating through the angle $-\pi$ the contour C_ε. Prove the Hankel integral representation

$$\frac{1}{\Gamma(z)} = \frac{1}{2\pi i} \int_{D_\varepsilon} e^{\zeta} \zeta^{-z} d\zeta.$$

Hint. (a) Consider the function $\zeta \to \zeta^{z-1}$ in the plane with a slit along the positive half-axis ($0 < \arg\zeta < 2\pi$). Observe that the integral on the right defines a holomorphic function of the argument z on the whole of the plane, which is zero when z is a positive integer. (b) Make a change of variable in the formula in (a) and use the reflection formula. In both cases it is convenient to restrict at first the range of z and then apply the rule of analytic continuation.

4. Show that for real x

(a) $$|\Gamma(ix)|^2 = \frac{\pi}{x \sinh \pi x}, \qquad x \neq 0,$$

(b) $$\left|\Gamma\left(\frac{1}{2} + ix\right)\right|^2 = \frac{\pi}{\cosh \pi x}.$$

5. What are the complex values z such that the following formulas hold

$$\int_0^\infty y^{z-1} \cos y \, dy = \cos \frac{\pi z}{2} \Gamma(z),$$

$$\int_0^\infty y^{z-1} \sin y \, dy = \sin \frac{\pi z}{2} \Gamma(z)?$$

6. Prove that for positive integers n

$$\Gamma\left(n + \frac{1}{2}\right) = \frac{(2n)!}{2^{2n} n!} \sqrt{\pi}.$$

7. Prove the formulas

(a) $B(u,v) = 2\int\limits_0^{\pi/2} \cos^{2u-1}\varphi \sin^{2v-1}\varphi \, d\varphi$,

(b) $B(u,v) = \int\limits_0^{\infty} \frac{x^{u-1}}{(1+x)^{u+v}} \, dx$,

(c) $B(u,v+1) = \frac{v}{u} B(u+1,v)$,

(d) $B\left(\frac{u+1}{2}, v-\frac{u+1}{2}\right) = 2\int\limits_0^{\infty} \frac{x^u}{(1+x^2)^v} \, dx$.

8. Compute the integrals

(a) $\int\limits_0^1 t^{u-1}(1-t)^{v-1}/(t+a)^{u+v} dt, \qquad a > 0$,

(b) $\int\limits_0^{\pi/2} \sin^{2k+1/2}\varphi \, d\varphi \qquad$ for $k \in \mathbf{Z}_+$

9. Let C be a closed contour on the complex plane, going twice around the points 1 and 0, first in the positive and then in the negative direction (such a contour, called a double loop around the points 1 and 0, is presented in Fig. 10). Show that

$$B(x,y) = \frac{1}{(1-e^{2\pi i x})(1-e^{2\pi i y})} \int\limits_C \zeta^{x-1}(1-\zeta)^{y-1} d\zeta .$$

Hint. Replace the loop C by a contour consisting of circular sectors and straight segments. Compute the integral under the assumption $\operatorname{Re} x > 0$, $\operatorname{Re} y > 0$ and apply the analytic continuation theorem. (A more general integral representation of this type is given in § 13.2.)

10. Prove that for $\operatorname{Re} z > 0$

$$(1-2^{1-z})\zeta(z) = \sum_{n=1}^{\infty} (1-)^{n+1} \frac{1}{n^z} = \frac{1}{\Gamma(z)} \int\limits_0^{\infty} \frac{x^{z-1}}{e^x+1} dx .$$

11. Let G be the group "$ax+b$", and A and B the subgroups of dilatations and translations, respectively. Let $\hat{B}$ be the group of unita-

ry characters of B and let $\hat{\nu}$ be the action of A on $\hat{B}$ given as $(\hat{\nu}(a)\chi)(b) = \chi(\nu(a^{-1})b)$.

(a) Show that $\hat{B}$ is the sum of three disjoint orbits of the group A. Under the isomorphism of $\hat{B}$ onto $\boldsymbol{R}^1$ given by the rule $\lambda \to \chi_{i\lambda}$, where $\chi_{i\lambda}(b) = e^{i\lambda b}$, those orbits are the following: 0, positive half-axis and negative half-axis.

(b) Prove that the unitary representations induced by the characters χ_λ, χ_μ of the group B are equivalent if, and only if, χ_λ and χ_μ belong to the same orbit of the group A.

Hint. (b) To prove the equivalence of U^{χ_λ} and U^{χ_μ} consider the operator T, such that $T\varphi(x) := \varphi(tx)$ where $\lambda = t\mu$ and $t > 0$.

12 (continued). (a) Prove that the differential of the representation U^{χ_λ} satisfies

$$dU^{\chi_\lambda}(X_1)\varphi(x) = \lambda x\varphi(x), \qquad dU^{\chi_\lambda}(X_2)\varphi(x) = x\varphi'(x),$$

for an arbitrary function $\varphi \in \mathcal{D}(\boldsymbol{R}_+)$. Here X_1 and X_2 are the generators of the subgroup B and A, respectively.

(b) Prove that the representations U^{χ_λ} are irreducible.

Hint. (b) Prove first that the only operators on $L^2(\boldsymbol{R}_+)$ which commute with the operators of multiplication by functions are the operators of multiplication by functions (i.e. the algebra of operators of multiplication by functions is a maximal abelian algebra in $L^2(\boldsymbol{R}_+)$).

Chapter 7

Bessel Functions

7.1. THE GROUP OF RIGID MOTIONS OF $\boldsymbol{R}^2$

Consider the space $\boldsymbol{R}^2$ with its natural unitary structure

$$(x|y) = x_1y_1 + x_2y_2, \qquad x = (x_1, x_2), \qquad y = (y_1, y_2).$$

The space of all affine isometries of $\boldsymbol{R}^2$ is called the *motion group of* $\boldsymbol{R}^2$. The mappings which preserve the orientation of the space constitute a subgroup called the *proper motion group* and denoted by $M(2)$.

To simplify notation we take into account the complex structure of $\boldsymbol{R}^2$. Thus, by identifying $x = (x_1, x_2)$ with $z = x_1 + ix_2$ and $y = (y_1, y_2)$ with $\zeta = y_1 + iy_2$, we write

$$(x|y) = \operatorname{Re}\bar{z}\zeta.$$

A proper motion is described by a pair (ζ, ξ) with $|\zeta| = 1$, $\xi \in \boldsymbol{C}$, which acts as follows

$$gz = \zeta z + \xi.$$

Given another motion g' defined by (ζ', ξ'), we have

$$(g'g)z = g'(gz) = \zeta'\xi z + \zeta'\xi + \xi'.$$

Hence to the product $g'g$ corresponds the pair $(\zeta'\zeta, \zeta'\xi + \xi')$.

We have thus established an isomorphism of $M(2)$ with the semi-direct product $\boldsymbol{T} \times_\tau \boldsymbol{C}$ where $\tau_\zeta(\xi) := \zeta\xi$. $M(2)$ is a connected group of dimension 3; however it is not simply connected.

The pair $\big(M(2), \boldsymbol{T}\big)$ is symmetric. An involutive automorphism for which $\boldsymbol{T}$ is the fixed point set, can be defined as follows:

$$(\zeta, \xi) = (\zeta, -\xi).$$

Owing to the decomposition $(\xi, \zeta) = (1, \xi)(\zeta, 0)$ we can identify the space $M(2)/\boldsymbol{T}$ with $\boldsymbol{C}$. Moreover, the action of $M(2)$ on the quotient space transforms under that identification into the natural action on G.

The invariant measure on $M(2)$ is the product of the invariant measures on $\boldsymbol{T}$ and $\boldsymbol{R}^2$.

$$\int_{M(2)} f \mathrm{d}\mu = \frac{i}{4\pi} \int_0^{2\pi} \int_{\boldsymbol{C}} f(\mathrm{e}^{i\varphi}, \xi) \mathrm{d}\varphi \mathrm{d}\xi \wedge \mathrm{d}\bar{\xi}$$

$$= \frac{1}{2\pi} \int_0^{2\pi} \int_{\boldsymbol{R}} \int_{\boldsymbol{R}} f(\mathrm{e}^{i\varphi}; x, y) \mathrm{d}\varphi \mathrm{d}x \mathrm{d}y. \tag{7.1.1}$$

This measure is actually bi-invariant. Hence $M(2)$ is a unimodular group. The Lebesgue measure on $\boldsymbol{R}^2$ is $M(2)$-invariant. The homogeneous space $M(2)/\boldsymbol{R}^2$ is isomorphic to $\boldsymbol{T}$ and also possesses an $M(2)$-invariant measure, namely the Haar measure for $\boldsymbol{T}$.

We shall now describe the Lie algebra of the group $M(2)$. For this purpose it is convenient to represent $M(2)$ as a subgroup of GL(2; $\boldsymbol{C}$) via the mapping

$$M(2) \ni (\zeta, \xi) \to \begin{bmatrix} \zeta & \xi \\ 0 & 1 \end{bmatrix} \in \mathrm{GL}(2; \boldsymbol{C})$$

The reader will verify that this is indeed a homomorphism. The algebra $\mathfrak{g}$ of the group $M(2)$ is then isomorphic to the algebra of the corresponding subgroup of GL(2; $\boldsymbol{C}$).

Define the following one-parameter subgroups:

$$a_1(t) = \begin{bmatrix} \mathrm{e}^{it} & 0 \\ 0 & 1 \end{bmatrix}, \qquad a_2(t) = \begin{bmatrix} 1 & t \\ 0 & 1 \end{bmatrix}, \qquad a_3(t) = \begin{bmatrix} 1 & it \\ 0 & 1 \end{bmatrix}.$$

The corresponding elements of the algebra $\mathfrak{gl}(2, \boldsymbol{C})$ are of the form

$$X_1 = \begin{bmatrix} i & 0 \\ 0 & 0 \end{bmatrix}, \qquad X_2 = \begin{bmatrix} 0 & 1 \\ 0 & 0 \end{bmatrix}, \qquad X_3 = \begin{bmatrix} 0 & i \\ 0 & 0 \end{bmatrix}.$$

Direct computation shows that the following commutation relations hold

$$[X_1, X_2] = X_3, \qquad [X_3, X_1] = X_2, \qquad [X_2, X_3] = 0. \tag{7.1.2}$$

7.2. SPHERICAL REPRESENTATIONS OF THE GROUP $M(2)$

The series of spherical representations of the group $M(2)$ is obtained by the operation of inducing from the one-dimensional representations of the normal subgroup $\boldsymbol{R}^2$. These representations will be realized in the multiplier form on the space $L^2(\boldsymbol{T})$ according to the technique worked out in §§ 2.2 and 2.4.

First of all we describe the characters of $\boldsymbol{R}^2$ (= $\boldsymbol{C}$):

$$\chi^\beta\colon z \to \mathrm{e}^{i \operatorname{Re} \beta \bar{z}}, \qquad \beta \in \boldsymbol{C}.$$

The transition from an induced representation in its usual form to the multiplier form is accomplished by means of the simplest imbedding of $\boldsymbol{T}$ into $M(2)$

$$S\colon \boldsymbol{T} \ni \alpha \to (\alpha, 0) \in M(2).$$

On the basis of § 2.4 we obtain the following expression for the induced representation U^{χ^β}:

$$\begin{aligned} T^\beta(g) f(\alpha) &:= \mathrm{e}^{i \operatorname{Re} \beta \bar{\alpha} \xi} f(\bar{\zeta}\alpha) \\ &=: \sigma_\beta(g^{-1}, \alpha) f(\bar{\zeta}\alpha), \qquad g = (\zeta, \xi). \end{aligned} \tag{7.2.1}$$

Thus the subgroup $\boldsymbol{T}$ acts on $L^2(\boldsymbol{T})$ as the left regular representation

$$T^\beta\big((\zeta, 0)\big) f(\alpha) = f(\bar{\zeta}\alpha).$$

On the other hand, the action of $\boldsymbol{R}^2$ has the form

$$T^\beta\big((1, \xi)\big) f(\alpha) = \mathrm{e}^{i \operatorname{Re} \bar{\beta} \bar{\alpha} \xi} f(\alpha).$$

The operator $T^\beta\big((1, \xi)\big)$ is a diagonal operator of multiplication by a continuous function. The representation space $L^2(\boldsymbol{T})$ admits a basis relative to which the operators $T^\beta\big((\zeta, 0)\big)$ are all diagonal. Obviously the basis consists of the characters of $\boldsymbol{T}$:

$$e_n(\alpha) := \alpha^n, \qquad n \in \boldsymbol{Z}.$$

Let us compute the matrix elements of the representations T^β with respect to this basis:

$$t_{mn}(g) = \big(e_m | T^\beta(g) e_n\big) = \int_{\boldsymbol{T}} \alpha^{n-m} \zeta^{-n} \mathrm{e}^{i \operatorname{Re} \bar{\beta} \bar{\alpha} \xi} \mathrm{d}\alpha. \tag{7.2.2}$$

for $g = (\zeta, \xi)$.

We immediately observe that owing to the invariance of the measure $\mathrm{d}\alpha$, the matrix element t^β_{00}, is identical with $t^{\beta^1}_{00}$ whenever $|\beta^1| = |\beta|$. Therefore the spherical functions corresponding to the representations T^β and T^{β^1} are identical in this case.

We shall utilize this observation in the proof of the following fact.

PROPOSITION 7.2.1. (1) *The representations* T^β *are irreducible for* $\beta \neq 0$.

(2) $T^\beta \cong T^{\beta'} \Leftrightarrow |\beta| = |\beta'|$.

Proof. (1) We shall prove that the representations T^β, $\beta \neq 0$ are spherical. Obviously, the function e_0 is the only (up to a constant factor) T-invariant vector in the representation space of T^β. Therefore it remains to prove that the T-invariant vector in $L^2(T)$ is cyclic. After calculating

$$T^\beta((1, \xi))e_0(\alpha) = e^{i \operatorname{Re} \beta \bar{\alpha} \xi}$$

we note that the set $T^\beta((1, \xi))e_0 \subset L^2(T)$ contains in particular all powers of the function $T \ni \alpha \to e^{\pm i \operatorname{Re} \beta \bar{\alpha}}$. Such functions span the algebra of continuous functions, and moreover, separate points in T if $\beta \neq 0$. It follows from the Stone theorem that this algebra is dense in the space $\mathscr{C}(T)$, which contains a basis of $L^2(T)$. The cyclicity of e_0 is therefore established. Thus it follows from Proposition 5.1.9 that the representations T^β, $\beta \neq 0$, are irreducible.

(2) $\Leftarrow$ The assertion follows by Theorem 5.1.4 since the representations T^β and $T^{|\beta|}$ have the same spherical function.

$\Rightarrow$ It can be assumed that $\beta, \beta' \in R_+$. When restricted to the subgroup T, all representations T^β are equivalent to the regular representation of T on $L^2(T)$. The unitary operator intertwining T^β with $T^{\beta'}$, is also an intertwining operator for the left regular representation of T. This operator is diagonal in the basis $\{e_n\}_{-\infty}^{\infty}$. So, if $U \in L_{M(2)}(T^\beta, T^{\beta'})$, then $Ue_m = \lambda_m e_m$, and the unitarity imposes the condition $|\lambda_m| = 1$. Then

$$\begin{aligned} t_{nn}^{\beta'}(g) &= (e_n | T^{\beta'}(g) e_n) = (Ue_n | UT^{\beta'}(g) e_n) = (Ue_n | T^\beta(g) Ue_n) \\ &= (e_n | T^\beta(g) e_n) = t_{nn}^\beta(g). \end{aligned}$$

In particular, for $n = 0$ and an arbitrary ξ, we have the identity

$$\int_T e^{i\beta \operatorname{Re} \bar{\alpha} \xi} d\alpha = \int_T e^{i\beta' \operatorname{Re} \bar{\alpha} \xi} d\alpha,$$

which yields, after differentiation with respect to ξ, the desired result

$$\beta = \beta'.$$ □

The set of equivalence classes of the irreducible representations T^β is parametrized, as we see, by the set R_+ or—from another point of view—by the set of orbits of the group $\boldsymbol{T}$ acting by rotations. It turns out that these are all infinite-dimensional representations of $M(2)$. A complete set of representations is obtained by adjoint the one-dimensional representations defined as $M(2) \ni (\zeta, \xi) \xrightarrow{\varkappa_n} \zeta^n$, $n \in Z$.

The structure of the set of irreducible representations of $M(2)$ is represented schematically in Fig. 5.

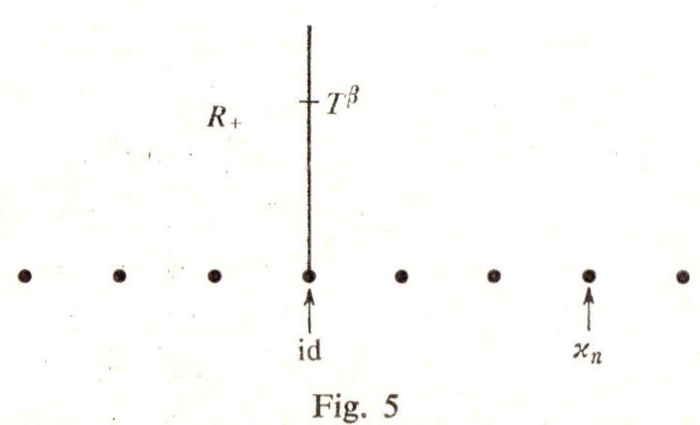

Fig. 5

The reason for placing the identity representation at the zero point of the real line is that if $\beta \to 0$, then the matrix elements of the representations T^β converge almost uniformly to the function 1.

Further transformations of formula (7.2.2) will be carried out for special values of the parameters.

Put $\beta > 0$ and $g = (e^{i\varphi}, re^{i\theta})$; then

$$t^\beta_{mn}(g) = \frac{e^{-in\psi}}{2\pi}\int_0^{2\pi} e^{i(n-m)\varphi + i\beta r\cos(\theta-\varphi)} d\varphi$$

$$= \frac{i^{m-n}}{2\pi} e^{i(n-m)\theta - in\psi}\int_0^{2\pi} e^{i\beta r\sin\varphi - i(m-n)\varphi} d\varphi \tag{7.2.3}$$

the last identity following after the substitution

$$\varphi \to \theta + \varphi - \tfrac{1}{2}\pi.$$

The function

$$J_n(z) = \frac{1}{2\pi}\int_0^{2\pi} e^{iz\sin\varphi - in\varphi} d\varphi \tag{7.2.4}$$

is called *Bessel function* of order n.

Thus we have reduced the matrix element to the form

$$t_{mn}(g) = i^{m-n}e^{i(n-m)\theta - in\varphi}J_{m-n}(\beta r). \tag{7.2.5}$$

The spherical functions have the form

$$t^{\beta}_{m0}(g) = i^m e^{-im\theta}J_m(\beta r), \tag{7.2.6}$$

and the zonal spherical functions have the form

$$t^{\beta}_{00}(g) = J_0(\beta r), \tag{7.2.7}$$

where $g = (e^{i\varphi}, re^{i\theta})$.

Upon identifying the left cosets gK with $re^{i\theta} \in \boldsymbol{C}$ we see that the spherical function can indeed be regarded as functions on $\boldsymbol{C}$ and the zonal spherical functions—as functions in the radius in $\boldsymbol{C}$.

Next we shall find the form of the differential of the representations T^{β} and of the representations of the group on the space $L^2(\boldsymbol{R}^2)$.

Assuming that $f \in \mathscr{E}(\boldsymbol{T})$ we differentiate the function $\boldsymbol{R} \ni t \to T^{\beta}(a_i(t))f \in L^2(\boldsymbol{T})$ and hence we find the form of the infinitesimal generators of the representation T^{β} corresponding to the one-parameter subgroups a_i

$$\mathrm{d}T^{\beta}(X_1)f(\alpha) = \frac{\partial}{\partial t}\bigg|_{t=0} f(e^{-it}\alpha) = -\tilde{X}_1 f(\alpha), \tag{7.2.8}$$

where $\tilde{X}_1$ denotes the invariant vector field on $\boldsymbol{T}$ corresponding to the one-parameter subgroup $t \to a_1(t)$.

Accordingly

$$\mathrm{d}T^{\beta}(X_2)f(\alpha) = i\beta \operatorname{Re}\alpha f(\alpha), \tag{7.2.9}$$

$$\mathrm{d}T^{\beta}(X_3)f(\alpha) = i\beta \operatorname{Im}\alpha f(\alpha). \tag{7.2.10}$$

In the basis $\{e_m\}^{\infty}_{-\infty}$ the action of the operators $T^{\beta}(X_1)$ takes the following simple diagonal form:

$$\mathrm{d}T^{\beta}(X_1)e_m = -ime_m. \tag{7.2.11}$$

The action of the operators

$$\begin{aligned} A^{\beta}_{+} &= \mathrm{d}T^{\beta}(X_2) + i\,\mathrm{d}T^{\beta}(X_3), \\ A^{\beta}_{-} &= \mathrm{d}T^{\beta}(X_2) - i\,\mathrm{d}T^{\beta}(X_3) \end{aligned} \tag{7.2.12}$$

with respect to the basis $\{e_m\}^{\infty}_{-\infty}$ is also of interest.

As can easily be seen from formulas (7.2.9) and (7.2.10) these operators raise and lower respectively, the index of the vector e_m

$$A^{\beta}_{+}e_m = i\beta e_{m+1}, \qquad A^{\beta}_{-}e_m = i\beta e_{m-1}, \tag{7.2.13}$$

and hence

$$A^{\beta}_{+}A^{\beta}_{-} = -\beta^2 \mathrm{id}.$$

The infinitesimal operators of the representation of $L^2(\boldsymbol{R})$ given by $U_g f(z) := L_g f(z) = f(\bar{\zeta}z - \bar{\zeta}\xi)$ for $g = (\zeta, \xi)$ are of the form

$$\mathrm{d}U(X_2) = -\left(\frac{\partial}{\partial z} + \frac{\partial}{\partial \bar{z}}\right), \qquad \mathrm{d}U(X_3) = i\left(\frac{\partial}{\partial \bar{z}} - \frac{\partial}{\partial z}\right),$$

$$\mathrm{d}U(X_1) = \frac{2}{i}\left(z\frac{\partial}{\partial \bar{z}} - \bar{z}\frac{\partial}{\partial z}\right).$$

The creation and annihilation operators have the form

$$A_+ = -2\frac{\partial}{\partial \bar{z}}, \qquad A_- = -2\frac{\partial}{\partial z}, \qquad A_+A_- = 4\frac{\partial}{\partial \bar{z}}\frac{\partial}{\partial z}. \tag{7.2.14}$$

The last operator is the Laplace operator on $\boldsymbol{C}$.

This ends the list of the properties of the representations of the group $M(2)$. These properties will be translated into the language of Bessel functions in the next chapter.

7.3. PROPERTIES OF THE BESSEL FUNCTIONS

In what follows we shall mainly deal with the representation T^1_g:

$$T^1_g f(\alpha) = e^{i\,\mathrm{Re}\,\bar{\alpha}\xi} f(\bar{\zeta}\alpha), \qquad g = (\zeta, \xi). \tag{7.3.1}$$

We can write

$$T^1_g e_0 = \sum_{-\infty}^{\infty} (e_m | T^1_g e_0) e_m = \sum_{-\infty}^{\infty} t^1_{m0}(g) e_m.$$

Hence, by applying formula (7.2.6) and putting $g = (1, r)$, we obtain

$$e^{ir\,\mathrm{Re}\,\bar{\alpha}} = \sum_{m=-\infty}^{\infty} iJ_m(r)\alpha^m.$$

The substitution $i\alpha = z$ gives

$$e^{\frac{1}{2}r(z-z^{-1})} = \sum_{m=-\infty}^{\infty} J_m(r) z^m. \tag{7.3.2}$$

Thus the Bessel function of order m is the coefficient of z^m in the expansion of the function

$$z \to e^{\frac{1}{2}r(z-z^{-1})},$$

which is called the *generating function* for Bessel functions.

By specializing the values of the parameter z $(z = i, -i, 1)$ we obtain

$$\sum_{-\infty}^{\infty} i^n J_n(x) = e^{ix}, \quad \sum_{-\infty}^{\infty} i^{-n} J_n(x) = e^{-ix}, \quad \sum_{-\infty}^{\infty} J_n(x) = 1. \tag{7.3.3}$$

Addition Formulas

One can derive these formulas from the identity

$$t^{\lambda}_{m0}(gh) = \sum_{k=-\infty}^{\infty} t^{\lambda}_{mk}(g)\, t^{\lambda}_{k0}(h)$$

by putting $g = (1, r_1)$, $h = (1, r_2 e^{i\theta_2})$. By substituting (7.2.5) and (7.2.6) to the above formula and noting that

$$gh = (1, r_1 + r_2 e^{i\theta_2}) =: (1, re^{i\theta})$$

we obtain

$$\sum_{k=-\infty}^{\infty} J_{m-k}(r_1) J_k(r_2) e^{-ik\theta_2} = J_m(r) e^{-im\theta}.$$

It follows directly from (7.2.4) that for real values of arguments the Bessel function also assume real values, and so, by complex conjugation, we arrive at the following formula, called *Graff's formula*:

$$\sum_{k=-\infty}^{\infty} J_{m-k}(r_1) J_k(r_2) e^{ik\theta_2} = J_m(r) e^{im\theta} \tag{7.3.4}$$

for the arguments r_1, r_2, θ_2; r related by the condition

$$r_1 + r_2 e^{i\theta_2} = re^{i\theta}.$$

By substituting special values of the parameters we obtain a series of classical formulas:

$\theta_2 = 0,\ r_1+r_2 = r$:

$$J_n(r_1+r_2) = \sum_{k=-\infty}^{\infty} J_{n-k}(r_1)J_k(r_2); \tag{7.3.5}$$

$\theta_2 = \pi,\ r_1 \geqslant r_2,\ r = r_1-r_2$:

$$J_n(r_1-r_2) = \sum_{k=-\infty}^{\infty} (-1)^k J_{n-k}(r_1)J_k(r_2). \tag{7.3.5'}$$

In particular, for $r_1 = r_2$ we get the identity

$$\sum_{k=-\infty}^{\infty} J_{n+k}(r)J_k(r) = J_n(0) = \begin{cases} 0, & n \neq 0, \\ 1, & n = 0, \end{cases} \tag{7.3.6}$$

known as the *Hansen formula.*

Next, putting $n = 0$, we obtain a decomposition of the Bessel function of order 0:

$$J_0(r) = \sum_{k=-\infty}^{\infty} (-1)^k e^{ik\theta} J_k(r_1)J_k(r_2). \tag{7.3.7}$$

A Mean Value Theorem

Formula (5.1.2) leads to a series of integral formulas for the Bessel functions. The more general are obtained by taking $v = e_l$, $x = e_m$, $y = e_n$ in the representation space of $\boldsymbol{T}^1$. We also assume that $g = (1, r_1)$, $k = (e^{i\varphi}, 0)$, $h = (1, r_2)$, and hence

$$gkh = (e^{i\varphi}, r_1+r_2 e^{i\varphi}).$$

Put $r_1+r_2 e^{i\varphi} =: re^{i\theta}$. Then the spherical equation takes the form

$$\frac{1}{2\pi}\int_0^{2\pi} t_{mn}(gkh)e^{il\varphi}\, d\varphi = t_{ml}(g)t_{ln}(h).$$

Hence, on the basis of (7.2.6),

$$\frac{1}{2\pi}\int_0^{2\pi} e^{i(n-m)\theta+i(l-n)\varphi} J_{m-n}(r)\, d\varphi = J_{m-l}(r_1)J_{l-n}(r_2).$$

Put $m-n=:p$, $l-n=:q$. We then obtain

$$\frac{1}{2\pi}\int\limits_0^{2\pi} e^{iq\psi-ip\theta}J_p(r)\,d\psi = J_{p-q}(r_1)J_q(r_2). \tag{7.3.8}$$

For $q=0$, this formula has the following clear interpretation: when ψ ranges over the interval $[0, 2\pi[$, the point $re^{i\theta}$ describes a circle with centre at r_1 and radius r_2. The integrand coincides with the spherical function $i^p t^1_{00}$. Formula (7.3.8), asserts that the mean value of this spherical function over the circle, is proportional to the value of the function at the centre of the circle and the factor of proportionality is equal to the value at r_2 of the zonal spherical function with which t^1_{00} is associated.

Therefore formula (7.3.8) can be regarded as a mean value theorem for spherical functions.

In some special cases formula (7.3.8) takes a considerably simpler form.

Let $r_1 = r_2 =: R$; then $\theta = \frac{1}{2}\psi$ and $r = 2R\cos\theta$,

$$\frac{1}{\pi}\int\limits_0^{2\pi} e^{i(q-2p)\theta}J_q(2R\cos\theta)\,d\theta = J_{q-p}(R)J_p(R). \tag{7.3.9}$$

Further consequences of formula (7.3.8) will be found in Problems.

Symmetry with Respect to the Index

The representation space of T admits a unitary operator which commutes with the representation operators $T(g)$ for g, belonging to the one-parameter subgroup, $a_2(t) = (1, t)$.

Such an operator $B\colon L^2(T) \to L^2(T)$ is defined by

$$Bf(\alpha) := f(\bar{\alpha}).$$

Plainly $T(a_2(t))Bf(\alpha) = e^{it\,\mathrm{Re}\,\bar{\alpha}}f(\bar{\alpha}) = e^{it\,\mathrm{Re}\,\alpha}f(\bar{\alpha}) = BT(a_2(t))f(\alpha)$.

Moreover, B is unipotent: $BB = \mathrm{id}$. Under the action of B the indices of the basis $\{e_n\}_{-\infty}^{\infty}$ change sign

$$Be_n = e_{-n}.$$

Hence it follows that for $g = a_1(t)$

$$\begin{aligned} t^1_{m,n}(g) &= (e_m|T^1_g e_{-n}) = (e_m|T^1_g Be_n) = (e_m|BT^1_g e_n) \\ &= (Be_m|T^1_g e_n) = t^1_{-m,n}(g). \end{aligned}$$

In particular, for $n = 0$, formula (7.2.6) gives

$$J_m = (-1)^m J_{-m}. \tag{7.3.10}$$

On the other hand, it follows directly from the definition (7.2.4) of the function J_n that

$$J_n(-z) = J_{-n}(z),$$

so that

$$J_n(-z) = (-1)^n J_n(z). \tag{7.3.11}$$

Recurrence Formulas and the Bessel Equation

The mapping $S\colon L^2(T) \ni x \to \overline{t^1_{x, e_0}} \in \mathscr{C}(G)$, intertwines the representation T^1 of the group $M(2)$, with the left regular representation of this group. The functions $\overline{t^1_{x,e}}$ are right-invariant under $\boldsymbol{T}$ and therefore can be identified with function on $M(2)/\boldsymbol{T} = \boldsymbol{C}$.

As we know (§ 2.8), the operator S also intertwines the representations of the Lie algebras. Hence

$$SA^\beta_\pm = A_\pm S \qquad \text{(cf. § 7.2).}$$

In view of formulas (7.2.13) and (7.2.14) we obtain

$$\begin{aligned} SA^1_+ e_m &= iSe_{m+1} = i\bar{t}_{m+1,0} = -2\frac{\partial}{\partial \bar{z}}\bar{t}_{m,0}, \\ SA^1_- e_m &= iSe_{m-1} = i\bar{t}_{m-1,0} = -2\frac{\partial}{\partial z}\bar{t}_{m,0}, \\ 4\frac{\partial}{\partial \bar{z}}\frac{\partial}{\partial z}\bar{t}_{m,0} &= \bar{t}_{m,0}. \end{aligned} \tag{7.3.12}$$

Passing to spherical coordinates on $\boldsymbol{C}$ and representing $t_{m,0}$ in form (7.2.6), we easily obtain the recurrence formulas

$$\begin{aligned} xJ_{m+1}(x) &= mJ_m(x) - xJ'_m(x), \\ xJ_{m-1}(x) &= mJ_m(x) + xJ'_m(x) \end{aligned} \tag{7.3.13}$$

and the Bessel equation

$$x^2J''_m(x) + xJ'_m(x) + (x^2 - m^2)J_m(x) = 0. \tag{7.3.14}$$

The last equation of (7.3.12), which leads to the Bessel equation, is a special case of formula (5.4.1) for a spherical function. The operator $\frac{\partial}{\partial z}\frac{\partial}{\partial \bar{z}}$ is indeed an $M(2)$-invariant operator on $\boldsymbol{C} = M(2)/\boldsymbol{T}$. Incidentally, we note that the above implies

$$\frac{\partial}{\partial z}\frac{\partial}{\partial \bar{z}} J_0(|z|)|_{z=0} = \frac{1}{4}.$$

7.4. HARMONIC ANALYSIS ON THE SYMMETRIC SPACE OF THE MOTION GROUP $M(2)$. THE FOURIER-BESSEL TRANSFORMATION

Viewing $\boldsymbol{C}$ as a homogeneous space of the motion group, we will now look for a decomposition into irreducible components of the representation U of the group $M(2)$ on $L^2(\boldsymbol{C})$. This will lead us to the Plancherel theorem and to the inversion formula for a new integral transformation, which commutes not only with translation, as the Fourier transformation does, but with the action of the whole group $M(2)$.

In seeking this transformation we follow the same way as that in which the Fourier transformation on $\boldsymbol{R}^n$ has been obtained.

We begin by constructing an operator which intertwines the action of $M(2)$ on $L^2(\boldsymbol{C})$, with the regular representation of the grup $M(2)$, on the space of matrix elements of the spherical representation T^β. Bearing in mind the case of a compact symmetric space (the Frobenius duality) we arrive at the idea that the totality of irreducible spherical representation should "suffice" for the decomposition of U.

We recall that $U_g f(z) = f(g^{-1} \circ z)$, $f \in L^2(\boldsymbol{C})$.

The representations U and T^β are intertwined by the operator

$$F_\beta \colon \mathscr{C}_0(\boldsymbol{C}) \ni f \to T^\beta(f \circ \pi) e_0 = \int_{M(2)} f((\pi(g)) T^\beta(g) e_0 \mathrm{d}g.$$

Substituting expression (7.1.1) for the measure $\mathrm{d}g$ and taking into account the form of T^β, we see that

$$F_\beta f(\alpha) = \frac{i}{2} \int_{\boldsymbol{C}} f(z) \mathrm{e}^{i\beta \operatorname{Re}(\alpha \bar{z})} \mathrm{d}z \wedge \mathrm{d}\bar{z} = \hat{f}(\beta\alpha), \tag{7.4.1}$$

where $\hat{f}$ denotes the classical Fourier transform on $\boldsymbol{R}^2$.

The restriction of the transform to the circle of radius β is the required operator from $L^2(G)$ into the representation space of T^β. Let us compose this mapping with the operation of forming a matrix element relative to $e_0 \in L^2(T)$:

$$\begin{aligned} SF_\beta f(g) &:= (T_g^\beta e_0 | F_\beta f) = (e_0 | T_{g^{-1}}^\beta F_\beta f) \\ &= (e_0 | F_\beta U_{g^{-1}} f) = \left(e_0 | T_{\hat{\cdot}}^\beta (L_{g^{-1}}(f \circ \pi)) e_0\right) \\ &= \int_{M(2)} f(gx \cdot 0)(e | T_x^\beta e_0) \mathrm{d}x \\ &= \frac{i}{2} \int_C f(z) J_0(\beta |g^{-1} \cdot z|) \mathrm{d}z \wedge \mathrm{d}\bar{z} \\ &= \frac{i}{2} \int_C f(z) J_0(\beta |z - \xi|) \mathrm{d}z \wedge \mathrm{d}\bar{z}, \end{aligned} \tag{7.4.2}$$

if $g = (\zeta, \xi)$.

It turns out that the matrix element $SF_\beta(f)$, being a right T-invariant function on $M(2)$, is actually a function of ξ. Thus we have obtained the integral transformation

$$\mathscr{C}_0(C) \ni f \to f^\beta \in \mathscr{C}(C),$$

where

$$f^\beta(\zeta) = \frac{i}{2} \int_C f(z) J_0(\beta |z - \zeta|) \mathrm{d}z \wedge \mathrm{d}\bar{z}.$$

The function f^β is the "component" of f in the space of matrix elements of the representation T^β. Observe an analogy with formula (7.2.3) from the analysis on compact symmetric spaces.

We have thus obtained the expected result:

THEOREM 7.4.1. *For every $f \in \mathscr{C}_0(C)$ the following decomposition holds*

$$f(\xi) = \frac{1}{2\pi} \int_0^\infty \beta f^\beta(\xi) \mathrm{d}\beta. \tag{7.4.3}$$

Moreover, we have the isometry

$$\frac{1}{2\pi} \int_0^\infty \|F_\beta f\|^2 \beta \mathrm{d}\beta = \|f\|_{L^2(C)}$$

($\|\cdot\|$ denoting the norm on $L^2(T)$).

Proof. We shall reduce both statements to well known facts in the classical Fourier analysis. The substitution $f^\beta(\xi) = (T_g^\beta e_0 | F_\beta f)$ for $g = (1, \xi)$ gives

$$\int_0^\infty \beta f^\beta(\xi) \mathrm{d}\beta = \int_0^\infty \beta \mathrm{d}\beta \int_T \mathrm{e}^{-i\beta \operatorname{Re} \bar{\alpha}\xi} \hat{f}(\alpha\beta) \mathrm{d}\alpha$$

$$= \frac{i}{4\pi} \int_C \hat{f}(\zeta) \mathrm{e}^{-i \operatorname{Re}(\zeta\xi)} \mathrm{d}\zeta \wedge \mathrm{d}\bar{\zeta} = 2\pi f(\xi).$$

In a similar way

$$\int_0^\infty \beta ||F_\beta f||^2 \mathrm{d}\beta = \int_0^\infty \beta \int_T |\hat{f}(\alpha\beta)|^2 \mathrm{d}\alpha \mathrm{d}\beta = 2\pi ||f||_{L^2(C)},$$

owing to the Plancherel formula. □

We shall state this result in the form of a theorem on the decomposition of the representation U on $L^2(C)$ into a direct integral.

THEOREM 7.4.2. *The representation U decomposes into the direct sum of spherical representations T^β*

$$U = \int_0^\infty T^\beta \beta \frac{\mathrm{d}\beta}{2\pi}.$$

The corresponding Fourier transformation is the operator F_β given by (7.4.1).

The component $f_m = SF_\beta f$ occurring in the decomposition of f into a direct integral (7.4.3) can also be expressed as a linear combination of matrix elements relative to the basis $\{e_m\}$. For $g = (1, \xi)$ we find

$$f^\beta(\xi) = (T^\beta e_0 | F_\beta f) = \sum_{k=-\infty}^{\infty} (T_g^\beta e_0 | e_k)(e_k | F_\beta f)$$

$$= \sum_{k=-\infty}^{\infty} \bar{t}_{k,0}^\beta(\xi) \frac{i}{2} \int_C f(z) t_{k0}^\beta(z) \mathrm{d}z \wedge \mathrm{d}\bar{z}$$

$$= \frac{i}{2} \sum_{k=-\infty}^{\infty} \left(\frac{\xi}{|\xi|}\right)^k J_k(\beta|\xi|) \int_C f(z) \left(\frac{z}{|z|}\right)^{-k} J_k(\beta|z|) \mathrm{d}z \wedge \mathrm{d}\bar{z}.$$

Put

$$Ff(k,\beta) := \frac{i}{4\pi}\int_{C} f(z)\left(\frac{z}{|z|}\right)^{-k} J_k(\beta|z|)\,dz\wedge d\bar{z}. \tag{7.4.5}$$

Then formula (7.4.3) takes the form

$$f(\xi) = \int_0^\infty \beta\,d\beta \sum_{k=-\infty}^{\infty}\left(\frac{\xi}{|\xi|}\right)^{k} J_k(\beta|\xi|)Ff(k,\beta). \tag{7.4.6}$$

This formula, just as formula (7.4.3), provides a decomposition of f into the eigenfunctions of the Laplace operator which, in addition, depend in a special way upon the argument ξ; namely, they are eigenfunctions of rotation operators.

Apply (7.4.6) to the function of the form $f(r)e^{ik\theta}$. Then, owing to the orthonormality of the power functions on the circle, we observe that the sum appearing in (7.4.6) consists only of a single summand and hence we obtain the inversion formula

$$f(r) = \int_0^\infty \beta\,d\beta J_k(\beta r)Ff(k,\beta), \tag{7.4.7}$$

with $Ff(k,\beta)$ being now defined as

$$Ff(k,\beta) = \int_0^\infty \varrho\,d\varrho J_k(\beta\varrho)f(\varrho). \tag{7.4.8}$$

The transformation $\mathscr{C}_0(\boldsymbol{R}_+)\ni f\to Fk(k,\beta)$ is called the *Fourier–Bessel transformation.* Expression (7.4.7) is the inversion formula for this transformation. Formula (7.4.4) now transforms into the Plancherel formula for the Fourier-Bessel transformation:

$$\int_0^\infty |f|^2(\varrho)\varrho\,d\varrho = \int_0^\infty |Ff(k,\beta)|^2\beta\,d\beta. \tag{7.4.9}$$

This formula implies that the transformation F admits an isometric extension to $L^2(\boldsymbol{R}_+,\varrho\,d\varrho)$.

PROBLEMS

1. Show that the group $M(2)$ is isomorphic to the subgroup of $\mathrm{GL}(3, \boldsymbol{R})$ consisting of all matrices of the form

$$\begin{bmatrix} \cos\varphi & -\sin\varphi & x_1 \\ \sin\varphi & \cos\varphi & x_2 \\ 0 & 0 & 1 \end{bmatrix}$$

and find a Lie subalgebra of $\mathfrak{gl}(3, \boldsymbol{R})$ isomorphic to the Lie algebra of the group $M(2)$. Prove that it is a 3-dimensional solvable Lie algebra isomorphic to the Lie subalgebra of vector fields on $\boldsymbol{R}^2$ spanned by the vector fields $\dfrac{\partial}{\partial x_1}, \dfrac{\partial}{\partial x_2}, x_1\dfrac{\partial}{\partial x_2} - x_2\dfrac{\partial}{\partial x_1}$.

2. (a) Find an explicit form of invariant vector fields on the group $M(2)$.

(b) Prove that the centre of the universal enveloping algebra for $M(2)$ is precisely the algebra of polynomials in the Laplace operator $\Delta = \dfrac{\partial^2}{\partial x^2} + \dfrac{\partial^2}{\partial y^2}$ regarded as an operator on $G \cong \boldsymbol{T} \times_\tau \boldsymbol{R}^2$.

(c) Show that every function on $M(2)$ which is bi-invariant under $\boldsymbol{T}$ and is an eigenfunction for the centre of the enveloping algebra of $M(2)$ satisfies Bessel's differential equation

$$\frac{\mathrm{d}^2 f}{\mathrm{d}r^2} + \frac{1}{r}\frac{\mathrm{d}f}{\mathrm{d}r} = \alpha f.$$

Hint. Introduce the coordinates (φ, x, y) on $M(2)$ where $g = (\mathrm{e}^{i\varphi}, (x, y))$ and express invariant vector fields on the group in terms of these coordinates. To prove (b) one can also apply the preceding problem.

3. (a) Prove that every finite-dimensional unitary representation of the group $M(2)$ contains a one-dimensional subrepresentation and consequently every finite-dimensional irreducible representation of $M(2)$ is one-dimensional.

(b) Suppose that $M(2)$ acts on the group of unitary characters of $\boldsymbol{C}$ according to the formula $g \cdot \chi(\xi) := \chi(g^{-1} \cdot \xi \cdot g)$, where $\xi \in \boldsymbol{C}$ is identified with the element $(1, \xi)$ of the group $M(2)$. Prove that the isotropy subgroup of a character is either the group of all translations or the whole group $M(2)$. The latter possibility occurs only in the case of the trivial character. Check that every irreducible representation of the group $M(2)$ described in Section 2, is equivalent to the representation induced by a unitary irreducible representation of the isotropy group of a certain character.

Hint. (a) By using the fact that the translation group is a normal subgroup in $M(2)$ show that every one-dimensional translation invariant subspace is invariant under the whole group $M(2)$.

Remark: (a) is a special case of a Lie theorem which asserts that irreducible (not necessarily unitary) representations of a solvable group are necessarily one-dimensional—see e.g. Chevalley, 1946.

4. (a) Prove that the integral

$$\int_0^{2\pi} e^{i(z\sin\varphi - n\varphi)} d\varphi$$

represents an entire function of the parameter z and hence derive the Taylor series expansion for the Bessel function of order $n, n \geqslant 0$

$$J_n(z) = \left(\frac{z}{2}\right)^n \sum_{k=0}^{\infty} \frac{(-1)^k}{(n+k)!k!} \left(\frac{z}{2}\right)^{2k}.$$

(b) Prove that

$$J_n(z) = \frac{1}{\pi} \int_0^{\pi} \cos(n\varphi - z\sin\varphi) d\varphi.$$

(c) Show that the series

$$S_\nu(z) = \sum_{k=0}^{\infty} \frac{(-1)^k}{k!\Gamma(\nu+k+1)} \left(\frac{z}{2}\right)^{2k}$$

is convergent for arbitrary complex values of ν and z and represents an entire function of both these variables. Prove that for every $\nu \in \boldsymbol{C}$ the function

$$J_\nu(z) := \left(\frac{z}{2}\right)^\nu S_\nu(z) = \left(\frac{z}{2}\right)^\nu \sum_{k=0}^{\infty} \frac{(-1)^k}{k!\Gamma(\nu+k+1)} \left(\frac{z}{2}\right)^{2k}$$

is a solution of Bessel's equation of order v:

$$\frac{d^2w}{dz^2}+\frac{1}{z}\frac{dw}{dz}+\left(1-\frac{v^2}{z^2}\right)w=0.$$

(d) Show that formula (7.3.2) remains valid for complex values of r, i.e. that

$$e^{\frac{1}{2}w(z-z^{-1})}=\sum_{m=-\infty}^{\infty}=J_m(w)z^m, \qquad z\in \boldsymbol{C}.$$

Hint. (d) Compare this formula with the Laurent expansion of the function on the left.

5. Derive recurrence relation (7.3.13) by applying the generating function (formula (7.3.2) and Problem 4(c)). In the same way derive also the relation between J_m and J_{-m} given by formula (7.3.11).

Hint. Differentiate formula (7.3.2) with respect to t.

6. Prove that

$$\left(\frac{z}{2}\right)^n=\sum_{r=0}^{\infty}\frac{(n+2r)\Gamma(n+r)}{r!}J_{n+2r}(z).$$

Hint. Make use of the expansion of $e^{\frac{1}{2}tz}$ obtained by applying Problem 4(d).

7. Prove that $z(z^2+1)=z\sum_{n=0}^{\infty}(2n+1)^3J_{2n+1}(z)$.

Hint. Apply Problem 6 or derive the formula directly from the form of the generating function.

8. Prove the formula

$$\frac{1}{2}z\cos z=\sum_{n=0}^{\infty}(-1)^n(2n+1)^2J_{2n+1}(z).$$

9. Show that the addition formula (7.3.5) is valid in the same form for complex values of r_1, r_2 and deduce hence the relation

$$1=J_0^2(z)+2\sum_{k=1}^{\infty}J_k^2(z).$$

10. Prove the recurrence formulas

$$\left(\frac{1}{z}\frac{d}{dz}\right)^n (z^\nu J_\nu(z)) = z^{\nu-n} J_{\nu-n}(z),$$

$$\left(\frac{1}{z}\frac{d}{dz}\right)^n (z^{-\nu} J_\nu(z)) = (-1)^n z^{-\nu-n} J_{\nu+n}(z).$$

11. Prove the inequality

$$|J_n(z)| \leqslant e^{|\operatorname{Im} z|}, \qquad n \in Z.$$

Hint. Apply (7.2.4).

12. (a) Prove, by expanding in a series the exponential factor in the integrand below and integrating term by term, that the following Poisson formula holds

$$J_n(z) = \frac{1}{\Gamma\left(n+\frac{1}{2}\right)\Gamma\left(\frac{1}{2}\right)}\left(\frac{z}{2}\right)^n \int_{-1}^{1} e^{izt}(1-t^2)^{n-\frac{1}{2}} dt, \qquad n \geqslant 0.$$

(b) Show that by replacing n in the Poisson formula by a complex parameter ν we obtain an integral representation of the Bessel function of order ν for ν satisfying the condition $\operatorname{Re}\nu > -\frac{1}{2}$.

(c) Show that

$$J_{\frac{1}{2}}(z) = \left(\frac{2}{\pi z}\right)^{\frac{1}{2}} \sin z, \qquad J_{\frac{3}{2}}(z) = \left(\frac{2}{\pi z}\right)^{\frac{1}{2}}\left(\frac{1}{z}\sin z - \cos z\right)$$

$$J_{-\frac{1}{2}}(z) = \left(\frac{2}{\pi z}\right)^{\frac{1}{2}} \cos z, \qquad J_{-\frac{3}{2}}(z) = \left(\frac{2}{\pi z}\right)^{\frac{1}{2}}\left(\frac{1}{z}\cos z + \sin z\right).$$

(d) Prove that for integer values of m the functions $J_{m+\frac{1}{2}}(z)$ can be expressed by elementary functions, or more precisely, that for certain polynomials, P_m, Q_m of degree not greater than m we have

$$J_{m+\frac{1}{2}}(z) = \left(\frac{2}{\pi z}\right)^{\frac{1}{2}}\left(P_m\left(\frac{1}{z}\right)\cos z + Q_m\left(\frac{1}{z}\right)\sin z\right).$$

Hint. (b) Proceed as in (a) by applying the integral formula for the Euler B-function given in Problem 7(a) in Chapter 6.

13. Let D be a contour (a single loop) going once the negative axis in the positive direction (cf. the Hankel representation for the Euler Γ-function—Problem 3(b), Chapter 6). Prove the following formula (known as the Schläfli integral for the Bessel function)

$$J_\nu(z) = \left(\frac{z}{2}\right)^\nu \frac{1}{2\pi i} \int_D e^{\left(t-\frac{z^2}{4t}\right)} \frac{dt}{t^{\nu+1}}.$$

Hint. Substitute the Hankel representation for the Γ-function in the series defining $J_\nu(z)$.

14. Derive the Sonine integral ($n \geqslant 0$)

$$J_n(z) = \left(\frac{z}{2}\right)^n \frac{2}{(n-1)!} \int_0^{\frac{1}{2}\pi} \sin\varphi(\cos\varphi)^{2n-1} J_0(z\sin\varphi)\,d\varphi.$$

Show that after appropriate modifications ($(n-1)! = \Gamma(n)$) the formula above makes sense for non-integer n and defines the same function as the Poisson integral in Problem 12.

Hint. Apply formula (7.3.9).

15. Prove that

$$\int_0^{\frac{1}{2}\pi} J_{2n}(2z\cos\varphi)\,d\varphi = \tfrac{1}{2}\pi(J_n(z))^2.$$

16. Prove by applying recurrence relations (7.3.13) and the Taylor expansion of the function $z \to J_0(\sqrt{z})$, that for a suitably chosen branch of the function $z \to \sqrt{z}$ the following formula, known as the *multiplication formula*, holds true

$$J_0(\lambda z) = \sum_{m=0}^{\infty} \frac{((1-\lambda^2)z)^m}{2^m m!} J_m(z),$$

and for $n > 0$,

$$J_n(\lambda z) = \lambda^n \sum_{m=0}^{\infty} \frac{((1-\lambda^2)z)^m}{2^m m!} J_{m+n}(z).$$

17. Prove the following formula, known as the Gegenbauer formula:

$$J_0(r) = \sum_{m=0}^{\infty} \varepsilon_m J_m(r_1) J_m(r_2) \cos\theta_2,$$

where $\varepsilon_m = \begin{cases} 1, & m = 0 \\ 2, & m > 0 \end{cases}$ and r, r_1, r_2 and have the same meaning as in (7.3.4).

18. Prove that in the case where ν is not an integer the Bessel equation of order ν

$$x^2w''+xw'+(x^2-\nu^2)w = 0$$

has two linearly independent solutions J_ν, $J_{-\nu}$ given by the formulas

$$J_\nu(x) = \left(\frac{x}{2}\right)^\nu \sum_{r=0}^{\infty} \frac{(-1)^r}{r!\Gamma(\nu+r+1)} \left(\frac{x}{2}\right)^{2r},$$

$$J_{-\nu}(x) = \left(\frac{x}{2}\right)^{-\nu} \sum_{r=0}^{\infty} \frac{(-1)^r}{\Gamma(-\nu+r+1)r!} \left(\frac{x}{2}\right)^{2r}.$$

Prove that the Wronskian of these solutions satisfies the identity

$$W(J_\nu, J_{-\nu}) = \frac{c}{x}$$

with $c = -\dfrac{2\sin\pi\nu}{\pi}$.

19. Show that for $a > 0$

(i) $\displaystyle\int_0^\infty J_0(t)\frac{\sin at}{t}\,dt = \begin{cases} \arcsin a, & 0 < a \leqslant 1, \\ \frac{1}{2}\pi, & 1 < a. \end{cases}$

(ii) $\displaystyle\int_0^\infty J_0(t)\cos at\,dt = \begin{cases} (1-a^2)^{-\frac{1}{2}}, & 0 < a < 1, \\ \infty & a = 1, \\ 0, & 1 < a. \end{cases}$

20. Show that

$$\sin z = z\left[J_0(z)+2\sum_{n=1}^{\infty}\frac{(-1)^n J_{2n}(z)}{1-4n^2}\right].$$

Hint. Use the Poisson integral (Problem (12)).

21. Prove that

$$J_0(z) = \frac{2^n n!}{(2n)!} z^n\left(1+\frac{d^2}{dz^2}\right)^n J_0(z).$$

Chapter 8

Theory of Jacobi and Legendre Polynomials

8.1. REPRESENTATIONS OF THE GROUP SL(2; *C*) ON A SPACE OF POLYNOMIALS

The classical Jacobi and Legendre polynomials are defined and investigated by menas of matrix elements of unitary irreducible representations of the matrix group SU(2).

We shall begin, however, by constructing a series of finite-dimensional representations of the group SL(2; *C*) which contains SU(2) as a compact subgroup. The matrix group SL(2; *C*) acts in a natural way on the space C^2:

$$\text{if } g = \begin{bmatrix} \alpha & \beta \\ \gamma & \delta \end{bmatrix} \text{ with } \alpha\delta - \beta\gamma = 1 \text{ and } x = \begin{bmatrix} z_1 \\ z_2 \end{bmatrix}$$

then gx denotes the element

$$\begin{bmatrix} \alpha & \beta \\ \gamma & \delta \end{bmatrix} \begin{bmatrix} z_1 \\ z_2 \end{bmatrix} = (\alpha z_1 + \beta z_2, \gamma z_1 + \delta z_2). \tag{8.1.1}$$

Denote by H^l, $l = 0, \frac{1}{2}, 1, \frac{3}{2}, \ldots$ the space of all homogeneous complex polynomials of degree $2l$, in two variables. This space is spanned by the polynomials

$$e_k(z_1, z_2) = \frac{z_1^{l+k} z_2^{l-k}}{[(l+k)!(l-k)!]^{\frac{1}{2}}},$$

where k ranges over the numbers $k = -l, -l+1, \ldots, l-1, l$.

The role of the numerical factor will be explained later. We shall define the action of the group on H^l by setting:

$$T_g^l w(x) = w(g^t x), \qquad g^t = \begin{bmatrix} \alpha & \gamma \\ \beta & \delta \end{bmatrix}.$$

The degree of the polynomials and the property of homogenity remain unchanged under the above action, which means that H^l are invariant subspaces.

Denote by $\{e^j\}$ the basis dual to $\{e_k\}$, i.e. verifying the condition $\langle e^j, e_k\rangle = \delta^j_k$.

We shall determine the matrix elements of the representation T^l_g relative to the vectors of the dual bases

$$t^l_{jk}(g) = \langle e^j, T^l_g e_k\rangle.$$

For $g = \begin{bmatrix}\alpha & \beta\\ \gamma & \delta\end{bmatrix}$ we find

$$T^l_g e_k(z_1, z_2) = \frac{(\alpha z_1+\gamma z_2)^{l+k}(\beta z_1+\delta z_2)^{l-k}}{[(l+i)!(l-i)!]^{\frac{1}{2}}}.$$

Hence the matrix element t^l_{jk} is the coefficient of the above polynomial appearing at the monomial $z_1^{l+j}z_2^{l-j}$ multiplied by the normalizing factor $[(l+\mathrm{i})!(l-j)!]^{\frac{1}{2}}$. We compute it by indicating $z_2 = 1$ (say) and applying the Taylor formula to the variable z_1:

$$t^l_{jk}(g)$$
$$= \sqrt{\frac{(l-j)!}{(l+k)!(l-k)!(l+j)!}}\,\frac{\mathrm{d}^{l+j}}{\mathrm{d}z^{l+j}}\,[(\alpha z+\gamma)^{l+k}(\beta z+\delta)^{l-k}|_{z=0}.$$

We transform this expression by making the substitution $\dfrac{y-1}{z} = (\alpha z+\gamma)\beta$. The unimodularity of g then implies that $\beta z+\delta = \dfrac{y+1}{2\alpha}$, and hence we obtain

$$t^l_{jk}(g) = \sqrt{\frac{(l-j)!}{(l+k)!(l-k)!(l+j)!}}\,(2\beta\alpha)^{l+j}$$
$$\times \frac{1}{(2\beta)^{l+k}}\,\frac{1}{(2\alpha)^{l-k}}\,\frac{\mathrm{d}^{l+j}}{\mathrm{d}y^{l+j}}[(y-1)^{l+k}(y+1)^{l-k}]|_{y=\gamma\beta+\frac{1}{2}}$$
$$= \sqrt{\frac{(l-j)!}{(l+k)!(l-k)!(l+j)!}}\,2^{j-l}\alpha^{j+k}\beta^{j-k}\,\frac{\mathrm{d}^{l+j}}{\mathrm{d}y^{l+j}}$$
$$\times [(y-1)^{l+k}(y+1)^{l-k}]|_{y=\gamma\beta+\frac{1}{2}}$$

Let us restrict the function t^1_{jk} to the one-parameter subgroup

$$g(\theta) = \begin{bmatrix}\cos\frac{1}{2}\theta & i\sin\frac{1}{2}\theta\\ i\sin\frac{1}{2}\theta & \cos\frac{1}{2}\theta\end{bmatrix}. \tag{8.1.3}$$

If $0 \leqslant \theta < \pi$, then the values $\alpha = \delta = \cos\frac{1}{2}\theta$ and $\gamma = \beta = i\sin\frac{1}{2}\theta$ are univalent functions of the variable $x = \cos\theta$. We define

$$P^l_{jk}(\cos\theta) := t^l_{jk}(g(\theta)) = \sqrt{\frac{(l-j)!}{(l+k)!(l-k)!(l+j)!}}$$
$$\times 2^{j-l} i^{j-k} \left(\cos\frac{1}{2}\theta\right)^{j+k} \left(\sin\frac{1}{2}\theta\right)^{j-k} \frac{\mathrm{d}^{l+j}}{\mathrm{d}y^{l+j}}$$
$$\times [(y-1)^{l+k}(y+1)^{l-k}]|_{y=\cos\theta}, \tag{8.1.4}$$

$$P^l_{jk}(x) = \sqrt{\frac{(l-j)!}{(l+k)!(l-k)!(l+j)!}}\,(1+x)^{\frac{j+k}{2}}(1-x)^{\frac{j-k}{2}}$$
$$\times 2^{-l-j} i^{j-k} \frac{\mathrm{d}^{l+j}}{\mathrm{d}x^{l+j}}[(x-1)^{l+k}(x+1)^{l-k}].$$

The functions P^l_{jk} differ by a factor from the classical *Jacobi polynomials*:

$$P^{(m,n)}_k(x) = \frac{(-1)^k}{2^k k!}(1-x)^{-m}(1+x)^{-n}\frac{\mathrm{d}^k}{\mathrm{d}x^k}$$
$$\times [(1-x)^{m+k}(1+x)^{n+k}]. \tag{8.1.5}$$

We also introduce the *Legendre polynomials*

$$P_k(x) = P^{(0,0)}_k(x) = \frac{(-1)^k}{2^k k!}\frac{\mathrm{d}^k}{\mathrm{d}x^k}(1-x^2)^k \tag{8.1.6}$$

and the *associated Legendre functions* defined for all integers k, m, with $m \geqslant 0$:

$$P^m_k(x) = \frac{(-1)^{m+k}}{2^k k!}(1-x^2)^{\frac{1}{2}m}\frac{\mathrm{d}^{m+k}}{\mathrm{d}x^{m+k}}(1-x^2)^k, \tag{8.1.7}$$

or equivalently

$$P^m_k(x) = \frac{2^m(k+m)!}{k!}(1-x^2)^{-\frac{1}{2}m}P^{(-m,-n)}_{k+m}(x). \tag{8.1.8}$$

The Jacobi polynomials are connected with functions P^l_{jk}, which are of fundamental importance for our subsequent considerations:

$$P^l_{jk}(x) = P^{(k-j,-k-j)}_{l+j}(x)(1-x)^{\frac{k-j}{2}}(1+x)^{\frac{-k-j}{2}} i^{k-j}$$
$$\times \sqrt{\frac{(l+k)!(l-k)!}{(l+j)!(l-j)!}} \tag{8.1.9}$$

Before we proceed to study the properties of the representations T^l and translating them into the properties of $P_k^{m,n}$ we shall first establish several new expressions for the functions P_{jk}^l.

Let us return to formula (8.1.2). The matrix elements t_{jk}^l can be computed directly in view of Newton's binomial formula

$$T_g^l e_k(z_1, z_2)$$

$$= \sum_{\substack{s \in I_1 \\ t \in I_2}} \frac{\binom{l+k}{s}\binom{l-k}{t}}{[(l+k)!(l-k)!]^{\frac{1}{2}}} (\alpha z_1)^s (\beta z_2)^{l+i-s} (\gamma z_1)^t (\delta z_2)^{l-k-t}$$

$$= \sum_{s,t} z_1^{s+t} z_2^{2l-s-t} \alpha^s \beta^{l+k-s} \gamma^t \delta^{l-k-t} A_{l,i,j,s,t},$$

where $I_1 = (0, 1, \dots, l+k)$, $I_2 = (0, 1, \dots, l-k)$.

We are concerned with the coefficient at $e_j = \dfrac{z_1^{l+j} z_2^{l-j}}{(l+j)!(l-j)!}$.

This monomial appears in the above sum whenever $s+t = l+j$. These are the summands for which $t \in I_2$ and $s = l+j-t \in I_1$, i.e. $t \in I_3 = \{p\colon 0 \leqslant p \leqslant l-k \text{ and } j-k \leqslant p \leqslant l+j\}$. Hence we get

$$t_{jk}^l(g) = \sum_{t \in I_3} \frac{[(l+k)!(l-k)!(l-j)!(l+j)!]^{\frac{1}{2}}}{(l+j-t)!(k-j+t)!t!(l-k-t)!}$$

$$\times \alpha^{l+j-t} \beta^{k-j+t} \delta^{l-k-t} \gamma^t = [(l+k)!(l-k)!(l+j)!(l-j)!]^{\frac{1}{2}}$$

$$\times \alpha^{l+j} \beta^{k-j} \delta^{l-k} \sum_{t \in I_3} \frac{1}{(l+j-t)!(k-j+t)!t!(l-k-t)!} \left(\frac{\alpha\delta - 1}{\alpha\delta}\right)^t. \tag{8.1.10}$$

Note that the sum is taken over the very set of indices for which the summands are well-defined. Upon substituting $g = g(\theta)$ it follows that

$$P_{jk}^l(\cos\theta) = [(l+k)!(l-k)!(l+j)!(l-j)!]^{\frac{1}{2}} (\cos\tfrac{1}{2}\theta)^{2l+j-k} i^{k-j}$$

$$\times (\sin\tfrac{1}{2}\theta)^{k-j} \sum_{t \in I_3} \frac{(-1)^t}{(l+j-t)!(k-j+t)!t!(l-k-t)!} (\tan\tfrac{1}{2}\theta)^{2t}.$$

As before, put $\cos\theta = z$:

$$P^l_{jk}(z) = [(l+k)!(l-k)!(l+j)!(l-j)!]^{\frac{1}{2}} i^{k-j}\left(\frac{1+z}{2}\right)^l\left(\frac{1-z}{1+z}\right)^{\frac{k-j}{2}}$$

$$\times \sum_{t\in I_3} \frac{(-1)^t}{(l+j-t)!(k-j+t)!t!(l-k-t)!}\left(\frac{1-z}{1+z}\right)^t. \tag{8.1.11}$$

Finally we can compute the matrix element t^l_{jk} from formula (8.1.2) in view of the orthonormality of the set of characters of the unit circle. We put $z_1 = e^{i\varphi}$, $z_2 = 1$ in formula (8.1.2) and integrate both sides with respect to $d\varphi$ with the factor $e^{-i(l+j)\varphi}$. Since

$$\frac{1}{2\pi}\int_0^{2\pi} e^{i(k-l)\varphi}d\varphi = \delta_{kl},$$

$$t^l_{jk}(g) = \left[\frac{(l+j)!(l-j)!}{(l+k)!(l-k)!}\right]^{\frac{1}{2}} \frac{1}{2\pi}\int_0^{2\pi} (\alpha e^{i\varphi}+\gamma)^{l+k}$$

$$\times (\beta e^{i\varphi}+\delta)^{l-k}e^{-i(l+j)\varphi}d\varphi$$

$$= \frac{1}{2\pi i}\left[\frac{(l+j)!(l-j)!}{(l+k)!(l-k)!}\right]^{\frac{1}{2}} \oint_\Gamma (\alpha\zeta+\delta)^{l+k}(\beta\zeta+\delta)^{l-k}\zeta^{-l-j-1}d\zeta, \tag{8.1.12}$$

where Γ is the unit circle centred at zero. The resulting expression for P^l_{jk} has the form

$$P^l_{jk}(z) = \frac{1}{2\pi i}\left[\frac{(l+j)!(l-j)!}{(l+k)!(l-k)!}\right]^{\frac{1}{2}},$$

$$\times \oint (\zeta\cos\tfrac{1}{2}\theta+i\sin\tfrac{1}{2}\theta)^{l-k}(i\zeta\sin\tfrac{1}{2}\theta+\cos\tfrac{1}{2}\theta)^{l+k}\zeta^{-l-j-1}d\zeta. \tag{8.1.13}$$

In particular, if $k = 0$ we obtain the following form of the associated Legendre functions:

$$P^m_l(z) = \frac{i^m(l+m)!}{2\pi l!}\int_0^{2\pi} (\cos\theta+i\sin\theta\cos\varphi)^l e^{im\varphi}d\varphi \tag{8.1.14}$$

(l, m—integers).

8.2. PROPERTIES OF THE REPRESENTATIONS T^l AND THEIR CONSEQUENCES

The representation T^l acts on the $(2l+1)$-dimensional linear space H^l over the field of complex numbers, l being a non-negative half-integer. We provide H^l with an inner product by putting

$$(e_k|e_j) = \delta_{kj} \tag{8.2.1}$$

for the basis $\{e_i\}$ defined previously.

PROPOSITION 8.2.1. *The representation T^l of the group* SU(2), *which is a subgroup of* SL(2, *C*), *is unitary with respect to the inner product so defined.*

Proof. We introduce a function f: $C^2 \times C^2 \to C$ according to the formula

$$f(z_1, z_2;\ w_1, w_2) = \sum_{j=-l}^{l} \bar{e}_j(z_1, z_2) e_j(w_1, w_2).$$

LEMMA 8.2.2. *The function f is invariant with respect to the natural action of* SU(2) *on $C^2 \times C^2$, i.e.*

$$f\big(g(z_1, z_2); g(w_1, w_2)\big) = f(z_1, z_2;\ w_1, w_2).$$

Proof. $f(z_1, z_2;\ w_1, w_2) = \displaystyle\sum_j \frac{\bar{z}_1^{l+j}\bar{z}_2^{l-j}w_1^{l+j}w_2^{l-j}}{(l+j)!(l-j)!} = \frac{(\bar{z}_1 w_1 + \bar{z}_2 w_2)^{2l}}{(2l)!}$.

Thus the function f is proportional to the $2l$-th power of a bilinear form which is SU(2) invariant, as can be seen directly from the definition of the group. This proves the lemma. □

The invariance of f is

$$\sum_j \left(\overline{T_g^l e_j(z_1, z_2)}\, T_g^l e_j(w_1, w_2) - \bar{e}_j(z_1, z_2) e_j(w_1, w_2)\right) = 0$$

and in the explicit form

$$\sum_j \Big(\sum_s \bar{t}_{sj}(g)\bar{e}_s(z_1, z_2) \sum_k t_{kj} e_k(w_1, w_2) - \bar{e}_j(z_1, z_2) e_j(w_1, w_2)\Big) = 0,$$

$$\sum_j \sum_{sk} \left(\bar{t}_{sj}(g)\, t_{kj}(g) - \delta_{kj}\delta_{sj}\right)\bar{e}_s(z_1, z_2) e_k(w_1, w_2)\big) = 0.$$

Interchanging the order of summation and using the fact that the system of functions $\bar{e}_j e_k$ is linearly independent, we see that

$$\sum_j \bar{t}_{sj}(g)\, t_{kj}(g) = \sum_j \delta_{kj}\delta_{sj} = \delta_{ks}. \tag{8.2.2}$$

This proves the unitarity of the matrix t_{sk} in the basis e_j, which is equivalent by Proposition 2.4.2, to the unitarity of T^l with respect to the inner product defined by identity (8.2.1). □

PROPOSITION 8.2.3. *The representation T^l of the group* SU(2) *is irreducible.*

Proof. We first observe that the vectors e_k are eigenvectors of the operators T_g^l for $g = g(t) = \begin{bmatrix} e^{\frac{1}{2}it} & 0 \\ 0 & e^{-\frac{1}{2}it} \end{bmatrix}$. More precisely, it follows from (8.1.2) that

$$T_{g(t)}^l e_k = e^{ikt} e_k.$$

Hence the vectors e_k have different eigenvalues. If the space were decomposable into a direct sum of invariant orthogonal subspaces, then every vector e_k would belong to one of them. Consequently, for every e_k there would be a matrix element $t_{kj} = 0$, namely that for which e_j belongs to the invariant subspace orthogonal to e_k. We compute

$$t_{lj}(g(\theta)) = P_{lj}^l(\cos\theta) = \frac{[(2l)!]^{\frac{1}{2}}}{(l+j)^{\frac{1}{2}}(l-j)^{\frac{1}{2}}}\, i^{l-j}(1+z)^{\frac{l+j}{2}}(1-z)^{\frac{l-j}{2}}$$

(in formula (8.1.11) only the summand corresponding to $t = 0$ is different from zero). Since the matrix elements t_{lj} are non-zero for arbitrary j, we get a contradiction of the assumption of reducibility of T^l. This concludes the proof. □

We shall now transfer onto the functions $P_{j,k}^l$ the properties of matrix elements of irreducible unitary representations which were established in § 8.1.

The equation

$$t_{jn}^l(sz) = \sum_k t_{jk}^l(s)\, t_{kn}^l(z)$$

when applied to the representation of the group SL(2, C), provides a new expression for the function P^l_{jk}. An element $g = \begin{bmatrix} \alpha & \beta \\ \gamma & \delta \end{bmatrix}$ with $\alpha \neq 0$ can be represented in the form

$$g = sz = \begin{bmatrix} 1 & 0 \\ \gamma/\alpha & 1 \end{bmatrix} \begin{bmatrix} \alpha & \beta \\ 0 & \alpha^{-1} \end{bmatrix}.$$

Hence the problem of computing $t_{ij}(g)$ reduces to the computation of matrix elements for triangular matrices. By formula (8.1.10) we find

$$t^l_{js}(s) = \begin{cases} 0, & s > j, \\ \left[\dfrac{(l-s)!(l+j)!}{(l+s)!(l-j)!}\right]^{\frac{1}{2}} \dfrac{1}{(j-s)!} \left(\dfrac{\gamma}{\alpha}\right)^{j-s}, & s \leqslant j, \end{cases}$$

$$t^l_{sk}z = \begin{cases} 0, & s > k, \\ \left[\dfrac{(l-s)!(l+k)!}{(l+s)!(l-k)!}\right]^{\frac{1}{2}} \dfrac{1}{(k-s)!} \alpha^{s+k}\beta^{k-s}, & s \leqslant k. \end{cases}$$

And finally

$$t^l_{jk}(g) = \left[\frac{(l+j)!(l+k)!}{(l-j)!(l-k)!}\right]^{\frac{1}{2}} \gamma^j \alpha^{k-j} \beta^k \sum_{s=\max(k,j)}^{l} \alpha^{-2s}(\beta\gamma)^s$$
$$\times \frac{(l+s)!}{(l-s)!(j+s)!(i+s)!}.$$

Hence

$$P^l_{jk}(z) = \left[\frac{(l+j)!(l+k)!}{(l-j)!(l-k)!}\right]^{\frac{1}{2}} i^{k+j} \left(\frac{1-z}{2}\right)^{\frac{k+j}{2}} \left(\frac{1+z}{2}\right)^{\frac{k-j}{2}}$$
$$\times \sum_{s=\max(k,j)}^{l} \frac{(l+s)!}{(l-s)!(j+s)!(i+s)!} \left(\frac{1-z}{1+z}\right)^s. \tag{8.2.3}$$

The advantage of this formula over formula (8.1.11), consists in the fact that the sum ranges here over a more convenient set of indices.

Now we shall make use of the identity

$$g(\pi)g(\theta) = g(\theta)g(\pi), \tag{8.2.4}$$

where $g(\theta)$ denotes as before

$$g(\theta) = \begin{bmatrix} \cos\frac{1}{2}\theta & i\sin\frac{1}{2}\theta \\ i\sin\frac{1}{2}\theta & \cos\frac{1}{2}\theta \end{bmatrix}.$$

The matrix element $t^l_{jk}(\pi)$ is computed from formula (8.1.11) since (8.2.3) applies to the case $\alpha \neq 0$

$$t^l_{kj}(\pi) = \begin{cases} 0 & k+j \neq 0 \\ i^{2l} & k+j = 0 \end{cases}$$

As a consequence of (8.2.4) we obtain the first symmetry rule:

$$\sum_n \delta_{n,-j} t^l_{nk}\big(g(\theta)\big) = \sum_n t^l_{jn}\big(g(\theta)\big)\delta_{k,-n}, \tag{8.2.5}$$

$$P^l_{-j,k} = P^l_{j,-k}.$$

Since $g(-\theta)$ is the element inverse to $g(\theta)$, we have $T^l\big(g(-\theta)\big) = \big(T^l(g(\theta))\big)^{-1}$. And hence the unitarity of the representation of SU(2) implies

$$t^l_{jk}\big(g(-\theta)\big) = \bar{P}^l_{jk}(\cos\theta).$$

On the basis of formula (8.1.10) we find directly that

$$t^l_{jk}\big(g(-\theta)\big) = (-1)^{k-j}P^l_{kj}(\cos\theta).$$

The same formula also implies that

$$\bar{P}^l_{jk} = (-1)^{k-j}P^l_{jk}.$$

Combining the above, we obtain the second symmetry rule:

$$P^l_{kj} = P^l_{jk}. \tag{8.2.6}$$

8.3. INTEGRAL EQUATIONS FOR THE FUNCTIONS P^l_{jk}

Throughout the present section G will denote the group SU(2) and K will be its subgroup composed of all matrices of the form $\begin{bmatrix} e^{i\varphi} & 0 \\ 0 & e^{-i\varphi} \end{bmatrix}$.

The pair (G, K) constitutes a symmetric space and all the assertions proved below, follow by regarding the matrix elements of the representation T^l as spherical functions.

On the underlying space of the group G we shall define the coordinates called *Euler's angles*. These coordinates will be given a geometrical meaning in the study of the transformation group G, G/K: the homogeneous space G/K is a sphere in $\boldsymbol{R}^3$ on which G acts by rotations.

The group SU(2) consisting of unimodular matrices of the form

$$g = \begin{bmatrix} \alpha & \beta \\ -\bar{\beta} & \bar{\alpha} \end{bmatrix}$$

is a three-dimensional manifold: the unit sphere $|\alpha|^2 + |\beta|^2 = 1$ in $\boldsymbol{C}^2$. For $\alpha\beta \neq 0$ we may select $0 < \theta < \pi$ so that $|\alpha| = \cos\frac{1}{2}\theta$. Then the angles $\operatorname{Arg}\alpha$ and $\operatorname{Arg}\beta$ uniquely determine the matrix g. We introduce new parameters φ, ψ by requiring

$$\operatorname{Arg}\alpha = \frac{\varphi+\psi}{2}, \qquad \operatorname{Arg}\beta = \frac{\varphi-\psi+\pi}{2}.$$

When φ and ψ run through the sets $[0, 2\pi[$ and $[-2\pi, 2\pi[$, respectively, $\operatorname{Arg}\alpha$ and $\operatorname{Arg}\beta$ run through the set $[0, 2\pi[$.

Therefore every element such that $\alpha\beta \neq 0$ (such elements form a dense subset of G) has the representation

$$\begin{aligned} g &= \begin{bmatrix} \cos\frac{1}{2}\theta\, e^{\frac{1}{2}i(\varphi+\psi)} & i\sin\frac{1}{2}\theta\, e^{\frac{1}{2}i(\varphi-\psi)} \\ i\sin\frac{1}{2}\theta\, e^{\frac{1}{2}i(\psi-\varphi)} & \cos\frac{1}{2}\theta\, e^{-\frac{1}{2}i(\varphi+\psi)} \end{bmatrix} \\ &= \begin{bmatrix} e^{\frac{1}{2}i\varphi} & 0 \\ 0 & e^{-\frac{1}{2}i\varphi} \end{bmatrix} \begin{bmatrix} \cos\frac{1}{2}\theta & i\sin\frac{1}{2}\theta \\ i\sin\frac{1}{2}\theta & \cos\frac{1}{2}\theta \end{bmatrix} \begin{bmatrix} e^{\frac{1}{2}i\psi} & 0 \\ 0 & e^{-\frac{1}{2}i\psi} \end{bmatrix} \\ &:= k(\varphi)a(\theta)k(\psi). \end{aligned} \tag{8.3.1}$$

The coordinates (φ, θ, ψ) are called *Euler's angles of the element g*.

The group SU(2) acts on the compactified complex plane by conformal mappings

$$g(z) = \frac{\alpha z+\beta}{-\bar{\beta}z+\bar{\alpha}}.$$

The subgroup K comprised of all elements of the form $k(\psi)$ is the isotropy group at the point $z = 0$ and its action on $\boldsymbol{C}$ consists in rotations of angle ψ about zero.

The subgroup A of all elements of the form $a(\theta)$ moves the point 0 along the imaginary axis in such a way that for $0 \leqslant \theta \leqslant \pi$ the point $a(\theta)\cdot 0$ runs through the positive part of the axis. Thus the parameters φ and θ correspond to spherical coordinates on the plane.

The Lie algebra of the group SU(2) is isomorphic to the algebra of anti-hermitian matrices of trace zero. Through this identification the adjoint representation acts by inner automorphisms

$$\operatorname{Ad}g(X) = gXg^{-1}, \quad \text{for } g \in \mathrm{SU}(2),\ X \in \mathfrak{g}.$$

We introduce coordinates in the space $\mathfrak{g}$ by putting

$$X = \begin{bmatrix} x_1 & x_2+ix_3 \\ x_2-ix_3 & -x_1 \end{bmatrix}.$$

Then $\det X = -(x_1^2+x_2^2+x_3^2)$. The obvious identity $\det \mathrm{Ad}g(X) = \det X$ means that $\mathrm{Ad}g \in \mathrm{O}(3)$.

The image of the mapping α: $\mathrm{SU}(2) \ni g \to \mathrm{Ad}g \in \mathrm{O}(3)$ being a connected subset of the identity, we see that $\mathrm{Ad}g$ is actually in $\mathrm{SO}(3)$.

Let us compute explicitly the result of the action of an element $g = \begin{bmatrix} \alpha & \beta \\ -\bar{\beta} & \bar{\alpha} \end{bmatrix}$ on $X = \begin{bmatrix} x & z \\ \bar{z} & -x \end{bmatrix}$:

$$\begin{bmatrix} \alpha & \beta \\ -\bar{\beta} & \bar{\alpha} \end{bmatrix}\begin{bmatrix} x_1 & z \\ \bar{z} & -x_1 \end{bmatrix}\begin{bmatrix} \bar{\alpha} & -\beta \\ \bar{\beta} & \alpha \end{bmatrix}$$

$$= \begin{bmatrix} (\alpha\bar{\alpha}-\beta\bar{\beta})x_1+\bar{\alpha}\beta z+\alpha\bar{\beta}\bar{z} & -2\alpha\beta x_1-\beta^2\bar{z}+\alpha^2 z \\ -2\bar{\alpha}\bar{\beta}x_1+\bar{\alpha}^2\bar{z}-\bar{\beta}^2 z & (\bar{\beta}\beta-\alpha\bar{\alpha})x_1-\bar{\alpha}\beta\bar{z}-\alpha\bar{\beta}\bar{z} \end{bmatrix}. \tag{8.3.2}$$

In particular we get the formula

$$\alpha(k(\varphi)) = \begin{bmatrix} 1 & 0 & 0 \\ 0 & \cos\varphi & -\sin\varphi \\ 0 & \sin\varphi & \cos\varphi \end{bmatrix},$$

which means that the operator $\alpha(k(\varphi))$ is the rotation through angle φ about the axis x_1. Accordingly $\alpha(a(\theta))$ agrees with the rotation through angle θ about x_2.

Rotations about two axes generate the whole group $\mathrm{SO}(3)$; hence it follows that α is surjective. The kernel of the homomorphism α consists of two elements:

$$\ker\alpha = \left\{\begin{bmatrix} 1 & 0 \\ 0 & 1 \end{bmatrix}, \begin{bmatrix} -1 & 0 \\ 0 & -1 \end{bmatrix}\right\}.$$

This being so, it follows that the group $\mathrm{SU}(2)$ is a two-fold covering of $\mathrm{SO}(3)$. Hence we have the diagram

$$\begin{array}{ccc} \mathrm{SU}(2) & \xrightarrow{\alpha} & \mathrm{SO}(3) \\ {\scriptstyle\pi_1}\downarrow & \sigma & \downarrow{\scriptstyle\pi_2} \\ P^1 & \leftarrow & S^2 \end{array}$$

Here π_1 is the projection onto the quotient space $SU(2)/K$ which is isomorphic to the compactified complex plane P^1 and π_2 is the projection onto $SO(3)/\alpha(K) = S^2$. The mapping σ is the stereographic projection of the unit sphere S^2 from the point $(1, 0, 0)$ onto the plane $x_1 = 0$ which is identified with the complex plane.

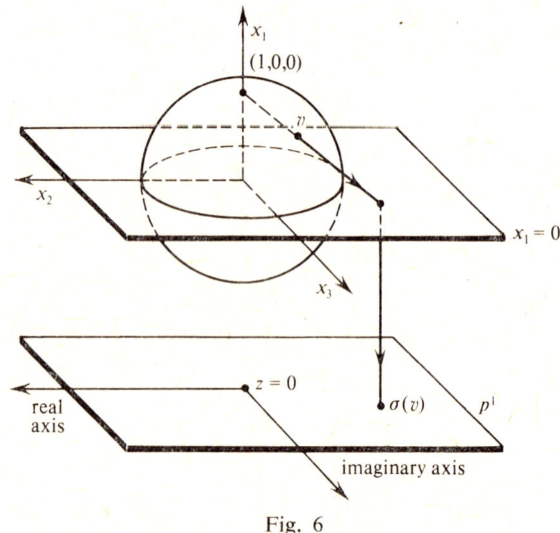

Fig. 6

In an explicit form

$$\sigma(x_1, x_2, x_3) := (x_2 + ix_3)(1 - x_1)^{-1},$$
$$\sigma(1, 0, 0) := \infty.$$

Applying formula (8.3.2) we find for $v = (x_1, x_2, x_3)$

$$\sigma(\alpha(g)v) = \frac{\alpha^2 z - \beta^2 \bar{z} - 2\alpha\beta x_1}{1 - 2(\alpha\bar{\alpha} - \beta\bar{\beta})x_1 + 2\bar{\alpha}\beta\bar{z} + 2\alpha\bar{\beta}z}$$
$$= \frac{\alpha z + \beta(1 - x_1)}{-\bar{\beta}z + \bar{\alpha}(1 - x_1)} = g\sigma(v),$$

where $z = x_2 + ix_3$.

Thus we obtain the formula

$$\sigma \circ (g) = g \circ \sigma \tag{8.3.3}$$

which denotes that the diagram represents a homomorphism of homogeneous spaces.

Now, the invariant measure on SU(2) can be expressed in terms of Euler's angles as

$$\int_G f\mathrm{d}g = \frac{1}{16\pi^2}\int_{-2\pi}^{2\pi}\int_0^{2\pi}\int_0^{\pi} f(\varphi, \theta, \psi)\sin\theta\mathrm{d}\varphi\mathrm{d}\psi\mathrm{d}\theta; \tag{8.3.4}$$

$\frac{1}{16\pi^2}$ is the normalizing factor. Further, we observe that in view of decomposition (8.3.1) the manifold $G/K \simeq S^2$ is parametrized by the angles φ and θ, which determine the longitude and the latitude on the sphere measured relative to the axis x_1 in the plane (x_1, x_2) and to the "south pole" respectively.

The measure

$$\int_K f\mathrm{d}k = \frac{1}{4\pi}\int_{-2\pi}^{2\pi} f(\mathrm{e}^{i\psi})\mathrm{d}\psi$$

is the normalized invariant measure on K and

$$\int_{S^2} f(x)\mathrm{d}m(x) := \frac{1}{4\pi}\int_0^{2\pi}\int_0^{\pi} f(\varphi, \theta)\sin\theta\mathrm{d}\varphi\mathrm{d}\theta \tag{8.3.5}$$

is the normalized invariant measure on S^2. Identity (8.3.3) serves as an example of a decomposition of the Haar measure on a group with respect to the invariant measure on a homogeneous space.

We end our preliminary discussion devoted to establishing the integral properties of the functions P^l_{ij} by expressing the matrix elements of the representation in terms of Euler's angles.

Let $g = k(\varphi)a(\theta)k(\psi)$ then

$$\begin{aligned} t^l_{mn}(g) &= \langle e^m, T^l_g e_n\rangle = (e_m | T^l_g e_n) = (e_m | T^l_{k(\varphi)} T^l_{a(\theta)} T^l_{k(\psi)} e_n) \\ &= (T^l_{k(-\varphi)} e_m | T^l_{a(\theta)} T^l_{k(\psi)} e_n) = \mathrm{e}^{i(m\varphi + n\psi)} t^l_{mn}(a(\theta)) \\ &= \mathrm{e}^{i(m\varphi + n\psi)} P(\cos\theta). \end{aligned} \tag{8.3.6}$$

Orthogonality Relations

We shall apply Schur's orthogonality relations to matrix elements in the form (8.3.4). Actually, the orthogonality of t^l_{mn} and $t^{l'}_{m'n'}$ for $m \neq m'$, $l \neq l'$ is evident in view of the form of the integral on the group and of the fact that the exponential factors appearing in (8.3.4), being characters of T^l, are mutually orthogonal. In these cases the orthogonality relations do not provide additional information about the functions P^l_{mm}. However, for $m = m'$ and $n = n'$ these relations imply

$$\frac{1}{2}\int_0^\pi P^l_{mn}(\cos\theta) P^{l'}_{mn}(\cos\theta)\sin\theta\,d\theta = \frac{1}{2l+1}\,\delta_{ll'}, \tag{8.3.7}$$

and by substituting $\cos\theta = x$ we obtain

$$\int_{-1}^{1} \overline{P^l_{mn}(x)} P^l_{mn}(x)\,dx = \frac{2}{2l+1}\,\delta_{ll'}. \tag{8.3.8}$$

This formula is applicable in constructing orthonormal systems in the space $L^2([-1, 1], dx)$ (see Problem 23).

A Mean Value Theorem (Spherical Equation)

We shall now use the fact that the matrix element t^l_{mj} is a $\varkappa_j$-spherical function with respect to the pair (G, K) $(\varkappa_j(k(\beta)) = e^{ij\gamma})$. Applying formula (5.1.2) to $g = a(\theta_1)$, $h = a(\theta_2)$ and denoting by φ, θ, ψ Euler's angles of the element gkh, we obtain

$$\frac{1}{2\pi}\int_{-\pi}^{\pi} P^l_{mn}(\cos\theta) e^{-i(j\gamma - m\varphi - n\psi)}d\gamma = P^l_{mj}(\cos\theta_1) P^l_{jn}(\cos\theta_2). \tag{8.3.9}$$

We shall find the dependence of the angles φ, ψ, θ upon β, θ_1, θ_2. By (8.3.1), we see that Euler's angles of the matrix $\begin{bmatrix} \alpha & \beta \\ -\bar\beta & \bar\alpha \end{bmatrix}$ satisfy the identities

$$\begin{aligned} \cos\theta &= 2|\alpha|^2 - 1, \\ e^{i\varphi} &= -\frac{\alpha\beta i}{|\alpha|\,|\beta|}, \\ e^{\frac{1}{2}i\psi} &= \frac{\alpha e^{-\frac{1}{2}i\varphi}}{|\alpha|}. \end{aligned} \tag{8.3.10}$$

By substituting the coefficients of the matrix $a(\theta_1)k(\gamma)a(\theta_2)$ in identities (8.3.9) we get

$$\cos\theta = \cos\theta_1\cos\theta_2 - \sin\theta_1\sin\theta_2\cos\gamma,$$

$$e^{i\varphi} = \frac{\sin\theta_1\cos\theta_2 + \cos\theta_1\sin\theta_2\cos\gamma + i\sin\theta_2\sin\gamma}{\sin\theta}, \tag{8.3.11}$$

$$e^{\frac{1}{2}i(\varphi+\psi)} = \frac{\cos\frac{1}{2}\theta_1\cos\frac{1}{2}\theta_2 e^{\frac{1}{2}i\gamma} - \sin\frac{1}{2}\theta_1\sin\frac{1}{2}\theta_2 e^{-\frac{1}{2}i\gamma}}{\cos\frac{1}{2}\theta}.$$

We shall consider a special case of formula (8.3.8) corresponding to the zonal spherical function t^l_{00} for an integer value of l:

$$t^l_{00}(g) = P^l_{00}(\cos\theta) = P_l(\cos\theta),$$

where θ is the second Euler angle of $g \in G$. We have

$$\frac{1}{2\pi}\int_0^{2\pi} P_l(\cos\theta_1\cos\theta_2 - \sin\theta_1\sin\theta_2\cos\gamma)\,d\gamma$$
$$= P_l(\cos\theta_1)P_l(\cos\theta_2). \tag{8.3.12}$$

This is the spherical equation for Legendre polynomials.

As in the case of Bessel functions, equation (8.3.11) can be interpreted as a mean value theorem for a spherical function. To this end we note that the curve defined on the sphere as

$$[-\pi, \pi[\ni \gamma \to k(\gamma)a(\theta_2)\cdot x_0 \in S^2$$

with $x_0 = (0, 0, 1)$ agrees with the circle given by $x_3 = -\cos\theta_2$. Next, the set $\{a(\theta_1)k(\gamma)a(\theta_2)\cdot x_0 \colon \gamma \in [-\pi, \pi[\}$ is a circle in S^2 with centre $(0, \sin\theta_1, -\cos\theta_1)$ of radius θ_2 (in the arc measure on S^2).

The left-hand side of (8.3.10), represents the mean value of the function P_l over this circle which is equal, as is shown by the right-hand side, to the product of the value of P_l at the centre of the circle and the value at a point lying at a distance of θ_2 from x_0.

Addition Formulas for Functions P^l_{mn}

These formulas are a consequence of identity (2.4.1)

$$t^l_{mn}(gh) = \sum_{k=-l}^{l} t^l_{mk}(g)t^l_{kn}(h).$$

We shall apply this identity to the special values $g = (0, \theta_1, 0)$, $h = (\gamma, \theta_2, 0)$. Keeping in mind that Euler's angles of the product gh are again given by (8.3.10), we arrive at

$$P^l_{mn}(\cos\theta)\mathrm{e}^{i(m\varphi+n\psi)} = \sum_{k=-l}^{l} P^l_{mk}(\cos\theta_1)P^l_{kn}(\cos\theta_2)\mathrm{e}^{ik\gamma}. \tag{8.3.13}$$

We shall consider a number of cases in which identities (8.3.10) can easily be solved.

Put $\gamma = 0$ and suppose $\theta_1+\theta_2 < \pi$; then $\theta = \theta_1+\theta_2$, $\varphi = \psi = 0$ and

$$P^l_{mn}(\cos(\theta_1+\theta_2)) = \sum_{k=-l}^{l} P^l_{mk}(\cos\theta_1)P^l_{kn}(\cos\theta_2). \tag{8.3.14}$$

Next, for $\gamma = \pi$, $\theta_1 \geqslant \theta_2$, we obtain $\theta = \theta_1-\theta_2$, $\varphi = 0$, $\psi = \pi$

$$P^l_{mn}(\cos(\theta_1-\theta_2)) = \sum_{k=-l}^{l} (-1)^{k-m}P^l_{mk}(\cos\theta_1)P^l_{kn}(\cos\theta_2). \tag{8.3.15}$$

For $\theta_1 = \theta_2$ we observe that formula (8.3.15) coincides with the orthogonality conditions for representation matrices:

$$\sum_{k=-l}^{l} \overline{P^l_{mk}}P^l_{nk} = \delta_{mn}.$$

8.4. THE DIFFERENTIAL OF THE REPRESENTATION T^l. RECURRENCE AND DIFFERENTIAL EQUATIONS FOR THE FUNCTIONS P^l_{mn}

The one-parameter subgroups of SU(2)

$$a_1(t) = \begin{bmatrix}\cos\frac{1}{2}t & i\sin\frac{1}{2}t\\ i\sin\frac{1}{2}t & \cos\frac{1}{2}t\end{bmatrix}, \quad a_2(t) = \begin{bmatrix}\cos\frac{1}{2}t & -\sin\frac{1}{2}t\\ \sin\frac{1}{2}t & \cos\frac{1}{2}t\end{bmatrix},$$

$$k(t) = a_3(t) = \begin{bmatrix}\mathrm{e}^{\frac{1}{2}it} & 0\\ 0 & \mathrm{e}^{-\frac{1}{2}it}\end{bmatrix} \tag{8.4.1}$$

correspond, via the homomorphism α: SU(2) → SO(3) established in § 8.3 to rotations about the respective axes in $\boldsymbol{R}^3$. The corresponding infinitesimal vectors form a basis for $\mathfrak{su}(2)$ and are known as *Pauli's matrices*. Direct calculation shows

$$X_1 = \frac{i}{2}\begin{bmatrix} 0 & 1 \\ 1 & 0 \end{bmatrix}, \quad X_2 = \frac{1}{2}\begin{bmatrix} 0 & -1 \\ 1 & 0 \end{bmatrix}, \quad X_3 = \frac{i}{2}\begin{bmatrix} 1 & 0 \\ 0 & -1 \end{bmatrix}, \tag{8.4.2}$$

and the commutation relations are the following:

$$[X_1, X_2] = X_3, \qquad [X_2, X_3] = X_1, \qquad [X_3, X_1] = X_2. \tag{8.4.3}$$

We shall now compute the differential of the representation T^l. The elements e_k of the basis for H^l are eigenvectors for the generators of the subgroup a_3

$$\mathrm{d}T^l(X_3)e_k = ike_k.$$

The vectors $\mathrm{d}T^l(X_1), \mathrm{d}T^l(X_2)$ are computed by differentiating at $s = 0$ the mapping

$$s \to (a_j(s))^t(z_1, z_2) \to e_k((a_j(s))^t(z_1, z_2)), \qquad j = 1, 2.$$

The derivative of the first mapping is given by (8.4.2) since the action of $a_j(s)$ on C^2 is a natural action of a 2×2 matrix. For $j = 1$ this derivative has the value $\frac{1}{2}i(z_2, z_1)$ and for $j = 2$—the value $\frac{1}{2}(z_2, -z_1)$.

The derivative of the mapping is computed directly:

$$\begin{bmatrix} \dfrac{\partial e_k}{\partial z_1} \\[2ex] \dfrac{\partial e_k}{\partial z_2} \end{bmatrix} = \begin{bmatrix} \dfrac{z_1^{l+k-1} z_2^{l-k}}{[(l+k)!\,(l-k)!]^{1/2}}(l+k) \\[2ex] \dfrac{z_1^{l+k} z_2^{l-k-1}}{[(l+k)!\,(l-k)!]^{1/2}}(l-k) \end{bmatrix}.$$

Finally, we find

$$\mathrm{d}T^l(X_1)e_k(z_1, z_2) = \frac{i}{2}\,\frac{z_1^{l+k-1} z_2^{l-k+1}}{[(l+k)!\,(l-k)!]^{1/2}}(l+k) + \frac{i}{2}\,\frac{z_1^{l+k+1} z_2^{l-k-1}}{[(l+k)!\,(l-k)!]^{1/2}}, \tag{8.4.4}$$

$$\mathrm{d}T^l(X_2)e_k(z_1, z_2) = \frac{1}{2}\,\frac{z_1^{l+k-1} z_2^{l-k+1}}{[(l+k)!\,(l-k)!]^{1/2}}(l+k) - \frac{1}{2}\,\frac{z_1^{l+k+1} z_2^{l-k-1}}{[(l+k)!\,(l-k)!]^{1/2}}(l-k).$$

It is useful to introduce the operators

$$\begin{aligned} H_+ &= i\,\mathrm{d}T^l(X_1)+\mathrm{d}T^l(X_2), \\ H_- &= i\,\mathrm{d}T^l(X_1)-\mathrm{d}T^l(X_2). \end{aligned} \tag{8.4.5}$$

Then with the help of (8.4.4) we obtain

$$\begin{cases} H_+ e_k = -\sqrt{(l-k)(l+k+1)}e_{k+1} \\ H_- e_k = -\sqrt{(l+k)(l-k+1)}e_{k-1} \\ \mathrm{d}T^l(X_3)e_k = ike_k. \end{cases} \tag{8.4.6}$$

Note that these formulas immediately imply the irreducibility of the representation T^l.

We shall use the fact that the operation

$$S_y\colon H^l \ni x \to \overline{t^l_{x,y}} = \check{t}^l_{y,x}$$

intertwines the representation T^l with the left regular representation of G on $\mathscr{E}(G)$. This gives the formulas

$$\begin{aligned} \mathrm{d}L(H_+)\check{t}^l_{m,k} &:= \big(i\,\mathrm{d}L(X_1)+\mathrm{d}L(X_2)\big)\check{t}^l_{m,k} \\ &= -\sqrt{(l-k)(l+k+1)}\;\check{t}^l_{m,k+1}, \\ \mathrm{d}L(H_-)\check{t}^l_{m,k} &:= \big(i\,\mathrm{d}L(X_1)-\mathrm{d}L(X_2)\big)\check{t}^l_{m,k} \\ &= -\sqrt{(l+k)(l-k+1)}\;t^l_{m,k-1}. \\ \mathrm{d}L(X_3)\check{t}^l_{m,k} &:= ik\check{t}^l_{m,k} \qquad (\text{where } \check{t}(g) := t(g^{-1})). \end{aligned} \tag{8.4.7}$$

These formulas will be a source of differential equations of the recurrence type for the system of functions $t^l_{m,k}$ and then for $P^l_{m,k}$ provided we know the operators $\mathrm{d}L(H_\pm)$.

In order to compute the operator $\mathrm{d}L(X_1)$ we apply formulas (8.3.10), which express in an implicit way the relations between Euler's angles of the element $a_1(t)k(\varphi_2)a_1(\theta_2)$ and the angles t, φ_2, θ_2. It follows from these formulas that Euler's angles φ, θ, ψ of the element $a_1(t)k(\varphi_2)a_1(\theta_2)k(\psi_2)$ are related with $t, \varphi_2, \theta_2, \psi_2$ in the following manner:

$$\begin{aligned} \cos\theta &= \cos t\cos\theta_2-\sin t\sin\theta_2\cos\varphi, \\ \sin\theta\,\mathrm{e}^{i\varphi} &= \sin t\cos\theta_2+\cos t\sin\theta_2\cos\varphi_2+i\sin\theta_2\sin\varphi_2, \\ \cos\tfrac{1}{2}\theta\,\mathrm{e}^{\frac{1}{2}i(\varphi+\psi)} &= \mathrm{e}^{\frac{1}{2}i\psi_2}\Big(\cos\tfrac{1}{2}\theta_1\cos\tfrac{1}{2}\theta_2\mathrm{e}^{i\varphi_2}-\sin\tfrac{1}{2}\theta_1\sin\tfrac{1}{2}\theta_2\mathrm{e}^{-\frac{1}{2}i\varphi_2}\Big). \end{aligned} \tag{8.4.8}$$

Let us differentiate (8.4.8) at the point $t = 0$. The resulting system is easily solved, and consequently

$$\mathrm{d}L(X_1) = -\cos\varphi\frac{\partial}{\partial\theta} - \frac{\sin\varphi}{\sin\theta}\frac{\partial}{\partial\psi} + \cot\theta\sin\varphi\frac{\partial}{\partial\varphi}.$$

A similar computation for the one-parameter subgroup a_2 leads to the formula

$$\mathrm{d}L(X_2) = \sin\varphi\frac{\partial}{\partial\theta} - \frac{\cos\varphi}{\sin\theta}\frac{\partial}{\partial\psi} + \cot\theta\cos\varphi\frac{\partial}{\partial\varphi}.$$

Hence the operators $\mathrm{d}L(H_+)$, $\mathrm{d}L(H_-)$ have the form

$$\begin{aligned} \mathrm{d}L(H_+) &= \mathrm{e}^{i\varphi}\left(-i\frac{\partial}{\partial\theta} - \frac{1}{\sin\theta}\frac{\partial}{\partial\psi} + \cot\theta\frac{\partial}{\partial\varphi}\right), \\ \mathrm{d}L(H_-) &= \mathrm{e}^{-i\varphi}\left(-i\frac{\partial}{\partial\theta} + \frac{1}{\sin\theta}\frac{\partial}{\partial\psi} - \cot\theta\frac{\partial}{\partial\varphi}\right). \end{aligned} \tag{8.4.9}$$

It remains to write formulas (8.4.7) bearing in mind that

$$\check{t}^l_{m,k}(g) = \mathrm{e}^{-l(m\psi+k\varphi)}P^l_{mk}(\cos\theta).$$

By performing some simple calculations and substituting $\cos\theta = z$ we arrive at

$$\sqrt{1-z^2}\frac{\mathrm{d}}{\mathrm{d}z}P^l_{mk} - \frac{m-kz}{\sqrt{1-z^2}}P^l_{mk} = -i\sqrt{(l-k)(l+k+1)}\,P^l_{m,k+1}, \tag{8.4.10}$$

$$\sqrt{1-z^2}\frac{\mathrm{d}}{\mathrm{d}z}P^l_{mk} + \frac{m-kz}{\sqrt{1-z^2}}P^l_{mk} = -i\sqrt{(l+k)(l-k+1)}\,P^l_{m,k+1}. \tag{8.4.11}$$

In formulas (8.4.8)–(8.4.11) i denotes the imaginary unit. Combining equation (8.4.10) with equation (8.4.11) in the form corresponding to the parameters l, m, $k+1$, we obtain the *associated Legendre equation*

$$\begin{aligned} &(1-z^2)\frac{\mathrm{d}^2}{\mathrm{d}z^2}P^l_{mk} - 2z\frac{\mathrm{d}}{\mathrm{d}z}P^l_{mk} - \frac{m^2+k^2-2mkz}{1-z^2}P^l_{mk} \\ &= -l(l+1)P^l_{mk}. \end{aligned} \tag{8.4.12}$$

By combining formulas (8.4.7). we easily see that this formula reflects the fact that the functions t^l_{mk} are eigenfunctions of the Laplace–Beltrami operator

$$\Delta := \mathrm{d}L(H_+)\mathrm{d}L(H_-) - \mathrm{d}L(X_3)^2,$$

corresponding to the eigenvalue $-l(l+1)$. The operator Δ commutes with right and left translations on SU(2). From (8.4.9) we see that

$$\Delta = \frac{\partial^2}{\partial\theta^2} + \cot\theta\frac{\partial}{\partial\theta} + \frac{1}{\sin^2\theta}\left(\frac{\partial^2}{\partial\varphi^2} - 2\cos\theta\frac{\partial^2}{\partial\varphi\,\partial\psi} + \frac{\partial^2}{\partial\psi^2}\right). \tag{8.4.13}$$

8.5. CHARACTERS OF IRREDUCIBLE REPRESENTATIONS AND NEW INTEGRAL FORMULAS FOR LEGENDRE FUNCTIONS

A convenient way of computing the character of the representation T^z is to apply the basis e_n of eigenvectors of the operators T^l_k, $k \in K$.

The set of the conjugacy classes of elements of the group SU(2) can be parametrized by the elements of the subgroup K. For every matrix $g \in \mathrm{SU}(2)$ there exists a unitary matrix u such that ugu^{-1} is diagonal, and hence $ugu^{-1} \in K$. A conjugacy class with respect to inner automorphisms is uniquely determined by the set of eigenvalues of this diagonal matrix. Consequently, elements $k_1 \neq k_2 \in K$ are conjugate with respect to an inner automorphism if and only if they are conjugate in the sense of complex conjugation.

In any case it is enough to find the values of the character on the elements of the subgroup K

$$\chi^l(k) = \sum_{j=-l}^{l} (e_j | T^l_k e_j) = \sum_{j=-l}^{l} \mathrm{e}^{i\varphi j} = \frac{\mathrm{e}^{i(l+1)\varphi} - \mathrm{e}^{-il\varphi}}{\mathrm{e}^{i\varphi} - 1}$$

$$= \frac{\sin(l+\frac{1}{2})\varphi}{\sin\frac{1}{2}\varphi}. \tag{8.5.1}$$

The value of $\chi^l(g)$ for

$$g = \begin{bmatrix} \mathrm{e}^{\frac{1}{2}i\varphi}\cos\frac{1}{2}\theta & i\mathrm{e}^{-\frac{1}{2}i\varphi}\sin\frac{1}{2}\theta \\ i\mathrm{e}^{\frac{1}{2}i\varphi}\sin\frac{1}{2}\theta & \mathrm{e}^{-\frac{1}{2}i\varphi}\cos\frac{1}{2}\theta \end{bmatrix}$$

can be computed by finding an element $k \in G$ which is conjugate to g.

We shall use the fact that the matrices k and g have the same eigenvalues. We thus find

$$\det(g-\lambda \mathrm{id}) = \lambda^2 - 2\lambda \cos\tfrac{1}{2}\varphi \cos\tfrac{1}{2}\theta + 1 .$$

Hence the eigenvalues are equal to

$$\lambda_{1,2} = \cos\tfrac{1}{2}\varphi \cos\tfrac{1}{2}\theta \pm i\sqrt{1-\cos^2\tfrac{1}{2}\varphi \cos^2\tfrac{1}{2}\theta} .$$

Since are we looking for a $k = \begin{bmatrix} e^{\frac{1}{2}it} & 0 \\ 0 & e^{-\frac{1}{2}it} \end{bmatrix}$, we must have

$$\cos\tfrac{1}{2}t = \cos\tfrac{1}{2}\varphi \cos\tfrac{1}{2}\theta . \tag{8.5.2}$$

The value of t is determined uniquely up to sign. In particular, for $\varphi = 0$ we may take $t = 0$ and so

$$\chi^l(a(\theta)) = \frac{\sin(l+\frac{1}{2})\theta}{\sin\frac{1}{2}\theta} . \tag{8.5.3}$$

We now apply formula (5.2.1), which establishes a relationship between the character of an irreducible representation and its spherical function. In our case this relation takes the form

$$\int_K \chi^l(a(\theta)k)\,\mathrm{d}k = \begin{cases} 0, & l \notin Z, \\ P_l(\cos\theta), & l \in Z. \end{cases} \tag{8.5.4}$$

On the basis of (8.5.1) we obtain for $l \in Z$

$$P_l(\cos\theta) = \frac{2}{\pi}\int_0^\pi \frac{\sin[(l+\frac{1}{2})t(\theta,\varphi)]}{\sin\frac{1}{2}t(\theta,\varphi)}\,\mathrm{d}\varphi ,$$

where $t(\theta, \varphi)$ is the implicit function defined by equation (8.5.2) (which is uniquely solved in the interval of integration). With the change of variable

$$t = t(\theta,\varphi), \qquad \sin\tfrac{1}{2}t\,\mathrm{d}t = \cos\tfrac{1}{2}\theta \sin\varphi\,\mathrm{d}\varphi ,$$

the above integral becomes

$$P_l(\cos\theta) = \frac{2}{\pi}\int_0^\pi \frac{\sin t(l+\frac{1}{2})}{\sqrt{\cos^2\frac{1}{2}\theta - \cos^2\frac{1}{2}t}}\,\mathrm{d}t$$

$$= \frac{2}{\pi}\int_0^\pi \frac{\sin t(l+\frac{1}{2})}{\sqrt{2(\cos\theta - \cos t)}}\,\mathrm{d}t . \tag{8.5.5}$$

The same formula can also be obtained by a suitable substitution in the integral expression (8.1.13).

Finally, we note that direct computation of the character χ^l leads to the formula

$$\sum_{j=-l}^{l} P^l_{jj}(\cos\theta) = \chi^l\left(a(\theta)\right) = \frac{\sin(l+\frac{1}{2})\theta}{\sin\frac{1}{2}\theta}. \tag{8.5.6}$$

8.6. HARMONIC ANALYSIS ON THE GROUP SU(2) AND THE SPHERE S^2

In the case of the group SU(2) and the symmetric pair $\left(\mathrm{SU}(2), K\right)$ we shall apply the theorem from Chapter 5 concerning the decomposition of functions with respect to the system of matrix elements of irreducible representations.

We know that the representations T^l are irreducible, and since $\dim H^l = 2l+1$, they are non-equivalent. However, we do not know whether the system of the representations

$$\mathscr{T} = \{[T^l]:\, l = 0, \tfrac{1}{2}, 1, \ldots\}$$

exhausts the set of irreducible representations of SU(2).

This is indeed the case and proof will be based on the following

PROPOSITION 8.6.1. *Matrix elements of the representations belonging to the set* $\mathscr{T}$ *constitute a dense subset in* $\mathscr{C}\left(\mathrm{SU}(2)\right)$.

Proof. We first note that, as remarked in § 8.3, the underlying space of the group SU(2), is a sphere in $\boldsymbol{C}^2$. We shall distinguish a set of matrix elements which separates points in the sphere and is closed under complex conjugation.

Let

$$f_1:\, H^l \ni e \to e(1, 0) \in \boldsymbol{C}^1,$$
$$f_2:\, H^l \ni e \to e(0, 1) \in \boldsymbol{C}^1.$$

For $g = \begin{bmatrix} \alpha & \beta \\ -\bar{\beta} & \bar{\alpha} \end{bmatrix}$ we find

$$f_1(T^l_g e_k) = A\alpha^{l+k}\beta^{l-k},$$
$$f_2(T^l_g e_k) = B(-\bar{\beta})^{l+k}\bar{\alpha}^{l-k} \qquad A, B \in \boldsymbol{C}^1.$$

The linear span of the set of matrix elements of the form $G \ni g \to \alpha^{l+k}\beta^{l-k} \ni C^1$ is the algebra of complex polynomials in variables α, β. The functions $G \ni g \to (\bar{\alpha})^{l-k}(\bar{\beta})^{l-k}$, on the other hand, span the algebra of polynomials in variables $\bar{\alpha}$ and $\bar{\beta}$. Linear combinations of these elements form the algebra of polynomials in variables $\mathrm{Re}\,\alpha$, $\mathrm{Im}\,\alpha$, $\mathrm{Re}\,\beta$, $\mathrm{Im}\,\beta$, with complex coefficients. Obviously, this algebra separates points in the sphere in C^2 and is closed under conjugation. Hence, since the assumptions of the Stone theorem are satisfied, the assertion follows. □

Now let us assume that there exists an irreducible representation (H, T) of the group SU(2) which is not contained in $\mathscr{T}$. Then the matrix elements of this representation are orthogonal in $L^2(G)$ to the matrix elements of the representations T^l. But in view of Proposition 8.1.6 they are then orthogonal to the set of continuous functions and therefore are zero. Thus we obtain

COROLLARY 8.6.2. *For every irreducible representation (H, T) of the group* SU(2) *there exists a half-integer l such that the representation T is equivalent to T^l.*

The following system of functions on the group

$$\sqrt{(2l+1)}\; t^l_{mn}(\varphi, \theta, \psi) = \sqrt{(2l+1)}\; e^{(m\varphi+n\psi)} P^l_{mn}(\cos\theta)$$

constitutes an orthonormal basis for the space $L^2(G)$. The decomposition of functions with respect to this basis is called the *trigonometric series expansion on the group* SU(2). For every function $f \in L^2(G)$ we have

$$f(\varphi, \theta, \psi) = \sum_{k=0}^{\infty} (k+1) \sum_{m,n=-\frac{1}{2}k}^{\frac{1}{2}k} C^k_{mn} e^{i(m\varphi+n\psi)} P^{\frac{1}{2}k}_{mn}(\cos\theta), \quad (8.6.1)$$

where

$$C^k_{mn} = \frac{(-1)^{m-n}}{16\pi^2} \int_{-2\pi}^{2\pi}\int_0^{2\pi}\int_0^{\pi} f(\varphi, \theta, \psi) e^{-i(m\varphi+n\psi)} P^{\frac{1}{2}}_{mn}(\cos\theta) \times \sin\theta \, d\varphi \, d\psi \, d\theta .$$

The series is convergent in $L^2(G)$.

The representations T^l contain a K-invariant vector only for the integer values of the parameter l. The pair (G, K), being symmetric, we can apply the results of § 5.2.

We find that the regular representation of the group G on $L^2(G/K) = L^2(S^2)$ is a direct sum of the representations T^l, $l = 0, 1, \ldots$ which occur once only in that sum. Hence follow decompositions of functions with respect to the matrix elements t^l_{j0}. We introduce the classical harmonics:

$$Y_{lj}(\varphi, \theta) := t^l_{-j0}(g) = e^{-i\varphi j} P^l_{j0}(\cos\theta). \tag{8.6.2}$$

The functions $\sqrt{2l+1}\; Y_{lj}(\varphi, \theta)$, $l = 0, 1, \ldots, \; -l \leqslant j \leqslant l$ constitute an orthonormal system in $L^2(S^2)$.

Formula (8.6.1) assumes the form

$$f(\varphi, \theta) = \sum_{l=0} (2l+1) \sum_{k=-l}^{l} C_{lk}\, Y_{lk}(\varphi, \theta) \tag{8.6.3}$$

with

$$C_{lk} = \frac{1}{4\pi} \int_0^{2\pi} \int_0^{\pi} f(\varphi, \theta) \overline{Y}_{lk}(\varphi, \theta) \sin\theta \, d\theta \, d\varphi,$$

the series being convergent in $L^2(S^2)$ with respect to the measure (8.3.5).

Moreover, the Plancherel formula

$$||f||^2_{L^2(S^2)} = \sum_{l=0}^{\infty} (2l+1) \sum_{k=-l}^{l} |C^2_{lk}| \tag{8.6.4}$$

also holds.

The Laplace–Beltrami operator which commutes with the action of the group on the sphere is obtained by neglecting that part of the operator Δ (8.4.13) which depends on the angle ψ. In terms of spherical coordinates we have

$$\Delta_0 = \frac{\partial^2}{\partial \theta^2} + \cot\theta \frac{\partial}{\partial \theta} + \frac{1}{\sin^2\theta} \frac{\partial^2}{\partial \varphi^2}. \tag{8.6.5}$$

As can easily be seen this is the angularpart of the Laplace operator in $\boldsymbol{R}^3$. The functions Y_{lk} are eigenfunctions of the operator Δ_0 corresponding to the eigenvalue $-l(l+1)$.

The decomposition of the space $L^2(S^2)$ into irreducible representations of the group SU(2) is identical with the decomposition into eigenspaces of the Laplace operator. Analogous statements are valid for arbitrary symmetric spaces.

From the point of view of the theory of spherical functions this phenomenon results from the fact that the eigenvalues of a spherical function, with respect to differential operators which commute with the action of the group, determine the function uniquely.

We shall now make use of formula (5.3.3) from the analysis on a compact and symmetric space (Chapter 5), which provides a decomposition of an arbitrary function in $L^2(G/K)$ into zonal spherical functions

$$f(g\cdot 0) = \sum_{l=0}^{\infty}(2l+1)f * t^l_{00}(g\cdot 0)$$

$$= \sum_{l=0}^{\infty}(2l+1)\int_G f(x\cdot 0)t^l_{00}(g^{-1}x\cdot 0)\mathrm{d}x. \tag{8.6.6}$$

Let $g = k(\varphi)a(\theta)k(\psi)$, $h = k(\varphi_1)a(\theta_1)k(\psi_1)$. The zonal spherical function t^l_{00}, being bi-invariant under K, depends only on the Euler angle θ:

$$t^l_{00}(g) = P_l(\cos\theta).$$

We thus have to determine this angle for the element

$$g^{-1}h = k(-\psi)a(-\theta)k(\varphi_1-\varphi)a(\theta_1)k(\psi_1).$$

It follows from (8.3.9) that the cosine of the required angle equals

$$\cos\theta\cos\theta_1+\sin\theta\sin\theta_1\cos(\varphi_1-\varphi).$$

Formula (8.6.6) in terms of spherical coordinates takes the form

$$f(\varphi,\theta) = \frac{1}{4\pi}\sum_{l=0}^{\infty}(2l+1)\int_0^{2\pi}\int_0^{\pi} f(\varphi_1,\theta_1)P_l\big(\cos\theta\cos\theta_1$$

$$+\sin\theta\sin\theta_1\cos(\varphi_1-\varphi)\big)\sin\theta_1\,\mathrm{d}\varphi_1\,\mathrm{d}\theta_1. \tag{8.6.7}$$

It follows from this formula or directly from decomposition (8.6.3) that for every function on the sphere which is spherically symmetric (i.e., independent of the angle φ) the following formula takes place:

$$f(\theta) = \frac{1}{2}\sum_{l=0}^{\infty}(2l+1)P_l(\cos\theta)\int_0^{\pi} f(\theta_1)P_l(\cos\theta_1)\sin\theta_1\, d\theta_1 .$$

Upon setting $\varphi(\cos\theta) := f(\theta)$ it follows that

$$\varphi(x) = \frac{1}{2}\sum_{l=0}^{\infty}(2l+1)P_l(x)\int_{-1}^{1}\varphi(x)P_l(x)\,dx.$$

This formula represents the decomposition of functions in $L^2([-1, 1])$ with respect to the orthonormal basis

$$e_l(x) = \sqrt{\frac{2l+1}{2}}\,P_l(x).$$

A number of classical formulas concerning decompositions of functions with respect to the Jacobi functions and to the associated Legendre polynomials will be found in the Problems.

8.7. DECOMPOSITION OF THE TENSOR PRODUCT OF REPRESENTATIONS T^l. THE CLEBSCH–GORDAN COEFFICIENTS

We shall carry out a decomposition into irreducible components of the tensor product of two representations T^l and $T^{l'}$. Let us assume for instance that $l \geqslant l'$. The character of the representation $T^l \otimes T^{l'}$ is the product of the characters of T^l and $T^{l'}$:

$$\chi^{ll'} = \chi^l \chi^{l'}.$$

In view of what has been said in the preceding section we know that the character χ^l can be expressed in terms of the eigenvalues of the argument on the form

$$\chi^l(g) = \frac{\omega^{2(l+1)} - \omega^{-2l}}{\omega^2 - 1}, \qquad (8.7.1)$$

where ω denotes an eigenvalue of the matrix g. Hence

$$
\begin{aligned}
\chi^{ll'}(g) &= \frac{(\omega^{2(l+1)}-\omega^{-2l})(\omega^{2(l'+1)}-\omega^{-2l'})}{(\omega^2-1)^2} \\
&= \frac{(\omega^{2(l+1-l')}-\omega^{-2(l+l')})(\omega^{2(2l'+1)}-1)}{(\omega^2-1)^2} \\
&= \frac{\omega^{2(l-l'+1)}-\omega^{-2(l+l')}}{\omega^2-1}\sum_{k=0}^{2l'}\omega^{2k} \\
&= \frac{1}{\omega^2-1}\sum_{k=0}^{2l'}\omega^{2(l-l'+k+1)}-\omega^{-2(l+l'-k)} \\
&= \frac{1}{\omega^2-1}\sum_{k=l-l'}^{l+l'}\omega^{2(k+1)}-\omega^{-k} = \sum_{l=l-l'}^{l+l'}\chi^k(g).
\end{aligned} \tag{8.7.2}
$$

It follows from (4.6.3) and Theorem 4.7.3 that the representation $T^l\otimes T^{l'}$ admits the decomposition

$$
T^l\otimes T^{l'} = \sum_{k=l-l'}^{l+l'} T^k. \tag{8.7.3}
$$

There are two distinguished bases in the representation space of $T^l\otimes T^{l'}$. One of them is the canonical basis for the tensor product

$$
f_{mn} = e^l_m \otimes e^{l'}_n ,
$$

where e^l_m denotes the canonical basis for the representation space of T^l (denoted previously by e_m since we dealt with a fixed representation T^l). The other basis consists of normalized eigenvectors of the subgroup K. We shall parametrize that basis by the indices $l-l' \leqslant k \leqslant l'+l$ and $-k \leqslant j \leqslant k$, in such a way that e^k_j will denote the vector in the representation space of T^k corresponding to the eigenvalue $e^{ij\varphi}$ of the operator $T^k(k)$ for

$$
k = \begin{bmatrix} e^{\frac{1}{2}i\varphi} & 0 \\ 0 & e^{-\frac{1}{2}i\varphi} \end{bmatrix}.
$$

This notation is consistent with that used previously, inasmuch the vectors e_j defined before, were required to verify the same conditions and therefore they differ from the present ones by at most a unitary factor.

Both bases being unitary, this implies that there exists a unitary operator C which sends one onto the other:

$$f_{mn} = \sum_{\substack{l-l' \leqslant k \leqslant l'+l \\ -k \leqslant j \leqslant k}} C(l, l', k;\ m, n, j) e_j^k. \qquad (8.7.4)$$

The numbers $C(l, l', k;\ m, n, j)$ are called the *Clebsch-Gordan coefficients* (C–G *coefficients*). The arbitrariness of the choice of the basis e_j^k results in the non-uniqueness of these coefficients.

The matrix elements of the representation $T^{ll'}$ in the basis f_{mn} are products of the respective matrix elements of the representations T^l and $T^{l'}$:

$$t^{ll'}_{(mn)(m'n')} = t^l_{mn'} t^{l'}_{mn'}. \qquad (8.7.5)$$

Next, the matrix elements $S^{ll'}_{(kj)(k'j')}$ in the basis e_j^k are zero for $k \neq k'$ and for $k = k'$ they are identical with the matrix elements of T^k with respect to the basis e_j^k

$$S^{ll'}_{(kj)(k'j')} = \delta_{kk'} t^k_{jj'}. \qquad (8.7.6)$$

It follows from formulas (8.7.4), (8.7.5) and (8.7.6) that

$$\begin{aligned} t^l_{mm'} t^{l'}_{mn'} &= \sum_{\substack{k, k' \\ j, j'}} \overline{C}(l, l', k; m, n, j) C(l, l', k'; m', n', j')\, \delta_{kk'} t^k_{jj'} \\ &= \sum_{k, j, j'} \overline{C}(l, l', k; m, n, j) C(l, l', k; m', n', j') t^k_{jj}. \qquad (8.7.7) \end{aligned}$$

The above formula is the main tool in computing the C-G coefficients.

Let us integrate both sides of (8.7.7) multiplied by t^k_{jj}. The orthogonality relations imply that

$$\begin{aligned} &\overline{C}(l, l', k;\ m, n, j) C(l, l', k;\ m', n', j') \\ &= (2k+1) \int_G t^l_{mm}(g) t^{l'}_{nn'}(g) \overline{t}^k_{jj'}(g) \mathrm{d}g. \qquad (8.7.8) \end{aligned}$$

To compute the integral on the right we substitute the expressions (8.3.6) for the matrix elements. The measure on the group is given in terms of Euler angles as

$$\frac{1}{16\pi^2} \sin\theta \,\mathrm{d}\theta \,\mathrm{d}\varphi \,\mathrm{d}\psi,$$

where integration is taken over $[0, 2\pi]$ with respect to φ and over $[-2\pi, 2\pi]$ with respect to ψ. As can be seen from (8.3.6) the integrand occurring in (8.7.8) contains the factors $e^{i\varphi(m+n-j)}$ and $e^{i\psi(m'+n'-k)}$ and a third one, which depends only on θ. Therefore the integral is equal to zero if $m+n \neq j$ and $m'+n' \neq k$. For the non-zero values of the coefficients we have the expression

$$\overline{C}(l, l', k;\ m, n, m+n)C(l, l', k;\ m', n', m'+n')$$
$$= \frac{2k+1}{2} \int_{-1}^{1} P^l_{mm'}(x)\overline{P^k_{n+m,\, n'+m'}}(x)P^{l'}_{nn'}(x)\,dx. \tag{8.7.9}$$

As a special case we get

$$\overline{C}(l, l', k;\ l, -l', l-l')C(l, l', k;\ m', n', m'+n')$$
$$= \frac{2k+1}{2} \int_{-1}^{1} P^l_{lm'}(x)P^{l'}_{-l'n'}(x)\overline{P^k_{l-l',\, m'+n'}}(x)\,dx. \tag{8.7.10}$$

From formula (8.1.11) we obtain

$$P^l_{lm'}(x) = \frac{(-1)^{\frac{1}{2}l-m}}{2^l} \sqrt{\frac{(2l)!}{(l-m')!(l+m')!}}$$
$$\times (1-x)^{\frac{1}{2}(l-m')}(1+x)^{\frac{1}{2}(l+m')},$$

$$P^l_{-l'n'}(x) = \frac{(-1)^{\frac{1}{2}l'+n'}}{2^l} \sqrt{\frac{(2l')!}{(l'+n')!(l'-n')!}}$$
$$\times (1-x)^{\frac{1}{2}(l'+n')}(1+x)^{\frac{1}{2}(l'-n')}.$$

Finally, by substituting the value of $P^k_{l-l',\, m'+n'}$ given directly by (8.1.11), we see that the integral in (7.8.10) equals

$$\sum_t A(l, l', k;\ m', n', t) \int_{-1}^{1} (1-x)^{l-m'+t}(1+x)^{l'+k-t+m'}\,dx,$$

where the index t runs over the set

$$0 \leqslant t \leqslant k-l+l', \qquad l-l'-m'-n' \leqslant t \leqslant k+m'+n'.$$

Before substituting the values of the coefficients $A(\cdot)$, we compute

$$\int_{-1}^{1} (1-x)^{\alpha}(1+x)^{\beta}\,dx = 2^{\alpha+\beta}\frac{\alpha!\beta!}{(\alpha+\beta+1)!}$$

for positive integers α, β. (The proof is established by induction upon making the change of variable $x = -\cos\varphi$.) Consequently, (8.7.10) takes the form

$$(2k+1)(-1)^{l-m'} \times \frac{[(2l)!(2l')!(k+l-l')!(k-l+l')!(k+m'+n')!(k-m'-n')!]^{\frac{1}{2}}}{[(l-m')!(l+m')!(l'+n')!(l'-n')!]^{\frac{1}{2}}} \times \sum_t \frac{(-1)^t(l-m'+t)!(l'+k-t+m')!}{(k+m'+n'-t)!(l-l'-m'-n'+t)!t!(k+l+l'-t)!}.$$

We shall compute the sum in the above expression for $m' = l$, $n' = -l'$ $m'+n' = l-l'$:

$$|C(l,l',k;l,-l',l-l')|^2 = (2k+1)\frac{(k+l-l')!(k-l+l')!}{(l+l'+k+1)!} \times \sum_{0\leqslant t\leqslant k+|l-l'|}(-1)^t\frac{(l+l'+k-t)!}{(l-l'+k-t)!t!(l'-l+k-t)!}.$$

The sum can be computed from the formula

$$\sum_{0\leqslant t\leqslant \min(\gamma,\beta)}(-1)^t\frac{(\alpha+\beta-t)!}{t!(\gamma-t)!(\beta-t)!} = \frac{\alpha!(\alpha+\beta-\gamma)!}{\beta!\gamma!(\alpha-\gamma)!}.$$

By setting $\alpha = 2l$, $\beta = l-l'+k$, $\gamma = l'-l+k$ we obtain

$$|C(l,l',k;l,-l',l-l')|^2 = (2k+1)\frac{(2l)!(2l')!(l-l'+k)!(k-l+l')}{(l'-l+k)!(l-l'+k)!(l+l'-k)!(l+l'+k+1)!} = \frac{(2l)!(2l')!(2k+1)}{(l+l'+k+1)!(l+l'-k)!}.$$

Thus we have found the absolute value of the C-G coefficient of the form $C(l,l,k;\ l,-l,l-l)$. Owing to the arbitrariness of the choice of the factors of absolute value 1 in the definition of the basis f_{mn} we may choose, for every l,l',k the phase of one of the corresponding C-G coefficients. We do so by indicating

$$C(l,l',k;l,-l',l-l') = \sqrt{\frac{(2k+1)(2l)(2l')!}{(l+l'+k+1)!(l+l'-k)!}}. \tag{8.7.13}$$

Now, in conjunction with formula (8.7.10) we can find the value of an arbitrary coefficient $C(l, l', k; m, n', m'+n':)$. Shown is one of the possible expressions

$$C(l, l', k; m', n', m'+n')$$
$$= \sqrt{2k+1}\,(-1)^{l-m}[(l+l'+k+1)!(l+l'-k)!]^{\frac{1}{2}}$$
$$\times \frac{[(k+l-l')!(k-l+l')!(k+m'+n')!(k-m'-n')!]^{\frac{1}{2}}}{[(l-m')!(l+m')!(l'+n')!(l'-n')!]^{\frac{1}{2}}}$$
$$\times \sum_{0\leqslant t\leqslant k+|l-l'|} \frac{(-1)^t(l-m'-t)!(l'+k-t+m')!}{(k+m'+n'-t)!(l-l'-m'-n'+t)!t!(k-l+l'-t)!}.$$

Other expressions for the C-G coefficients can be obtained by applying the remaining formulas for the functions P^l_{mn} computed in § 8.1. A number can be found in the Problems.

PROBLEMS

1. (a) Prove that the Killing form of the Lie algebra $\mathfrak{su}(2)$ is given by the formula $B(X, Y) = 4\,\mathrm{tr}(XY)$ and satisfies the identity $B(X, X) = -8\det X$. Check that the Pauli matrices X_i, $i = 1, 2, 3$, constitute an orthogonal basis for $\mathfrak{su}(2)$ relative to the Killing form.

(b) Prove that every matrix $g \in \mathrm{SU}(2)$ is conjugate (in SU(2)) to the matrix $k(t)$ where $t = 2\arccos\frac{1}{2}\mathrm{tr}\,g$, and that matrices g, g^{-1} are conjugate to each other.

Hint. (b) Prove that a unitary matrix which reduces g to a diagonal form can be modified so that its determinant is equal to 1.

2. (a) Show that the set H of the matrices of the form

$$\begin{bmatrix} a & b \\ -\bar{b} & \bar{a} \end{bmatrix}, \qquad a, b \in \boldsymbol{C},$$

forms, under usual matrix operations, an associative algebra over the field $\boldsymbol{R}$, such that for every non-zero element there exists a miltiplicative inverse. Check that the matrices

$$\tau_0 = \begin{bmatrix} 1 & 0 \\ 0 & 1 \end{bmatrix}, \quad \tau_1 = \begin{bmatrix} i & 0 \\ 0 & -i \end{bmatrix}, \quad \tau_2 = \begin{bmatrix} 0 & 1 \\ -1 & 0 \end{bmatrix}, \quad \tau_3 = \begin{bmatrix} 0 & i \\ i & 0 \end{bmatrix}$$

form a basis (over $\boldsymbol{R}$) for H and satisfy the relations

$$\tau_k^2 = -\tau_0, \qquad \tau_0 \tau_k = \tau_k \tau_0, \qquad k = 1, 2, 3,$$

$$\tau_k \tau_l = \operatorname{sgn}\begin{pmatrix} 1 & 2 & 3 \\ k & l & m \end{pmatrix} \tau_m$$

(as usual, sgn( ) denotes the sign of a permutation), and derive hence that H is isomorphic to the algebra of quaternions.

(b) Show that $x \to \det x$ is an Euclidean norm on H (i.e. arises from a real inner product) and the basis $\{\tau_\alpha\}_{\alpha=0}^3$ is an orthonormal basis relative to this norm.

(c) Show that every $H \ni x \neq 0$ admits a unique decomposition $x = \lambda g$ $g \in \mathrm{SU}(2)$, $\lambda \in \boldsymbol{R}_+$.

(d) For $g \in \mathrm{SU}(2)$ we define a mapping $R(g)$ of H (identified with the help of the basis $\{\tau_\alpha\}_{\alpha=0}^3$ with $\boldsymbol{R}^4$ equipped with a natural inner product) by putting $R(g)x := gx$. Prove that $R(g)$ is a proper rotation (i.e. $\det Rg = 1$) and the rule $g \to R(g)$ is an injective homomorphism of SU(2) into SO(4).

(e) Show that the invariant measure on SU(2) is, up to a normalization factor, the measure on the unit sphere in H induced by the inner product in (b).

Hint. (d) To prove $\det R(g) = 1$ apply (b) of the preceding problem.

3. (a) Prove that $\mathfrak{sl}(2, \boldsymbol{C})$ is a complexification of the algebra $\mathfrak{su}(2)$ (cf. § 17.2) (in other words every element $Z \in \mathfrak{sl}(\in 2, \boldsymbol{C})$ is uniquely represented as $Z = X + iY$, $X, Y \in \mathfrak{su}(2)$ and $\mathfrak{su}(2)$ is a Lie subalgebra of the algebra $\mathfrak{sl}(2, \boldsymbol{C})$ (regarded as an algebra over the field $\boldsymbol{R}$)) and deduce hence that every representation of the Lie algebra $\mathfrak{su}(2)$ on a complex vector space extends to a representation of the algebra $\mathfrak{sl}(2, \boldsymbol{C})$ and the latter representation is uniquely determined by the initial representation of $\mathfrak{su}(2)$.

(b) Let X_j, $j = 1, 2, 3$ be the Pauli matrices (8.4.2). Check that the matrices $X^+ = X_2 + iX_1$, $X^- = -X_2 + iX_1$, $X^3 = -iX_3$ satisfy the following commutation relation

$$[X^3, X^\pm] = \pm X^\pm, \qquad [X^+, X^-] = 2X^3.$$

(c) Let J^+, J^-, J^3 be linear operators on a complex vector space W satisfying the commutation relation

$$[J^3, J^\pm] = \pm J^\pm, \qquad [J^+, J^-] = 2J^3.$$

Prove that there exists a unique representation T of the Lie algebra $\mathfrak{sl}(2, \boldsymbol{C})$ on the space W such that

$$J^+ = T(X^+), \qquad J^- = T(X^-), \qquad J_3 = T(X^3).$$

(d) Let W be the space of polynomials of one complex variable (of any degree) and let J^+, J^-, J^3 be the operators on W defined as

$$J^+ = -2uz + z^2 \frac{\mathrm{d}}{\mathrm{d}z}, \qquad J^- = -\frac{\mathrm{d}}{\mathrm{d}z}, \qquad J^3 = -u + z\frac{\mathrm{d}}{\mathrm{d}z},$$

where $u \in \boldsymbol{C}$ is a parameter. Prove that if $2u \in \boldsymbol{Z}_+$ then the representation of the algebra $\mathfrak{sl}(2, \boldsymbol{C})$ defined by these operators on some subspace is the differential of an irreducible representation of the group $\mathrm{SL}(2, \boldsymbol{C})$. Prove that in the remaining cases the operators J^+, J^-, J^3 define on W an irreducible, infinite-dimensional representation of the Lie algebra $\mathfrak{sl}(2, \boldsymbol{C})$ which cannot be integrated to a representation of the group $\mathrm{SL}(2, \boldsymbol{C})$. Check that the operators J^+, J^-, J^3 act on the monomials $f_j(z) := z^j$, $j = 0, 1, \ldots$ according to the formulas

$$J^+ f_j = (j - 2u) f_{j+1}, \qquad j = 0, 1, \ldots,$$
$$J^- f_j = (-j) f_{j-1}, \qquad j = 1, 2, \ldots \qquad \text{and} \qquad J^- f_0 = 0,$$
$$J^3 f_j = (j - u) f_j, \qquad j = 0, 1, \ldots$$

Hint. (d) The operator iJ^3 would have to be the generator of the one-parameter subgroup $a_3(\cdot)$ of $\mathrm{SU}(2)$.

4. (a) Let $l = 0, \frac{1}{2}, 1, \ldots, g = \begin{bmatrix} \alpha & \beta \\ \gamma & \delta \end{bmatrix} \in \mathrm{SL}(2, \boldsymbol{C})$. Show that the formula

$$(D^l(g)f)(z) := (\beta z + \delta)^{2l} f\left(\frac{\alpha z + \gamma}{\beta z + \delta}\right)$$

for polynomials f of degree $\leqslant 2l$ of variable z defines a representation of $\mathrm{SL}(2, \boldsymbol{C})$ equivalent to the representation T^l. Construct an operator intertwining T^l with D^l. Indicating $h_j(z) = z^j$, $0 \leqslant j \leqslant 2l$, prove that

$$\mathrm{d}D^l(X^3) h_j = (j - l) h_j, \qquad \mathrm{d}D^l(X^+) h_j = (j - 2l) h_{j+1},$$
$$\mathrm{d}D^l(X^-) h_j = j h_{j-1},$$

where $\mathrm{d}D^l$ denotes the differential of the representation D^l and $X^\pm$, X^3 are defined as in Problem 3.

(b) Let K^l be the space of all trigonometric polynomials of degree not greater than l, i.e. $K^l = \left\{ \sum_{n=-l}^{l} a_n e^{-in\varphi} \colon a_n \in \boldsymbol{C} \right\}$. Show that U^l defined as

$$(U^l(g)f)e^{i\varphi} := e^{-il\varphi}(\alpha e^{i\varphi}+\gamma)^l(\beta e^{i\varphi}+\delta)^l f\left(\frac{\alpha e^{i\varphi}+\gamma}{\beta e^{i\varphi}+\delta}\right)$$

is a representation equivalent to T^l.

5. Let (T, E) be an irreducible finite-dimensional representation of the Lie algebra $\mathfrak{sl}(2, \boldsymbol{C})$. Put $J^+ = T(X^+)$, $J^- = T(X^-)$, $J^3 = T(X^3)$, where X^+, X^-, X^3 are the elements of $\mathfrak{sl}(2, \boldsymbol{C})$ defined in Problem 3(b).

(a) Show that there exists a basis for E consisting of eigenvectors for J^3 and derive hence that the representation (T, E) is equivalent to the representation T^l for l such that $2l = \dim E - 1$.

(b) Derive from (a) a different proof of Corollary 8.6.2.

Hint. (a) Deduce from the commutation relations for J^+, J^-, J^3 that if $v \in E$ is an eigenvector for J^3 then non-zero vectors $(J^+)^k v$, $(J^-)^l v$, $k, l = 1, 2, \ldots$ are also eigenvectors for J^3. Apply the finiteness of the dimension and the irreducibility, to prove that those of the above vectors which are non-zero, constitute a basis for E and reduce the action of the operators J^+, J^-, J^3 to the form given in Problem 4(a) by a suitable normalizations of these vectors. (b) Since SU(2) is simply connected, every finite-dimensional representation of the Lie algebra $\mathfrak{su}(2)$ is the differential of a certain representation of the group.

6. Let F be the space spanned by the functions $f_n(x, t) := P_n(x)t^{n+\frac{1}{2}}$ for $n = 0, 1, \ldots$, where P_n is the Legendre polynomial. On the space F we define the operators

$$J^{\pm} := t^{\pm 1}\left((x^2-1)\frac{\partial}{\partial x} \pm xt\frac{\partial}{\partial t} + \frac{x}{2}\right), \qquad J^3 := t\frac{\partial}{\partial t}.$$

Prove that $J^{\pm}$, J^3 define an irreducible representation of $\mathfrak{sl}(2, \boldsymbol{C})$ on the space F. Compute $J^{\pm}f_n$, $J^3 f_n$ and examine the possibility of extending it to a representation of the group.

Hint. Apply the recurrence formulas for the Legendre polynomials (cf. Problem 16).

7. (a) Prove that the generators of the rotations about the axes x_1, x_2, x_3 are the matrices

$$R_1 = \begin{bmatrix} 0 & 0 & 0 \\ 0 & 0 & -1 \\ 0 & 1 & 0 \end{bmatrix}, \quad R_2 = \begin{bmatrix} 0 & 0 & 1 \\ 0 & 0 & 0 \\ -1 & 0 & 0 \end{bmatrix}, \quad R_3 = \begin{bmatrix} 0 & -1 & 0 \\ 1 & 0 & 0 \\ 0 & 0 & 0 \end{bmatrix}$$

respectively.

(b) Prove that the differential of the left regular representation of the group SO(3) on the sphere S^2 is given by θ, φ — polar coordinates on the sphere

$$\mathrm{d}L(R_1) = \left(\sin\varphi\frac{\partial}{\partial\theta} + \cot\theta\cos\varphi\frac{\partial}{\partial\varphi}\right),$$

$$\mathrm{d}L(R_2) = \left(-\cos\varphi\frac{\partial}{\partial\theta} + \cot\theta\sin\varphi\frac{\partial}{\partial\varphi}\right),$$

$$\mathrm{d}L(R_3) = -\frac{\partial}{\partial\varphi}.$$

(c) Show that $C = \sum_{i=1}^{3} [dL(R_i)]^2 = \left(\frac{\partial^2}{\partial\theta^2} + \cot\theta\frac{\partial}{\partial\theta} + \frac{1}{\sin^2\theta}\frac{\partial^2}{\partial\varphi^2}\right)$ is the spherical part of the Laplace operator $\Delta = \frac{\partial^2}{\partial x^2} + \frac{\partial^2}{\partial y^2} + \frac{\partial^2}{\partial z^2}$.

Hint. (b) To compute $\mathrm{d}L(R_i)$ find the tangent vector to the curve $t \to \exp(tR_i)\cdot p$, $p \in s^2$ on the sphere S^2.

8.(a) Let α: SU(2) → SO(3) be the surjective homomorphism constructed in § 8.3. Prove that in order that the representation T^l of the group SU(2) defines a representation S^l of the group SO(3) by means of the formula $S^l \circ \alpha = T^l$ it is necessary and sufficient that l be an integer. Prove also that all irreducible finite-dimensional representations of the group SO(3) arise in this way.

(b) Show that every irreducible unitary representation T of the group SO(3) has a basis $\{e_{-l}, \ldots, e_l\}$ in which $\mathrm{d}T(R_i)$, $i = 1, 2, 3$ are of the form

$$\mathrm{d}T(R_1)e_m = \tfrac{1}{2}i(\alpha_{m+1}e_{m+1} + \alpha_m e_{m-1}),$$

$$\mathrm{d}T(R_2)e_m = -\tfrac{1}{2}i(\alpha_{m+1}e_{m+1} - \alpha_m e_{m-1}),$$

where $\alpha_m = \sqrt{(l+m)(l-m+1)}$ and moreover

$$-\sum_{i=1}^{3}[\mathrm{d}\,T(R_i)]^2 e_m = l(l+1)e_m, \qquad m = -l, \ldots, l.$$

(c) Prove without using the results of § 8.6 that if T is an irreducible subrepresentation of the rugular representation of SO(3) on S^2 then, up to normalization, the set $\{Y_{lm}\colon m = -l, \ldots, l\}$ is the basis mentioned above.

Hint. The condition $\mathrm{d}T(R_3)e_m = ime_m$ implies that $e_m(\theta, \varphi) = \mathrm{e}^{im}F_m(\theta)$. The identity $Ce_m = l(l+1)e_m$ leads to an equation for the function F_m which transforms into the associated Legendre equation.

9. Prove that

$$P^l_{rs}(z) = (-1)^{\frac{-r-s}{2}}\left[\frac{(l-s)!(l-r)!}{(l+s)!(l+r)!}\right]^{\frac{1}{2}}\left(\frac{1+z}{1-z}\right)^{\frac{r+s}{2}}$$

$$\times \sum_{k=\max(r,s)}^{1}\frac{(l+k)!}{(l-k)!(k-s)!(k-r)!}\left(\frac{1-z}{2}\right)^k.$$

10. Prove the following formulas:

$$P^l_{ll}(z) = \left(\frac{1+z}{2}\right)^l, \qquad P^l_{l,-l}(z) = (-1)^l\left(\frac{1-z}{2}\right)^l,$$

$$P^l_{l0}z = (-1)^{\frac{l}{2}}\frac{\sqrt{(2l)!}}{2^l l!}(1-z^2)^{\frac{l}{2}}.$$

11. (a) Check that the following are the first four Legendre polynomials:

$$P_0(x) = 1, \qquad P_1(x) = x, \qquad P_2(x) = \tfrac{1}{2}(3x^2-1),$$
$$P_3(x) = \tfrac{1}{2}(5x^3-3x).$$

(b) Prove that the Legendre polynomials arise, up to a constant factor, by the Gramm–Schmidt orthogonalization in the space $L^2([-1,1])$ of the sequence of polynomials t^k, $k = 0, 1, \ldots$

12. Show that

$$P_n(x) = \sum_{k=0}^{E(\frac{1}{2}n)}(-1)^k\frac{(2n-2k)!}{2^n k!(n-k)!(n-2k)!}x^{n-2k},$$

where $E(\frac{1}{2}n)$ is the greatest integer not greater than $\frac{1}{2}n$.

Hint. $P_n(x) = P_n^0(x) = P_{00}^n(x)$. Apply the binomial formula to the integrand in formula (8.1.14) or apply the same device to formula (8.1.6).

13. Prove that for $m \geqslant 0$

$$P_l^m(x) = i^m \sqrt{\frac{(l+m)!}{(l-m)!}}\, P_{m0}^l(x).$$

Show that by defining P_l^m for $m < 0$ by the same formula we have

$$P_l^{-m}(x) = (-1)^m \frac{(l-m)!}{(l+m)!}\, P_l^m(x)$$

and then prove that the classical spherical harmonics Y_{lk} are given by the formula

$$Y_{lk}(\theta, \varphi) = (-i)^k \sqrt{\frac{(l-k)!}{(l+k)!}}\, P_l^k(\cos\theta) e^{-ik\varphi}$$

for $-l \leqslant k \leqslant l$, $l = 0, 1, 2, \ldots$

14. Apply Problem 12 and formulas (8.1.6) and (8.1.7) to show that

$$P_l^m(x) = \frac{(-1)^{l+m}(l+m)!}{2^l(l-m)!}(1-x^2)^{\frac{1}{2}m} \sum_k \frac{(-1)^k(2k)!\,x^{2k-l+m}}{k!(l-k)!(2k-l+m)!},$$

where the sum is taken over the set of all integers k such that $0 \leqslant k \leqslant l$ and $2k \geqslant l+m$.

15. Prove that

$$P_l^m(z) = \left(\frac{1+z}{1-z}\right)^{\frac{1}{2}m} \sum_{j=\max(m,0)}^{l} \frac{(-1)^j(l+j)!}{(l-j)!(j-m)!j!}\left(\frac{1-z}{2}\right)^j.$$

Hint. Apply Problem 9.

16. (a) Prove the following integral formula for the Legendre polynomials

$$P_l(x) = \frac{1}{2\pi i}\int_\Gamma \frac{z^l dz}{(z^2-2zx+1)^{\frac{1}{2}}}, \qquad x = \cos\varphi,\ \varphi \text{ real},$$

where Γ is a circle of radius $r > 1$ about $z = 0$.

(b) Show that $g(z,t) = (1-2zt+z^2)^{-\frac{1}{2}}$ is the generating function for the Legendre polynomials, i.e. that

$$(1-2zt+z^2)^{-\frac{1}{2}} = \sum_{n=0}^{\infty} P_n(t)z^n.$$

Discuss the convergence of the series on the right.

(c) Prove the following recurrence relations for the Legendre polynomials (here $P_l'(z) := \frac{\mathrm{d}}{\mathrm{d}z} P_l(z)$):

(1) $zP_l'(z) = lP_l(z) + P_{l-1}'(z)$,
(2) $(2l+1)P_l(z) = P_{l+1}'(z) - P_{l-1}'(z)$,
(3) $(1-z^2)P_l'(z) + lzP_l(z) = lP_{l-1}(z)$,
(4) $(1-z^2)P_l'(z) - (l+1)zP_l(z) = -(l+1)P_{l+1}(z)$,
(5) $(2l+1)zP_l(z) = (l+1)P_{l+1}(z) - lP_{l-1}(z)$.

(d) Show that P_l satisfies the Legendre differential equation

$$\begin{aligned}&((1-z^2)P_l'(z)) + l(l+1)P_l(z)\\ &= (1-z^2)P_l''(z) - 2zP_l'(z) + l(l+1)P_l(z) = 0.\end{aligned}$$

(e) Show that $P_l(z) = \sum_{k=0}^{l} \binom{l}{k}^2 \left(\frac{z-1}{2}\right)^{l-k} \left(\frac{z+1}{2}\right)^k$.

(f) Show that (the Schläfli integral)

$$P_l(z) = \frac{1}{2^l}\frac{1}{2\pi i}\int_\Gamma \frac{(t^2-1)^l}{(t-z)^{l+1}}\,\mathrm{d}t,$$

Γ—the unit circle, $|z| < 1$.

(g) Prove the following relation with the hypergeometric function

$$P_l(z) = F(-l, l+1, 1; \tfrac{1}{2}(1-z)).$$

Hint. (a) Transform formula (8.1.14). (b) Apply (a) and the Cauchy integral formula for the coefficients of the Taylor series. (c) By direct differentiation of the generating identity (b) we get

$$(z^2-2zt+1)^{-\frac{3}{2}} = \sum_{n=1}^{\infty} P_n'(t)z^n,$$

$$(t-z)(z^2-2zt+1)^{-\frac{3}{2}} = \sum_{n=1}^{\infty} nP_n(t)z^{n-1}.$$

Since $1-z^2-2z(t-z) = 1-2zt+z^2$ by multiplying the first identity by $(1-z^2)$ and the second by $2z$ and subtracting them we get (2). The other relations are obtained similarly.

17. The associated Legendre functions.

(a) (the Schläfli integral) Show that

$$P_l^m(z) = \frac{(-1)^m(1-z^2)^{\frac{1}{2}m}}{2^l l!} \frac{(l-m)!}{2\pi i} \oint_\Gamma \frac{(t^2-1)^l}{(t-z)^{l+m+1}} \, dt, \quad |z|<1.$$

(b) Show the following relation with the hypergeometric function:

$$P_l^m(z) = \left(-\frac{1}{2}\right)^m \frac{(l+m)!}{(l-m)!m!} (1-z^2)^{\frac{1}{2}m} \times F\left(m-l, m+l+1, m+l; \tfrac{1}{2}(1-z)\right).$$

(c) Prove

$$P_l^m(z) = \left(\frac{1+z}{1-z}\right)^{\frac{1}{2}m} \sum_{j=\max(m,0)}^{l} \frac{(-1)^j(l+j)!}{(l-j)!(j-m)!j!} \left(\frac{1-z}{2}\right)^j.$$

(d) Prove that the function P_l^m is the only solution of the associated Legendre equation

$$\frac{d}{dz}\left[(1-z^2)\frac{du}{dz}\right] + \left[l(l+1) - \frac{m^2}{1-z^2}\right] u = 0$$

bounded at the points -1, 1.

(e) Prove the following recurrence formulas

$$\frac{d}{dz} P_l^m(z) = (1-z^2)^{-\frac{1}{2}} P_l^{m+1}(z) - \frac{mz}{1-z^2} P_l^m(z),$$

$$(2l+1)(1-z^2)^{\frac{1}{2}} P_l^m(z) = P_{l+1}^{m+1}(z) - P_{l-1}^{m+1}(z).$$

(f) Show that $P_l^l(\cos\theta) = (-1)^l \dfrac{(2l)!}{2^l l!} \sin^l\theta$.

(g) Prove that for an arbitrary $l>0$ the functions

$$\left[\frac{(2l+2k+1)k!}{2(2l+k)!}\right]^{\frac{1}{2}} P_{l+k}^l(x), \qquad k = 0, 1, \dots$$

constitute an orthonormal system of functions in $L^2([-1, 1]), dx)$.

Hint. All the formulas can be obtained either as special cases of the formulas for P^l_{nk} or from the formulas for the Legendre polynomials.

18. Let w be a non-negative function on $[a, b]$, $a < b$, and φ_n a polynomial of degree n precisely. Prove that in order that $\{\varphi_n\}$ be an orthogonal system relative to the weight w, i.e. that

$$\int_a^b w(x)\varphi_n(x)\overline{\varphi_m(x)}\mathrm{d}x = \delta_{nm}, \qquad m, n = 0, 1, \dots$$

it is necessary and sufficient that $\int_a^b w(x)\varphi_n(x)x^k\mathrm{d}x = 0$ for $k = 0, 1, \dots$ $\dots, n-1$. A sequence $\{\varphi_n\}_{n=0}^{\infty}$ of polynomials φ_n such that the degree of φ_n equals n and φ_n are mutually orthogonal relative to weight w, is called an orthogonal system of polynomials with weight w.

Show that if $[a, b]$ is a bounded interval then the orthogonal system of the polynomials $\{\varphi_n\}$ is a basis for the Hilbert space $L^2([a, b], w(x)\mathrm{d}x)$

Hint. Expand φ_n with respect to the powers x^k. Apply the Stone–Weierstrass theorem.

19. Show that if $\{\varphi_n\}$ is an orthogonal sequence of real polynomials in $L^2([a, b], w(x)\mathrm{d}x)$ with $w(x) > 0$ and φ_n is of degree n precisely, then all roots of φ_n are different and lie in the open interval $[a, b]$.

Hint. Consider $\int_a^b w(x)\varphi_n(x)\mathrm{d}x = 0$. If $\alpha_1, \dots, \alpha_t$ are the points in $]a, b[$ at which φ_n changes sign, then $s < n$ and for $\psi(x) = \prod_{i=1}^{s}(x-\alpha_i)$ we have

$$\int_a^b w(x)\varphi_n(x)\psi(x)\mathrm{d}x \doteq 0$$

(Problem 18). Derive hence that $\varphi_n(x) = C\psi(x)$.

20. Show that for $\alpha, \beta > -1$ the Jacobi polynomials $P_n^{(\alpha,\beta)}$ defined by

$$P_n^{(\alpha,\beta)}(x) = \frac{(-1)^n}{2^n n!}(1-x)^{-\alpha}(1-x)^{-\beta}\frac{\mathrm{d}^n}{\mathrm{d}x^n}(1-x)^{\alpha+n}(1+x)^{\beta+n}$$

form an orthogonal system of polynomials with the weight

$$w(x) = (1-x)^{\alpha}(1+x)^{\beta}$$

(α, β are not assumed to be integers).

Remark: Besides the Legendre polynomials corresponding to $\alpha = \beta = 0$ we obtain for $\alpha = \beta = -\frac{1}{2}$ the Tchebyshev polynomials and for $\gamma > -\frac{1}{2}$ and $\alpha = \beta = \gamma - \frac{1}{2}$ the Gegenbauer polynomials.

Hint. Write $P_n^{(\alpha,\beta)}$ in the form $P_n^{(\alpha,\beta)} x = \frac{c_n}{w(x)} \frac{d^n}{dx^n} [w(x)(1-x^2)^n]$. Show that $P_n^{(\alpha,\beta)}$ is a polynomial of degree n precisely and apply the criterion from Problem 18. For integer values of α, β one can also apply the orthogonality of functions P_{mm}^l proved already and their relation with the Jacobi polynomials (Problem 19).

21. Prove that for fixed l, k such that $l \geqslant k \geqslant 0$, the functions $\sqrt{\frac{2l+2n+1}{2}}\, P_{lk}^{l+n}(\cdot)$, $n = 0, 1, 2, \ldots$, constitute an orthonormal basis on the interval $[-1, 1]$, with respect to the Lebesgue measure. As a special case, once again derive the orthonormality of the system of functions

$$\sqrt{\frac{(2l+2n+1)n!}{2(2l+n)!}}\, P_{l+n}^l(\cdot),$$

$n = 0, 1, 2, \ldots$ on the interval $[-1, 1]$.

Hint. Either apply the orthogonality relations and the results of Problem 18 or the result of the preceding problem.

22. (a) Show that if $t \to g(t) \in SU(2)$ is a one-parameter subgroup with generator $\sum_{i=1}^{3} a_i X_i$, where X_i are the Pauli matrices then the subgroup $g_0 g(t) g_0^{-1}$ has generator $\sum_{i=1}^{3} b_i X_i$ where $b = \alpha(g_0)a$. Here a, $b \in \boldsymbol{R}^3$ are vectors with coordinates a_i, b_i resp.

(b) Apply (a) to show

$$\begin{aligned} T^l(g)\mathrm{d}T^l(X_1) &= (\cos\varphi\cos\psi - \sin\varphi\sin\psi\cos\theta)\mathrm{d}T^l(X_1)T^l(g) \\ &\quad + (\sin\varphi\cos\psi + \cos\varphi\sin\psi\cos\theta)\mathrm{d}T^l(X_2)T^l(g) \\ &\quad + \sin\varphi\sin\theta\,\mathrm{d}T^l(X_3)T^l(g) \end{aligned}$$

(φ, ψ, θ) being the Euler angles of the matrix g.

(c) Show by applying (b) to the vectors e_l the following recurrence relation

$$\begin{aligned} &\sqrt{(l-n)(l+n+1)}\, P_{m,n+1}^l(z) \\ &= \tfrac{1}{2}\sqrt{(l+m)(l-m+1)}\,(1+z) P_{m-1,n}^l(z) \\ &\quad + \tfrac{1}{2}\sqrt{(l-m)(l+m+1)}\,(1-z)\, P_{m+1,n}^l(z) - im\sqrt{1-z^2}\, P_{mn}^l(z). \end{aligned}$$

(d) Prove that

$$T^l(g)\mathrm{d}T^l(X_3) = \sin\varphi\sin\theta\,\mathrm{d}T^l(X_1)T^l(g) - \cos\varphi\sin\theta\,\mathrm{d}T^l(X_2)T^l(g) + \cos\theta\,\mathrm{d}T^l(X_3)T^l(g).$$

Deduce hence the recurrence formula

$$2i\frac{m-nz}{\sqrt{1-z^2}}P^l_{mn}(z) = \sqrt{(l+n)(l-n+1)}\,P^l_{m,n-1}(z) - \sqrt{(l-n)(l+m+1)}\,P^l_{m,n+1}(z).$$

23. Show that

(a) $$e^{i-(m\varphi+\psi n)}P^l_{mn}(\cos\theta) = \sum_{k=-l}^{l} i^{-k}P^l_{mk}(\cos\theta_1)P^l_{kn}(\cos\theta_2),$$

where

$$\cos\theta = \cos\theta_1\cos\theta_2, \qquad \tan\varphi = \frac{\sin\varphi_2}{\sin\theta_1\cos\theta_2},$$

$$\tan\psi = \frac{\sin\theta_1}{\cos\theta_1\sin\theta_2}.$$

(b) $$P_l(\cos\theta_1\cos\theta_2 - \sin\theta_1\sin\theta_2\cos\varphi_2) = \sum_{k=-l}^{l} e^{-ik\varphi_2}P^k_l(\cos\theta_1)P^{-k}_l(\cos\theta_2).$$

Hint. (a) Apply the addition formula (8.3.11) to the case $\varphi_2 = \frac{1}{2}\pi$. (b) Put $m = n = 0$ in formula (8.3.11). Make use of the symmetry of P^l_{m0}.

24. (a) Apply the addition formula (8.3.12) to derive the recurrence relation

$$2i\sqrt{1-z^2}\,\frac{\mathrm{d}}{\mathrm{d}z}P^l_{mn}(z) = \sqrt{(l-n)(l+n+1)}\,P^l_{m,n+1}(z) + \sqrt{(l+n)(l-n+1)}\,P^l_{m,n-1}(z).$$

(b) Apply 23(a) to obtain the recurrence from 22(d).

Hint. (a) Differentiate (8.3.12) with respect to θ_2. To compute the derivative of the function P^l_{mn} make use of integral formula (8.1.13).

25. Prove that the Clebsch–Gordan coefficients can be represented as

(a) $$C(l, l', k;\ m, n, m+n) = \frac{(-1)^{-k+l'+n}}{2^{k+l+l'+1}}$$
$$\times\left[\frac{(2k+l)(k+m+n)!(l+l'-k)!(l+l'+k+1)!}{(l-m)!(l+m)!(l-n)!(l'+n)!(k-m-n)!(k+l-l')!(k-l+l')!}\right]^{\frac{1}{2}}$$
$$\times\int_{-1}^{1}(1-x)^{l'-m}(1+x)^{l'-n}\frac{\mathrm{d}^{k-m-n}}{\mathrm{d}x^{k-m-n}}[(1-x)^{k-l+l'}(1+x)^{k+l-l'}]\mathrm{d}x.$$

(b) $$C(l, l', k;\ m, n, m+n) = (-1)^{l-m}$$
$$\times\left[\frac{(2k+l)(k+m+n)!(k-m-n)!(l-m)!(l'-n)!(l+l'-k)!}{(l+m)!(l'+n)!(k+l-l')!(k-l+l')!(k+l+l'+1)!}\right]^{\frac{1}{2}}$$
$$\times\sum_t\frac{(-1)^t(l+m+t)!(k+l'-m-t)!}{t!(k-m-n-t)!(l-m-t)!(l'-k+m+t)!}.$$

Hint. Substitute formula (8.7.13) into (8.7.10) and expression (8.1.4) for $P^k_{l-l',m+n}$ taking into account the properties of changing the sign of the coefficients (cf. (8.2.5)). To derive (b) apply the Leibniz formula to (a) and integrate by parts.

26. Show that for $l' = \frac{1}{2}$ the Clebsch–Gordan decomposition has the form

$$T^l\otimes T^{\frac{1}{2}} = T^{l-\frac{1}{2}}\oplus T^{l+\frac{1}{2}}$$

and the Clebsch–Gordan coefficients for the non-zero terms are given by

	$n = -\frac{1}{2}$	$n = \frac{1}{2}$
$k = l-\frac{1}{2}$	$\left[\frac{l+m+n+\frac{1}{2}}{2l+1}\right]^{\frac{1}{2}}$	$-\left[\frac{l-m-n+\frac{1}{2}}{2l+1}\right]^{\frac{1}{2}}$
$k = l+\frac{1}{2}$	$\left[\frac{l-m-n+\frac{1}{2}}{2l+1}\right]^{\frac{1}{2}}$	$\left[\frac{l+m+n+\frac{1}{2}}{2l+1}\right]^{\frac{1}{2}}$

27. Compute the Clebsch–Gordan coefficients for $l' = 1$.

28. Prove the following formulas called the *symmetry formulas*

$$C(l, l'; k; m, n, m+n) = (-1)^{k-l-l'}C(l, l', k;\ -m,\ -n,\ -m-n),$$
$$C(l, l', k; m, n, m+n) = (-1)^{k-l-l'}C(l'. l, k; n, m, m+n).$$

29. Prove that

$$\mathrm{e}^{r\cos\theta}J_0(r\sin\theta) = \sum_{n=0}^{\infty}\frac{r^n}{n!}P_n(\cos\theta).$$

Chapter 9

Gegenbauer Polynomials

9.1. INFORMATION ABOUT THE GROUP SO(n) AND THE HOMOGENEOUS SPACE S^{n-1}

In the present chapter we study the class of orthogonal polynomials known as the Gegenbauer polynomials, which appear as spherical functions of irreducible representations of the group SO(n).

In 1929 E. Cartan computed the spherical functions of the group SO(n), thus giving rise to the application of the method of representations theory to the study of special functions.

The Gegenbauer polynomials were the first class of special functions investigated systematically by means of this method.

In fact, it is the fundamental text "Higher Transcendental Functions" by Erdelyi *et al.*, 1953 which already presented the Gegenbauer polynomials, although in an implicit way, from the point of view of representation theory.

Our exposition is mainly based on N. J. Vilenkin, 1965.

Let $(E_n, (\cdot \mid \cdot))$ be an n-dimensional real Euclidean space. SO(n) is the Lie group of orientation-preserving linear isomorphisms of the space E_n.

Choosing an orthonormal basis for E_n allows us to identify the space E_n with $\boldsymbol{R}^n$ and the group SO(n) with the group of all orthogonal matrices with determinant 1.

We introduce spherical coordinates in $\boldsymbol{R}^n$ according to the formula

$$\begin{aligned}
x_1 &= r\sin\theta_{n-1} \dots \sin\theta_2 \sin\theta_1, \\
x_2 &= r\sin\theta_{n-1} \dots \sin\theta_2 \cos\theta_1, \\
x_3 &= r\sin\theta_{n-1} \dots \cos\theta_2 \qquad\qquad (9.1.1) \\
&\dots\dots\dots\dots \\
x_{n-1} &= r\sin\theta_{n-1}\cos\theta_{n-2}, \\
x_n &= r\cos\theta_{n-1}
\end{aligned}$$

where $0 \leqslant r < \infty$, $0 \leqslant \theta_1 < 2\pi$, $0 \leqslant \theta_k < \pi$, $1 < k \leqslant n-1$. Clearly, the Cartesian coordinates determine the spherical coordinates in the following way:

$$\cos\theta_k = \frac{x_{k+1}}{r_{k+1}}, \qquad \sin\theta_k = \frac{r_k}{r_{k+1}}, \tag{9.1.2}$$

where $r_n = r$ and $r_k^2 = \sum_{i=1}^{k} x_i^2$.

For a fixed coordinate r, (9.1.1) defines a parametrization of the sphere of radius r in E_n in terms of the angles $(\theta_k)_{k=1,\ldots,n-1}$.

Obviously, the invariant measure on E_n is the Lebesgue measure $\mathrm{d}\lambda^n$. As regards the SO(n)-invariant normalized measure on the unit sphere S^{n-1}, it can be expressed in terms of the coordinates $(\theta_1, \theta_2, \ldots, \theta_{n-1})$ by the formula

$$\mathrm{d}m(s) = \frac{\Gamma(\frac{1}{2}n)}{2\pi^{\frac{1}{2}n}} \sin^{n-2}\theta_{n-1} \ldots \sin\theta_2 \,\mathrm{d}\theta_1 \ldots \mathrm{d}\theta_{n-1}. \tag{9.1.3}$$

The sphere S^{n-1} is a homogeneous space of the group SO(n) with the isotropy subgroup at a point isomorphic to SO($n-1$).

The isotropy group at the point $(0, \ldots, 1)$ (in Cartesian coordinates) is the set K of all matrices (u_{ij}) such that $u_{nj} = u_{jn} = \delta_{nj}$ and $(u_{ij})_{i,j=1,\ldots,n-1} \in \mathrm{SO}(n-1)$.

9.2. SPHERICAL REPRESENTATIONS OF THE GROUP SO(n)

The pair $(\mathrm{SO}(n), K)$ is symmetric (cf. Example 1(f), § 5.3).

On the basis of the results of Chapter 5 we can derive a decomposition of a unitary spherical representation U of the group SO(n) on $L^2(S^{n-1})$ and a description of irreducible components of the representation.

As we know,

$$U = \bigoplus_{l=0}^{\infty} U^l, \qquad H = L^2(S^{n-1}) = \bigoplus_{l=0}^{\infty} H^l,$$

where each of the representations (H^l, U^l) is finite-dimensional and spherical, i.e. the subspace of all K-invariant vectors is one-dimensional. Moreover, the representations U^l are non-equivalent.

We shall describe the representations (H^l, U^l) below. Let J_n^l be the space of homogeneous polynomial functions of n variables of degree l over the field $\boldsymbol{C}$. The dimension of the space J_n^l equals $\frac{(n+l-1)!}{(n-1)!l!}$.

For $g \in \mathrm{SO}(n)$ the mappings $T_g^l\colon J_n^l \ni w \to w \circ g^{-1}$ transform J_m^l into itself.

The pair (J_n^l, T^l) is a representation of SO(n).

This representation is in general, reducible since for $l \geqslant 2$ the space J_n^l contains the subspace of polynomials of the form $J_n^{l-2}r^2$ where $r^2 = \sum_{i=1}^{n} x_i^2$.

We define two sequences of mappings:

$$\begin{gathered} J_n^0 \xrightarrow{\tau} J_n^2 \xrightarrow{\tau} J_n^4 \xrightarrow{\tau} \dots \xrightarrow{\tau} J_n^{2(j-1)} \xrightarrow{\tau} J_n^{2j} \xrightarrow{\tau} \dots, \\ J_n^1 \xrightarrow{\tau} J_n^3 \xrightarrow{\tau} J_n^5 \xrightarrow{\tau} \dots \xrightarrow{\tau} J_n^{2j-1} \xrightarrow{\tau} J_n^{2j+1} \xrightarrow{\tau} \dots, \end{gathered} \tag{9.2.1}$$

where $\tau(w) = r^2 w$.

We note that τ intertwines T^l with T^{l+2} since the function r is rotation invariant. Being a finite-dimensional representation of a compact group, T^l decomposes into a direct sum of irreducible components.

For the time being we shall only give some rather general information concerning the components of the representation (J_n^l, T^l).

Lemma 9.2.1. *Let $E \subset J^l$ be an invariant subspace having the property that T^l is irreducible on E. Then (E, T^l) is a spherical representation.*

Proof. Homogeneous polynomials are uniquely determined by the restrictions to the sphere S^{n-1}. The mapping $R\colon w \to w|_{S^{n-1}}$ intertwines $T^l|_E$ with the representation U. Since irreducible components of the representation U are spherical (Theorem 5.2.2), the desired result follows. □

Corollary 9.2.2. *Let E be an invariant subspace of J_n^l. $T^l|_E$ is an irreducible representation if and only if the subspace of K-fixed vectors in E is* 1*-dimensional.*

It turns out that, apart from the case of $n = 2$, all irreducible components of the representation U are found among the spaces J_n^l, more precisely: among the complementary spaces to $\tau(J_n^{l-2})$ in J_n^l. It is now possible to describe such representations in an explicit and convenient form.

In the forthcoming considerations it will be useful to apply the dimension argument. For this purpose we find the dimension of the quotient space $J_n^l/\tau(J_n^{l-2})$:

$$\dim J_n^l/\tau(J_n^{l-2}) = \dim J_n^l - \dim \tau(J_n^{l-2}) = \frac{(2l-n-2)(n+l-3)!}{(n-2)!\,l!}. \tag{9.2.2}$$

A complementary subspace for $\tau(J_n^{l-2})$ in J_n^l is also easily determined. Denote by $K_n^l \subset J_n^l$ the subspace of all those polynomial whose terms do not contain x_n of power greater than 1. Clearly, every polynomial $w \in K_n^l$ is of the form

$$w(x) = x_n w_1(x) + w_2(x),$$

where $w_1 \in J_{n-1}^{l-1}$ and $w_2 \in J_{n-1}^l$. Hence follows the dimension estimate:

$$\begin{aligned}\dim K_n^l &= \dim J_{n-1}^{l-1} + \dim J_{n-1}^l = \frac{(2l-n-2)(n+l-3)!}{(n-2)!\,l!}\\ &= \dim J_n^l/\tau(J_n^{l-2}).\end{aligned}$$

Obviously $K_n^l \cap \tau(J_n^{l-2}) = 0$ and so

$$J_n^l = K_n^l \oplus \tau(J_n^{l-2}). \tag{9.2.3}$$

Unfortunately K_n^l is not invariant under the action of the group. To overcome this difficulty we shall define an operator A on J_n^l which annihilates $\tau(J_n^{l-2})$ and is injective on K_n^l.

To this end we view the polynomials in J_n^l as functions on $\boldsymbol{R}^{n-1} \times \boldsymbol{C}$ and define a submanifold $\mathscr{C} \subset \boldsymbol{R}^{n-1} \times \boldsymbol{C}$ by the condition

$$x_n = i\Big(\sum_{j=1}^{n-1} x_j^2\Big)^{\frac{1}{2}} =: i r_{n-1}, \qquad i = \sqrt{-1}.$$

To every polynomial $w \in J_n^l$ we now assign a function of $n-1$ variables by restricting w to the set $\mathscr{C}$. In other words,

$$Aw(x_1, \ldots, x_{n-1}) := w(x_1, \ldots, x_{n-1}, i r_{n-1}).$$

Since $r^2|_{\mathscr{C}} = 0$ we have $A\tau(J_n^{l-2}) = 0$. Obviously $A|_{K_n^l}$ is injective. The image of A consists of positive homogeneous functions of order l (i.e. $f(rx) = r^l f(x)$, $r \in \boldsymbol{R}_+$).

Now we can define an action of the group on $F_n^l := A(J_n^l)$ so as to ensure that

$$Q^l(g)A = AT^l(g). \tag{9.2.4}$$

Clearly, for every $g \in SO(n-1)$, i.e. rotation about x_n, we have

$$Q^l(g)Aw(x) = Aw(g^{-1}x).$$

In order to complete the definition of the representation Q^l we must define $Q^l(g(\varphi))$ for the rotations $g(\varphi)$ of the plane x_{n-1}, x_n through the angle φ. Showing $F = Aw$, we obtain

$$\begin{aligned} Q^l(g(\varphi))F(x_1, \dots, x_{n-1}) &= AT^l(g(\varphi))w(x_1, \dots, x_{n-1}) \\ &= w(x_1, \dots, x_{n-2}, x_{n-1}\cos\varphi + ir_{n-1}\sin\varphi, -x_{n-1}\sin\varphi \\ &\quad + ir_{n-1}\cos\varphi). \end{aligned} \tag{9.2.5}$$

The representation Q^l is equivalent to the subrepresentation T^l on the complement of $\tau(J_n^l)$. The realization of the representation thus obtained, immadiately leads to the following

COROLLARY 9.2.3. *For $n > 2$ the representations Q^l, $l = 0, 1, \dots$ are irreducible.*

Proof. All irreducible components of the representation Q^l are spherical (Lemma 9.2.1). Owing to Corollary 9.2.2 it suffices to show that the subspace of K-invariant vectors in F_n^l is one-dimensional. Excluding the case of $n \leqslant 2$, the group $SO(n-1)$ acts transitively on S^{n-2}. This being so, it follows that every $SO(n-1)$-invariant positively homogeneous function of order l on $\boldsymbol{R}^{n-1}$ is proportional to r_{n-1}^l, which concludes the proof. □

The above realizations of spherical representations of the group $SO(n)$ are due to Vilenkin.

We shall now proceed to a more thorough study of the series of spherical representations. But let us first verify that the above representations exhaust the set of spherical representations of the pair $(SO(n), K)$.

THEOREM 9.2.4. *Let $n > 2$. Every spherical representation of the group $SO(n)$ is equivalent to the representation (F_n^l, Q^l) for a certain l.*

Proof. Let us examine the sequences (9.2.1). The initial elements J_n^0 and J_n^1 are irreducible. The same is true of the representations on $J_n^l/\tau(J_n^{l-2})$ since they are equivalent to Q^l. Hence every representation (J_n^l, T^l) decomposes into a direct sum whose components are of the form $Q^{l_1} \dots Q^{l_p}$. The assertion will follow if we prove that the representation U on $L^2(S^{n-1})$ decomposes into representations (J_n^l, T^l), $l = 0, 1, \dots, \infty$.

We see that the operator R of the restriction of a polynomial $w \in J_n^l$ to the sphere maps the representation (J_n^l, T^l) into $(U, L^2(S^{n-1}))$. Thus we only have to prove that the linear span of the set $\bigcup_{l=0}^{\infty} RJ_n^l$ is dense in $L^2(S^{n-1})$.

We observe that the ring W of all polynomials on $\boldsymbol{R}^n$ separates points in the sphere and the operator R is an isomorphism of the ring W into $\mathscr{C}(S^{n-1})$. Thus, in view of the Stone–Weierstrass theorem, the set RW is dense in $\mathscr{C}(S^{n-1})$. Hence RW is dense in $L^2(S^{n-1})$. Since W is the linear span of the set $\bigcup_{l=0}^{\infty} J_n^l$, the assertion follows. □

The results that we have obtained so far allow us to derive integral formulas for the spherical functions related to the group SO(n) and the corresponding expansions of SO($n-1$)—invariant continuous functions on the sphere. This will be done in §§ 9.4 and 9.5.

Now we shall define another realization of the representation of the group on $J_n^l/\tau(J_n^{l-2})$ by distinguishing an invariant complement of the space $\tau(J_n^{l-2})$ in J_n^l. It is that realization which establishes links between the theory of spherical functions and the Gegenbauer polynomials. Obviously, one can define a complementary subspace as the orthogonal complement of $\tau(J_n^{l-2})$ with respect to a G-invariant inner product on J_n^l. This method, however, is not the simplest one. It is more convenient to resort to the classical method of determining $\tau(J_n^{l-2})$ as an eigenspace of the Laplace operator

$$\Delta = \sum_{i=1}^{n} \frac{\partial^2}{\partial x_i^2}.$$

This operator commutes with the action of G on $\boldsymbol{R}^n$, and therefore its kernel, $H^l = \{w \in J_n^l : \Delta w = 0\}$ is a G-invariant subspace of J_n^l. As an operator from J_n^l into a lower dimensional space J_n^{l-2} it has a non-zero kernel.

THEOREM 9.2.5. $J_n^l = H^l + r^2 J_n^{l-2}$.

Proof. Since the representation on $J_n^l/\tau(J_n^{l-2})$ is irreducible, it is enough to prove the following statement.

LEMMA 9.2.6. *Δ acts injectively on $\tau(J_n^{l-2})$.*

Proof. Represent $f \in \tau(J_n^{l-2})$ in the form $f = r^{2k} f_1$ where f_1 is a polynomial not divisible by r^2. Let us compute Δf. First, the Leibnitz rule gives

$$\Delta(r^2 f_1) = 2nf_1 + 4 \sum_{j=1}^{n} x_j \frac{\partial f}{\partial x_j} + r^2 \Delta f_1.$$

The homogeneity of f_1 implies that

$$\sum_{j=0}^{n} x_j \frac{\partial f_1}{\partial x_j} = (l-2) f_1.$$

Hence

$$\Delta(r^2 f_1) = (2n+4l-8) f_1 + r^2 \Delta f_1.$$

Inductively, we obtain

$$\Delta(r^{2k} f_1) = 2k(n+2l-2k-1) r^{2k-2} f_1 + r^{2k} \Delta f_1. \tag{9.2.6}$$

Let us suppose now that $\Delta(r^{2k} f_1) = 0$. Then from (9.2.6) we get after reductions

$$2k(n+2l-2k-2) f_1 - r^2 \Delta f_1 = 0.$$

The coefficient of f_1 is different from zero since $l \geqslant 2k \geqslant 2$. Consequently, f_1 is divisible by r^2, which contradicts our assumption. This proves the lemma. It can now be seen that $\ker \Delta$ is the complementary subspace for $\tau(J_n^{l-2})$; this ends the proof of the theorem. □

The operator $\Delta \colon J_n^l \to J_n^{l-2}$ is related with the operator τ by means of the formula

$$\operatorname{im} \tau + \ker \Delta = J_n^l. \tag{9.2.7}$$

Thus we have obtained a realization of the irreducible spherical representations of the group $\mathrm{SO}(n)$ on the spaces H^l of homogeneous harmonic polynomials on $\boldsymbol{R}^n$. It is this kind of realization of a spherical representation that will be denoted by (H^l, U^l).

The imbedding of the representations U^l into $L^2(S^{n-1})$ is accomplished via the operator R of the restriction of harmonic polynomials to the unit sphere.

Now we can carry out the decomposition of the representations T^l on J_n^l. Let $2k \leqslant l$. Then the operator $\Delta^k(= \Delta \circ \Delta \circ \ldots \circ \Delta)$ intertwines T^l with T^{l-2k} and $A^k := r^k \circ \Delta^k$ intertwines T^l with T^l. The image of the operator A^k is an invariant subspace of J_n^l on which the representation T^l is equivalent to T_n^{l-2k}. The following decomposition holds:

$$J_n^l = r^{2k} J_n^{l-2k} + \sum_{i=0}^{k-1} H^i, \qquad T^l = T^{l-2k} + \bigoplus_{i=0}^{k-1} U^l.$$

It follows by induction that the space of operators intertwining T^l with T^l is spanned by the operators A^k, $2k \leqslant l$. These observations serve to determine the intertwining projection $P^l \colon J_n^l \to H^l$. It should be of the form

$$P^l f = \sum_{k=0}^{E(\frac{1}{2}l)} \alpha_k r^{2k} \Delta^k f, \qquad f \in J_n^l$$

with the constants α_k verifying the condition $\Delta P^l f = 0$. By applying formula (9.2.6) we find that the coefficients satisfy the reccurrence re lation

$$\alpha_k + (2k+2)(n+2l-2k-4)\alpha_{k+1} = 0,$$

and hence we obtain the final expression

$$P^l f = \sum_{k=0}^{E(\frac{1}{2}l)} \frac{(-1)^k r^{2k} \Delta^k f}{2^k k!(n+2l-4)(n+2l-6)\ldots(n+2l-2k-2)}. \tag{9.2.8}$$

The zonal spherical function of the representation U^l, regarded as a function on the sphere S^{n-1}, is a restriction to the sphere of a harmonic polynomial in J^l, which is symmetric with respect to rotations about x_n and moreover assumes the value 1 at $x = (0, \ldots, 1)$.

We know that these properties determine the polynomial uniquely, and moreover, that its form can be obtained by acting with P^l on the K-invariant polynomial $w_0 \colon x \to x_n^l$. It remains to apply formula (9.2.8)

$$P^l w_0(x) = \sum_{k=0}^{E(\frac{1}{2}l)} \frac{(-1)^k l! r^{2k} x_n^{l-2k}}{k! 2^k (l-2k)!(n+2l-4)\ldots(n+2l-2k+1)}$$

for $x \in S^{n-1}$.

The above sum can also be written in terms of the Gegenbauer polynomials defined as follows:

$$C_j^\alpha(t) = \frac{2^j\Gamma(\alpha+j)}{j!}\left[t^j - \frac{j(j-1)}{2^2(\alpha+j-1)}\,t^{j-2}\right.$$
$$+\frac{j(j-1)(j-2)(j-3)}{2^4\cdot 1\cdot 2(\alpha+j-1)(\alpha+j-2)}\,t^{j-4}+\dots$$
$$\left.+\frac{(-1)^l j!\Gamma(\alpha+j-1)}{(j-2l)!2^{2l}l!\Gamma(\alpha+j)}\,t^{j-2l}+\dots\right], \tag{9.2.9}$$

$$P^l w_0(x) = \frac{l!\Gamma\left(\frac{n-2}{2}\right)}{2^l\Gamma\left(l+\frac{n-2}{2}\right)}\,r^l C_l^{\frac{n-2}{2}}\left(\frac{x_n}{r}\right). \tag{9.2.10}$$

Thus the zonal spherical function of the representation U^l has the form

$$\omega_n^l(x) = \frac{C_l^{\frac{n-2}{2}}(x_n)}{C_l^{\frac{n-2}{2}}(1)}, \qquad x\in S^{n-1}. \tag{9.2.11}$$

The spherical functions of the representations H^l are real.

9.3. GEGENBAUER'S EQUATION AND BASIC RECURRENCES

The polynomial $P^l w_0$ is a harmonic function, i.e. satisfies the differential equation $\Delta P^l w_0 = 0$. We shall write this equation in terms of spherical coordinates in $\boldsymbol{R}^n$ by applying (9.2.10).

The function under consideration depends only upon the radius r and the angle θ_{n-1}, and therefore we may neglect those terms occurring in the expression of Δ in spherical coordinates which do not contain differentiations with respect to the above variables. Thus

$$P^l w_0(r,\theta_{n-1}) = \left[r^{1-n}\frac{\partial}{\partial r}\left(r\frac{\partial}{\partial r}\right)+r^{-2}\sin^{2-n}\theta_{n-1}\right.$$
$$\left.+\frac{\partial}{\partial\theta_{n-1}}\left(\sin^{n-2}\theta_{n-1}\frac{\partial}{\partial\theta_{n-1}}\right)\right]P^l w_0(r,\theta_{n-1}) = 0.$$

Making the substitution (9.2.10) and performing the differentiations with respect to r we obtain

$$\frac{\partial^2}{\partial\theta^2} C_l^p(\cos\theta)+2p\cot\theta\,\frac{\partial}{\partial\theta}\,C_l^p(\cos\theta)+l(2p+1)\,C_l^p(\cos\theta) = 0,$$

$$p=\frac{n-2}{2}, \qquad n\in N,$$

with the change of variable $\cos\theta = t$ this becomes

$$(1-t^2)\frac{\partial^2}{\partial t^2}\,C_l^p(t)-(2p+1)t\,\frac{\partial}{\partial t}\,C_l^p(t)+l(2p+l)C_l^p(t) = 0, \tag{9.3.1}$$

which is called *Gegenbauer's equation*

Direct calculation shows that the Gegenbauer polynomials defined by the series (9.2.9) satisfy the identity

$$\frac{\mathrm{d}}{\mathrm{d}t}\,C_m^\alpha(t) = 2\alpha C_{m-1}^{\alpha+1}(t). \tag{9.3.2}$$

An important recurrence formula can be deduced by comparing the result of the action of the projection operators P^l and P^{l+1}. By (9.2.10)

$$P^{l+1}x_n^{l+1} = \frac{(l+1)!\,\Gamma\left(\dfrac{n-2}{2}\right)}{2^{l+1}\Gamma\left(l+\dfrac{n}{2}\right)}\,r^{l+1}C_{l+1}^{\frac{n-2}{2}}\left(\frac{x_n}{r}\right).$$

On the other hand, one can decompose the polynomial x_n^{l+1} into a harmonic part and a polynomial of the form $r^2w_1(x)$ on the basis of the following observation:

LEMMA 9.3.1. *If w is a homogeneous harmonic polynomial of degree l, then the polynomial w' of degree $l+1$:*

$$x\to x_n w(x)-\frac{1}{n+2l-2}\,r^2\,\frac{\partial w}{\partial x_n}\,(x)$$

is a homogeneous harmonic polynomial of degree $l+1$.

The proof consists in direct computation of $\Delta w'$. Since $\Delta w = 0$, we have $\Delta(x_n w) = 2\frac{\partial w}{\partial x_n}$. On the other hand, by (9.2.1) we get

$$\Delta\left(r^2 \frac{\partial w}{\partial x_n}\right) = (2n+4l-4)\frac{\partial w}{\partial x_n} + r^2 \Delta \frac{\partial w}{\partial x_n} = 2(n+2l-2)\frac{\partial w}{\partial x_n},$$

because $\frac{\partial w}{\partial x_n}$ is also a harmonic polynomial. □

It follows from the lemma that if $x_n^l = P^l x_n + r^2 w_1$ then the resulting decomposition

$$x_n^{l+1} = x_n P^l x_n^l - \frac{1}{n+2l-2} r^2 \frac{\partial P^l x_n^l}{\partial x_n} + r^2 \left(\frac{1}{n+2l-2}\frac{\partial P^l x_n^l}{\partial x_n} + x_n w_1\right)$$

is the required decomposition into a harmonic component and an element of the space $\tau(J^{n-1})$. The uniqueness of the decomposition ensures that

$$P^{l+1} x_n^{l+1} = x_n P^l x_n^l - \frac{1}{n+2l-2} r^2 \frac{\partial P^l x_n^l}{\partial x_n}. \tag{9.3.3}$$

By substituting (9.2.10) and (9.3.2) into this formula we obtain the recurrence relation

$$(l+1) C_{l+1}^p(t) = (2p+l) t C_l^p(t) - 2p(1-t^2) C_{l-1}^{p+1}(t). \tag{9.3.4}$$

Corollary 9.3.2. $C_l^p(1) = \dfrac{\Gamma(2p+l)}{l!\Gamma(2p)}$.

Proof. Putting $t = 1$ in (9.3.4), we find

$$C_{l+1}^p(1) = \frac{2p+1}{l+1} C_l^p(1).$$

From the definition of C_l^p we conclude that $C_1^p(1) = 2p$. The assertion now follows by induction. □

9.4. Integral Formulas for the Gegenbauer Polynomials

We shall recall the basic facts from the general theory of spherical representations (§ 5.1).

The operator of orthogonal projection onto the space of K-fixed vectors has the form $P\xi = \int U_k \xi \mathrm{d}k$, where $\mathrm{d}k$ is the normalized Haar measure on K and $\xi \in H$. Thus, for an arbitrary $\xi \in H$, the vector $P\xi$ is proportional to ξ_0, where ξ_0 is a fixed K-fixed vector spanning the space PH. Taking $\xi = U_g \xi_0$, we obtain

$$PU_g \xi_0 = \omega(g)\xi_0 \qquad \text{(Corollary 5.1.2)}, \tag{9.4.1}$$

where ω is the zonal spherical function of the representation (U, H).

Let $\xi^* \in H$ be a functional on the representation space which is not zero at ξ_0. Then (9.4.1) leads to

$$\langle \xi^*, PU_g \xi_0 \rangle / \langle \xi^*, \xi_0 \rangle = \omega(g). \tag{9.4.2}$$

We shall apply this formula to the representation (Q^l, F^l_n) of the group SO(n). As we know, the SO($n-1$)-invariant function in F^l_n is the image of the polynomial $w_0(x) = x^l_n$, i.e. is the function $(ir_{n-1})^l$. We choose the functional ξ^* in the form

$$\langle \xi^*, f \rangle := f(0, 0, \ldots, 1).$$

Then $\langle \xi^*, Aw_0 \rangle = i^l$.

We shall compute the value of $\langle \xi^*, PU_g Aw_0 \rangle$ only for a rotation g in the plane (x_{n-1}, x_n) since the spherical function is completely determined by its values at the elements of this one-parameter subgroup:

$$\begin{aligned} \langle \xi^*, PQ_g Aw_0 \rangle &= \int_K \langle \xi^*, Q^l_{kg} Aw_0 \rangle dk = \int_K \langle \xi^*, AU^l_{kg} w_0 \rangle dk \\ &= \int_K w_0(g^{-1}k^{-1}x_0) dk, \end{aligned} \tag{9.4.3}$$

where $x_0 = (0, \ldots, 1, i)$ and the action of the group on C^n is a natural action of matrices.

$K = \mathrm{SO}(n-1)$ being a unimodular group, we can transform (9.4.3) into the form

$$\int_K w_0(g^{-1}k \cdot x_0) dk.$$

Owing to the fact that x_0 is invariant under SO($n-2$) (rotations in the space $(x_1, \ldots, x_{n-2})$, the function

$$K \ni k \to w_0(g^{-1}k \cdot x_0)$$

is right invariant under SO($n-2$) and therefore may be regarded as a function on $SO(n-1)/SO(n-2) = S^{n-2}$. Consequently, the expression (9.4.3) takes the form

$$\int_{S^{n-2}} w_0(g^{-1}y(x))dm(x), \tag{9.4.4}$$

where for $x \in S^{n-2}$ the vector $y(x) \in C^n$ is of the form (x, i). In terms of spherical coordinates on S^{n-2} we have

$$w_0(g^{-1}y(x)) = \cos\theta_{n-2}\sin\varphi + i\cos\varphi,$$

where g is the rotation through angle φ in the plane (x_{n-1}, x_n). Finally, by substituting in (9.4.4) the form of the invariant integral on S^{n-2} given in § 9.1, we find

$$\omega_n^l(g) = \frac{\Gamma\left(\frac{n-1}{2}\right)}{\Gamma\left(\frac{n-2}{2}\right)} \int_0^\pi (\cos\varphi - i\sin\varphi\cos\theta)^l \sin^{n-3}\theta\, d\theta. \tag{9.4.5}$$

Combining this formula with (9.2.11) and Corollary 9.3.2, we get

$$C_l^p(\cos\varphi) = \frac{\Gamma(2p+l)}{2^{2p-1}\Gamma(p)^2 l!} \int_0^\pi (\cos\varphi - i\sin\varphi\cos\theta)^l \sin^{n-3}\theta\, d\theta. \tag{9.4.6}$$

Let us return to the realizations of the spherical representations of the rotation group on the finite-dimensional subspaces H^l of $L^2(S^{n-1})$.

The inner product on $L^2(S^{n-1})$ induces an invariant product on H^l. The spherical function

$$\omega_n^l(x) = C_l^{\frac{n-2}{2}}(x_n) \frac{l!\Gamma(n-2)}{\Gamma(n+l-2)}, \qquad x \in S^{n-1} \tag{9.4.7}$$

is a normalized K-invariant vector in H^l.

Thus from the definition of a spherical function as a matrix element of the representation we obtain the integral relation

$$\omega_n^l(gx_0) = \int_{S^{n-1}} \omega_n^l(g^{-1}x)\omega_n^l(x)dm(x). \tag{9.4.8}$$

We shall find an explicit expression for (9.4.8) when g is a rotation through angle φ in the space of variables (x_{n-1}, x_n).

We apply spherical coordinates (9.1.1) and formula (9.1.3). Then the value of the n-th coordinate of the vector $g^{-1}x$ is equal to

$$\sin\theta_{n-1}\cos\theta_{n-2}\sin\varphi+\cos\varphi\cos\theta_{n-1}.$$

Substituting (9.4.7) in (9.4.8), we obtain after reductions

$$C_l^p(\cos\varphi) = \frac{l+p}{\pi}\int_0^{\pi}\int_0^{\pi} C_l^p(\cos\theta\cos\varphi+\sin\theta\sin\varphi\cos\theta_1) \times C_l^p(\cos\theta)\sin^{2p}\theta\sin^{2p-1}\theta_1\,d\theta\,d\theta_1 \tag{9.4.9}$$

for $p = \frac{n-2}{2}$, n being a positive integer.

Let us now apply the orthogonality relations for ω_n^l in order to obtain new integral relations for the Gegenbauer polynomials. As matrix elements of irreducible representations, the functions ω_n^l fulfil the following relations:

$$\int_G \omega_n^l(gx_0)\omega_n^{l'}(gx_0)\,dg = \int_{S^{n-1}} \omega_n^l(x)\omega_n^{l'}(x)\,dm(x) = \frac{\delta_{ll'}}{\dim H^l}. \tag{9.4.10}$$

The dimension of H^l equals $\frac{(2l+n-2)(n+l-3)}{(n-2)!l!}$. Substituting (9.4.7) and (9.1.3), we arrive at the orthogonality relations for the Gegenbauer polynomials:

$$\int_0^{\pi} C_l^p(\cos\theta)\,C_{l'}^p(\cos\theta)\sin^{2p}\theta\,d\theta = \delta_{ll'}\frac{\pi\Gamma(2p+l)}{2^{2p-1}l!(l+p)\Gamma^2(p)}. \tag{9.4.11}$$

From (9.4.11) we obtain, in view of previous considerations,

THEOREM 9.4.1. *The system of functions*

$$[-1,1]\ni t\to f_l(t) = 2^{p-1}\Gamma(p)\left[\frac{2(l+p)l!}{\Gamma(2p+l)\pi}\right]^{\frac{1}{2}} C_l^p(t)$$

is a complete orthonormal system in the space $L^2([-1,1],d\mu(t))$, *where* $d\mu(t) = (1-t^2)^{p-\frac{1}{2}}dt$.

Proof. It is only the completeness of this system that has to be proved. In the proof we shall make use of formula (5.2.4), which implies that spherical functions constitute a complete system in the space $L^2(K\backslash G/K)$ of functions in $L^2(G/K)$ which are invariant under rotations in K.

In the case under consideration $L^2(K\backslash G/K)$ is isomorpnic to the space $L^2([-1, 1], d\mu)$ under the assignment $L^2([-1, 1], d\mu) \ni f \to \tilde{f}$ where $\tilde{f}(x) = f(x_n)$ and $\tilde{f}_l$ are proportional to ω_n^l. □

In order to complete these considerations we shall find the form of formula (9.1.9) and of the Parseval formula in the present case. For $f \in L^2([-1, 1], d\mu)$ we thus have

$$f = \sum_{l=0}^{\infty}{}' a_l C_l^p, \tag{9.4.12}$$

where

$$a_l = \frac{2^{2p-1} l!(l+p)\Gamma^2(p)}{\pi\Gamma(2p+1)} \int_{-1}^{1} f(x) C_l^p(x) (1-x^2)^{p-\frac{1}{2}} dx$$

(the series being convergent in $L^2([-1, 1], d\mu)$), and

$$\int_{-1}^{1} |f|^2(t)(1-t^2)^{p-\frac{1}{2}} dt = \sum_{l=0}^{\infty} \frac{\pi\Gamma(2p+1)}{2^{2p-1} l!(l+p)\Gamma^2(p)} |a_l|^2. \tag{9.4.13}$$

9.5. THE MEAN VALUE THEOREM FOR A SPHERICAL FUNCTION

The sphere S^{n-1} in $\boldsymbol{R}^n$ can be given a metric structure which is invariant under the transitive action of the rotation group. Namely, an invariant distance between points $x, x' \in S^{n-1}$ can be defined either as the Cartesian distance between these points or as the arc distance in the plane passing through x, x' and 0. Thus we see that the condition of invariance does not determine the metric uniquely; still it is worth noting that for both metrics such geometrical objects as a sphere $S(x) = \{x' | d(x, x') = \text{const}\}$ and a geodesic, i.e. the shortest line connecting two given points, are identical.

In both cases the sphere with centre $(0, \ldots, 1)$ passing through a given point x consist of those points in S^{n-1} whose n-th coordinate equals x_n. Equivalently, this sphere can be defined as the orbit of x with respect to the action of $SO(n-1)$ on the space of the coordinates $x_1, \ldots, x_{n-1}$.

Thus the invariance of the metric implies that the set gKx is a sphere with centre gx_0 passing through the point gx and therefore of radius $d(x, x_0)$. It follows from formula (5.1.2) of that the function ω_n^l satisfies the spherical equation

$$\int_K \omega_n^l(gkx)\mathrm{d}k = \omega_n^l(gx_0)\omega_n^l(x). \tag{9.5.1}$$

This formula can be given the following geometrical interpretation: the mean value of a spherical function over the sphere of radius $d(x, x_0)$ with centre gx_0 is the product of the value of the function at the centre of the sphere and at the point x.

For special values of x and g we obtain the well-known integral relations for the Gegenbauer polynomials. To see this we put $x = (0,\ldots$ $\ldots, \sin\theta, \cos\theta)$ and suppose that g is the rotation group of the coordinates $(x_1, \ldots, x_{n-1})$. Then the vector kx runs through the subset of the sphere S^{n-1} of vectors of the form $(s, \sin\theta)$ where s belongs to the sphere S^{n-2} of radius $\cos\theta$. In spherical coordinates the last coordinate of the vector $s \in S^{n-2}$ has the form $\cos\theta\cos\theta_1$. Consequently, the n-th coordinate of the vector gkx is equal to

$$\sin\varphi\sin\theta\cos\theta_1 + \cos\varphi\cos\theta$$

and formula (9.5.1) takes the form

$$\int_0^\pi C_l^p(\sin\varphi\sin\theta\cos\theta_1 + \cos\varphi\cos\theta)\sin^{2p-1}\theta_1\,\mathrm{d}\theta_1$$

$$= \frac{2^{2p-1}l!\,\Gamma(p)^2}{\Gamma(2p+l)}\, C_l^p(\cos\varphi)C_l^p(\cos\theta). \tag{9.5.2}$$

PROBLEMS

1. For $\alpha = (\alpha_1, \ldots, \alpha_n) \in \boldsymbol{Z}_+^n$ ($\boldsymbol{Z}_+$—the set of non-negative integers) and $x \in \boldsymbol{R}^n$ we write as usual $|\alpha| = \sum_{k=1}^n \alpha_k$, $x^\alpha = x_1^{\alpha_1} \ldots x_n^{\alpha_n}$. Prove that the monomials x^α for α such that $|\alpha| = l$ constitute a basis for the space J_n^l and prove that $\dim J_n^l = \binom{n+l-1}{l}$.

Hint. Prove that the number of multi-indices α such that $|\alpha| = l$ equals the number of choices of $n-1$ elements from a set of $n+l-1$ elements.

2. Denote by C_n^l the vector space of all functions on $\boldsymbol{C}^n$ of the form

$$p(z) = \sum_{|\alpha|=l} p_\alpha z^\alpha, \qquad \text{where } z^\alpha = z_1^{\alpha_1} \ldots z_n^{\alpha_n}.$$

(Obviously C_n^l is isomorphic to J_n^l by the restriction of functions to $\boldsymbol{R}^n$.)

Define a representation of the group $U(n)$ on C_n^l by putting

$$T_g^l p(z) = p(g^t z).$$

(a) Prove that the representation T^l on C_n^l is irreducible.

(b) Prove that the restriction of T^l to the group $\mathrm{SO}(n)$ is (up to the isomorphism of C_n^l with J_n^l mentioned above) the representation T^l of the group $\mathrm{SO}(n)$ defined in § 9.2.

(c) Prove that the restriction of T^l to the group $\mathrm{SU}(n)$ is also an irreducible representation on C_n^l.

Hint. (a) If P is an operator on C_n^l such that $T_g^l P = P T_g^l$ then by applying this formula to a diagonal matrix g, prove that P is represented in the basis $\{z^\alpha\}$ by a diagonal matrix. Now find $g \in U(n)$ permuting the elements of the basis. (c) Modify the proof of (a).

3. For $\alpha = (\alpha_1, \ldots, \alpha_n)$ let D^α denote, as usual, $\left(\frac{\partial}{\partial x_1}\right)^{\alpha_1} \ldots$ $\ldots \left(\frac{\partial}{\partial x_n}\right)^{\alpha_n}$ and, for every polynomial $p(x) = \sum_{|\alpha|=l} p_\alpha x^\alpha$, put $p(D) = \sum_{|\alpha|=l} p_\alpha D^\alpha$. Denote by $\bar{p}$ the polynomial with the coefficients conjugate to the coefficients of p (i.e. $\bar{p}(x) = \sum_{|\alpha|=l} \bar{p}_\alpha x^\alpha$ if $p(x) = \sum_{|\alpha|=l} p_\alpha x^\alpha$).

(a) Prove that the mapping

$$J_n^l \times J_n^l \ni (p, q) \to (p|q) = \frac{1}{l!} \bar{p}(D) q$$

is an inner product on J_n^l and the monomials $\frac{l!}{\alpha_1! \ldots \alpha_n!} x^\alpha$ form an orthonormal basis.

(b) Show that J_n^l is an orthogonal sum relative to this inner product of the subspace H^l and $\tau(J_n^{l-2})$. Do not use the decomposition of J_n^l proved in Theorem 9.2.5.

(c) Prove that the same formula (with obvious modifications) defines an inner product on C_n^l and that the representation T^l of the group $U(n)$ is unitary relative to that product.

Hint. (b) If $p \in \tau(J_n^{l-2})$ then $p(D) = q(D)\Delta = \Delta q(D)$ for a certain $q \in J_n^{l-2}$. Further, observe that any polynomial orthogonal to the space $\tau(J_n^{l-2})$ must be harmonic.

4. Prove that for an arbitrary polynomial function on $\boldsymbol{R}^n$ of degree $l-2$, $f(x) = \sum_{|\alpha|=0}^{l-2} f_\alpha x^\alpha$ there exists a unique polynomial function (of degree l) of the form $g(x) = (r^2-1)\sum_{|\alpha|=0}^{l-2} g_\alpha x^\alpha$ such that $\Delta g = f$ and deduce hence that $\Delta: J_n^l \to J_n^{l-2}$ is a surjective mapping.

Hint. To prove the uniqueness of g make use of the uniqueness of solutions of the Dirichlet problem for the unit ball.

5. For a function f on the unit sphere we define its radial extension f^* to the spherical shell $\{x: a < r < b\}$, where $0 < a < 1 < b$ by putting $f^*(x) = f\left(\dfrac{x}{r}\right)$.

(a) Prove that f is differentiable on the sphere (with respect to the natural manifold structure) if and only if f^* is differentiable in $\boldsymbol{R}^n$.

(b) Define the Laplace–Beltrami operator Δ_0 on the sphere by a decomposition of the n-dimensional Laplacian Δ with respect to spherical coordinates:

$$\Delta = \frac{\partial^2}{\partial x^2} + \frac{n-1}{r}\frac{\partial}{\partial r} + \frac{1}{r^2}\Delta_0.$$

Prove the formula

$$\Delta_0 = \sum_{j=1}^{n-1} \frac{1}{q_j \sin^{j-1}\theta_j}\frac{\partial}{\partial\theta_j}\sin^{j-1}\theta_j\frac{\partial}{\partial\theta_j},$$

where

$$q_j = \begin{cases} 1, & j = n-1, \\ \sin^2\theta_{j+1}\ldots\sin^2\theta_{n-1}, & 1 \leqslant j \leqslant n-2. \end{cases}$$

(c) Show that Δ_0 is a symmetric negative operator on $L^2(S^{n-1})$, i.e. that $(\Delta_0 f, h) = (f, \Delta_0 h)$ and $(\Delta_0 f, f) \leqslant 0$ for $f, h \in C^\infty(S^{n-1})$.

(d) Prove without using the decomposition of $L^2(S^{n-1})$ into a direct sum of the spaces H^l that any two spherical harmonics of degrees k, l with $k \neq l$ (i.e. the restrictions to the unit sphere of harmonic homogeneous polynomials of degrees k and l) are orthogonal in $L^2(S^{n-1})$.

(e) Let $Y^{(k)}$ be a spherical harmonic of degree k. Find all solutions of the Laplace operator $\Delta u = 0$ in $\boldsymbol{R}^n \backslash \{0\}$ of the form $u(x) = \varphi(r) Y^{(k)}\left(\frac{x}{r}\right)$ and prove that the function $r^k Y^{(k)}\left(\frac{x}{r}\right)$ is a unique solution admitting an extension to the whole of $\boldsymbol{R}^n$.

Hint: (c) Observe that $\Delta_0 f = \Delta f^*|_{r=1}$, which implies

$$(\Delta_0 f, h) - (f, \Delta_0 h) = \int_{S^{n-1}} (h^* \overline{\Delta f^*} - \bar{f}^* \Delta h^*) \mathrm{d}\sigma(x).$$

Integrate this identity over r varying from a to b and apply the Green formula

$$\int_V (u \Delta v - v \Delta u) \mathrm{d}x = \int_{\partial V} \left(u \frac{\partial v}{\partial n} - v \frac{\partial u}{\partial n}\right) \mathrm{d}\sigma(x),$$

where $\frac{\partial}{\partial n}$ denotes differentiation in the direction of the outer normal and $\mathrm{d}\sigma(x)$ an element of the surface measure. For the other part use the formula

$$\int_V v \Delta v \,\mathrm{d}x = -\int_V ||\operatorname{grad} v||^2 \mathrm{d}x + \int_{\partial V} v \frac{\partial v}{\partial v} \mathrm{d}\sigma.$$

(d) Apply the Green formula as in the proof of symmetricity in (c).

6. Prove that every real harmonic polynomial function on $\boldsymbol{R}^2$, homogeneous of degree k, $k > 0$, is of the form $u_k(x, y) = \operatorname{Re}(c_k(x+iy)^k)$ $c_k \in \boldsymbol{C}$. Further, prove that $L^2(S^1)$ is an orthogonal direct sum of the spaces H^k spanned by the restrictions of these function to the unit circle S^1.

Hint. Find a conjugate harmonic function to u_k, i.e. a function v_k such that $u_k + iv_k$ is a holomorphic function. Then reduce the problem to a theorem on decomposition into Fourier series.

7. Prove that for every $z = (z_1, \ldots, z_n) \in \boldsymbol{C}^n$ such that $\sum_{j=1}^{n} z_j^2 = 0$ and every natural number p the functions $\boldsymbol{R}^n \ni x \to (x|z)^p \in \boldsymbol{C}$ belong

to H^p. Prove also that the functions of this form span the space of harmonic polynomials in $\boldsymbol{R}^n$.

8. (a) Prove that the zonal spherical function of the representation (U^l, H^l) (regarded as a subrepresentation of the representation U on $L^2(S^{n-1})$) is characterized by the conditions

(i) $\omega_n^l(0, \ldots, 0, 1) = 1$,

(ii) if $h \in H^l$ and $h(0, \ldots, 0, 1) = 0$, then $(\omega_n^l | h) = 0$.

Deduce hence that for every function $h \in H^l$ the following identity holds:

$$h(0, \ldots, 0, 1) = \dim H^l (\omega_n^l | h).$$

(b) Prove that for arbitrary $y \in S^{n-1}$ and $h \in H^l$ we have

$$h(y) = \dim H^l (\varphi_y^l | h)$$

for a certain function $\varphi_y^l \in H^l$ and then show that

$$\varphi_y^l(x) = \left[\frac{\Gamma(2p+l)}{l!\,\Gamma(2p)}\right]^{-1} C_l^p((x|y)), \qquad p = \frac{n-2}{2}.$$

(c) Let $\{Y_j^l\}_{j=1}^{\dim H^l}$ be a basis for H^l orthogonal with respect to the inner product on $L^2(S^{n-1})$. Prove the addition formula

$$\sum_j \overline{Y_j^l}(y) Y_j^l(x) = \dim H^l \varphi_y^l(x).$$

Check that for $n = 3$ we get the addition formula for the Legendre polynomial:

$$P_l(\cos\theta\cos\theta' + \sin\theta\sin\theta'\cos(\varphi-\varphi')) = P_l(\cos\theta) P_l(\cos\theta')$$

$$+2\sum_{j=1}^{l} \frac{(l-j)!}{(l+j)!} P_l^j(\cos\theta) P_l^j(\cos\theta')\cos(\varphi-\varphi').$$

Hint. (a) Consider the functional on H^l given by evaluation at $P = (0, \ldots, 0, 1) \in S^{n-1}$, i.e. $H^l \ni h \to h(P)$. (b) Exploit the transitivity of SO(n) on the sphere and apply (a).

Remark: All the facts considered in this problem follow, as special cases, from Problem 10, Chapter 5. Cf. also (9.4.8).

9. Derive the following recurrence formula for the Gegenbauer polynomials

$$(2p+1) C_l^p(t) = 2p\left(C_l^{p+1}(t) - t C_{l-1}^{p+1}(t)\right).$$

Hint. Replacing p by $p+1$ and l by $l+1$ in equation (9.3.4), we obtain

$lC_l^{p+1}(t) = (2p+l+1)tC_{l-1}^{p+1}(t) - (2p+2)(1-t^2)C_{l-2}^{p+2}(t)$. By (9.3.1) and (9.3.2) it follows that

$$4p(p+1)(1-t^2)C_{l-2}^{p+2}(t) - 2p(2p+1)tC_{l-1}^{p+2}(t) + l(2p+l)C_l^p(t) = 0.$$

Eliminating $(1-t^2)C_{l-2}^{p+2}(t)$, we obtain the required result.

10. (a) Prove that for odd $n = 2s+3$, $s = 0, 1, \ldots$

$$C_l^p(t) = C_l^{s+\frac{1}{2}}(t) = \frac{2^s s!}{(2s)!}\frac{d^s}{dt^s}P_{l+s}(t),$$

where $P_k(t)$ is the Legendre polynomial.

Prove that for even $n = 2s+2$, $s = 0, 1, \ldots$

$$C_l^p(t) = C_l^s(t) = \frac{1}{2^{s-1}(s-1)!(l+s)}\frac{d^s}{dt^s}T_{l+s}(t),$$

where $T_k(t)$ is a Tchebyshev polynomial, i.e. $T_k(t) = \cos(k \arccos t)$.

(b) Prove that for $p = 0, \frac{1}{2}, 1, \ldots$ and $-1 \leqslant t \leqslant 1$

$$C_l^p(t) = \frac{(-2)^l \Gamma(p+l)\Gamma(2p+l)}{l!\Gamma(p)\Gamma(2p+2l)}(1-t^2)^{\frac{1}{2}-p}\frac{d^l}{dt^l}(1-t^2)^{l+p-\frac{1}{2}}.$$

(c) Prove that for any real φ and $|z| \leqslant 1$

$$(1-2\cos\varphi z+z^2)^{-p} = \sum_{l=0}^{\infty} C_l^p(\cos\varphi)z^l.$$

Hint. (a) Compute $C_l^{\frac{1}{2}}(t)$ and $C_l^1(t)$ by applying integral representation (9.4.6) and then make use of recurrence formula (9.3.2). (b) Make use of the Rodrigues formula for the Legendre polynomials

$$P_l(t) = \frac{(-1)^l}{2^l l!}\left(\frac{d}{dt}\right)^l (1-t^2)^l$$

and for the Tchebyshev polynomials

$$T_l(t) = \frac{(-1)^l \Gamma(\frac{1}{2})}{\Gamma(1+\frac{1}{2})2^l}(1-t^2)^{\frac{1}{2}}\left(\frac{d}{dt}\right)^l [(1-t^2)^{l-\frac{1}{2}}].$$

(c) Expand in series $(1-2\cos\varphi z+z^2)^{-p}$ for $p = \frac{1}{2}$ and for $p = 1$ and differentiate those series.

11. (In this problem we assume that certain facts from Chapter 13 concerning the hypergeometric function and the Legendre function are already known.)

Let $\lambda \in \boldsymbol{R}$ satisfy the condition $\lambda > -\frac{1}{2}$. Let $C_l^\lambda(t)$ be defined as the l-th coefficient in the Taylor expansion of the function $(1-2tz+z^2)^{-\lambda}$, i.e.

$$(1-2tz+z^2)^{-\lambda} = \sum_{l=0}^{\infty} C_l^\lambda(t) z^l.$$

(a) Show that C_l^λ is a polynomial, of degree l precisely, given by the formula

$$C_l^\lambda(z) = \sum_{n=0}^{l} \frac{(-1)^n \Gamma(\lambda+n)\Gamma(l+2\lambda+n)}{n!\Gamma(\lambda)\Gamma(2n+2\lambda)(l-n)!} \left(\frac{1-z}{2}\right)^n$$

and that

$$C_l^\lambda(z) = \frac{\Gamma(l+2\lambda)}{\Gamma(l+1)\Gamma(2\lambda)} F\left(l+2\lambda,\ -l;\ \lambda+\frac{1}{2};\ \frac{1-z}{2}\right).$$

(b) Prove the following relation with the Legendre functions P_ν^μ:

$$C_l^\lambda(z) = 2^{\lambda-\frac{1}{2}} \frac{\Gamma(l+2\lambda)\Gamma(\lambda+\frac{1}{2})}{\Gamma(2\lambda)\Gamma(l+1)} (z^2-1)^{\frac{1}{4}-\frac{\lambda}{2}} P_{l+\lambda-\frac{1}{2}}^{\frac{1}{2}-\lambda}(z);$$

derive hence the Rodrigues formula

$$C_l^\lambda(z) = \frac{(-2)^l \Gamma(\lambda+l)\Gamma(2\lambda+l)}{\Gamma(l+1)\Gamma(\lambda)\Gamma(2\lambda+2l)} (1-z^2)^{\frac{1}{2}-\lambda} \frac{d^l}{dz^l} (1-z^2)^{l+\lambda-\frac{1}{2}}.$$

(c) Prove that C_l^λ constitute an orthogonal system of polynomials on $[-1, 1]$ with respect to the weight $(1-t^2)^{\lambda-\frac{1}{2}}$ and that

$$\int_{-1}^{1} (C_l^\lambda(t))^2 (1-t^2)^{\lambda-\frac{1}{2}} dt = \pi \frac{2^{1-2\lambda}\Gamma(l+2\lambda)}{\Gamma(l+1)(l+\lambda)(\Gamma(\lambda))^2}.$$

(d) Prove that C_l^λ satisfy the differential equation

$$(1-t^2)\frac{d^2x}{dt^2} - (2\lambda+1)t\frac{dx}{dt} + l(l+2\lambda)x = 0.$$

Chapter 10

Jacobi and Legendre Functions

10.1. STRUCTURE OF THE GROUP SL(2, $\boldsymbol{R}$) AND ITS HOMOGENEOUS SPACES

Some classes of special functions are connected with the theory of representations of the group SL(2, $\boldsymbol{R}$) of real-valued unimodular 2×2 matrices. This group acts transitively on such manifolds as the open unit disc in $\boldsymbol{C}$ and the half-space conformal to it, on the boundary of the disc (or, respectively, on the real line), on the space $\boldsymbol{C}^2-\{0\}$ and on the projective space P^1. This is the cause of the particular role of the group SL(2, $\boldsymbol{R}$) in the theory of special functions.

In the present chapter, devoted to Jacobi and Legendre functions, we treat SL(2, $\boldsymbol{R}$) first of all as a group of isometries of the unit disc

$$D := \{z \in \boldsymbol{C}\colon |z| < 1\}$$

and as a transformation group of its boundary

$$B = \{z \in \boldsymbol{C}\colon |z| = 1\}.$$

The space $\Xi = \boldsymbol{C}^2-\{0\}$ will appear as a set of "hypersurfaces" in D, called *horocycles*. Thus we begin with a list of properties of these homogeneous spaces and of the group itself.

The group $G :=$ SL(2, $\boldsymbol{R}$) is a 3-dimensional Lie group of real matrices

$$g = \begin{bmatrix} \alpha & \beta \\ \gamma & \delta \end{bmatrix}, \qquad \alpha\delta-\beta\gamma = 1.$$

The group acts on $\boldsymbol{C}$ by homographies:

$$g\cdot z = \frac{\alpha z+\beta}{\gamma z+\delta}. \tag{10.1.1}$$

There are three orbits with respect to the above action of the group on $\boldsymbol{C}$:

$$\boldsymbol{H}_+ = \{z \in \boldsymbol{C}: \operatorname{Im} z > 0\},$$

the real axis and the lower half-space $\boldsymbol{H}_- = \{z: \bar{z} \in \boldsymbol{H}_+\}$.

The point $z = i \in M$ has the isotropy subgroup

$$K = \left\{\begin{bmatrix} \cos\frac{\varphi}{2} & \sin\frac{\varphi}{2} \\ -\sin\frac{\varphi}{2} & \cos\frac{\varphi}{2} \end{bmatrix} : \varphi \in [-2\pi, 2\pi]\right\},$$

which is compact, commutative and isomorphic to $\boldsymbol{T}$.

From Theorem 1.4.2, it follows that $\boldsymbol{H}_+ \cong G/K$, where the isomorphism refers to the structure of a homogeneous space.

The half-space $\boldsymbol{H}_+$ is the image of the unit disc under the conformal transformation

$$\tau: D \ni z \to \frac{z+i}{iz+1} \in \boldsymbol{H}_+.$$

The diameter $]-1, 1[$ is then transformed onto the imaginary half-line $\boldsymbol{R}_+ i$.

We can thus define the action of G on D by the formula

$$\mathfrak{t}(g) \cdot z = \tau^{-1}(g \cdot \tau(z)) = \frac{az+b}{\bar{b}z+\bar{a}}, \tag{10.1.2}$$

where

$$\begin{bmatrix} a & b \\ \bar{b} & \bar{a} \end{bmatrix} = \begin{bmatrix} 1 & i \\ i & 1 \end{bmatrix}^{-1} \begin{bmatrix} \alpha & \beta \\ \gamma & \delta \end{bmatrix} \begin{bmatrix} 1 & i \\ i & 1 \end{bmatrix} =: \mathfrak{t}(g);$$

in precise terms:

$$a = \tfrac{1}{2}\left(\alpha+\beta+i(\beta-\gamma)\right),$$
$$b = \tfrac{1}{2}\left(\beta+\gamma+i(\alpha-\delta)\right).$$

The condition of unimodularity takes the form $|a|^2 - |b|^2 = 1$. The mapping $G \ni g \to \mathfrak{t}(g)$ proves to be an isomorphism onto the group $\mathrm{SU}(1;1) = \left\{\begin{bmatrix} a & b \\ \bar{b} & \bar{a} \end{bmatrix} : |a|^2 - |b|^2 = 1\right\}$, and this group acts transitively on the unit disc. Hence we have the following commutative diagram:

$$\begin{array}{ccc} \mathrm{SL}(2, \boldsymbol{R}) & \xrightarrow{\mathfrak{t}} & \mathrm{SU}(1,1) \\ \downarrow & & \downarrow \\ \boldsymbol{H}_+ & \xleftarrow{\tau} & D \end{array}$$

The perpendicular arrows are projections onto homogeneous spaces: the groups act on them by homographies.

The image of the group K under the mapping $\mathfrak{t}$ is the isotropy group at the point $\tau^{-1}(i) = 0$, i.e.

$$\mathscr{K} = \left\{\begin{bmatrix} e^{i\frac{\varphi}{2}} & 0 \\ 0 & e^{-i\frac{\varphi}{2}} \end{bmatrix} : \varphi \in [-2\pi, 2\pi[\right\}.$$

The action of the group SU(1, 1) on the disc can be extended to the boundary B directly by formula (10.1.2) and the action of the subgroup $\mathscr{K}$ is transitive on B. More precisely,

$$B = \mathscr{K} \Big/ \left\{\begin{bmatrix} 1 & 0 \\ 0 & 1 \end{bmatrix}, \begin{bmatrix} -1 & 0 \\ 0 & -1 \end{bmatrix}\right\} =: \mathscr{K}/\mathscr{M}.$$

The two-element subgroup $\mathscr{M}$ is the stability subgroup at the point $1 \in B$.

Similarly, the 1-point compactification of the real axis $\boldsymbol{R}$ is subject to the transitive action of the group SL(2, $\boldsymbol{R}$) and the isotropy subgroup at the point ∞ is formed by the group of triangular matrices

$$P = \{g \in G: \gamma = 0\}.$$

The group P is a semi-direct product of its subgroups

$$A_1 = \left\{\begin{bmatrix} \alpha & 0 \\ 0 & \alpha^{-1} \end{bmatrix} : \alpha \in \boldsymbol{R}_*\right\} \quad \text{and} \quad N = \left\{\begin{bmatrix} 1 & t \\ 0 & 1 \end{bmatrix} : t \in \boldsymbol{R}^1\right\}.$$

The action of A_1 on N is given by the formula

$$\nu(a)\begin{bmatrix} 1 & \beta \\ 0 & 1 \end{bmatrix} := \begin{bmatrix} 1 & \beta\alpha^2 \\ 0 & 1 \end{bmatrix},$$

and then $P = A_1 \times_\nu N$.

Further, let us distinguish the subgroup

$$A = \left\{\begin{bmatrix} \alpha & 0 \\ 0 & \alpha^{-1} \end{bmatrix} : \alpha > 0\right\}.$$

Then $A_1 = M \times_\nu A$.

PROPOSITION 10.1.1 *Every element $g \in G$ can be uniquely represented in the form*

$$g = kan, \tag{10.1.3}$$

where $k \in K$, $a \in A$, $n \in N$.

Proof. The group A acts transitively and effectively on every half-line in $\boldsymbol{C}$ passing through zero:

$$\begin{bmatrix} \alpha & 0 \\ 0 & \alpha^{-1} \end{bmatrix} \cdot z = \alpha^2 z;$$

on the other hand, the action of N is the following:

$$\begin{bmatrix} 1 & \beta \\ 0 & 1 \end{bmatrix} \cdot z = z + \beta.$$

Hence for every $z = g^{-1} \cdot i \in \boldsymbol{H}_+$ there exist a unique $a \in A$ and a unique $n \in N$ such that $z = n^{-1}a^{-1} \cdot i$. The element ang^{-1} belongs to the isotropy group at the point $i \in \boldsymbol{H}_+$, which equals K. Setting $k^{-1} = ang^{-1}$, we obtain the assertion. □

Decomposition (10.1.3) is called the *Iwasawa decomposition* of the group SL(2, $\boldsymbol{R}$). It plays a key role in the theory of representations of this group. The corresponding decomposition of group SU(1, 1) is given by

$$\mathrm{SU}(1, 1) = \mathscr{K}\mathscr{A}\mathscr{N},$$

where

$$\mathscr{K} = \mathfrak{t}(K),$$

$$\mathscr{A} := \mathfrak{t}(A) = \left\{ \begin{bmatrix} \cosh\frac{t}{2} & \sinh\frac{t}{2} \\ -i\sinh\frac{t}{2} & \cosh\frac{t}{2} \end{bmatrix} : t \in \boldsymbol{R} \right\}, \tag{10.1.4}$$

$$\mathscr{N} := \mathfrak{t}(N) = \left\{ \begin{bmatrix} 1+\frac{t}{2}i & \frac{t}{2} \\ -\frac{t}{2} & 1-\frac{t}{2}i \end{bmatrix} : t \in \boldsymbol{R} \right\}.$$

In terms of the coordinates of the Iwasawa decomposition the Haar measure for the group SL(2, $\boldsymbol{R}$) assumes the form

$$\int_G f(g)\mathrm{d}g = \frac{1}{4\pi} \int_{-\infty}^{+\infty} \int_{-\infty}^{+\infty} \int_{-2\pi}^{2\pi} f(k(\varphi)a(t)n(s))\mathrm{e}^t \mathrm{d}\varphi \mathrm{d}t \mathrm{d}s,$$

where

$$k(\varphi) = \begin{bmatrix} \cos\frac{\varphi}{2} & \sin\frac{\varphi}{2} \\ -\sin\frac{\varphi}{2} & \cos\frac{\varphi}{2} \end{bmatrix}, \qquad a(t) = \begin{bmatrix} e^{\frac{t}{2}} & 0 \\ 0 & e^{\frac{t}{2}} \end{bmatrix},$$

$$n(s) = \begin{bmatrix} 1 & s \\ 0 & 1 \end{bmatrix}.$$

Obviously this formula applies also to SU(1, 1) if we replace k, a, n by $\mathfrak{k}(k)$, $\mathfrak{k}(a)$, $\mathfrak{k}(n)$, respectively. Symbolically we write

$$dg = \frac{e^t}{4\pi} d\varphi dt ds.$$

The orbit of the point zero in the disc with respect to the action of $\mathscr{A}$ is the diameter $J =]-i, i[$:

$$R \ni t \to a(t) \cdot 0 = i \tanh\frac{t}{2}.$$

The subgroup $\mathscr{K}$ acts on the unit disc by rotations:

$$]-2\pi, 2\pi[\ni \varphi \to \mathfrak{k}(\varphi) \cdot z = e^{i\varphi} z.$$

Every element $z \in D - \{0\}$ can be represented in the form $z = ka \cdot 0$. This is not a unique representation. The set $\mathscr{K}\mathscr{A}$ covers $D - \{0\}$ four times with respect to the projection $G \ni g \to g \cdot 0$. The uniqueness is obtained after restricting the domain to $0 \leqslant \varphi < 2\pi$ and $\tau > 0$. If $ka \cdot 0 = g \cdot 0$, then $a^{-1}k^{-1}g \in \mathscr{K}$. Therefore every element $g \in$ SU(1, 1) has the representation

$$g = k_1 a k_2, \qquad k_i \in \mathscr{K}, \quad a \in \mathscr{A}.$$

THEOREM 10.1.2. *The mapping*

$$[0, 2\pi[\times]-\infty, \infty[\times [-2\pi, 2\pi[\ni (\varphi, \tau, \psi)$$

$$\to \begin{bmatrix} \cosh\frac{\tau}{2} e^{i\frac{\varphi+\psi}{2}} & \sinh\frac{\tau}{2} e^{i\frac{\varphi-\psi}{2}} \\ \sinh\frac{\tau}{2} e^{i\frac{\varphi-\psi}{2}} & \cosh\frac{\tau}{2} e^{i\frac{\varphi+\psi}{2}} \end{bmatrix} \in \mathrm{SU}(1, 1)$$

is a continuous bijection onto SU(1, 1)$-$ $\{\mathscr{K}\}$. *When restricted to the interior of the domain, it is a diffeomorphism onto an open dense subset of the group.*

The parameters $(\varphi, -i\tau, \psi)$ are called the *Euler angles* of an element $g \in SU(1, 1)$.

With the help of the isomorphism t, the parametrization in terms of the Euler angles, is transported to the group SL(2, **R**).

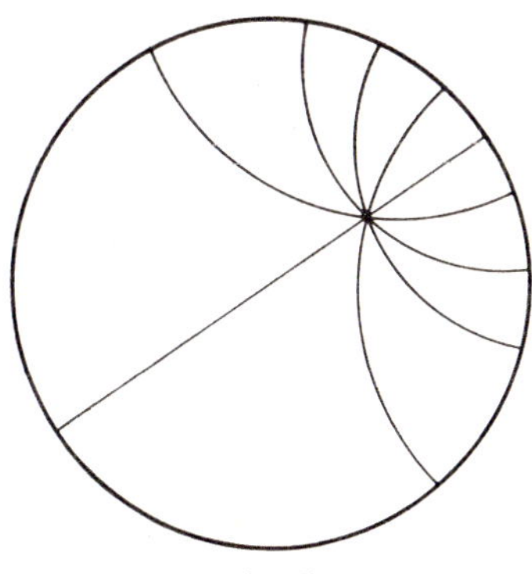

Fig. 7

The invariant integral on the group has the following form in terms of the coordinates φ, τ, ψ:

$$dg = \frac{1}{8\pi} \sinh \tau \, d\tau \, d\varphi \, d\psi . \qquad (10.1.5)$$

The orbits of the subgroups $\mathscr{A}$ and $\mathscr{K}$ in D (the level sets of the coordinates (φ, τ, ψ)) have important geometrical interpretations in terms of the invariant Riemannian structure on the unit disc.

The length element on D has the form

$$ds^2 = \frac{dx^2 + dy^2}{(1-(x^2+y^2))^2} \qquad (10.1.6)$$

in Cartesian coordinates $(z = x+iy)$.

Geodesics passing through the point 0 are straight lines. The action of the group SU(1, 1) preserves the Riemannian structure and therefore it maps geodesics onto geodesics. Hence, if L is a geodesic passing through $z = g \cdot 0$ then $g^{-1}L$ is an interval on a straight line passing through 0, so $g^{-1}L = k\mathscr{A} \cdot 0$, where $k \in \mathscr{K}$. Therefore $L = gk\mathscr{A} \cdot 0 = gk\mathscr{A}k^{-1}g^{-1}g \cdot 0 = h\mathscr{A}h^{-1} \cdot z$.

The geodesics are the orbits in D of subgroups conjugate to $\mathscr{A}$. The pencil of geodesics issuing from a point $z \in D$ is a pencil of circles passing through z and normal to the boundary of the disc (Fig. 7).

The distance from z to zero, calculated directly from (10.1.6) is

$$d(0, z) = \frac{1}{2}\log\frac{1+|z|}{1-|z|} = \frac{1}{2}\log\frac{0-\dfrac{z}{|z|}}{0+\dfrac{z}{|z|}} : \frac{z-\dfrac{z}{|z|}}{z+\dfrac{z}{|z|}}.$$

As is well known, homographies preserve the double ratio. Writing out the formula $d(g\cdot 0, g\cdot z) = d(0, z)$ and substituting $g0 = z_1$, $gz = z_2$, we obtain

$$d(z_1, z_2) = \frac{1}{2}\log\left(\frac{z_1-b_2}{z_1-b_1} : \frac{z_2-b_2}{z_2-b_1}\right),$$

where b_1 and b_2 are the points at which the geodesics passing through z_1 and z_2 intersect the boundary of the unit disc (Fig. 8).

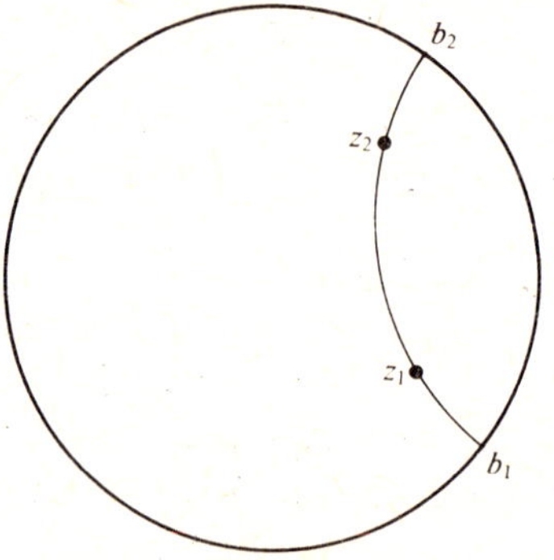

Fig. 8

The orbits of the group $\mathscr{K}$ are concentric circles with centre at zero. The distance being invariant under the action of the group, we conclude that the orbits of the subgroup $g\mathscr{K}g^{-1}$ are spheres in the sense of the Riemannian metric (10.1.6) with centre at the point $g\cdot 0$.

Let us now pass to the orbits of the subgroups conjugate to $\mathscr{N}$. The orbit of the point 0 with respect to the action of the group $\mathscr{N}$ itself consists of points of the form

$$\zeta(t) = \frac{t}{2-ti}, \qquad t \in \boldsymbol{R}.$$

We find

$$\left|\zeta(t)-\frac{i}{2}\right| = \frac{1}{2}.$$

Hence the orbit is a circle with centre at $\frac{1}{2}i$ and diameter 1, the point of contact with the boundary of the disc being removed.

In the same way we check that the orbit of the point xi, $x \in]-1,1[$, is the circle of diameter $1-x$ tangent to the boundary at the point i (Fig. 9, p. 368).

The orbits of the subgroup $\mathcal{N}$ form a pencil of circles, tangent to the boundary at the point i, and therefore intersecting the geodesic $]-i, i[$ at right angles. The subgroup $\mathcal{N}$ is invariant under the inner automorphisms induced by the elements of the subgroup $\mathcal{P}$. An element k of the subgroup $\mathcal{K}$ maps the pencil of circles onto a pencil of circles tangent to the boundary at the point ki. But $k\mathcal{N}z = k\mathcal{N}k^{-1}kz = \alpha_k\mathcal{N}\cdot z_1$. Thus the orbits of the subgroup $\alpha_k\mathcal{N}$ constitute the pencil of circles tangent to B at the point $k\cdot i$.

The orbits of the subgroups conjugate to $\mathcal{N}$ are called *horocycles*. Horocycles are said to be *parallel* when they are orbits of the same subgroup. Parallel horocycles do not intersect each other and are tangent to the disc at the same point, hence they intersect the same diameter of the disc at right angles. The pencil of parallel horocycles plays, in the geometry of the disc a similar role to the pencil of hyperplanes in $\boldsymbol{R}^n$ normal to a fixed direction.

The space of horocycles, denoted by Ξ, has the structure of a homogeneous space of the group. We have already seen that the group acts transitively on Ξ. The stability subgroup of the horocycle $\xi_0 = \mathcal{N}\cdot 0$ is $\mathcal{M}\mathcal{N}$. Hence $\Xi = \mathrm{SU}(1,1)/\mathcal{M}\mathcal{N} = \mathrm{SL}(2,\boldsymbol{R})/MN$.

A manifold structure on Ξ can be represented by means of the following model:

The group SL(2, $\boldsymbol{R}$) acts transitively on $\boldsymbol{R}^2-\{0\}$:

$$g\begin{bmatrix}x\\y\end{bmatrix} = \begin{bmatrix}\alpha & \beta\\ \gamma & \delta\end{bmatrix}\begin{bmatrix}x\\y\end{bmatrix} = \begin{bmatrix}\alpha x+\beta y\\ \gamma x+\delta y\end{bmatrix}.$$

The stability subgroup at the point $(1, 0)$ is just N. The subgroup M on the other hand acts on the space as a central symmetry. The space $G/MN = \Xi$ arises through the identification of the points x and $-x$ in $R^2 - \{0\}$. Hence it is the Cartesian product of the radius and the projective line P^1.

10.2. Induced Representations of the Group SL(2, R)

We are concerned with representations of the group G induced by means of the representations of the subgroup $P = MAN = A_1N$.

We shall first give a description of the finite-dimensional irreducible representations of this subgroup.

Let χ be the character of such a representation. For $n \in N$ and $a \in A_1$ we have $ana^{-1} \in N$, and the action of A on $N \cong R$ consists in multiplication by numbers different from zero. Because $\chi(n) = \chi(ana^{-1})$, the character is constant on N. Thus every finite-dimensional representation of the group P is trivial on N and hence uniquely determined by a representation of $A_1 = P/N$.

Since A_1 is a commutative group, all its irreducible representations are 1-dimensional. Representations of P are parametrized by the set of homomorphisms of $A_1 \cong R_*$ into C_*, and these are of the form

$$A_1 \ni \begin{bmatrix} \alpha & 0 \\ 0 & \alpha^{-1} \end{bmatrix} \to (\operatorname{sgn}\alpha)^\varepsilon |\alpha|^l, \qquad \varepsilon = 0, 1,\ l \in C.$$

We introduce the notation $\sigma = (\varepsilon, l)$. Then

$$\chi_\sigma\left(\begin{bmatrix} \alpha & \beta \\ 0 & \alpha^{-1} \end{bmatrix}\right) = (\operatorname{sgn}\alpha)^\varepsilon |\alpha|^l, \qquad \varepsilon = 0, 1,\ l \in C. \tag{10.2.1}$$

The induced representations are numbered by the same indices.

$$U^\sigma = U^{(\varepsilon, l)} := U^{\chi_\sigma}. \tag{10.2.2}$$

Let us recall that an induced representation acts by left translations on the space $\mathscr{E}^\sigma$ of smooth functions on G satisfying the condition

$$f(gp) = \delta^{\frac{1}{2}}(p)\chi_\sigma^{-1}(p)f(g). \tag{10.2.3}$$

The modular function $\mathscr{E}^\sigma$ has the form

$$\delta\left(\begin{bmatrix} \alpha & \beta \\ 0 & \alpha^{-1} \end{bmatrix}\right) = |\alpha|^{-2}. \tag{10.2.4}$$

We now pass to the multiplier form of this representation. On account of the Iwasawa decomposition the values of a function $f \in \mathscr{E}^\sigma$ are uniquely determined by its restriction to the subgroup K. The function $f|K$ satisfies

$$f|K(km) = \chi_\sigma(m)\,(f|K)\,(k). \tag{10.2.5}$$

For $\varepsilon = 0$ the set $\mathscr{E}^\sigma|K$ consists of symmetric functions on the circle, for $\varepsilon = 1$ it is the set of antisymmetric functions on the circle. We seek the form of the operator $\mathscr{U}_g^\sigma$ acting on $\mathscr{E}(K)$ for which $(U_g^\sigma f)|K = \mathscr{U}_g^\sigma(f|K)$.

Let $h(g)$ stand for an element of the group A such that the Iwasawa decomposition of the element $g \in G$ has the form

$$g = kh(g)n.$$

The following property of the function $h\colon G \to A$ is immediate:

$$h(kgn) = h(g), \qquad k \in K, \quad n \in N.$$

Then, according to (10.2.3)

$$(\mathscr{U}_g^\sigma f)(k) = f(g^{-1}k) = f(g^{-1}\cdot k)\,\delta^{\frac{1}{2}}\chi_\sigma^{-1}\big(h(g^{-1}k)\big).$$

By $g \cdot k$ we denote an element of the group K in the Iwasawa decomposition of the product gk, which can be identified with $gkAN \in G/AN$.

Introducing $S_\sigma(g, k) := \delta^{\frac{1}{2}}\chi_\sigma^{-1}\big(h(gk)\big)$ we pass to the multiplier form

$$\mathscr{U}_g^\sigma f(k) := S_\sigma(g^{-1}, k) f(g^{-1}\cdot k). \tag{10.2.6}$$

We shall determine the form of the function h and the action of G on K in the coordinates introduced earlier.

A convenient interpretation of the function h can be obtained by using the action of the group $\boldsymbol{R}^2 - \{0\} \cong G/N$. Namely, whenever

$$g = \begin{bmatrix} \cos\frac{\varphi}{2} & \sin\frac{\varphi}{2} \\ -\sin\frac{\varphi}{2} & \cos\frac{\varphi}{2} \end{bmatrix} \begin{bmatrix} e^{\frac{t}{2}} & 0 \\ 0 & e^{\frac{t}{2}} \end{bmatrix} \begin{bmatrix} 1 & s \\ 0 & 1 \end{bmatrix},$$

then

$$g \cdot \begin{bmatrix} 1 \\ 0 \end{bmatrix} = e^{\frac{t}{2}} \begin{bmatrix} \cos\frac{\varphi}{2} \\ -\sin\frac{\varphi}{2} \end{bmatrix}, \tag{10.2.7}$$

which means that

$$h(g) = \begin{bmatrix} \left\| g \cdot \begin{bmatrix} 1 \\ 0 \end{bmatrix} \right\| & 0 \\ 0 & \left\| g \cdot \begin{bmatrix} 1 \\ 0 \end{bmatrix} \right\|^{-1} \end{bmatrix}. \tag{10.2.8}$$

Using this observation, we find $h(g^{-1}k)$ for g and k expressed in terms of the Euler angles. Let us begin with the case

$$g = a = \begin{bmatrix} e^{\frac{\tau}{2}} & 0 \\ 0 & e^{-\frac{\tau}{2}} \end{bmatrix}, \qquad k = \begin{bmatrix} \cos\frac{\theta}{2} & \sin\frac{\theta}{2} \\ -\sin\frac{\theta}{2} & \cos\frac{\theta}{2} \end{bmatrix}.$$

Computing

$$g^{-1}k\begin{bmatrix} 1 \\ 0 \end{bmatrix} = \begin{bmatrix} e^{-\frac{\tau}{2}}\cos\frac{\theta}{2} \\ -e^{\frac{\tau}{2}}\sin\frac{\theta}{2} \end{bmatrix},$$

we find

$$\left\| g^{-1}k\begin{bmatrix} 1 \\ 0 \end{bmatrix} \right\| = \left(e^{-\tau}\cos^2\frac{\theta}{2} + e^{\tau}\sin^2\frac{\theta}{2}\right)^{\frac{1}{2}} = (\cosh\tau - \sinh\tau\cos\theta)^{\frac{1}{2}}. \tag{10.2.9}$$

The element $g \in SL(2, \boldsymbol{R})$ corresponding to the Euler angles $(\varphi, -i\tau, \psi)$ is of the form $g = k_1 a k_2$, where a is as before, and

$$k_j = \begin{bmatrix} \cos\frac{\varphi_j}{2} & \sin\frac{\varphi_j}{2} \\ -\sin\frac{\varphi_j}{2} & \cos\frac{\varphi_j}{2} \end{bmatrix}, \qquad \text{where } \varphi_1 = \varphi - \frac{\pi}{2},\ \varphi_2 = \psi + \frac{\pi}{2}.$$

Then

$$\begin{aligned} \left\| g^{-1}k \cdot \begin{bmatrix} 1 \\ 0 \end{bmatrix} \right\| &= \left\| a^{-1}k_1^{-1}k\begin{bmatrix} 1 \\ 0 \end{bmatrix} \right\| \\ &= \left(\cosh\tau - \sinh\tau\cos\left(\theta - \varphi + \frac{\pi}{2}\right)\right)^{\frac{1}{2}} \\ &= (\cosh\tau + \sinh\tau\sin(\theta - \varphi))^{\frac{1}{2}} =: e_{-1}(g, k). \end{aligned} \tag{10.2.10}$$

The function thus obtained, appearing here as a multiplier defining a representation of the group on $\mathscr{E}^\sigma$, plays a central role in harmonic analysis on G/K.

For the present, just to attract the reader's attention, we note a connection between this function and the theory of harmonic functions on the disc.

The function $G\times K\ni(g,k)\to e_{-1}(g,k)$ is right K-invariant with respect to the first variable, and so it is in fact a function on $G/K\times\times K\cong D\times\mathscr{K}$. Introducing spherical variables on the disc: $r=\tanh\frac{t}{2}$, we obtain

$$e_{-1}(g\mathscr{K},k)=\left[\frac{1-2r\cos(\varphi-\theta)+r^2}{1-r^2}\right]^{\frac{1}{2}}$$

for $g\mathscr{K}=re^{i\varphi}\in D$, $k=e^{\frac{i\theta}{2}}$.

In the function $(e_{-1})^2$ we recognize the reciprocal of the *Poisson kernel*, i.e. the elementary harmonic function inside the unit disc.

Let us note yet another property of the function e_{-1}, resembling the properties of a plane wave on R^n. Let us find

$$e_{-1}(knk^{-1}gk,k)\quad\text{for } n\in N.$$

Since $h((knk^{-1}g)^{-1}k)=h(g^{-1}kn)=h(g^{-1}k)$ we have

$$e_{-1}(knk^{-1}g\mathscr{K},k)=e_{-1}(k\mathscr{K},k)t$$

This means that the function $e_{-1}(\cdot\,,k)$ is constant on every horocycle from the pencil associated with the subgroup $k\mathscr{N}k^{-1}$. This is the pencil of tangents to the boundary of the disc at the point $k\cdot i$.

We shall also make use of the action of the group on $R^2-\{0\}$ in order to determine its action on $K\cong G/AN$. The group K can be bijectively mapped onto S^1:

$$K\ni k\to\begin{bmatrix}\cos\frac{\varphi}{2}\\ -\sin\frac{\varphi}{2}\end{bmatrix},\qquad \varphi\in]-2\pi,2\pi[.$$

After this identification we note that by formula (10.2.7) the compact constituent of the Iwasawa decomposition of an element g has the form

$$g\cdot\begin{bmatrix}1\\0\end{bmatrix}\Big/\left\|g\cdot\begin{bmatrix}1\\0\end{bmatrix}\right\|.$$

From the above we find for $g = a$

$$g^{-1} \cdot k = \begin{bmatrix} e^{-\frac{\tau}{2}} \cos \frac{1}{2}\theta \\ -e^{\frac{\tau}{2}} \sin \frac{1}{2}\theta \end{bmatrix} \cdot (\cosh \tau - \sinh \tau \cos \theta)^{-\frac{1}{2}}. \qquad (10.2.11)$$

We proceed to determine the matrix elements of $\mathscr{U}^\sigma$.

The representation $\mathscr{U}^{(\varepsilon, l)}$ acts on the space $\mathscr{E}^\sigma$ of all functions on the group K, which are symmetric $(f(km) = f(k))$ for $\varepsilon = 0$ and anti-symmetric $(f(km) = -f(k))$ for $\varepsilon = 1$, $m = \begin{bmatrix} -1 & 0 \\ 0 & -1 \end{bmatrix}$.

We distinguish the following system of functions on K:

$$e_n^\varepsilon(k(\varphi)) = e^{i(2n+\varepsilon)\frac{\varphi}{2}}, \qquad -\infty < n < \infty$$

The set $\{e_n^\varepsilon\}_{n=-\infty}^{\infty}$ belongs to $\mathscr{E}^\sigma$. By virtue of (10.2.6), (10.2.9), (10.2.11) we have

$$\mathscr{U}_a^\sigma e_n^\varepsilon(k(\theta)) = \left(\cos\frac{\theta}{2} e^{-\frac{\tau}{2}} + i \sin\frac{\theta}{2} e^{\frac{\tau}{2}}\right)^{2n+\varepsilon}$$

$$\times (\cosh \tau - \sinh \tau \cos \theta)^{-n-\frac{\varepsilon+l+1}{2}}$$

On account of the identities

$$\cos \tau - \sinh \tau \cos \theta = \left(e^{-\frac{\tau}{2}} \cos\frac{\theta}{2} + i e^{\frac{\tau}{2}} \sin\frac{\theta}{2}\right)$$

$$\times \left(e^{-\frac{\tau}{2}} \cos\frac{\theta}{2} - i e^{\frac{\tau}{2}} \sin\frac{\theta}{2}\right)$$

and

$$e^{-\frac{\tau}{2}} \cos\frac{\theta}{2} - i e^{\frac{\tau}{2}} \sin\frac{\theta}{2} = \cosh\frac{\tau}{2} e^{-\frac{\theta}{2}i} - \sinh\frac{\tau}{2} e^{i\frac{\theta}{2}}$$

we obtain the formula

$$\mathscr{U}_a^\sigma e_n(k(\theta)) = \left(\cosh \tfrac{1}{2}\tau e^{\frac{1}{2}i\theta} - \sinh \tfrac{1}{2}\tau e^{-\frac{1}{2}i\theta}\right)^{n-\frac{1}{2}(l+1-\varepsilon)}$$

$$\times \left(\cosh \tfrac{1}{2}\tau e^{-\frac{1}{2}i\theta} - \sinh \tfrac{1}{2}\tau e^{\frac{1}{2}i\theta}\right)^{-n-\frac{1}{2}(l+1+\varepsilon)}. \qquad (10.2.12)$$

We shall calculate the matrix elements relative to the inner product in $L^2(K)$.

$$u^\sigma_{mn}(a) := (e_m|\mathscr{U}^\sigma_a e_n) = \frac{1}{4\pi}\int_0^{4\pi} e^{-\frac{1}{2}i(2m+\varepsilon)\theta}\left(\cosh\tfrac{1}{2}\tau e^{\frac{1}{2}i\theta}\right.$$

$$\left.-\sinh\tfrac{1}{2}\tau e^{-\frac{1}{2}i\theta}\right)^{n-\frac{1}{2}(l+1-\varepsilon)}\left(\cosh\tfrac{1}{2}\tau e^{\frac{1}{2}i\theta}\right.$$

$$\left.-\sinh\tfrac{1}{2}\tau e^{\frac{1}{2}i\theta}\right)^{-n-\frac{1}{2}(l+1+\varepsilon)}d\theta$$

$$= \frac{1}{2\pi}\int_0^{2\pi} (\cosh\tfrac{1}{2}\tau - \sinh\tfrac{1}{2}\tau e^{-i\theta})^{n-\frac{1}{2}(l+1-\varepsilon)}(\cosh\tfrac{1}{2}\tau$$

$$-\sinh\tfrac{1}{2}\tau e^{i\theta})^{-n-\frac{1}{2}(l+1+\varepsilon)}e^{i(n-m)\theta}d\theta = \frac{(-1)^{m-n}}{2\pi}\int_0^{2\pi}(\cosh\tfrac{1}{2}\tau$$

$$+\sinh\tfrac{1}{2}\tau e^{i\theta})^{n-\frac{1}{2}(l+1-\varepsilon)}(\cosh\tfrac{1}{2}\tau + \sinh\tfrac{1}{2}\tau e^{-i\theta})^{-n-\frac{1}{2}(l+1+\varepsilon)}$$

$$\times e^{i(m-n)\theta}d\theta \tag{10.2.13}$$

after the substitution $\theta \to \theta+\pi$. The Jacobi function is defined by the formula

$$\mathfrak{P}^\zeta_{mn}(\cosh\tau) := \frac{1}{2\pi}\int_0^{2\pi}(\cosh\tfrac{1}{2}\tau + \sinh\tfrac{1}{2}\tau e^{i\theta})^{\zeta+n}(\cosh\tfrac{1}{2}\tau$$

$$+\sinh\tfrac{1}{2}\tau e^{-i\theta})^{\zeta-n}e^{i(m-n)\theta}d\theta \tag{10.2.14}$$

for $0 \leqslant \tau < \infty$, $\zeta \in \boldsymbol{C}$, m and n being simultanously either integers or half-integers. By comparing the two formulas we conclude that

$$u^\sigma_{mn}(a) = (-1)^{m-n}\mathfrak{P}^{-\frac{1}{2}(l+1)}_{m+\frac{1}{2}\varepsilon,\, n+\frac{1}{2}\varepsilon}(\cosh\tau). \tag{10.2.15}$$

The Jacobi function with $p = q = 0$ is called the *Legendre function* and denoted by

$$\mathfrak{P}_\zeta(\cosh\tau) = \mathfrak{P}^\zeta_{00}(\cosh\tau) = \frac{1}{2\pi}\int_0^{2\pi}(\cosh\tau + \sinh\tau\cos\theta)^\zeta d\theta.$$

The functions with indices $n = 0$, $m \in \boldsymbol{Z}$ define the associated Legendre functions.

$$\mathfrak{P}_\zeta^m(\cosh\tau) = \frac{\Gamma(\zeta+m+1)}{\Gamma(\zeta+1)}\mathfrak{P}_{m0}^\zeta(\cosh\tau)$$

$$= \frac{1}{2\pi}\frac{\Gamma(\zeta+m+1)}{\Gamma(\zeta+1)}\int_{-\pi}^{\pi}(\cosh\tau+\sinh\tau\cos\theta)^\zeta e^{im\theta}d\theta. \tag{10.2.16}$$

Finally let us find the form of the matrix element for an arbitrary argument $g \in G$. Suppose that $g = k_1 a k_2$.

$$u_{mn}^\sigma(g) = (e_m^\varepsilon|\mathscr{U}_g e_n^\varepsilon) = (\mathscr{U}_{k_1^{-1}}^\sigma e_m^\varepsilon|\mathscr{U}_a^\sigma\mathscr{U}_{k_2}^\sigma e_n^\varepsilon).$$

The vector e_m^ε is an eigenvector of the operator $\mathscr{U}_k^\sigma$, $k \in K$:

$$\mathscr{U}_{k(\varphi)}^\sigma e_m^\varepsilon = e^{-i(m+\frac{1}{2}\varepsilon)\varphi}e_m^\varepsilon, \tag{10.2.17}$$

$$u_{mn}^\sigma(g) = e^{-i(m+\frac{1}{2}\varepsilon)\varphi_1-i(n+\frac{1}{2}\varepsilon)\varphi_2}u_{mn}^\sigma(a).$$

The matrix element depends on Euler angles

$$\varphi = \varphi_1+\frac{\pi}{2}, \qquad \psi = \varphi_2-\frac{\pi}{2}, \qquad -i\tau \tag{10.2.18}$$

in the following manner:

$$u_{mn}^\sigma(g) = i^{n-m}e^{-i(m+\frac{1}{2}\varepsilon)\varphi-i(n+\frac{1}{2}\varepsilon)\psi}\mathfrak{P}_{m+\frac{1}{2}\varepsilon,n+\frac{1}{2}\varepsilon}^{-\frac{1}{2}(l+1)}(\cosh\tau). \tag{10.2.19}$$

We shall write the case $\varepsilon = 0$, $n = 0$ separately:

$$u_{m0}^\sigma(g) = i^{-m}e^{-i(m+\frac{1}{2}\varepsilon)\varphi}\mathfrak{P}_{m0}^{-\frac{1}{2}(l+1)}(\cosh\tau) \tag{10.2.20}$$

$$= i^{-m}e^{-i(m+\frac{1}{2}\varepsilon)\varphi}\frac{\Gamma(\frac{1}{2}(1-l))}{\Gamma(m+\frac{1}{2}(1-l))}\mathfrak{P}_{-\frac{1}{2}(l+1)}^m(\cosh\tau).$$

Finally

$$u_{00}^\sigma(g) = \mathfrak{P}_{-\frac{1}{2}(l+1)}(\cosh\tau). \tag{10.2.21}$$

10.3. PROPERTIES OF THE REPRESENTATION $\mathscr{U}^\sigma$ AND THE FUNCTION $\mathfrak{P}_{mn}^l$

The invariant measure dk on the group K is quasi-invariant when regarded as a measure on the homogeneous space $K = G/AN$. Knowing the form of the Haar measure in terms of the coordinates of the Iwasawa decomposition, we can find the ϱ-function corresponding to this measure and the Radon–Nikodym derivative. Formula (10.1.5) can be written in the form

$$\int_G f(g)\mathrm{d}g = \int_K\int_A\int_N \delta^{-1}(a)f(kan)\mathrm{d}k\,s\,\mathrm{d}a\,\mathrm{d}n,$$

or equivalently

$$\int_G f(g)\,\delta(h(g))\mathrm{d}g = \int_K\int_A\int_N f(kan)\mathrm{d}k\,\mathrm{d}a\,\mathrm{d}n. \qquad (10.3.1)$$

The measure da dn on AN is the left Haar measure for this subgroup. Decomposition (10.3.1) is a special case of the decomposition described in Theorem 1.7.3. Comparing (10.3.1) with formula (10.7.5) in this chapter, we observe that the ϱ-function of the measure dk is the function $g \to \delta(h(g))$. Formula (1.7.6) gives the form of the derivative of the measure $\mathrm{d}g\cdot k$ with respect to dk:

$$S(g, k) = \delta(h(gk)). \qquad (10.3.2)$$

Thus we have the identity

$$\int_K f(g\cdot k)\mathrm{d}k = \int_K \delta(h(g^{-1}k))f(k)\mathrm{d}k. \qquad (10.3.3)$$

PROPOSITION 10.3.1. *For* $f_1, f_2 \in \mathscr{E}^\sigma$ *we have*

$$\langle U^\sigma_g f_1, U^{-\sigma}_g f_2\rangle = \langle f_1, f_2\rangle. \qquad (10.3.4)$$

Proof. The left-hand side of the expression has the form

$$\int_K \delta^{\frac{1}{2}}\chi_\sigma^{-1}(h(g^{-1}k))f_1(g^{-1}\cdot k)\,\delta^{\frac{1}{2}}\chi_{-\sigma}^{-1}(h(g^{-1}k))\,f_2(g^{-1}\cdot k)\mathrm{d}k$$

$$= \int_K \delta(h(g^{-1}k))f_1(g^{-1}\cdot k)f_2(g^{-1}\cdot k)\mathrm{d}k = \int_K f_1(k)f_2(k)\mathrm{d}k. \qquad \square$$

REMARK. During the proof we found the expression for the function $S_\varrho(g^{-1}, k) = \delta(h(g^{-1}k)) = (e_{-1}(g, k))^{-2}$ (formulas (10.2.4), (10.2.8), and (10.2.10)). Thus, the Radon–Nikodym derivative of the measure $\mathrm{d}g\cdot k$ with respect to dk is identical with the Poisson kernel on the disc. This fact was noted by H. Furstenberg, who used it to create the theory of harmonic functions on symmetric spaces. Furstenberg's results are presented in § 18.2.

COROLLARY 10.3.2. *The inner product* $(\cdot\,|\,\cdot)$ *in* $L^2(K)$ *has the following invariance property*

$$(\mathcal{U}_g f_1 | \mathcal{U}_g^{-\bar{\sigma}} f_2) = (f_1 | f_2). \tag{10.3.5}$$

Indeed, since the complex conjugation of the multiplier S_σ determining the representation $\mathcal{U}^\sigma$ gives the multiplier $S_{\bar{\sigma}}$, the corollary follows by Proposition 10.3.1.

COROLLARY 10.3.3. *The representation* $\mathcal{U}^\sigma$ *is unitary if and only if* l *is a pure imaginary number.*

We proceed to prove certain properties of the functions $\mathfrak{P}^\zeta_{mn}$ under transformations of indices.

First, by a change of variable in formulas (10.2.14), we get

$$\mathfrak{P}^\zeta_{mn} = \mathfrak{P}^\zeta_{-m,-n}. \tag{10.3.6}$$

The matrix elements with respect to $\langle\cdot,\cdot\rangle$ and $(\cdot\,|\,\cdot)$ are related by the following formula:

$$(e^\varepsilon_m | \mathcal{U}_g e_n) = \langle \overline{e^\varepsilon_m}, \mathcal{U}^\sigma_g e^\varepsilon_n \rangle = \langle e^\varepsilon_{-m-\varepsilon}, \mathcal{U}^\sigma_g e^\varepsilon_n \rangle.$$

Applying the symmetry of the form $\langle\cdot,\cdot\rangle$ and Proposition 10.3.1 we obtain

$$\begin{aligned} u^\sigma_{mn}(g) &= \langle e^\varepsilon_n, \mathcal{U}^{-\sigma}_{g^{-1}} e^\varepsilon_{-m-\varepsilon} \rangle = \langle \overline{e^\varepsilon_{-n-\varepsilon}}, \mathcal{U}^{-\sigma}_{g^{-1}} e^\varepsilon_{-m-\varepsilon} \rangle \\ &= (e^\varepsilon_{-n-\varepsilon} | \mathcal{U}^\sigma_{g^{-1}} e^\varepsilon_{-m-\varepsilon}) = u^{-\sigma}_{-n-\varepsilon,-m-\varepsilon}(g^{-1}). \end{aligned}$$

Let $g = a(\tau)$, and thus $a(\tau)^{-1} = k(\pi)a(\tau)k(-\pi)$. Applying formula (10.2.19), we get

$$\mathfrak{P}^{-\frac{l+1}{2}}_{m+\frac{\varepsilon}{2},\, n+\frac{\varepsilon}{2}} = (-1)^{(m-n)} \mathfrak{P}^{\frac{l+1}{2}}_{-n-\frac{\varepsilon}{2},\, -m-\frac{\varepsilon}{2}},$$

The numbers $l \in \mathbf{C}$, $m, n \in \frac{1}{2}Z$, being arbitrary we can write in view of formula (10.3.3)

$$\mathfrak{P}^\zeta_{n,m} = (-1)^{m-n} \mathfrak{P}^{-\zeta-1}_{m,n}. \tag{10.3.7}$$

The invariance of the form $(\cdot\,|\,\cdot)$ given by (10.3.2) yields the following identities

$$u^\sigma_{mn}(g) = (e_m | \mathcal{U}^\sigma_g e_n) = \overline{(\mathcal{U}^\sigma_g e_n | e_m)} = \overline{(e_n | \mathcal{U}^{-\bar{\sigma}}_{g^{-1}} e_m)} = \overline{u^{-\bar{\sigma}}_{nm}(g^{-1})}.$$

Substituting $g = a(\tau)$ and using formula (10.3.7) we obtain

$$\overline{\mathfrak{P}^\zeta_{n,m}} = \mathfrak{P}^{\bar{\zeta}}_{n,m}. \tag{10.3.8}$$

It follows directly from (10.3.8) that for real ζ the Jacobi functions $\mathfrak{P}^\zeta_{m,n}$ are real-valued. We note that also for imaginary l the function $\mathfrak{P}^{\frac{l-1}{2}}_{nn}$ is real-valued.

$$\overline{\mathfrak{P}^{\frac{l-1}{2}}_{nn}} = \mathfrak{P}^{-\frac{1}{2}l-\frac{1}{2}}_{nn} = \mathfrak{P}^{\frac{1}{2}l+\frac{1}{2}-1}_{nn} = \mathfrak{P}^{\frac{l-1}{2}}_{nn}.$$

In particular, the Legendre functions $\mathfrak{P}_{i\nu-\frac{1}{2}}$ are real-valued. We shall formulate this fact in a form which will be significant from the point of view of representation theory.

$$\mathfrak{P}^{i\nu-\frac{1}{2}}_{nn} = \mathfrak{P}^{-i\nu-\frac{1}{2}}_{nn}, \qquad \nu \in \boldsymbol{R} \tag{10.3.9}$$

Addition Formulas

These formulas are derived from the equation

$$u^\sigma_{mn}(g_1 g_2) = \sum_{k=-\infty}^{\infty} u^\sigma_{mk}(g_1) u^\sigma_{kn}(g_2).$$

We take g_1 with the Euler angles $0, -i\tau_1, 0$, and g_2 with the Euler angles $\varphi_2, -i\tau_2, 0$. The Euler angles $\varphi, -i\tau, \psi$ of the product $g_1 g_2$ can be found from the relations

$$\begin{bmatrix} \cosh\frac{1}{2}\tau_1 & \sin\frac{1}{2}\tau_1 \\ \sinh\frac{1}{2}\tau_1 & \cosh\frac{1}{2}\tau_1 \end{bmatrix} \begin{bmatrix} \cosh\frac{1}{2}\tau_2 e^{\frac{1}{2}i\varphi_2} & \sinh\frac{1}{2}\tau_2 e^{\frac{1}{2}i\varphi_2} \\ \sinh\frac{1}{2}\tau_2 e^{-\frac{1}{2}\varphi_2} & \cosh\frac{1}{2}\tau_2 e^{-\frac{1}{2}\varphi_2} \end{bmatrix}$$

$$= \begin{bmatrix} \cos\frac{1}{2}\tau e^{\frac{1}{2}i(\varphi+\psi)} & \sinh\frac{1}{2}\tau e^{\frac{1}{2}(i\varphi-\psi)} \\ \sinh\frac{1}{2}\tau e^{\frac{1}{2}i(\psi-\varphi)} & \cosh\frac{1}{2}\tau e^{-\frac{1}{2}i(\varphi+\psi)} \end{bmatrix},$$

i.e., after obvious transformations,

$$\cosh\tau = \cosh\tau_1\cosh\tau_2 + \sinh\tau_1\sinh\tau_2\cos\varphi_2, \tag{10.3.10}$$

$$e^{i\varphi} = \frac{\sinh\tau_1\cos\tau_2 + \cosh\tau_1\sinh\tau_2\cos\varphi_2 + i\sinh\tau_2\sin\tau_2}{\sinh\frac{1}{2}\tau},$$

$$e^{\frac{1}{2}i(\varphi+\psi)} = \frac{\cosh\frac{1}{2}\tau_1\cosh\frac{1}{2}\tau_2 e^{\frac{1}{2}i\varphi_2} + \sinh\frac{1}{2}\tau_1\sinh\frac{1}{2}\tau_2 e^{-\frac{1}{2}i\varphi_2}}{\cosh\frac{1}{2}\tau}.$$

Using the description of the matrix elements in terms of the functions $\mathfrak{P}_{mn}$ given by (10.2.19), we get

$$e^{-i(m\varphi+n\psi)}\mathfrak{P}^{\zeta}_{mn}(\cosh\tau) = \sum_{k=-\infty}^{\infty} e^{-ik\varphi_2}\mathfrak{P}^{\zeta}_{mk}(\cosh\tau_1)\mathfrak{P}^{\zeta}_{kn}(\cosh\tau_2). \tag{10.3.11}$$

Summation with respect to k is taken over the set of integers if m, n are integers and over the set $\{l+\frac{1}{2}\}_{l=-\infty}^{\infty}$ if m, n are half integers. Special cases of this formula will be found in the problems.

Mean Values of Jacobi Functions

Each of the representations $\mathscr{U}^{\sigma}$, when restricted to K, decomposes into a direct sum of non-equivalent 1-dimensional representations. The basis $\{e^{\varepsilon}_n\}_{n=-\infty}$ is the set of eigenvectors of the operators $\mathscr{U}^{\sigma}_k$, $k \in K$. Thus, the results of § 5.1 apply if we set $v = e^{\varepsilon}_j$ for an arbitrary j.

$$\int_K u_{mn}(g_1 k g_2)\bar{\varkappa}_j(k)\,\mathrm{d}k = u^{\sigma}_{mj}(g_1)u^{\sigma}_{jn}(g_2), \tag{10.3.12}$$

where $\varkappa_j(k)$ is the eigenvalue of $\mathscr{U}^{\sigma}_k$ on e^{ε}_j, given by formula (10.2.17). Substituting $g_1 = a(\tau_1)$, $g_2 = a(\tau_2)$ and $k = k(\varphi_2)$, we have on account of formulas (10.2.19),

$$\frac{1}{2\pi}\int_0^{2\pi} e^{i(j\varphi_2-m\varphi-n\psi)}\mathfrak{P}^{\zeta}_{mn}(\cosh\tau)\,\mathrm{d}\varphi_2 = \mathfrak{P}^{\zeta}_{mj}(\cosh\tau_1)\mathfrak{P}^{\zeta}_{jn}(\cosh\tau_2), \tag{10.3.13}$$

where φ, τ, ψ are related with φ_2, τ_1, τ_2 by relations (10.3.10).

The curve in the disc D

$$\varphi \to g_1 k(\varphi) g_2 \cdot 0 \in D, \qquad g_1, g_2 \in \mathrm{SU}(1,1)$$

has the interpretation of a sphere with centre at $g_1 \cdot 0$ and radius $\cosh\tau_2$, if $h(g_2) = a(\tau_2)$. The sphere is to be understood in the sense of an SU(1, 1)-invariant Riemannian structure on D. For $n = 0$ the function u^{σ}_{mn} is right K-invariant. Formula (10.3.12) expresses the mean value of the integrand over a sphere in D by the value of u^{σ}_{mj} at $g_1 \cdot 0$ and the value of u^{σ}_{j0} at $g_2 \cdot 0$.

If $l = n = 0$ the formula takes the form of a mean value theorem for the function u^{σ}_{m0}, which states that the mean value is equal to the product of the value of the function at the centre of the sphere and the value at $g_2 \cdot 0$ of the zonal spherical function u^σ_{00}. Special cases of formulas (10.3.13) can be found in Problem 14.

10.4. DIFFERENTIALS OF THE REPRESENTATIONS $\mathcal{U}^\sigma$ RECURRENCE RELATIONS. IRREDUCIBILITY

One-parameter subgroups of the group SL(2, $\boldsymbol{R}$) defined by

$$a_1(t) = \begin{bmatrix} e^{\frac{t}{2}} & 0 \\ 0 & e^{-\frac{t}{2}} \end{bmatrix}, \qquad a_2(\tau) = \begin{bmatrix} \cosh\frac{\tau}{2} & \sinh\frac{\tau}{2} \\ \sinh\frac{\tau}{2} & \cosh\frac{\tau}{2} \end{bmatrix},$$

$$k(\varphi) = \begin{bmatrix} \cos\frac{\varphi}{2} & \sin\frac{\varphi}{2} \\ -\sin\frac{\varphi}{2} & \cos\frac{\varphi}{2} \end{bmatrix}$$

have the infinitesimal generators

$$X_1 = \frac{1}{2}\begin{bmatrix} 1 & 0 \\ 0 & -1 \end{bmatrix}, \qquad X_2 = \frac{1}{2}\begin{bmatrix} 0 & 1 \\ 1 & 0 \end{bmatrix}, \qquad X_3 = \frac{1}{2}\begin{bmatrix} 0 & 1 \\ -1 & 0 \end{bmatrix}$$

which satisfy the following commutation relations

$$[X_1, X_2] = X_3, \qquad [X_1, X_3] = X_2, \qquad [X_2, X_3] = -X_1. \tag{10.4.1}$$

The operators of the representation of the Lie algebra SL(2, $\boldsymbol{R}$) associated with the representation $\mathcal{U}^\sigma$ are denoted by

$$A_i := d\mathcal{U}^\sigma(X_i).$$

The vectors of the basis $\{e^\varepsilon_n\}_{n=-\infty}^{n=\infty}$ are eigenvectors of the operators $\mathcal{U}^\sigma_k$, and hence also eigenvectors of the operators $d\mathcal{U}^\sigma(X_3)$. More precisely, we have

$$A_3 e^\varepsilon_n = -i(n+\tfrac{1}{2}\varepsilon)e^\varepsilon_n. \tag{10.4.2}$$

The operators A_i satisfy commutation relations (10.4.1). We introduce the following operators:

$$H_+ = A_1 - iA_2, \qquad H_- = A_1 + iA_2, \qquad H = iA_3. \tag{10.4.3}$$

PROPOSITION 10.4.1. *The operators H_+, H_-, H act on the basis vectors in the following way*:

$$H_+ e_n^\varepsilon = \left(n+\frac{l+1+\varepsilon}{2}\right)e_{n+1},$$
$$H_- e_n = \left(-n+\frac{l+1-\varepsilon}{2}\right)e_{n-1}^\varepsilon, \tag{10.4.4}$$
$$He_n = \left(n+\frac{\varepsilon}{2}\right)e_n^\varepsilon.$$

Proof. Direct computation is to some extent replaced by a standard application of the commutation rules and of the third formula, which obviously follows from (10.4.3).

$$\begin{aligned} HH_+ e_n &= A_3(iA_1+A_2)e_n^\varepsilon = (iA_3A_1+A_3A_2)e_n^\varepsilon \\ &= [-i(A_2-A_1A_3)+(A_1+A_2A_3)]e_n^\varepsilon \\ &= (H_+e_n+H_+iA_3)e_n^\varepsilon = \left(1+n+\frac{\varepsilon}{2}\right)e_n^\varepsilon. \end{aligned}$$

A similar computation for H_- leads to the relation

$$HH_- e_n^\varepsilon = \left(-1+n+\frac{\varepsilon}{2}\right)e_n^\varepsilon.$$

Thus the vector $H_\pm e_n^\varepsilon$ is proportional to $e_{n\pm 1}^\varepsilon$. The value of the coefficient will be found directly from the definition of the differential of a representation by means of formula (10.2.12) and the observation that $a_2(t) = k_0^{-1}a_1(t)k_0$, where $k_0 = \frac{1}{\sqrt{2}}\begin{bmatrix} 1 & 1 \\ -1 & 1\end{bmatrix} = k\left(\frac{\pi}{2}\right)$. To simplify the calculations we shall find $d\mathscr{U}^\sigma(X_i)e_n|_{\theta=0}$

$$d\mathscr{U}^\sigma(X_1)e_n(0) = \frac{d}{dt}\left(e^{-\frac{1}{2}t(2n+\varepsilon)+t\left(n+\frac{1}{2}(\varepsilon+1+l)\right)}\right)\Big|_{t=0} = \tfrac{1}{2}(l+1),$$

$$\begin{aligned} d\mathscr{U}^\sigma(X_2)e_n(0) &= e^{-\frac{1}{2}i\pi\left(n+\frac{1}{2}\varepsilon\right)}\frac{d}{dt}\left[\left(\frac{1}{\sqrt{2}}e^{-\frac{1}{2}\tau}\right.\right. \\ &\qquad \left.\left. +i\frac{1}{\sqrt{2}}e^{\frac{1}{2}\tau}\right)^{2n+\varepsilon}(\cosh\tau)^{-n-\frac{1}{2}(\varepsilon+l+1)}\right]\Big|_{t=0} \\ &= \tfrac{1}{2}i(2n+\varepsilon), \end{aligned}$$ □

which is the desired result.

We shall use formulas (10.4.4) first of all to investigate the reducibility of the representations $\mathscr{U}^\sigma$. These representations are reducible only for σ such that $l+\varepsilon$ is odd. Thus, setting $\Omega = \{\sigma = (\varepsilon, l)\colon l+\varepsilon$ not an odd integer$\}$, we have the following

THEOREM 10.4.2. *The representations $\mathscr{U}^\sigma$ for $\sigma \in \Omega$ are irreducible.*

Proof. If $\sigma \in \Omega$, then $\frac{l+1\pm\varepsilon}{2}$ is not an integer. The operators H_+, H_- are not zero on any vector in the basis. Every non-zero invariant closed subspace E of the representation space of $\mathscr{U}^\sigma$ decomposes into 1-dimensional eigensubspaces of the operators $\mathscr{U}^\sigma_k$, $k \in K$. If e^ε_n belongs to such a subspace, then by (10.4.4) every vector in the basis $\{e^\varepsilon_n\}$ also belongs to it. Therefore E is identical with the representation space. □

Problems 18, and 19 show that the representations $\mathscr{U}^\sigma$ for $\sigma \notin \Omega$ are in fact reducible and have implications for the theory of Jacobi functions.

Only the representations corresponding to $\varepsilon = 0$ contain a vector invariant with respect to the action of the subgroup K. Thus spherical representations are obtained for $\sigma = (0, l)$ where l is an odd integer. The spherical function of such a representation is determined by the Legendre function

$$u^\sigma_{00}(a) = \mathfrak{P}_{00}^{\frac{-l+1}{2}}(\cosh\tau) = \mathfrak{P}_{-\frac{l+1}{2}}(\cosh\tau).$$

Thus formula (10.3.6) indicates that the representation corresponding to the parameter $l = i\nu$ has the same spherical function as the representation related to $l = -i\nu$. Since representations are uniquely determined by spherical functions, we conclude that

$$\mathscr{U}^{(0,i\nu)} \simeq \mathscr{U}^{(0,-i\nu)}.$$

Recurrence Relations

The way in which the recurrence relations for Jacobi functions can be obtained is precisely the same as that applied in Chapter 8 for obtaining the recurrence relations for P^l_{mn}.

Thus we use relations (10.4.4) and the convolutive properties of the linear operator

$$\mathscr{E}^\sigma \ni e_n \to \check{u}^\sigma_{mn} \in \mathscr{E}(G), \qquad \text{where } \check{u}^\sigma_{mn} = u^\sigma_{mn}(g^{-1}).$$

Formulas (10.4.4) become

$$\begin{aligned}\tilde{H}_+\check{u}^\sigma_{mn} &= \left(n+\frac{l+1+\varepsilon}{2}\right)\check{u}^\sigma_{m,n+1},\\ \tilde{H}_-\check{u}^\sigma_{m,n} &= \left(-n+\frac{l+1-\varepsilon}{2}\right)\check{u}^\sigma_{m,n-1},\end{aligned} \tag{10.4.5}$$

where $\tilde{H}_\pm$ are linear combinations of the infinitesimal operators of the regular representation:

$$\tilde{H}_+ = \mathrm{d}L(X_1)-i\mathrm{d}L(X_2),$$

$$\tilde{H}_- = \mathrm{d}L(X_1)+i\mathrm{d}L(X_2).$$

The explicit form of these operators in the Euler coordinates is obtained from formulas (10.3.10) by differentiation. The calculations similar to those performed in Chapter 8 are omitted.

The final formula takes the form

$$\begin{aligned}\tilde{H}_+ &= \mathrm{e}^{i\varphi}\left(-\coth\tau\frac{\partial}{\partial\varphi}+i\frac{\partial}{\partial\tau}+\frac{1}{\sinh\tau}\frac{\partial}{\partial\psi}\right),\\ \tilde{H}_- &= \mathrm{e}^{-i\varphi}\left(-\coth\tau\frac{\partial}{\partial\varphi}-i\frac{\partial}{\partial\tau}+\frac{1}{\sinh\tau}\frac{\partial}{\partial\psi}\right).\end{aligned} \tag{10.4.6}$$

Substitution of the value of $H_\pm$ and $\check{u}^\sigma_{m,n}$ given by (10.2.19) in formulas (10.4.5) leads, after some trivial transformations, to the equations

$$\begin{aligned}\sqrt{z^2-1}\,\frac{\mathrm{d}}{\mathrm{d}z}\mathfrak{P}^\zeta_{mn}+\frac{m-nz}{\sqrt{z^2-1}}\mathfrak{P}^\zeta_{mn} &= (\zeta-n)\mathfrak{P}^\zeta_{m,n+1},\\ \sqrt{z^2-1}\,\frac{\mathrm{d}}{\mathrm{d}z}\mathfrak{P}^\zeta_{mn}+\frac{nz-m}{\sqrt{z^2-1}}\mathfrak{P}^\zeta_{mn} &= (\zeta+n)\mathfrak{P}^\zeta_{m,n-1},\end{aligned} \tag{10.4.7}$$

where m and n are both integers or half-integers.

Many of the well-known formulas for $\mathfrak{P}^\zeta_{mn}$, $\mathfrak{P}^\zeta_n$, $\mathfrak{P}^\zeta$, namely those given in the list of relations, follow from the above equations by linear combinations and the rules for a change of indices.

A successive application of formulas (10.4.7) leads to the second order differential equation (called the *associated Legendre equation*)

$$(z^2-1)\frac{\mathrm{d}^2}{\mathrm{d}z^2}\mathfrak{P}^\zeta_{mn}+2z\frac{\mathrm{d}}{\mathrm{d}z}\mathfrak{P}^\zeta_{mn}-\frac{m^2+n^2-2mnz}{z^2-1}\mathfrak{P}^\zeta_{mn}$$

$$= \zeta(\zeta+1)\mathfrak{P}^\zeta_{mn}. \tag{10.4.8}$$

This equation corresponds to an important property of matrix elements, which is seen directly from (10.4.5). Namely they are eigenfunctions of the operator

$$-\mathrm{d}L(X_3)^2+\tilde{H}_+\tilde{H}_- = [\mathrm{d}L(X_1)]^2+[\mathrm{d}L(X_2)^2-[\mathrm{d}L(X_3)]^2$$
$$= \frac{1}{\sinh\tau}\frac{\partial}{\partial\tau}\sinh\tau\frac{\partial}{\partial\tau}+\frac{1}{\sinh^2\tau}\left(\frac{\partial^2}{\partial\varphi^2}-2\cosh\tau\frac{\partial^2}{\partial\varphi\,\partial\psi}+\frac{\partial^2}{\partial\varphi^2}\right). \tag{10.4.9}$$

This is the Laplace–Beltrami operator of the group SL(2, $\boldsymbol{R}$), which commutes with left and right translations on G.

Let us compare formula (10.4.8) with the differential equation for the function P^l_{mn} (8.4.12). The form of the two equations is identical and the only difference is that formula (10.4.12) is defined exclusively for $l\in\frac{1}{2}\boldsymbol{Z}$. The parameter ζ is related to the parameter $\sigma = (\varepsilon, l)$, numbering the representations by means of the formula $\zeta = -l+\frac{1}{2}$ (see (10.2.15)).

Note that the equation for the function P^l_{mn} appears in (10.4.8) when ζ corresponds to a $\sigma\notin\Omega$, i.e. to a reducible representation $\mathscr{U}^\sigma$.

10.5. HARMONIC ANALYSIS ON THE DISC SU(1,1)/$\mathscr{K}$

The space D interests us as a homogeneous manifold of the group SU(1, 1). Since in the description of the space we apply coordinates originating from the parameters of the group, all formulas can be related to the other model of that homogeneous space, i.e. to the upper complex half-plane.

The coordinates of the homogeneous space which we shall use arise from the parameters of the subgroups of SU(1, 1):

$$\mathscr{K}:\ k(\theta) = \begin{bmatrix} \mathrm{e}^{i\frac{\theta}{2}} & 0\\ 0 & \mathrm{e}^{-i\frac{\theta}{2}}\end{bmatrix},$$

$$\mathscr{A}:\ a(t) = \begin{bmatrix} \cosh\frac{t}{2} & i\sinh\frac{t}{2}\\ -i\sinh\frac{t}{2} & \cosh\frac{t}{2}\end{bmatrix},$$

$$\mathcal{N}: n(\xi) = \begin{bmatrix} 1+\frac{\xi}{2}i & \frac{\xi}{2} \\ \frac{\xi}{2} & 1-\frac{\xi}{2}i \end{bmatrix}.$$

Besides the spherical coordinates introduced by means of Theorem 10.1.2, which assign to a pair (θ, τ) the point $k(\theta)a(\tau)\cdot 0 = i \tanh\frac{\tau}{2}e^{i\theta}$, we shall introduce in D the so-called horocyclic coordinates, which are associated with the Iwasawa decomposition of the group. Namely, for every $z \in D$ there is a unique pair (t, ξ) such that $z = a(t)n(\xi)\cdot 0$. This pair is called the *horocyclic coordinates of the point* z. An explanation of this term is given in the description of the level lines of these coordinates. From § 10.1 we know that the curves $t =$ const are horocycles tangent to the boundary of the unit disc at the point i and the curves $\xi =$ const are geodesics intersecting every horocycle from this pencil at right angles (Fig. 9).

We begin the study of decompositions of functions on the disc by investigating the class of functions in $\mathscr{D}(D)$ which are rotation invariant, i.e. invariant with respect to the action of the group $\mathscr{K}$. In other words, we are concerned with functions depending only on the modulus of z. We denote the space of these functions by $\mathscr{D}(\mathscr{K}\backslash D)$.

First of all we shall look for a connection between the horocyclic coordinates of a point z and the modulus of z. Put $z(t, \xi) = a(t)n(\xi)\cdot 0$.

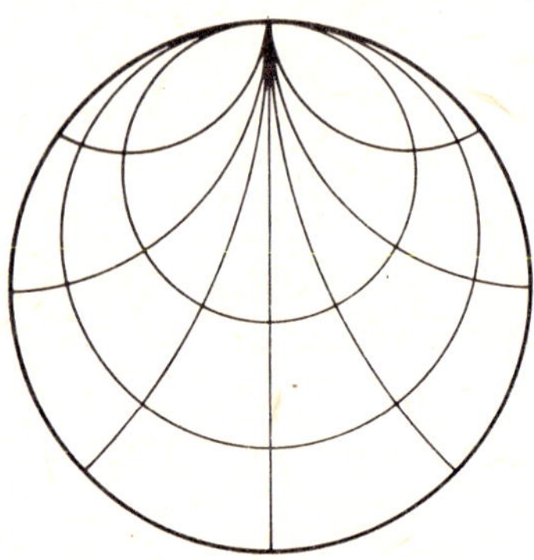

Fig. 9

LEMMA 10.5.1. *If* $|z(t,\xi)| = \tanh\frac{1}{2}\tau$, *then*

$$\cosh\tau = \cosh t + \tfrac{1}{2}\xi^2 e^t. \tag{10.5.1}$$

Proof. We see that the entry a_{11} of the matrix $a(t)n(\xi)$ has the form:

$$a_{11} = \cosh\frac{t}{2} + i\frac{\xi}{2}e^{\frac{t}{2}}$$

Comparing the above with the decomposition from Theorem 10.1.2, we find that $2\cosh^2\frac{\tau}{2} = 2\cosh^2\frac{t}{2} + \frac{\xi^2}{2}e^t$, hence we obtain (10.5.1) by subtracting unity from both sides. □

The invariant measure on the group can be expressed in terms of the coordinates θ, t, ξ as follows:

$$dg = \frac{1}{4\pi}e^t d\theta\, dt\, d\xi. \tag{10.5.2}$$

DEFINITION. The *spherical Fourier transformation on D* is the mapping which to $f \in \mathscr{D}(K\backslash D)$ assigns the function $\hat{f}$ on $\boldsymbol{R}$ given by the formula

$$\hat{f}(\nu) = \int_D f(z)\omega_\nu(z)\,dm(z), \tag{10.5.3}$$

ω_ν being the spherical function of the representation $\mathscr{U}^{(0,2\nu i)}$, which is unitary. (The measure dm is determined by the condition $\int_D \psi(z)\,dm(z) = \int_G \psi(g\cdot 0)\,dg$.)

Let us recall that in terms of the coordinates τ, θ on the disc, the spherical function ω_ν has the form

$$\omega_\nu(a(\tau)\cdot 0) = \omega_\nu\left(\tanh\frac{\tau}{2}\right) = \mathfrak{P}_{i\nu-\frac{1}{2}}(\cosh\tau)$$

$$= \frac{1}{2\pi}\int_0^{2\pi}(\cosh\tau - \sinh\tau\cos\theta)^{i\nu-\frac{1}{2}}\,d\theta,$$

which in the notation of § 10.2 can be written as

$$\omega_\nu(a(\tau)\cdot 0) = \int_K \delta^{-i\nu+\frac{1}{2}}\big(h(a(\tau)k)\big)\,dk.$$

The functions ω_ν are real-valued and symmetric. Therefore the Fourier transform can be written in the form

$$\hat{f}(\nu) = \int_G f(g\cdot 0)\,\omega_\nu(g\cdot 0)\,dg = \int_G f(g\cdot 0)\int_K \delta^{-i+\frac{1}{2}}(h(gk))\,dk\,dg$$

$$= \int_G f(g\cdot 0)\,\delta^{-i\nu+\frac{1}{2}}(h(g))\,dg \tag{10.5.4}$$

as a result of changing the order of integration and using the K-invariance of f.

Finally let us write the right-hand side in terms of the horocycle coordinates.

For simplicity we write:

$$f[\cosh t] := f\left(\tanh\frac{t}{2}\right).$$

Then by means of (10.5.2) we obtain

$$\hat{f}(\nu) = \int_{-\infty}^{\infty}\int_{-\infty}^{\infty} f\left[\cosh t + e^t\frac{\xi^2}{2}\right] e^{\frac{t}{2}(2i\nu-1)}\,e^t\,dt\,d\xi.$$

Set

$$F_f(\cosh t) = e^{\frac{t}{2}}\int_{-\infty}^{\infty} f\left[\cosh t + e^t\frac{\xi^2}{2}\right]d\xi = \int_{-\infty}^{\infty} f\left[\cosh t + \frac{x^2}{2}\right]dx. \tag{10.5.5}$$

Then

$$\hat{f}(\nu) = \int_{-\infty}^{\infty} e^{i\nu t}F_f(\cosh t)\,dt. \tag{10.5.6}$$

Thus the function $\hat{f}(\nu)$ is represented in the form of the "usual" Fourier transform of the function F_f, which is called the *Abel transform* of the function f.

The terminology is transferred from general theory. The cases of the Abel transform, which are investigated in the present chapter and the next, were introduced by Gelfand and Graev [53].

The Abel transform at a point t is the integral of the function f over the horocycle $a(t)\mathcal{N}\cdot 0$ with respect to the measure $e^{\frac{t}{2}}d\xi$.

By means of the inversion rule for the Fourier transform on $\boldsymbol{R}$ we can express F_f in terms of $\hat{f}$:

$$F_f(\cosh\tau) = \frac{1}{2\pi}\int_{-\infty}^{\infty} \hat{f}(\nu)e^{-i\nu\tau}d\nu$$

and

$$F_f'(\cosh\tau) = \frac{i}{2\pi\sinh\tau}\int_{-\infty}^{\infty} \hat{f}(\nu)\nu e^{-i\nu\tau}d\nu.$$

The admissibility of the differentiation of the integral with respect to the parameter is confirmed by the following fact, which is very important for further considerations.

LEMMA 10.5.2. *The transform* $\hat{f}$ *is a function of class* $\mathscr{S}(\boldsymbol{R}^1)$.

Proof. The definition of the Abel transform F_f implies that it is a function in the class $\mathscr{D}(\boldsymbol{R}^1)$. Thus $\hat{f}$, as the Fourier transform of a smooth, compactly supported function, belongs to $\mathscr{S}(\boldsymbol{R}^1)$. □

A decisive step in the proof of the inversion formula for the spherical Fourier transform has been made by R. Takahasi (1952) and is based upon the following observation.

PROPOSITION 10.5.3.

$$f[t] = \frac{-1}{2\pi}\int_{-\infty}^{\infty} F_f'\left(t+\frac{x^2}{2}\,dx\right) \quad \textit{for} \quad t>0.$$

Proof. We compute

$$\int_{-\infty}^{\infty} F_f'\left(t+\frac{x^2}{2}\right)dx = \int_{-\infty}^{\infty}\int_{-\infty}^{\infty} f'\left[t+\frac{x^2+v^2}{2}\right]dv\,dx$$

$$= \int_0^{\infty}\int_0^{2\pi} f'\left[t+\frac{\varrho^2}{2}\right]\varrho\,d\varrho\,d\theta = -2\pi f[t]. \quad \square$$

Substituting the previously obtained formula, for F'_f, we get

$$f(0) = f[1] = \frac{-1}{2\pi}\int_{-\infty}^{\infty} F'_f\left(1+\frac{x^2}{2}\right)dx$$

$$= \frac{-1}{2\pi}\int_{-\infty}^{\infty} F'_f(\cosh\tau)\cosh\frac{\tau}{2}d\tau = \frac{-i}{8\pi^2}\int_{-\infty}^{\infty}\int_{-\infty}^{\infty}\hat{f}(\nu)\nu\frac{e^{-i\nu\tau}}{\sinh\frac{\tau}{2}}d\tau d\nu$$

$$= \frac{-i}{8\pi^2}\int_{-\infty}^{\infty}\hat{f}(\nu)\nu\int_{-\infty}^{\infty}\frac{e^{-i\nu\tau}}{\sinh\frac{\tau}{2}}d\tau d\nu = \frac{1}{4\pi}\int_{-\infty}^{\infty}\hat{f}(\nu)\nu\tanh(\pi\nu)d\nu$$

$$= \frac{1}{2\pi}\int_{0}^{\infty}\hat{f}(\nu)\nu\tanh(\pi\nu)d\nu, \tag{10.5.7}$$

using the evenness of the function $\hat{f}$, which results from the evenness of F_f.

The evenness of the function $\hat{f}$ reflects the fact that the representations corresponding to the parameters $(0, i\nu)$ and $(0, -i\nu)$ are equivalent.

Formula (10.5.7) leads to the fundamental relations for analysis on the space D.

THEOREM 10.5.4. *The following formulas hold for* $f \in \mathscr{D}(\mathscr{K}\backslash D)$:

$$\int_D |f|^2(z)dm(z) = \frac{1}{2\pi}\int_0^\infty |\hat{f}|^2(\nu)\nu\tanh(\pi\nu)d\nu, \tag{10.5.8}$$

$$f(z) = \frac{1}{2\pi}\int_0^\infty \hat{f}(\nu)\omega_\nu(z)\nu\tanh(\pi\nu)d\nu. \tag{10.5.9}$$

Proof. We check (10.5.9) by applying (10.5.7) to the function

$$F_g(z) := \int_K f(gk\cdot z)dk, \tag{10.5.10}$$

$$f(g\cdot 0) = F_g(0) = \frac{1}{2\pi}\int_0^\infty \hat{F}_g(\nu)\nu\tanh(\pi\nu)d\nu.$$

On the other hand, we have

$$\hat{F}_g(\nu) = \int_D \int_{\mathscr{K}} f(gk \cdot z) dk \omega_\nu(z) dm(z) = \int_D f(g \cdot z) \omega_\nu(z) dm(z)$$

$$= \omega_\nu(g^{-1}) \hat{f}(\nu) = \bar{\omega}_\nu(g) \hat{f}(\nu)$$

in virtue of formula (5.1.9) and of ω_ν being positive definite for real values of ν. We know however, that ω_ν is a real function. This ends the proof of formula (10.5.9).

Now, substituting $f_1 = f^* * f$ into (10.5.7), we obtain

$$f^* * f(0) = \int_G \bar{f}(g^{-1} \cdot 0) f(g^{-1} \cdot 0) dg = \int_D |f|^2(z) dm(z)$$

$$= \frac{1}{2\pi} \int_0^\infty (f^* * f)^\wedge (\nu) \nu \tanh(\pi\nu) d\nu .$$

We know from the theory of spherical function that the mapping $\mathscr{D}(\mathscr{K} \backslash D) \ni f \to \hat{f}$ is a multiplicative functional on the convolution algebra. It is unitary when ω_ν is a positive definite spherical function. This means that $(f^* * f)^\wedge(\nu) = |\hat{f}|^2(\nu)$, which concludes the proof. □

The spherical Fourier transformation on D, when expressed in terms of the spherical coordinates (cf. (10.1.5)), assumes the form of the classical Mehler–Fok transformation

$$\hat{f}(\nu) = \pi \int_0^\infty f\left(\tanh \frac{\tau}{2}\right) \sinh \tau \mathfrak{P}_{i\nu - \frac{1}{2}} (\cosh \tau) d\tau .$$

Put $F(\tau) := f\left(\tanh \dfrac{\tau}{2}\right)$, and

$$\tilde{F}(\nu) = \int_0^\infty F(\tau) \mathfrak{P}_{i\nu - \frac{1}{2}} (\cosh \tau) \sinh \tau d\tau . \tag{10.5.11}$$

From (10.5.9) we obtain the well-known inversion formula for the Mehler–Fok transformation

$$F(\tau) = \frac{1}{2\pi^2} \int_0^\infty \tilde{F}(\nu) \mathfrak{P}_{-\frac{1}{2} - i\nu} (\cosh \tau) \nu \tanh(\pi\nu) d\nu \tag{10.5.12}$$

and from (10.5.8) the Plancherel formula

$$\int_0^\infty |F|^2(\tau)\sinh\tau\,\mathrm{d}\tau = \frac{1}{2\pi^2}\int_0^\infty \tilde{F}(\nu)\nu\tanh(\pi\nu)\mathrm{d}\nu. \qquad (10.5.13)$$

We proceed to generalize Fourier analysis to arbitrary functions of class $\mathscr{D}(D)$. We set

$${}^1f(z) = \int_K f(k\cdot z)\mathrm{d}k.$$

The function ${}^1f(z)$ is already K-invariant and we note that

$${}^1f(0) = f(0).$$

By applying (10.5.7) we find

$$f(0) = \frac{1}{2\pi}\int_0^\infty {}^1\hat{f}(\nu)\nu\tanh(\pi\nu)\mathrm{d}\nu.$$

The function ω_ν being K-invariant, we have the identity ${}^1\hat{f}(\nu) = \hat{f}(\nu)$, and thus formula (10.5.7) applies to an arbitrary function in $\mathscr{D}(D)$.

As in the proof of Theorem 10.5.4 we now apply this formula to the function $F_g(z)$ defined by (10.5.10):

$$f(g\cdot 0) = \frac{1}{2\pi}\int_0^\infty \tilde{F}_g(\nu)\nu\tanh\pi\nu\,\mathrm{d}\nu,$$

$$\hat{F}_g(\nu) = \int_D f(g\cdot z)\omega_\nu(z)\mathrm{d}m(z) = \int_D f(z)\omega_\nu(g^{-1}\cdot z)\mathrm{d}m(z).$$

We return to the definition of the spherical function by means of the multiplier of the representation $\mathscr{U}^\sigma$

$$\omega_\nu(y\cdot 0) = \int_K \delta^{-i\nu+\frac{1}{2}}(h(y^{-1}k))\mathrm{d}k,$$

$$\omega_\nu(g^{-1}y\cdot 0) = \int_K \delta^{-i\nu+\frac{1}{2}}(h(y^{-1}(gk))\mathrm{d}k$$

$$= \int_K \delta^{-i\nu+\frac{1}{2}}(h(y^{-1}(g\cdot k))\,\delta^{-i\nu+\frac{1}{2}}(h(gk))\mathrm{d}k,$$

in view of the properties of a multiplier.

In the remark which follows Proposition 10.3.1 we established the form of the Radon–Nikodym derivative of the measure dgk with respect to dk. It follows from this formula that

$$\omega_\nu(g^{-1}y\cdot 0) = \int_K \delta^{-i\nu+\frac{1}{2}}(h(y^{-1}k))\,\delta^{-i\nu+\frac{1}{2}}(g(g^{-1}\cdot k))\,\delta(h(g^{-1}k))\,\mathrm{d}k$$

$$= \int_K \delta^{-i\nu+\frac{1}{2}}(h(y^{-1}k))\,\delta^{i\nu+\frac{1}{2}}(h(g^{-1}k))\,\mathrm{d}k.$$

Hence we get the following relation:

$$f(g\cdot 0) = \frac{1}{2\pi}\int_0^\infty \int_K \delta^{i\nu-\frac{1}{2}}(h(g^{-1}k))\int_G f(y\cdot 0)$$

$$\times\,\delta^{-i\nu+\frac{1}{2}}(h(y^{-1}k))\,\mathrm{d}m(y)\,\nu\tanh(\pi\nu)\,\mathrm{d}\nu. \tag{10.5.14}$$

We shall interpret this relation as the inversion formula for a suitable integral transformation.

DEFINITION. We introduce a family of functions on $D\times B$ by the formula

$$e_\lambda(gK, kM) := \delta^{-\frac{1}{2}}(h(g^{-1}k)). \tag{10.5.15}$$

The admissibility of the definition of e_λ, or more precisely, the possibility of dropping the function $(g, h)\to h(g^{-1}k)$ onto $G/K\times G/MAN$, is not obvious. Let us recall however, that this function was investigated in § 10.2. We found that in spherical coordinates on the disc,

$$e_{-1}(gK, kM) = \left(\frac{1-2r\cos(\varphi-\theta)+r^2}{1-r}\right)^{\frac{1}{2}}. \tag{10.5.16}$$

On the other hand, in the coordinates arising from the Euler angles on the group we have

$$e_{-1}(gK, kM) = (\cosh\tau+\sinh\tau\sin(\theta-\varphi))^{\frac{1}{2}}.$$

Finally, it follows that

$$e_\lambda(z, \mathrm{e}^{i\theta}) = (\cosh\tau+\sinh\tau\sin(\theta-\varphi))^{-\frac{1}{2}\lambda}$$

if $z = \tanh\frac{\tau}{2}\mathrm{e}^{i\varphi}$.

The function $e_\lambda(\cdot, e^{i\theta})$ is constant on every horocycle in the pencil tangent to the boundary at the point $e^{i\theta}$.

Definition. The function defined on $\boldsymbol{R}_+ \times B$ by the formula

$$\tilde{f}(\nu, b) = \int_D f(z) e_{1-2i\nu}(z, b) dm(z)$$

is called the *Fourier transform of the function* $f \in \mathscr{D}(D)$.

Theorem 10.5.5. *The following formulas hold: the inversion formula*

$$f(z) = \frac{1}{2\pi} \int_0^\infty \int_B e_{2i\nu+1}(z, b) \tilde{f}(\lambda, b) db \nu \tanh(\pi\nu) d\nu \tag{10.5.17}$$

$$\left(\int_B f(b) db = \frac{1}{2\pi} \int_0^{2\pi} f(e^{i\varphi}) d\varphi \right),$$

and the Plancherel formula

$$\int_D |f|^2(z) dm(z) = \frac{1}{2\pi} \int_0^\infty \int_B |\tilde{f}|^2(\nu, b) db \nu \tanh(\pi\nu) d\nu. \tag{10.5.18}$$

The first formula follows immediately from (10.5.14) and the second can be obtained from (10.5.17) in a way similar to that in which (10.5.8) was obtained from (10.5.9).

Remarks. For a fixed parameter ν the operator $\mathscr{D}(D) \ni f \to \tilde{f}(\nu, \cdot) \in \mathscr{E}(B)$ intertwines the regular representation with the representation $\mathscr{U}^{(0,2i\nu)}$ (functions on B are identified in a natural way with symmetric functions on $\mathscr{K}$):

$$\begin{aligned} (L_g f)^\wedge(\nu, b) &= \int_D f(g^{-1}z) e_{1-2i\nu}(z, b) dm(z) \\ &= \int_D f(z) e_{1-2i\nu}(g \cdot z, b) dm(z) \\ &= e_{1-2i\nu}(g^{-1}k, b) \int_D f(z) e_{1-2i\nu}(z, g^{-1} \cdot b) dm(z) \\ &= e_{1-2i\nu}(g^{-1}k, b) \tilde{f}(\nu, g^{-1} \cdot b) \\ &= \delta^{\frac{1}{2}} \chi_{2i\nu}(h(g^{-1}k)) \tilde{f}(\nu, g^{-1} \cdot b) \end{aligned}$$

for $b = kM$.

The function $\tilde{f}(\nu, b)$ can be regarded as an element of the direct integral

$$H = \int_0^\infty H^{2\nu i}\left(\frac{1}{2\pi}\nu\tanh(\pi\nu)\right)\mathrm{d}\nu,$$

where $(H^{2\nu i}, \mathscr{U}^{0,2i\nu})$ is an irreducible spherical representation of the group SL(2, $\boldsymbol{R}$).

The isometry $f \to \tilde{f}$ is an abstract Fourier transformation realizing the decomposition of the representation of the group in $L^2(D)$ into irreducible components.

$$U = \frac{1}{2\pi}\int_0^\infty \mathscr{U}^{(0,2i\nu)}\nu\tanh(\pi\nu)\mathrm{d}\nu.$$

This representation consists only of the spherical representations induced from the group MAN. Each of them appears in the decomposition only once.

Finally, we note that the functions $e_\lambda(\cdot\,, b)$ defined on D as K-multipliers associated with a spherical function are eigenfunctions of the Laplace operator on D, and in virtue of Theorem 5.4.2 we have

$$\Delta e_\lambda = \lambda(\lambda-2)e_\lambda$$

PROBLEMS

1. Let $\mathscr{K}$, $\mathscr{A}$, $\mathscr{N}$ be the subgroups of the group SU(1, 1) defined by formulas (10.1.4).

(a) Express the constituents $k(g)$, $h(g)$ of the Iwasawa decomposition of an element $g = k(g)h(g)n$ in terms of the coefficients of the matrix $g = \begin{bmatrix} a & b \\ \bar{b} & \bar{a}\end{bmatrix} \in \mathrm{SU}(1, 1)$. Prove that the function $t(g)$ defined by the formula $h(g) = \exp(it(g)X_3)$, $X_3 = \frac{1}{2}\begin{bmatrix} 0 & 1 \\ -1 & 0\end{bmatrix}$ has the form $t(g) = 2\log|a-ib|$.

(b) Prove that the function

$$G/\mathscr{K} \times B \ni (g\mathscr{K}, k\cdot 1) \to P(g\mathscr{K}, k\cdot 1) = \mathrm{e}^{-t}(g^{-1}k)$$

equals the Poisson kernel for $D \simeq G/\mathscr{K}$ rotated by $-\frac{1}{2}\pi$

$$P(g\mathscr{K}, k \cdot 1) = \frac{1-|z|^2}{|e^{i\varphi}-iz|^2},$$

where gK is identified with the element $z = g \cdot 0 \in D$ and $e^{i\varphi} = k \cdot 1 \in B$.

(c) Put

$$V = \left\{\begin{bmatrix} 1+it & -t \\ -t & 1-it \end{bmatrix} : t \in \boldsymbol{R}\right\}.$$

Compute $k(v)$ and $t(v)$ for $v \in V$, defined by the Iwasawa decomposition of the element v. Prove that

(i) $V \ni v \to k(v) \cdot 1 \in B$ is a diffeomorphism onto $B \backslash \{-1\}$;

(ii) $$\frac{1}{2}\int_0^{2\pi} f(e^{i\varphi})d\varphi = \frac{1}{\pi}\int_V f(k(v) \cdot 1)e^{-t(v)}dv,$$

dv being the (suitably normalized) Haar measure on V.

2. Prove that the measure $dm(z)$ on D determined by the condition

$$\int_D f(z)dm(z) = \int_G f(g \cdot 0)dg$$

is given in terms of the polar coordinates $z = \tanh \tau e^{i\theta}$ by the formula

$$dm(z) = \tfrac{1}{2}\sinh 2\tau \, d\tau \, d\theta$$

and in terms of the Cartesian coordinates $z = x+iy$ by

$$dm(z) = (1-(x^2+y^2))^{-2}dx\,dy.$$

3. Let

$$X = \begin{bmatrix} 0 & 1 \\ 0 & 0 \end{bmatrix}, \quad Y = \begin{bmatrix} 0 & 0 \\ 1 & 0 \end{bmatrix}, \quad H = \begin{bmatrix} 1 & 0 \\ 0 & -1 \end{bmatrix}.$$

(a) Check that $\{X, Y, H\}$ is a basis for the Lie algebra $\mathfrak{sl}(2, \boldsymbol{R})$ satisfying the commutation relations

$$[X, Y] = H, \qquad [H, X] = 2X, \qquad [H, Y] = -2Y.$$

(b) Let B be the Killing form of the Lie algebra $\mathfrak{sl}(2, \boldsymbol{R})$. Prove that $B(X, X) = -8 \det X$ and that the adjoint representation $g \to \operatorname{Ad} g$ sets up an isomorphism of $SL(2, \boldsymbol{R})/\{I, -I\}$ with $SO_0(2, 1)$, the connected component of identity of the group of all automorphisms of $\boldsymbol{R}^3$ preserving a quadratic form of signature $(+\,+\,-)$ (cf. Ex. 3, § 5.1).

(c) Check that the operator

$$\omega = X_1^2+X_2^2-X_3^2 = \tfrac{1}{4}H^2-\tfrac{1}{2}H+XY$$

belongs to the centre of the enveloping algebra of the group SL(2, $\boldsymbol{R}$) (multiplication in the above formula is multiplication in the enveloping algebra and not matrix multiplication!).

4. Let V be a subgroup in G = SL(2, $\boldsymbol{R}$) given as

$$V = \left\{\begin{bmatrix}1 & 0\\ x & 1\end{bmatrix} : x \in \boldsymbol{R}\right\},$$

and let $P = \left\{\begin{bmatrix}a & b\\ 0 & a^{-1}\end{bmatrix} : a \neq 0\right\}$.

(a) Prove that the mapping

$$V\times P \ni (v, p) \to vp \in G$$

is a diffeomorphism onto a dense open subset of G whose complement is the set sP, $s = \begin{bmatrix}0 & 1\\ -1 & 0\end{bmatrix}$.

(b) Prove that the left Haar measure on P is given by

$$\int_G f(p)\mathrm{d}p = \int_{R_*}\int_R f\left(\begin{bmatrix}a & b\\ 0 & a^{-1}\end{bmatrix}\right)a^{-2}\mathrm{d}a\,\mathrm{d}b$$

and the modular function δ is given by formula (10.2.4).

(c) Prove that for the Haar measure dg on G the following decomposition holds:

$$\int_G f(g)\mathrm{d}g = \int_V\int_P f(vp)\,\delta(p)^{-1}\mathrm{d}p\,\mathrm{d}v = \int_P\int_V f(pv)\mathrm{d}p\,\mathrm{d}v,$$

dp, dv being the left Haar measures on V and P respectively.

(d) Let $\mathscr{E}^\sigma$, U^σ be the representation induced from P. Find a multiplier realization of that representation on the space of functions on the subgroup V, and determine the form of the multiplier. Find the form of a measure μ on V such that for the $\sigma = (\varepsilon, l)$ corresponding to a unitary representation, the multiplier representation thus obtained on the space $L^2(V, \mathrm{d}\mu)$ is unitarily equivalent to U.

Hint. (c) Cf. Example 1(d), § 1.6. (d) To determine the measure μ consider the mapping $V \ni v \to vP \in G/P \simeq P^1\mathbf{R}$. (Cf. Problem 1, § 1.6.)

5. Let H_+ be equipped with the Riemannian metric

$$ds^2 = y^{-2}(dx^2+dy^2)$$

(here, as usual, x, y denote the Cartesian coordinates of a point $z \in H_+$).

(a) Prove that the homographic mapping $\tau(z) = \dfrac{z+i}{iz+1}$ is up to the factor 2 an isometry of D onto H_+.

(b) Show that the geodesics in H_+ are identical with the Euclidean semicircles in H_+ perpendicular to the line $\operatorname{Im} z = 0$ (including the half-lines in H_+ perpendicular to $\operatorname{Im} z = 0$).

(c) Prove that

$$d(z_1, z_2) = \log \frac{1+\left|\dfrac{z_1-z_2}{z_1-\bar{z}_2}\right|}{1-\left|\dfrac{z_1-z_2}{z_1-\bar{z}_2}\right|}$$

and find an analogous formula for the metric on D. Prove that

$$d(i, a(t)i) = |t| \quad \text{for } a(t) = \begin{bmatrix} e^{t/2} & 0 \\ 0 & e^{-t/2} \end{bmatrix}.$$

(d) Prove that for an arbitrary $r > 0$ the orbit $V \cdot ri$ is a geodesic (see Problem 4 for the definition of the subgroup V). Compute the non-Euclidean distance between two arbitrary points in this orbit and prove that it satisfies the relation

$$\cosh \varrho(z, z_1) = 1 + \frac{1}{2} \frac{\varrho(z, z_1)^2}{\operatorname{Im}(z)\operatorname{Im}(z_1)},$$

$\varrho(z, z_1)$ denoting the Euclidean distance.

(e) Two geodesics in H_+ are said to be *parallel* provided they pass through the same point on the boundary (more precisely provided the corresponding Euclidean circles pass through the same point on the line $\operatorname{Im} z = 0$). Prove that any pencil of parallel horocycles is the set of trajectories orthogonal to a pencil of parallel geodesics.

(f) Prove that every geodesic in H_+ belonging to the pencil of parallels determined by the point $0 \in \boldsymbol{R}$ is of the form

$$t \to va(t)i, \qquad \text{where } v = \begin{bmatrix} 1 & 0 \\ x & 1 \end{bmatrix}, \ a(t) = \begin{bmatrix} e^{t/2} & 0 \\ 0 & 2e^{-t/2} \end{bmatrix}.$$

6. (a) Prove that every homographic mapping $z \to g \cdot z$ for $g \in \mathrm{SL}(2, \boldsymbol{C})$ has either one or two fixed points and moreover if $g \in \mathrm{SL}(2, \boldsymbol{R})$ then the fixed points are either both real or conjugate. Find an analogous condition for the fixed points of the homographic mappings corresponding to the elements of SU(1, 1).

(b) Prove that every homographic mapping corresponding to a $g \in \mathrm{SU}(1, 1)$ can be represented in the form

$$z \to w = e^{i\varphi} \frac{z+a}{1+\bar{a}z}, \qquad a \in D, \ \varphi \in \boldsymbol{R}.$$

(c) An element $g \in \mathrm{SL}(2, \boldsymbol{R})$ ($g \in \mathrm{SU}(1, 1)$) different from identity is said to be elliptic, hyperbolic or parabolic if it is conjugate to an element in K, A_1, N (in $\mathscr{K}, t(\mathscr{A}_1), \mathscr{N}$), respectively.

For $g \in \mathrm{SL}(2, \boldsymbol{R})$ prove that

(i) (g is hyperbolic) $\Leftrightarrow$ ($|\mathrm{tr}\, g| > 2$),

(ii) (g is elliptic) $\Leftrightarrow$ ($|\mathrm{tr}\, g| < 2$),

(iii) (g is parabolic) $\Leftrightarrow$ ($|\mathrm{tr}\, g| = 2$).

(d) Prove that every homographic mapping of disc D onto itself defined by an elliptic element is of the form $z \to w$, where w is given by the identity

$$\frac{w-z_0}{w-\bar{z}_0^{-1}} = e^{i\varphi} \frac{z-z_0}{z-\bar{z}_0^{-1}}, \qquad z_0 \in D.$$

Derive hence the equation of the non-Euclidean circle through the point $z \in D$ with centre z_0.

(e) Prove that the homographic mapping of the disc D given by a hyperbolic element is of the form $z \to w$ where

$$\frac{w-z_1}{w-z_2} = e^t \frac{z-z_1}{z-z_2}, \qquad |z_1| = |z_2| = 1, \ t \in \boldsymbol{R}.$$

Deduce hence that the subgroup $g\mathscr{A}g^{-1}$ preserves the unique geodesic in D and that its orbits are lines of constant distance from that geodesic (the so-called *equidistants*).

(f) Prove that the homographic mappings of the disc corresponding to parabolic elements, and only those mappings, have a unique fixed point which belongs to the circle $|z| = 1$. Find an expression for such mappings analogous to that given above.

(g) Given $z_0 \in D$ let L be the diameter of D through z_0 and denote by ζ one of the points of the intersection of L with the boundary of the disc. For $z \in L$ denote by $S(z)$ the non-Euclidean circle with centre z through the point z_0. Prove that the circles $S(z)$ converge as $z \to \zeta$ to the horocycle passing through z_0 and tangent at ζ to the boundary B of the disc.

7. Prove that $|\ln e_{-1}(g, k)|$ is the distance of the point 0 from the horocycle tangent to B at the point $k \cdot i$ and passing through $g \cdot 0$.

Hint. Prove that the distance is equal to the length of the geodesic passing through 0 and orthogonal to the horocycle.

8. Let Δ be the Laplace–Beltrami operator on D.

(a) Prove that in terms of the polar coordinates $z = \tanh \tau e^{i\theta}$ on D

$$\Delta = \frac{\partial^2}{\partial \tau^2} + \coth \tau \frac{\partial}{\partial \tau} + \frac{1}{\sinh^2 \tau} \frac{\partial^2}{\partial \theta^2}$$

and show that

$$\Delta \mathfrak{P}_{i\nu-\frac{1}{2}}(\cosh \tau) = -\left(\frac{1}{4} + \nu\right)^2 \mathfrak{P}_{i\nu-\frac{1}{2}}(\cosh \tau).$$

(b) Let $f \in \mathscr{C}_0^\infty(D)$. Prove that

$$(\Delta f, f) \leqslant -\tfrac{1}{4}\|f\|^2,$$

where the inner product $(\cdot\,,\,\cdot)$ and the norm $\|\cdot\|$ are those of the space $L^2(D, \mathrm{d}m(z))$.

Hint. (b) Apply the polar coordinates and integrate by parts with respect to τ.

9. (a) Prove that $\mathfrak{P}_\zeta$ is a solution of the Legendre equation

$$(z^2-1)\frac{\mathrm{d}^2 u}{\mathrm{d}z^2} + 2z\frac{\mathrm{d}u}{\mathrm{d}z} - \zeta(\zeta+1)u = 0.$$

(b) Investigate the analiticity of the function

$$z \to \frac{1}{2\pi} \int_{-\pi}^{\pi} [z + (z^2-1)^{\frac{1}{2}} \cos \varphi]^\zeta \mathrm{d}\varphi, \qquad \text{where} \quad \zeta \in C$$

and prove that after a suitable choice of the branch of the function $z \to (z^2-1)^{1/2}$ the integral above gives an analytic extension of the Legendre function $\mathfrak{P}_\zeta$ onto the half-plane $\operatorname{Re} z > 0$.

(c) Prove by choosing a suitable contour of integration for the integral in (b) that $\mathfrak{P}_\zeta$ is given by the Schläfli integral formula

$$\mathfrak{P}_\zeta(z) = \frac{1}{2\pi i}\int_C \frac{(t^2-1)^\zeta \mathrm{d}t}{2^\zeta (t-z)^{\zeta+1}}.$$

Prove that one can take as C an arbitrary contour enclosing the points z and 1 inside and leaving the point -1 outside.

(d) Prove the Murphy formula

$$\mathfrak{P}_\zeta(z) = F(\zeta+1, -\zeta; 1; \tfrac{1}{2}(1-z)),$$

where F is the hypergeometric function (cf. Chapter 13).

(e) Prove the following addition formula for the Legendre functions (the values τ, τ_1, τ_2, φ φ_2 being related by formulas (10.3.7))

$$\mathfrak{P}_\zeta(\cosh\tau) = \sum_{k=-\infty}^{\infty} \mathrm{e}^{-ik\varphi}\mathfrak{P}_\zeta^k(\cosh\tau_1)\mathfrak{P}_\zeta^{-k}(\cosh\tau_2).$$

10. (a) Prove that

$$\mathfrak{P}_\zeta(\cosh\tau) = \frac{2}{\pi}\mathrm{e}^{\zeta\tau}\int_0^{\frac{\pi}{2}} (\cos^2\varphi + \mathrm{e}^{-2\tau}\sin^2\varphi)\,\mathrm{d}\varphi$$

and deduce hence that, for $\operatorname{Re}\zeta > 0$, P_ζ is assymptotically equivalent

$$\mathfrak{P}_\zeta(\cosh\tau) \sim \frac{\Gamma\left(\zeta+\frac{1}{2}\right)}{\pi^{\frac{1}{2}}\Gamma(\zeta+1)}\mathrm{e}^{\zeta\tau}\,(\tau\to\infty).^1$$

(b) Let ω_l be the zonal spherical function of the representation U^σ for $\sigma = (0, l)$. Prove that ω_l is bounded if and only if $-1 \leqslant \operatorname{Re} l \leqslant 1$.

Hint. (a) Apply the Lebesgue dominated convergence theorem. (b) Deduce from (a) the necessity of this condition. To prove its sufficiency consider $\int f(g)\omega_l(g)\mathrm{d}g$ and prove that this is a continuous functional on $L^1(G)$. ω_l being K-invariant it suffices to consider this integral

---

[1] $f(x) \sim g(x)$ $(x \to a)$ if $\lim\limits_{x\to a}\dfrac{f(x)}{g(x)} = 1.$

only for the functions f bi-invariant under K and the required estimate can be obtained by expressing the integral in terms of the horocyclic coordinates and comparing it with the expression for the norm of f in those coordinates. Both norms can be expressed by means of the Abel transform (see Helgason and Johnson, 1969).

11. (a) Prove that (10.2.16) can be written in the form

$$\mathfrak{P}_\zeta^m(\cosh\tau) = \frac{\Gamma(\zeta+m+1)}{\pi\Gamma(\zeta+1)}\int_0^\pi(\cosh\tau+\sinh\tau\cos\varphi)\zeta\cos m\varphi\,\mathrm{d}\varphi.$$

(b) Prove that $\mathfrak{P}_\zeta^m$ are solutions of the associated Legendre equation for $n = 0$.

(c) Show that an analytic continuation of the associated Legendre function $\mathfrak{P}_\zeta^n$ is given by the Schläfli integral

$$\mathfrak{P}_\zeta^n(z) = \frac{\Gamma(\zeta+n+1)}{\Gamma(\zeta+1)}(z^2-1)^{\frac{1}{2}n}\frac{1}{2\pi i}\int_C\frac{(t^2-1)^\zeta}{2^\zeta(t-z)^{\zeta+n+1}}\,\mathrm{d}t,$$

where the contour C satisfies the same conditions as in (c) in the previous problem.

(d) Prove

$$\mathfrak{P}_\zeta^n(z) = \frac{\Gamma(\zeta+n+1)}{n!\,\Gamma(\zeta-n+1)}\left(\frac{z-1}{z+1}\right)^{\frac{1}{2}n}F(\zeta+1,\,-\zeta;\,1+n;\,\tfrac{1}{2}(1-z))$$

(F—the hypergeometric function) (cf. Chapter 13).

(e) Prove the addition formula for the associated Legendre functions

$$\mathrm{e}^{-im\varphi}P_\zeta^m(\cosh\tau) = \sum_{k=-\infty}^{\infty}\frac{\Gamma(\zeta+m+1)}{\Gamma(\zeta+k+1)}\mathrm{e}^{-ik\varphi_2}\mathfrak{P}_\zeta^{mk}(\cosh\tau_1)x$$

$$\times\,\mathfrak{P}_\zeta^k(\cosh\tau_2).$$

12. Define Tchebyshev polynomials by the formula $T_n(x) = \cos(n\arccos x)$. Prove the following formulas:

$$\mathfrak{P}_\zeta^n(\cosh\tau) = \frac{\Gamma(\zeta+n+1)}{\Gamma(\zeta+1)}$$

$$\times\,\frac{1}{2\pi}x\int_{-\tau}^{\tau}\frac{\exp((\zeta+\frac{1}{2})t)\,T_n\!\left(\dfrac{e^t-\cosh\tau}{\sinh\tau}\right)}{(\cosh^2\frac{1}{2}\tau-\cosh^2\frac{1}{2}t)^{\frac{1}{2}}}\,dt,$$

$$\mathfrak{P}_\zeta(\cosh\tau) = \frac{1}{\pi}\int_0^\tau \frac{\cosh(\zeta+\frac{1}{2})t}{(\cosh^2\frac{1}{2}\tau-\cosh^2\frac{1}{2}t)^{\frac{1}{2}}}\,dt.$$

Hint. The function $\mathfrak{P}_\zeta^n$ being analytic it follows that

$$\mathfrak{P}_\zeta^n(\cosh\tau) = \frac{1}{2\pi i}\int_C \left(\cosh\tau + \frac{z^2+1}{2z}\sinh\tau\right)^\zeta z^{n-1}\,dz,$$

where C is the circle of radius a; $1 < a < \cot\frac{1}{2}\tau$. Make the substitution $e^t = \cosh\tau + \dfrac{z^2-1}{2z}\sinh\tau$ and deform the contour thus obtained to the interval $[-\tau, \tau]$ ran over twice.

13. (a) Prove the following integral representation for the associated Legendre functions

$$\mathfrak{P}_\zeta^{-m}(\cosh\tau) = \frac{\sinh^m\tau}{2^m\pi^{\frac{1}{2}}\Gamma(\frac{1}{2}+m)} \times \int_0^\pi (\sin\varphi)^{2m}(\cosh\tau+\sinh\tau\cos\varphi)^{\zeta-m}\,d\varphi$$

valid for arbitrary complex ζ and positive integer m.

(b) For complex ζ and μ with $\operatorname{Re}\mu < \frac{1}{2}$ define

$$\mathfrak{P}_\zeta^\mu(z) = \frac{2^\mu(z^2-1)^{-\frac{\mu}{2}}}{\pi^{\frac{1}{2}}\Gamma(\frac{1}{2}-\mu)}\int_0^\pi (\sin\varphi)^{-2\mu}\left(z+\sqrt{(z^2-1)}\cos\varphi\right)^{\zeta+\mu}d\varphi$$

provided z is not a real number between -1 and 1. Expanding the integrand into power series in $\tanh\tau$ and integrating termwise prove that

$$\mathfrak{P}_\zeta^\mu(\cosh\tau) = \frac{2^\mu\cosh^{\zeta+\mu}\tau}{\Gamma(1-\mu)(\sinh^2\tau)^{\frac{\mu}{2}}} xF\left(-\frac{\zeta+\mu}{2}, \frac{1}{2}-\frac{\zeta+\mu}{2}; 1-\mu;\ \tanh^2\tau\right)$$

and thus infer that $\mathfrak{P}_\zeta^\mu(z)$ is a solution of associated Legendre differential equation

(c) For integer $n \geqslant 2$ and arbitrary complex ν put

$$f_\nu^n(\tau) = \frac{\Gamma\left(\frac{n}{2}\right)}{\pi^{\frac{1}{2}}\Gamma\left(\frac{n-1}{2}\right)} \int_0^\pi \sin^{n-2}\varphi(\cosh\tau - \sinh\tau\cos\varphi)^{-i\nu-\frac{n-1}{2}} \, d\varphi.$$

Prove that

$$\left(-\frac{1}{\sinh\tau}\frac{d}{d\tau}\right) f_\nu^n(\tau) = \frac{\left(\frac{n-1}{2}\right)^2 + \nu^2}{n} f_\nu^{n+2}(\tau)$$

and hence deduce formulas

$$f_\nu^{2m}(\tau) = \frac{2^{m-1}(m-1)!}{(\nu^2+(\frac{1}{2})^2)(\nu^2+(\frac{3}{2})^2)\dots(\nu^2+(m-\frac{3}{2})^2)} \times \left(\frac{-1}{\sinh\tau}\frac{d}{d\tau}\right)^{m-1} \mathfrak{P}_{\frac{1}{2}+i\nu}(\cosh\tau),$$

$$f_\nu^{2m+1}(\tau) = \frac{1 \cdot 3 \dots (2m-1)}{\nu^2(\nu^2+1^2)\dots(\nu^2+(m-1)^2)} \left(\frac{-1}{\sinh\tau}\frac{d}{d\tau}\right)^m \cos\nu\tau.$$

Hint. (a) In the formula of Problem 11(a) substitute $\cos(m\varphi) = T_m(\cos\varphi)$ where T_m is the m-th Tchebyshev polynomial. Use Rodrigues formula

$$T_m(x) = \frac{(-1)^m(1-x^2)^{\frac{1}{2}}\Gamma(\frac{1}{2})}{2^m\Gamma(\frac{1}{2}+m)} \frac{d^m}{dx^m}(1-x^2)^{m-\frac{1}{2}}$$

for integration by parts. (c) To prove the recurrence formula differen tiate the integral and integrate by parts.

14. Apply the addition formula (10.3.8) to prove that

(a) $\mathfrak{P}_{mn}^{\xi}(\cosh(\tau_1+\tau_2)) = \sum_{k=-\infty}^{\infty} \mathfrak{P}_{mk}^{\xi}(\cosh\tau_1)\mathfrak{P}_{kn}^{\xi}(\cosh\tau_2)$,

(b) $\mathfrak{P}_{mn}^{\xi}(\cosh(\tau_1-\tau_2)) = \sum_{k=-\infty}^{\infty} (-1)^{n-k}\mathfrak{P}_{mk}^{\xi}(\cosh\tau_1)\mathfrak{P}_{kn}^{\xi}(\cosh\tau_2)$,

(c) $\sum_{k=-\infty}^{\infty} \overline{\mathfrak{P}_{mk}^{-\frac{1}{2}+i\nu}(\cosh\tau)}\mathfrak{P}_{kn}^{-\frac{1}{2}-i\nu}(\cosh\tau) = \delta_{mn}$.

15. Derive from the addition formula (10.3.8) the mean value formula (10.3.10) for the Jacobi function.

16. Check that the Legendre functions P satisfy the equation

$$\frac{1}{2\pi}\int_0^{2\pi} \mathfrak{P}_\zeta(\cosh\tau_1\cosh\tau_2+\sinh\tau_1\sinh\tau_2\cos\varphi)e^{ik\varphi}d\varphi$$

$$= \mathfrak{P}_\zeta^k(\cosh\tau_1)\mathfrak{P}_\zeta^{-k}(\cosh\tau_2)$$

and deduce hence

$$\mathfrak{P}_\zeta^k(\cosh\tau_1)\mathfrak{P}_\zeta^{-k}(\cosh\tau_2)$$

$$= \frac{1}{\pi}\int_{|\tau_1-\tau_2|}^{\tau_1+\tau_2} \frac{\mathfrak{P}_\zeta(\cosh\tau)\, T_k\left(\dfrac{\cosh\tau-\cosh\tau_1\cosh\tau_2}{\sinh\tau_1\sinh\tau_2}\right)\sinh\tau\, d\tau}{\left[\left(\cosh(\tau_1+\tau_2)-\cosh\tau\right)\left(\cosh\tau-\cosh(\tau_1-\tau_2)\right)\right]^{\frac{1}{2}}}$$

17. Prove

(a) $(z\cosh\frac{1}{2}\tau+\sinh\frac{1}{2}\tau)^{\zeta+n}(z\sinh\frac{1}{2}\tau+\cosh\frac{1}{2}\tau)^{\zeta-n}$

$$= \sum_{m=-\infty}^{\infty} \mathfrak{P}_{mn}^\zeta(\cosh\tau)z^{\zeta+m},$$

(b) $\displaystyle\sum_{m=-\infty}^{+\infty} \mathfrak{P}_{mn}^\zeta(\cosh\tau) = e^{\zeta\tau},$

(c) $\displaystyle\frac{\cosh^\zeta\tau}{\Gamma(\zeta+1)} = \sum_{m=-\infty}^{\infty}\frac{i^m}{\Gamma(\zeta+m+1)}\mathfrak{P}_\zeta^m(\cosh\tau).$

18. Prove the following recurrence formulas by applying Problem 17:

(a) $\mathfrak{P}^{\zeta+\frac{1}{2}}_{m+\frac{1}{2},\,n+\frac{1}{2}}(\cosh\tau) = \cosh\frac{1}{2}\tau\,\mathfrak{P}_{mn}^\zeta(\cosh\tau)$
$+\sinh\frac{1}{2}\tau\,\mathfrak{P}_{m+1,n}^\zeta(\cosh\tau),$

(b) $\mathfrak{P}^{\zeta+\frac{1}{2}}_{m+\frac{1}{2},\,n-\frac{1}{2}}(\cosh\tau) = \sinh\frac{1}{2}\tau\,\mathfrak{P}_{mn}^\zeta(\cosh\tau)$
$+\cosh\frac{1}{2}\tau\,\mathfrak{P}_{m+1,n}(\cosh\tau),$

(c) $(\zeta+n)\cosh\frac{1}{2}\tau\,\mathfrak{P}^{\zeta-\frac{1}{2}}_{m-\frac{1}{2},\,n-\frac{1}{2}}(\cosh\tau)+(\zeta-n)\sinh\frac{1}{2}\tau\,\mathfrak{P}^{\zeta-\frac{1}{2}}_{m-\frac{1}{2},\,n-\frac{1}{2}}(\cosh\tau)$
$= (\zeta+m)\mathfrak{P}_{mn}^\zeta(\cosh\tau).$

Hint. To obtain (a) ((b), respectively) multiply both sides of identity (a) in Problem 17 by $z\cosh\frac{1}{2}\tau+\sinh\frac{1}{2}\tau z\sinh\frac{1}{2}\tau+\cosh\frac{1}{2}\tau,$

respectively) and expand in a series the left-hand side of the identity thus obtained.

To derive (c) differentiate both sides of (a) from Problem 17 with respect to z and apply a suitably modified form of the decomposition (a).

19. Let $\sigma = (\varepsilon, l)$ be such that $l+\varepsilon$ is an odd integer, say, $l+\varepsilon = 2k-1$. Further let $\mathscr{F}_+^\sigma$, $\mathscr{F}_-^\sigma$ be a subspace of E^σ determined by the condition

$$\mathscr{F}_+^\sigma = \left\{f \in \mathscr{E}^\sigma : f = \sum_{m=-\infty}^{-k} a_n e_n^\varepsilon\right\},$$

$$\mathscr{F}_-^\sigma = \left\{f \in \mathscr{E}^\sigma : f = \sum_{n=k-\varepsilon}^{\infty} a_n e_n^\varepsilon\right\}.$$

Check that $\mathscr{F}_+$ and $\mathscr{F}_-$ are invariant under $\mathscr{U}^\sigma$ and that $\mathscr{F}_+^\sigma \cap \mathscr{F}_-^\sigma = \{0\}$ if and only if $l+1 > 0$. Prove also that if $\sigma = (\varepsilon, l)$ satisfies the condition $l+1 > 0$ then the restriction of $\mathscr{U}^\sigma$ to $\mathscr{F}_+^\sigma$ and $\mathscr{F}_-^\sigma$ are irreducible. If $\mathscr{F}_+^\sigma \cap \mathscr{F}_-^\sigma \neq 0$ then the restriction of $\mathscr{U}^\sigma$ to $\mathscr{F}_+^\sigma \cap \mathscr{F}_-^\sigma$ is irreducible and $\mathscr{U}^\sigma$ induces an irreducible representation on both spaces $\mathscr{F}_+^\sigma / \mathscr{F}_+^\sigma \cap \mathscr{F}_-^\sigma$ and $\mathscr{F}_-^\sigma / \mathscr{F}_+^\sigma \cap \mathscr{F}_-^\sigma$.

Hint. To prove the invariance of the subspaces $\mathscr{F}_\pm^\sigma$ apply the explicit form of the operator $\mathscr{U}$ in the basis consisting of eigenvectors e_n^ε (§ 10.2). The proof of irreducibility follows the method presented in Section 4.

20. Let l be an integer and let $\varepsilon = 0$ if l is odd and $\varepsilon = 1$ if l is even. Prove that

(a) $\mathfrak{P}^{-\frac{1}{2}(l-1)}_{m+\frac{1}{2}\varepsilon, n+\frac{1}{2}\varepsilon}(z) \equiv 0$ if $l \geqslant 0$ and either $2n < -l-\varepsilon-1$ and $2m > l+1-\varepsilon$ or $l \geqslant 0$ or $2n > l+1-\varepsilon$ and $2m < -l-\varepsilon-1$.

(b) Let $l < 0$; impose conditions on m and n under which the function $\mathfrak{P}^{-\frac{1}{2}(l+1)}_{m+\frac{1}{2}\varepsilon, n+\frac{1}{2}}$ equals identically zero.

(c) The function $\mathfrak{P}^{-\frac{1}{2}(l+1)}_{m+\frac{1}{2}\varepsilon, n+\frac{1}{2}\varepsilon}(\cosh\tau)$ is a polynomial in $\sinh\frac{1}{2}\tau$ and $\tanh\frac{1}{2}\tau$ (for those values of parameters m and n for which the function is not identically zero). Find that polynomial.

Hint. (a) and (b). Apply Problem 19. Compute the integral in (10.2.14) by means of residua.

Chapter 11

Harmonic Analysis on the Lobatschevsky space

11.1. THE GROUP $\mathrm{SL}(2,\boldsymbol{C})$. INDUCED SPHERICAL REPRESENTATIONS

Throughout this chapter G will denote the group $\mathrm{SL}(2, \boldsymbol{C})$ of unimodular complex 2×2 matrices. It is a Lie group of real dimension 6.

The group acts transitively on the projective plane P^1 (i.e. on the compactified plane $\boldsymbol{C}^1$) by homographic transformations:

$$g \cdot z = \frac{\alpha z + \beta}{\gamma z + \delta}, \qquad g = \begin{bmatrix} \alpha & \beta \\ \gamma & \delta \end{bmatrix}.$$

The isotropy subgroup at the point ∞ is the subgroup of all triangular matrices

$$P = \left\{ \begin{bmatrix} \alpha & \beta \\ 0 & \alpha^{-1} \end{bmatrix}; \ \alpha \in \boldsymbol{C}_*, \quad \beta \in \boldsymbol{C} \right\}.$$

The group P is a semi-direct product of the multiplicative group $\boldsymbol{C}_*$ and the additive group $\boldsymbol{C}$ with the action of $\boldsymbol{C}_*$ on $\boldsymbol{C}$ by the automorphisms $\boldsymbol{R} \ni \beta \to \alpha^2 \beta$.

We also write $A_1 = \left\{ \begin{bmatrix} \alpha & 0 \\ 0 & \alpha^{-1} \end{bmatrix} : \alpha \in \boldsymbol{C}_* \right\}$, and $A = \left\{ \begin{bmatrix} e^{\frac{t}{2}} & 0 \\ 0 & e^{-\frac{t}{2}} \end{bmatrix} : t \in \boldsymbol{R} \right\}$. Put $B := G/P = P^1$. $\mathrm{SU}(2)$ is a compact subgroup of G which also acts transitively on $\boldsymbol{R}$ by homographies. The isotropy subgroup at zero equals $M = \mathrm{SU}(2) \cap P = \left\{ \begin{bmatrix} e^{\frac{1}{2} i\varphi} & 0 \\ 0 & e^{-\frac{1}{2} i\varphi} \end{bmatrix} \right\}$. Hence we also have $P^1 \cong \mathrm{SU}(2)/M \cong S^2$.

The group G acts on $\boldsymbol{C}^2$ in a natural way:

$$g \begin{bmatrix} z_1 \\ z_2 \end{bmatrix} = \begin{bmatrix} \alpha z_1 + \beta z_2 \\ \gamma z_1 + \delta z_2 \end{bmatrix}.$$

The action so defined is transitive on $C^2 - \{0\}$. The corresponding homogeneous space will be denoted by Ξ_0. By computing the isotropy group at the point $\begin{bmatrix}1\\0\end{bmatrix}$ we check that Ξ_0 is isomorphic to the homogeneous space G/N, where $N = \left\{\begin{bmatrix}1 & \beta\\ 0 & 1\end{bmatrix} : \beta \in C\right\}$.

The subgroup N is normal in P. Every element $\xi \in \Xi_0$ can be uniquely represented in the form $ka \cdot \begin{bmatrix}1\\0\end{bmatrix}$, where $k \in \mathrm{SU}(2)$ and a is a matrix of the form

$$a(t) = \begin{bmatrix} e^{\frac{1}{2}t} & 0 \\ 0 & e^{-\frac{1}{2}t}\end{bmatrix}. \tag{11.1.1}$$

Namely, if $\xi = \begin{bmatrix}\alpha\\ \beta\end{bmatrix}$ and $r := (|\alpha|^2 + |\beta|^2)^{\frac{1}{2}}$ then we take

$$a = \begin{bmatrix} r & 0\\ 0 & r^{-1}\end{bmatrix} \quad \text{and} \quad k = \frac{1}{r}\begin{bmatrix}\alpha & \beta\\ -\bar{\beta} & \bar{\alpha}\end{bmatrix}.$$

This leads to the following

PROPOSITION 11.1.1. *The group G admits the Iwasawa decomposition $G = KAN$.*

The proof is identical to that of Proposition 10.1.1.

As in the case of the group $\mathrm{SL}(2; \boldsymbol{R})$ we shall examine the representations of G induced by the character of the group MAN.

Since, at present, our main concern is analysis on the symmetric space of the group G, we shall restrict ourselves to a description of that series of spherical representation which are induced by identity characters on M. The subgroup N is the commutator of AN, which means that it is spanned by the elements of the form $aba^{-1}b^{-1}$, a, $b \in AN$. Therefore every character assumes value 1 on N. The characters which induce representations will be parametrized by a complex parameter l:

$$\chi^l(p) := |\alpha|^l \quad \text{for } p = \begin{bmatrix}\alpha & \beta\\ 0 & \alpha^{-1}\end{bmatrix}.$$

The modular function δ of the subgroup P is of the form

$$P \ni \begin{bmatrix} \alpha & \beta \\ 0 & \alpha^{-1} \end{bmatrix} \to |\alpha|^{-4}.$$

The representation space of the induced representation H^l is comprised of smooth functions on G such that

$$f(gman) = \delta^{\frac{1}{2}}\chi^{-l}(a)f(g) = |\alpha|^{-l-2}f(g). \tag{11.1.2}$$

The function f being N-invariant it can be dropped onto $\Xi_0 = G/N$:

$$\tilde{f}\left(g \cdot \begin{bmatrix} 1 \\ 0 \end{bmatrix}\right) := f(g).$$

Condition (11.1.2) can be written in the following form:

$$\tilde{f}(\alpha z_1, \alpha z_2) = |\alpha|^{-l-2}\tilde{f}(z_1, z_2). \tag{11.1.3}$$

Thus the space H^l consists of positively homogeneous functions on $\boldsymbol{C}^2$ of order $-(l+2)$. Such functions are completely determined by their values on the sphere S^3 in $\boldsymbol{C}^2$ and moreover satisfy on S^3 the condition

$$f(\alpha z_1, \alpha z_2) = f(z_1, z_2) \qquad \text{for } |\alpha| = 1.$$

The induced representation U^{χ^l} is intertwined by the operator $f \to f|S^3$ with the following representation

$$\begin{aligned} U^l_g f(z_1, z_2) &= f(\delta z_1 - \beta z_2, -\gamma z_1 + \alpha z_2) \\ &= r(g^{-1}; z_1, z_2)^{-l-2}\tilde{f}\left(\frac{\delta z_1 - \beta z_2}{r}, \frac{-\gamma z_1 + \alpha z_2}{r}\right), \end{aligned} \tag{11.1.4}$$

where

$$r(g^{-1}; z_1, z_2) := (|\delta z_1 - \beta z_2|^2 + |\gamma z_1 - \alpha z_2|^{1/2}),$$

$r(\cdot; \cdot)$ is a multiplier on $G \times S^3$ with the action of G on S^3 corresponding to the action on G/AN, i.e. given by the formula

$$g \cdot (z_1, z_2) = \frac{1}{r}(\alpha z_1 + \beta z_2, \gamma z_1 + \delta z_2).$$

For the time being we shall compute only the zonal spherical functions of the representations U^l by applying the inner product which is the transport of that on the space $L^2(\mathrm{SU}(2))$.

$$(\tilde{f}_1, \tilde{f}_2) = \int_{\mathrm{SU}(2)} \bar{\tilde{f}}_1(k\cdot(1,0))\tilde{f}_2(k\cdot(1,0))\,\mathrm{d}k$$

$$= \frac{1}{4\pi}\int_0^{2\pi}\int_0^{\pi} \bar{\tilde{f}}_1\tilde{f}_2\left(\cos\tfrac{1}{2}\theta\,\mathrm{e}^{\frac{1}{2}i\varphi}, i\sin\tfrac{1}{2}\theta\,\mathrm{e}^{-\frac{1}{2}i\varphi}\right)\sin\theta\,\mathrm{d}\varphi\,\mathrm{d}\theta. \tag{11.1.5}$$

For $a(t) = \begin{bmatrix} \mathrm{e}^{\frac{1}{2}t} & 0 \\ 0 & \mathrm{e}^{-\frac{1}{2}t} \end{bmatrix}$ we shall compute the matrix element $u^l_{00}(a(t))$ $:= (f_0, U^l_{a(t)} f_0)$, where $f_0 \equiv 1$. We find

$$u^l_{00}(a(t)) = \frac{1}{4\pi}\int_0^{2\pi}\int_0^{\pi}(\mathrm{e}^{-t}\cos^2\tfrac{1}{2}\theta + \mathrm{e}^{t}\sin^2\tfrac{1}{2}\theta)^{\frac{1}{2}(-l-2)}\sin\theta\,\mathrm{d}\varphi\,\mathrm{d}\theta$$

$$= \frac{1}{2}\int_0^{\pi}(\cosh t - \sinh t\cos\theta)^{-\frac{1}{2}l-1}\sin\theta\,\mathrm{d}\theta$$

$$= \frac{1}{2}\int_{-1}^{1}(\cosh t + y\sinh t)^{-\frac{1}{2}l-1}\,\mathrm{d}y = \frac{2\sinh\frac{l}{2}t}{l\sinh t}.$$

In particular, for $l = iv$, $v \in \boldsymbol{R}$ we obtain

$$u^{iv}_{00}(a(t)) = \frac{2\sin\frac{t}{2}v}{v\sinh t}.$$

Obviously for every representation U^l the inner product (11.1.5) is invariant under the action of the group K. Moreover, the following relations hold identically with respect to l and g:

PROPOSITION 11.1.2. $(U^l_g f_1 | U^{-\bar{l}}_g f_2) = (f_1 | f_2)$.

PROPOSITION 11.1.3. $\langle U^l_g f_1 | U^{-l}_g f_2\rangle = \langle f_1 | f_2\rangle$, *where* $\langle\cdot|\cdot\rangle$ *is a symmetric bilinear from defined as* $\langle f_1 | f_2\rangle := (\bar{f}_1, f_2)$.

The proof of both these statements is a repetition of the reasoning followed in the proof of Proposition 9.3.1 and thus reduces to verifying that

$$S_\varrho(g, k) = \delta(h(gk))$$

is the Radon–Nikodym derivative of the measure $dg^{-1} \cdot k$ with respect to dk. By $h(g)$ we denote, as in the previous section, the constituent a in the Iwasawa decomposition $g = kan$. The form of the Haar measure on the group G is provided by formula (10.2.14).

COROLLARY 11.1.4. *The representation U^l is unitary with respect to the product* $(\cdot \mid \cdot) \Leftrightarrow l = i\nu,\ \nu \in \boldsymbol{R}$.

Anticipating a little the later course of our exposition we claim that the representations $U^{i\nu}$, $\nu \in \boldsymbol{R}$ are irreducible. Since unitary spherical representations are in a 1–1 correspondence with positive definite spherical functions, we obtain, in view of what has been said above.

COROLLARY 11.1.5. *The representation $U^{i\nu}$ and $U^{-i\nu}$ are equivalent.*

Indeed, we have $u^{i\nu}_{00} = \overline{u^{i\nu}_{00}} = u^{-i\nu}_{00}$.

11.2. ON THE STRUCTURE OF THE LOBATSCHEVSKY SPACE

We shall present a number of models of the symmetric space of the group $G = \mathrm{SL}(2; \boldsymbol{C})$. Let $\mathfrak{X}$ be the set of complex unimodular positive definite 2×2 matrices. Every element of the space $\mathfrak{X}$ is of the form

$$x = \begin{bmatrix} a & b \\ \bar{b} & c \end{bmatrix}, \qquad a, c > 0, \qquad ac - |b|^2 = 1. \tag{11.2.1}$$

We define an action of the group on $\mathfrak{X}$:

$$g \circ x := gxg^*, \qquad \text{where } \begin{bmatrix} \alpha & \beta \\ \gamma & \delta \end{bmatrix}^* := \begin{bmatrix} \bar{\alpha} & \bar{\gamma} \\ \bar{\beta} & \bar{\delta} \end{bmatrix}.$$

The group G act transitively on $\mathfrak{X}$. The solution of the equation

$$gxg^* = e = \begin{bmatrix} 1 & 0 \\ 0 & 1 \end{bmatrix}$$

is the matrix $g = x^{-\frac{1}{2}}$. The isotropy subgroup at the element $e \in P$ is the set of all matrices satisfying the condition $gg^* = e$, i.e. of all unitary matrices. Thus $\mathfrak{X} = G/K$.

Next, let us imbed $\mathfrak{X}$ in the 4-dimensional real space by means of the mapping

$$\mathfrak{X} \ni x \to \left(\frac{a+c}{2}, \ -\mathrm{Im}\, b, \ \mathrm{Re}\, b, \ \frac{c-a}{2}\right)$$
$$=: p(x) =: (p_0, p_1, p_2, p_3) \in \boldsymbol{R}^4. \tag{11.2.2}$$

The unimodularity condition for a matrix x now takes the form

$$p_0^2 - p_1^2 - p_2^2 - p_3^2 = 1 \tag{11.2.3}$$

and a, $c > 0$ means that $p_0 > 0$.

The image of $\mathfrak{X}$ under this mapping is the upper hyperboloid $\mathscr{H}$ in $\boldsymbol{R}^4$.

The quotient manifold $G/K \simeq \mathfrak{X} \simeq \mathscr{H}$ is called the *Lobatschevsky space*.

Formula (11.2.2) can be regarded as a mapping of the space of all hermitian 2×2 matrices onto $\boldsymbol{R}^4$. The canonical basis for $\boldsymbol{R}^4$ is obtained as the image of the Pauli matrices:

$$\frac{1}{2}\begin{bmatrix} 1 & 0 \\ 0 & 1 \end{bmatrix}, \quad \frac{1}{2}\begin{bmatrix} 0 & -i \\ i & 0 \end{bmatrix}, \quad \frac{1}{2}\begin{bmatrix} 0 & 1 \\ 1 & 0 \end{bmatrix}, \quad \frac{1}{2}\begin{bmatrix} -1 & 0 \\ 0 & 1 \end{bmatrix}.$$

Under the action of the mapping p representation (11.2.1) of the group SL(2, $\boldsymbol{C}$) on the space of hermitian matrices transforms onto the following representation of the group on the space $\boldsymbol{R}^4$:

$$\tau(g) = \begin{bmatrix} \frac{1}{2}(|\alpha|^2+|\beta|^2+|\gamma|^2+|\delta|^2) & \mathrm{Im}(\alpha\bar{\beta}+\gamma\bar{\delta}) & \mathrm{Re}(\alpha\bar{\beta}+\gamma\bar{\delta}) & \frac{1}{2}(|\beta|^2+|\delta|^2-|\alpha|^2-|\gamma|^2) \\ \mathrm{Im}(\bar{\alpha}\gamma+\bar{\beta}\delta) & \mathrm{Re}(\bar{\alpha}\delta-\bar{\beta}\gamma) & -\mathrm{Im}(\beta\bar{\gamma}+\delta\bar{\alpha}) & \mathrm{Im}(\bar{\alpha}\gamma-\bar{\beta}\delta) \\ \mathrm{Re}(\bar{\alpha}\gamma+\bar{\beta}\delta) & \mathrm{Im}(\bar{\beta}\gamma-\bar{\alpha}\delta) & \mathrm{Re}(\beta\bar{\gamma}+\delta\bar{\alpha}) & \mathrm{Re}(\bar{\beta}\delta-\alpha\bar{\gamma}) \\ \frac{1}{2}(|\gamma|^2+|\delta|^2-|\alpha|^2-|\beta|^2) & \mathrm{Im}(\bar{\gamma}\delta-\bar{\alpha}\beta) & \mathrm{Re}(\bar{\gamma}\delta-\bar{\alpha}\beta) & \frac{1}{2}(|\alpha|^2+|\delta|^2-|\beta|^2-|\gamma|^2) \end{bmatrix},$$

We prove by direct calculation that

$$\tau(g)p(x) = p(g \circ x). \tag{11.2.4}$$

The kernel of the homomorphism τ consists of two elements

$$\left\{\begin{bmatrix}1 & 0\\ 0 & 1\end{bmatrix}, \begin{bmatrix}-1 & 0\\ 0 & 1\end{bmatrix}\right\}.$$

Thus the image of τ is a 6-dimensional connected subgroup of GL(4; $\boldsymbol{R}$) consisting of all matrices preserving the bilinear symmetric form on $\boldsymbol{R}^4$

$$[p|q] := p_0 q_0 - p_1 q_1 - p_2 q_1 - p_3 q_3, \tag{11.2.5}$$

and such that the sign of p_0 remains unchanged. So the image of τ coincides with Λ_+—*the proper Lorentz group*. We have thus constructed a commutative diagram

$$\begin{array}{ccc} SL(2, \boldsymbol{C}) & \xrightarrow{\tau} & \Lambda_+ \\ \pi\downarrow & & \downarrow\pi \\ \mathfrak{X} & \xrightarrow{p} & \mathscr{H} \end{array}$$

where π denotes natural projections onto homogeneous spaces. From the point of view of harmonic analysis on homogeneous spaces the pairs (SL(2; $\boldsymbol{C}$), SU(2)) and (Λ_+, SO(3)) are equivalent. The kernel of the homomorphism τ being contained in SU(2) implies that the mapping p: $\mathfrak{X} \to \mathscr{H}$ is a diffeomorphism which can be applied to transport functions from $\mathscr{H}$ to $\mathfrak{X}$:

$$F(x) := f(p(x))$$

for an arbitrary function f on $\mathscr{H}$.

A vector-function on the group SL(2, $\boldsymbol{C}$) can be transported into Λ_+ provided the function is constant on the cosets with respect to ker τ. This holds, in particular, for the representations U^l, and therefore we can define a representation of the group Λ_+ by setting

$$\mathscr{U}^l_{\tau(g)} := U^l_g.$$

The zonal spherical functions u^l_{00} are bi-invariant under the subgroup K and consequently can be transported to the quotient space G/K. The transport of these functions to Λ_+ leads to functions which are matrix elements of the representations $\mathscr{U}^l$. To simplify notation we shall denote all these functions by the same symbol (§ 11.3). Regarded as a representation of the group Λ_+, $\mathscr{U}^l$ can be given a simple interpretation. Namely, the multiplier corresponding to this representation plays the role of a plane wave on the Lobatschevsky space.

We shall find the images of the subgroups K, A_1, N of the group $SL(2; \boldsymbol{C})$ under the homomorphism τ. We have

$$\tau\left(\begin{bmatrix} \alpha & 0 \\ 0 & \alpha^{-1} \end{bmatrix}\right) = \begin{bmatrix} \frac{1}{2}(|\alpha|^2+|\alpha|^{-2}) & 0 & 0 & \frac{1}{2}(|\alpha|^{-2}-|\alpha|^2) \\ 0 & \operatorname{Re}\frac{\bar{\alpha}}{\alpha} & -\operatorname{Im}\frac{\bar{\alpha}}{\alpha} & 0 \\ 0 & -\operatorname{Im}\frac{\bar{\alpha}}{\alpha} & \operatorname{Re}\frac{\alpha}{\bar{\alpha}} & 0 \\ \frac{1}{2}(|\alpha|^{-2}-|\alpha|^2) & 0 & 0 & \frac{1}{2}(|\alpha|^2+|\alpha|^{-2}) \end{bmatrix}, \tag{11.2.6}$$

$$\tau\left(\begin{bmatrix} 1 & \beta \\ 0 & 1 \end{bmatrix}\right) = \begin{bmatrix} 1+\frac{1}{2}|\beta|^2 & -\operatorname{Im}\beta & \operatorname{Re}\beta & \frac{1}{2}|\beta|^2 \\ -\operatorname{Im}\beta & 1 & 0 & -\operatorname{Im}\beta \\ \operatorname{Re}\beta & 0 & 1 & \operatorname{Re}\beta \\ -\frac{1}{2}|\beta|^2 & -\operatorname{Im}\beta & -\operatorname{Re}\beta & 1-\frac{1}{2}|\beta|^2 \end{bmatrix}, \tag{11.2.7}$$

$$\tau\left(\begin{bmatrix} \alpha & \beta \\ -\bar{\beta} & \bar{\alpha} \end{bmatrix}\right) = \begin{bmatrix} 1 & 0 & 0 & 0 \\ 0 & \operatorname{Re}(\bar{\alpha}^2+\beta^2) & \operatorname{Im}(\beta^2-\bar{\alpha}^2) & 2\operatorname{Im}\alpha\beta \\ 0 & -\operatorname{Im}(\bar{\beta}^2+\bar{\alpha}^2) & \operatorname{Re}(\alpha^2-\beta^2) & 2\operatorname{Re}\alpha\beta \\ 0 & -2\operatorname{Im}\beta\bar{\alpha} & -2\operatorname{Re}\bar{\alpha}\beta & |\alpha|^2-|\beta|^2 \end{bmatrix}. \tag{11.2.8}$$

The group $\mathscr{K} = \tau(K)$ is isomorphic to the group of rotations in the hyperplane $p_0 = 0$. It also coincides with the isotropy group at the point $w = (1, 0, 0, 0)$ in the hyperboloid $\mathscr{H}$.

The group acts transitively on the submanifold $S_r(w) = \{p \in \mathscr{H}: p_0 = \cosh r\}$, which is isomorphic to a sphere in $\boldsymbol{R}^3$.

It follows from formula (11.2.6) that the orbit of the point w under the action of the group $\mathscr{A} := \tau(A)$ is a branch of the hyperbola $p_1 = p_2 = 0$, $p_0^2 - p_3^2 = 1$.

Every point $x \in \mathscr{H}$ can be represented as the image of the vertex w of the hyperboloid under the action of ka, $k \in \tau(K) =: \mathscr{K}$, $a \in \mathscr{A}$. Namely, to a point of the form $p = (\cosh t, p_1, p_2, p_3)$, we assign $a(t) = \tau\left(\begin{bmatrix} e^{\frac{1}{2}t} & 0 \\ 0 & e^{-\frac{1}{2}t} \end{bmatrix}\right)$ and an element $k \in \mathscr{H}$ which maps the point

$(\cosh t, 0, 0, -\sinh t)$ onto $(\cosh t, p_1, p_2, p_3)$. Such an element k is not uniquely determined: as follows from (11.2.8) or (11.2.6), the elements $k = \tau\left(\begin{bmatrix} \alpha & 0 \\ 0 & \bar{\alpha} \end{bmatrix}\right)$ for $|\alpha| = 1$ constitute the isotropy subgroup at the point $(\cosh t, 0, 0, -\sinh t)$. If we apply the desrciption of $K = SU(2)$ in terms of the Euler angles (φ, θ, ψ), we see that the result of the action of k on $a(t) \cdot w$ is independent of ψ.

We shall now parametrize $\mathscr{H}$ by applying the "geographical" coordinates on the sphere.

PROPOSITION 11.2.1. *The assignment*

$$R_+ \times]0, \pi[\times [0, 2\pi[\ni (t, \theta, \varphi) \to \begin{bmatrix} \cosh t \\ \sinh t \sin\theta \cos\varphi \\ \sinh t \sin\theta \cos\varphi \\ \sinh t \cos\theta \end{bmatrix} \in \mathscr{H}$$

defines a continuous bijective mapping which is a diffeomorphism when restricted to the interior of the domain. The image equals $\mathscr{H} \setminus \{\mathscr{A} w\}$.

The coordinates (φ, θ, t) will be referred to as the polar coordinates. We introduce a distance on the space $\mathscr{H}$ by putting

$$\cosh d(p, q) = [p|q].$$

The function d is well defined since the points in the hyperboloid satisfy the inequality $[x|y] \geqslant 1$. On the basis of the Schwarz inequality

$$1 + p_1 q_1 + p_2 q_2 + p_3 q_3 \leqslant \left(1 + \sum_1^3 p_j^2\right)^{\frac{1}{2}} \left(1 + \sum_1^3 q_j^2\right)^{\frac{1}{2}} = p_0 q_0 .$$

Hence if $[p|p] = [q|q] = 1$ then $[p|q] \geqslant 1$.

The Schwarz inequality passes into equality if and only if p is colinear with q. Since the lines passing through zero in $\boldsymbol{R}^4$ intersect $\mathscr{H}$ at most at one point we conclude that the equality $d(p, q) = 0$ implies $p = q$.

The distance thus defined is invariant under group action. By a sphere of radius r with centre p we shall understand the set $S_r(p) = \{q \in \mathscr{H}: d(p, q) = r\}$, $r > 0$. Spheres with centre at a point $x \in \mathscr{H}$ coincide with the intersections of $\mathscr{H}$ by the planes $[x|q] = \text{const}$.

Proposition 11.2.2. *Spheres with centre at $g \cdot x$ are orbits of the subgroup $g\mathscr{K}g^{-1}$.*

Proof. We know that $S_r(w)$ is an orbit of the subgroup $\mathscr{K}$. The distance being invariant we have

$$S_r(g \cdot w) = \{y \colon d(y, g \cdot w) = r = \{y \colon d(g^{-1}y, w) = r\}.$$

Hence there exists a $p \in \mathscr{H}$ such that $g^{-1}S_r(g \cdot w) = p$, i.e.

$$S_r(g \cdot w) = g\mathscr{K}p = g\mathscr{K}g^{-1}g \cdot p = g\mathscr{K}g^{-1} \cdot q. \qquad \square$$

Thus we can equivalently define spheres in $\mathscr{H}$ in the group-theoretic sense as the orbits of the maximal compact subgroups of the group Λ_+. Similarly, geodesics in the manifold $\mathscr{H}$ can be viewed as the orbits of subgroups conjugate to $\mathscr{A} = \tau(A)$.

Proposition 11.2.3. *A geodesic passing through a $p \in \mathscr{H}$ intersects every sphere with centre p at two points p_1, p_2. The diameter of the sphere is equal to $d(p_1, p_2)$.*

Proof. The distance being invariant under group action, it suffices to consider the orbit under the action of $\mathscr{A}$ of the point $p = w$. By (11.2.6) we obtain $a(t)w = [\cosh t, 0, 0, -\sinh t]$. The geodesic intersects the sphere of radius r with centre w at the points $p' = a(r)w$ and $p'' = a(-r)w$. We also find that

$$[a(t)w|a(t')w] = \cosh t \cosh t' - \sinh t \sinh t' = \cosh(t-t'), \tag{11.2.9}$$

which proves that $d(p', p'') = 2r$. $\square$

From Proposition 11.2.3 (and formula (11.2.9)) we infer that the geometry of the space $\mathscr{H}$ induces on the geodesics $g\mathscr{A}g^{-1}p$ the "usual" Euclidean geometry.

Now we shall introduce geometric objects in $\mathscr{H}$ which correspond to hyperplanes in the Euclidean geometry.

By a *horocycle* in $\mathscr{H}$ we mean the orbit of a point under the action of an arbitrary subgroup conjugate to $\mathscr{N} = \tau(N)$. The set of horocycles corresponding to a fixed subgroup $g\mathscr{N}g^{-1}$ is called the *pencil of parallel horocycles.*

We shall first investigate the family of horocycles corresponding to the subgroup $\mathscr{N}$. On the basis of Proposition 11.1.1 we infer that every element $p \in \mathscr{H}$ is uniquely defined by a pair $(n, a) \in \mathscr{N} \times \mathscr{A}$ such that

$$p = na \cdot w. \tag{11.2.10}$$

It follows that parallel horocycles are either disjoint or identical. The family of parallel horocycles is parametrized by the elements of the group $\mathscr{A}$.

We can introduce new coordinates in $\mathscr{H}$, this time global ones.

PROPOSITION 11.2.4. *The mapping*

$$R \times C \ni (t, \beta) \to \tau\left(\begin{bmatrix} e^{\frac{1}{2}t} & 0 \\ 0 & e^{-\frac{1}{2}t} \end{bmatrix}\right) \cdot w \in \mathscr{H}$$

is a diffeomorphism.

We shall find the form of the horocycle $\xi_0 := \mathscr{N} \cdot w$ by applying formula (11.2.7). We have

$$C^1 \ni \beta \to n(\beta) \cdot w = (1 + \tfrac{1}{2}|\beta|^2, \ -\mathrm{Im}\beta, \mathrm{Re}\beta, \ -\tfrac{1}{2}|\beta|^2).$$

Hence this set is a section of the hyperboloid $\mathscr{H}$ by a plane given by the equation $p_0 + p_3 = 1$. This equation can be written in the form

$$[p|s_0] = 1, \tag{11.2.11}$$

where $s_0 = (1, 0, 0, -1)$ is a vector of the "light" cone $[p|p] = 0$. The set $g \cdot \xi_0$, which is the orbit of the point $g \cdot w$ under the group $g\mathscr{N}g^{-1}$, coincides with the section of $\mathscr{H}$ by the hyperplane

$$[p|g^{-1} \cdot s_0] = 1. \tag{11.2.12}$$

For instance, the family of parallel horocycles of the form $a \cdot \xi_0$, $a \in \mathscr{A}$ has an equation of the form

$$x_0 + x_3 = \text{const}.$$

The group Λ_+ acts transitively on $S_+ = \{p\colon [p|p] = 0, p_0 > 0\}$, i.e. on the upper part of the "light" cone. Theerfore we can identify the space of horocycles with the set S_+.

PROPOSITION 11.2.5. $\Xi = S_+ \simeq \Lambda_+/\mathscr{M}\mathscr{N}$, *where* $\mathscr{M} = \tau(M)$.

Proof. It is enough to verify that the stability subgroup at a certain point in the cone is equal to $\mathscr{M}\mathscr{N}$. Let $s_0 = (1, 0, 0, -1)$. We already know that $\mathscr{N}s_0 = s_0$.

It remains to describe pairs (k, a) such that $ka \cdot s_0 = s_0$. The subgroup $\mathscr{A}$ moves every element $s \in S_+$ along the radius, and so the elements of the required stability group must be rotations. But only rotations about the axis x_3 which are of the form $\tau(m)$, $m \in M$ leave the point s_0 fixed. This ends the proof. □

The description of the space of horocycles provided by the proposition resembles a description of the set of hypersurfaces in $\boldsymbol{R}^4$. By choosing an element $b = k\mathscr{M} \in \mathscr{K}/\mathscr{M} = S^2$ we determine the family of parallel horocycles which contains $\xi = a(t)k\xi_0$. The parameter a defines the position of a horocycle in the pencil. In particular $|t|$ is the distance between the horocycle and the point w, equal to the distance between ξ_0 and ξ.

We shall express the invariant integral on $\mathscr{H}$ in polar coordinates

$$\int_{\mathscr{H}} f\mathrm{d}m = \frac{1}{4\pi}\int_0^\infty\int_0^\pi\int_0^{2\pi} f(\varphi, \theta, t)\sin\theta\sinh^2 t\,\mathrm{d}\varphi\,\mathrm{d}\theta\,\mathrm{d}t$$

$$= \frac{1}{2}i\int_C\int_R f(a(t)n(\beta)\cdot w)\,\mathrm{d}\beta\wedge\mathrm{d}\bar{\beta}\,\mathrm{d}t. \tag{11.2.13}$$

Both these formulas can be derived from the following expression for the measure $\mathrm{d}m$:

$$\mathrm{d}m(p) = \frac{\mathrm{d}p_1\mathrm{d}p_2\mathrm{d}p_3}{p_0} = \frac{\mathrm{d}p_1\mathrm{d}p_2\mathrm{d}p_3}{\sqrt{1+p_1^2+p_2^2+p_3^2}}.$$

The invariant measure on G expressed in terms of the coordinates of the Iwasawa decomposition has the form

$$\mathrm{d}g = \frac{i}{16\pi^2}\mathrm{e}^{2t}\mathrm{d}\varphi\,\mathrm{d}\theta\,\mathrm{d}\psi\,\mathrm{d}t\,\mathrm{d}\beta\wedge\mathrm{d}\bar{\beta}, \tag{11.2.14}$$

which also implies (11.2.13).

11.3. The Spherical Fourier Transformation on $\mathscr{H}$

This section is devoted to the study of integral operators on the space $\mathscr{D}(\mathscr{K}\backslash\mathscr{H})$ of functions of class $\mathscr{D}(\mathscr{H})$, having the property that $f(k\cdot p) = f(p)$, $k \in \mathscr{K}$, $p \in \mathscr{H}$. Such functions are invariant under rotations about the axis x_0 and depend only upon the distance $d(p, w)$.

In terms of polar coordinates a function $f \in \mathscr{D}(\mathscr{K}\backslash\mathscr{H})$ is a symmetric function of t.

By ω_ν, $\nu \in C^1$ we denote the spherical function of the representation $U^{2i\nu}$. ω_ν is a bi-invariant function on the group and so it projects onto $\mathscr{H} \simeq G/\mathscr{K}$. The function so obtained is denoted by the same symbol,

$$\omega_\nu(g \cdot \mathscr{K}) = (x_0 | U_g^{2i\nu} x_0), \qquad \omega_\nu\big(a(t)K\big) = \frac{\sin t\nu}{\nu \sinh t}, \tag{11.3.1}$$

where x_0 is a normalized $\mathscr{K}$-fixed vector in the representation space of $U^{2i\nu}$.

Definition. By the *spherical Fourier transform of a function* $f \in \mathscr{D}(\mathscr{K}\backslash\mathscr{H})$ we mean the function defined on C by the formula

$$\hat{f}(\nu) = \int_{\mathscr{H}} f(p)\omega_\nu(p)\mathrm{d}m(p). \tag{11.3.2}$$

In terms of polar coordinates, (11.3.2) assumes the form

$$\begin{aligned}\hat{f}(\nu) &- \frac{1}{\nu}\int_0^\infty \sinh t \sin t\nu f\big(a(t)\mathscr{K}\big)\mathrm{d}t \\ &= \frac{1}{2\nu}\int_{\infty -}^{\infty} \sinh t f\big(a(t)\mathscr{K}\big)\mathrm{e}^{i\nu t}\mathrm{d}t\end{aligned} \tag{11.3.3}$$

owing to the symmetry of the function $t \to f\big(a(t)\mathscr{K}\big)$.

By applying the inversion formula for the classical Fourier transformation, we get

$$f\big(a(t)K\big)\sinh t = \frac{1}{\pi}\int_{-\infty}^{\infty} \nu\hat{f}(\nu)\mathrm{e}^{-i\nu t}\mathrm{d}\nu,$$

which leads to the formula

$$f\big(a(t)K\big) = \frac{2}{\pi}\int_0^\infty \nu^2\hat{f}(\nu)\omega_\nu\big(a(t)\mathscr{K}\big)\mathrm{d}\nu. \tag{11.3.4}$$

This is the inversion formula for the spherical Fourier transformation.

To complete the above results we shall prove

THEOREM 11.3.1. *Let $f \in \mathscr{D}(\mathscr{K}\backslash\mathscr{H})$. Then*

$$f(p) = \frac{2}{\pi}\int_0^\infty \hat{f}(\nu)\omega_\nu(p)\nu^2 \mathrm{d}\nu. \tag{11.3.5}$$

Moreover, for arbitrary $f_1, f_2 \in \mathscr{D}(\mathscr{K}\backslash\mathscr{H})$ the following Plancherel formula holds:

$$(f_1|f_2)_{L^2(\mathscr{H})} = \frac{2}{\pi}\int_0^\infty \nu^2\hat{f}_1(\nu)\hat{f}_2(\nu)\mathrm{d}\nu. \tag{11.3.6}$$

Proof. Formula (11.3.5) follows from (11.3.4). Next, keeping in mind that $\omega_\nu(f_1^* * f_2) = \overline{\omega_\nu(f_1)}\omega_\nu(f_2) = \hat{f}_1$, we obtain in view of (11.3.5) the Plancherel formula (11.3.6).

11.4. DECOMPOSITION INTO PLANE WAVES ON $\mathscr{H}$

The spherical Fourier transformation on $\mathscr{H}$ applies to a decomposition of a spherically symmetric function into "elementary" functions of this type, i.e. zonal spherical functions of the representations $\mathscr{U}_{i\nu}$.

We are now concerned with decompositions of "arbitrary" functions on $\mathscr{H}$ into elementary components of various types.

We shall obtain, consecutively, decompositions with resepct to the family of translates of spherical functions, decompositions into plane waves on $\mathscr{H}$, and finally decompositions into functions which are analogues of the classical harmonics on the sphere.

The first of the above-mentioned decompositions, which resembles formula (7.4.3) for Bessel functions, is a direct consequence of formula (11.3.5).

Define $F_g(p) = \int_K f(gk \cdot p)\mathrm{d}k$. The function F_g is of class $\mathscr{D}(\mathscr{K}\backslash\mathscr{H})$ for an arbitrary $f \in \mathscr{D}(\mathscr{H})$ and moreover $F_g(w) = f(gw)$. It follows by (11.3.5) that

$$f(p) = \int_0^\infty F_g(\nu)\omega_\nu(p)\nu^2\mathrm{d}\nu. \tag{11.4.1}$$

On the other hand,

$$\hat{F}_g(\nu) = \int_{\mathscr{H}} f(p)\omega_\nu(g^{-1}p)\mathrm{d}m(p).$$

Let us view f and ω_ν as funotions on G. Then, owing to the symmetry of the function ω_ν $(\omega_\nu(g) = \omega_\nu(g^{-1}) = \omega_{-\nu}(g))$, we obtain

$$\int_G f(x)\,\omega_\nu(g^{-1}x)\mathrm{d}x = \int_G f(x)\omega_\nu(x^{-1}g)\mathrm{d}x = f*\omega_\nu(g).$$

Then formula (11.4.1) becomes

$$f(p) = \frac{2}{\pi}\int_0^\infty \nu^2(f*\omega_\nu)\,(p)\mathrm{d}\nu. \tag{11.4.2}$$

This is the first of the decompositions mentioned above.

DEFINITION. By a *plane wave of direction* $s \in S_+$ we shall understand a function of the form

$$\mathscr{H} \ni p \to [p|s]^{-\mu} =: e_\mu(p, s), \qquad \mu \in C. \tag{11.4.3}$$

The parameter μ is the "frequency" of the plane wave. The function $e_\mu(\cdot\,,\cdot)$ is smooth on $\mathscr{H} \times S_+$. By finding the form of the plane wave in spherical coordinates we shall discover its connection with the known multiplier of the induced representation. Let $s_1 := (1, 0, 0, 1)$.

The invariance of the form $[\cdot\,|\,\cdot]$ under the action of the Lorentz group implies that

$$e_\mu(p, k\cdot s_1) = e_\mu(k^{-1}p, s_1). \tag{11.4.4}$$

Hence the restriction of the wave to $\mathscr{H} \times \{s_1\}$ determines the wave uniquely. By expressing p in hyperbolic coordinates (Proposition 11.2.1) we find

$$e_\mu(p, s_1) = (\cosh t - \sinh t\cos\theta)^{-\mu} = \delta\big(h(g^{-1})\big)^{\frac{\mu}{2}} \tag{11.4.5}$$

if $p = \tau(g)\cdot w$.

We have $e_\mu e_\nu = e_{\mu+\nu}$. On account of (11.4.5) we observe that the plane wave e_μ is a multiplier which defines the representation $\mathscr{U}^{2\mu-2}$

Applying (11.4.4), we obtain

$$\int_B e_\mu(p, s)\mathrm{d}s = \int_{\mathscr{H}} e_\mu(p, k\cdot s_1)\mathrm{d}k = \int_{\mathscr{K}} e_\mu(k^{-1}p, s_1)\mathrm{d}k$$

$$= \int_{\mathscr{K}} e_\mu(k^{-1}a(t)\cdot w, s_1)\mathrm{d}k,$$

where $p = k_1 a(t) \cdot w$. Consequently

$$\int_B e_\mu(p, s)\mathrm{d}s = \frac{1}{2}\int_0^\pi (\cosh t - \sinh t \cos\theta)^{-\mu} \sin\theta \, \mathrm{d}\theta .$$

By comparing (11.1.6) with (11.3.1) we find, owing to the symmetry of a spherical function,

$$\omega_\nu(p) = u_{00}^{2i\nu}(p) = \int_B e_{i\nu+1}(p, s)\mathrm{d}s . \tag{11.4.6}$$

For every $s \in B$ the function $\mathscr{H} \ni p \to e_{i\nu+1}(p, s)$ is a spherical function associated with ω_ν.

As we know (cf. § 11.1), the function $G \times B \ni (g, b) \to e_\mu(g^{-1}w, b)$ is a multiplier. Hence by substituting $B = \mathscr{K}s_1$ in definition (11.4.3) we obtain

$$e_\mu(gp, s) = e_\mu(p, g^{-1}s) e_\mu(gw, s) . \tag{11.4.7}$$

We shall apply representation (11.4.6) of a spherical function to obtain a decomposition of any function $f \in \mathscr{D}(\mathscr{H})$ into plane waves. Namely, we have

$$\omega_\nu(g^{-1}p) = \int_B e_{i\nu+1}(g^{-1}p, s)\,\mathrm{d}s = \int_B e_{i\nu+1}(p, gs) e_{i\nu+1}(g^{-1}w, s)\mathrm{d}s$$

$$= \int_B e_{i\nu+1}(p, s) e_{i\nu+1}(g^{-1}w, g^{-1}s) S_\varrho(g^{-1}, s)\mathrm{d}s ,$$

where S_ϱ is the Radon–Nikodym derivative of $\mathrm{d}g^{-1}s$ with respect to $\mathrm{d}s$, and can be written as

$$S_\varrho(g, ks_0) = \delta\big(h(gk)\big) = e_2(g^{-1}w, ks_0) .$$

Again on the basis of (11.4.7) we find

$$\omega_\nu(g^{-1}p) = \int_B e_{i\nu+1}(p, s) e_{-i\nu-1}(gw, s) e_2(gw, s)\mathrm{d}s$$

$$= \int_B e_{i\nu+1}(p, s) e_{-i\nu+1}(gw, s)\mathrm{d}s . \tag{11.4.8}$$

Finally, by substituting (11.4.8) into (11.4.3) we arrive at the desired decomposition into plane waves

$$f(p) = \frac{2}{\pi}\int_0^\infty \nu^2 \int_B \int_{\mathscr{H}} f(q) e_{i\nu+1}(q, s) e_{-i\nu+1}(p, s) \mathrm{d}s \mathrm{d}m(q) \mathrm{d}\nu$$

$$= \frac{2}{\pi}\int_0^\infty \nu^2 \int_B e_{-i\nu+1}(p, s)\tilde{f}(\nu, s) \mathrm{d}s \mathrm{d}\nu, \tag{11.4.9}$$

where

$$\tilde{f}(\nu, s) := \int_{\mathscr{H}} f(q) e_{i\nu+1}(q, s) \mathrm{d}m(q). \tag{11.4.10}$$

For every $f \in \mathscr{D}(\mathscr{H})$ the function $\tilde{f}$ on $C \times B$ defined by (11.4.10) is called the *Fourier transform of the function f*.

THEOREM 11.4.1. *The Fourier transformation on $\mathscr{H}$ has the following properties*:

$$(f_1 | f_2)_{L^2(\mathscr{H})} = \frac{2}{\pi}\int_0^\infty \nu^2 \int_B \bar{\tilde{f}}_1(\nu, s) \tilde{f}_2(\nu, s) \mathrm{d}s \mathrm{d}\nu, \tag{11.4.11}$$

$$f(p) = \frac{2}{\pi}\int_0^\infty \nu^2 \int_B e_{-i\nu+1}(p, s)\tilde{f}(\nu, s) \mathrm{d}s \mathrm{d}\nu, \tag{11.4.12}$$

the integral in (11.4.12) *is absolutely convergent.*

Formula (11.4.11) is proved by substituting in (11.4.12) the convolution $f_1^* * f_2$. This is a standard argument and therefore omitted. The absolute convergence of the integral (11.4.12) follows from the fact that the spherical Fourier transform of a function $f \in \mathscr{D}(\mathscr{K} \setminus \mathscr{H})$ is a function of class $\mathscr{S}(\boldsymbol{R}^1)$ and integral (11.4.11) is absolutely convergent.

By (11.4.11) the Fourier transformations extends to an isometry of $L^2(\mathscr{H})$ into $L^2\left(\boldsymbol{R}_+ \times B, \frac{2\nu^2}{\pi} \mathrm{d}\nu \otimes \mathrm{d}s\right)$.

PROPOSITION 11.4.2. *The operator $\mathscr{D}(\mathscr{H}) \ni f \to \tilde{f}(\nu, \cdot) \in \mathscr{E}(B)$ intertwines the regular representation on $\mathscr{D}(\mathscr{H})$ with the induced representation $\mathscr{U}^{2i\nu}$.*

Proof. $\widetilde{L_g f}(\nu, s) = \int_{\mathscr{H}} f(g^{-1}\cdot q)e_{i\nu+1}(q, s)\mathrm{d}m(q) = \int_{\mathscr{H}} f(q)e_{i\nu+1}(g\cdot q, s)\mathrm{d}m(q)$
$= \left(\int_{\mathscr{H}} f(q)e_{i\nu+1}(q, g^{-1}s)\mathrm{d}g\right)e_{i\nu+1}(gw, s) = \left(\mathscr{U}^{2i\nu}(g)\tilde{f}\right)(\nu, s)$.

COROLLARY 11.4.3. *The Fourier transformation is an isometry intertwining the representation $(L, L^2(\mathscr{H}))$ with the direct integral of the representations*

$$L = \int_0^\infty \mathscr{U}^{2i\nu}\frac{2\nu^2}{\pi}\mathrm{d}\nu,$$
$$L^2(\mathscr{H}) = \int_0^\infty H^{2i\nu}\frac{2\nu^2}{\pi}\mathrm{d}\nu. \tag{11.4.13}$$

On account of Corollary 11.4.3 we shall derive a new type of decomposition in the following manner. We shall use the fact that all representations $\mathscr{U}^{i\nu}$ are realized on the same space $L^2(B)$. We choose an orthonormal basis for $L^2(B)$ by means of the classical harmonics on the sphere. For every function $f \in L^2(B)$ we have the decomposition

$$f = \sum_{j=1}^\infty (h_j|f)h_j.$$

In particular

$$\tilde{f}(\nu, s) = \sum_j (h_j|\tilde{f}(\nu, \cdot))h_j(s)$$

and from formula (11.4.12) we get

$$f(p) = \frac{1}{\pi^2}\int_0^\infty \nu^2\mathrm{d}\nu \sum_j \tilde{f}_{\nu,j}\Phi_{\nu j}(p), \tag{11.4.14}$$

where $\Phi_{\nu,j}(p) := \int_B \bar{h}_j(s)e_{i\nu+1}(p, s)ds$ and

$$\tilde{f}_{\nu,j} = \int_B \bar{h}_j(s)\tilde{f}(\nu, s)\mathrm{d}s = \int_B \bar{h}_j(s)\int_{\mathscr{H}} f(q)e_{i\nu+1}(q, s)\mathrm{d}m(q)\mathrm{d}s. \tag{11.4.15}$$

The functions $\Phi_{\nu,j}$ are simply spherical functions associated with the zonal spherical function ω_ν. They are matrix elements of the type $\Phi_{\nu,j} = (h_j|U^{2i\nu}(g)h_0)$ where h_0 is a K-fixed vector. Now let us chose a basis for $L^2(B)$ consisting of normalized classical harmonics. We take

$$h_j := i^m \sqrt{2l+1}\, Y_{lm}, \tag{11.4.16}$$

where the index j runs over all pairs (l, m) such that $l = 0, 1, 2, \dots;$ $|m| \leqslant l$.

In terms of the coordinates θ, φ on thc sphere we have

$$h_j(\varphi, \theta) = P_l^m(\cos\theta)\mathrm{e}^{-im\varphi}\sqrt{\frac{(2l+1)(l-k)!}{(l+k)!}}. \tag{11.4.17}$$

We shall compute the function $\Phi_{\nu,j}$ beginning with the case $q = a(t)\cdot w$.

$$\Phi_{\nu,j}(q) = \int_{\mathscr{K}} \bar{h}_j(ks_1)[a(t)w|k\cdot s_1]^{-i\nu-1}\mathrm{d}k$$

$$= \int_{\mathscr{K}} \bar{h}_j(ks_1)[k^{-1}a(t)w|s_1]^{-i\nu-1}\mathrm{d}k.$$

Hence denoting by φ, θ, ψ the Euler angles of k, we obtain

$$\Phi_{\nu,j}(q) = \frac{1}{4\pi}\int_0^{2\pi}\int_0^{\pi} \bar{h}_j(\varphi, \theta)(\cosh t - \sinh t\cos\theta)^{-i\nu-1}\sin\theta\,\mathrm{d}\theta\,\mathrm{d}\varphi.$$

By comparing this formula with (11.4.17) we observe that for $m \neq 0$ the integral is zero. However, for $j = (l, 0)$ we get

$$\Phi_{\nu,j}(q) = \tfrac{1}{2}\sqrt{2l+1}\int_0^{\pi} P_l(\cos\theta)(\cosh t - \sinh t\cos\theta)^{-i\nu-1}\sin\theta\,\mathrm{d}\theta$$

$$= \tfrac{1}{2}\sqrt{2l+1}\int_{-1}^{1} P_l(y)(\cosh t - y\sinh t)^{-\nu i-1}\mathrm{d}y.$$

Next, we represent the polynomial P_l in the form

$$\frac{\mathrm{d}^l}{\mathrm{d}y^l}(1-y^2)^l\frac{(-1)^l}{2^l l!}$$

(Rodrigues' formula (8.1.6)). Integration by parts leads to

$$\Phi_{\nu,j}(q) = \frac{(-1)^l\sqrt{2l+1}}{2^{l+1}l!}\int\limits_{-1}^{1}\frac{\mathrm{d}^l}{\mathrm{d}y^l}(1-y^2)^l(\cosh t - y\sinh t)^{-i\nu-1}\mathrm{d}y$$

$$= \frac{\Gamma(-i\nu)\sqrt{2l+1}}{\Gamma(-i\nu-l)2^{l+1}l!}\sinh^l t\int\limits_{-1}^{1}(1-y^2)^l(\cosh t - a\sinh t)^{-i\nu-l-1}\mathrm{d}y$$

$$= \frac{\Gamma(-i\nu)}{\Gamma(-i\nu-l)}\frac{\sqrt{2l+1}}{2^{l+1}l!}\sinh^l t$$

$$\times\int\limits_0^{\pi}\sin^{2l+1}\theta(\cosh t - \cos\theta\sinh t)^{-i\nu-l-1}\mathrm{d}\theta. \tag{11.4.18}$$

By expanding the integrand in a series with respect to tanhθ, performing the integration and then comparing the coefficients of the expansion (Problem 13, Ch. 10) we discover a connection with the associated Legendre functions

$$\int\limits_0^{\pi}\sin^{2l+1}\theta(\cosh t - \cos\theta\sinh t)^{-i\nu-l-1}\mathrm{d}\theta$$

$$= (\cosh t)^{-i\nu-l-1}\int\limits_0^{\pi}\sin^{2l+1}\theta(1-\cosh t\tanh t)^{-i\nu-l-1}\mathrm{d}\theta$$

$$= (\cosh t)^{-i\nu-l-1}\sum_{k=0}^{\infty}\frac{\Gamma(-i\nu-l-1)\tanh^k t}{\Gamma(-i\nu-l-k)\Gamma(k+1)}(-1)^k$$

$$\times\int\limits_0^{\pi}\cos^k\theta\sin^{2l+1}\theta\,\mathrm{d}\theta$$

$$= (\cosh t)^{-i\nu-l-1}\sum_{j=0}^{\infty}\frac{\Gamma(-i\nu-l-1)\tanh^{2j}t}{\Gamma(-i\nu-l-2j)\Gamma(2j+1)}\frac{\Gamma(j+\frac{1}{2})l!}{\Gamma(j+\frac{3}{2}+l}$$

$$= \sqrt{\pi}\,2^{\frac{1}{2}(2l+1)}(2l+1)!(\sinh t)^{-\frac{2l+1}{2}}\mathfrak{P}_{-i\nu-\frac{1}{2}}^{-\frac{1}{2}(2l+1)}(\cosh t). \tag{11.4.19}$$

It follows that

$$\Phi_{\nu,j}q = \frac{\Gamma(-i\nu)}{\Gamma(-i\nu-l)}\frac{(2l+1)!}{l!}\sqrt{\frac{\pi(2l+1)}{2}} \times \sinh^{\frac{1}{2}}t\,\mathfrak{p}^{-\frac{1}{2}}_{-i\nu-\frac{1}{2}}(\cosh t) \tag{11.4.20}$$

for $q = a(t)w$. We now face the problem of determining the function $\Phi_{\nu,j}$ for an arbitrary $q \in \mathscr{H}$. So we put $q = k_1 a(t)\cdot w$

$$\begin{aligned}\Phi_{\nu,j}(q) &= \int_{\mathscr{K}} \overline{h}_j(ks_1)[k_1 a(t)w|ks_1]^{-i\nu-1}\mathrm{d}k\\ &= \int_{\mathscr{K}} \overline{h}_j(ks_1)[a(t)w|k_1^{-1}ks_1]^{-i\nu-1}\mathrm{d}k\\ &= \int_{\mathscr{K}} \overline{h}_j(k_1 ks_1)[a(t)w|ks_1]^{-i\nu-1}\mathrm{d}k.\end{aligned}$$

The translates of the function h_j for $j = (l, m)$ span, as we know, a $(2l+1)$-dimensional subspace on which the representation is irreducible. Since the vectors $h_{j'}$, $j' = (l, m')$, $|m'| \leq l$ constitute a basis for that space, we get

$$L_{k_1}h_j = \sum_{|m'|\leq l} (h_{l,m'}|L_{k_1}h_j)h_{(l,m')}.$$

Consequently

$$\Phi_{\nu,j}(q) = \sum_{|m'|\leq l} (L_{k_1}^{-1}h_j|h_{(l,m')})\Phi_{\nu,j'}(a(t)w).$$

We know, however, that it is only for $m' = 0$ that the function $\Phi_{\nu,j'}$ does not vanish at $a(t)w$. Therefore the above sum reduces to

$$\Phi_{\nu,j}(q) = (h_j|L_{k_1}h_{j'})\Phi_{\nu,j'}(a(t)w).$$

The first factor is a matrix element of the irreducible representation $\mathscr{U}^l$ of the group K with respect to the basis consisting of the classical harmonics. The addition formulas for those functions imply that

$$(h_j|L_k h_{j'}) = Y_{lm}(k_1 s_1),$$

where $j = (l, m)$, $j' = (l, 0)$.

Finally, we get the following simple expression:

$$\Phi_{\nu,j}(q) = Y_{lm}(k_1 s_1)\Phi_{\nu,j}(a(t)w) \tag{11.4.21}$$

if $q = k_1 a(t) w$; $j = (l, m)$. Equivalently, in terms of Euler angles,

$$\Phi_{\nu,j}(k(\varphi, \theta)\, a(t) w) = \frac{\Gamma(-i\nu)\,(2l+1)!}{\Gamma(-i\nu-l)\, l!} \sqrt{\frac{\pi(2l+1)}{2}}$$

$$\times \sinh^{-\frac{1}{2}} t\, \mathfrak{P}^{-l-\frac{1}{2}}_{-i\nu-\frac{1}{2}}(\cosh t)\, i^{-m} P_l^m(\cos\theta) \mathrm{e}^{-im\varphi}. \tag{11.4.22}$$

Let us list the consequences of Theorem 11.4.1 concerning the present decomposition.

THEOREM 11.4.2. *The integral transformation* $\mathscr{D}(\mathscr{H}) \ni f \to f_{\nu,j}$ *where* $f_{\nu,j} := \int_B \bar{h}_j(s) \tilde{f}(\nu, s)$, *has the following property of an isometry*

$$\|f\|_{L^2(\mathscr{H})} = \frac{2}{\pi} \sum_{l=0}^{\infty} \sum_{|m| \leqslant l} \int_0^{\infty} \nu^2 \mathrm{d}\nu |\tilde{f}_{\nu,(l,m)}|^2. \tag{11.4.23}$$

Moreover, there exists an inverse transformation given by

$$f(p) = \frac{2}{\pi} \sum_{l=0}^{\infty} \sum_{|m| \leqslant l} \int_0^{\infty} \nu^2 \mathrm{d}\nu\, \tilde{f}_{\nu,(l,m)} \bar{\Phi}_{\nu,(l,m)}(p). \tag{11.4.24}$$

Since the integrals under consideration are absolutely convergent, it is indeed permissible to interchange the order of integration and summation in formula (11.4.13). Formulas (11.4.23) and (11.4.24) are known as expansions into trigonometric series on $\mathscr{H}$.

The three types of expansions described above have their counterparts in Fourier analysis on the plane. The transformation on $\mathscr{H}$ called the *Fourier traansformation* is indeed the closest counterpart of the classical Fourier transformation. To underline the formal resemblance we represent the latter transformation in terms of spherical coordinates on $\boldsymbol{R}^2$: for $f \in \mathscr{D}(\boldsymbol{R}^2)$, $\varrho \in \boldsymbol{R}_+$, $s \in S^1$

$$\hat{f}(\varrho s) = \int_{R^2} f(x) \mathrm{e}^{\varrho i (x|s)} \mathrm{d}x,$$

$$f(x) = \frac{1}{(2\pi)^2} \int_{R_+} \int_{S^1} \hat{f}(\varrho s) \mathrm{e}^{-i\varrho(x,s)} \lambda \mathrm{d}\lambda \mathrm{d}s.$$

Formally these expressions differ from formulas (11.4.10) and (11.4.12) not only in the different forms of the Plancherel measures but also in the occurrence of a non-unitary factor $e_1(\cdot, \cdot)$ in the plane waves on $\mathscr{H}$.

We have already observed this phenomenon in harmonic analysis on symmetric spaces. Namely, it is not by accident that the function $e_2(g^{-1}, \cdot)$ is the Radon–Nikodym derivative of the measure dgs with respect to ds.

The appearance of the factor $e_1(\cdot, \cdot)$ results from the fact that in the decomposition $\Lambda_+ = KAN$ the group AN is not normal and therefore its action on $K = \Lambda_+/AN$ is non-trivial. An analogous decomposition for the motion group $M(2)$ in $\boldsymbol{R}^2$ is of the form $K \times_r \boldsymbol{R}^2$ with $\boldsymbol{R}^2$ denoting the translation group which in this case is normal.

11.5. DIFFERENTIAL PROPERTIES OF SPHERICAL FUNCTIONS

When looking for a Λ_+-invariant operator on $\mathscr{H}$, it is convenient to imbed this manifold into the Minkovsky space $(\boldsymbol{R}^4, [\cdot|\cdot])$, the form $[\cdot|\cdot]$ being invariant under the action of Λ_+.

A Λ_+-invariant differential operator on $\boldsymbol{R}^4$ is a power of the d'Alembertian

$$\square = \sum_{i=1}^{3} \frac{\partial^2}{\partial x_i^2} - \frac{\partial^2}{\partial x_0^2}. \tag{11.5.1}$$

In terms of hyperbolic coordinates $(r^2 = [x|x], t, \theta, \varphi)$ this operator decomposes into the radial part

$$D_r = \frac{1}{r^3} \frac{\partial}{\partial r} r^3 \frac{\partial}{\partial r} \tag{11.5.2}$$

and the "angular" part

$$D = \frac{1}{\sinh^2 t} \left[\frac{\partial}{\partial t} \sinh^2 t \frac{\partial}{\partial t} + \frac{1}{\sin\theta} \frac{\partial}{\partial\theta} \sin\theta \frac{\partial}{\partial\theta} + \frac{1}{\sin^2\theta \sin^2\varphi} \frac{\partial^2}{\partial\varphi^2} \right]. \tag{11.5.3}$$

D is the required invariant operator on the hyperboloid $\mathscr{H}$. We shall call it the *Laplace–Beltrami operator* for the manifold $\mathscr{H}$.

Plane waves, being spherical functions associated with zonal spherical function, are eigenfunctions of the operator D. More precisely, the plane wave $e_{i\nu+1}(\cdot, s)$ is a spherical function associated with ω_ν (equation (11.4.6)).

In virtue of Theorem 5.4.2 we obtain

$$De_{i\nu+1} = D\omega_\nu(w) e_{i\nu+1}. \tag{11.5.4}$$

The computation of $De_{i\nu+1}(w)$ will be equivalent to finding $D\omega_\nu(w)$, which will allow us to determine the differential properties of an arbitrary spherical function associated with ω_ν, and thus, in particular, besides $e_{i\nu+1}$ also of the function $\phi_{\nu,j}$ and of the function ω_ν itself. The computation of $De_\mu(w)$ is based on the following simple device. Let $f_s^\mu(x) = [x, s]^{-\mu}$, $x \in \boldsymbol{R}^4$, $s \in S$. We observe that $\Box f^\mu(w) = 0$. Since $f_s^\mu | \mathscr{H} = e_\mu(\cdot, s)$ we get $D_r f_s^\mu(w) + De_\mu(w, s) = 0$. The valuc of $D_r f_s$ is now easily computed. Since the radial component of the function f^μ is equal to $r^{-\mu}$ it follows from (11.5.2) that

$$De_\mu(w, s) = \mu(\mu-2).$$

THEOREM 11.5.1. *The elementary functions of the Fourier analysis on $\mathscr{H}$ satisfy the differential equations*

$$D\Phi_{\nu,j} = -(\nu^2+1)\Phi_{\nu,j}, \tag{11.5.5}$$

$$De_\mu = \mu(\mu-2)e_\mu. \tag{11.5.6}$$

In particular, the identity (11.5.5) is satisfied by the zonal spherical function

$$\omega_\nu = \Phi_{\nu,(0,0)}.$$

As regards other cases that of $\mu = 2$ is of special interest. The functions on $\mathscr{H}$ of the form

$$f(p) = \int_B e_2(p, s)\, d\sigma(s), \tag{11.5.7}$$

where σ is a finite measure on B, prove to be solutions of the equation

$$Df = 0. \tag{11.5.8}$$

Solutions of equation (11.5.8) are called *harmonic functions on* $\mathscr{H}$.

Just as in the case of the analysis on the unit disc the function $e_2(p, s)$ which as we know is related to the Radon–Nikodym derivative of the measure $dg^{-1}s$ with respect to ds plays the role of the Poisson kernel. In Part III we shall prove, following Furstenberg, 1963 that every bounded harmonic function is of form (11.5.7) for a certain uniquely determined measure $d\sigma$ on B. Theorem 11.5.1 strengthens the analogy between the functions $e_\mu(\cdot, s)$ and the exponential functions on $\boldsymbol{R}$, also constituting solutions of the Laplace operator constant on hyperplanes.

Finally, we note that the decomposition of the regular representation on $L^2(\mathscr{H})$ has turned out to be a decomposition into eigenspaces of the Laplace operator which correspond to different eigenvalues.

11.6. THE GELFAND-GRAEV TRANSFORMATION

Harmonic analysis on the symmetric spaces corresponding to the groups of hyperbolic rotations in $\boldsymbol{R}^n$ is due to Gelfand and Graev. We shall study only the cases of $n = 3$ (Chapter 10) and $n = 4$ (the present chapter). In the original papers of these authors Fourier analysis is a consequence of the theory of yet another integral transformation, which can be regarded as the Radon transformation on a symmetric space.

The transform of a function $f \in \mathscr{D}(\mathscr{H})$ is a function on the space of horocycles and its value at a point $\xi \in \Xi$ is given by the integral of f over the horocycle ξ with respect to the $\mathscr{N}$-invariant measure.

In general theory, presented briefly in Part III, we shall call that transform the *generalized Radon transformation*. In this section, however, we call it the *Gelfand–Graev transformation*. When restricted to functions in $\mathscr{D}(\mathscr{K}\backslash\mathscr{H})$, this transformation is closely related to the Abel transformation. In our exposition of harmonic analysis on $\mathscr{H}$, based on the methods of Harish-Chandra and Helgason, we proceed from a particular case of the Gelfand–Graev transformation through the theory of the spherical and the general Fourier transformation to the general Gelfand–Graev transformation. This approach in our opinion is clearer than original one.

Let $\xi \in \Xi$ be of the form $\xi = ka \cdot \mathscr{N} \cdot w_0$. We define a measure $d\xi$ on ξ as the transport of the invariant measure dn on $\xi_0 = \mathscr{N} \cdot w$ via the mapping

$$\xi_0 \ni n \cdot w \to kan \cdot w \in \xi .$$

DEFINITION. By the *Gelfand–Graev transformation on* $\mathscr{H}$ we shall understand the mapping

$$\mathscr{D}(\mathscr{H}) \ni f \to \hat{f} \in \mathscr{D}(\Xi),$$

given by the formula

$$\hat{f}(\xi) = \int_{\xi} f \mathrm{d}\xi. \tag{11.6.1}$$

A horocycle being a closed subset of $\mathscr{H}$, it follows that the restriction of f to ξ is a compactly supported function and the value of $\hat{f}(\xi)$ is well defined.

First of all we establish a connection between the Fourier and the Gelfand–Graev transformations.

The invariant measure on $\mathscr{H}$ can be represented as

$$\int_{\mathscr{H}} f \mathrm{d}m = \frac{1}{2} i \int_{C^1} \int_{R^1} f(a(t)n(t) \cdot w) \mathrm{d}t \mathrm{d}\beta \wedge \mathrm{d}\bar{\beta}$$

(formula(11.2.13)). Then for $s = k \cdot s_1$

$$\begin{aligned}
\tilde{f}(\nu, s) &= \int_{\mathscr{H}} f(x) e_{i\nu+1}(x, s) \mathrm{d}m(x) \\
&= \int_{\mathscr{H}} f(x) [x | k \cdot s_1]^{-i\nu-1} \mathrm{d}m(x) \\
&= \int_{\mathscr{H}} f(x) [k^{-1}x | s_1]^{-i\nu-1} \mathrm{d}m(x) = \int_{\mathscr{H}} f(k \cdot x) [x | s_1]^{-i\nu-1} \mathrm{d}m(x) \\
&= \frac{1}{2} i \int_{R^1} \int_{C^1} f(ka(t)n(\beta)w) \mathrm{e}^{(i\nu+1)t} \mathrm{d}\beta \wedge \mathrm{d}\bar{\beta} \\
&= \int_{R^1} \int_{\mathscr{N}} \hat{f}(ka(t)n) \mathrm{e}^{(i\nu+1)t} \mathrm{d}n \mathrm{d}t \\
&= \int_{R^1} \hat{f}(ka(t) \cdot \xi_0) \mathrm{e}^{(i\nu+1)t} \mathrm{d}t.
\end{aligned} \tag{11.6.2}$$

We shall use this relation to transform into the language of $\hat{f}$ the properties of the Fourier transformation. To this end we observe that $\tilde{f}$, as a function of ν, is equal to the Fourier transform of the function $t \to \mathrm{e}^t \hat{f}(ka(t) \cdot \xi_0)$.

PROPOSITION 11.6.1. *The Gelfand–Graev transformation is a continuous linear injection of $\mathscr{D}(\mathscr{H})$ into $\mathscr{D}(\Xi)$ intertwining the regular representations on these spaces.*

Proof. Linearity, continuity and the intertwining property follow directly from the definition. To prove injectiveness we suppose that $\hat{f} \equiv 0$; then by (11.6.2) we get $\hat{f} \equiv 0$, which means that $f \equiv 0$. □

Let us note that the Gelfand–Graev transformation is not a surjection onto $\mathscr{D}(\Xi)$. For the proof let $\varphi \in \mathscr{D}(\Xi)$ be a K-invariant function of the space of horocycles, i.e. on the upper half of the "light" cone in $\boldsymbol{R}^4$. If the mapping $f \to \hat{f}$ were surjective, then a function $f \in \mathscr{D}(\mathscr{H})$ such that $\hat{f} = \varphi$ would also be K-invariant.

In §3 we observed that the Abel transformation of a K-invariant function is a function symmetric in the argument t. Since

$$F_f(a(t)) = \mathrm{e}^{-t}\hat{f}(a(t)\xi_0) = \mathrm{e}^{-t}\varphi(a(t)\xi_0)$$

it turns out that the condition of K-invariance of a function on Ξ implies the symmetry condition

$$\mathrm{e}^{-2t}\varphi(a(t)\xi_0) = \varphi(a(-t)\xi_0).$$

Now it is enough to choose a function φ supported in a small neighbourhood of the vertex of the light cone to obtain an evident contradiction which excludes the surjectivity of the mapping $f \to \hat{f}$.

$$(L_k f)^\wedge = L_k\hat{f} = L_k\varphi = \varphi = \hat{f}.$$

Now we proceed to construct a transformation on $\mathscr{D}(\Xi)$ which is inverse to the Gelfand–Graev transformation.

We recall that the identification of the space of horocycles with the points of the light cone consisted in assigning to a point $s \in S_+$ the horocycle $\xi = \{p \in H\colon [p|s] = 1\}$.

Now to the points of the space $\mathscr{H}$ we shall assign geometric objects in $S_+ \cong \Xi$. Namely, for a point $p \in \mathscr{H}$ we define

$$\check{p} = \{s \in S_+\colon [p|s] = 1\}.$$

Thus the set $\check{p} \subset \Xi$ consists of the pencil of horocycles passing through $p \in \mathscr{H}$. In particular, $\check{w}$ is the set of horocycles passing through the vertex of the hyperboloid, and this set, as we remember, is comprised of the horocycles $\mathscr{K} \cdot \xi_0$. The invariance of the form $[\cdot\,|\,\cdot]$ now yields

$$\begin{aligned}(g \cdot w)^\vee &= \{s \in S_+\colon [w|g^{-1}s] = 1\} = \{s \in S_+ : g^{-1}s \in \check{w}\}\\ &= g \cdot \check{w} = g\mathscr{K}g^{-1}(g \cdot w).\end{aligned}$$

Thus every set p is the orbit in Ξ of a group conjugate to $\mathscr{K}$ under an inner automorphism. In particular, $\breve{p}$ is compact, and we may equip it with the measure $d\breve{p}$, which is the transport of the $\mathscr{K}$-invariant measure dk.

DEFINITION. The mapping

$$\mathscr{E}(\Xi) \ni \varphi \to \breve{\varphi} \in \mathscr{E}(\mathscr{H})$$

defined as

$$\breve{\varphi}(p) = \int_{\breve{p}} \varphi(\xi) d\breve{p}(\xi) \tag{11.6.3}$$

is called the *dual Gelfand–Graev transformation.*

Even for functions φ in $\mathscr{D}(\Xi)$ the transform $\breve{\varphi}$ is not supported by a compact set in $\mathscr{H}$. Therefore the connection between a function $f \in \mathscr{D}(\mathscr{H})$ and its double transform $\hat{f}^{\smile}$ must be of another type than in the case of the Fourier transformation.

It is seen directly from formula (11.6.3) that the operation $\varphi \to \breve{\varphi}$ intertwines the regular representations.

THEOREM 11.6.2. *For every function* $f \in \mathscr{D}(\mathscr{H})$ *we have*

$$f = 2(-D-1)\hat{f}^{\smile}, \tag{11.6.4}$$

where D is the Laplace–Beltrani operator on $\mathscr{H}$ given by (11.5.3).

Proof. The inversion formula for the classical Fourier transformation, when applied to formula (11.6.2), yields the relation

$$\hat{f}(ka(t)\xi_0) = \frac{1}{2\pi} \int_{-\infty}^{\infty} e^{-(i\nu+1)t} \tilde{f}(\nu, ks_1) d\nu. \tag{11.6.5}$$

Hence the double transform $\hat{f}^{\smile}$ can be expressed in terms of the Fourier transformation as

$$(\hat{f})^{\smile}(gw) = (L_{g^{-1}}f)^{\wedge\smile}(w) = \frac{1}{2\pi} \int_B \int_{-\infty}^{\infty} (L_g^{-1}f)^{\sim}(\nu, s) d\nu ds. \tag{11.6.6}$$

In view of the group properties of a multiplier

$$(\hat{f})^{\smile}(gw) = \frac{1}{2\pi} \int_B \int_{-\infty}^{\infty} e_{i\nu+1}(g^{-1}w, s)\tilde{f}(\nu, gs) d\nu ds$$

$$= \frac{1}{2\pi}\int_B \int_{-\infty}^{\infty} e_{-i\nu-1}(gw, gs)\tilde{f}(\nu, gs)\,d\nu\,ds$$

$$= \frac{1}{2\pi}\int_B \int_{-\infty}^{\infty} e_{-i\nu+1}(gw, s)\tilde{f}(\nu, s)\,d\nu\,ds, \qquad (11.6.7)$$

where we have used the expression for the function S_ϱ (the Radon–Nikodym derivative of the measure $dg^{-1}s$ with respect to ds) established in § 11.4.

It follows from (11.6.2) and (11.6.6) that for an arbitrary $g \in G$ the integrand is of class $\mathscr{S}(\boldsymbol{R}^1)$. The operator D acts on the integrand as described in Theorem 11.5.1 and preserves the class $\mathscr{S}(\boldsymbol{R}^1)$. Therefore we conclude that

$$D(\hat{f})^{\vee}(gw) = -\frac{1}{2\pi}\int_B \int_{-\infty}^{\infty} (\nu^2+1)e_{-i\nu+1}(gw, s)\tilde{f}(\nu, s)\,ds\,d\nu$$

$$= -\frac{1}{\pi}\int_0^{\infty} \nu^2 \int_B e_{-i\nu+1}(gw, s)\tilde{f}(\nu, s)\,ds\,d\nu$$

$$-\frac{1}{2\pi}\int_B \int_{-\infty}^{\infty} e_{-i\nu+1}(gw, s)\tilde{f}(\nu, s)\,ds\,ds$$

$$= -\frac{1}{2}f(gw) - (\hat{f})^{\vee}(gw).$$

This completes the proof of the theorem. □

As is seen from the above proof, the simplicity of formula (11.6.4) results from the fact that the Plancherel measure on $\mathscr{H}$ differs only by a constant from the "Fourier transform" of the operator D. This time it is a "lucky" coincidence although the theory of the Radon transformation can actually be extended to arbitrary symmetric Riemannian spaces, as will be done in Part III.

Just as in the case of the classical Radon transformation on $\boldsymbol{R}^n$, where the even- and the odd-dimensional cases are essentially different, symmetric spaces possess a characteristic which decides whether the inversion formula for the Radon transformation contains only an invariant operator on the initial space. This problem will be dealt with more thoroughly later on. Now we only note that the group SL(2, ***R***), apparently "easier" than SL(2, ***C***), is the more complicated one from the point of view of the Radon transformation. The reason lies in the form of the Plancherel measure on SL(2, ***R***), more precisely: in the fact that this measure is not a polynomial function.

It was lucky for the developement of the theory that the case of the low-dimensional group SL(2, ***R***) already exemplified all the problems and complexities of the general theory. It has even become customary to say that "what is true of the group SL(2, ***R***) is true of an arbitrary non-compact semi-simple group".

11.7. IRREDUCIBILITY PROBLEMS OF THE REPRESENTATIONS $\mathscr{U}^l$

Let us consider again the representations $\mathscr{U}^l$ induced from the subgroup $P \subset \mathrm{SL}(2, \boldsymbol{C})$ as described in § 11.1. In view of the subsequent notation it will be convenient to number these representations by an index $\nu \in \boldsymbol{C}$ which is related to l in such a way that $l = 2i\nu$. The representation space is $L^2(B)$ and the action is defined by means of the multiplier

$$(\mathscr{U}^{2i\nu}(g)f)(s) = e_{i\nu+1}(gw, s)f(g^{-1}s). \tag{11.7.1}$$

The representations $\mathscr{U}^{2i\nu}$ are unitary for real values of ν. Still when restricted to the subgroup K, $\mathscr{U}^{2i\nu}$ is unitary for all ν since it coincides with the regular representation on $L^2(K/M)$.

Suppose that the representation $\mathscr{U}^{2i\nu}$ is reducible and let $L^2(B) = H_1 + H_2$ be a decomposition into non-trivial orthogonal K-invariant subspaces such that H_1 is G-invariant. Then it follows from Proposition 11.1.1 that the space H_2 is invariant under the action of the representation $\mathscr{U}^{2i\bar{\nu}}$. Thus reducible representations occur in pairs: the representation corresponding to the parameter ν is reducible if and only if the representation corresponding to $\bar{\nu}$ is reducible. Shortly we say that ν is reducible if the corresponding representation is so.

The representation of K on $L^2(B)$ decomposes into irreducible spherical representation; every such representation occurs only once in the decomposition. In particular a K-fixed vector f_0 in $L^2(B)$ belongs either to H_1 or H_2. We may assume that $f_0 \in H_1$; otherwise we consider the representation $\mathscr{U}^{2i\bar{\nu}}$ and the space H_2 instead of $\mathscr{U}^{2i\nu}$ and H_1.

Let $\{h_j\}$ be the basis for $L^2(B)$ introduced in § 11.4. The index j varies over all pairs (l, k); $l \in N$, $k \in \boldsymbol{Z}$, $|k| \leqslant 1$.

For a fixed l the set $h_{(l,k)}$, $|k| \leqslant l$ spans the representation space of the irreducible representation corresponding to the zonal spherical function of the group K determined by the Legendre polynomial P_l.

Since the space H_2 is non-trivial there exists a $j_0 = (l_0, k_0)$ such that $h_{(l_0,k_0)} \in H_2$. The spaces H_1 and H_2 are mutually orthogonal and thus

$$(h_{j_0} | \mathscr{U}^{2i\nu}(g) f_0) = 0$$

identically for $g \in G$.

In the notation of § 11.4 we have $\Phi_{\nu,j_0}(g) \equiv 0$. Next, it follows from formula (11.4.21) that the function $\Phi_{\nu,j}$ is zero on G if and only if it is zero on the subset $\mathscr{A} \cdot w \in \mathscr{H}$. The values of the function on the set $\mathscr{A} \cdot w$ are given by formula (11.4.18). We choose the form

$$\Phi_{\nu,j}(a(t)w) = \frac{\Gamma(-i\nu)}{\Gamma(-i\nu-l)} \frac{\sqrt{2l+1}}{2^{l+1}l!} \sinh^l t \int_{-1}^{1} (1-y^2)^l (\cosh t - y \sinh t)^{-i\nu-l-1} dy.$$

The integral occurring in this expression is a continuous function with respect to the parameter t and at $t = 0$ assumes a positive value. Thus the vanishing of $\Phi_{\nu,j}$ is equivalent to the vanishing of the factor

$$n(\nu, l) := \frac{\Gamma(-i\nu)}{\Gamma(-i\nu-l)} = (-i\nu-1)(-i\nu-2) \ldots (-i\nu-1-l).$$

In the case where f_0 belongs to the space H_2, which is $\mathscr{U}^{2i\nu}$-invariant, we conclude by following the same argument that $\Phi_{\bar{\nu},j} = 0$ if and only if $n(\bar{\nu}, l) = 0$.

Finally, we obtain

THEOREM 11.7.1. *For $\nu \neq iZ$ the representations $\mathscr{U}^{2i\nu}$ are irreducible.*

Thus all representations occurring in the decomposition of the representation of the group into a direct integral described in Corollary 11.4.3 are irreducible.

PROBLEMS

1. Let $\mathfrak{g}$ denote the Lie algebra $\mathfrak{sl}(2, C)$, i.e. the set of complex 2×2 traceless matrices, and let $\mathfrak{k} = \mathfrak{su}(2)$ be the algebra of anti-hermitean matrices. If, as usual, $\exp A$ denotes the exponential of a matrix

A, show that $\mathfrak{X} = \exp(i\mathfrak{k})$ and that the mapping $K \times \mathfrak{X} \ni (k, x) \to kx \in G$ is a diffeomorphism.

Hint. Use the spectral theorem for positive definite hermitean matrices and the observation that, for any $g \in SL(2, \boldsymbol{C})$, $gg^* \in X$. The decomposition $g = kx$ is known as a *polar decomposition* of a matrix.

2.(a) Express the Iwasawa decomposition of $g \in SL(2, \boldsymbol{C})$ in terms of the matrix coefficients of g.

(b) Let $H = \frac{1}{2}\begin{bmatrix} 1 & 0 \\ 0 & -1 \end{bmatrix}$ and define a function $G \ni g \to t(g) \in \boldsymbol{R}$ by the Iwasawa decomposition $g = k(g)\exp(t(g)H)n$. Show that $t(g) = \log(|\alpha|^2 + |\gamma|^2)$ and express the modular function $\delta(\cdot)$ and characters $\chi^l(\cdot)$ of the subgroup P in terms of $t(\cdot)$. Also express the multiplier defining the representation U^l in terms of $t(\cdot)$.

3. Let $H(2)$ be the space of 2×2 hermitean matrices and let

$$\sigma^0 = \begin{bmatrix} 1 & 0 \\ 0 & 1 \end{bmatrix}, \quad \sigma^1 = \begin{bmatrix} 0 & -i \\ i & 0 \end{bmatrix}, \quad \sigma^2 = \begin{bmatrix} 0 & 1 \\ 1 & 0 \end{bmatrix}, \quad \sigma^3 = \begin{bmatrix} -1 & 0 \\ 0 & 1 \end{bmatrix}$$

be the hermitean Pauli matrices.

(a) Show that $\{\sigma^\alpha\}_{\alpha=0}^3$ is a basis in $H(2)$ and that $\det\left(\sum_{\mu=0}^{3} x_\mu \sigma^\mu\right) = x_0^2 - x_1^2 - x_2^2 - x_3^2$. Show that the coordinates of a point $x \in H(2)$ with respect to that basis are given by the formula $x_\mu = \frac{1}{2}\operatorname{tr}(x\sigma^\mu)$ and that the mapping defined by formula (11.2.2) is precisely the restriction to $\mathfrak{X}$ of the coordinate mapping $H(2) \ni x \to (x_0, x_1, x_2, x_3) \in \boldsymbol{R}^4$.

(b) For $g \in SL(2, \boldsymbol{C})$ define a mapping $\tau(g)\colon H(2) \to H(2)$ by $\tau(g)x := gxg^*$. Show that $\tau(g)$ is represented in the basis $\{\sigma^\alpha\}_0^3$ by the matrix $[M_{\mu\nu}]$ with entries $M_{\mu\nu} = \frac{1}{2}\operatorname{tr}(\sigma^\mu g \sigma^\nu g^*)$, which coincides with the matrix $\tau(g)$ given in the text.

(c) Let $O(1, 3)$ denote the subgroup of $GL(4, \boldsymbol{R})$ consisting of the matrices which leave the form $[\cdot|\cdot]$ invariant. Show that $O(1, 3)$ has four connected components and that its identity component Λ_+ is given as $\Lambda_+ = \{M \in O(1, 3) : \det M = 1,\ M_{00} \geqslant 1\}$.

(d) Apply (b) to show that for $g \in SU(2)$, $\tau(g)$ is a rotation in the hyperplane $x_0 = 0$ and that for g hermitean and positive definite $\tau(g)$ is a hyperbolic rotation. The latter fact means that, in an appropriate basis of $H(2)$, $\tau(g)$ has the form

$$\tau(g)(y_0, y_1, y_2, y_3)$$
$$= (y_0\cosh t + y_1\sinh t,\ y_0\sinh t + y_1\cosh t,\ y_2, y_3)$$

with t real.

4. (continuation) Let $C^+ \subset H(2)$ be the open cone of positive definite matrices.

(a) Show that C^+ is invariant under $\tau(g)$ for every $g \in \mathrm{SL}(2, \boldsymbol{C})$ and that the Lebesgue measure on C^+ is invariant under $\tau(g)$, $g \in \mathrm{SL}(2, \boldsymbol{C})$. Further, show that the mapping

$$\boldsymbol{R}_+ \times \mathscr{H} \ni (r, p) \to r(p_0, p_1, p_2, p_3) \in C^+,\ p = (p_0, p_1, p_2, p_3)$$

is a diffeomorphism and, by expressing the measure on C^+ in terms of the above parameters, prove the following formula for the invariant measure $\mathrm{d}m$ on $\mathscr{H}$; $\mathrm{d}m(p) = \dfrac{\mathrm{d}p_1\,\mathrm{d}p_2\,\mathrm{d}p_3}{p_0}$.

(b) Modify the method of (a) to show that the invariant measure on the "future light cone" S_+ is given as $\mathrm{d}m(p) = \dfrac{\mathrm{d}p_1\,\mathrm{d}p_2\,\mathrm{d}p_3}{p_0}$.

(c) Prove that every positive finite measure μ on $H(2)$ invariant under τ $\mathrm{SL}(2, \boldsymbol{C})$ (the finiteness means $\mu(H(2)) < \infty$) is a multiple of the Dirac delta supported at the origin $O \in H(2)$.

5. (a) Let $\mathfrak{l}$ be the Lie algebra of the group Λ_+. Show that $\mathfrak{l}$ consists of matrices of the form

$$\begin{bmatrix} 0 & \beta_1 & \beta_2 & \beta_3 \\ \beta_1 & 0 & -\alpha_3 & \alpha_2 \\ \beta_2 & \alpha_3 & 0 & -\alpha_1 \\ \beta_3 & -\alpha_2 & \alpha_1 & 0 \end{bmatrix},$$

where α_i, β_i are real.

(b) Define R_i, $i = 1, 2, 3$, by substituting 1 for α_i and zero for the remaining entries, and define H_j, $j = 1, 2, 3$, by substituting 1 for β_j and zero for the remaining entries. Check the commutation relations

$$[R_j, R_k] = \sum_{l=1}^{3} c_{jk}^l R_l, \qquad [H_j, R_k] = \sum_{l=1}^{3} c_{jk}^l H_l,$$

$$[H_j, H_k] = -\sum_{l=1}^{3} c_{jk}^l H_l,$$

where

$$c_{jk}^{l} = \begin{cases} 0, & \text{not all } j, k, l \text{ are distinct,} \\ \operatorname{sgn}\begin{pmatrix} 1 & 2 & 3 \\ j & k & l \end{pmatrix}, & \text{otherwise.} \end{cases}$$

(c) Show that R_i is a generator of rotations in the plane $p_0 = p_i = 0$ and H_j is a generator of hyperbolic rotations in the plane (p_0, p_j).

(d) Show that the complexification $\mathfrak{l}_c$ of $\mathfrak{l}$ (cf. the definition in Chapter 17) is isomorphic with the direct sum $\mathfrak{sl}(2, C) \oplus \mathfrak{sl}(2, C)$. (Hence a complexification of a simple Lie algebra need not be simple, cf. Ch. 17.)

Hint. (a) Compare § 1.8. (d) Take as a basis of λ_c the elements $R^{\pm} = \pm R_2 + iR_1$, $R = -Ri_3$, $H^{\pm} = \pm H_2 + iH_1$, $H = -iH_3$ and compute the commutation relations.

6. (a) Show that the left Haar measure and the modular function for the group P of the unimodular triangular matrices are given in the coordinates α, β, defined by indicating $p = \begin{bmatrix} \alpha & \beta \\ 0 & \alpha^{-1} \end{bmatrix}$ by the formulas

$$\int_P f(p) \mathrm{d}p = -\frac{1}{4} \int_{C_*} \int_C f(\alpha, \beta), \alpha|^{-3} \mathrm{d}\alpha \mathrm{d}\bar{\alpha} \mathrm{d}\beta \mathrm{d}\bar{\beta},$$

$$\delta(p) = |\alpha|^{-4}.$$

(b) Compute the left Haar measure and the modular function δ_S on the group $S = AN$ and show that the left Haar measure on $G = \mathrm{SL}(2, C)$ is given by

$$\int_G f(g) \mathrm{d}g = \int_K \int_S f(ks) \delta_S(s^{-1}) \mathrm{d}k \mathrm{d}s,$$

where $\mathrm{d}k$, $\mathrm{d}s$ are the left Haar measures on K and S, respectively. Check that this is equivalent to (11.2.14).

Hint. Note that P is a semidirect product of M with S. Use Problem 4 § 1.7.

7. Show that the horocycles in $\mathscr{H}$ are precisely the sets $\{p \in \mathscr{H}: [p|s] = 1\}$ for $s \in S_+$ and non-Euclidean spheres are precisely $\{p \in \mathscr{H}: [p|q] = l\}$ for q inside the light cone, i.e. $[q|q] > 0$. What can be said about sets of the form $\{p \in \mathscr{H}: [p|v] = 1\}$ where $[v|v] < 0$, i.e. v is outside the light cone?

Hint. A pair of such sets is an equidistant surface with respect to a geodesic on $\mathcal{H}$.

8. Prove that the distance on $\mathcal{H}$ defined by $\cosh d(p, q) = [p|q]$ satisfies the triangle inequality, i.e. $d(p, q) \leqslant d(p, r) + d(r, q)$ for all p, q, r.

9. By exhibiting an invariant subspace of H^{-2k} prove that the representations $\mathcal{U}^{2k}$ are reducible.

10. Let $f \in \mathcal{D}(G/K)$, and let ω_ν be the spherical function corresponding to the representation $\mathcal{U}^{2i\nu}$ with real ν. Show that the integral $\int\limits_G \omega_\nu(g) f * f^*(g)\,dg$ depends only on the Fourier transform $\tilde{f}(\nu, b)$ of f and thus obtain a new proof of the fact that ω_ν is positive definite for real ν.

Hint. Cf. the computations leading to the proof of (11.4.9)

11. Let $T\colon L^2(G/K) \to L^2(G/K)$ be an integral operator $(Tf)(x) = \int\limits_{G/K} t(x, y) f(y)\,dm(y)$ with a continuous kernel t.

(a) Show that T intertwines the natural action of G on $L^2(G/K)$ if and only if $t(gw, hw) = k(h^{-1}g)$ for a certain function k bi-invariant under K.

(b) Under the condition of (a) show that

$$(Tf)^{\sim}(\nu, s) = \hat{k}(\nu)\tilde{f}(\nu, s),$$

i.e. that by means of the Fourier transform, T is reduced to the multiplication by the function $\hat{k}(\cdot)$—the Fourier transform of the kernel k.

(c) Show that $\|T\| = \sup\limits_{\nu \in R} |\hat{k}(\nu)|$, where $\|T\| = \sup\limits_{\|f\| \leqslant 1} \|Tf\|$ is the norm of the operator T.

12. (a) Show that for any continuous function φ on B the function $x \to \int\limits_B e_{i\nu+1}(x, s)\varphi(s)\,ds$ is an eigenfunction of the Laplace–Beltrami operator $\tilde{\mathcal{D}}$ on $\mathcal{H}$.

(b) Compute $(\widehat{Df})$ for $f \in \mathcal{D}(\mathcal{H})$.

Chapter 12

The Laguerre Polynomials

12.1. THE GROUP, THE REPRESENTATION, MATRIX ELEMENTS

The present chapter is devoted to a parallel study of representations of two groups, G_1 and G_2 of similar structures.

The group G_1 consists of triangular 3×3 matrices of the form

$$g(z,x)=\begin{bmatrix}1 & \bar{z} & \frac{1}{2}\bar{z}z+ix\\ 0 & 1 & z\\ 0 & 0 & 1\end{bmatrix},\qquad z\in\boldsymbol{C},\ x\in\boldsymbol{R}. \tag{12.1.1}$$

G_2 is the group of all triangular 4×4 matrices of the form

$$g(z_1,z_2,x)=\begin{bmatrix}1 & \bar{z}_1 & \bar{z}_2 & \frac{1}{2}(z|z)+ix\\ 0 & 1 & 0 & z_1\\ 0 & 0 & 1 & z_2\\ 0 & 0 & 0 & 1\end{bmatrix}, \tag{12.1.2}$$

where $x\in\boldsymbol{R}$, $z=(z_1,z_2)$, $z_n\in\boldsymbol{C}$, $n=1,2$, $(z|z):=|z_1|^2+|z_2|^2$.

We shall denote the elements of both groups in a uniform way by setting

$$g(z,x):=\begin{cases}g(z,x) & \text{for } G_1,\\ g(z_1,z_2,x) & \text{for } G_2\end{cases} \tag{12.1.3}$$

and defining the group operation by the formula

$$g(z,x)g(w,y):=g\big(z+w,x+y+\mathrm{Im}(z|w)\big), \tag{12.1.4}$$

where

$$(z|w):=\begin{cases}\bar{z}\,w & \text{for } G_1,\\ \bar{z}_1w_1+\bar{z}_2w_2 & \text{for } G_2.\end{cases}$$

Note that $g(z,x)^{-1}=g(-z,-x)$.

The subgroup H_n consisting of all elements of the form $g(0,x)$,

$x \in \boldsymbol{R}$ is normal in G_n and the quotient group G_n/H_n is isomorphic to $\boldsymbol{C}^n$ ($n = 1, 2$). The invariant measure on $\boldsymbol{C}^n$ has the form

$$\mathrm{d}z := \begin{cases} \mathrm{d}x\,\mathrm{d}y & \text{for } n = 1, \\ \mathrm{d}x_1\,\mathrm{d}x_2\,\mathrm{d}y_1\,\mathrm{d}y_2 & \text{for } n = 2. \end{cases} \tag{12.1.5}$$

There exists a continuous and even smooth mapping $s\colon \boldsymbol{C}^n \to G_n$ satisfying the condition $\pi\,(s(z)) = z\,(\pi\colon G_n \to G_n/H_n$ being a natural projection). This mapping is given by the formula

$$s(z) := g(z, 0).$$

This implies that there is a 1-1 correspondence between the reperesentations induced from the subgroup H_n and the multiplier representations on $\boldsymbol{C}^n$. This being so, we shall introduce a series of representations of the group G_n already in the multiplier form

$$\begin{aligned} &H_\lambda := L^2\left(\boldsymbol{C}^n, \frac{\lambda^n}{\pi^n}\exp\left(-\lambda(z|z)\mathrm{d}z\right)\right), \\ &(U^\lambda_{g(w,x)}\varphi)(z) := \exp\lambda(ix - \tfrac{1}{2}(w|w) - (w|z))\,\varphi(z+w), \end{aligned} \tag{12.1.6}$$

where λ is assumed to be greater than zero.

The representation U^λ is the multiplier form of the representation of the group G_n induced by the character $g(0, x) \to \mathrm{e}^{i\lambda x}$ of the subgroup H_n. We call it the *Bargmann representation.*

One can verify by direct calculation, without referring to the above fact, that U^λ is a unitary representation of the group G_n.

By way of example we compute for $g = g(w, x)$:

$$\begin{aligned} (U^\lambda_g\varphi|U^\lambda_g\varphi) &= \frac{\lambda^{2n}}{\pi^{2n}}\int\limits_{\boldsymbol{C}^n} \exp\big(-\lambda((z|z) + (w|w) \\ &\qquad + 2\,\mathrm{Re}(w|z))\big)|\varphi(z+w)|^2\mathrm{d}z \\ &= \frac{\lambda^{2n}}{\pi^{2n}}\int\limits_{\boldsymbol{C}^n} \exp\big(-\lambda(z+w|z+w)\big)|\varphi(z+w)|^2\mathrm{d}z \\ &= \frac{\lambda^{2n}}{\pi^{2n}}\int\limits_{\boldsymbol{C}^n} \exp\big(-\lambda(z|z)\big)|\varphi(z)|^2\mathrm{d}z = (\varphi|\varphi)_\lambda, \end{aligned}$$

which means that U^λ_g is unitary.

In order to check that U^λ is a representation it is enough to verify the multiplier identity for the function

$$G_n \times \boldsymbol{C}^n \ni ((w, x), z) \to \exp\big(-\lambda(ix - \tfrac{1}{2}(w|w) - (w|z))\big).$$

This identity means that

$$\exp\left(\lambda\left(ix'-\tfrac{1}{2}(w'|w')-(w'|z)\right)\right)\exp\left(\lambda\left(ix-\tfrac{1}{2}(w|w)\right.\right.$$
$$\left.\left.-(w|z+w')\right)\right)$$
$$=\exp\left(\lambda\left(i(x+x')+\mathrm{Im}(w'|w)-\tfrac{1}{2}(w+w'|w+w')-(w+w'|z)\right)\right),$$

which is obviously the case.

The space H_λ contains an invariant close subspace

$$\mathscr{H}_\lambda := \{f \in H : f \text{ is entire}\}.$$

All polynomials belong to the space H_λ and so it is non-empty. To prove that $\mathscr{H}_\lambda$ is closed in H_λ we observe that the theory of holomorphic functions (see e.g. Maurin, 1980) supplies us with an inequality which states that for every compact set $K \subset C^n$ there exists a constant $C > 0$ so that

$$\sup_K |f| \leqslant C \int_K |f|\,\mathrm{d}z \tag{12.1.7}$$

for every entire function f. By transforming the right-hand side of (12.1.7) we find that

$$\int_K |f|\,\mathrm{d}z = \int_K \exp\tfrac{1}{2}\lambda(z|z)\exp(-\tfrac{1}{2}\lambda(z|z))|f(z)|\,\mathrm{d}z$$
$$\leqslant C' \int_K \exp(-\lambda(z|z))|f(z)|^2\mathrm{d}z \leqslant C'\|f\|^2$$

in view of the Schwarz inequality in the space $L^2(K, \mathrm{d}z)$.

It follows from inequality (11.1.7) that the convergence of a sequence of functions in the sense of $\mathscr{H}_\lambda$ implies almost uniform convergence and the space of entire functions is complete relative to the latter. Therefore $\mathscr{H}_\lambda$ is closed in H_λ as claimed.

The restriction of the representation U^λ to the space $\mathscr{H}_\lambda$ will be denoted by $\mathscr{U}_\lambda$.

We shall fix a polynomial basis for the space $\mathscr{H}_\lambda$

$$e_m z := \begin{cases} \left[\dfrac{\lambda^m}{m!}\right]^{\frac{1}{2}} z^m & \text{for } G_1, \\[2ex] \left[\dfrac{\lambda^{m_1+m_2}}{m_1!m_2!}\right]^{\frac{1}{2}} z_1^{m_1} z_2^{m_2} & \text{for } G_2, \end{cases}$$

the index m denoting a non-negative integer in the case of the group G_1 and a pair (m_1, m_2) of such numbers in the case of G_2.

Let us also introduce the notation: $|m| := m_1+m_2$, $m! := m_1!\,m_2!$.

LEMMA 12.1.1. *The basis* $\{e_m\}$ *is orthonormal in* $\mathscr{H}_\lambda$.

Proof. The Lebesgue measure on C^n can be represented in terms of spherical coordinates on this space in the form

$$\int_{C^n} f(z)\mathrm{d}z = \int_0^\infty r^{2n-1}\mathrm{d}r \int_{S^{2n-1}} f(rs)\mathrm{d}s, \qquad n = 1, 2. \tag{12.1.8}$$

The sphere S^{2n-1} is diffeomorphic to the group SU(n) (§ 8.3) and the measure ds, being invariant under rotations in R^{2n}, is also invariant with respect to group multiplication, which is realized as a restriction of a certain rotation in R^{2n} to the sphere. It follows that ds is the Haar measure for SU(n). From Chapter 8 we know that, when restricted to SU(n), the polynomials e_m are mutually orthogonal in $L^2(\mathrm{SU}(n))$. From (12.1.8) we read that the vectors are also mutually orthogonal in $\mathscr{H}_\lambda$.

The normalization of the basis results from the identity

$$\int_{C^n} \exp(-(z|z))\mathrm{d}z = \pi^n$$

upon integration by parts.

The operators of complex differentiation on C^n will be written in a uniform way by applying the following convention: for $m = (m_1, m_2)$

$$\frac{\mathrm{d}^m}{\mathrm{d}z^m} := \frac{\mathrm{d}^{m_1}}{\mathrm{d}z_1^{m_1}}\frac{\mathrm{d}^{m_2}}{\mathrm{d}z_2^{m_2}}.$$

Now we proceed to the investigation of matrix elements of the representation $\mathscr{U}^\lambda$. In view of the special structure of the space $\mathscr{H}_\lambda$ two possibilities suggest themselves.

The Hilbert space structure allows us to compute in a standard way

$$u^\lambda_{nm}(g) = (e_n|\mathscr{U}^\lambda_g e_m).$$

On the other hand the decomposition of the vector $\mathscr{U}^\lambda_g e_m$ in the basis $\{e_j\}$ has the form

$$\mathscr{U}^\lambda_g e_m(z) = \sum_{|n|=0}^\infty (e_n|\mathscr{U}^\lambda_g e_m)e_n(z) = \sum_{|n|=0}^\infty \left[\frac{\lambda^{|n|}}{n!}\right]^{\frac{1}{2}} (e_n|\mathscr{U}^\lambda_g e_m)z^n$$

and therefore can be regarded as the power series expansion of an entire function. Indeed, as was shown above, convergence with respect to the norm on $\mathscr{H}_\lambda$ implies almost uniform convergence.

Using the well known formulas for the coefficients of the power expansion of a holomorphic function, we find

$$\left[\frac{\lambda^{|n|}}{n!}\right]^{\frac{1}{2}} (e_n|\mathscr{U}_g^\lambda e_m) = \frac{1}{n!}\frac{\mathrm{d}^n}{\mathrm{d}z^n}(\mathscr{U}_g^\lambda e_m)\bigg|_{z=0}. \tag{12.1.10}$$

First we shall compute the right-hand side of (12.1.10) for $\mathscr{U}_g := \mathscr{U}_g^1$, where $g = (w, 0)$:

$$u_{nm}(g) = \frac{1}{(n!m!)^{\frac{1}{2}}}\frac{\mathrm{d}^n}{\mathrm{d}z^n}\left(\exp\left(-\tfrac{1}{2}(w|w+z)\right)(z+w)^m\right)\Big|_{z=0}.$$

Since translations commute with differentiation, we get

$$u_{nm}(g) = \frac{1}{(n!m!)^{\frac{1}{2}}}\exp\tfrac{1}{2}(w|w)\frac{\mathrm{d}^n}{\mathrm{d}z^n}\left(\exp\left(-(w|z)\right)z^m\right)\Big|_{z=w}.$$

After the substitution $s_i := \bar{w}_i z_i$, $s = (s_1, s_2)$, this becomes

$$u_{nm}(g) = \frac{1}{(n!m!)^{\frac{1}{2}}}\exp\tfrac{1}{2}(w|w)\,\bar{w}^{n-m}\frac{\mathrm{d}^n}{\mathrm{d}s^n}\left(\exp(-s)s^m\right)\Big|_{s=(w|w)}. \tag{12.1.11}$$

The matrix elements prove to be closely related with the Laguerre polynomials, which are defined by the formula

$$L_n^\alpha(x) = \mathrm{e}^x\frac{x^{-\alpha}}{n!}\frac{\mathrm{d}^n}{\mathrm{d}x^n}(\mathrm{e}^{-x}x^{n+\alpha}), \qquad n = 0, 1, 2, \ldots;\ \alpha > -1. \tag{12.1.12}$$

Clearly

$$n!\mathrm{e}^{-x}x^\alpha L_n^\alpha(x) = \frac{\mathrm{d}^n}{\mathrm{d}x^n}(\mathrm{e}^{-x}x^{n+\alpha}).$$

If we confine ourselves to integer values of the parameter α, then both sides of this identity extend to entire functions on $\boldsymbol{C}$. By putting $m = n+\alpha$ we have for $n \leqslant m$

$$\frac{\mathrm{d}^n}{\mathrm{d}z^n}(\mathrm{e}^{-z}z^m) = n!\mathrm{e}^{-z}z^{m-n}L_n^{m-n}(z).$$

We adopt this formula as a definition of the function L_n^{m-n} for $n > m$. Then (12.1.11) can be written as

$$u_{nm}(g) = \left(\frac{n!}{m!}\right)^{\frac{1}{2}} \exp(-\tfrac{1}{2}(w|w))\,\overline{w}^{m-n} L_{n_1}^{m_1-n_1}(|w_1|^2) L_{n_2}^{m_2-n_2}(|w_2|^2)$$

and symbolically

$$u_{nm}(g) = \left(\frac{n!}{m!}\right)^{\frac{1}{2}} (\exp -\tfrac{1}{2}(w|w))\;\overline{w^{m-n}} L_n^{m-n}((w|w)).$$

Returning to the general case, we note that

$$u_{nm}^{\lambda}((w,0))e_m(z) = u_{nm}^{1}((\lambda^{\frac{1}{2}}w, 0))e_m(\lambda^{\frac{1}{2}}z)$$

and therefore

$$u_{nm}^{\lambda}(g) = \left(\frac{n!}{m!}\right)^{\frac{1}{2}} \exp\left(-\frac{1}{2}\lambda(w|w)\right) \overline{w}^{m-n} \lambda^{\frac{m-n}{2}} L_n^{m-n}(\lambda(w|w)). \tag{12.1.13}$$

12.2. BASIC PROPERTIES OF THE LAGUERRE POLYNOMIALS

We shall now apply formula (12.1.13) to represent expansion (12.1.9) of the representation of the group G_1. We have

$$\left(\frac{\lambda^m}{m!}\right)^{\frac{1}{2}} \exp(-\lambda(\tfrac{1}{2}w\overline{w} - \overline{w}z))(z+w)^m$$

$$= \left(\frac{\lambda^m}{m!}\right)^{\frac{1}{2}} \exp(-\tfrac{1}{2}\lambda\overline{w}w)w^m \sum_{n=0}^{\infty} \overline{w}^{-n} L_n^{m-n}(\lambda|w|^2)z^n.$$

Putting $w = 1$, we obtain after obvious simplifications

$$\exp(-\lambda z)(z+1)^m = \sum_{n=0}^{\infty} L_n^{m-n}(\lambda)z^n. \tag{12.2.1}$$

Thus we have a generating function for the family of the polynomials L_n^{m-n}. It can be expected that by computing the matrix elements directly from the definition we shall obtain integral expressions for the functions L_n^{m-n}. So we multiply both sides of identity (12.2.1) by $\overline{z}^k$

and integrate the expression with respect to the measure $\exp(-\lambda|z|^2)\mathrm{d}z$. In view of the orthogonality of the system z^n we obtain

$$\frac{\lambda^{k+1}}{\pi k!}\int_C \exp(-\lambda(z+|z|^2)(z+1)^m\bar{z}^k \mathrm{d}z = L_k^{m-k}(\lambda).$$

Assuming $m-k =: l \geqslant 0$, we obtain

$$L_k^l(\lambda) = \frac{\lambda^{k+1}}{\pi k!}\int_C \exp(-\lambda(z+|z|)^2)(z+1)^{k+1}\bar{z}^k \mathrm{d}z. \tag{12.2.2}$$

Next we shall derive in a standard way the addition formulas for the Laguerre polynomials. To this end we set $g_1 = g(r_1, 0), g_2 = g(r_2 e^{i\varphi}, 0)$. Then $g_1 g_2 = g(r_1 + r_2 e^{i\varphi}, r_1 r_2 \sin\varphi)$. Now we apply the formula

$$u_{nm}(g_1 g_2) = \sum_{k=0}^{\infty} u_{nk}(g_1) u_{km}(g_2).$$

The definition of the representation $\mathscr{U}$ gives the identity

$$u_{nm}(g(r_1+r_2 e^{i\varphi}, r_1 r_2 \sin\varphi)) = \exp(ir_1 r_2 \sin\varphi) u_{nm}(g(r_1+r_2 e^{i\varphi}, 0)).$$

Write $r_1 + r_2 e^{i\varphi} = re^{i\psi}$. Then

$$\begin{aligned} r &= (r_1^2+r_2^2+2r_1 r_2\cos\varphi)^{\frac{1}{2}}, \\ e^{2i\psi} &= \frac{r_1+r_2 e^{i\varphi}}{r_1+r_2 e^{-i\varphi}}. \end{aligned} \tag{12.2.3}$$

Combining the above with formula (12.1.13) for $\lambda = 1$, we obtain after simple manipulation

$$\begin{aligned} &\exp(r_1 r_2 e^{-i\varphi}) \sum_{k=0}^{\infty} r_1^{k-n} L_n^{k-n}(r_1^2)(r_2 e^{i\varphi})^{m-k} L_k^{m-k}(r_2^2) \\ &= (re^{i\psi})^{m-n} L_n^{m-n}(r^2). \end{aligned} \tag{12.2.4}$$

12.3. DIFFERENTIAL PROPERTIES OF THE LAGUERRE POLYNOMIALS

The Lie algebra $\mathfrak{g}_1$ of the group G_1 is 3-dimensional. We define a basis for $\mathfrak{g}_1$ comprised of the vectors X_i corresponding to the one-parameter subgroups a_i:

$$a_1(t) := g\left(\frac{t}{\sqrt{2}}, 0\right), \quad a_2(t) = g\left(\frac{it}{\sqrt{2}}, 0\right), \quad a_3(t) = g(0, t). \tag{12.3.1}$$

The following commutation relations hold among the vectors X_i:

$$[X_1, X_3] = [X_2, X_3] = 0, \qquad [X_1, X_2] = X_3. \tag{12.3.2}$$

Representations of the algebra $\mathfrak{g}_1$ such that X_3 is represented by a non-zero scalar, are called *classical commutation relations*. The representation of X_2 and X_1 as the position operator $\hat{x}: f \to xf$ and the momentum $\hat{p}: f \to \frac{\partial}{\partial x} f$, respectively, serves as a fundamental example of such a situation.

We shall consider the representations of the algebra $\mathfrak{g}_1$ related to the regular representation of the group and to the representations $\mathcal{U}^\lambda$.

By differentiating (12.1.6) we easily find

$$\begin{aligned} X_1 &= \mathrm{d}\mathcal{U}^\lambda(X_1) = \frac{1}{\sqrt{2}}\left(\frac{\mathrm{d}}{\mathrm{d}z} - \lambda z\right), \\ X_2 &= \mathrm{d}\mathcal{U}^\lambda(X_2) = \frac{i}{\sqrt{2}}\left(\frac{\mathrm{d}}{\mathrm{d}z} + \lambda z\right), \\ X_3 &= \mathrm{d}\mathcal{U}^\lambda(X_3) = i\lambda. \end{aligned} \tag{12.3.3}$$

The elements of the basis e_m belong to the domains of these operators and satisfy the relations

$$X_1 e_m = \sqrt{\tfrac{1}{2}\lambda}\left(\sqrt{m}\, e_{m-1} - \sqrt{m+1}\, e_{m+1}\right), \tag{12.3.4}$$

$$X_2 e_m = i\sqrt{\tfrac{1}{2}\lambda}\left(\sqrt{m}\, e_{m-1} + \sqrt{m+1}\, e_{m+1}\right). \tag{12.3.5}$$

Put

$$F_{nm}(t) := u^\lambda_{nm}\big(g(t, 0)\big).$$

Owing to the intertwining property of the operation of forming a matrix element we get by (12.3.4)

$$\frac{\mathrm{d}F_{nm}}{\mathrm{d}t} = \sqrt{m}\, F_{n,m-1} - \sqrt{m+1}\, GF_{n,m+1}. \tag{12.3.6}$$

On the other hand, in view of (12.3.5) and the identity

$$\left(e_n \middle| \mathcal{U}^\lambda_{g\left(w+i\frac{t}{2},\, 0\right)} e_m\right) e^{iw\frac{t}{\sqrt{2}}} = (e_n | \mathcal{U}^\lambda_{g(w,\,0)}) \mathcal{U}^\lambda_{g\left(i\frac{t}{\sqrt{2}},\, 0\right)} e_m), \quad w, t \in \boldsymbol{R}^1,$$

we get the relation

$$t\frac{\mathrm{d}}{\mathrm{d}t} F_{nm} + \frac{m-n}{t} F_{nm} = \sqrt{m}\, F_{n,m-1} + \sqrt{m+1}\, F_{n,m+1}. \tag{12.3.7}$$

Thanks to identities (12.3.4) and (12.3.5) it is easy to define on the representation space of $\mathcal{U}^\lambda$ operators of "raising" and "lowering" the indices:

$$E_+ = -X_1 - iX_2, \qquad E_- = X_1 - iX_2. \tag{12.3.8}$$

These operators verify the relations

$$E_+ e_m = \sqrt{2\lambda(m+1)}\, e_{m+1}, \qquad E_- e_m = \sqrt{2\lambda m}\, e_{m-1}. \tag{12.3.9}$$

The composite $E_+ E_-$ is diagonal relative to the basis $\{e_m\}$:

$$E_+ E_- e_m = 2\lambda m e_m.$$

Because of this identity (for $\lambda = 1$) we obtain a second order differential equation for the function F_{nm}.

Simple computations which consist in combining formulas (12.3.5) and (12.3.7) lead to the relation

$$F''_{nm} + \frac{1}{t} F'_{nm} - \left[t^2 - 2(m+n+1) + \frac{(m-n)^2}{t^2} \right] F_{nm} = 0. \tag{12.3.10}$$

Formulas (12.3.5), (12.3.7) and (12.3.10) also express the recurrence and differential properties of the Laguerre polynomials. Representing the matrix element in the form (12.1.13) we obtain after transformations

$$\begin{gathered} \frac{d}{dx} L_n^k(x) = -L_n^{k+1}(x), \\ x \frac{d}{dx} L_n^k(x) + (k-x) L_n^k(x) = (n+1) L_{n+1}^{k-1}(x) \end{gathered} \tag{12.3.11}$$

and

$$x \frac{d^2}{dx^2} L_n^k(x) + (k+1-x) \frac{d}{dx} L_n^k(x) + n L_n^k(x) = 0. \tag{12.3.12}$$

Moreover, by combining the identities in (12.3.11) we obtain yet another relation

$$L_n^k = L_n^{k+1} - L_{n-1}^{k+1}.$$

The above properties can also be derived directly from the definition of the Laguerre polynomial in terms of the generating function (12.2.1). We have chosen a different approach, not only to preserve the connection with representation theory but because in many physical applications it is exactly functions, F_{mn}, i.e. the matrix elements of representations, that we have to deal with. We shall now present some connections between identity (12.3.10) and the equations of quantum physics. Namely, if we put $M = m-n$ and $E = m+n+1$ in (12.3.10), we obtain

$$-\frac{1}{2r}\frac{\mathrm{d}}{\mathrm{d}r}r\frac{\mathrm{d}\psi}{\mathrm{d}r}+\left[\frac{M^2}{2r^2}+\frac{1}{2}r^2\right]\psi = E\psi. \tag{12.3.13}$$

This is the radial part of the 2-dimensional isotropic harmonic oscillator (relative to the system of units in which the Planck constant and the coefficient of elasticity equal 1). A solution of the general equation of the harmonic oscillator is given in polar coordinates in $\boldsymbol{R}^2$ in the form $\psi(r)\mathrm{e}^{iM\varphi}$. It follows from equation (12.3.10) that the functions $F_{\frac{1}{2}(E-M-1),\frac{1}{2}(E+M-1)}$ are the radial parts of solutions of the equation of the oscillator with energy E; E and M assume such values that the parameters of the function F are integer numbers. In particular, for a given value of $E \in N$ there exists an M such that $F_{\frac{1}{2}(E-M-1),\frac{1}{2}(E+M-1)}$ is the radial part of a solution of the equation of the harmonic oscillator. Equation (12.3.10) is also connected with the wave function of the hydrogen atom. Let us introduce the variable $r = \frac{m+n+1}{4}t^2$ and the function $w(r) = t^{-1}F_{nm}(\mathrm{t})$. Let $\nu = \frac{n+m+1}{2}$ and $l = \frac{m-n-1}{2}$.

It follows from (12.3.10) that the function w is a solution of the equation which is the radial part of the Schrödinger equation for the hydrogen atom

$$\frac{1}{2}\frac{1}{r^2}\frac{\mathrm{d}}{\mathrm{d}r}r^2\frac{\mathrm{d}w}{\mathrm{d}r}+\frac{l(l+1)}{2r^2}w-\frac{1}{r}w = -\frac{1}{2n^2}w. \tag{12.3.14}$$

Thus the function

$$R_{\nu l}(r) := \frac{2}{\nu^2}\left[\frac{(\nu-l-1)!}{(\nu+l)!}\right]^{\frac{1}{2}}\left(\frac{2r}{\nu}\right)^l \mathrm{e}^{-\frac{r}{\nu}}L_{\nu-l-1}^{2l+1}\left(\frac{2r}{\nu}\right)$$

is a solution of equation (12.3.14) and, moreover, owing to the choice of the coefficient, is a normalized vector in $L^2(\boldsymbol{R}_+, r^2\mathrm{d}r)$.

12.4. ONE-DIMENSIONAL HARMONIC OSCILLATOR AND THE HERMITE POLYNOMIALS

The basis $\{e_m\}$ introduced in order to compute the matrix elements of the representation $\mathscr{U}^\lambda$ consists, as we proved in § 12.3 of eigenvectors of the operator E_+E_- and, moreover, the image under the operator E_+ E_- of an element e_m is proportional to the element e_{m+1}, e_{m-1}, respectively. From these observations we have already derived the recurrence formulas and the differential equations for functions L_n^k. It would be interesting to find a basis for which $\boldsymbol{X}_1, \boldsymbol{X}_2$ would be the operators of raising and lowering the indices. This would amount to finding a family of solutions of the eigenproblem for the operator $\boldsymbol{X}_1\boldsymbol{X}_2$, which is closely related to the Hamiltonian of the harmonic oscillator. We need some new symbols:

$$J_+ := \frac{1}{\sqrt{2}}\left(\frac{\mathrm{d}}{\mathrm{d}s} - z\right) = X_1, \qquad J_- := \frac{1}{\sqrt{2}}\left(-\frac{\mathrm{d}}{\mathrm{d}z} - z\right) = iX_2. \tag{12.4.1}$$

Then

$$J_+J_- = -\frac{1}{2}\frac{\mathrm{d}^2}{\mathrm{d}z^2} + \frac{1}{2}z^2 - \frac{1}{2}id =: H - \frac{1}{2}E,$$

where E denotes the identity matrix and H is the Hamiltonian of the harmonic oscillator

$$H = -\frac{1}{2}\left(\frac{\mathrm{d}^2}{\mathrm{d}z^2} - z^2\right). \tag{12.4.2}$$

The operators $J_\pm$, H, E span a 4-dimensional Lie algebra with the commutation relations

$$[H, J_\pm] = \pm J_\pm, \qquad [J_+, J_-] = -E, \qquad [H, E] = [J_\pm, E] = 0.$$

The commutation relations imply that if ψ is an eigenfunction of the operator H with the eigenvalue λ, then

$$H(J_\pm\psi) = (\lambda \pm 1)J_\pm,$$

i.e. $J_\pm$ ψ is also an eigenvector.

The considerations which follow are based on the observation that there exists an entire function on $\boldsymbol{C}$ annihilated by the operator J namely the function $\psi_0(z) = \pi^{-\frac{1}{4}}\exp(-\frac{1}{2}z^2)$. It is an eigenfunction of the operator H with the eigenvalue $\frac{1}{2}$.

Next we define reccurrently

$$(n+1)^{\frac{1}{2}}\psi_{n+1} = J_+\psi_n, \qquad n = 1, 2, 3, \ldots \tag{12.4.3}$$

The vector ψ_n is an eigenvector of the operator H with the eigenvalue $n+\frac{1}{2}$. We observe that ψ_0 and also ψ_n are not elements of the space H_λ; nevertheless they are entire functions. Therefore we shall consider the operators $J_\pm$ and the representation operators of $\mathscr{U}^\lambda$ on the larger space $\mathscr{E}$ of all entire functions. By inequality (12.1.7) the operators $J_\pm$ are continuous on $\mathscr{E}$ in the topology of almost uniform convergence. In particular, it follows that the family of operators

$$U(t) = \exp tJ_+ = \sum_{n=0}^{\infty} \frac{t^n}{n!} J_+^n$$

is well defined on $\mathscr{E}$.

The infinitesimal generator of this one-parameter subgroup equals J_+. As we remember, the operator J_+ was defined as the infinitesimal generator of the one-parameter subgroup $\mathscr{U}\left(g\left(\frac{t}{\sqrt{2}},0\right)\right)$. These subgroups must therefore be identical:

$$\exp\left(-\tfrac{1}{4}t^2 - \frac{t}{\sqrt{2}}z\right)\psi\left(z+\frac{t}{\sqrt{2}}\right)$$

$$= \sum_{n=0}^{\infty} \frac{t^n}{n!}(J_+^n\psi)(z) \qquad \text{for} \quad \psi \in \mathscr{E}, \quad t \in \boldsymbol{R}. \tag{12.4.4}$$

By applying (12.4.4) to the function ψ_0 and substituting $\alpha = \frac{t}{\sqrt{2}}$ we obtain

$$\pi^{-\frac{1}{4}}\exp(\alpha^2 - 2z\alpha - \tfrac{1}{2}z^2) = \sum_{n=0}^{\infty} \frac{2^{\frac{1}{2}n}}{(n!)^{\frac{1}{2}}} \alpha^n \psi_n(z). \tag{12 4.5}$$

Thus we have found a generating function for the family ψ_n. Now it is easy to establish a connection between these functions and the Hermite polynomials. Namely, if we define functions H_n by means of the generating function

$$\exp(-\beta^2+2\beta z) = \sum_{n=0}^{\infty} \beta^n H_n(z)/n!, \tag{12.4.6}$$

then by expanding the left-hand side of the identity in the power series with respect to β we find

$$H_n(z) = (-1)^n e^{z^2} \frac{d^n}{dz^n} e^{-z^2}, \qquad n = 0, 1, 2, \ldots$$

Thus the functions H_n agree with the Hermite polynomials, which were defined by the above formula in Chapter 3. Comparing (12.4.5) with (12.4.6), we see that

$$\psi_n(z) = \pi^{-\frac{1}{4}}(n!)^{-\frac{1}{2}}(-1)^n 2^{-\frac{1}{2}n} \exp(-\tfrac{1}{2}z^2) H_n(z). \tag{12.4.7}$$

Recalling (see Chapter 3) that

$$\int_{-\infty}^{\infty} H_n(x) H_{n'}(x) \exp(-x^2) dx = \pi^{\frac{1}{2}} 2^n n!\, \delta_{nn'},$$

we observe that the functions ψ_n when restricted to the real axis, constitute an orthonormal system in $L^2(\boldsymbol{R})$. Since the *Hermite functions* $H_n e^{-\frac{1}{2}x^2}$ constitute a complete system in this space, $\{\psi_n\}_0^\infty$ is also a basis for $L^2(\boldsymbol{R})$.

We shall formulate the results of these considerations in the following manner.

THEOREM 12.4.1. *The system of functions* $\{\psi_n\}$ *defined by formula* (12.4.7) *is an orthonormal basis for the space* $L^2(\boldsymbol{R})$, *which consists of eigenvectors of the operator*

$$H = -\frac{1}{2}\left(\frac{d^2}{dz^2} - z^2\right)$$

with eigenvalues $n+\frac{1}{2}$, $n = 0, 1, 2, \ldots$

12.5. CONNECTION BETWEEN THE LAGUERRE POLYNOMIALS AND THE JACOBI FUNCTIONS

We shall imbed every group G_j, $j = 1, 2$ in the semi-direct product Γ_j of $\mathrm{SU}(j)$ and G_j.

The elements of the group Γ_j will be written as $g(k, z, x)$, $k \in \mathrm{SU}(j)$, $g(z, x) \in G_j$. Group multiplication in Γ_j is given by the formula

$$\begin{aligned} &g(k_1, z_1, x_1) g(k_2, z_2, x_2) \\ &:= g\big(k_1 k_2, {}^t k_2 \cdot z_1 + z_2, x_1 + x_2 + \mathrm{Im}({}^t k_2 \cdot z_1 | z_2)\big), \end{aligned} \tag{12.5.1}$$

where ${}^t k \cdot z$ denotes a natural action on C^j of the transpose of the matrix k.

The representation $\mathscr{U}^\lambda$ extends to a representation of the group Γ_j thanks to the property that the action of $\mathrm{SU}(j)$ on C^j is holomorphic. Namely, we define

$$\begin{aligned} T^\lambda_{g(k,w,x)} \varphi(z) &:= (\mathscr{U}^\lambda_{g(w,x)} \varphi)({}^t k \cdot z) \\ &= \exp\Big(\lambda\big(ix - \tfrac{1}{2}(w|w) - (w|{}^t k \cdot z)\big)\Big) \varphi({}^t k \cdot z + w). \end{aligned} \tag{12.5.2}$$

We shall confine ourselves to the study of the representation $T := T^1$ of the group G_2. The restriction of T to the subgroup $\mathrm{SU}(2)$ coincides with the action of the group on the space of polynomials on C^2. In Chapter 8 this action was used to construct irreducible representations of the group $\mathrm{SU}(2)$ and to determine the corresponding matrix elements, which, as we remember, are connected with the Jacobi functions.

Retaining the notation introduced in Chapter 8 we shall express an element e_m, $m = (m_1, m_2)$ of the basis for H_1 as a vector in $H^{\frac{1}{2}|m|}$:

$$e_m(z) = e_{\frac{1}{2}(m_1 - m_2)}(z_1, z_2).$$

Thus the matrix elements of the representation T are also determined by the matrix elements of the representation T^1 of the group $\mathrm{SU}(2)$. Formula (8.1.2) can now be written as

$$T_k e_m = \sum_{|n| = |m|} t^{\frac{1}{2}|m|}_{\frac{1}{2}(n_1 - n_2), \frac{1}{2}(m_1 - m_2)}(k) e_n = \sum_{n=0}^{\infty} T_{nm}(k) e_n. \tag{12.5.3}$$

We see that $T_{nm}(k)$ is a block matrix. In the special case where

$$k(\theta) = \begin{bmatrix} \cos\frac{1}{2}\theta & i\sin\frac{1}{2}\theta \\ -i\sin\frac{1}{2}\theta & \cos\frac{1}{2}\theta \end{bmatrix}$$

we obtain

$$T_{k(\theta)}e_m = \sum_{|n|=|m|} P^{\frac{1}{2}|m|}_{\frac{1}{2}(n_1-n_2),\frac{1}{2}(m_1-m_2)}(\cos\theta)\,e_n. \tag{12.5.4}$$

We shall make use of the identity

$$kg(0,z,0)k^{-1} = g(0,\bar{k}\cdot z,0) \tag{12.5.5}$$

($\bar{k}$ denoting complex conjugation of matrices), which is valid for arbitrary $k \in \mathrm{SU}(2)$ and $z \in \boldsymbol{C}^2$.

In the language of matrix elements this identity means that

$$\sum_{k,m} T_{jk}(k)\mathcal{U}_{km}(g(z,0))T_{mn}(k^{-1})$$
$$= \sum_{k,m} T_{jk}(k)\mathcal{U}_{km}(g(z,0))\overline{T_{nm}(k)} = \mathcal{U}_{jn}(\bar{k}\cdot z). \tag{12.5.6}$$

The coordinates of the vector $\bar{k}\cdot z$ are $(z_1\cos\frac{1}{2}\theta,\ z_2\sin\frac{1}{2}\theta)$. We substitute in (12.5.6) the matrix elements of the representation and then apply formula (12.5.4). Writing $l_1 := \frac{1}{2}|j|$, $l_2 := \frac{1}{2}|m|$, $j := j_1$, $m := m_1$, we obtain

$$\sum_{k=0}^{\min(2l_1,2l_2)} (-1)^{k-j}\sqrt{\frac{(2l_1-k)!}{(2l_2-k)!}}\,P^{l_1}_{l_1-j,l_1-k}(\cos\theta)P^{l_2}_{l_2-m,l_2-k}(\cos\theta)$$
$$\times L^{2(l_2-l_1)}_{2l_1-k}(|z|^2) = \sqrt{\frac{j!(2l_1-j)!}{m!(2l_2-m)!}}(i\sin\tfrac{1}{2}\theta)^{m-j}$$
$$\times(\cos\tfrac{1}{2}\theta)^{2l_2-2l_1-m+j}L^{m-j}_j(|z|^2\sin^2\tfrac{1}{2}\theta)$$
$$\times L^{2l_2-2l_1-m-j}_{2l_1-1}(|z|^2\cos^2\tfrac{1}{2}\theta). \tag{12.5.7}$$

In the special case where $l_1 = l_2 = 1$, $j = m$ we get

$$\sum_{k=0}^{2l}(P^l_{l-j,l-k}(\cos\theta))^2L^0_{2l-k}(r) = L^0_j(r\sin^2\tfrac{1}{2}\theta)L^0_{2l-j}(r\cos^2\tfrac{1}{2}\theta). \tag{12.5.8}$$

Next, by putting $j = m = 0$ we obtain

$$\sum_{k=0}^{m} \frac{(p+m)!}{k!(p+m-k)!}(1-y)^k y^{m-k} L_{m-k}^p(x) = L_m^p(xy) \qquad (12.5.9)$$

for $x > 0$, $0 \leqslant y < 1$.

Using (12.5.8), we can derive a "nice" integral formula for the Laguerre polynomials. Multiplying both sides of (12.5.8) by $\sin\theta$ and integrating over $[0, \pi]$ with respect to θ, we find owing to the orthogonality relations for P_{mn}^l:

$$\int_0^\pi L_j^0(r\sin^2\tfrac{1}{2}\theta)\, L_{2l-j}^0(r\cos^2\tfrac{1}{2}\theta)\sin\theta\, d\theta$$

$$= \frac{2}{2l+1}\sum_{k=0}^{2l} L_{2l-k}^0(r). \qquad (12.5.10)$$

Further implications of formula (12.5.7) will be given in the problems.

12.6. ORTHOGONALITY RELATIONS FOR THE LAGUERRE POLYNOMIALS

The representation $\boldsymbol{T}$ is irreducible. We are again concerned with the representation of the group Γ_1 and therefore we shall prove this statement for Γ_1 only; the case of Γ_2 can be treated analogously.

The basis $\{e_m\} \subset \mathscr{H}_1$ consists of eigenvectors of the subgroup SU(1) which correspond to different eigenvalues.

A Γ_1-invariant subspace $E \subset \mathscr{H}_1$ decomposes into a direct sum of eigenspaces of the operators $T(k)$. Thus E contains an element of the basis—a monomial e_{i_0}. On the other hand, the space E is also invariant under the operator $E_\pm$ defined by (12.3.6). Thus it follows from recurrence formulas (12.3.7) that E contains all elements of the basis. Consequently $E = \mathscr{H}_1$, which proves the irreducibility of $\boldsymbol{T}$.

The representation $\boldsymbol{T}$ assumes scalar values on the elements of the subgroup $\mathscr{H}_1 \subset \Gamma_1$ of the form $g(1, 0, x)$. The elements of the form $g(1, 0, 2\pi n)$ for integer values of n are represented by the identity operator.

The subgroup which constitutes the kernel of the representation $\boldsymbol{T}$ is normal in Γ_1 and will be denoted by Z.

In what follows, $\boldsymbol{T}$ will denote the representation of the group $\Gamma := \Gamma_1/Z$ obtained by projecting onto Γ the representation of the group Γ_1. The representation $\boldsymbol{T}$ acts on the space $\mathscr{H}_1$ of entire functions on $\boldsymbol{C}$. In this space convergence in the sense of the inner product on $L^2(\boldsymbol{C}^1, \mathrm{e}^{-(z|z)}\,\mathrm{d}z)$ implies almost uniform convergence. Therefore the functional on $\mathscr{H}_1$ which to a function $f \in \mathscr{H}_1$ assigns its value at $z = 0$ is continuous on $\mathscr{H}_1$.

For $f \in \mathscr{H}_1$ let

$$\eta(f) = f(0) \qquad \text{and} \qquad S(f)(g) = \eta(T_g f).$$

Substituting the value of the operator T_g we have

$$S(f)(g) = \exp\left(ix - \tfrac{1}{2}(w|w|)\right) f(w)$$

if $g = g(k, w, xZ)$.

The operator S, being by definition a matrix element, intertwines $\boldsymbol{T}$ with the right regular representation of the group on the space $\mathscr{C}(\Gamma)$.

LEMMA 12.6.1. *Subject to a suitable normalization of the Haar measure, the operator S is an isometry on $L^2(\Gamma)$.*

Proof. It is natural to fix the Haar measure on Γ in the form $\mathrm{d}g(k, w, x) = \mathrm{d}k\,\mathrm{d}w\,\mathrm{d}x$, where $\mathrm{d}k$ and $\mathrm{d}x$ are normalized measures on SU(1). Then

$$\int_{\Gamma} |Sf(g)|^2 \mathrm{d}g = \int_{\boldsymbol{C}^1} \mathrm{e}^{-(w|w)} |f(w)|^2 \mathrm{d}w = \|f\|^2_{H_1}. \qquad \square$$

We see that $\boldsymbol{T}$ has proved to be an irreducible subrepresentation of the regular representation of the group Γ on $L^2(\Gamma)$ and hence a representation of the L^2-type (cf. § 4.5).

The Peter–Weyl theory of compact group representations also applies, subject to pertinent changes to representations of the L^2-type. An essential difference is that L^2-representations of a non-compact group are infinite-dimensional. We recall the following theorem, proved in Chapter 4.

THEOREM 12.6.2 (R. Godement, 1947, see Dixmier, 1964). *For every irreducible L^2-representation $\boldsymbol{T}$ of a locally compact group G there exists* *a positive number d_T, called the formal dimension of the representation T, such that the following generalized orthogonality relations hold:*

$$\int_G \overline{(T_g a|b)}(T_g a_1|b_1)\mathrm{d}g = \frac{1}{d_T}(a_1|a)(b|b_1). \tag{12.6.1}$$

This theorem leads in particular to the orthogonality relations for the Laguerre polynomials.

Leaving temporarily open the problem of the formal dimension of the representation in question, we shall apply formula (12.6.1) to the elements of the basis $\{e_n\}$ of H_1

$$\int_G T_{nm}(g)\,\overline{T}_{jl}(g)\mathrm{d}g = \frac{1}{d_T}\delta_{ml}\,\delta_{nj}. \tag{12.6.2}$$

Put $m = l$, $n = j$ and note that

$$T_{nm}(g(k, w, x)) = k^n \exp(ix) u_{nm}(g(w, 0)).$$

By substituting the value of u_{nm} given by (12.1.13) we arrive at

$$\frac{1}{d_T} = \frac{1}{\pi}\frac{n!}{m!}\int_{C^1} \exp(-|w|^2)|w|^{2(m-n)}L_n^{m-n}(|w|^2)\mathrm{d}w$$

$$= \frac{n!}{m!}\int_0^\infty \mathrm{e}^{-r} r^{m-n} L_n^{m-n}(r)\mathrm{d}r.$$

Writing $m-n = k$, we have

$$\frac{1}{d_T} = \frac{n!}{(n+k)!}\int_0^\infty \mathrm{e}^{-r} r^k L_n^k(r)\mathrm{d}r.$$

Now it is enough to put $k = n = 0$ ($L_0^0 \equiv 1$) in order to obtain an integral which can be computed by elementary means. As a result we get $\mathrm{d}_T = 1$. The general orthogonality relationship for the Laguerre polynomial which follow from (12.6.2) have the form

$$\int_0^\infty \mathrm{e}^{-r} r^k L_m^k(r) L_n^k(r)\mathrm{d}r = \frac{n!}{(n+k)!}\delta_{nm}.$$

PROBLEMS

1. Prove that the group Γ_2 is isomorphic to the group of matrices of the form

$$g(k,z,x) = \begin{bmatrix} 1 & \bar{z}_1 & \bar{z}_2 & \frac{1}{2}(z|z)+ix \\ 0 & \bar{k}_{11} & \bar{k}_{12} & z_1 \\ 0 & \bar{k}_{21} & \bar{k}_{22} & z_2 \\ 0 & 0 & 0 & 1 \end{bmatrix}, \qquad k = [k_{ij}] \in \mathrm{SU}(2).$$

2. Prove that the group G_1 is isomorphic to the Heisenberg-Weyl group of matrices of the form

$$\begin{bmatrix} 1 & x & z \\ 0 & 1 & y \\ 0 & 0 & 1 \end{bmatrix}, \qquad \text{where } x,\ y,\ z \in \boldsymbol{R}^1.$$

3. Prove that the subgroup H_i constitutes the centre of the group G_i.

4. Prove the identity

$$L_k^{m-k}(x) = (-1)^{k-m}\frac{\Gamma(m+1)}{\Gamma(k+1)}x^{k-m}L_m^{k-m}(x) \qquad \text{for } k > m.$$

5. Apply addition formula (12.2.4) to prove the identities

$$\sum_l \left(\frac{x}{y}\right)^l L_l^{m-l}(x^2)L_n^{l-n}(y^2)$$

$$= \left(1+\frac{y}{x}\right)^m\left(1+\frac{x}{y}\right)^{-n} e^{-xy}L_n^{m-n}((x+y^2))$$

and

$$\sum_l \left(\frac{-iy}{x}\right)^l L_l^{m-l}(x^2)L_n^{l-n}(y^2)$$

$$= \left[1-\frac{iy}{x}\right]^m\left[1+\frac{ix}{y}\right]^{-n} e^{ixy}L_n^{m-n}(x^2+y^2).$$

6. Show that

$$L_n^l x = \sum_{k=0}^{n} \frac{(n+l)!}{(k+l+1)!}\,\frac{(-x)^k}{k!(n-k)!}.$$

7. Prove that the generating function for the family of functions $L_n^\alpha(x)$, $n = 0, 1, 2$ is the function $w(x, z) = (1-z)^{-\alpha-1}\exp\dfrac{-x^z}{1-z}$, i.e. that

$$w(x, z) = \sum_{n=0}^{\infty} L_n^\alpha(x)z^n.$$

Hint. Compute the coefficients of the expansion of the function $w(x, z)$ with respect to $z = 0$ by the method of residua.

8. Prove that

$$H_n(x) = \sum_{k=0}^{E(\frac{n}{2})} \frac{(-1)^k n!}{k!(n-2k)!}(2x)^{n-2k}.$$

9. Prove by applying the generating function for the Hermite polynomials the recurrence relations

$$H_{n+1}(x)-2xH_n(x)+2nH_{n-1}(x) = 0,$$

$$H_n'(x) = 2nH_{n-1}(x)$$

and the differential equation

$$H_n''(x)-2xH'(x)+2nH_n(x) = 0.$$

Deduce the same from Theorem 12.4.1.

10. Verify that the expansion of the function e^x in a series with respect to the Hermite polynomials has the form

$$e^x = e^{\frac{1}{2}}\sum_{n=0}^{\infty} H_n(x)/2^n n!.$$

11. Prove the following generalization of orthogonality relations for the Laguerre polynomials

$$\int_0^\infty x^p e^{-x} L_m^p(xy)\, L_{m-s}^p(x)\,dx = \frac{(p+m)!}{s!(m-s)!}(1-y)^s y^{m-s}.$$

Hint. Multiply both sides of addition formula (12.2.4) by $r^{2l_2-2l_1}$ $e^{-r}L_{2l_1-s}^{2l_2-2l_1}(r)$ and integrate over $0 \leqslant r < \infty$ by applying the orthogonality relations.

12. Apply the identity

$$T(k)\,T(z) = T(\bar{k}z)\,T(k)$$

to prove the formula

$$\sqrt{\frac{(2l_1-k)!}{(2l_2-k)!}}\,P^{l_1}_{l_1-j,\,l_1-k}(\cos\theta)\,L^{2l_2-2l_1}_{2l_1-k}(r)$$

$$=\sum_{m=0}^{2l_2}\sqrt{\frac{j!(2l_1-j)!}{m!(2l_2-m)}}\,(i\sin\tfrac{1}{2}\theta)^{m-j}(\cos\tfrac{1}{2}\theta)^{2l_2-2l_1-m+j}$$

$$\times\, L^{m-j}_{j}(r\sin^2\tfrac{1}{2}\theta)\,L^{2l_2-2l_1-m+j}_{2l_1-j}(r\cos^2\tfrac{1}{2}\theta)\,P^{l_2}_{l_2-m,\,l_2-k}(\cos\theta).$$

Chapter 13

The Hypergeometric Equation

The present chapter differs in character from the remaining parts of the book. The method we employed to investigate the different classes of special functions emphasised their integral properties and led to expansions into eigenfunctions of certain differential operators.

These operators, however, appeared in the background of our considerations and methods of differential equations were not employed in the slightest degree. Although in the domain of expansions into special functions, the elimination of methods of spectral analysis of differential operators proves advantageous, the fact remains that it is the differential properties that determine the particular role of these functions in analysis, and the methods of differential equations are the most universal.

In the present chapter we intend to throw more light on the position in the theory of differential equations for the particular equations and solutions which appeared in the preceding chapters. By investigating the possibilities of analytic extensions of those special solutions and the relations between them we considerably extend the application range of the results obtained so far. The exposition is elementary and completely traditional.

13.1. THE SECOND ORDER HOMOGENEOUS LINEAR DIFFERENTIAL EQUATION ON $\boldsymbol{C}$

Suppose that we are given an equation

$$\frac{\mathrm{d}^2u}{\mathrm{d}z^2}+p(z)\frac{\mathrm{d}u}{\mathrm{d}z}+q(z)u = 0, \tag{13.1.1}$$

where the functions p and q are holomorphic outside a finite subset of $\boldsymbol{C}$. A point at which either of the functions p, q is singular is called a *singular point* of equation (13.1.1). Singularity at a point $a \in \boldsymbol{C}$ is regular if the function $(z-a)p$ and $(z-a)^2q$ are holomorphic in a neighbourhood of the point a. The role of this condition will become clear in the course of our exposition.

As long as we deal with the problem of the existence and regularity of solutions in a neighbourhood of a fixed point z_0 we can assume without loss of generality that $z_0 = 0$. Namely, the transformation $z \to (z-z_0)$ maps z_0 into 0 and preserves the regularity (singularity) type of the point z_0.

The method presented below is due to Frobenius.

According to the foregoing let us assume that the point zero is a non-singular point of equation (13.1.1). Then in a neighbourhood with radius, say, $R > 0$ the series

$$p(z) = \sum_{n=0}^{\infty} p_n z^n, \qquad q(z) = \sum_{n=0}^{\infty} q_n z^n$$

are both convergent.

Suppose that there exists a holomorphic solution of equation (13.1.1) on this neighbourhood, of the form

$$u(z) = \sum_{n=0}^{\infty} c_n z^n. \tag{13.1.2}$$

By substituting the above expansions in (13.1.1) we obtain a system of linear equations with respect to c_n:

$$0 = (n+2)(n+1)c_{n+2} + \sum_{l=0}^{n} \{(l+1)p_{n-l}c_{l+1} + q_{n-l}c_l\}, \tag{13.1.3}$$

$n = 0, 1, 2, \ldots$ It follows that the values of c_{n+2} for $n = 0, 1, 2, \ldots$ can be computed one after another provided we are given the complex numbers c_0 and c_1. It follows from the form of equation (13.1.3) that c_n depends linearly upon c_0 and c_2 for $n \geqslant 2$. Thus

$$w(z) = \sum_{n=0}^{\infty} (\lambda_n c_0 + \mu_n c_1) z^n,$$

where $\lambda_0 = \mu_1 = 1$, $\lambda_1 = \mu_0 = 0$.

THEOREM 13.1.1 (Spain, Smith, 1970 and Kratzer, Franz, 1960). *There exists a regular solution of equation* (13.1.1) *in a neighbourhood* *of every non-singular point. Given $a, b \in \mathbf{C}$, there exists a unique solution w with the property that $w(z_0) = a$, $w'(z_0) = b$.*

The coefficients of the Taylor expansion of the solution w are given by recurrence formula (13.1.3).

Now we suppose that 0 is a regular singular point of the equation. We investigate the existence of solutions of the form

$$u(z) = z^{\alpha} \sum_{s=0} c_s z^s. \tag{13.1.4}$$

The functions $zp(z)$ and $z^2q(z)$ can be expanded in a neighbourhood of 0 into the Taylor series

$$zp(z) = \sum_{n=0}^{\infty} p_n z^n, \qquad z^2 q(z) = \sum_{n=0}^{\infty} q_n z^n,$$

which are simultaneously convergent for a certain $R > 0$.

Substituting the above series into (13.1.1) we obtain

$$\sum_{s=0}^{\infty} (s+\alpha)(s+\alpha-1) c_s z^{s+\alpha} + \sum_{n=0}^{\infty} p_n z^n \sum_{s=0}^{\infty} (s+\alpha) c_s z^{s+\alpha}$$

$$+ \sum_{n=0}^{\infty} q_n z^n \sum_{s=0}^{\infty} c_s z^{s+\alpha} = 0. \tag{13.1.5}$$

Equating with zero the coefficient of z we have the relation

$$c_0 F(\alpha) := c_0 \big(\alpha(\alpha-1) + p_0 \alpha + q_0\big) = 0.$$

The equation $F(\alpha) = 0$ is called the *indicial equation of the singularity*. Numbers ϱ_1, ϱ_2, which satisfy this square equation, are called the *exponents of the singularity*. The exponents will be indexed in such a way that $\operatorname{Re} \varrho_1 \geqslant \operatorname{Re} \varrho_2$.

The requirement that the coefficients of the remaining powers of z in expression (13.1.5) be zero takes the form of the recurrence equation

$$0 = c_s F(\alpha + s) + \sum_{n=0}^{s-1} c_n \{(n+\alpha) p_{s-n} + q_{s-n}\}. \tag{13.1.6}$$

The possibility of using equations (13.1.6) to construct solutions for $\alpha = \varrho_1$ depends on the properties of the indicial equation. We consider three separate cases:

(A) $\varrho_1 - \varrho_2$ is not an integer.

The coefficients $F(\alpha+s)$ are all different from zero for $s > 0$, and the successive constants c_s in decomposition (13.1.4) are uniquely determined for the two roots ϱ_1, ϱ_2.

The series $\sum_{s=0}^{\infty} c_s z^s$ proves to be convergent at least in the disc $|z| < R$ (cf. Kratzer, Franz, 1960 and Spain, Smith, 1970).

PROPOSITION 13.1.2. *In case A the two-dimensional space of solutions is spanned by the functions*

$$u_i(z) = z^{\varrho_i} \sum_{s=0}^{\infty} c_s^i z^s, \qquad i = 1, 2,$$

which are regular functions at least on the disc of radius $|z| < R$, with the exclusion of a curve joining the centre with the boundary to ensure uniqueness of the function z^{ϱ_i} if $\varrho_i \notin Z$.

(B) $\varrho_1 = \varrho_2$.

By solving the recurrence equations we obtain only one solution having the same properties as in case (A).

Suppose now that $u_1(z)$ is a function of the form (13.1.4) for $\alpha \neq \varrho_1 = \varrho_2$ whereas the coefficients c_s satisfy the recurrence equations (13.1.6). Let us compute the value on the function u_1 of the differential operator on the left-hand side of (13.1.1)

$$L(u_1) = \frac{d^2}{dz^2} u_1 + p(z) \frac{d}{dz} u_1 + q(z) u_1 = c_0 z^\alpha F(\alpha),$$

because the coefficients of $z^{\alpha+s}$ in the expansion of the left-hand side are zero for $s \geqslant 1$. Thus

$$L\left(\frac{\partial u_1}{\partial \alpha}\right) = \frac{\partial}{\partial \alpha} L(u_1) = c_0 z^\alpha \left(F'(\alpha) + F(\alpha) \log z\right).$$

Since $\varrho_1 = \varrho_2$ we have $F'(\varrho_1) = 0$ and consequently

$$L\left(\frac{\partial u_1}{\partial \alpha}\right)_{\alpha=\varrho_1} = 0.$$

Let us find the form of the solution so obtained, bearing in mind that the coefficients c_s are functions of α

$$u_2(z) = \left.\frac{\partial u_1}{\partial \alpha}\right|_{\alpha=\varrho_1}(z) = u_1(z)\log z + z^{\varrho_1}\sum_{s=1}^{\infty}\left.\frac{\partial c_s}{\partial \alpha}\right|_{\alpha=\varrho_1} z^s. \qquad (13.1.7)$$

PROPOSITION 13.1.3. *In case* (*B*) *there exist two linearly independent solutions of equation* (13.1.1), *one given by* (13.1.4) *with the coefficients defined by* (13.1.6) *and the other given by formula* (13.1.7). *Both have the same regularity properties as in the case* (A).

(C) $\varrho_1 - \varrho_2 = n \in N$.

The recurrence formulas (13.1.6) define a solution of the form (13.1.4) for $\alpha = \varrho_1$ since $F(\alpha+s) \neq 0$ for $s > 0$.

For $\alpha = \varrho_2$ in the equation for c_n the coefficient of the unknown $F(\alpha_2+n) = F(\varrho_1)$ is zero and therefore $c_n(\varrho_2)$ cannot be determined.

A modification of the method applied in case (B) allows us to overcome these difficulties. We set $c_0(\alpha) = \alpha - \varrho_2$ and determine the successive values of c_s from the recurrence formulas (13.1.6). For $s < n$ we have $c_s = 0$; the value of $c_n(\varrho_2)$ can be chosen arbitrarily, and then it determines uniquely the remaining numbers $c_s(\varrho_2)$ for $s > n$. The function

$$z^{\varrho_2}\sum_{s=n}^{\infty} c_s(\varrho_2)z^s = z^{\varrho_2+n}\sum_{s=0}^{\infty} c_{s+n}(\varrho_2)z^s = z^{\varrho_1}\sum_{s=0}^{\infty} c_{s+n}(\varrho_2)z^s \qquad (13.1.8)$$

has the form (13.1.4) with the parameters (α, c_s) satisfying equations (13.1.6) and the indicial equation and therefore is a solution of equation (13.1.1). However, the way in which the coefficients $c_s(\varrho_2)$ are computed shows that this solution is proportional to $u_1(z)$.

Let

$$u_1(z, \alpha) := z^{\alpha}\sum_{s=0}^{\infty} c_s(\alpha)z^s,$$

and compute

$$L(u_1)(z, \alpha) = c_0(\alpha)F(\alpha)z^{\alpha} = (\alpha-\varrho_2)^2(\alpha-\varrho_1)z^{\alpha}.$$

As in case (B), we find

$$L\left(\frac{\partial u_1}{\partial \alpha}\bigg|_{\alpha=\varrho_2}\right) = 0,$$

which means that the function

$$\frac{\partial u_1}{\partial \alpha}(z, \alpha) = z^{\alpha}\log z \sum_{s=0}^{\infty} c_s(\alpha)z^s + z \sum_{s=0}^{\infty} \frac{\partial c_s}{\partial \alpha}(\alpha)z^s$$

for $\alpha = \varrho_2$ is a solution of the equation in question. Thus we write

$$u_2(z) = z^{\varrho_2} \sum_{s=0}^{\infty} \frac{\partial c_s}{\partial \alpha}(\varrho_2)z^s + cu_1(z)\log z,$$

where c is the factor of proportionality between the function given by (13.1.8) and u_1.

PROPOSITION 13.1.4. *In case* (C) *there exist two solutions of equation* (13.1.1) *on a disc around the singular point*: u_1*—of the form* (13.1.4) *with the coefficients defined by the recurrence formula* (13.1.6) *and* u_2 *with the coefficients defined above. They are regular in the domain obtained by excluding from the disc a segment joining the singular point with the circle of convergence of the series defining* u_1.

Summing up the result of these considerations, we can say that if z_0 is a regular singularity of equation (13.1.1) and inside the disc $D(z_0, R)$ the functions $(z-z_0)p(z)$, $(z-z_0)^2q(z)$ are holomorphic, then on every simply connected subset of the set $D(z_0, R)-\{z_0\}$ equation (13.1.1) has a 2-dimensional space of holomorphic solutions. The singularity of solutions at the point z_0 is of the ramification type. The singularity type is determined by the characteristic equation. If the singularity is non-regular then the characteristic equation is at most of first order. We shall not discuss the case since we do not need it in our applications. General results concerning this case were formulated by Forsyth. Regular solutions of equation (13.1.1) need not then exist or there may exist only a single solution.

Properties of equation (13.1.1) at the point ∞ are defined by means of inversion of the complex plane. The change of variable $z = 1/\zeta$ in (13.1.1) leads to the equation

$$\zeta^4 \frac{d^2u}{d\zeta^2} + \left\{2\zeta^3 - \zeta^2 p\left(\frac{1}{\zeta}\right)\right\} \frac{du}{d\zeta} + q\left(\frac{1}{\zeta}\right) u = 0. \tag{13.1.9}$$

By saying that equation (13.1.1) has certain properties at the point ∞ we mean that these properties are true of equation (13.1.9) at the point $\zeta = 0$.

In particular we say that equation (13.1.1) is non-singular (singular) at the point $z = \infty$ if equation (13.1.9) is non-singular (singular) at $\zeta = 0$. In an analogous way we define the regularity of the point ∞ and the value of the exponents of the regular point. A homogeneous second-order ordinary differential equation having precisely three regular singularities on $C \cup \{\infty\}$ is called a *Riemann P-equation.* Its general form was given by Papperitz, see Whittaker, Watson (1963).

Let ξ, η, ζ be regular singular points with the corresponding exponents (α, α'), (β, β'), (γ, γ'), respectively. It follows that p is a meromorphic function having three simple poles at ξ, η, ζ with the residua $1-\alpha-\alpha'$, $1-\beta-\beta'$, $1-\gamma-\gamma'$. Hence

$$p(z) = \frac{1-\alpha-\alpha'}{z-\xi} + \frac{1-\beta-\beta'}{z-\eta} + \frac{1-\gamma-\gamma'}{z-\zeta}.$$

The regularity condition at $z = \infty$ leads to the following relation between the exponents of the points

$$\alpha+\alpha'+\beta+\beta'+\gamma+\gamma' = 1.$$

Accordingly, the function q which satisfies the regularity conditions is of the form

$$q(z) = \left\{\frac{\alpha\alpha'(\xi-\eta)(\xi-\zeta)}{z-\xi} + \frac{\beta\beta'(\eta-\zeta)(\eta-\xi)}{z-\eta} + \frac{\gamma\gamma'(\zeta-\xi)(\zeta-\eta)}{z-\zeta}\right\} \frac{1}{(z-\zeta)(z-\eta)(z-\zeta)}.$$

Following Riemann, we denote a solution of the equation corresponding to such p and q by

$$u = P\begin{Bmatrix} \zeta, & \eta, & \zeta & \\ \alpha, & \beta, & \gamma, & z \\ \alpha', & \beta', & \gamma' & \end{Bmatrix}.$$

The Riemann P-equation with the parameters $\xi = 0$, $\eta = \infty$, $\zeta = 1$, $\alpha = 0$, $\alpha' = 1-c$, $\beta = a$, $\beta' = b$, $\gamma = 0$, $\gamma' = c-a-1$ is called the *hypergeometric equation* and is given in the form

$$z(1-z)\frac{d^2u}{dz^2}+\{c-(a+b+1)z\}\frac{du}{dz}-abu = 0. \tag{13.1.10}$$

The problem of solving a Riemann P-equation may be reduced to that of solving the hypergeometric equation. The method consists in transforming the compactified complex plane in such a way that the type of the equation is preserved and only the system of parameters is changed. Let

$$W(\alpha_1, \beta_1, \gamma_1)(z) := (z-\zeta)^{-\alpha_1}(z-\eta)^{-\beta_1}(z-\zeta)^{-\gamma_1}.$$

Whenever $\alpha_1+\beta_1+\gamma_1 = 0$, the product of the function $W(\alpha_1, \beta_1, \gamma_1)$ and

$$P\begin{Bmatrix} \xi, & \eta, & \zeta & \\ \alpha, & \beta, & \gamma, & z \\ \alpha', & \beta', & \gamma' & \end{Bmatrix}$$

is a solution of the equation

$$P\begin{Bmatrix} \xi, & \eta, & \zeta & \\ \alpha+\alpha_1, & \beta+\beta_1, & \gamma+\gamma_1, & z \\ \alpha'+\alpha_1, & \beta'+\beta_1, & \gamma'+\gamma_1 & \end{Bmatrix}$$

with translated parameters.

It follows that the mappings $W(\alpha_1, \beta_1, \gamma_1)$ preserve the values $\alpha-\alpha'$, $\beta-\beta'$, $\gamma-\gamma'$ and $\alpha+\beta+\gamma$ and allow us to choose arbitrarily the remaining parameters. Therefore we may assume that two of the poles have one of the exponents equal to zero. This implies that we can describe the functions

$$P\begin{Bmatrix} \xi, & \eta, & \zeta & \\ \alpha, & \beta, & \gamma, & z \\ \alpha, & \beta', & \gamma' & \end{Bmatrix}$$

provided we know the position of the poles and the functions

$$P\begin{Bmatrix} \xi, & \eta, & \zeta & \\ 0, & 0, & \gamma, & z \\ \alpha', & 1-\alpha'-\gamma', & \gamma' & \end{Bmatrix}.$$

Thus the set of functions

$$P\begin{Bmatrix} 0, & 1, & \infty & \\ 0, & 0, & a, & z \\ 1-c, & c-a-b, & b & \end{Bmatrix}$$

is sufficient to recreate the set

$$P\begin{Bmatrix} \xi, & \eta, & \zeta & \\ \alpha, & \beta, & \gamma, & z \\ \alpha', & \beta', & \gamma' & \end{Bmatrix}.$$

By the *hypergeometric function* we mean the solution of the hypergeometric equation corresponding to the pole 0 and the exponent 0 with free term 1, under the assumption that c is not a negative integer. Then, by virtue of Proposition 13.1.1, recurrence relations (13.1.6), which now become

$$s(c-1+s)c_s = (a+s-1)(b+s-1)c_{s-1},$$

define a series whose radius of convergence equals 1. The solution with free term 1 has the form of the hypergeometric series

$$F(a,b,c;z) = 1+\frac{ab}{c}z+\frac{a(a+1)\,b(b+1)}{c(c+1)2!}z^2$$
$$+\frac{a(a+1)(a+2)b(b+1)(b+2)}{c(c+1)(c+2)3!}z^3+\dots.$$

The expansion of the function F can also be written in the following convenient form:

$$\frac{\Gamma(a)\Gamma(b)}{\Gamma(c)}F(a,b,c;z) = \sum_{s=0}^{\infty}\frac{\Gamma(a+s)\Gamma(b+s)}{\Gamma(c+s)s!}z^s.$$

If c is a negative integer, then we can obtain in a neighbourhood of zero a solution of the hypergeometric equation by applying recurrence relations (13.1.6) corresponding to $1-c$. The solution so obtained can also be expressed by means of the hypergeometric function. Namely, it is given by the function $z^{1-c}F(1+a-c,\ 1+b-c,\ 2-c;\ z)$. In both cases considered above, Propositions 13.1.1–13.1.4 provide a method of determining the other solution, which may contain the logarithmic function.

The methods described above can also be used to obtain solutions of the hypergeometric equation on other domains, which are neighbourhoods of the remaining singularities, and also to find solutions of the general Riemann P-equation. Special cases will be found in the problems. A detailed description of all solutions of the hypergeometric equations which can be expressed by means of the hypergeometric function can be found in Whittaker and Watson (1963).

The following chapter is devoted to the methods of analytic continuation of the hypergeometric series. This will give us an opportunity to describe more general methods of finding hypergeometric solutions in the form of contour integrals. In those integrals we shall recognize a series of integral representations of the special function investigated earlier.

Example 1

The Legendre differential equation

$$(1-z^2)\frac{\mathrm{d}^2u}{\mathrm{d}z^2} - 2z\frac{\mathrm{d}u}{\mathrm{d}z} + \nu(\nu+1)u = 0$$

has ± 1 and x as its singular points. They are all regular. The change of independent variable $z = 1-2\xi$ shifts the finite singular points to 0 and 1, respectively, and leaves ∞ invariant. The equation is thus transformed to (a special case of the hypergeometric equation)

$$\xi(1-\xi)\frac{\mathrm{d}^2u}{\mathrm{d}z^2} + (1-2\xi)\frac{\mathrm{d}u}{\mathrm{d}\xi} + \nu(\nu+1)u = 0$$

and it is immediately seen that this equation has the characteristic exponents at $\xi = 0$ both equal to zero.

Now, (13.1.6) reduces to

$$s^2c_s = (s+\nu)(s-\nu-1)c_{s-1}, \qquad s \geqslant 1,$$

so by induction

$$(s!)^2c_s = (\nu+1)\ldots(s+\nu)(-\nu)\ldots(s-\nu-1)c_0.$$

Therefore, unless ν is an integer, in which the case c_s vanishes from a certain value on, we obtain

$$c_s = \frac{\Gamma(\nu+s+1)\Gamma(-\nu+s)}{\Gamma(\nu+1)\Gamma(-\nu)(s!)^2}c_0, \qquad s \geqslant 1.$$

Noting that the power series $\sum_{s=0}^{\infty} c_s z^s$ has radius of convergence equal to one, we see that for ν different from an integer, the function

$$u_1(z) = \frac{1}{\Gamma(-\nu)\Gamma(\nu+1)} \sum_{s=0}^{\infty} \frac{\Gamma(-\nu+s)\Gamma(\nu+s+1)}{(s!)^2} \left(\frac{1-z}{2}\right)^s,$$

defined in the circle $|z-1| < 2$, is a solution of the Legendre equation. Another independent solution is obtained by using the method of case (B). We write down the resulting solution leaving out the details for the reader.

$$u_2(z) = u_1(z)\log\frac{1-z}{2} + \frac{1}{\Gamma(-\nu)\Gamma(\nu+1)} \sum_{s=1}^{\infty} \frac{\Gamma(-\nu+s)\Gamma(\nu+s+1)}{(s!)^2}$$

$$\times \left[\sum_{r=1}^{s} \frac{1}{r+\nu} + \frac{1}{r-\nu-1} - \frac{2}{r}\right]\left(\frac{1-z}{2}\right)^s.$$

When ν is a positive integer, say $\nu = n \geqslant 0$ then $c_s = 0$ for $s \geqslant n+1$ and

$$c_s(n) = \frac{(-1)^s(n+s)!}{(s!)^2(n-s)!} c_0, \qquad 1 \leqslant s \leqslant n.$$

When ν is a negative integer, $\nu = -n$, $n > 0$, the coefficients c_s vanish for $s \geqslant n$ and

$$c_s(-n) = \frac{(-1)^{s-1}(n+s-1)!}{(s!)^2(n-s-1)!} c_0 = (-1)c_s(n-1), \quad 1 \leqslant s < n.$$

Accordingly, defining a polynomial $P_n(z)$ of degree n by

$$P_n(z) = \sum_{s=0}^{n} \frac{(-1)^s(n+s)!}{(s!)^2(n-s)!} \left(\frac{1-z}{2}\right),$$

we see that $P_n(z)$ is a solution of the Legendre equation with $\nu = n$ and $P_{n-1}(z)$ is a solution of the same equation corresponding to $\nu = -n$.

Example 2

The Bessel equation provides an illustration to all possible cases of the relation between characteristic exponents (cases (A), (B), (C)). Recall that the Bessel equation has the form

$$\frac{d^2u}{dz^2} + \frac{1}{z}\frac{du}{dz} + \left(1 - \frac{\nu^2}{z^2}\right)u = 0$$

and the complex parameter ν is caled the *order of the Bessel equation* (cf. Ch. 7, Problem 4(c)). The equation has two singular points: $z = 0$ which is regular and $z = \infty$ which is not. The characteristic equation at $z = 0$ is

$$(\alpha+\nu)(\alpha-\nu) = 0$$

and the recurrence equation (13.1.6) takes the form

$$\text{(a)} \quad c_1[(\alpha+1)^2-\nu^2] = 0,$$

$$\text{(b)} \quad c_s[(\alpha+s)^2-\nu^2]+c_{s-2} = 0, \qquad s \geqslant 2.$$

These recurrence relations enable us to construct a basis of solutions of the Bessel equation provided the difference of characteristic exponents is not an even integer, i.e. ν is not an integer. In fact, if 2ν is not an integer (case (A)), then for $\alpha = \pm\nu$ (a) and (b) have a solution depending on a single arbitrary constant and it is given by the formulas

$$c_{2s}(\nu) = \frac{(-1)^s}{2^{2s}s!(\nu+1)\dots(\nu+s)}c_0,$$

$$c_{2s}(-\nu) = \frac{(-1)^s}{2^{2s}s!(-\nu+1)\dots(-\nu+s)}c_0,$$

$$c_{2s+1}(\nu) = c_{2s+1}(-\nu) = 0.$$

If 2ν is an odd integer, say $\nu = k+\frac{1}{2}, k \geqslant 0$ then $c_r(\nu)$ are given by the same formulas as those given above. A solution of (a) and (b) for $\alpha = -\nu$ depends however on two arbitrary constants c_0 and $c_{2\nu}$. Putting $c_{2\nu} = 0$ we get the same formulas for $c_s(-\nu)$ as above. For $c_0 = 0$ we obtain however

$$c_{2s}(-\nu) = 0, \qquad s = 1, 2, \dots,$$

$$c_1(-\nu) = \dots = c_{2\nu-2}(-\nu) = 0, \qquad \text{provided} \quad 2\nu > 1,$$

and

$$c_{2\nu+2s}(-\nu) = \frac{(-1)^s}{2^{2s}s!(\nu+1)\dots(\nu+k)}c_{2\nu}, \qquad s = 1, 2, \dots$$

Therefore unless ν is an integer, there are two solutions $J_\nu(z)$, $J_{-\nu}(z)$, called the *Bessel functions of the first kind and of order* ν *and* $-\nu$ respectively given by

$$J_\nu(z) = \left(\frac{z}{2}\right)^\nu \sum_{s=0}^{\infty} \frac{(-1)^s}{s!\Gamma(\nu+s+1)} \left(\frac{z}{2}\right)^{2s},$$

$$J_{-\nu}(z) = \left(\frac{z}{2}\right)^{-\nu} \sum_{s=0}^{\infty} \frac{(-1)^s}{s!\Gamma(-\nu+s+1)} \left(\frac{z}{2}\right)^{2s}$$

and they are independent.

Let us note that another choice of arbitrary constants in the case ν is a half of an odd integer would lead to a solution proportional to J_ν, or more generally, to a combination of J_ν and $J_{-\nu}$. The formulas given above for J_ν and $J_{-\nu}$ in the case of integer ν lead to the identity

$$J_{-\nu}(z) = (-1)^\nu J_\nu(z).$$

To construct a solution of the Bessel equation of integer order n, independent from $J_n(z)$ we proceed differently according whether $n = 0$ or $n > 0$.

If $n = 0$ we follow the method of case B, so from (a) and (b) by putting $\nu = 0$ and $c_0 = 1$ we get

$$c_{2s}(\alpha) = \frac{(-1)^s}{(\alpha+2)^2 \ldots (\alpha+2s)^2}, \qquad c_{2s+1}(\alpha) = 0.$$

Therefore

$$c_{2s}'(0) = -c_{2s}(0) \sum_{n=1}^{s} \frac{1}{n}$$

and so the second solution is

$$Y^0(z) = J_0(z) \log z - \sum_{s=1}^{\infty} \frac{(-1)^s}{(s!)^2} \left(1 + \frac{1}{2} + \ldots + \frac{1}{n}\right) \left(\frac{z}{2}\right)^{2s}.$$

The function $Y^{(0)}(z)$ is called the *Bessel function of the second kind and of order zero*, or the *Neumann function*.

For $n > 0$ a direct construction of a solution using the method of the case (C) leads to rather burdensome computations and is left as an exercise (Problem 21). A simpler approach is based on the use of recurrence relations satisfied by Bessel functions.

Recalling the formula (cf. Ch. 7, Problem 10)

$$J_{n+1}(z) = -z^n \frac{\mathrm{d}}{\mathrm{d}z}\left(z^{-n} J_n(z)\right)$$

we define the Neumann function of order n for an integer $n \geqslant 0$, by the recurrence formula

$$Y^{(n+1)}(z) = -z^n \frac{\mathrm{d}}{\mathrm{d}z}\left(z^{-n} Y^{(n)}(z)\right), \qquad n = 0, 1, \ldots,$$

where for $n = 0$ we take the Neumann function $Y^{(0)}$ defined above. It is left to the reader to verify that $Y^{(n+1)}(z)$ provides another independent solution of the Bessel equation.

13.2. SOLUTIONS OF THE HYPERGEOMETRIC EQUATION IN THE FORM OF EULER INTEGRALS

We shall seek solutions of the hypergeometric equation in the form of an integral with a parameter

$$u(z) = \int_{\mathscr{C}} w(t)(t-z)^k \mathrm{d}t. \tag{13.2.1}$$

Our objective is to determine the function w, the contour $\mathscr{C}$ in $\boldsymbol{C}$ and the value of the parameter k for which the function u is a special solution of the hypergeometric equation (h.e.). The mapping $w \to u$ is termed the *Euler transformation.*

The method presented below allows us to determine only some special solutions, just as the method applied in the preceding section. In the course of our discussion we shall impose certain simple conditions on the contour $\mathscr{C}$, ensuring that u is a solution of the h.e. It will be possible to obtain many integral representations of a given solution at the same time. In particular, we shall be able to determine analytic continuations of the function F beyond its domain, i.e. the unit disc.

Suppose that integral (13.2.1) is absolutely convergent. Then differentiation under the integral sign is legitimate and the condition

$$Lu = z(1-z)\frac{\mathrm{d}^2 u}{\mathrm{d}z^2} + \{c-(a+b+1)z\}\frac{\mathrm{d}u}{\mathrm{d}z} - abu = 0$$

assumes the form

$$\int_{\mathscr{C}} w[z(1-z)(k-1)k(t-z)^{k-2}-k(c-(a+b+1)z)(t-z)^{k-1}$$
$$-ab]\mathrm{d}t = 0.$$

Substituting $v(t) = (t-z)^k$ and applying the identities

$$z(1-z) = t(1-t)-(1-2t)(t-z)-(t-z)^2$$

and

$$c-(a+b+1)z = c-(a+b+1)t+(a+b+1)(t-z),$$

we arrive at

$$\int_{\mathscr{C}} \big[t(1-t)v''+\big((k-1)(2t-1)-c+(a+b+1)t\big)v'$$
$$-\big(k(k-1)+(a+b+1)k+ab\big)v\big]w\,\mathrm{d}t = 0 \tag{13.2.2}$$

($'$ denotes differentiation with respect to t).

We shall try to select a function $\Omega(w, v)$ and w in such a manner that the integrand will be the differential of the function $t \to \Omega(w(t), v(t))$. To this end we first represent the left-hand side of (13.2.2) symbolically as

$$\int_{\mathscr{C}} [pv''+qv'+rv]w\,\mathrm{d}t$$
$$= \int_{\mathscr{C}} \left\{\frac{\mathrm{d}}{\mathrm{d}t}(wpv+wqv-vw'p-vwp')+v\left(\frac{\mathrm{d}^2}{\mathrm{d}t^2}(wp)\right.\right.$$
$$\left.\left.-\frac{\mathrm{d}}{\mathrm{d}t}(wq)+wr\right)\right\}\mathrm{d}t.$$

Our requirements will be satisfied if we take for w a function fulfilling the differential equation

$$\frac{\mathrm{d}^2}{\mathrm{d}t^2}(wp)-\frac{\mathrm{d}}{\mathrm{d}t}(wq)+wr = 0. \tag{13.2.3}$$

Recalling the form of r, we observe that it is convenient to put $r = 0$, which can be done by choosing $k = -a$ or $k = -b$. Thus we look for a solution of the equation

$$\frac{\mathrm{d}}{\mathrm{d}t}\left(\frac{\mathrm{d}}{\mathrm{d}t}(wp)-wq\right) = 0.$$

The solution, up to an unimportant factor, is given by the function

$$w(t) = t^{b-c}(t-1)^{c-a-1}.$$

Hence (13.2.2) takes the form

$$\int_{\mathscr{C}} \frac{\mathrm{d}}{\mathrm{d}t}\{t^{b-c+1}(t-1)^{c-a}(t-z)^{-b-1}\}\mathrm{d}t = 0.$$

Denote by $\Delta(t, z)$ the function in braces. The equation will be satisfied if we choose the contour $\mathscr{C}$ in such a way that the function Δ assumes the same value at the end-points of $\mathscr{C}$.

Proposition 13.2.1. *The function*

$$u(z) = \int_{\mathscr{C}} t^{b-c}(t-1)^{c-a-1}(t-z)^{-b}\mathrm{d}t \tag{13.2.4}$$

is a solution of the hypergeometric equation for an arbitrary contour $\mathscr{C}$ *such that the integral is absolutely convergent and the value of the function* Δ *at both end-points is the same.*

If all exponents appearing in the definition of the function Δ have positive real parts, then Δ is zero at the points 1, 0, z and ∞. Then it is most natural to choose for $\mathscr{C}$ the contour which joins an arbitrary pair of these points. However, all assumptions concerning the values of the parameters will become redundant if we choose a contour called the *double loop*, corresponding to a pair of ramification points of the function Δ.

Such a contour, corresponding to a pair of points z_1, z_2, is presented in Figure 10. We shall denote it by $\mathscr{D}(z_1, z_2)$.

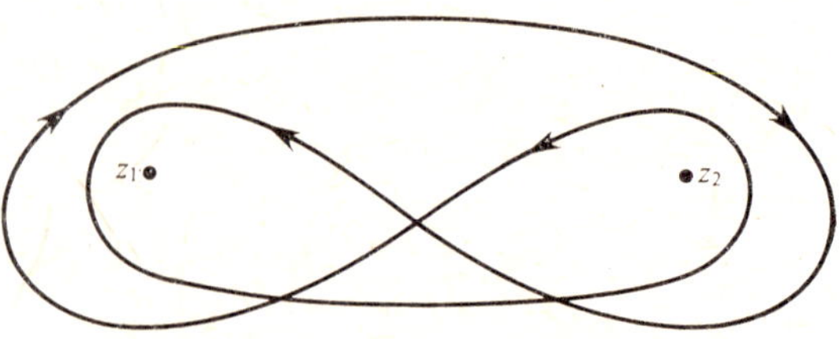

Fig. 10

Let us see how the function $(z-z_1)^p(z-z_2)^q$ behaves along the contour. We fix a starting point z_0. If the argument of the function at the point z equals γ, then after passing around z_2 it changes to $\gamma-2\pi q$, next after turning around z_1 in the positive direction it assumes the value $\gamma-2\pi q+2\pi p$ and after turning around z_2 in the negative direction the value $\gamma+2\pi q$. Finally, returning to the point z it again assumes the value γ.

It follows that $\mathscr{D}(z_1, z_2)$ defines a closed contour on the Riemann surface corresponding to the function $(z-z_1)^p(z-z_2)^q$.

We shall now investigate the basic properties of integrals along the double loop of functions of the form $s^{p-1}(1-s)^{q-1}f(s)$, where f is a function holomorphic in a domain containing the loop $\mathscr{D}$.

We assume that p and q have positive real parts. Let us choose the family of loops presented in Figure 11.

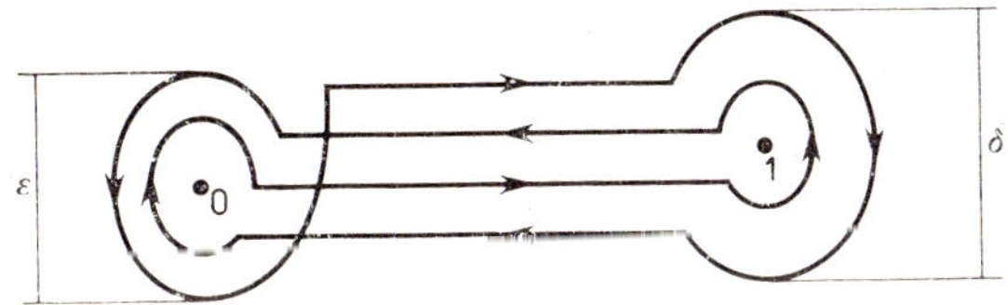

Fig. 11

On the basis of the Cauchy theorem integrals along arbitrary double loops lying in the domain of analyticity of the function f have the same value. If the parameters ε and δ converge to zero, then the integrals along the arcs around the singular point also converge to zero.

For instance, the integral over the arc around zero can be estimated by

$$\frac{1}{2\pi}\left(\frac{1}{2}\varepsilon\right)^{\operatorname{Re} p}\int_0^{2\pi}|1-\varepsilon e^{i\varphi}|^{q-1}d\varphi \underset{\varepsilon\to 0}{\to} 0.$$

Next, the integrals along the parallel lines which in the limit coincide with the interval $[0, 1]$ are convergent to

$$I_1 = -\int_0^1 s^{p-1}(1-s)^{q-1}f(s)ds, \qquad I_2 = -e^{-2\pi i(q-1)}I_1,$$

$$I_3 = -e^{2\pi i(p-1)}I_1, \qquad I_4 = -e^{-2\pi i(q-1)}I_3.$$

It follows that the integral along the contour has the value

$$I = \sum_{i=1}^{4} I_i$$

$$= (1 - e^{-2\pi i(q-1)})(e^{2\pi i(p-1)} - 1)\int_0^1 s^{p-1}(1-s)^{q-1} f(s)\,ds. \tag{13.2.5}$$

THEOREM 13.2.2. *Let $\mathscr{D}$ be a fixed double loop around the points* 0 *and* 1 *and let Ω be the set comprised of those $z \in \mathbf{C}$ for which $\frac{1}{z}$ lies outside $\mathscr{D}$. Under these assumptions the function*

$$\mathbf{C} \ni z \to \int_{\mathscr{D}} s^{a-1}(1-s)^{c-a-1}(1-zs)^{-b}ds\left[\int_{\mathscr{D}} s^{a-1}(1-s)^{c-a-1}ds\right]^{-1} \tag{13.2.6}$$

is an analytic extension of the function F to Ω.

REMARK. Computation of integrals in (13.2.6) should involve the same sheet of the multivalued integrand. Thus for instance we may fix for $s \in]0, 1[$: $\arg(s) = 0$, $\arg(1-s) = 0$, $\lim_{s \to 0} \arg(1-zs) = 0$.

Proof. The change of variable $s \to 1/t$ in (13.2.6) transforms the numerator into an integral of the form (13.2.4). The image $\mathscr{D}'$ of the contour $\mathscr{D}$ satisfies the assumption of Proposition 13.1.2. Thus the proposition implies that the function

$$\mathbf{C} \ni z \to \int_{\mathscr{D}} s^{a-1}(1-s)^{c-a-1}(1-zs)^{-b}ds$$

$$= \int_{\mathscr{D}'} t^{b-c}(t-1)^{c-a-1}(t-z)^{-b}dt \tag{13.2.7}$$

is a solution of the hypergeometric equation in a neighbourhood of the point $z = 0$. This solution is regular at zero since for z small enough the contour $\mathscr{D}'$ does not touch the singular points of the integrand, and hence the theorem on the differentiation of an integral with a parameter applies.

Since in the definition of the function F, c was to be different from $0, -1, -2, \ldots$, it follows that the only regular solution of the h.e. in a neighbourood of zero is the series corresponding to the exponent zero and therefore proportional to the hypergeometric series. The function defined by formula (13.2.6) assumes the value 1 at $z = 0$ and so it agrees with this series in a neighbourhood of zero. □

Theorem 13.2.2 provides convenient methods of constructing analytic extensions of the hypergeometric function. By choosing a suitable contour $\mathscr{C}$ we can obtain an extension of the function F to an arbitrary simply connected domain which does not contain the point $z = 1$. Still, it is useful to have at hand a formula in which arbitrariness of the contour does not appear. We shall present such a formula under the assumption that the conditions given in the Remark following Theorem 13.2.2 are fulfilled.

COROLLARY 13.2.3. *For* $\operatorname{Re} c > \operatorname{Re} a > 0$ *the following function is an analytic extension of the hypergeometric series to the domain* $\Omega = \boldsymbol{C} - \{[1, \infty]\}$

$$F(a, b, c; z) = B(c-a, a)^{-1} \int_0^1 s^{a-1}(1-s)^{c-a-1}(1-zs)^{-b} ds. \tag{13.2.8}$$

Proof. Fix a $z \in \Omega$ and choose a contour $\mathscr{D}$ such that $1/z$ lies outside the contour. An extension of the hypergeometric function which reaches the point z is obtained by using Theorem 13.2.2. If $z \notin [1, \infty]$, then the integrals appearing in (13.2.6) can be computed by means of formula (13.2.5). Since the denominator is equal to $B(c-a, a)$, formula (13.2.8) follows. □

Suppose in addition that $\operatorname{Re}(c-a-b) > 0$. Then function (13.2.8) has a limit at the point 1 (by Lebesgue's theorem), which equals

$$F(a, b, c; 1) = B(b, c-b-a)/B(c-a, a). \tag{13.2.9}$$

13.3. THE HYPERGEOMETRIC FUNCTION FOR SOME SPECIAL VALUES OF THE PARAMETERS

By comparing the Taylor expansions of suitable functions we observe that

$$(1-z)^{-a} = \frac{1}{\Gamma(a)} \sum_{s=0}^{\infty} \frac{\Gamma(a+s)}{s!} z^s = F(a, b, b; z), \tag{13.3.1}$$

$$\log(1+z) = z\sum_{s=0}^{\infty}(-1)^s\frac{\Gamma(s+1)}{\Gamma(s+2)}z^s = zF(1,1,2;-z)$$

for $|z|<1$, (13.3.2)

$$\sin^{-1}(z) = \int_0^z(1-t^2)^{-\frac{1}{2}}\,dt = \sum_{s=0}^{\infty}\frac{\Gamma(s+\frac{1}{2})}{s!\Gamma(\frac{1}{2})}\int_0^z t^{2s}dt$$

$$= zF(\tfrac{1}{2},\tfrac{1}{2},\tfrac{3}{2};z^2). \tag{13.3.3}$$

Except the Bessel functions and the Laguerre polynomials, all special functions described previously can be represented by means of the hypergeometric series:

Gegenbauer Polynomials:

$$C_v^p(z) = \frac{\Gamma(v+2p)}{\Gamma(v+1)\Gamma(2p)}F(v+2p,\ -v,\ p+\tfrac{1}{2};\tfrac{1}{2}(1-z)). \tag{13.3.4}$$

Jacobi functions and polynomials:

$$P_n^{\alpha,\beta}(z) = (-1)^n\binom{n+\beta}{n}F(-n,n+\alpha+\beta+1,\beta+1,\tfrac{1}{2}(1+z)), \tag{13.3.5}$$

$$\mathfrak{P}_{mn}^l(z) = \frac{\Gamma(l+n+1)}{2\Gamma(l+m+1)(n-m)!}(z-1)^{\frac{1}{2}(n-m)}(z+1)^{l-\frac{1}{2}(n-m)}$$

$$\times F\left(-l-m,n-l,n-m+1;\frac{z-1}{z+1}\right)$$

$$= \frac{\Gamma(l+n+1)}{2^n\Gamma(l+m+1)(n-m)!}(z-1)^{\frac{1}{2}(n-m)}(z+1)^{\frac{1}{2}(n+m)}$$

$$\times F(n+l+1,n-l,n-m+1;\tfrac{1}{2}(1-z)). \tag{13.3.6}$$

As special cases of the above relations we have *associated Legendre functions*:

$$\mathfrak{P}_l^n(z) = \frac{\Gamma(l+n+1)}{\Gamma(l-n+1)n!}\left(\frac{z-1}{z+1}\right)^{\frac{n}{2}}F(l+1,\ -l,n+1;\tfrac{1}{2}(1-z)). \tag{13.3.7}$$

If the parameters a, b are negative integers, then the hypergeometric series terminates and for $m \geqslant n$ we have

$$P^l_{mn}(z) = \frac{i^{m-n}}{2^m(m-n)!}\sqrt{\frac{(l-n)!(l+m)!}{(l-m)!(l+n)!}}(1+z)^{\frac{1}{2}(m+n)}(1-z)^{\frac{1}{2}(m-n)}$$
$$\times F(l+m+1, m-l, m-n+1; \tfrac{1}{2}(1-z)). \qquad (13.3.8)$$

13.4. THE CONFLUENT HYPERGEOMETRIC EQUATION AND THE CONFLUENT HYPERGEOMETRIC FUNCTION

By the *confluent hypergeometric equation* we shall understand the ordinary homogeneous second order differential equation

$$z\frac{d^2}{dz^2}u + (c-z)\frac{du}{dz} - au = 0. \qquad (13.4.1)$$

Next the power series

$${}_1F_1(a, c, z) = \frac{\Gamma(c)}{\Gamma(a)}\sum_{n=0}^{\infty}\frac{\Gamma(a+n)z^n}{\Gamma(c+n)n!} \qquad (13.4.2)$$

which is convergent for $|z| < 1$ will be called the *confluent hypergeometric series* (*confluent hypergeometric function*). Direct calculation shows that ${}_1F_1$ is a solution of equation (13.4.1); we shall prove this later without resorting to computation.

We shall define a limit passage which transform the hypergeometric series into the confluent series and the hypergeometric equation into the confluent equation. The nature of that passage explains the term "confluent", i.e. flowing together or uniting.

The Taylor series of the function $F\left(a, b, c; \frac{z}{b}\right)$, i.e.

$$\frac{\Gamma(c)}{\Gamma(a)}\left\{\frac{\Gamma(a)}{\Gamma(c)} + \frac{\Gamma(a+1)bz}{\Gamma(c+1)b} + \frac{\Gamma(a+2)}{\Gamma(c+2)}\frac{b(b+1)z^2}{b^2 2!} + \dots\right.$$
$$\left.\dots + \frac{\Gamma(a+n)}{\Gamma(c+n)}\frac{b(b+1)\dots(b+n-1)}{b^n}\frac{z^n}{n!} + \dots\right\},$$

has a radius of convergence equal to $|b|$.

Fixing arbitrarily a radius R, and applying the dominated convergence theorem for series, we see that for $|z| < R$

$$\lim_{|b|\to\infty} F\left(a, b, c; \frac{z}{b}\right) = \frac{\Gamma(c)}{\Gamma(a)} \sum_{n=0}^{\infty} \frac{\Gamma(a+n)}{\Gamma(c+n)} \frac{z^n}{n!}$$

$$\times \lim_{|b|\to\infty} \frac{b(b+1)\dots(b+n-1)}{b^n} = {}_1F_1(a, c; z).$$

R being arbitrary, this implies that the radius of convergence of series (13.4.2) is equal to ∞. The function $F\left(a, b, c; \frac{z}{b}\right)$ satisfies the differential equation arising from the hypergeometric equation by the substitution $z \to \frac{z}{b}$, i.e. the equation

$$z(b-z)\frac{\mathrm{d}^2u}{\mathrm{d}z^2} + \{bc-(a-b-1)z\}\frac{\mathrm{d}u}{\mathrm{d}z} - abu = 0. \tag{13.4.3}$$

Dividing equation (13.4.3) by b and passing to the limit as $|b| \to \infty$, we arrive at the confluent equation (13.4.1), having as a solution the function ${}_1F_1$. Equation (13.4.3) has three singular points: 0, b and ∞. Thus the transition from the hypergeometric equation to equation (13.4.1) can be interpreted as "bringing together" the regular singular points b and ∞.

The resulting equation has singularities at the points 0 and ∞; the first one is regular while the other one, obtained by "bringing together" two regular singularities, is irregular.

Moreover, we can obtain an integral expression for the function ${}_1F_1$ by passing to the limit in an analytic extension of the function F.

For a fixed contour $\mathscr{D}$, a fixed z_n and b sufficiently large we have

$$F\left(a, b, c; \frac{z}{b}\right) = \frac{\Gamma(c)}{\Gamma(c-a)\Gamma(a)} \int_{\mathscr{D}} t^{a-1}(1-t)^{c-a-1}\left(1-\frac{zt}{b}\right)^{-b} \mathrm{d}t. \tag{13.4.4}$$

When $|b|$ tends to infinity, the last component of the integrand converges uniformly, with respect to $t \in \mathscr{D}$, to the function e^{zt}. Thus the limit value of the integral gives the following representation of the confluent function:

$$ {}_1F_1(a, c; z) = \frac{\Gamma(c)}{\Gamma(c-a)\Gamma(a)} \int_{\mathscr{D}} t^{a-1}(1-t)^{c-a-1} e^{zt} \mathrm{d}t. \tag{13.4.5} $$

Under the assumption of $\operatorname{Re} c > \operatorname{Re} a > 0$ integration over $\mathscr{D}$ can be replaced by integration along the interval $[0, 1]$:

$$ {}_1F_1(a, c; z) = \frac{\Gamma(c)}{\Gamma(c-a)\Gamma(a)} \int_0^1 t^{a-1}(1-t)^{c-a-1} e^{zt} \mathrm{d}t. \tag{13.4.6} $$

Finally we note that the Laguerre polynomials are a special case of the confluent hypergeometric function

$$ L_n^{\alpha}(z) = \frac{\Gamma(n+\alpha+1)}{n!\Gamma(\alpha+1)} {}_1F_1(-n, \alpha+1; z). \tag{13.4.7} $$

For Bessel functions we have

$$ J_\nu(z) = \frac{1}{\Gamma(\nu+1)} \left(\frac{z}{2}\right)^{\nu} e^{-iz} {}_1F_1(\nu+\tfrac{1}{2}, 2\nu+1; 2iz). \tag{13.4.8} $$

PROBLEMS

1. Let z_0 be a regular singular point of equation (13.1.1). Show that expanding $p(z)$ and $q(z)$ in the power series about the point z_0 and neglecting all terms except the first one we get the Euler equation

$$ \frac{\mathrm{d}^2u}{\mathrm{d}z^2} + \frac{p_0}{z-z_0}\frac{\mathrm{d}u}{\mathrm{d}z} + \frac{q_0}{(z-z_0)^2} u = 0. $$

Show that the Euler equation has solutions of the form $u(z) = (z-z_0)^{\alpha}$, where α is the exponent of the singularity z_0 of equation (13.1.1).

2. Find a basis of solutions in a neighbourhood of $z = 0$ of the Weber differential equation

$$ \frac{\mathrm{d}^2u}{\mathrm{d}z^2} = (\tfrac{1}{4}z^2+\alpha)u. $$

3. Find the general form of the differential equation (13.1.1) with at most two regular singular points in the extended plane $\boldsymbol{C} \cup \{\infty\}$.

Hint. Take ∞ as one of the singular points. Then the equation is the Euler equation of Problem 1 in the case where z_0 is another singular point or $\frac{d^2u}{dz^2} = 0$ when there are no other singularities.

4. (a) Show that for the Hermite equation

$$\frac{d^2u}{dz^2} - 2z\frac{du}{dz} + \lambda u = 0$$

the functions $u_1(z)$, $u_2(z)$ given by

$$u_1(z) = \Gamma(\tfrac{1}{4}\lambda+1)\sum_{r=0}^{\infty}\frac{(-1)^r(2z)^{2r}}{\Gamma(\frac{1}{4}\lambda-r+1)(2r)!},$$

$$u_2(z) = \frac{1}{2}\Gamma(\tfrac{1}{4}\lambda+1)\sum_{r=0}^{\infty}\frac{(-1)^r(2z)^{2r+1}}{\Gamma(\frac{1}{4}\lambda-r+\frac{1}{2})(2r+1)!}$$

form a basis of solutions. Determine the cases when there are polynomial solutions.

(b) Show that the substitution $u(z) = e^{-\frac{1}{2}z^2}y(z)$ transforms the Hermite equation into the equation for eigenfunctions of the harmonic oscillator (cf. § 12.4)

$$\left(\frac{d^2}{dz^2} - z^2\right)y = -(\lambda+1)y.$$

(c) Let u denote either of the functions $u_1(z)$, $u_2(z)$. Show that for real x $u(x)e^{-\frac{1}{2}x^2} \to 0$, when $|x| \to \infty$ if and only if u reduces to a polynomial, proportional to the Hermite polynomial and that this happens only when $\lambda = 2n$, n a non-negative integer.

5. Show that

$$u_1(z) = \frac{\Gamma(\frac{1}{2}\nu+1)}{\Gamma(\frac{1}{2}\nu+\frac{1}{2})}\sum_{r=0}^{\infty}(-1)^r\frac{\Gamma(\frac{1}{2}\nu+r+\frac{1}{2})}{\Gamma(\frac{1}{2}\nu-r+1)(2r)!}(2z)^{2r},$$

$$u_2(z) = \frac{\Gamma(\frac{1}{2}\nu+\frac{1}{2})}{\Gamma(\frac{1}{2}\nu+1)}\sum_{r=0}^{\infty}(-1)^r\frac{\Gamma(\frac{1}{2}\nu+r+1)}{\Gamma(\frac{1}{2}\nu-r+\frac{1}{2})(2r+1)!}(2z)^{2r+1}$$

constitute a basis of solutions in the unit circle $|z| < 1$ of the Legendre equation

$$(1-z^2)\frac{d^2u}{dz^2}-2z\frac{du}{dz}+\nu(\nu+1)u = 0.$$

Determine when any of $u_1(z)$, $u_2(z)$ is a polynomial.

6. (a) For each of the following equations find all its singular points and determine whether or not they are regular.

$$(1-z^2)\frac{d^2u}{dz^2}-z\frac{du}{dz}+n^2u = 0 \quad \text{(Tchebyshev)},$$

$$z\frac{d^2u}{dz^2}+(\beta+1-z)\frac{du}{dz}+\nu u = 0 \quad \text{(Laguerre)},$$

$$(1-z^2)\frac{d^2u}{dz^2}-2z\frac{du}{dz}+\left[n(n+1)-\frac{m^2}{1-z^2}\right]u = 0 \quad \text{(associated Legendre)},$$

$$(1-z^2)\frac{d^2u}{dz^2}+[b-a-(a+b+2)z]\frac{du}{dz}+n(n+a+b+1)u = 0 \quad \text{(Jacobi)}.$$

(b) For the equations of Tchebyshev, Laguerre and associated Legendre find bases of solutions in a neighbourhood of $z = 0$. Discuss the existence of polynomial solutions.

7. Show that for the equation

$$z\frac{d^2u}{dz^2}+(2-z)\frac{du}{dz}-u = 0$$

(a special case of the confluent h.e.) the characteristic exponents differ by an integer and yet there is a basis of solutions with no logarithmic term.

8. (a) Derive the Schläfli integral (Problem 9(c), Ch. 10) for the Legendre function from Proposition 13.2.1.

(b) Define the Legendre function of the second kind $Q_\nu(z)$ by

$$Q_\nu(z) = \frac{1}{4i\sin\nu\pi}\int_{\mathscr{C}}\frac{(t^2-1)^\nu}{(z-t)^{\nu+1}}\,dt$$

with the contour $\mathscr{C}$ being a figure of eight which encircles $t = 1$ and $t = -1$ in opposite directions and such that z lies outside the contour. Show that

$$Q_\nu(z) = 2^{-\nu-1} \int_{-1}^{1} \frac{(1-t^2)^\nu}{(z-t)^{\nu+1}} \, dt.$$

(c) Derive the power series expansion for $Q_\nu(z)$

$$Q_\nu(z) = (2z)^{-\nu-1} \sum_{r=1}^{\infty} \frac{\Gamma(r+\frac{1}{2})\Gamma(\nu+1+2r)}{(2r)!\,\Gamma(\nu+r+\frac{3}{2})} z^{-2r}$$

and show by using the duplication formula for $\Gamma(z)$ that

$$Q_\nu(z) = \frac{\pi^{\frac{1}{2}}\Gamma(\nu+1)}{\Gamma(\nu+\frac{3}{2})} (2z)^{-(\nu+1)} F\left(\tfrac{1}{2}(\nu+1), \tfrac{1}{2}\nu+1, \nu+\tfrac{3}{2}; z^{-2}\right).$$

9. Let h be a homographic mapping of the extended complex plane $\boldsymbol{C} \cup \{\infty\}$. Show that the effect of h on solutions of the Riemann's P-equation can be expressed as

$$P\begin{Bmatrix} \xi, & \eta, & \zeta & \\ \alpha_1, & \beta_1, & \gamma_1, & z \\ \alpha_2, & \beta_2, & \gamma_2 & \end{Bmatrix} = P\begin{Bmatrix} h(\xi), & h(\eta), & h(\zeta) & \\ \alpha_1, & \beta_1, & \gamma_1, & h(z) \\ \alpha_2, & \beta_2, & \gamma_2 & \end{Bmatrix}.$$

10. Show that

$$\text{(a) } P\begin{Bmatrix} 0, & 1, & \infty & \\ \alpha_1, & \beta_1, & \gamma_1, & z \\ \alpha_2, & \beta_2, & \gamma_2 & \end{Bmatrix} = z^\alpha (z-1)^\beta P\begin{Bmatrix} 0, & 1, & \infty & \\ \alpha_1-\alpha, & \beta_1-\beta, & \gamma_1+\alpha+\beta, & z \\ \alpha_2-\alpha, & \beta_2-\beta, & \gamma_2+\alpha+\beta & \end{Bmatrix},$$

$$\text{(b) } P\begin{Bmatrix} 0, & 1, & \infty & \\ \alpha_1, & \beta_1, & \gamma_1, & z \\ \alpha_2, & \beta_2, & \gamma_2 & \end{Bmatrix} = P\begin{Bmatrix} 0, & 1, & \infty & \\ \beta_1, & \alpha_1, & \gamma_1, & 1-z \\ \beta_2, & \alpha_2, & \gamma_2 & \end{Bmatrix}.$$

11. Prove the identities

$$F(a, b, c; z) = (1-z)^{c-a-b} F(c-a, c-b, c; z),$$

$$F(a, b, c; z) = (1-z)^{-b} F\left(b, c-a, c; \frac{z}{z-1}\right).$$

12. Show that

$$2F(-\tfrac{1}{2}a,\ -\tfrac{1}{2}a+\tfrac{1}{2},\tfrac{1}{2};z^2) = (1+z)^a+(1-z)^a,$$

$$F(\tfrac{1}{2}a,\ -\tfrac{1}{2}a,\tfrac{1}{2}\sin^2 z) = \cos az,$$

$$F(\tfrac{1}{2}(a+1),\tfrac{1}{2}(1-a),\tfrac{3}{2};\ \sin^2 z) = \frac{\sin az}{a\sin z},$$

$$F(\tfrac{1}{2}(a+1),\tfrac{1}{2}(1-a),\tfrac{1}{2};\ \sin^2 z) = \frac{\cos az}{\cos z},$$

$$F(\tfrac{1}{2}(n+1),\tfrac{1}{2}n,n+1;z) = 2^n(1+\sqrt{1-z})^{-n},$$

$$F(\tfrac{1}{2},1,\tfrac{3}{2};z^2) = \frac{1}{2z}\log\frac{1+z}{1-z}.$$

13. Prove the following recurrence relations for the hypergeometric function

$$F(a,b+1,c;z)-F(a,b,c;z) = \frac{az}{c}F(a+1,b+1,c+1;z),$$

$$F(a+1,b+1,c;z)-F(a,b,c;z) = \frac{abz}{c}F(a+1,b+1,c+1;z),$$

$$\frac{\mathrm{d}}{\mathrm{d}z}F(a,b,c;z) = \frac{ab}{c}F(a+1,b+1;c+1;z).$$

14. (a) Prove that the radius of convergence of the hypergeometric series is 1 except the case when a or b is 0 or a negative integer, in which case the series reduces to a polynomial.

(b) Prove under the assumption $\mathrm{Re}(c-a-b) > 0$ that the hypergeometric series converges also on the circle $|z| = 1$.

Hint for (b). Use the Gauss criterion convergence.

15. Show that the function

$$u_2(z) = z^{1-c}F(a+1-c,b+1-c,2-c;z)$$

satisfies the hypergeometric equation in a neighbourhood of $z = 0$ with zero deleted. Moreover prove by computing the Wronskian that if $c \neq 1$ then $u_1(z) = F(a,b,c;z)$ and $u_2(z)$ are independent solutions.

16. (a) Show that

$$u_1(z) = F(a,b,a+b+1-c;1-z),$$

$$u_2(z) = (1-z)^{c-a-b}F(c-b,c-a,c+1-a-b;1-z)$$

are independent solutions of h.e. in a neighbourhood of $z = 1$ (with 1 deleted), provided $a-b-c$ is not an integer.

(b) Show that

$$u_1(z) = z^{-a}F\left(a, a-c+1, a-b+1; \frac{1}{z}\right),$$

$$u_2(z) = z^{-b}F\left(b, b-c+1, b-a+1; \frac{1}{z}\right)$$

are independent solutions around $z = \infty$ provided $a-b$ is not an integer.

Hint. Combine Problems 9 and 10.

17. Let G denote the six-element group of homographic mappings permuting among themselves the singular points of the h.e. Using the Problem 9 investigate what solutions of the h.e. may be obtained by superposition of the hypergeometric function with elements of G. For each of the solutions so obtained describe its domain of definition.

18. Prove

$$(1-t)^{b-c}(1-t+zt)^{-b} = \sum_{n=0}^{\infty} \frac{(c)_n}{n!} F(-n, b, c; z)t^n,$$

where $(c)_n := c(c+1) \dots (c+n-1)$.

19. Obtain the integral (13.2.8) for the hypergeometric function from its power series expansion by using the integral formula for the Euler B-function given in (6.3.15).

20. Show that the integral representation (13.4.6) of the confluent hypergeometric function reduces to the Poisson integral for the Bessel function (cf. Problem 12, Ch. 7).

21. Find the second independent solution of the Bessel equation of order n, n a positive integer, by the method of case (C), § 13.1. Compare it with the Neumann function $Y^{(n)}(z)$.

Hint. The function $u_1(z, \alpha) = z^\alpha \sum_{s=0}^{\infty} c_s(\alpha)z^s$ has the coefficients

$$c_{2s}(\alpha) = \frac{(-1)^s(\alpha+n)}{(\alpha-n+2) \dots (\alpha-n+2s)(\alpha+n+2) \dots (\alpha+n+2s)},$$

$$c_{2s+1}(\alpha) = 0.$$

Rewrite u_1 as the sum of a polynomial and a function of the form

$$cz^{\alpha+2n}\left[1+\sum_{s=1}^{\infty} b_s z^{2s}\right]$$

with appropriately chosen constants c and coefficients b_s. The answer is:

$$-2^{2n-1}(n-1)!\frac{\partial u_1}{\partial\alpha}\bigg|_{\alpha=-n}(z)$$

$$= J_n(z)\log z - \frac{1}{2}\sum_{s=0}^{n-1}\frac{(n-s-1)!}{s!}\left(\frac{z}{2}\right)^{-n+2s}$$

$$-\frac{1}{2}\sum_{s=0}^{\infty}\frac{(-1)^s}{s!(n+s)!}\left(\frac{z}{2}\right)^{n+2s}\left\{\sum_{r=1}^{s}\frac{1}{r}+\sum_{r=0}^{s}\frac{1}{n+r}\right\}.$$

By subtracting $\frac{1}{2}\sum_{r=1}^{n-1}\left(\frac{1}{r}\right)J_n(z)$ one obtains the following formula for the Neumann function $Y^{(n)}(z)$,

$$Y^{(n)}(z) = J_n(z)\log z - \frac{1}{2}\sum_{s=0}^{n-1}\frac{(n-s-1)!}{s!}\left(\frac{z}{2}\right)^{-n+2s}$$

$$-\frac{1}{2}\sum_{s=0}^{\infty}\frac{(-1)^s}{s!(n+s)!}\left(\frac{z}{2}\right)^{n+2s}\left\{\sum_{r=1}^{s}\frac{1}{r}+\sum_{r=1}^{n+s}\frac{1}{r}\right\}.$$

PART III

Introduction

This part of the book is devoted to the general theory. It presents the theorems and relations whose special cases have been described in the previous part. Clearly, the presentation of the general theory opens the possibility of infinitely multiplying the examples of spherical functions of formulas describing their properties and of expansions of functions with respect to various orthonormal systems. However, this is not the only reason why it is worthwhile to get acquainted in a more extensive theory than that presented before. The strongest inducement is the chance of observing the relations between the simple geometric properties of the manifolds here presented and the abundance and complexity of the theory of functions on these domains.

The symmetric structure which will be dealt with denotes that to every two points $x \neq y$ in a connected manifold M corresponds a unique point $S_x y$ symmetric to y with respect to x.

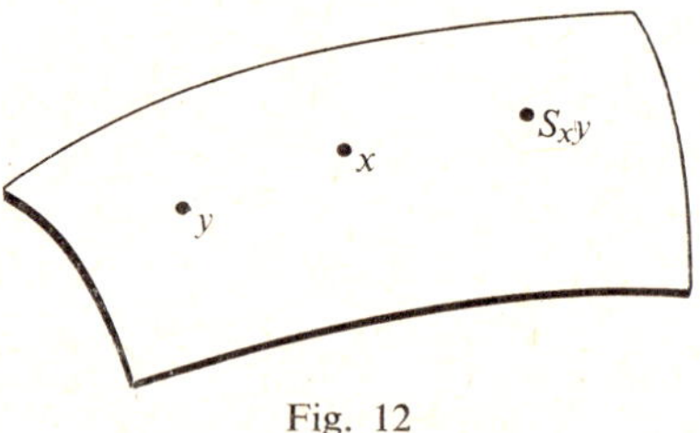

Fig. 12

The axioms of the symmetry operation S_x consist of a number of main properties of central symmetry in a Euclidean space. This is in fact the basic example of a symmetric structure.

The existence of a symmetric structure has profound consequences for the geometry of the underlying manifold. For every two fixed points $x \neq y$ one can find a unique central point o, i.e. such a point that x is symmetric to y with respect to o. Repeating this procedure, one gets a set whose closure is a segment of a smooth curve joining x and y. The curve can be prolonged to infinity, so that the symmetry with respect to an arbitrary point on the curve maps the curve into itself. In the geometry of symmetric spaces such curves play the role of straight lines. All straight lines which go through a fixed point cover the symmetric manifold.

To apply the methods of group theory, one considers the group of those diffeomorphisms of M which map lines into lines (group Aff M). The group acts transitively on the symmetric manifold. Straight lines arise through the action on M of 1-parameter subgroups of Aff M. It turns out that Aff M has the structure of a Lie group with M as a homogeneous space. The description of M as a homogeneous space reduces the problem of the classification of symmetric manifolds to the study of pairs of Lie groups (G, H) such that $H \subset G$ and $M = G/H$ is a symmetric space. The last condition is equivalent to the requirement that there should be an involutive automorphism σ of G which leaves the points of the subgroup H fixed.

The geometry of a symmetric space may differ considerably from the Euclidean geometry or the geometry of a sphere, which are the simplest models of a symmetric space. In both these spaces one can introduce polar coordinates: the isotropy subgroup H of the point $o \in M$ acts transitively in the set of all lines through o, which allows us to define for any $x \in M$ "the radial coordinate" and "the angular coordinates" corresponding to the parametrization of the subgroup H. Among general symmetric spaces this situation is quite exceptional. In order to define "generalized polar coordinates" one usually has to consider pencils of lines through a fixed point. The elements of the pencil cannot be mapped one onto another under the action of diffeomorphisms from the subgroup H. Consequently the "radius" of a point is a system of numbers. The number of the elements of the system is called the rank of the space M.

The method of description of a symmetric manifold as a homogeneous space of a Lie group as well as the final classification of Riemannian symmetric spaces were worked out by E. Cartan in the years 1926–1929.

There are three distinct types of symmetric Riemannian spaces, from which, by taking Cartesian products, one can obtain all the remaining symmetric spaces. The fundamental types differ in the sign of the sectional curvature of the underlying manifold.

Spaces of the Euclidean type (curvature zero) and of the compact type (non-negative curvature) are simpler cases from the point of view of harmonic analysis. The first case is covered by the classical Fourier transform and the Fourier-Bessel transform while in the second the Peter–Weyl theorem, the Frobenius–Weil theorem and the theory of spherical functions supply the basic information concerning expansions of functions on those spaces. The third type of symmetric manifolds is characterized by a non-positive sectional curvature and is called a symmetric space of the non-compact type. The group Aff$M$ of such a manifold has no finite-dimensional unitary representations, and consequently completely new phenomena take place in the harmonic analysis on manifolds of the non-compact type.

The theory of expansions of functions on those spaces originated only in the late forties in the period of a particularly fast development of the abstract harmonic analysis. One has the impression, however, that rather than supplying technical methods the general theory of decomposition of unitary representations brought the realization that the problems of harmonic analysis on homogeneous spaces were ripe for solving. The contribution of the general harmonic analysis to this domain is described in Chapter 16. The complete and detailed solution required a substantial increase in the knowledge of the structure of semi-simple groups themselves and the geometry of symmetric spaces. In the forties Harish-Chandra started the most systematic and completely general investigations of representations of semi-simple groups. He began with numerous deep results in the theory of Lie groups and algebras laying particular stress on semi-simple groups. His achievements in this domain can only be compared with the work of E. Cartan.

In the early fifties the theory of spherical functions was formed, first by Gelfand and Neumark for the models of the classical groups and then in its general form by Harish-Chandra, Godement and Gelfand. Among other things, Gelfand reduced the problem of the description of the Plancherel measure for the representation of G on $L^2(G/H)$ to the description of expansions of functions of class $\mathscr{D}(H\backslash G/H)$ with respect to the family of zonal spherical functions on G/H (1950). This is an essential step, but in the case of a manifold of the non-compact type it marks only the beginning of a long way, which we outline in

Chapter 18. The next important result was the general formula for a spherical function given by Harish-Chandra in 1953. In 1958 Harish-Chandra again described the Plancherel measure associated with the homogeneous space of the non-compact type. The knowledge of this measure is clearly of fundamental importance. Yet it fails to supply a tool as convenient as the Fourier transform in the classical analysis. Seeking an integral operation with the properties of the Fourier transform, we have to choose counterparts of the plane waves from the analysis on a Euclidean space. This involves the geometrical problems of finding counterparts of hypersurfaces in $\boldsymbol{R}^n$ and selecting from among them the families of parallel hypersurfaces which should constitute the system of levels of a plane wave. In the case of a symmetric space of the non-compact type the role of hypersurfaces is played by horocycles, which are submanifolds of dimension dim M—rank M. The geometrical considerations aimed at justifying the adopted definition of plane waves and the Fourier transform on M occupy a major part of § 18.1 and § 18.2. Getting familiarized with them will allow one relate the differences which appear between the classical Fourier analysis and the analysis on symmetric spaces with the corresponding properties of plane geometry and non-Euclidean geometry.

Chapter 14

Affine Transformations

14.1. ASSOCIATED VECTOR BUNDLES

The construction that we are going to describe now is a natural extension of the construction of the bundle associated with the frame bundle of a differentiable manifold. Recall that a principal bundle is a system $\mathrm{P} = (P, M, \pi, G)$ where $\pi\colon P \to M$ is a smooth surjection of a differentiable manifold P onto a manifolds M; the group G acts on the manifold P to the right without fixed points and the action is transitive on every fibre $\pi^{-1}(m)$ for $m \in M$. Moreover, we assume that there exists a neighbourhood $O \ni m$ and a diffeomorphism

$$f_0\colon \pi^{-1}(O) \ni x \to (\pi(x), \eta(x)) \in O \times G,$$

which satisfies the condition

$$f_0(x \cdot g) = (\pi(x), \eta(x) \cdot g) \qquad \text{for an arbitrary } g \in G.$$

Among the basic examples of principal bundles known from Chapter 1 we have: the trivial bundle $M \times G$, the frame bundle, G-structures and the bundle $\pi\colon G \to G/H$ determined by an arbitrary closed subgroup H of the Lie group G.

DEFINITION. Let P be a principal bundle over M. We put

$$V(x) = \ker \mathrm{d}\pi_x \qquad \text{for } x \in P.$$

The elements of the space $V(x)$ are referred to as *vertical vectors*. The dimension of the space $V(x)$ equals the dimension of the structural group G.

The action of the structural group on the space of the principal bundle P defines a system of fields on P called the *system of fundamental vector fields*.

Denote by O_x the mapping $G \ni g \to r_g x \in P$. Let $\mathfrak{g}$ be the Lie algebra of the group G. For every $X \in \mathfrak{g}$ we define a fundamental vector field on P as follows:

$$X^*(x) = (\mathrm{d}O_x)_e(X).$$

The mapping $\mathfrak{g} \ni X \to X^* \in \mathscr{T}(P)$ is injective since the group acts freely on P.

Fundamental fields are all vertical as a consequence of the fact that $\pi \circ O_x$ is a constant mapping

$$0 = \mathrm{d}(\pi \circ O_x)_e(X) = \mathrm{d}\pi_x \circ (dO_x)_e(X) = \mathrm{d}\pi_x(X^*(x)).$$

Moreover, we note that the values of fundamental fields at a fixed point form the space of vertical vectors:

$$V(x) = \ker \mathrm{d}\pi_x = \{X^*(x) : X \in \mathfrak{g}\}.$$

Given an arbitrary finite-dimensional representation of the structural group, we can define a bundle whose fibres are linearly isomorphic to that respresentation. This method leads in particular to tensor bundles on manifolds and to a series of other important constructions.

Suppose $\mathrm{P} = (P, M, \pi, G)$ is a principal bundle and let the structural group G act to the left on a manifold W of dimension k over $\boldsymbol{R}$. We define a left-hand action of the group G on the Cartesian product $P \times W$ by setting

$$g(x, v) = (xg^{-1}, gv). \tag{14.1.1}$$

The space of orbits of the group G in $P \times W$ will constitute the space of a new bundle, which we shall denote by $E(\mathrm{P}, W)$ or simply $E(W)$.

We define a projection $p \colon E(W) \to M$ by the formula

$$p([(x, v)]) := \pi(x), \tag{14.1.2}$$

where $[(x, v)]$ is the orbit of the point $(x, v) \in P \times W$.

The definition is correct since the choice of a representative of the orbit $[(x, v)]$ does not influence the right-hand side of formula (14.1.2).

Next, we introduce an atlas for the space $E(W)$ in the following manner:

For an open subset $O \subset M$ such that there exists a diffeomorphism

$$\pi^{-1}(O) \ni x \to (\pi(x), \eta(x)) \in M \times G,$$

we define the mapping θ_0

$$p^{-1}(O) \ni [(x, v)] \to (\pi(x), \eta(x)v) \in M \times W.$$

We leave it to the reader to verify that the mapping θ_0 is one-to-one.

Finally, charts for $E(W)$ are defined by composing θ_0 with the charts for the manifolds M and W.

Definition. The system $(E(\mathrm{P}, W), M, p)$ is termed a *bundle*, with *typical fibre W* and *base M, associated with the principal bundle* P.

By a *section* of the bundle we mean a mapping s of the base M into the bundle space such that

$$p\big(s(m)\big) = m. \tag{14.1.3}$$

14.2. Operations on differential forms

The exterior multiplication is a bilinear operation of the space $\vartheta_k \times \vartheta_l$ into ϑ_{k+l}. To begin with, we define the exterior product of elements $\omega_1(m) \in \Lambda^k T_m^*$ and $\omega_2(m) \in \Lambda^l T_m^*$ belonging to fibres over the same point.

Let j_m denote the isomorphism of the typical fibre onto the fibre over $m \in M$

$$W \ni v \to [(m, v)] \in p^{-1}(m).$$

We define

$$\omega_1(m) \wedge \omega_2(m) := j_m\Big(j_m^{-1}\omega_1(m) \wedge j_m^{-1}\big(\omega_2(m)\big)\Big), \tag{14.2.1}$$

where "$\wedge$" on the right denotes the exterior product of elements of the typical fibres.

We now define the exterior product of forms $\omega_1 \in \vartheta_k$ and $\omega_2 \in \vartheta_l$ as

$$\omega_1 \wedge \omega_2(m) := \omega_1(m) \wedge \omega_2(m). \tag{14.2.2}$$

The following formulas result directly from the definition of the exterior product of finite-dimensional spaces:

$$\begin{aligned} \omega_1 \wedge \omega_2 &= (-1)^{kl} \omega_2 \wedge \omega_1, \\ (\omega_1 \wedge \omega_2)\omega_3 &= \omega_1 \wedge (\omega_2 \wedge \omega_3). \end{aligned} \tag{14.2.3}$$

The exterior differentiation is a linear operation of the type $\mathrm{d}\colon \vartheta_k(M) \to \vartheta_{k+1}$. It is uniquely determined by the system of conditions:

(1) $\omega \in \vartheta_0 = \mathscr{E}(M) \Rightarrow \mathrm{d}\omega(X) = X\omega, \qquad X \in \mathscr{T}(M),$

(2) $\mathrm{d} \cdot \mathrm{d} = 0,$

(3) $\mathrm{d}(\omega_1 \wedge \omega_2) = \mathrm{d}\omega_1 \wedge \omega_2 + (-1)^r \omega_1 \wedge \mathrm{d}\omega_2,$

where r is the degree of the form ω_1.

One can prove that the operator d determined by these conditions can be written in the form

$$(k+1)\mathrm{d}\omega(X_0, \ldots, X_k) = \sum_{i=0}^{k} (-1)^i X_i \omega(X_0, \ldots, \hat{X}_i, \ldots, X_k)$$

$$+\sum_{i<j} (-1)^{i+j} \omega([X_i, X_j], X_0, \hat{X}_i, \ldots, \hat{X}_j, \ldots, X_k) \tag{14.2.4}$$

(the symbol ^ over a field denotes omission of the field).

Analogous operations can also be defined for vector-valued forms. To this end let $\omega \in \vartheta_k^W(M)$ be a W-valued form on M. By fixing a basis $\{e_i\}$ for W we can write $\omega = \sum \omega_i \otimes e_i$ where $\omega_i \in \vartheta_k(M)$.

Now we put

$$\mathrm{d}\omega := \sum \mathrm{d}\omega_i \otimes e_i$$

and owing to the linearity of d the value of $\mathrm{d}\omega$ is independent of the choice of a basis. Next, suppose that W_j, $i = 1, 2$ are finite-dimensional vector spaces over $\boldsymbol{R}$ and $\varrho: W_1 \times W_2 \to W_3$ is a bilinear form. Let $\{e_i\}$ and $\{f_j\}$ be bases for W_1 and W_2, respectively. We can define the exterior multiplication

$$\varrho_{\#} : \vartheta_k^{W_1} \times \vartheta_l^{W_2} \to \vartheta_{k+l}^{W_3} \tag{14.2.5}$$

by setting

$$\varrho_{\#}(\omega^1, \omega^2) = \sum \omega_i^1 \wedge \omega_j^2 \varrho(e_i, f_j) \tag{14.2.6}$$

for

$$\omega^1 = \sum_i \omega_i^1 \otimes e_i, \qquad \omega^2 = \sum_j \omega_j^2 \otimes f_j.$$

Condition (3) for the exterior differentiation is also valid in this case. Namely, we have

$$\mathrm{d}\varrho_{\#}(\omega^1, \omega^2) = \varrho_{\#}(\mathrm{d}\omega^1, \omega^2) + (-1)^r \varrho_{\#}(\omega^1, \mathrm{d}\omega^2), \tag{14.2.7}$$

where r is the degree of the form ω^1.

The following two examples of the operation $\varrho_{\#}$ are of special importance:

(1) $W_1 = W_2 = W_3 = \mathfrak{g}$ is a Lie algebra and $\varrho = [\cdot, \cdot]$ is the commutator of $\mathfrak{g}$. In this case we usually write $[\omega^1 \wedge \omega^2]$ for $\varrho_{\#}(\omega^1, \omega^2)$.

(2) $W_2 = W_3 = V$, $W_1 = V \otimes V^*$ and ϱ is defined by the canonical isomorphism of the space W_1 onto the space $L(V)$ of linear operators on V:

$$\varrho\Big(\sum_{ij} \lambda_{ij}\, v_i \otimes v_j^*\Big)\, v = \sum_{ij} \lambda_{ij} v_i \langle v, v_j^* \rangle .$$

In this case $\varrho_{\#}(\omega^1, \omega^2)$ is denoted simply by $\omega_1 \wedge \omega_2$.

Example 1

Let G denote a Lie group and $\mathfrak{g}$ be its Lie algebra (regarded as the algebra of left invariant vector fields on G). We shall describe in detail the case of $\mathfrak{g}$-valued differential forms on G. As in the case of scalar-valued forms, we say that a form ω is invariant if, for every $g \in G$, $l_g^* \omega = \omega$ where $l_g^* \omega(X_1, \ldots, X_r) = \omega(\mathrm{d}l_g X_1, \ldots, \mathrm{d}l_g X_r)$, $X_1, \ldots, X_r \in \mathfrak{g}$. For an arbitrary $g \in G$ we define a $\mathfrak{g}$-valued form $\theta\colon T_g G \to \mathfrak{g}$ so that for every $Y \in T_g G$ the value $\theta(Y)$ is the only vector field in $\mathfrak{g}$ having the property that $\theta(Y)(g) = Y$. It is easily verified that the mapping defined in this way for every $g \in G$ depends smoothly upon the point g, i.e., represents a differential form with values in the Lie algebra $\mathfrak{g}$. This form will be denoted by θ and we shall refer to it as the canonical 1-form on G. By fixing a basis $\{X_i\}_1^n$ for $\mathfrak{g}$ we can write

$$\theta = \sum_{i=1}^{n} \theta^i \otimes X_i ,$$

where $\theta^i \in \vartheta_1(G)$ for $i = 1, \ldots, n$.

It is easy to observe that the form θ and the scalar valued forms θ^i are all left invariant. As a matter of fact, $\{\theta^i\}_1^n$ is a basis for the space of all invariant (scalar valued) forms on G. This space can be viewed as the dual space of $\mathfrak{g}$ under the duality $\langle \varphi, X \rangle = \varphi(X)$ (note that the right-hand side of this identity is a constant function on G).

We shall verify the following properties of the forms θ, θ^i:

(i) For arbitrary $X, Y \in T_g G$

$$[\theta \wedge \theta](X, Y) = [\theta(X), \theta(Y)],$$

(ii) The Maurer–Cartan equations

$$\mathrm{d}\theta^k = -\tfrac{1}{2} \sum_{i,j=1}^{n} c_{ij}^k \theta^i \wedge \theta^j , \qquad k = 1, \ldots, N,$$

where c_{ij}^k are the structural constants given by the condition

$$[X_i, X_j] = \sum_{k=1}^{n} c_{ij}^k X_k, \qquad i, j = 1, \ldots, n.$$

(iii) The Maurer–Cartan equations in the intrinsic form

$$\mathrm{d}\theta = -\tfrac{1}{2}[\theta \wedge \theta].$$

To prove (i) we note that according to (14.2.6)

$$[\theta \wedge \theta](X, Y) = \sum_{i,j=1}^{n} \theta^i \wedge \theta^j(X, Y)[X_i, X_j]$$

$$= \tfrac{1}{2} \sum_{i,j=1}^{n} \big(\theta^i(X)\theta^j(Y) - \theta^i(Y)\theta^j(X)\big)[X_i, X_j]$$

$$= \tfrac{1}{2}[\theta(X)], \theta(Y)] - \tfrac{1}{2}[\theta(Y), \theta(X)] = [\theta(X), \theta(Y)].$$

Concerning (ii) we observe that the fact of the functions $\theta^i(X)$ being constant for an arbitrary $X \in \mathfrak{g}$ implies that (14.2.4) becomes

$$\mathrm{d}\theta^k(X, Y) = -\tfrac{1}{2}\theta^k([X, Y]), \qquad X, Y \in \mathfrak{g}.$$

Further, since $X = \sum_{i=1}^{n} \theta^i(X)X_i$, $Y = \sum_{i=1}^{n} \theta^i(Y)X_i$ we have

$$\theta^k([X, Y]) = \sum_{i,j=1}^{n} \theta^i(X)\theta^j(Y)\theta^k([X_i, X_j])$$

$$= \sum_{i,j=1}^{n} c_{ij}^k \theta^i(X)\theta^j(Y) = \sum_{i,j=1}^{n} c_{ij}^k \theta^i \wedge \theta^j(X, Y).$$

The last identity follows in view of the relation $c_{ij}^k = -c_{ji}^k$.

The verification that (iii) is indeed a coordinate free expression of the Maurer-Cartan equation is left to the reader.

14.3. AFFINE CONNECTIONS

We now proceed to define the basic concept of differential geometry, i.e. the concept of an affine connection. The definition given below is due to Nomizu (1954). The definition itself is rather abstract, whereas the derived notion of the parallel translation along a curve in a manifold has a very clear interpretation.

One can differentiate a function on a manifold in the direction of a curve, but there is no natural way of differentiating a vector field or a tensor field.

This results from the fact that a tensor field is not a function with values in a vector space since its values at different points belong to different vector space. Although such spaces are linearly isomorphic there is no cannonical isomorphism between them arising from the manifold structure itself. Defining an affine connection on a manifold is tantamount to establishing a method of identification in tensor bundles of fibres, over different points of a curve in the manifold.

As a result an affine connection allows us to develop the differential calculus on vector fields and tensor fields on a manifold.

The notion of an affine connection has been introduced by Levi-Civita.

DEFINITION. Let P $= (P, M, \pi, G)$ be a principal bundle over an n-dimensional manifold M. Denote the action on P of the structural group as follows:

$$r_g p = p \cdot g, \qquad p \in P, \qquad g \in G.$$

By a *connection on the principal bundle* P we shall understand a distribution H on the manifold P satisfying the conditions

$$H(p) + V(p) = T_p(P), \qquad p \in P, \tag{14.3.1}$$

$$\mathrm{d}r_g(H(p)) = H(r_g p), \qquad p \in P, \qquad g \in G. \tag{14.3.2}$$

Condition (14.3.1) means that at every point in the manifold P the distribution H determines a complementary subspace to the space of vertical vectors. Elements of the space $H(p)$ are called the *horizontal vectors*.

Covariance condition (14.3.2) means that the value of a connection at a point p determines the values on the whole fibre $p \cdot G$.

Given an $X \in T_p(P)$, we shall denote by HX and VX the horizontal and the vertical component of a vector X, respectively. The tangent mapping $\mathrm{d}\pi_p$ to the projection in the bundle P establishes an isomorphism of the space $H(p)$ onto $T_{\pi(p)}(M)$. The inverse operation $l_p\colon T_{\pi(p)}(M) \to H(p)$, i.e. the operation satisfying the condition

$$\mathrm{d}\pi_p \circ l_p = \mathrm{id}, \tag{14.3.3}$$

is called the *lifting* to the point $p \in P$ of tangent vectors at the point $\pi(p) \in M$. We shall also write

$$l_p(X) =: \hat{X}_p \qquad \text{for } X \in T_{\pi(p)}(M).$$

Given a vector field X on M, we shall denote by $\hat{X}$ a vector field on P such that $\hat{X}(p) = (X(\pi(p)))_p^\wedge$ and we shall call it the *lift* of the vector field X.

The operation of lifting satisfies the identities

$$dr_g(\hat{X}) = \hat{X} \qquad \text{(invariance of the field)}, \tag{14.3.4}$$

$$H[\hat{X}, \hat{Y}] = [X, Y]^\wedge. \tag{14.3.5}$$

The first identity is a straightforward consequence of property (14.3.2) of a connection. To prove the second one we act with the operator $d\pi$ on the left-hand side of identity (14.3.5). We have

$$d\pi(H[\hat{X}, \hat{Y}]) = d\pi([\hat{X}, \hat{Y}]) = [d\pi\hat{X}, d\pi\hat{Y}] = [X, Y],$$

which gives the desired result.

Example 1 (*Connection on a trivializable bundle*).

Suppose that a principal bundle $\mathrm{P} = (P, M, \pi, G)$ admits a global smooth section s. We then say that P is a trivializable bundle since the mapping $\psi_s : M \times G \ni (m, g) \to s(m)g \in P$ defines an isomorphism of P onto the trivial bundle $M \times G$. The mapping $ds_m : T_m M \to T_{s(m)}P$ is injective owing to the condition $d\pi_{s(m)} \circ ds_m = \mathrm{id}$. We define a distribution H on P by putting for $p \in \pi^{-1}(m)$

$$H(p) := dr_g(T_{s(m)}P) \qquad \text{if} \quad p = s(m)g.$$

Clearly, H satisfies conditions (14.3.1) and (14.3.2) and therefore defines a connection in P. Also it is easily seen that the space $H(p)$ of horizontal vectors and the space $V(p)$ of vertical vectors are the images under the mapping $(d\psi_s)_{(m,g)}$ of the space of the tangent vectors to the manifolds $M \times \{g\}$ and $\{m\} \times G$, respectively.

Later we shall consider special cases of the above situation, where P is the frame bundle of a manifold M.

DEFINITION. As before, let P be a principal bundle over M with the structural group G whose Lie algebra is denoted by $\mathfrak{g}$. Let H be a connection on P. By a *connection form* of the connection H we mean the 1-form on the space P given by the equation

$$\varphi_p(X_p)^* = VX_p, \qquad X_p \in T_p(P). \tag{14.3.6}$$

As we know, the mapping $\mathfrak{g} \ni X \to X^*(p) \in V(p)$ is an isomorphism and therefore equation (14.3.6) determines uniquely the form φ.

Below we list the basic properties of the form φ, which are direct consequences of the respective properties of the connection H:

$$\varphi(HX) = 0 \qquad \text{(verticality of } \varphi\text{)}, \tag{14.3.7}$$

$$r_g^*(\varphi) = \mathrm{Ad}\, g^{-1} \circ \varphi \qquad \text{(covariance of } \varphi\text{)}, \tag{14.3.8}$$

$$\varphi(X^*) = X \qquad \text{for} \quad X \in \mathfrak{g}. \tag{14.3.9}$$

For a fixed $p \in P$ the function $\varphi_p\colon V(p) \to \mathfrak{g}$ is an isomorphism. The connection form determines the connection uniquely. Namely, we have

$$H(p) = \ker \varphi_p. \tag{14.3.10}$$

It can be proved that every form having properties (14.3.7)-(14.3.9) defines a connection in P by formula (14.3.10).

Definition. By the *curvature form* of a connection H corresponding to the connection form φ we mean the $\mathfrak{g}$-valued 2-form given by the identity

$$\Phi(X, Y) = \mathrm{d}\varphi(HX, HY).$$

Symbolically $\Phi = \mathrm{d}\varphi \circ H$.

Theorem 14.3.1. (1) *The connection form and the curvature form are connected by the structural equation*

$$\mathrm{d}\varphi = -\tfrac{1}{2}[\varphi \wedge \varphi] + \Phi.$$

(2) *The Bianchi identity holds*:

$$\mathrm{d}\Phi(HX, HY] = 0.$$

For the proof of this theorem the reader is recommended to consult one of fundamental monographs: Bishop, Crittenden (1964), Kobayashi, Nomizu (1963, 1969), Helgason (1968) and Sternberg (1964).

Example 2

As an illustration we shall find the connection form and the curvature form on a trivializable bundle P. Let ϱ denote the projection onto the second factor in $M \times G$ and let $p = \psi_s(m, g) \in P$. Then for $X_p \in T_p(P)$ we have

$$\varphi_p(X_p) = \theta \circ \mathrm{d}\varrho_{(m,g)}(\mathrm{d}\psi_s^{-1})_p(X_p),$$

where θ is the canonical 1-form on G defined in Example 1. Shortly

$$\varphi = \psi_s^{*-1} \circ \varrho^*\theta.$$

Using the Maurer–Cartan equations and the properties of the pull-over operation on forms, we get

$$\mathrm{d}\varphi = (\psi_s^{-1})^* \circ \varrho^*\mathrm{d}\theta = -\tfrac{1}{2}(\psi_s^{-1})^* \circ \varrho^*[\theta \wedge \theta] = -\tfrac{1}{2}[\varphi \wedge \varphi],$$

which means in view of Theorem 14.3.1 that the curvature form Φ vanishes identically.

Example 3

Suppose that G is a connected Lie group and let $K \subset G$ be a closed subgroup. Denote by $\mathfrak{g}$, $\mathfrak{k}$ the Lie algebra of G and K, respectively. In addition let us assume that there exists a subspace $\mathfrak{m}$ of $\mathfrak{g}$ such that $\mathfrak{g} = \mathfrak{k} \oplus \mathfrak{m}$, and $\mathrm{Ad}(k)\mathfrak{m} = \mathfrak{m}$ for every $k \in K$ (the homogeneous space G/K is then said to be *reductive*). We define a distribution H on G by putting

$$H(g) := \mathfrak{m}(g) = \{X(g) : X \in \mathfrak{m}\}.$$

We note that the distribution H is left invariant, more precisely

$$H(g) = \mathrm{d}l_g H(e), \qquad g \in G.$$

It will present no difficulty to the reader to verify that H satisfies conditions (14.3.1) and (14.3.4) and thus is a connection on the principal bundle $\mathrm{G} = (G, G/K, \pi, K)$ (e.g. condition (14.3.2) reduces to showing that $\mathrm{d}l_g \circ \mathrm{d}l_k H(e) = \mathrm{d}r_k \circ \mathrm{d}l_g H(e)$, which is immediate since $\mathrm{d}l_k \circ \mathrm{Ad}(k^{-1}) = \mathrm{d}r_k$ and because left translations commute with right translations).

The connection H being invariant, it follows that an element X of the Lie algebra $\mathfrak{g}$, i.e. a left invariant vector field X on G, is a horizontal field if and only if $X \in \mathfrak{m}$.

Let θ denote the canonical $\mathfrak{g}$-valued 1-form on G and let φ be the $\mathfrak{k}$-valued 1-form on G obtained by composing θ with the projection on $\mathfrak{k}$ along $\mathfrak{m}$. Then φ is the connection form corresponding to the connection H defined above. (The left invariance of the connection H translates to the left invariance of the form φ.)

In order to describe the curvature form Φ we note that it is also invariant under left translations and hence determined by its values on the Lie algebra $\mathfrak{g}$ (on the left invariant vector fields) and even by the values on the horizontal vector-fields belonging to $\mathfrak{m}$. The Maurer–Cartan equation implies that for $X, Y \in \mathfrak{m}$

$$\Phi[X, Y] = -\tfrac{1}{2}[X, Y]_{\mathfrak{k}},$$

where the index $\mathfrak{k}$ denotes the projection onto $\mathfrak{k}$ along $\mathfrak{m}$.

Affine Connections

DEFINITION. A connection on the frame bundle $\boldsymbol{B}(M)$ of a manifold M is called an *affine connection on M*.

Every basis $b \in \boldsymbol{B}(M)$ defines a mapping

$$i_b\colon \boldsymbol{R}^n \to T_{\pi(b)}(M)$$

by the formula

$$\boldsymbol{R}^n \ni (t^1, \ldots, t^n) \to \sum_1^n t^j X_j \qquad \text{where} \quad b = (\pi(b); X_1, \ldots, X_n).$$

By the canonical form on the frame bundle $\boldsymbol{B}(M)$ we mean the vertical 1-form with values in $\boldsymbol{R}^n$ defined as follows:

$$\omega(T) = i_b^{-1}(\mathrm{d}\pi(T)), \qquad b \in \boldsymbol{B}(M), \qquad T \in T_b(\boldsymbol{B}(M)) \quad (14.3.11)$$

(the mapping i_b is defined by (14.1.3)). This form satisfies the following covariance condition

$$\omega \circ \mathrm{d}r_g = g^{-1} \circ \omega,$$

where the action of $g \in \mathrm{GL}(n)$ on $\boldsymbol{R}^n$ is a natural action of a matrix. For a fixed $b \in \boldsymbol{B}(M)$ the value of the form ω at this point defines an isomorphism of the space H_b onto $\boldsymbol{R}^n$;

$$H_b \ni T \to i_b^{-1}(\mathrm{d}\pi_b(T)) = \omega_b(T) \in \boldsymbol{R}^n.$$

The form ω itself is independent of the choice of a connection and so are the fundamental fields. By means of a connection, however, we define the horizontal vector fields on $\boldsymbol{B}(M)$, called the *basis vector fields*. Namely, for $t \in \boldsymbol{R}^n$ we define

$$E(t)_b := l_b \circ i_b(t) = \hat{i}_b(t). \tag{14.3.12}$$

The vector fields $E(t)$ satisfy

$$\mathrm{d}r_g\big(E(t)\big) = E(g^{-1}t). \tag{14.3.13}$$

Let us select a basis $\{t_i\}_{i=1}^n$ for $\boldsymbol{R}^n$ and a basis $\{X_j\}_{j=1}^{n^2}$ for $\mathfrak{gl}\,(n)$.

The system of the basic vector fields $E_i := E(t_i)$, $i = 1, \ldots, n$ and the fundamental vector fields $X_j^*, j = 1, \ldots, n^2$ has the property that for a fixed $b \in \boldsymbol{B}(M)$ the set of values at the point b of the vector fields in this system constitutes a basis for the space $T_b\big(\boldsymbol{B}(M)\big)$. It follows that an affine connection on M defines an $\{e\}$-structure on $\boldsymbol{B}(M)$. The canonical form ω is actually a left inverse of the operation $\boldsymbol{R}^n \ni t \to E(t)$, as is seen from the computation

$$\omega\big(E(t)\big) = i_b^{-1}\Big(\mathrm{d}\pi\big(E(t)\big)\Big) = i_b^{-1} \circ l_b^{-1} \circ l_b \circ i_0(t) = t. \tag{14.3.14}$$

Definition. The 2-form on $\boldsymbol{B}(M)$ defined as

$$\Omega := \mathrm{d}\omega \circ H \tag{14.3.15}$$

is called the *torsion form* of the connection H.

The form Ω is horizontal and satisfies the condition

$$\Omega \circ \mathrm{d}r_g = g^{-1} \circ \Omega. \tag{14.3.16}$$

A good illustration of the role of the curvature form and the torsion form is given by the following

Proposition 14.3.1. *For every t, $s \in \boldsymbol{R}^n$ we have*

$$[E(t), E(s)] = -\Big(\Phi\big(E(t), E(s)\big)\Big)^* - E\Big(\Omega\big(E(t), E(s)\big)\Big).$$

Remark. Thus the vertical and the horizontal components of the commutator of basic vector fields are determined by the values of the curvature form and the torsion form on these fields.

Proof. If $\varphi_b(T)$ and $\omega_b(T)$ are zero for a $T \in T_b(\boldsymbol{B}(M))$, then $T = 0$ since φ_b is injective on $V_b(b)$ and ω_b on $H(b)$. Moreover,

$$\begin{aligned}
&\omega[(E(t), E(s)] + \big(\Phi(E(t), E(s))\big)^* + E\big(\Omega(E(t), E(s))\big) \\
&= \varphi([E(t), E(s)]) + \Phi(E(t), E(s)) \\
&= -E(t)\varphi(E(s)) + E(s)\varphi(E(t)) + \varphi([E(t), E(s)]) \\
&\quad + \Phi(E(t), E(s)) = -\mathrm{d}\varphi(E(t), E(s)) + \Phi(E(t), E(s)) = 0.
\end{aligned}$$

In the computations we have applied the identity $\varphi \circ E = 0$, property(3) of a connection form and the definition of the exterior derivative.

On the other hand,

$$\begin{aligned}
&\omega\Big([E(t), E(s)] + \big(\Phi(E(t), E(s))\big)^* + E\big(\Omega(E(t), E(s))\big)\Big) \\
&= \omega([E(t), E(s)]) + E\big(\Omega(E(t), E(s))\big) \\
&= \omega([E(t), E(s)]) + \Omega(E(t), E(s)). \qquad (14.3.17)
\end{aligned}$$

Since $\omega(E(t)) = t$ is a constant mapping and $E(s)\omega(E(t)) = 0$, we can represent expression (14.3.17) in the form

$$\begin{aligned}
&\omega(E(t), E(s)) - E(t)\omega(E(s)) + E(s)\omega(E(t)) + \Omega(E(t), E(s)) \\
&= -\mathrm{d}\omega(E(t), E(s)) + \Omega(E(t), E(s)).
\end{aligned}$$

Finally, it follows directly from the definition of Ω that the right-hand side of this formula is zero. □

To the torsion form of an affine connection there correspond new structural equations and Bianchi's identities. For the convenience of the reader we shall present them together with the formulas given in Theorem 14.3.1.

THEOREM 14.3.3. *Let H be an affine connection on a manifold M and let φ, Φ, Ω be the corresponding connection, curvature and torsion forms, respectively. Then the following equations hold:*

the first structural equation

$$\mathrm{d}\omega = -\varphi \wedge \omega + \Omega,$$

the second structural equation

$$\mathrm{d}\varphi = -\tfrac{1}{2}[\varphi \wedge \varphi] + \Phi,$$

Bianchi's identities

$$\mathrm{d}\Phi(HX, HY) = 0, \qquad \mathrm{d}\Omega(HX, HY) = \Phi \wedge \omega(X, Y).$$

The proofs can be found in the monographs quoted earlier.

Example 4

A global section $s\colon M \to B(M)$ of the frame bundle is actually a system of $n = \dim M$ vector fields $S_1, \ldots, S_n \in \mathscr{T}(M)$ whose values at every point in M constitute a basis for the tangent space at that point. A manifold which admits a global section (a global basis field) is said to be parallelizable. Following the construction described in Example 2 we arrive at an affine connection on M determined by the given basis field. By applying local charts we can give an explicit description of the spaces of horizontal vectors $H(b) \subset T_b(B(M))$. Let $m = \pi(b)$ and $b = s(m)g$ for a certain $g \in G$. Let us choose a chart (χ, U) on M around the point m and let $X_1, \ldots, X_n$ be the basis field determined by the chart, i.e. let X_i be the vector field tangent to the curve γ_i obtained by fixing all but the i-th coordinate. The vectors $\tilde{X}_i(b) := \mathrm{d}r_g \circ \mathrm{d}s_m X_i(m)$, $b = s(m)g$ are in fact the tangent vectors to the curves $r_g \circ s \circ \gamma_i$ and, moreover, coincide with the horizontal lifts of the vectors $X_i(m)$. Thus for every $X \in T_m(M)$, $X = \sum_{i=1}^{n} x^i X_i(m)$ we have the identity $\hat{X}_b = \sum_{i=1}^{n} x^i \tilde{X}_i(b)$.

14.4. PARALLEL TRANSLATION. GEODESICS. THE EXPONENTIAL MAPPING

Let $\mathrm{P} = (P, M, \pi, G)$ be a principal bundle with a connection H. Let γ be a smooth curve on the manifold M. By a horizontal lift of the curve γ to the bundle P we shall understand a curve $\hat{\gamma}$ on P which satisfies $\pi \circ \hat{\gamma} = \gamma$ and whose tangent vectors are all horizontal. There exist many lifts of the same curve but if we fix a point $b \in \pi^{-1}(x)$ then there is only one lift $\hat{\gamma}$ passing through b.

THEOREM 14.4.1. *For every smooth curve (I, γ) on M and $b \in \pi^{-1}(\gamma)$ there exists a unique curve $\hat{\gamma}$ in P satisfying the conditions*:

(1) $\pi(\hat{\gamma}) = \gamma$,

(2) $\hat{\gamma}'(t) = (\gamma'(t))^{\wedge}$,

(3) $\bigvee_{t \in I} \hat{\gamma}(t) = b$.

Proof. We shall reduce the problem to finding an integral curve of a vector field. By horizontal lifting the vectors $d\gamma(t)\left(\frac{\partial}{\partial t}\right) = \gamma'(t)$ to points $b \in \pi^{-1}(\gamma(t))$ we obtain a unique vector field on $\pi^{-1}(\gamma) \subset P$. We shall define an auxiliary manifold:

$$N = \{(t, b) \in I \times P \colon \gamma(t) = \pi(b)\}.$$

The set N acquires a bundle structure under the projection

$$p \colon (t, b) \to t \in I.$$

Denote by θ the mapping $N \ni (t, b) \to b \in \pi^{-1}(\gamma)$ and let $\varphi' := \theta^*\varphi$ be the transport of the connection from P to N. Then there exists precisely one vector field X on N satisfying the conditions:

(1) $\varphi'(X) = 0$,

(2) $dp(X) = \frac{\partial}{\partial t} \in \mathscr{T}(I)$,

which immediately follows from the decomposition $T_n(N) = \ker d\pi_n + \ker \varphi'_n$.

Let γ_0 be the integral curve of the vector field X passing through $n = \theta^{-1}(b) \in N$ (the existence of such a curve follows from Th. 1.2.3). Then the curve $\hat{\gamma}_0 \colon t \to \theta(\gamma_0(t)) \in P$ is a lift of the curve γ and satisfies conditions (1)–(3). The only trouble is that it is defined on a subinterval of the interval I. We have to prove that the lift so defined extends to the whole of the interval I. Let $]t_-, t_+[$ be the maximal interval to which the curve $\hat{\gamma}_0$ can be extended so that conditions (1)–(3) remain valid.

Suppose $t_+ \neq \sup I$. Then by choosing $b_+ \in \pi^{-1}(t_+)$ we can obtain a lift γ_1 of the curve γ passing through b_+ and defined in a neighbourhood of t_+. Thus there exists a $t \in I$ for which both $\hat{\gamma}_0(t)$ and $\hat{\gamma}_1(t)$ are defined. The action of the structural group being transitive on the fibre over $\gamma(t)$, it follows that there exists a $g \in G$ such that $\hat{\gamma}_1(t)g = \hat{\gamma}_0(t)$. Owing to the covariance of an affine connection the curve $\hat{\gamma}_1 g$ is a lift of γ satisfying (1)–(3) and hence an extension of $\hat{\gamma}_0$ beyond t_+. Thus $t_+ = \sup I$ and arguing in a similar way we get $t_- = \inf I$. □

An affine connection on a manifold M determines the way of transporting an arbitrary basis b of the tangent space to M at a point m

along any curve which passes through the point m. The transport of a basis determines the way of transporting an arbitrary vector $X \in T_m(M)$.

DEFINITION. Let $X \in T_{m_1}(M)$ and suppose that γ is a smooth curve such that $\gamma(t_1) = m_1$, $\gamma(t_2) = m_2$. Let $t_1, \dots, t_n$ be the coordinates of the vector X in a basis b_1 and $\hat{\gamma}$ the lift of γ such that $\hat{\gamma}(t_1) = b_1$. Then the vector $Y \in T_p(M)$ with the coordinates $t_1, \dots, t_n$ in the basis $\hat{\gamma}(t_2)$ is called the *parallel translate of the vector X along the curve* γ from m_1 to m_2.

Put

$$\tau_\gamma(X) := Y = i_{b_2} \circ i_{b_1}^{-1}(X), \tag{14.4.1}$$

where $b_2 = \hat{\gamma}(t_2)$. The value $\tau_\gamma(X)$ is independent of the choice of $b_1 \in \pi^{-1}(m_1)$.

For every bundle $E(\mathrm{P}, W)$ associated with the bundle P and every $q \in P$ we have the mapping

$$V \ni v \to l_q(v) := (q, v)G \in p^{-1}(\pi(q)) \in \Gamma.$$

We define the parallel translate of an element $u \in p^{-1}(m_1)$ along γ to $m_2 = \gamma(t_2)$ by setting

$$\tau_\gamma(u) = i_{q_2} \circ i_{q_1}^{-1}(u), \qquad \text{where} \quad q_i = \hat{\gamma}(t_i). \tag{14.4.2}$$

Example 1 (*continued from* § 14.3)

The connection constructed in Example 4 for a parallelizable manifold can be characterized by the following property. Let ϱ be an $\boldsymbol{R}^n$-valued 1-form on M defined by assigning to a vector field its coordinates relative to a given basis field $S_1, \dots, S_n$. Then a vector field is parallel if and only if ϱ is constant on that field. In other words, the translate of a vector X from a point p to a point q does not depend upon the curve joining these points and is given by

$$\sum_{i=1}^{n} x^i S_i(p) \to \sum_{i=1}^{n} x^i S_i(q), \qquad (x^1, \dots, x^n) \in \boldsymbol{R}^n.$$

This is a direct conclusion from the observation that if γ is curve in M then $s \circ \gamma$ is a horizontal lift of γ.

DEFINITION. Let (γ, I) be a smooth curve on M. For $s, t \in I$ we put $\gamma_{s,t} := \gamma|[s, t]$. The curve γ is called a *geodesic relative to the affine connection H on M* provided that for arbitrary $s, t \in I$, we have

$$\gamma'(t) = \tau_{\gamma_{s,t}}(\gamma'(s)).$$

PROPOSITION 14.4.2. *Let $\hat{\gamma}$ be a lift of a curve γ to the frame bundle. Then the following conditions are equivalent:*

(1) *the curve γ is a geodesic,*

(2) *$\hat{\gamma}$ is an integral curve of the basic field $E(x)$ for a certain $x \in \boldsymbol{R}^n$,*

(3) $\omega(\hat{\gamma}') = \text{const}$.

Proof. (1) $\Rightarrow$ (2). We know that $d\pi(\hat{\gamma}') = \gamma'$. Since $\hat{\gamma}'$ is a horizontal vector, we have the relations

$$\hat{\gamma}'(t) = l_{\hat{\gamma}(t)}\gamma'(t) = l_{\hat{\gamma}(t)} \circ i_{\hat{\gamma}(t)} \circ i^{-1}_{\hat{\gamma}(t_1)}(\gamma'(t_1)).$$

Finally, by putting $x = i^{-1}_{\hat{\gamma}(t_1)}(\gamma'(t_1))$ in the above identity, we obtain

$$\hat{\gamma}'(t) = E(x)_{\gamma(t)}.$$

(2) $\Rightarrow$ (1). The proof is obtained by reading the above transformations in the opposite direction.

(2) $\Rightarrow$ (3). This is obtained by applying the identity $\omega(E(x)) = x$.

(3) $\Rightarrow$ (2). It is enough to note that ω_b is an isomorphism onto the subspace $H_b \subset T_b(\boldsymbol{B}(M))$. □

THEOREM 14.4.3. *For every $X \in T_m(M)$ there exists a unique geodesic γ_X in M satisfying the condition*

$$\gamma_X(0) = m, \qquad \gamma_X'(0) = X. \tag{14.4.3}$$

Proof. Let us form the vector field $E(x)$ for $x = i_b^{-1}(X)$ where b is an arbitrary point in $\pi^{-1}(m)$. Let $\hat{\gamma}$ be the integral curve of the vector field $E(x)$, satisfying $\hat{\gamma}(0) = b$. Then Proposition 14.4.2 implies that $\pi \circ \hat{\gamma}$ is the required geodesic. □

If a geodesic γ_X is defined on the interval $[0, \varepsilon]$, then γ_{sX}, $s > 0$ is defined on the interval $\left[0, \frac{\varepsilon}{s}\right]$. This follows from the identity

$$\gamma_{sX}(u) = \gamma_X(su), \tag{14.4.4}$$

which is verified by computing the corresponding tangent vectors and applying the uniqueness of a geodesics with a prescribed tangent vector.

A connection H is said to be complete if the vector field $E(x)$ is complete for every $x \in \boldsymbol{R}^n$. Geodesics of a complete connection are defined on the whole of the axis $\boldsymbol{R}$.

DEFINITION. By the *exponential mapping at a point* $m \in M$ *(relative to the connection* H on M) we mean the mapping

$$T_m(M) \ni X \to \mathrm{Exp}_m X := \gamma_X(1)$$

for those X for which the right-hand side is defined.

The following theorem concerning the exponential mapping plays a fundamental rôle in differential geometry.

THEOREM 14.4.4. *For every point* $m \in M$ *there exists a neighbourhood* O *of zero in* $T_m(M)$ *such that* Exp_m *is a diffeomorphism of* O *onto an open neighbourhood of the point* m.

14.5. COVARIANT DIFFERENTIATION

Let $E(\mathrm{P}, W)$ be a bundle associated with a principal bundle $p\colon \mathrm{P} \to M$ over a manifold M and let (γ, I) be a curve in P. Denote by γ_0 the curve obtained by projecting γ onto M:

$$\gamma_0(t) = p \circ \gamma(t).$$

We define

$$\frac{D}{\mathrm{d}t}\gamma(t_0) = \lim_{t \to t_0} \frac{\tau_{t_0,t}^{-1}(\gamma(t)) - \gamma(t_0)}{t - t_0}, \tag{14.5.1}$$

where $\tau_{t_0,t}$ is the parallel translation from $\gamma_0(t_0)$ to $\gamma_0(t)$ along γ_0. The curve $\dfrac{D}{\mathrm{d}t}\gamma$ in $E(\mathrm{P}, W)$ is called the *covariant derivative* of the curve γ along γ_0.

Example 1 *(continued from* § 14.4)
For a given curve (γ, I) in $T(M)$ and the basis field $S_1, \ldots, S_n$ on M defined before we shall write $\gamma(t) = \sum_{i=1}^{n} x^i(t) S_i \circ \gamma_0(t)$. Then

$$\frac{D}{\mathrm{d}t}\gamma(t) = \sum_{i=1}^{n} (x^i)'(t) S_i \circ \gamma_0(t).$$

An analogous formula is valid for a tensor field of an arbitrary type for a field of differential forms, etc.

For every $X \in \mathscr{T}(M)$ and a section Y of the bundle $E(\mathrm{P}, W)$ we define the covariant derivative of the section Y with respect to the vector field X as a section of the bundle $E(\mathrm{P}, W)$ given by the formula

$$(\nabla_X Y)(m) = \frac{DY \circ \gamma}{\mathrm{d}t}(0) \in p^{-1}(m), \tag{14.5.2}$$

where γ is an integral curve of the vector field X such that $\gamma(0) = m$. For a fixed $X \in \mathscr{T}(M)$ the operator ∇_X maps the space $C^\infty(E(\mathrm{P}, W))$ of smooth sections of the bundle linearly into itself in such a way that the following conditions hold:

$$\nabla_{fX+Y} = f\nabla_X + \nabla_Y \quad \text{for } Y \in \mathscr{T}(M),\ f \in \mathscr{E}(M) \tag{14.5.3}$$

if $E(\mathrm{P}, W) = E(\mathrm{P}, W_1) \otimes E(\mathrm{P}, W_2)$ then

$$\nabla_X(Y_1 \otimes Y_2) = (\nabla_X Y_1) \otimes Y_2 + Y_1 \otimes (\nabla_X Y_2). \tag{14.5.4}$$

In particular, for every $f \in \mathscr{E}(M)$:

$$\nabla_X fY = XfY + f\nabla_X Y \qquad (\text{obviously } \nabla_X f = Xf). \tag{14.5.5}$$

The above identities are verified by a simple computation.

It is more convenient to deal with ∇ than with the distribution of the connection H or with the connection form φ, since ∇ is defined on the manifold M itself and not on the frame bundle. It can be proved that every R-bilinear operator

$$\nabla\colon \mathscr{T}(M) \times \mathscr{T}(M) \to \mathscr{T}(M)$$

satisfying axioms (14.5.5) and (14.5.3) gives rise to a connection on M.

We shall present here an outline of the proof of this statement. In the proof we use the coordinates on the frame bundle introduced in § 1.2 to represent the horizontal lift of a tangent vector field. Given a chart $(U, \varkappa)$ on M we defined there a section D of the bundle $\boldsymbol{B}(U)$ by putting $D_m := (D_1(m), \ldots, D_n(m))$ where

$$D_i(m) := (\mathrm{d}\varkappa^{-1})_m \left(\frac{\partial}{\partial x^i}\right). \tag{14.5.6}$$

An arbitrary basis $b = (m, X_1, \ldots, X_n)$ can be represented in the form

$$X_i = \sum_{j=1}^{n} g_i^j D_j.$$

Finally, to an element $b \in \boldsymbol{B}(M)$ we assign the coordinates

$$(\varkappa^1(m), \ldots, \varkappa^n(m), g_i^j(b)).$$

These coordinates determine, according to a general procedure, a system of vector fields on $\pi^{-1}(U) \subset \boldsymbol{B}(M)$.

Denote by Y_i the differentation with respect to the coordinate $\varkappa^i$, and by Y_j^i the differentiation with respect to the variable g_i^j on $\boldsymbol{B}(M)$. We write

$$\nabla_{D_i} D_j = \sum_{k=1}^{n} \Gamma_{ij}^k D_k. \tag{14.5.7}$$

The functions Γ_{ij}^k are called the *Christoffel symbols of the connection.*

Now we can define the horizontal lift of a vector $D_j \in T_m(M)$ by the formula

$$\hat{D}_j = Y_j - \sum_{i,k,l} \Gamma_{ji}^k g_l^i Y_k^l. \tag{14.5.8}$$

Obviously, the knowledge of a horizontal lift allows us to determine the distribution of the connection and the connection form.

Example 2

Let $M \subset \boldsymbol{R}^n$ be an $(n-1)$-dimensional manifold. Denote by N a normal vector field on M (i.e. N_p is normal to the subspace $T_p(M) \subset T_p(\boldsymbol{R}^n)$). Let D denote the natural covariant differentiation on $\boldsymbol{R}^n$. We shall define a covariant differentiation ∇ on M by

$$\nabla_X Y = D_X Y + (D_X N | Y) N,$$

where $(\cdot|\cdot)$ is the standard inner product on $\boldsymbol{R}^n$. The reader will easily note that the left-hand side of this identity is the orthogonal projection of $D_X Y$ onto the tangent subspace to M, and that properties (14.5.3) and (14.5.5) are satisfied. Moreover, for arbitrary vector fields X, Y, Z on M we have

$$X(Y|Z) = (\nabla_X Y | Z) + (Y | \nabla_X Z).$$

Simple examples referring to the above situation (e.g. the sphere S^2) show that it may well be the case that a vector field X on M is parallel on M and is not parallel on $\boldsymbol{R}^n$. Consequently geodesics in M need not be geodesics in $\boldsymbol{R}^n$. Concerning the relation between the geodesics on M and on $\boldsymbol{R}^n$ (see Problem 14).

Note that if M does not admit a global normal vector field then the above construction can always be carried out locally and the resulting "local" connexions can be pieced together by applying a partition of unity.

It is also possible to assign to the torsion form and the curvature form certain objects on M, namely the torsion tensor and the curvature tensor. To this end let X, Y, $Z \in T_m(M)$ and suppose that b is a basis for the space $T_m(M)$. We define the curvature tensor

$$R(X, Y)Z := i_b\big(\Phi(\hat{X}, \hat{Y})i_b^{-1}Z\big) \tag{14.5.9}$$

($\mathfrak{gl}(n)$ is assumed to act in a natural way on $\boldsymbol{R}^n$).

The curvature tensor R is 3-covariant and 1-contravariant. Next we define the torsion tensor

$$T(X, Y) = i_b\big(\Phi(\hat{X}, \hat{Y})\big). \tag{14.5.10}$$

The torsion tensor T is of type (2,1).

First of all one should verify the correctness of the above definition, i.e. the independence of R and T from the choice of the basis $b \in \pi^{-1}(m)$. By way of example we shall prove this for the tensor T.

Let $b' = b \cdot g$. Then $\hat{X}_i' = \mathrm{d}r_g \hat{X}_i$. The right-hand side of identity (14.5.7) assumes the form

$$\begin{aligned} i_{b\cdot g}\Omega(\hat{X}_1', \hat{X}_2') &= i_b\big(g\Omega(\mathrm{d}r_g\hat{X}_1, \mathrm{d}r_g\hat{X}_2)\big) = i_b(g \circ g^{-1}\Omega(\hat{X}_1, \hat{X}_2)) \\ &= i_b\Omega(\hat{X}_1, \hat{X}_2) \end{aligned}$$

owing to property (14.3.16) of the form Ω.

The structural equations supply the relations between the tensors R, T and the operation ∇. We shall present them in the final form

$$R(X, Y) = [\nabla_X, \nabla_Y] - \nabla_{[X,Y]}, \tag{14.5.11}$$

$$T(X, Y) = \nabla_X Y - \nabla_Y X - [X, Y]. \tag{14.5.12}$$

We shall also give a description of geodesics in terms of the covariant differentiation.

Proposition 14.5.1. *A curve* (γ, I) *is a geodesic if and only if*

$$\frac{D\gamma'}{dt} = 0. \tag{14.5.13}$$

The proof follows directly from the definition of a geodesic and that of the covariant differentiation.

14.6. AFFINE MAPPINGS

Let (M, ∇), (M', ∇') be manifolds with connnections. A smooth mapping $\theta\colon M \to M'$ is said to be an affine mapping if it preserves the covariant differentiation, i.e. if

$$d\theta \circ \frac{D}{dt}\gamma = \frac{D' d\theta \circ \gamma}{dt} \tag{14.6.1}$$

for every curve γ in $\mathcal{T}(M)$. The above property can be expressed by means of the operator ∇ in the form

$$\nabla'_{X'} Y' = d\theta \circ \nabla_X Y \tag{14.6.2}$$

if X, X' and Y, Y' are pairs of θ-compatible vector fields on M and M', respectively.

On the basis of formula (14.5.10) we infer that affine mappings send geodesics into geodesics. Hence we have the identity

$$\theta \circ \mathrm{Exp}_m = \mathrm{Exp}_{\theta(m)} \circ d\theta_m \tag{14.6.3}$$

for every affine mapping $\theta\colon M \to M'$.

It turns out that an affine mapping on a connected manifold is completely determined by the differential at an arbitrary point.

Lemma 14.6.1. *Let* θ, ψ *be affine mappings from a connected manifold* (M, ∇) *into* (M', ∇') *and let* $d\theta_m = d\psi_m$ *for a certain* $m \in M$. *Then* $\theta = \psi$.

Proof. The set $M_1 = \{m \in M\colon d\theta_m = d\psi_m\}$ is non-empty and closed in M. On the other hand,

$$\begin{aligned}\theta(\mathrm{Exp}_m X) &= \mathrm{Exp}_{\theta(m)} \circ d\theta_m(X) = \mathrm{Exp}_{\psi(m)} \circ d\psi_m(X) \\ &= \psi(\mathrm{Exp}_m X) \qquad \text{for } X \in T_m(M).\end{aligned}$$

Since in view of Theorem 14.4.4 the set $\mathrm{Exp}_m(T_m(M))$ contains a neighbourhood of the point m, it follows that M_1 is open. Now the connectedness of M implies that $\psi = \theta$.

A vector field $X \in \mathscr{T}(M)$ is called an affine vector field if the corresponding one-parameter subgroup (Theorem 1.2.3) consists of affine mappings.

The vector space (over $\boldsymbol{R}$) of all affine vector fields on a manifold (M, ∇) will be denoted by $\mathfrak{A}_\nabla$ or simply by $\mathfrak{A}$.

THEOREM 14.6.2. *Let M be a connected manifold with a connection ∇. The space $\mathfrak{A}_\nabla$ is a finite-dimensional subalgebra of the Lie algebra $\mathscr{T}(M)$.*

Proof. With every vector field $X \in \mathscr{T}(M)$ and a point $m \in M$ we relate the following operator: $A_X^m : T_m(M) \ni Y \to A_X^m(Y) \in \mathscr{E}(M)$ where $A_X^m(Y)f = Y(Xf)(m)$, $f \in \mathscr{E}(M)$.

For a fixed m the totality of functionals $A_X^m(Y)$ span a finite-dimensional subspace, since in a local chart every such functional is a linear combination of $n(n+1)$ functionals of the form

$$\left.\frac{\partial^2}{\partial x_i \partial x_j}\right|_{\varkappa(m)} \quad \text{and} \quad \left.\frac{\partial}{\partial x_i}\right|_{\varkappa(m)}.$$

Moreover, the image of the mapping $\mathscr{T}(M) \ni X \to A_X^m$ is also finite dimensional since it is contained in the space of linear operators between finite-dimensional spaces.

It turns out that the operator

$$\mathfrak{A}_\nabla \ni X \to A_X$$

is an injection. For the proof we put $M' := \{m \in M : A_X^m = 0\}$. Then for $m \in M'$ we have $Y(Xf) = 0$ identically with respect to $Y \in T_m(M)$ and $f \in \mathscr{E}(M)$. Hence $X_m = 0$. The one-parameter group of affine transformations generated by the vector field X has the property $\varphi_t(m) = m$ for $m \in M$ because both sides are integral curves through m of the field X. The mapping $t \to (\mathrm{d}\varphi_t)_m$ is a one-parameter transformation group on $T_m(M)$ and hence a one-parameter subgroup of $\mathrm{GL}(T_m(M))$. Thus there exists a $B \in \mathfrak{gl}((T_m(M))$ such that $(\mathrm{d}\varphi_t)_m = \mathrm{e}^{tB}$. On the other hand,

$$B(Y)f = \frac{\mathrm{d}}{\mathrm{d}t}(\mathrm{d}\varphi_t(Y))|_{t=0} = \frac{\mathrm{d}}{\mathrm{d}t} Yf(\varphi_t)|_{t=0} = Y\frac{\mathrm{d}}{\mathrm{d}t}f(\varphi_t)|_{t=0}$$

$$= Y(Xf) = A_X^m(Y)f = 0, \qquad f \in \mathscr{E}(M),\ Y \in T_m(M).$$

It follows that $(\mathrm{d}\varphi_t)_m \underset{t}{\equiv} \mathrm{id}$ and by Lemma 14.6.1 we infer that $\varphi_t(m) \underset{t}{\equiv} \mathrm{id}$. Consequently $X = 0$. Thus we have established the existence of an injection from $\mathfrak{A}_\nabla$ into a finite-dimensional space. It remains to prove that $\mathfrak{A}_\nabla$ is closed under commutator. To this end we shall apply

LEMMA 14.6.3. *The following conditions are equivalent*: (1) $X \in \mathfrak{A}_\nabla$, *and* (2) $[X, \nabla_Y Z] = \nabla_Y [X, Z] + \nabla_{[X, Z]} Z$ *for any* $Y, Z \in \mathcal{T}(M)$.

Proof. Let φ_t be the one-parameter transformation group generated by X. We define a one-parameter transformation group on $\mathcal{T}(M)$ by setting

$$\varphi_t\colon\ Y \to \mathrm{d}\varphi_{-t}(Y).$$

Direct computation shows that $\dfrac{\mathrm{d}}{\mathrm{d}t}\tilde{\varphi}_t(Y)|_{t=0} = [X, Y]$. Thus, if X is an affine vector field, we must have

$$\begin{aligned}[X, \nabla_Y Z] &= \frac{\mathrm{d}}{\mathrm{d}t}(\tilde{\varphi}_t \nabla_Y Z)|_{t=0} = \frac{\mathrm{d}}{\mathrm{d}t}(\nabla_{\tilde{\varphi}_t Y}\tilde{\varphi}_t Z)|_{t=0}\\ &= \frac{\mathrm{d}}{\mathrm{d}t}(\nabla_{\tilde{\varphi}_t Y} Z)|_{t=0} + \frac{\mathrm{d}}{\mathrm{d}t}(\nabla_Y \tilde{\varphi}_t Z)|_{t=0}\\ &= \nabla_{[X, Y]} Z + \nabla_Y [X, Z],\end{aligned}$$

which is precisely condition (2).

Suppose now that (2) is satisfied. We consider two one-parameter transformation groups on $\mathcal{T}(M)$:

$$W_1(t) = \tilde{\varphi}_t \nabla_Y Z, \qquad W_2(t) = \nabla_{\tilde{\varphi}_t Y}\tilde{\varphi}_t Z.$$

Clearly, $W_1(0) = W_2(0) = \nabla_Y Z$. Next, condition (2) states that

$$\left.\frac{\mathrm{d}_1 W}{\mathrm{d}t}\right|_{t=0} = \left.\frac{\mathrm{d}W_2}{\mathrm{d}t}\right|_{t=0}.$$

Hence $W_1(t) = W_2(t)$ for an arbitrary t, which means that φ_t is affine. This concludes the proof of the lemma. □

It follows that the proof of Theorem 14.6.2 will be complete if we verify that the space of vector fields satisfying (2) is a subalgebra in $\mathcal{T}(M)$. This is proved by direct computation which uses only the Jacobi identity a few times. □

In our further investigations of the space of affine mappings it will be convenient to view a connection again as a distribution on the frame bundle. Every diffeomorphism θ of a manifold M into itself can be lifted to the frame bundle $\boldsymbol{B}(M)$. Namely, for $b = (m; X_1, \ldots, X_n)$ we define

$$\hat{\theta}_b := \big(\theta(m), \mathrm{d}\theta_m X_1, \ldots, \mathrm{d}\theta_m X_n\big).$$

$\hat{\theta}\colon \boldsymbol{B}(M) \to \boldsymbol{B}(M)$ is a smooth fibre preserving mapping. Thus

$$\mathrm{d}\hat{\theta}_b V_b \subset V_{\hat{\theta}_b}. \tag{14.6.5}$$

The condition of affinity of the diffeomorphism θ can now be formulated as

$$\mathrm{d}\hat{\theta}_b H_b \subset H_{\hat{\theta}_b} \tag{14.6.6}$$

or equivalently in the language of the operation of the horizontal lift:

$$(\mathrm{d}\theta(X))^\wedge = \mathrm{d}\hat{\theta}(\hat{X}), \tag{14.6.7}$$

By way of example we shall verify that (14.6.6) implies (14.6.7). Both sides of identity (14.6.7) being horizontal vectors, it is enough to prove that $\mathrm{d}\pi$ projects them onto the same vector:

$$\begin{aligned}\mathrm{d}\pi(\mathrm{d}\hat{\theta}\hat{X}) &= \mathrm{d}(\pi \circ \hat{\theta})\hat{X} = \mathrm{d}(\theta \circ \pi)\hat{X} = \mathrm{d}\theta\big(\mathrm{d}\pi(\hat{X})\big) = \mathrm{d}\theta(X)\\ &= \mathrm{d}\pi\Big(\big(d\theta(X)\big)^\wedge\Big).\end{aligned}$$

The basis fields prove to be invariant under $\hat{\theta}$:

$$\mathrm{d}\hat{\theta}_b(E_b(X)) = \mathrm{d}\hat{\theta}_b(i_b(x))^\wedge = \big(d\theta(i_b(x))\big)^\wedge = (i_{\hat{\theta}_b}(x))^\wedge = E(x)_{\hat{\theta}_b}.$$

Thus

$$\mathrm{d}\hat{\theta}\big(E(x)\big) = E(x). \tag{14.6.8}$$

Let $\Phi_t(x)$ be the local one-parameter transformation group generated by $E(x)$. We have

$$\hat{\theta}\Phi_t(x) = \Phi_t(x)\hat{\theta}, \tag{14.6.9}$$

because, by (14.6.8), $E(x)$ is also the infinitesimal generator of the one-parameter subgroup $t \to \hat{\theta}\Phi_t\hat{\theta}^{-1}$.

It follows directly from the definition of $\Phi_t x$ that the curve $\gamma(t) := \pi \circ \Phi_t(x)(b)$, for every $b \in \boldsymbol{B}(M)$, is a geodesic in M such that

$$\gamma(0 = \pi(b) \qquad \text{and} \qquad \gamma'(0) = i_b(x).$$

Putting $i_b(x) = X$ we finally obtain

$$\pi(\Phi_t(x)b) = \operatorname{Exp}_{\pi(b)} tX. \tag{14.6.10}$$

The following theorem play a fundamental rôle in our further considerations.

THEOREM 14.6.4 (Kobayashi, 1972). *If a connection H on M is complete, then every affine vector field on M is complete.*

Proof. We may assume that the manifold is connected since every continuous curve in M lies wholly in one of the connected components of the manifold.

Let X be an affine vector field and θ_t the corresponding local one-parameter transformation group with domain $D \subset \boldsymbol{R} \times M$. By $\hat{\theta}_t$ we shall denote the local one-parameter transformation group of the frame bundle $\boldsymbol{B}(M)$ determined by θ_t. For a fixed $m_0 \in M$ there exist an $\varepsilon > 0$ and a neighbourhood O of the point m such that $]-\varepsilon, \varepsilon[\times O \subset D$. Let γ be a geodesic emanating from m_0. The completeness of the connection means that γ is defined on $\boldsymbol{R}$ and that for every $p \subset \gamma$ there exists a $Y \in T_{m_0}(M)$ such that $p = \operatorname{Exp} Y$. Next, it follows from formula (14.6.10) that one can select $b \in \pi^{-1}(m_0)$ and $x \in \boldsymbol{R}^n$ so that

$$p = \pi(\Phi_1(x)b).$$

We shall prove that the integral curve of the field X issuing from the point p is also defined on $]-\varepsilon, \varepsilon[$, i.e. that

$$]-\varepsilon, \varepsilon[\times \pi(\Phi_1(x)\pi^{-1}(O)) \subset D.$$

For $b' \in \Phi_1(x)\pi^{-1}(O)$ we define the mappings

$$\psi(t)b' = \Phi_1(x)\hat{\theta}_t\Phi_{-1}(x)b',$$

which are diffeomorphisms defined on $]-\varepsilon, \varepsilon[\times\Phi_1\pi^{-1}(O)$.

Now we shall apply formula (14.6.9). For those t for which $\hat{\theta}_t$ is defined on $\Phi_1(x)\pi^{-1}(O)$ we have

$$\psi_t = \hat{\theta}_t.$$

Therefore, for $t \in]-\varepsilon, \varepsilon[$ the family of transformations

$$t \to \pi(\psi_t)$$

is an extension of the transformations θ_t which just means that

$$]-\varepsilon, \varepsilon[\times \pi(\Phi_1(x)\pi^{-1}(O)) \subset D.$$

Thus we proved that the integral curves of the field X passing through points which can be joined by a geodesic are all defined for $t \in]-\varepsilon, \varepsilon[$. In virture of Theorem 14.4.4 the set of points having this property is both closed and open and therefore coincides with M owing to the connectedness of M.

Now we can easily define θ_t for arbitrary $t \in \boldsymbol{R}^1$. Namely, let $0 < t = \frac{1}{2}m\varepsilon + s$, $0 \leqslant s \leqslant \frac{1}{2}\varepsilon$; we put

$$\theta_t := \theta^m_{\frac{1}{2}\varepsilon} \circ \theta_s .$$

In a similar way we define θ_t for $t < 0$. This completes the proof.

From the above theorem we derive

COROLLARY 14.6.5. *Let M, H be a connected manifold with a complete connexion. The group* Aff M *of all affine diffeomorphisms of M has a Lie group structure such that* (Aff M, M) *is a Lie transformation group with respect to a natural action of diffeomorphisms on M.*

The corollary follows directly from Theorem 1.4.4 and Theorems 14.6.2, 14.6.4.

Corollary 14.6.5 is the main tool in the study of different types of transformation groups of differentiable manifolds. Such structures in a manifold as the Riemannian symplectic or symmetric (this last type will be defined in the next chapter) determine a connection which is invariant under the group of transformations preserving that structure. The group imbeds in the Lie group Aff M and acquires a Lie group structure induced by that imbedding.

14.7. THE RIEMANNIAN CONNEXION. SECTIONAL CURVATURE

Suppose now that a manifold M possesses an $O(n)$ structure, i.e. a Riemannian structure. Denote by g the metric tensor and by $d(\cdot, \cdot)$ the Riemannian metric. The pair (M, g) is called a Riemannian manifold.

A diffeomorphism $S: M \to M$ which preserves the tensor g, i.e., for an arbitrary $m \in M$ satisfies the condition

$$g_{Sm}(\mathrm{d}S_m(X), \mathrm{d}S_m(Y)) = g_m(X, Y) \quad \text{for } X, Y \in T_m(M), \tag{14.7.1}$$

is called an *isometry*.

Obviously, an isometry preserves the distance on M and the length of curves. The converse statement is also true: a diffeomorphism which preserves distances is an isometry.

A connection ∇ on the manifold (M, g) is said to be a metric connection provided the parallel translations relative to ∇ are isometries. There exist many metric connections on a Riemannian manifold. However, there is a significant one among them. Namely, we have

THEOREM 14.7.1. *Every Riemannian manifold admits precisely one metric connection with zero torsion tensor. Isometries of the manifold are affine mappings.*

The connection distinguished by Theorem 14.7.1 is called the Riemannian connection on the Riemannian manifold (M, g).

An outline of the proof of the above theorem is contained in the problems. At present we shall describe the Riemannian connection by noting that its Christoffel symbols satisfy, in a local chart, the system of equations

$$\sum_l g_{lk}\Gamma^l_{ij} = \frac{1}{2}\left[\frac{\partial g_{ki}}{\partial x^j} + \frac{\partial g_{jk}}{\partial x^i} - \frac{\partial g_{ji}}{\partial x^k}\right], \tag{14.7.2}$$

where g_{ij} denotes the coefficients of the tensor g in the chart $(x^1, \ldots, x^n)$.

By a complete Riemannian manifold we shall understand a Riemannian manifold which is complete with respect to the natural metric.

THEOREM 14.7.2. *A Riemannian manifold is complete if and only if its Riemannian connection is such.*

Thus, the completness of a Riemannian manifold means precisely that all maximal geodesics are defined on the whole of the real line.

THEOREM 14.7.3. *Any pair of points in a complete Riemannian manifold can be joined by a geodesic of length equal to the distance between those points.*

The proofs of the theorems quoted above can be found in Helgason, 1968.

Example 1

Let us consider a model of a non-Euclidean space in the form of the upper half-plane $H_+ = \{(x, y): y > 0\}$ with the Poincaré–Klein metric tensor $g = y^{-2}g_0$, where g_0 is the canonical metric tensor induced from $\boldsymbol{R}^2$ ($g_0 = \mathrm{d}x\otimes\mathrm{d}x+\mathrm{d}y\otimes\mathrm{d}y$). Using (14.7.2), we find by direct computation the Christoffel symbols Γ^i_{jk} of the Riemannian connection determined by g (with respect to the identity chart on H_+). We have

$$\Gamma^1_{11} = \Gamma^1_{22} = \Gamma^2_{12} = \Gamma^2_{21} = 0, \qquad \Gamma^2_{11} = y^{-1},$$
$$\Gamma^1_{12} = \Gamma^1_{21} = \Gamma^2_{22} = -y^{-1}.$$

The equations of geodesics (see Problem 11) assume the form

$$\text{(a) } x''-2y^{-1}x'y' - 0, \qquad \text{(b) } y''+y^{-1}[(x')^2-(y')^2]=0.$$

Owing to the condition $y > 0$, equation (a) is equivalent to

$$\text{(a}'\text{) } y^2(x''-2y^{-1}x'y') = (y^{-2}x')' = 0,$$

which shows that either x' is identically zero or $y^{-2}x' = \text{const} \neq 0$. In the second case we put $y^{-2}x' = 1/\alpha$ and substitute this into (b). It follows after simple computations that

$$y^{-1}y'+\frac{1}{\alpha}x = \text{const.} = \frac{\beta}{\alpha}.$$

Thus system (a), (b) splits into two systems,

$$\text{(I)} \begin{cases} x' = \dfrac{1}{\alpha}y_2, \\ y' = -\dfrac{1}{\alpha}xy+\dfrac{\beta}{\alpha}y, \end{cases} \qquad \text{or} \qquad \text{(II)} \begin{cases} x' = 0, \\ y' = \delta y. \end{cases}$$

System II has solutions $x = \gamma$, $y = \mathrm{e}^t$ (in the parametrization given by the arc length along a geodesic). As regards system I, on applying the normalization condition for a tangent vector $y^{-2}[(x')^2+(y')^2] = 1$ we obtain

$$x' = \alpha\left[1-\frac{1}{\alpha^2}(x-\beta)^2\right].$$

Hence

$$x = \alpha\tanh t+\beta, \qquad y = \alpha(\coth t)^{-1}.$$

Thus we obtain a description already given in Chapter 10 of geodesics as semi-circles (including the semi-circles passing through the point

∞, i.e. the half-lines) perpendicular to the line $y = 0$. Moreover, it follows from the above parametrization of geodesics that the non-Euclidean plane is a complete Riemannian space (with respect to the Poincaré–Klein metric).

Sectional Curvature

Let M be a Riemannian manifold with the Riemannian form g and the curvature tensor R.

Let $X, Y \in T_p(M)$, $p \in M$ be linearly independent vectors. Denote by S the two-dimensional subspace spanned by the vectors X, Y. By the sectional curvature at p along the plane S we understand the number

$$K(S) = -\frac{g(R_p(X, Y)X, Y)}{|X \vee Y|^2}, \tag{14.7.3}$$

where $|X \vee Y|^2 = \|X\|^2\|Y\|^2 - g(X, Y)^2$ is the area of the parallelogram spanned by the vectors X, Y.

If the manifold M is two-dimensional, then the linearly independent vectors X, Y span the tangent space $T_p(M)$ and $K(S)$ coincides in this case with the curvature considered by Gauss in the study of surfaces in the 3-dimensional space.

The definition of the sectional curvature, as a function of a point and of the subspace S, is correct since the value of $K(S)$ is independent of the choice of vectors spanning the subspace S. Thus the sectional curvature is a mapping from the space of 2-dimensional dustributions on M into $\mathscr{E}(M)$.

The values of the curvature $K(S)$ for every point p and for every subspace S uniquely determine the curvature tensor for the manifold. The proof of these fundamental properties of the sectional curvature can be found in every book on differential geometry.

We say that a manifold M is of positive (negative) curvature if $K(S) \geqslant 0$ ($\leqslant 0$) for an arbitrary 2-dimensional distribution on M, and is not identically zero. If $K = 0$, then the manifold is said to be flat.

A Riemannian manifold whose sectional curvature depends exclusively upon the point m and is independent of the plane S, is called a space of constant curvature. This terminology is justified by Shur's theorem, which states that the sectional curvature of a space of constant curvature, is a scalar which is also independent of the point in the manifold.

The sign of the sectional curvature supplies important information concerning the topology of the manifold. For instance the curvature of a simply connected compact manifold cannot be non-positive. Next a non-compact complete manifold cannot have positive curvature separated from zero (i.e. greater than a fixed positive constant). Below we quote a more detailed result which will be useful in the study of symmetric manifolds of the non-compact type.

THEOREM 14.7.4. *The exponential mapping on a Riemannian manifold of negative curvature is, at any point, a local diffeomorphism. If M is simply connected, then for every $p \in M$ the mapping* Exp_p *is a diffeomorphism of $T_p(M)$ onto M.*

As a complementary literature to this section we recommend Helgason, 1968, Kobayashi, Nomizu, 1963, 1969 and Wolf, 1967.

PROBLEMS

1. Show that defining a k-dimensional distribution on a manifold M ($\dim M > k$) is equivalent to defining for each $p \in M$ a k-dimensional subspace $V_p \subset T_p(M)$ with the following smoothness condition: every point in M possesses a neighbourhood U such that there exist k smooth vector fields in U which constitute a basis for the space V_q for $q \in U$.

2. Let φ be an $L(V)$-valued 1-form on M and ω a V-valued 1-form on M. Prove that

$$\varphi \wedge \omega(X, Y) = \frac{1}{2}\big(\varphi(X)\omega(Y) - \varphi(Y)\omega(X)\big).$$

Assuming $V = \boldsymbol{R}^n$, $L(V) = \mathfrak{gl}(n, \boldsymbol{R})$ write $\varphi = [\varphi_{ij}]_{i,j=1}^n$, $\omega = (\omega)_{i=1}^n$ and prove now that

$$\varphi \wedge \omega = \Big(\sum_{j=1}^{n} \varphi_{ij} \wedge \omega_j\Big)_{i=1}^{n}.$$

3. Let $G = \mathrm{GL}(n, \boldsymbol{R})$ and let $\varkappa = [\varkappa_i^j]$ be a global chart for G provided by a natural inclusion mapping i: $\mathrm{GL}(n, \boldsymbol{R}) \to \boldsymbol{R}^{n^2}$. Denote by

X_j^i the basis field determined by this chart (then $X_j^i \varkappa_l^k = \delta_l^i \delta_j^k$) and let $\{d\varkappa_l^j\}$ denote the basis field conjugate to the basis field $\{X_j^i\}$.

(a) The form $\omega = \sum_{i,j=1}^{n} a_j^i d\varkappa_l^j$ is left invariant if and only if $a_j^i = \sum_k b_j^k (g^{-1})_k^i$, where $[(g^{-1})_j^i]$ is the matrix inverse to $[g_j^i]$.

(b) The canonical form θ defined in Example 1 is given by the formula $\theta = g^{-1}dg$, where $dg = [d\varkappa_j^i]$ is a $\mathfrak{gl}(n, \boldsymbol{R})$-valued 1-form.

(c) The forms $\theta_j^i = \sum_k (g^{-1})_j^k d\varkappa_k^i$ form a basis for the space of all left invariant 1-forms on G, which is conjugate to the basis $\{\tilde{X}_j^i\}$ for the space of all left invariant vector fields on G, which consists of vector fields such that $\tilde{X}_j^i(I) = X_j^i(I)$.

(d) Check that the Maurer–Cartan equations assume in this case the form

$$d\theta_j^i = -\sum_{k=1}^{n} \theta_j^i \wedge \theta_j^k .$$

4. Let $\mathrm{P} = (P, M, \pi, G)$ be a trivializable bundle and suppose that s_1 and s_2 are smooth sections of this bundle, defined on M. Prove that there exists a smooth mapping $\psi: M \to G$ such that $s_1(m) = s_2(m)\psi(m)$ and that the connections determined by s_1 and s_2 as in Example 4 in § 14.3 are identical if and only if ψ is a constant mapping.

5. Suppose that G and K are given the same meaning as in Example 4 in § 14.3 and let H be the connection on G/K defined in it. Put $o = eK$. Prove that for an arbitrary $X \in \mathfrak{m}$ the parallel translation of tangent vectors at 0 along the curve $\gamma(t) = \exp tX \cdot o$ is the differential of the mapping $X \to \exp tX \cdot x$. Deduce hence that the curves $\gamma(t) = \exp tX \cdot o$ are geodesics and H is a complete connection.

6. Let G be a connected Lie group. Put $K = G \times G$ and let K act on G by $(x, y) \cdot g := xgy^{-1}$.

(a) Prove that the isotropy subgroup at the identity $e \in G$ is $L := \Delta(G \times G) = \{(x, x): x \in G\}$ and that G is isomorphic (as a homogeneous space of the group K) to the reductive space K/L (cf. Example 3).

(b) Denote by $\mathfrak{g}$, $\mathfrak{k}$, $\mathfrak{l}$ the Lie algebras of the group G, K, L, respectively, and write $\mathfrak{m}_+ := \{(0, X): X \in \mathfrak{g}\}$, $\mathfrak{m}_- := \{(X, 0): X \in \mathfrak{g}\}$, $\mathfrak{m}_0 := \{(X, -X): X \in \mathfrak{g}\}$. Prove that each of these spaces is $\mathrm{ad}(L)$-invariant and is a complementary space to $\mathfrak{l}$ in the Lie algebra $\mathfrak{k}$.

(c) Denote by H_+, H_-, H_0 the K-invariant connections on the bundle $\mathrm{K} = (K, G, \pi, L)$ constructed according to Example 3 for the decompositions $\mathfrak{k} = \mathfrak{l}+\mathfrak{m}_+$, $\mathfrak{k} = \mathfrak{l}+\mathfrak{m}_-$, $\mathfrak{k} = \mathfrak{l}+\mathfrak{m}_0$, respectively. Show that the torsion tensor T and the curvature tensor R are:

(α) for H_+: $T(X, Y) = [X, Y]$, $R = 0$,

(β) for H_-: $T(X, Y) = -[X, Y]$, $R = 0$,

(γ) for H_0: $T = 0$, $R(X, Y)Z = -\frac{1}{4}[[X, Y], Z]$.

(d) Prove that the connection H_- is precisely the connection considered in Problem 7(c).

(e) Prove that the geodesics of the connections H_+, H_-, H_0 are identical.

7. (a) Let M be a manifold with an affine connection H such that the parallel translation is independent of the curve (such a connection is said to be flat). Prove that then there exists a global smooth basis field on M, i.e. that M is a parallelizable manifold. Conversely, if M is a parallelizable manifold and $S_1, \ldots, S_n$ is a global basis field on M, then the connection determined by this basis field (see Example 4) is flat. Prove also that the curvature tensor of a flat connection is zero.

(b) Let $S_1, \ldots, S_n$ be a smooth basis field on M and ϱ an $\boldsymbol{R}^n$-valued 1-form defined by $\varrho\left(\sum_{i=1}^{n} x^i S_i\right) = (x^1, \ldots, x^n)$. Further, let Ω be the torsion form of the connection H determined by the basis field. Prove that the structural equations for H reduce to the single equation

$$\mathrm{d}\varrho = s^*\Omega,$$

where $s\colon M \to \boldsymbol{B}(M)$ is the section of the frame bundle corresponding to the basis field $S_1, \ldots, S_n$.

(c) Let G be a Lie group and $\{X_i\}_1^n$ a basis for the Lie algebra of left-invariant vector fields on G. Prove that for the affine connection determined by this basis, left invariant vector fields are parallel fields, left translations of one-parameter subgroup coincide with geodesics and left translations are affine mappings. Show also that $T(X, Y) = -[X, Y]$.

(d) Prove that the following manifolds are parallelizable: S^1, open submanifolds of $\boldsymbol{R}^n$, the so-called flat n-dimensional torus $T^n := S^1 \times \ldots \times S^1$, with the manifold structure of a product of n circles S^1.

Note that the flat torus T^2 is not the same manifold as the two-dimensional torus with the manifold structure given by a natural imbedding in $\boldsymbol{R}^3$.

8. (a) Let φ, ψ be two connection forms on a bundle P and f a smooth function on the base M of the bundle. Write $\tilde{f} = f \circ \pi$. Show that $\tilde{f}\varphi + (1-\tilde{f})\psi$ is a connection form on P. (In particular, connection forms constitute a convex set in $\vartheta^1_{\mathfrak{g}}(\mathrm{P})$.)

(b) Let φ be a connection form on $\boldsymbol{B}(M)$ and τ a $\mathfrak{gl}(n, R)$-valued smooth 1-form on $\boldsymbol{B}(M)$ such that: (i) τ is horizontal, (ii) $r_g^*\tau = \mathrm{Ad}(g^{-1})\circ\tau$. Prove that $\psi = \varphi + \tau$ is a connection form on $\boldsymbol{B}(M)$ and conversely, that for every pair of connection forms φ, ψ on $\boldsymbol{B}(M)$ the form $\tau = \varphi - \psi$ has properties (i) and (ii).

(c) Prove that two affine connections with the connection forms φ, ψ have the same geodesics if and only if the form $\tau = \varphi - \psi$ satisfies the condition $\tau \wedge \omega = 0$.

(d) Let φ be a connection form on $\boldsymbol{B}(M)$. Prove that there exists a form τ satisfying conditions (i), (ii), in (b) and such that $\varphi + \tau$ is a connection form with zero torsion form.

Hint. (d) Define τ by the formula $\tau(X)(v) = \frac{1}{2}\Omega(X, E(v)b)$ where $X \in T_b(\boldsymbol{B}(M))$ and apply the first structural equation.

9. (a) Let M be a manifold with a connection H and γ a curve in M. Suppose that X is a vector field on M tangent to γ at the point $\gamma(0) = m$. Let $E^1(t), \ldots, E^n(t)$ be a basis for $T_{\gamma(t)}(M)$ such that $E^i(t)$ is obtained by parallel translating $E^i(0)$ along γ to the point $\gamma(t)$. For a vector field Y on M define functions y_i by the condition $Y(\gamma(t)) = \sum_{i=1}^{n} y_i(t)E^i(t)$. Show that

$$\nabla_X Y(m) = \sum_{i=1}^{n} y_i'(0)\, E^i(0).$$

(b) Let $M = \boldsymbol{R}^n$ and let H be the natural connection on $\boldsymbol{R}^n$ determined by the basis field $X_1, \ldots, X_n$ where $X_i = \dfrac{\partial}{\partial x^i}$. Show that for an arbitrary vector field $Y = \sum_{i=1}^{n} y^i X_i$,

$$\nabla_X Y = \sum_{i=1}^{n} X y^i X_i.$$

10. Let ∇ be an operator of covariant differentiation on M. Prove that

(a) ∇ has a local character, i.e. whenever either of the fields X and Y is zero on an open submanifold $U \subset M$, then so is $\nabla_X Y$.

(b) If a vector field X is zero at a point $m \in M$, then $\nabla_X Y(m)$ is also zero.

(c) Prove that ∇ defines an operator of covariant differentiation on every open submanifold on M.

Hint. (a) Use property (14.5.5) of the operator ∇ for a suitably chosen function f. (b) Apply (a) and (14.5.3), (14.5.5). (c) Define $\nabla|_U$ as the covariant differentiation of the extensions to M of vector fields defined on U.

11. Let ∇ be an operator of covariant differentiation on a manifold M.

(a) Prove that a vector field X is parallel along a smooth curve γ on M if and only if $\nabla_T X = 0$ identically on γ. (T denotes here the tangent vector field to γ.)

(b) Consider a vector field X defined on the domain of a chart (χ, U). Suppose that γ is a smooth curve in U and let $X_1, \ldots, X_n$ be the basis field on U determined by the chart. Write $X \circ \gamma(t) = \sum_{i=1}^{n} x^i(t) \times X_i \circ \gamma(t)$. Prove that the field X is parallel along γ if and only if $\frac{\mathrm{d}x^k}{\mathrm{d}t} + \sum_{i,j=1}^{n} x^i \frac{\mathrm{d}\gamma^j}{\mathrm{d}t} \Gamma^k_{ij} = 0$, where γ^j denotes the j-th coordinate of γ in the chart χ.

(c) Retaining the notation of (b), prove that γ is a geodesic if and only if

$$\frac{\mathrm{d}^2\gamma^k}{\mathrm{d}t^2} + \sum_{i,j=1}^{n} \Gamma^k_{ij} \frac{\mathrm{d}\gamma^i}{\mathrm{d}t} \frac{\mathrm{d}\gamma^j}{\mathrm{d}t} = 0.$$

(d) Let ∇ denote natural covariant differentiation on $\boldsymbol{R}^n$ (or on an open subset $O \subset \boldsymbol{R}^n$ (cf. Problem 9b)). Prove that a curve γ is a geodesic if and only if γ is a linear parametrization of an interval on a straight line.

12. Let (M, g) be a Riemannian manifold.

(a) Show that for every smooth 1-form ω on M there exists a unique vector field X on M such that

$$\omega(Y) = g(X, Y)$$

for every vector field Y on M.

(b) Prove that there exists a unique affine connection on M such that for arbitrary vector fields X, Y, Z on M

(1) $Zg(X, Y) = g(\nabla_Z X, Y) + g(X, \nabla_Z Y)$,

(2) $\nabla_X Y - \nabla_Y X = [X, Y]$, i.e. $T(X, Y) = 0$.

(c) Show that a connection ∇ on M satisfies equation (1) in (b) if and only if parallel translations along an arbitrary curve are isometries of tangent spaces equipped with the inner product given by the Riemannian structure g.

Hint. (a) Prove the existence of such a field on the domain of an arbitrary chart. (b) By permuting cyclically the fields X, Y, Z in equation (1), prove that a connection ∇ satisfying (1) and (2) is uniquely determined by the formula

$$g(\nabla_X Y, Z) = \frac{1}{2}\{Xg(X, Y) + Yg(Z, X) - Zg(X, Y) + \\ + g(Z, [X, Y]) + g(Y, [Z, X]) - g(X, [Y, Z]\}.$$

(d) Show that condition (1) implies the identity

$$\frac{\mathrm{d}}{\mathrm{d}t} g(X, Y) = g\left(\frac{D}{\mathrm{d}t} X, Y\right) + g\left(X, \frac{D}{\mathrm{d}t} Y\right),$$

which is valid for arbitrary vector fields along an (arbitrary) curve.

13. Let (M, g) be a Riemannian manifold and denote by R the curvature tensor of the Riemannian connection on M.

(a) Prove that arbitrary vector fields X, Y, Z, W on M satisfy the following identities

(α) $R(X, Y)Z = -R(Y, X)Z$,

(β) $g(R(X, Y)Z, W) = -g(R(X, Y)W, Z)$,

(γ) $g(R(X, Y)Z, W) = g(R(Z, W)X, Y)$,

(δ) $R(X, Y)Z + R(Z, X)Y + R(Y, Z)X = 0$ (Bianchi's identity).

(b) Prove that a tensor field R_1 of type (3,1) defined by the formula

$$R_1(X, Y)Z = g(Y, Z)X - g(X, Z)Y$$

fulfils identities (α)–(γ). Prove also that if $X(p)$ and $Y(p)$ are linearly independent then $r_1(X(p), Y(p)) := g(R_1(X, Y)Y, X)(p) > 0$.

(c) Show that if $\{X, Y\}$ and $\{X', Y'\}$ are two bases of a two-dimensional space $S \subset T_p(M)$ and

$$X' = \alpha_{11}X + \alpha_{21}Y, \qquad Y' = \alpha_{12}X + \alpha_{22}Y,$$

then

$$r_1(X', Y') = [\det(\alpha)]^2 r_1(X, Y)$$

and for $r(X, Y) := g(R(X, Y)Y, X)$ we also have

$$r(X', Y') = [\det(\alpha)]^2 r(X, Y).$$

(d) Prove that a mapping $k\colon \mathscr{T}(M) \times \mathscr{T}(M) \to \mathscr{E}(M)$ defined as $k(X, Y) = g(R(X, Y)Y, X)$ determines uniquely the curvature tensor R.

Hint. For the proof of (δ) perform a direct calculation using the identity $O = T(X, Y) = \nabla_X Y - \nabla_Y X - [X, Y]$ and the Jacobi identity for vector fields, (α) follows directly from definition and is true of an arbitrary, not necessarily Riemannian connection; (β) is obtained by applying the identity $Xg(Y, Z) = g(\nabla_X Y, Z) + g(Y, \nabla_X Z)$ (cf. Problem 10); we infer (γ) from (α), (β) and (δ) on remarking that the identity

$$g(R(Y, X)Z, W) + g(R(X, Z)Y, W) + g(R(Z, Y)X, W) = 0$$

is a consequence of (δ) and (α). Permuting cyclically the arguments in the first summand and adding the formulas thus obtained, we arrive at (γ). (d) Note that $k(X + Y, Y) = k(X, Y) = k(Y, X)$. Next, by applying ($\alpha$)–($\delta$), show that

$$3R(Y, X)Z = R(X, Y+Z)(Y+Z) - R(X, Y)Y - R(X, Z)Z$$
$$- R(Y, X+Z)(X+Z) + R(Y, X)X + R(Y, Z)Z,$$

and make use of the polarization formula for the symmetric bilinear form $(X, Y) \to g(R(X, Z)Z, Y)$.

14. Denote by D the natural covariant differentiation on $\boldsymbol{R}^n$. Let $M \subset \boldsymbol{R}^n$ be an $(n-1)$-dimensional submanifold. Define covariant differentiation ∇ on M as in Example 2, § 14.5 $\big(T_p(M)$ is identified in a natural way with a subspace of $T_p(\boldsymbol{R}^n)\big)$ by setting

$$\nabla_X Y := D_X Y + (D_X N | Y) N,$$

where N is a unit normal vector field on M (it suffices that the vector field N be defined locally on M).

(a) Prove that ∇ is a Riemannian connection on M if M is equipped with the Riemannian metric induced by the inclusion mapping $i: M \to \boldsymbol{R}^n$.

(b) Prove that the formula $L(X) := D_X N$ defines a linear mapping

$$L\colon\ T_p(M) \to T_p(M) \qquad \text{for an arbitrary } p \in M.$$

(c) Prove that $L(X) = \mathrm{d}N_p(X)$ where the unit normal vector field N is regarded as a local mapping of M into S^{n-1} the $(n-1)$-dimensional unit sphere in $\boldsymbol{R}^n$.

(d) Prove that the operator L is symmetric with respect to the inner product on $T_p(M)$, i.e. that

$$\big(L(X)|Y\big) = \big(X|L(Y)\big) \qquad \text{for } X, Y \in T_p(M).$$

(e) Show that a curve γ in M is a geodesic in $\boldsymbol{R}^n$ if and only if it is a geodesic in M and

$$\big(L(\gamma')|\gamma'\big) = 0.$$

(f) Show that a curve which is not a geodesic in $\boldsymbol{R}^n$ is a geodesic in M if and only if its tangent vector field T satisfies the condition $(D_T X) = 0$ for every $X \in T_p(M)$.

REMARK. The form $T_p(M) \times T_p(M) \ni (X, Y) \to \big(L(X)|Y\big)$ is called the *second fundamental form* for the manifold M.

Hint to (d). Prove first that every vector field X on M admits an extension $\tilde{X}$ to $\boldsymbol{R}^n$ with the property that $(\tilde{X}f)|_U = X(f|_U)$ and $[\tilde{X}, \tilde{Y}] = [X, Y]^{\sim}$.

15. Let M be a manifold with a connection H and the corresponding operator ∇. Let φ be a diffeomorphism of M into itself. Denote by $\hat{\varphi}$

the mapping of the frame bundle $\boldsymbol{B}(M)$ into itself, induced by φ: $\hat{\varphi}(b) = (\varphi(m), \mathrm{d}\varphi_m(X^1), \ldots, \mathrm{d}\varphi_m(X^n))$, where $b = (m, X^1, \ldots, X^n)$.

(a) Prove that $\hat{\varphi}$ is a diffeomorphism of $\boldsymbol{B}(M)$ into itself.

(b) Show equivalence of the conditions (1) and (2) below:

(1) $\mathrm{d}\hat{\varphi}_b H_b \subset H_{\varphi(b)}$,

(2) $(\mathrm{d}\varphi_m(X_m))^\wedge = \mathrm{d}\hat{\varphi}_b(\hat{X}_b)$.

(c) Prove that φ is affine if and only if it satisfies the conditions in (b).

16. Let M be a manifold with a connection ∇. Assume that the diffeomorphism S: $M \to M$ satisfies $S^2 = \mathrm{id}_M$. Prove that on M there exists a connection invariant with respect to the mapping S.

Hint. Prove first that the mapping

$$(X, Y) \to \nabla^S_X Y := \mathrm{d}S\nabla_{\mathrm{d}S(X)}(\mathrm{d}S(Y))$$

defines a connection on M and consider the connection

$$\tilde{\nabla}_X Y := 1/2\,(\nabla_X Y + \nabla^S_X Y).$$

Chapter 15

Symmetric Spaces

15.1. DEFINITIONS AND EXAMPLES

Let $\mathfrak{M}$ be a differentiable manifold and Diff $\mathfrak{M}$ the group of all diffeomorphisms of $\mathfrak{M}$.

DEFINITION. By a *symmetric space* we shall understand a pair $(\mathfrak{M}, S)$ where S is a mapping

$$\mathfrak{M} \ni m \to S_m \in \operatorname{Diff}\mathfrak{M}$$

satisfying the following axioms:

$$S_m \circ S_m = \mathrm{id}_{\mathfrak{M}} \qquad \text{for every } m \in \mathfrak{M}; \tag{15.1.1}$$

$$S_m(m) = m; \tag{15.1.2}$$

$$S_x \circ S_m \circ S_x = S_{S_x m}, \qquad x, m \in \mathfrak{M}; \tag{15.1.3}$$

for every $m \in \mathfrak{M}$ there exists a neighbourhood $\theta(m)$ such that m is a unique fixed point of S_m in $\theta(m)$. (15.1.4)

The mapping S_m is called *symmetry with respect to m*. Property (15.1.2) means that m is a fixed point of the symmetry S_m and condition (15.1.4) denotes that this point is isolated.

The meaning of the remaining algebraic conditions (15.1.1)–(15.1.3) is explained by the following fundamental examples.

Example 1

Let G be a Lie group. We define the symmetry S_g with respect to the point $g \in G$ by the formula

$$S_g x := g x^{-1} g.$$

The verification of conditions (15.1.1)–(15.1.3) is trivial. Property (15.1.4) follows from the existence of the exponential mapping.

A vector space is a special case of this general example. In that case the symmetry with respect to a point x takes the form $S_x y = 2x - y$.

Example 2

Let S^n be a sphere in $\boldsymbol{R}^{n+1}$ with centre at zero. We define a symmetry by the formula

$$S_x y := 2\frac{(x|y)}{(x|x)}x - y.$$

The verification of conditions (15.1.1)–(15.1.3) presents no difficulty. Since $y = \pm x$ are the only solutions of the equation $S_x y = y$, condition (15.1.4) is also valid.

By a homomorphism of symmetric spaces $(\mathfrak{M}, S)$ and $(\mathfrak{M}', S')$ we shall mean a smooth mapping $\varphi\colon \mathfrak{M} \to \mathfrak{M}'$ such that

$$\varphi(S_x y) = S_{\varphi(x)}\varphi(y), \qquad x, y \in \mathfrak{M}. \tag{15.1.5}$$

Condition (15.1.3) denotes that every symmetry S_m is a homomorphism of $\mathfrak{M}$ into itself, and property (15.1.1) implies even that it is an automorphism, i.e. an invertible homomorphism.

The group of all automorphisms of a symmetric space will be denoted by Aut $\mathfrak{M}$. We shall now distinguish a subgroup of Aut $\mathfrak{M}$ called the *group of displacements* of the symmetric space $\mathfrak{M}$ and denoted by $G(\mathfrak{M})$. Namely, $G(\mathfrak{M})$ is the group generated by the automorphisms of the form $S_x \circ S_y$ for $x, y \in \mathfrak{M}$. Writing (15.1.5) in the form $S_x = \varphi^{-1}S_{\varphi(x)}\varphi$, we observe that $G(\mathfrak{M})$ is a normal subgroup of Aut $\mathfrak{M}$. The term "group of displacements" is motivated by the case of $\mathfrak{M} = \boldsymbol{R}^n$, where $G(\mathfrak{M})$ consists of displacements of the space $\boldsymbol{R}^n$.

Fix a point o in $\mathfrak{M}$. We define inductively the power of x with respect to o:

$$x^0 := o, \qquad x^1 := x, \qquad x^{n+2} = S_x S_o x^n, \qquad x^{-n} = S_o x^n.$$

Examples 1, 2 (*continued*)

(a) In the case of $\mathfrak{M} = \boldsymbol{R}^n$ we easily verify that $x^n = nx$.

(b) Let $\mathfrak{M}$ be the unit sphere in $\boldsymbol{R}^{n+1}$. Take $o = (1, 0, \ldots, 0)$ and $x = (\cos\varphi, \sin\varphi, 0, \ldots, 0)$. Then we have $x^2 = S_x S_o o = S_x o = (\cos 2\varphi, \sin 2\varphi, 0, \ldots, 0)$ and generally $x^n = (\cos n\varphi, \sin n\varphi, 0, \ldots, 0)$.

The above examples admit a generalization, see Problem 7. Denote by $Q(x)$ the translation $S_x S_o$. The mapping

$$\mathfrak{M} \ni x \to Q(x) \in G(\mathfrak{M})$$

called the *quadratic representation* has the following properties:

Proposition 15.1.1. (1) *The elements of the form* $Q(x)$ *generate the group* $G(\mathfrak{M})$.

(2) $Q(x^n) = Q(x)^n$.

(3) *If* φ *is a homomorphism of* $(\mathfrak{M}, S)$ *into* $(\mathfrak{M}', S')$ *and* $\varphi(o) = o'$ *then* φ *preserves raising to a power with respect to* o, *i.e.*

$$\varphi(x^n) = \varphi(x)^n. \tag{15.1.6}$$

Proof. (1) $S_x S_y = S_x S_o S_o S_y = Q(x)Q(y)^{-1}$.

Property (2) is valid for $n = 0$ and $n = 1$, and by induction we find

$$\begin{aligned} Q(x^{n+2}) &= Q(Q(x)x^n) = Q(S_x S_o x^n) = S_x S_o x^n S_x S_o \\ &= S_x S_o S_{x^n} S_o S_x S_o = Q(x)Q(x^n)Q(x) = Q(x)^{n+2}. \end{aligned}$$

Property (3) follows directly from the definition of a power and that of a homomorphism. □

The examples described above are special cases of the following situation: Let G be a connected Lie group and σ an automorphism of G with period 2 (i.e. $\sigma^2 = \mathrm{id}$). Denote by G_σ the subgroup of all fixed points of the automorphism σ and by $(G_\sigma)_o$ the connected component of the identity of this subgroup. Finally, let K be an arbitrary closed subgroup of G such that

$$(G_\sigma)_o \subset K \subset G. \tag{15.1.7}$$

We take $\mathfrak{M} = G/K$ and define

$$S_{[g]}[h] = [g\sigma(g)^{-1}\sigma(h)] \tag{15.1.8}$$

(as usual $[g]$ denotes the class $gK \in \mathfrak{M}$).

We shall prove that $\mathfrak{M}$ is a symmetric space. The verification of algebraic conditions (15.1.1)–(15.1.3) is left to the reader. To prove that condition (15.1.4) also holds we first consider the case of the symmetry S_o, $o := [e]$.

Let O be a neighbourhood of zero in $\mathfrak{g}$ on which exp is a diffeomorphism. Since

$$S_o[\exp X] = [\sigma \exp Y] = [\exp (\mathrm{d}\sigma)_e X],$$

it follows that for a $X \in O$ the element $\exp X$ belongs to G if and only if $(\mathrm{d}\sigma)_e X = X$. We write

$$\mathfrak{g}_\pm = \{X \in \mathfrak{g} \colon (\mathrm{d}\sigma)_e X = \pm X\}.$$

It follows from what has been said above that the space $\mathfrak{g}_+$ is identical with the Lie algebra $\mathfrak{k}$ of the group K, regarded as a subset of $\mathfrak{g}$. Since $\mathfrak{g} = \mathfrak{g}_+ + \mathfrak{g}_-$ (because $X = \frac{1}{2}(X + \mathrm{d}\sigma X) + \frac{1}{2}(X - \mathrm{d}\sigma X)$), we see that $\mathfrak{g}_- \cap O$ is mapped homeomorphically onto a neighbourhood U of the point $o \in U$ under the mapping

$$\mathfrak{g}_- \cap O \ni X \to [\exp X].$$

For $X \ni \mathfrak{g}_-$ we have

$$S_o[\exp X] = [\exp(\mathrm{d}\sigma)_e X] = [\exp(-X)],$$

which shows that S_o has a unique fixed point in U, namely the point $o = [e]$.

For an arbitrary $y = [g]$ the condition

$$S_{[g]}[h] = [h]$$

leads to the identity $[g\sigma(g^{-1})\sigma(h)] = [h]$, and hence to

$$[\sigma(g^{-1}h)] = [g^{-1}h].$$

As we have proven above, the last identity has a unique solution $[g^{-1}h] = [e]$ in the neighbourhood $[g]$ of the point gU.

The assertion follows in view of the transitivity of the action of the group on $\mathfrak{M}$.

It turns out that the last example is of a general nature. The proof of this fact is the objective of § 15.3. For an arbitrary connected symmetric space $\mathfrak{M}$, we shall define a Lie transformation group G of $\mathfrak{M}$, which acts transitively on $\mathfrak{M}$. The structure of the symmetric space will determine an involutive automorphism σ whose fixed point set contains the isotropy subgroup of a point in $\mathfrak{M}$. To this end we shall apply the natural affine connection which is determined by the symmetric structure on $\mathfrak{M}$.

15.2. AFFINE CONNECTION ON A SYMMETRIC SPACE

One can define an affine connection on a symmetric space $\mathfrak{M}$ such that the symmetries S_m, $m \in \mathfrak{M}$ are all affine mappings.

We shall first construct that connection (formula (15.2.1)) and then prove its uniqueness.

To begin with we introduce some algebraic operations determined by the symmetric structure on the space of differential operators.

The value of a differential operator D at a point $p \in \mathfrak{M}$ is regarded as a functional on $\mathscr{E}(\mathfrak{M})$:

$$D_p f := (Df)(p).$$

Let D be a differential operator on $\mathfrak{M}$ and D' a differential operator on $\mathfrak{N}$ and let μ be a surjection $\mathfrak{M} \times \mathfrak{N} \to \mathfrak{M}$. For arbitrary $m \in \mathfrak{M}$, $n \in \mathfrak{N}$ we denote by $D_m \cdot D'_n$ the functional on $\mathfrak{M}$ defined by the formula

$$(D_m \cdot D'_n)f = (D \otimes D')(f \circ \mu)(m, n),$$

where $D \otimes D'$ is the differential operator on $\mathfrak{M} \times \mathfrak{N}$ defined by composing the operator D which acts on the variable m and the operator D' which acts on the variable n. For μ we take the mapping $\mathfrak{M} \times \mathfrak{M} \ni (m, m) \to S_m m' \in \mathfrak{M}$. Then given a pair of differential operators D, D' on $\mathfrak{M}$ we can define an operator from $\mathscr{E}(\mathfrak{M})$ into $\mathscr{E}(\mathfrak{M})$ according to the formula

$$(D \cdot D')f(m) := (D_m \circ D'_m)f. \tag{15.2.1}$$

The condition $S_m m = m$ implies that the right-hand side is a functional with support at the point m. Thus the linear operation $D \cdot D'$ does not extend the support of a function and this means that $D \cdot D'$ is indeed a differential operator on $\mathfrak{M}$ (the Peetre theorem in Schwartz, 1957).

The action of the operator $D \cdot D'$ can be written in a convenient way by applying the operation $\Delta: \mathfrak{M} \ni m \to \Delta(m) := (m, m) \in \mathfrak{M} \times \mathfrak{M}$:

$$(D \cdot D')f = [(D \otimes D')(f \circ \mu)] \circ \Delta.$$

If one of the operators D and D' is identity (e.g. $D = 1$), we shall write $m \cdot D'_x := 1_m \cdot D'_x$. In this notation we have $x \cdot y = S_x y$.

The "dot" calculus has been introduced and developed by O. Loos (1969). Properties (15.1.1)–(15.1.4) of a symmetry operation imply the following identities:

LEMMA 15.2.1. (1) *For any* $x, y, z \in \mathfrak{M}$ *and* $X \in T_x(\mathfrak{M})$ *we have* $x \cdot X = -X$,

(2) $X \cdot x = 2X$,

(3) $X \cdot (x \cdot y) + x \cdot (X \cdot y) = 0$,

(4) $X \cdot (y \cdot z) = (X \cdot y) \cdot (x \cdot z) + (x \cdot y) \cdot (X \cdot z)$.

Proof. The vector $x \cdot X$ is a different expression of the vector $\mathrm{d}S_x X$. It follows from property (15.1.2) that $(\mathrm{d}S_x)^2 = \mathrm{id}$. We shall prove that $\mathrm{d}S_x$ has no fixed points (except zero) in T_x. Let O be a neighbourhood of x such that x is the only fixed point of S_x in this neighbourhood. Then the same property is true of the S_x-invariant set $\mathscr{U} := (S_x O) \cap O$. There exists a connection on $\mathscr{U}$ such that $S_x|_U$ is an affine mapping (cf. Problem 16. Ch. 14).

Let Exp be the exponential mapping determined by that connection and let $X \in T_x(\mathfrak{M})$ be a vector in the domain of Exp satisfying the relation $\mathrm{d}S_x X = X$. Moreover, we may assume that $\mathrm{Exp} X \in \mathscr{U}$. Then $S_x \mathrm{Exp} X = \mathrm{Exp}\, \mathrm{d}S_x X = \mathrm{Exp} X$, which means that the symmetry S_x leaves $\mathrm{Exp} X$ fixed. Therefore we must have $\mathrm{Exp} X = x$ and hence $X = 0$. The vectors of the form $X + \mathrm{d}S_x X$ being $\mathrm{d}S_x$-invariant, it follows that $\mathrm{d}S_x = -\mathrm{id}$, which proves (1).

Property (2) is a consequence of (1) and (15.1.2).

Ad (3). For a fixed function $f \in \mathscr{E}(\mathfrak{M})$ we define a function on $\mathfrak{M} \times \mathfrak{M} \times \mathfrak{M}$ by the formula

$$\psi_f(x, y, z) = f(S_x S_y z).$$

Obviously $\psi_f(x, x, y) = f(y)$, and so the function $\varphi_f(x, y) := \psi_f(x, x, y)$ is constant in the first variable. Therefore $(X \otimes y)\varphi_f(x, y) = 0$ and the assertion follows in view of

$$\begin{aligned}(X \otimes y)\varphi_f(x, y) &= (X \otimes x \otimes y)\psi_f(x, x, y) + (x \otimes X \otimes y)\psi_f(x, x, y) \\ &= X \cdot (x \cdot y) f + x \cdot (X \cdot y) f.\end{aligned}$$

(4) is proved in an analogous way by applying (15.1.3). □

PROPOSITION 15.2.2. *If φ is a homomorphism of a symmetric space $(\mathfrak{M}, S)$ into $(\mathfrak{M}', S')$ then $\mathrm{d}\varphi(X \cdot Y) = \mathrm{d}\varphi(X) \cdot \mathrm{d}\varphi(Y)$ for $X, Y \in T(\mathfrak{M})$.*

Proof. Denote by μ' the symmetry mapping on M. Then

$$\begin{aligned}\mathrm{d}\varphi(X \cdot Y) f(\varphi(x)) &= [(X \otimes Y) f \circ \varphi \circ \mu)(x, x)] \\ &= [(X \otimes Y)(f \circ \mu' \circ \varphi)(x, x) \\ &= [(\mathrm{d}\varphi X) \otimes (\mathrm{d}\varphi Y)(f \circ \mu')](\varphi(x), \varphi(x)) \\ &= (\mathrm{d}\varphi(X) \cdot \mathrm{d}\varphi(Y)) f(\varphi(x))\end{aligned}$$

for every $f \in \mathscr{E}(\mathfrak{M}')$. □

Definition. If X and Y are in $\mathscr{T}(\mathfrak{M})$, then we put

$$\nabla_X Y := XY + \tfrac{1}{2} X \cdot Y. \tag{15.2.2}$$

The value $\nabla_X Y$ is a differential operator on $\mathfrak{M}$ of order not greater than 2.

Theorem 15.2.3. *The operation* $\nabla\colon \mathscr{T}(\mathfrak{M}) \times \mathscr{T}(\mathfrak{M}) \ni (X, Y) \to \nabla_X Y$ *has its image contained in* $\mathscr{T}(\mathfrak{M})$ *and is a connection on* $\mathfrak{M}$.

The curvature tensor of the connection ∇ *has the form*

$$R(X, Y)Z = \tfrac{1}{4}(X \cdot (Y \cdot Z) - Y \cdot (X \cdot Z)).$$

Proof. The verification that $\nabla_X Y \in \mathscr{T}(\mathfrak{M})$ is the most important thing. We shall prove by direct computation that

$$\nabla_X Y(fg) = g\nabla_X Y(f) + f\nabla_X Y(g). \tag{15.2.3}$$

We begin by computing

$$\begin{aligned}
X \cdot Y(fg)(x) &= (X \otimes Y)(fg \circ \mu)(x, x)\\
&= (X \otimes 1)(1 \otimes Y)(fg \circ \mu)(x, x)\\
&= \{(X \otimes 1)(1 \otimes Y)[(f \circ \mu)(g \circ \mu)]\}(x, x)\\
&= \{(X \otimes 1)[(1 \otimes Y)(f \circ \mu)](g \circ \mu)\\
&\quad + [(1 \otimes Y)(g \circ \mu)](f \circ \mu)\}(x, x)\\
&= \{\big((X \otimes Y)(f \circ \mu)\big)(g \circ \mu)\\
&\quad + \big((1 \otimes Y)(f \circ \mu)\big)\big((X \otimes 1)(g \circ \mu)\big)\\
&\quad + \big((X \otimes Y)(g \circ \mu)\big)(f \circ \mu)\\
&\quad + \big((1 \otimes Y)(g \circ \mu)\big)\big((X \otimes 1)(f \circ \mu)\big)\}(x, x)\\
&= g\big((X \cdot Y)f\big)(x) + f\big((X \cdot Y)g\big)(x)\\
&\quad + \big((x \cdot Y)f\big)(x)\big((X \cdot x)g\big)(x)\\
&\quad + \big((X \cdot x)f\big)(x)\big((x \cdot Y)g\big)(x)\\
&= g\big((X \cdot Y)f\big)(x) + f\big((X \cdot Y)g\big)(x)\\
&\quad - 2(Yf)(x)(Xg)(x) - 2(Xf)(x)(Yg)(x).
\end{aligned}$$

On the other hand,

$$(XY)(fg) = (XYf)g + f(XY)g + XfYg + YfXg.$$

By combining the two expressions we immediately obtain identity (15.2.3). The verification of identities (15.5.3), (15.5.4) of the preceding chapter presents no difficulty.

To prove the second part of the assertion we need the following

LEMMA 15.2.4. $Z(X \cdot Y) = (ZX) \cdot Y + X \cdot (ZY)$.

Proof. $Z(X \cdot Y)f(x) = (Z\otimes 1 + 1\otimes Z)(X\otimes Y)(f \circ \mu)(x, x) = (ZX\otimes Y + X\otimes ZY)(f \circ \mu)(x, x) = ((ZX) \cdot Y + X \cdot (ZY))f(x)$, which is the desired result. □

Returning to the proof of the theorem, we apply expression (14.5.11).

$$\begin{aligned} R(X, Y)Z &= X(YZ + \tfrac{1}{2}X \cdot Z) + \tfrac{1}{2}X \cdot (YZ + \tfrac{1}{2}Y \cdot Z) \\ &\quad - Y(XZ + \tfrac{1}{2}X \cdot Z) - \tfrac{1}{2}Y \cdot (XZ + \tfrac{1}{2}X \cdot Z) \\ &\quad - [X, Y]Z - \tfrac{1}{2}(XY - YX) \cdot Z \\ &= \tfrac{1}{4}(X \cdot (Y \cdot Z) - Y \cdot (X \cdot Z)). \end{aligned}$$

The following property accounts for the exceptional role of the connection defined above.

THEOREM 15.2.5. *Let $\varphi: \mathfrak{M} \to \mathfrak{M}'$ be a homomorphism of symmetric spaces. Then φ is an affine mapping relative to the connections ∇, ∇' determined by the symmetric structure on $\mathfrak{M}$ and $\mathfrak{M}'$, respectively.*

Proof. We have

$$\begin{aligned} d\varphi \nabla_X Y &= d\varphi(XY + \tfrac{1}{2}X \cdot Y) = d\varphi(X)\, d\varphi(Y) + \tfrac{1}{2}d\varphi(X) d\varphi(Y) \\ &= \nabla'_{d\varphi(X)} d\varphi(Y). \end{aligned}$$ □

In particular, the symmetries S_x are affine transformations of the manifold $\mathfrak{M}$.

The connection on a symmetric space given by formula (15.2.2) will be called the *canonical connection* on that space.

PROPOSITION 15.2.6. *The canonical connection is the only connection invariant under all symmetries on $\mathfrak{M}$.*

Proof. Suppose there are two connections ∇ and ∇' on $\mathfrak{M}$ invariant under all symmetries, i.e. such that S_x, $x \in \mathfrak{M}$ are affine mappings. Then the difference $\nabla - \nabla'$ is an $\mathscr{E}(\mathfrak{M})$-bilinear mapping of $\mathscr{T}(\mathfrak{M})$

into $\mathscr{T}(\mathfrak{M}')$. Write $B(X, Y) := \nabla_X Y - \nabla'_X Y$. The invariance of the both connections implies that

$$\mathrm{d}S_x B(X,Y)_x = B(\mathrm{d}S_x X_x, \mathrm{d}S_x Y_x).$$

Further, by Lemma 15.2.1 we have

$$\mathrm{d}S_x B(X, Y)_x = -B(X, Y)_x,$$

and since

$$\mathrm{d}S_x B(X, Y)_x = B(-X_x, -Y_x) = B(X, Y)_x,$$

we conclude that $B \equiv 0$.

Example 3

Let us find the canonical connection for the symmetric structure on the space $\boldsymbol{R}^n$ defined before. We represent vector fields X, Y on $\boldsymbol{R}^n$ in the natural basis $X_i = \dfrac{\partial}{\partial x^i}$:

$$X = \sum_{i=1}^{n} x^i X_i, \qquad Y = \sum_{i=1}^{n} y^i X_i, \qquad x^i, y^i \in \mathscr{E}(\boldsymbol{R}^n).$$

Directly from definition (15.2.1) we have

$$(X \cdot Y) f = -2 \sum_{i=1}^{n} y^i X(X_i f).$$

Substituting this into (15.2.2), we obtain

$$\nabla_X Y = \sum_{i=1}^{n} X y^i X_i,$$

which is precisely the expression for the natural (Riemannian) connection on $\boldsymbol{R}^n$ (cf Problem 9(c) in Chapter 14).

For connected symmetric manifolds the converse of Theorem 15.2.5 is also valid.

THEOREM 15.2.7. *If $\mathfrak{M}$, $\mathfrak{M}'$ are symmetric spaces, $\mathfrak{M}$ is connected and $\varphi \colon \mathfrak{M} \to \mathfrak{M}'$ is an affine mapping with respect to the canonical connections then φ is a homomorphism of the symmetric spaces.*

Proof. We have to prove that $\varphi \circ S_x = S_{\varphi(x)} \circ \varphi$ for every $x \in \mathfrak{M}$.

To this end we compute the differentials of both mappings at a point x. On the basis of Lemma 15.2.1 we find

$$d(\varphi \circ S_x)_x = (d\varphi)_x \circ (dS_x)_x = -(d\varphi)_x$$

and

$$d(S_{\varphi(x)} \circ \varphi)_x = (dS_{\varphi(x)})_{(\varphi)x} \circ (d\varphi)_x = -(d\varphi)_x.$$

Since the differentials are identical, the assertion now follows in view of Lemma 14.6.1. □

The theorems presented above link up the purely algebraic properties of mappings (since the rules of preserving the symmetric structure are of algebraic character) with the properties of the type of differential identities.

The role of these statements will be made clear in further parts of our exposition. For the time being let us consider more closely a very simple case, namely that of curves on a symmetric manifold. The real axis $\boldsymbol{R}$ carries a natural structure of a symmetric space:

$$S'_x y = 2x - y.$$

A mapping $\varphi: (\boldsymbol{R}, S') \to (\mathfrak{M}, S)$ is affine if (see (14.6.1))

$$\frac{D}{dt} \circ d\varphi(X_t) = d\varphi \circ \frac{D'}{dt}(X_t)$$

for every vector field X_t on $\boldsymbol{R}$. Replacing X_t by the field d/dt we obtain the following characterization of an affine mapping

$$\frac{D}{dt}\varphi' = 0.$$

This is the equation of a geodesic in $\mathfrak{M}$.

From Theorems 15.2.5 and 15.2.7 we derive a new description of geodesics on a symmetric space. A mapping $\varphi: \boldsymbol{R} \to \mathfrak{M}$ is a geodesic of the canonical connection on M if and only if

$$\varphi(2t - t_1) = S_{\varphi(t)}\varphi(t_1). \tag{15.2.4}$$

As a consequence of this observation we have

COROLLARY 15.2.8. *The canonical connection on a symmetric space is complete.*

Proof. Suppose γ is a geodesic in $\mathfrak{M}$ defined on an interval $[a, b]$ and let $p = \gamma(a)$, $q = \gamma(b)$. We define

$$\bar{\gamma}(b+t) = S_q\gamma(b-t) \qquad \text{for } 0 \leqslant t < b-a.$$

S_q being an affine mapping this implies that the curve $t \to \bar{\gamma}(b+t)$ is a geodesic and the tangent vector at the point $\bar{\gamma}(b)$ equals $-(\mathrm{d}S_q \circ \circ \gamma')(b) = \gamma'(b)$. Owing to the uniqueness of a geodesic with a fixed tangent vector we see that $\bar{\gamma}$ is an extension of the curve γ to the interval $[a, 2b-a]$. In this way every geodesic can be extended by means of symmetries to the whole of $\boldsymbol{R}$. □

We know from Theorems 14.6.2 and 14.6.4 that affine vector fields are complete and form a finite-dimensional subalgebra $\mathfrak{A}$ of $\mathscr{T}(\mathfrak{M})$.

Further, it follows from Theorem 1.4.4 that the group Aff $\mathfrak{M}$ can be given a unique Lie group structure such that the Lie algebra is isomorphic in a natural way to $\mathfrak{A}$ and such that the identity component is generated by the one-parameter subgroups corresponding to the vector fields in $\mathfrak{A}$.

15.3 STRUCTURE OF THE GROUP OF DISPLACEMENTS OF A SYMMETRIC SPACE

We shall now study the group $G(\mathfrak{M})$ of displacements of a symmetric space. We shall prove that $G(\mathfrak{M})$ has a Lie group structure and acts transitively on $\mathfrak{M}$. We begin by proving

PROPOSITION 15.3.1. *Let o be a fixed point in a symmetric manifold $\mathfrak{M}$ and let $X \in T_o(\mathfrak{M})$ and $\gamma(t) := \mathrm{Exp}_o tX$. Then the curve in $G(\mathfrak{M})$ defined by the equation*

$$\varphi_X(t) = S_{\gamma(t)}S_o = Q(\gamma(t))$$

is a one-parameter subgroup.

Proof. The curve γ is a geodesic in $\mathfrak{M}$, and hence a homomorphism of the symmetric space $\boldsymbol{R}$ into $\mathfrak{M}$. By Proposition 15.1.1(3) we have $\gamma(kt) = \gamma(t)^k$ if the raising to a power is performed with respect to $o = \gamma(0)$. Let

$$s = \frac{s_1}{s_2}, \qquad t = \frac{t_1}{t_2}, \qquad s_i, t_i \in \boldsymbol{Z}.$$

Then

$$s = \alpha s_1 t_2, \qquad t = \alpha s_2 t_1, \qquad \text{where } \frac{1}{\alpha} = s_2 t_2.$$

It follows from Proposition 15.1.1(2) that

$$\begin{aligned} Q(\gamma(s))Q(\gamma(t)) &= Q(\gamma(\alpha))^{s_1 t_2} Q(\gamma(\alpha))^{s_2 t_1} \\ &= Q(\gamma(\alpha))^{s_1 t_2 + s_2 t_1} = Q(\gamma(s+t)). \end{aligned}$$

Thus the mapping $\boldsymbol{R} \ni t \to \varphi_X(t) \in G(\mathfrak{M})$ has turned out to be a homomorphism for rational values of the arguments. The continuity of the mapping implies that the property

$$\varphi(t+s) = \varphi(t)\varphi(s)$$

extends to the whole of $\boldsymbol{R}$.

THEOREM 15.3.2. *Let $\mathfrak{M}$ be a connected symmetric manifold. Then the group $G(\mathfrak{M})$ is generated by the one-parameter subgroups*

$$\boldsymbol{R} \ni t \to Q(\mathrm{Exp}_o tX), \qquad X \in T_o(\mathfrak{M}), \qquad o \in \mathfrak{M}.$$

The group $G(\mathfrak{M})$ acts transitively on $\mathfrak{M}$.

Proof. Arbitrary points p, q in a connected manifold $\mathfrak{M}$ can be joined by a broken line consisting of geodesics in such a way that if $\{p_i\}_{i=0}^{n}$ are the end-points of the respective segments and $p_0 = p$, $p_n = q$, then $p_i = \mathrm{Exp}_{p_{i+1}} X_{i+1}$, where $X_i \in T_{p_i}(\mathfrak{M})$, $i = 0, \dots, n-1$ (see Fig. 13). Then we have

$$\begin{aligned} S_p S_q &= S_{p_0} S_{p_1} S_{p_1} \dots S_{p_{n-1}} S_{p_{n-1}} S_{p_n} = (S_{\mathrm{Exp} X_1} S_{p_1}) \dots (S_{\mathrm{Exp} X_n} S_{p_n}) \\ &= \varphi_{X_1}(1) \dots \varphi_{X_n}(1). \end{aligned}$$

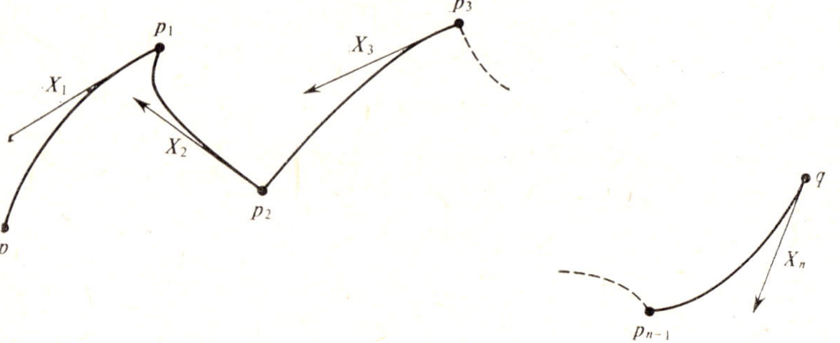

Fig. 13

Thus it remains to prove transitivity of the action of $G(\mathfrak{M})$ on $\mathfrak{M}$. We again exploit the existence of the broken line. It is enough to verify that for every pair of points p_i, p_{i+1} lying on the same geodesic there exists an element $g \in G(\mathfrak{M})$ which sends p_i onto p_{i+1}. Define $g = S_p S_{p_i}$ where $p = \mathrm{Exp}_{p+1} \frac{1}{2} X_{i+1}$. Then by (15.2.4) we get

$$gp_i = S_p S_{p_i} p_i = S_p p_i = S_{\mathrm{Exp}_{p_{i+1}} \frac{1}{2} X_{i+1}} \mathrm{Exp}_{p_{i+1}} X_{i+1}$$
$$= \mathrm{Exp}_{p_{i+1}} 0 = p_{i+1}. \qquad \square$$

The group of displacements $G(\mathfrak{M})$ constitutes a subgroup of $\mathrm{Aff}\,\mathfrak{M}$ generated by the one-parameter subgroups described in Theorem 15.3.2. Owing to Theorem 1.4.4 we know that $G(\mathfrak{M})$ has the structure of a Lie group which turns $(G(\mathfrak{M}), \mathfrak{M})$ into a Lie transformation group. Since the action of $G(\mathfrak{M})$ is transitive, we obtain an isomorphism of the homogeneous spaces $\mathfrak{M} = G/H$ where $H = \{g \in G(\mathfrak{M}) \colon g \cdot o = o\}$, $o \in \mathfrak{M}$.

The Lie algebra of the group $G(\mathfrak{M})$ is isomorphic to the algebra of the affine vector fields generated by the vector fields induced by the one-parameter transformation groups $\varphi_X(t) = Q(\mathrm{Exp}_o tX)$, $X \in T_o(\mathfrak{M})$.

The tangent mapping to the inclusion mapping $G(\mathfrak{M}) \to \mathrm{Aff}\,\mathfrak{M}$ is a natural injection of the Lie algebra of $G(\mathfrak{M})$ into $\mathfrak{A}$. Thus $G(\mathfrak{M})$ constitutes a connected Lie subgroup of the group $\mathrm{Aff}\,\mathfrak{M}$.

We shall now give a description of the symmetric structure on $\mathfrak{M}$ in terms of the structure of a Lie group acting transitively on $\mathfrak{M}$.

THEOREM 15.3.3. *Let* $(\mathfrak{M}, S)$ *be a connected symmetric manifold and* G *a subgroup of* $\mathrm{Aff}\,\mathfrak{M}$ *acting on* $\mathfrak{M}$ *transitively and such that* $S_o G S_o \subset G$. *Let* H *be the stability group at the point* $o \in \mathfrak{M}$. *Then*

(a) $\mathfrak{M} = G/H$ (*isomorphism of homogeneous spaces*).

(b) *The mapping* $\sigma \colon G \ni g \to S_o g S_o$ *is an involutive automorphism of* G *such that*

$$(G_\sigma)_0 \subset H \subset G_\sigma,$$

($(G_\sigma)_0$ *denoting the connected component of the identity*)

(c) *If* $\mathfrak{g}$ *is the Lie algebra of the group* G, *then*

$$\mathfrak{g} = \mathfrak{h} + \mathfrak{p},$$

where $\mathfrak{h}$ *is the Lie algebra of the group* H *imbedded in a natural way in* $\mathfrak{g}$, *and*

$$\mathfrak{p} = \{X \in \mathfrak{g} \colon (d\sigma)_e X = -X\}.$$

Moreover the following inclusions hold:

$$[\mathfrak{h}, \mathfrak{p}] \subset \mathfrak{p}, \qquad [\mathfrak{p}, \mathfrak{p}] \subset \mathfrak{h}.$$

Proof. Property (a) follows from the transitivity of G in view of Theorem 1.4.2. The involutiveness of the mapping σ is a consequence of axiom (15.1.2) in the definition of a symmetric space. Similarly, the fact that σ is a group automorphism results from the involutiveness of the symmetry S_o:

$$\sigma(gg') = S_o g g' S_o = S_o g S_o S_o g' S_o = \sigma(g)\sigma(g').$$

Let $h \in H$, i.e. $h \cdot o = o$. Let us see how the element $\sigma(h)$ acts on $\mathfrak{M}$:

$$S_o h S_o x = S_o S_{h\cdot o} h \cdot x = h \cdot x$$

(after applying formula (15.1.5)), which means that $\sigma(h) = h$. Thus we have the inclusion $H \subset G_\sigma$.

Next, for every one-parameter subgroup of G,

$$S_o\big((\exp tX)\cdot o\big) = S_o(\exp tX)S_o \cdot o = \sigma(\exp tX)\cdot o = \exp tX \cdot o.$$

For small values of t the element $\exp tX \cdot o$ is contained in a neighbourhood of o which does not contain S_o-fixed points besides o. Thus the above identity implies that $\exp tX \cdot o = o$ for an arbitrary $t \in \boldsymbol{R}$.

Since the connected component of the identity of G_σ, is generated by one-parameter subgroups, it follows that $(G_\sigma)_0 \subset H$.

Ad (c). The identity $\exp(\mathrm{d}\sigma)_e X = \sigma(\exp X)$ implies that

$$\exp tX \in G_\sigma \quad \text{if and only if} \quad (\mathrm{d}\sigma)_e X = X.$$

The Lie algebra of the subgroup H regarded as a subalgebra of $\mathfrak{g}$ consists of $(\mathrm{d}\sigma)_e$-fixed vectors. Since an arbitrary vector $X \in \mathfrak{g}$ can be represented as

$$X = \tfrac{1}{2}(X + (\mathrm{d}\sigma)_e X) + \tfrac{1}{2}(X - (\mathrm{d}\sigma)_e X),$$

we get the desired decomposition $\mathfrak{g} = \mathfrak{h} + \mathfrak{p}$.

Finally, the commutation relations can be proved by using the fact that $(\mathrm{d}\sigma)_e$ is an automorphism of the Lie algebra $\mathfrak{g}$. For instance, if $X \in \mathfrak{h}$, $Y \in \mathfrak{p}$ then

$$(\mathrm{d}\sigma)_e[X, Y] = [(\mathrm{d}\sigma)_e X, (\mathrm{d}\sigma)_e Y] = -[X, Y]$$

and hence $[X, Y] \in \mathfrak{p}$. □

The mapping which to a symmetric space $\mathfrak{M}$ assigns a pair (G, H) such that $\mathfrak{M} = G/H$ is obviously not univalent. In fact, the group $\operatorname{Aff}\mathfrak{M}$ and $G(\mathfrak{M})$ both verify the assumptions of Theorem 15.3.3. The special case of $\mathfrak{M} = \boldsymbol{R}^n$ shows that the groups $G(\mathfrak{M})$ (= group of displacements) and $\operatorname{Aff}\mathfrak{M}$ (= group of affine motions) may indeed differ.

On the basis of Theorem 15.3.3 we arrive at the conclusion that for an arbitrary subgroup $G \subset \operatorname{Aff}\mathfrak{M}$ which acts transitively on $\mathfrak{M}$ we get the same subspace (when regarded as imbedded in the Lie algebra of the group $\operatorname{Aff}\mathfrak{M}$). In any case this space has a dimension equal to $\dim T_0(\mathfrak{M})$ and consists of all vectors satisfying the condition $(\mathrm{d}\sigma)_e X = -X$. Moreover, we observe that the group $G(\mathfrak{M})$, being generated by one-parameter subgroups corresponding to the elements of $\mathfrak{p}$, is the smallest subgroup of $G(\mathfrak{M})$ which acts transitively on $\mathfrak{M}$. A detailed description of the affine vector fields corresponding to the space $\mathfrak{p}$ will be given in the next section.

Example 1

Consider the n-dimensional unit sphere as a symmetric space under the symmetries $S_x y = 2(x|y)x - y$. Take $o = (1, 0, \ldots, 0) \in \boldsymbol{R}^{n+1}$ and $G = \mathrm{SO}(n+1)$. (It can be proved that the orthogonal transformations of $\boldsymbol{R}^{n+1}$ are affine mappings relative to the canonical connection on S^n—in particular, this fact will follow from the discussion of Riemannian symmetric spaces in § 15.5.) S_0 is then given by the matrix $\begin{bmatrix} 1 & 0 \\ 0 & -I_n \end{bmatrix}$ where I_n is the unit matrix of degree n, and the set of σ-fixed points is the set of matrices $\begin{bmatrix} \det A^{-1} & 0 \\ 0 & A \end{bmatrix}$, $A \in O(n)$. The isotropy group at the point $o \in S^n$ is the group $\begin{bmatrix} 1 & 0 \\ 0 & A \end{bmatrix}$, $A \in \mathrm{SO}(n)$, which is the connected component of identity of the group G_σ of all σ-fixed points. Its Lie algebra consists of matrices $\begin{bmatrix} 0 & 0 \\ 0 & B \end{bmatrix}$, $B \in \mathfrak{o}(n)$ and the complementary subspace $\mathfrak{p}$ from the decomposition $\mathfrak{g} = \mathfrak{h} + \mathfrak{p}$ in Theorem 15.3.3(c) consists of matrices $\begin{bmatrix} 0 & -\xi_t \\ \xi & 0 \end{bmatrix}$ where ξ is a (column) vector in $\boldsymbol{R}^n$. The reader will note that the quotient space G/G_σ corresponding to the same group $G = SO(n+1)$ and the same involutive automorphism σ is the n-dimensional projective space $P_n(\boldsymbol{R})$ (cf. also Problem 5).

Owing to Theorem 15.3.3 the study of connected symmetric manifolds reduces to the study of group structures and Lie algebra structures. The example in § 15.1 showed that a Lie group equipped with an involutive automorphism defines a symmetric space; now it turns out that every connected symmetric space arises in this way.

The method of studying the geometry of symmetric spaces by considering the corresponding translation group was originated by E. Cartan, who worked out in this way a classification of symmetric Riemannian spaces. These spaces will be described later in this chapter.

The section which follows presents the basic facts resulting from the general description of symmetric spaces given above.

15.4. GEOMETRY OF SYMMETRIC SPACES

We consider a connected symmetric space $\mathfrak{M}$ as a homogeneous space of a group G. We retain the notation of Theorem 15.3.3. The Lie algebra of the group G constitutes a subalgebra of the space $\mathfrak{A}$ of all affine vector fields. We shall give a detailed description of those vector fields which correspond to the elements of $\mathfrak{p}$.

We define a mapping

$$T_o(\mathfrak{M}) \ni X \to X^{\#} \in \mathscr{T}(\mathfrak{M}),$$

where $X^{\#}(m) := X \cdot (o \cdot m)$.

PROPOSITION 15.4.1. (1) *The (affine) vector field* $X^{\#}$ *is the generator of the one-parameter group* $\varphi_X(t) = Q(\mathrm{Exp}_o tX)$.

(2) *The integral curve of the vector field* $X^{\#}$ *passing through the point* o *is the geodesic* $\mathrm{Exp}_o 2tX$.

(3) $(d\sigma)_e X = -X$.

Proof. Ad (1) We compute the tangent vector field to φ_X:

$$\frac{d}{dt}\varphi_X(t) \cdot m|_{t=o} = \frac{d}{dt}(\mathrm{Exp}\, tX) \cdot (o \cdot m)|_{t=o}$$
$$= X \cdot (o \cdot m) = X^{\#}(m).$$

Moreover, this proves that $X^{\#}$ is an affine vector field.

Ad (2). The geodesic $\gamma(t) := \mathrm{Exp}_o 2tX$ is a homomorphism of the symmetric space $\boldsymbol{R}$ into $\mathfrak{M}$. Hence

$$\gamma(s)\gamma(t) = \gamma(2s-t)$$

or equivalently

$$\gamma(2s+t) = \gamma(s)\gamma(-t).$$

The differentiation $\left.\frac{\mathrm{d}}{\mathrm{d}s}\right|_{s=0}$, when applied to both sides of this formula, leads to the relation

$$2X \cdot \gamma(-t) = 2X \cdot \big(o \cdot \gamma(t)\big) = 2\gamma'(t),$$

i.e.

$$\gamma'(t) = X^{\#}\big(\gamma(t)\big).$$

Ad (3). We find

$$S_o Q(\mathrm{Exp}_o tX) S_o = S_o S_{\mathrm{Exp}\, tX} S_o = S_{S_o \mathrm{Exp}\, tX} S_o - S_{\mathrm{Exp}(-tX)} S_o$$
$$= Q\big(\mathrm{Exp}(-tY)\big),$$

so that $\sigma\big(Q(\mathrm{Exp}\, tX)\big) = Q\big(\mathrm{Exp}(-tX)\big)$. Assertion (3) follows in view of (1) after differentiating with respect to t. □

Assertion (1) of Proposition 15.4.1 can also be formulated as follows:

$$\exp tX^{\#} = Q(\mathrm{Exp}_o tX) \qquad \text{for } X \in T_o(\mathfrak{M}), \tag{15.4.1}$$

where exp is the exponential mapping on the group G and the vector field $X^{\#}$ is regarded as an element of $\mathfrak{p} \subset \mathfrak{g}$. Writing $\pi\colon G \to G/H$, $\pi(g) := g \cdot o$ we obtain on account of (15.4.1) and Proposition 15.4.1(2)

$$\pi(\exp tX^{\#}) = \varphi_X(t) \cdot o = \mathrm{Exp}_o 2tX. \tag{15.4.2}$$

This relation is of fundamental importance. It shows that the geodesics emanating from the point $o \in M$ are orbits of the one-parameter subgroups of the group G.

Example 1 (*continuation of Example* 1 *in* § 15.3)

Under the notation of Example 1 in § 15.3, for every $X \in \mathfrak{p}$, X

$= \begin{bmatrix} 0 & -\xi^t \\ \xi & 0 \end{bmatrix}$ with $\xi \neq 0$ there exists an orthogonal matrix B of degree n

such that $B\xi = \begin{bmatrix} c \\ 0 \\ \vdots \\ 0 \end{bmatrix}$, where $c = \|\xi\|$. Then

$$X' = \begin{bmatrix} 0 & -(B\xi)^t \\ B\xi & 0 \end{bmatrix} = \mathrm{Ad}\begin{bmatrix} 1 & 0 \\ 0 & B \end{bmatrix} X$$

and

$$\exp tX' = \begin{bmatrix} 1 & 0 \\ 0 & B \end{bmatrix} \exp tX \begin{bmatrix} 1 & 0 \\ 0 & B^t \end{bmatrix}$$

which allows us to reduce the problem of constructing geodesics to the case where

$$\xi = e_1 = \begin{bmatrix} 1 \\ 0 \\ \vdots \\ 0 \end{bmatrix}.$$

Then

$$\exp tX = \begin{bmatrix} \cos t & -\sin t & 0 \dots 0 \\ \sin t & \cos t & 0 \dots 0 \\ 0 & 0 & \\ \vdots & \vdots & I_{n-1} \\ 0 & 0 & \end{bmatrix} \quad \text{and} \quad \pi(\exp tX) = \begin{bmatrix} \cos t \\ \sin t \\ 0 \\ \vdots \\ 0 \end{bmatrix}$$

Any other geodesic starting at the point o can be obtained from the one given above by applying a rotation about the axis O_{X_1}, i.e. by acting with a matrix $\begin{bmatrix} 1 & 0 \\ 0 & B \end{bmatrix}$, $B \in \mathrm{SO}(n)$.

The geometric operation of parallel translation along a geodesic emanating from the point $o \in \mathfrak{M}$ can also be described by means of the action of the group G.

Proposition 15.4.2. *The parallel translation along the geodesic* $\gamma(t) = \mathrm{Exp}_o tX$ *is identical with the mapping* $d\varphi_X(\frac{1}{2}t)$.

Proof. We have to prove that for an arbitrary $Y \in T_o(\mathfrak{M})$ the curve $Y_t\colon \boldsymbol{R} \ni t \to \mathrm{d}\varphi_X(\frac{1}{2}t)\,(Y) \in T_{\mathrm{Exp}_o tX}(M)$ is parallel along $\gamma(t)$, i.e. that $\dfrac{DY_t}{\mathrm{d}t} \equiv 0.$

We compute

$$Y_t f = Y(f \circ \varphi_X(\tfrac{1}{2}t)) = \mathrm{Exp}_o(\tfrac{1}{2}Xt) \cdot (o \cdot Y)(f)$$
$$= -(\mathrm{Exp}_o(\tfrac{1}{2}tX)) \cdot Y(f) = -(\gamma(\tfrac{1}{2}t) \cdot Y) f.$$

Let $\tilde{Y}$ be a vector field on $\mathfrak{M}$ which coincides with Y_t for $t \in]a, b[$. We have

$$\nabla_{X^\#} \tilde{Y}(\gamma(t_0)) = \frac{DY}{\mathrm{d}t}(t_0).$$

It follows from the definition of the connection ∇ that

$$\nabla_{X^\#} \tilde{Y}(\gamma(t)) = X^\# \tilde{Y}(\gamma(t_0)) + \tfrac{1}{2} X^\#(\gamma(t_0)) \cdot Y_{t_0}.$$

The geodesic γ is an integral curve of the field $X^\#$ and therefore $X^\#(\gamma(t_0)) = \gamma'(t_0)$ (recall that $\gamma'(t_0) f = \dfrac{\mathrm{d}}{\mathrm{d}t} f(\gamma(t))|_{t=t_0}$), and

$$X^\#(\gamma(t_0)) Y_{t_0} = -\gamma'(t_0) \cdot (\gamma(\tfrac{1}{2}t_0) \cdot Y).$$

For an arbitrary $m \in \mathfrak{M}$ we have the identity

$$\gamma(t-s) \cdot (o \cdot m) = \varphi_X(t-s) m = \varphi_X(t) \varphi(-s) m$$
$$= \gamma(t)\,(o \cdot \gamma(-s)) \cdot (o \cdot m) = \gamma(t) \cdot (\gamma(s) \cdot m).$$

Differentiation with respect to t at the point $s = \frac{1}{2}t$ leads to the identity

$$\gamma'(\tfrac{1}{2}t) \cdot (o \cdot m) = \gamma'(t) \cdot (\gamma(\tfrac{1}{2}t) \cdot m).$$

This being so, we have

$$-\gamma'(\tfrac{1}{2}t) \cdot Y = \gamma'(t) \cdot (\gamma(\tfrac{1}{2}t) \cdot Y),$$

and hence

$$X^\#(\gamma(t_0)) Y_{t_0} = \gamma'(\tfrac{1}{2}t_0) \cdot Y.$$

On the other hand,

$$X^\# \tilde{Y} f(\gamma(t_0)) = \frac{\mathrm{d}}{\mathrm{d}t}\bigg|_{t=0} \tilde{Y} f(\gamma(t+t_0)) = -\frac{\mathrm{d}}{\mathrm{d}t}\bigg|_{t=0} \gamma\left(\frac{1}{2}(t+t_0)\right) \tilde{Y}$$
$$= -\tfrac{1}{2}\gamma'(\tfrac{1}{2}t_0) \cdot Y.$$

Finally, we arrive at the desired identity

$$\nabla_X \tilde{Y}(\gamma(t)) = 0.$$

THEOREM 15.4.3. *The curvature and the torsion of the canonical connection has the following properties*:

(1) $T \equiv 0$

(2) $\nabla R = 0$.

Proof. In both cases we follow the method used to establish the uniqueness of a connection invariant under all symmetries.

By way of example we present the details concerning case (1). The mapping

$$T\colon \mathscr{T}(\mathfrak{M}) \times \mathscr{T}(\mathfrak{M}) \to \mathscr{T}(\mathfrak{M})$$

is invariant under symmetries

$$\mathrm{d}S_x\big(T(X, Y)\big) = T(\mathrm{d}S_x X, \mathrm{d}S_x Y).$$

In particular, at a point $x \in \mathfrak{M}$ we have

$$\begin{aligned} -T_x(X, Y) = \mathrm{d}S_x T_x(X, Y) &= T_x(\mathrm{d}S_x X_x, \mathrm{d}S_x Y_x) \\ &= T_x(-X_x, -Y_x) = T_x(X, Y), \end{aligned}$$

which establishes assertion (1). □

The properties of symmetric spaces formulated above form the definition of locally symmetric spaces, which were investigated by E. Cartan. Before classifying these spaces by means of the above methods Cartan used the notion of the holonomy group of a point in a symmetric space.

A closed curve γ in a space with an affine connection defines an isomorphism of the tangent space at a point p in the curve; namely, the parallel translation τ_γ from p to p along the curve γ.

We define a set H_p of isomorphisms of $T_p(\mathfrak{M})$:

$$H_p = \{g \in \mathrm{GL}\big(T_p(\mathfrak{M})\big)\colon g = \tau_\gamma \text{ for a certain closed curve } \gamma\}.$$

This set is a subgroup of $\mathrm{GL}\big(T_p(\mathfrak{M})\big)$.

To the trivial curve $\gamma(t) = p$ corresponds the identity element. The inverse of a τ_γ is given by the curve obtained by changing the orientation in γ. The product $\tau_{\gamma_1} \circ \tau_{\gamma_2}$ is the isomorphism corresponding to the curve obtained by extending γ_2 by means of γ_1.

The group H_p is called the *holonomy group of the point p*. We present a theorem which explains the connection between the two methods used by E. Cartan.

THEOREM 15.4.4 (Loos, 1969). *Let γ be a broken curve in $\mathfrak{M}$ joining m to n. Then there exists a $g \in G(\mathfrak{M})$ such that $gm = n$ and $\tau_\gamma = (\mathrm{d}l_g)_m$.*

COROLLARY 15.4.5. *The group H_m is isomorphic to the group of linear transformations of the form $(\mathrm{d}l_g)_m$ where $g \in G_m$ (G_m is the isotropy group at the point m). Since the isotropy groups of arbitrary points are isomorphic, the same is also true of the holonomy groups H_m.*

15.5. RIEMANNIAN SYMMETRIC SPACES. RIEMANN PAIRS

A Riemannian manifold $(\mathfrak{M}, g)$ which is also a symmetric space is called a *Riemannian symmetric space* provided all symmetries S_m are isometries:

$$g_{S_m x}(\mathrm{d}S_m X, \mathrm{d}S_m Y) = g_x(X, Y) \quad \text{for} \quad X, Y \in T_x(\mathfrak{M}), m \in \mathfrak{M}. \tag{15.5.1}$$

Then the group of translations $G(\mathfrak{M})$ imbeds in the group of all isometries of the space $\mathfrak{M}$.

In § 15.3 we gave a description of symmetric manifolds as homogeneous spaces G/H. The structure of a connected symmetric space turned out to be completely determined by a triple (G, σ, H) where G is a connected group, H—a subgroup and $\sigma \in \mathrm{Aut}\, G$ an involution which defines a symmetry with respect to $o = [e]$. We shall give a condition which from among all triples (G, σ, H) (traditionally called symmetric pairs) singles out the Riemannian symmetric spaces.

DEFINITION. If (G, σ, H) is a symmetric pair and $\mathrm{Ad}_G(H)$ is a compact subgroup of $\mathrm{GL}(\mathfrak{g})$, then such a system (G, σ, H) is called a *Riemann pair*.

THEOREM 15.5.1. *If (G, σ, H) is a Riemann pair, then the symmetric space G/H has a unique Riemannian structure invariant with respect to G and to all symmetries of the space G/H.*

Proof. A G-invariant Riemannian structure on $\mathfrak{M} = G/H$ can be defined on the basis of Proposition 1.4.7. If $\mathrm{Ad}_G(H)$ is a compact group then so is the group $\mathrm{Ad}_{\mathfrak{g}/\mathfrak{h}}(H)$ which acts on $T_o(\mathfrak{M})$. Thus $T_o(\mathfrak{M})$ admits an inner product Q invariant under all operators in $\mathrm{Ad}_{\mathfrak{g}/\mathfrak{h}}(H)$.

A G-invariant Riemannian tensor on $\mathfrak{M}$ is now defined as follows:

$$g_{x\cdot o}\big((\mathrm{d}l_x)_o X_1, (\mathrm{d}l_x)_o Y_1\big) := Q(X_1, Y_1).$$

We only have to verify that $\mathrm{d}S_m$ leaves the tensor invariant. We first check that this property is true of S_0. For the proof let X, Y be in $T_m(\mathfrak{M})$, $m = x \cdot o$. Then there exist $X_1, Y_1 \in T_o(\mathfrak{M})$ with the property that $X = (\mathrm{d}l_x)_o X_1$, $Y = (\mathrm{d}l_x)_o Y_1$. Hence

$$g_{S_o m}(\mathrm{d}S_o X, \mathrm{d}S_o Y) = g_{S_o m}\big(\mathrm{d}S_o(\mathrm{d}l_x)_o X_1, \mathrm{d}S_o(\mathrm{d}l_x)_o Y_1\big).$$

Since by definition

$$S_o(x \cdot m) = \sigma(x) \cdot m,$$

we infer that

$$\mathrm{d}S_o(\mathrm{d}l_x)_o = (\mathrm{d}l_{\sigma(x)})_o.$$

Consequently

$$\begin{aligned} g_{S_o m}(\mathrm{d}S_o X, \mathrm{d}S_o Y) &= g_{\sigma(x)\cdot o}\big((\mathrm{d}l_{\sigma(x)})_o X_1, (\mathrm{d}l_{\sigma(x)})_o Y_1\big) \\ &= Q(X_1, Y_1) = g_m(X, Y). \end{aligned}$$

To prove that an arbitrary S_m, $m \in \mathfrak{M}$ is an isometry it suffices to observe that $S_m = l_x \cdot S_o \cdot l_{x^{-1}}$ for $m = x \cdot o$.

Given a symmetric structure $(\mathfrak{M}, S)$ and a Riemannian structure $(\mathfrak{M}, g)$ on the same space $\mathfrak{M}$, we may speak of the corresponding canonical connections, which we shall denote by ∇^s and ∇^g, respectively. The torsions of both connections are zero. As we know, the connection ∇^s is the unique connection for which symmetries are affine mappings. Next, when we deal with a Riemannian symetric space all symmetries prove to be isometries, and hence owing to Theorem 14.7.1 affine mappings.

The above discussion proves the following

THEOREM 15.5.2. *For any Riemannian symmetric space* $(\mathfrak{M}, g, S)$ *the connections* ∇^s and ∇^g *are identical.*

As a straightforward consequence we have

THEOREM 15.5.3. *Suppose that* $\mathfrak{M}$ *is a Riemannian symmetric space and* G *the connected component of the identity of the group of isometries of the manifold* $\mathfrak{M}$. *Let* $H \subset G$ *be the isotropy subgroup at a point* $m \in G$ *and* σ *an involution on* G *given by the condition*

$$\sigma(g) = S_o g S_o.$$

Then (G, σ, H) *is a Riemann pair.*

Proof. In view of Theorem 15.3.3 it remains to prove that the group $\mathrm{Ad}_G H$ is compact. This is stated by the following

LEMMA 15.5.4 (see e.g. Helgason, 1969). *The group of all isometric isotropies of a fixed point in a Riemannian manifold is compact.* □

COROLLARY 15.5.5. (1) *Any Riemannian symmetric space* $\mathfrak{M}$ *is complete.*

(2) *For an arbitrary* $m \in \mathfrak{M}$ *the mapping* Exp_m *is a surjection of* $T_m(\mathfrak{M})$ *onto* $\mathfrak{M}$.

Proof. The canonical connection on a symmetric space is complete (Corollary 15.2.8). Since it coincides with the Riemannian connection, point (1) follows by Theorem 15.5.4. We note that the exponential mapping is defined on the whole of $T_m(\mathfrak{M})$. Every point $p \in \mathfrak{M}$ can be joined to m by a geodesic of the form $\mathrm{Exp}_m tX$, $X \in T_m(\mathfrak{M})$. This ends the proof of the corollary. □

In a further part of our exposition, we shall characterize the symmetric spaces for which Exp_m is a bijection. Riemannian symmetric spaces are classified by means of properties of the corresponding symmetric pairs $(G, \mathfrak{M})$, more precisely: by means of properties of the respective pairs of Lie algebras $(\mathfrak{g}, \mathfrak{h})$. Such pairs are said to be orthogonal.

We retain the notation of Theorem 15.3.3.

DEFINITION. We say that a pair $(\mathfrak{g}, \mathfrak{h})$ is

(1) *of the compact type* if the Killing form of the Lie algebra $\mathfrak{g}$ is negative definite (thus $\mathfrak{g}$ is semi-simple);

(2) *of the Euclidean type* if the subspace $\mathfrak{p}$ in the decomposition $\mathfrak{g} = \mathfrak{h} + \mathfrak{p}$ is an abelian ideal;

(3) *of the non-compact type* if $\mathfrak{g}$ is a semi-simple algebra, the Killing form being negative definite on $\mathfrak{h}$ and positive definite on $\mathfrak{p}$.

The same terminology also applies to the corresponding Riemannian symmetric space $\mathfrak{M}$; in case (1) the space $\mathfrak{M}$ is said to be of the *compact type*, in case (2) (resp. (3)) of the *euclidean* (resp. *noncompact*) type.

An arbitrary pair of Lie algebras $(\mathfrak{g}, \mathfrak{h})$ arising from a Riemannian symmetric space can be described in terms of pairs of the above types.

THEOREM 15.5.6 (E. Cartan). *Let* $(\mathfrak{g}, \mathfrak{h})$ *be a pair of Lie algebras corresponding to a Riemannian symmetric space. Then there exist ideals* $\mathfrak{g}_+, \mathfrak{g}_0, \mathfrak{g}_-$ in $\mathfrak{g}$ *with the following properties*:

(1) $\mathfrak{g} = \mathfrak{g}_+ + \mathfrak{g}_0 + \mathfrak{g}_-$;

(2) $\mathfrak{g}_\alpha$, $\alpha = +, 0, -$ *are invariant under the automorphism* $d\sigma$ *of the algebra* $\mathfrak{g}$;

(3) $(\mathfrak{g}_+, \mathfrak{h} \cap \mathfrak{g}_+)$ *is of the non-compact type,*
$(\mathfrak{g}_0, \mathfrak{h} \cap \mathfrak{g}_0)$ *is of the Euclidean type,*
$(\mathfrak{g}_-, \mathfrak{h} \cap \mathfrak{g}_-)$ *is of the compact type.*

We do not give the proof of this theorem. It can be found in Helgason (1968), Kobayashi, Nomizu (1969). The above results lead to fundamental theorems on the structure of the manifold $\mathfrak{M}$ corresponding to a pair (G, σ, H). We present them below, again without proof.

THEOREM 15.5.7 (E. Cartan). *Let* $\mathfrak{M}$ *be a simply connected Riemannian symmetric manifold. Then there exist manifolds* $\mathfrak{M}_+, \mathfrak{M}_0, \mathfrak{M}_-$ *of the non-compact, Euclidean and compact types, respectively, such that*

$$\mathfrak{M} = \mathfrak{M}_+ \times \mathfrak{M}_0 \times \mathfrak{M}_-$$

where the equality sign denotes an isomorphism of the symmetric and the Riemannian structure.

Example 1

Let $G = \mathrm{SL}(n, \boldsymbol{R})$ and σ be defined as $\sigma(g) = (g^t)^{-1}$. The group of σ-fixed points coincides with $\mathrm{SO}(n)$, which is compact, and consequently $(\mathrm{SL}(n, \boldsymbol{R}), \sigma, \mathrm{SO}(n))$ is a Riemannian symmetric pair. The corresponding symmetric space is the set $\mathfrak{M}$ of all positive definite matrices of degree n with determinant 1. To see this it is enough to define an action of G on $\mathfrak{M}$ by the formula $A \to BAB^t$ where $A \in \mathfrak{M}$ and $B \in \mathrm{SL}(n, \boldsymbol{R})$. The isotropy subgroup at the identity matrix I is precisely $\mathrm{SO}(n)$ and the transitivity of this action follows from the well-known theorem in linear algebra concerning the diagonalization of symmetric positive definite matrices. Decomposition (15.3.3) of the Lie algebra $\mathfrak{g} = \mathfrak{sl}(n, \boldsymbol{R})$

has the form $\mathfrak{sl}(n, \boldsymbol{R}) = \mathfrak{o}(n)+\mathfrak{p}$ where $\mathfrak{p}$ is the space of symmetric matrices of trace 0. Under the identification of $\mathfrak{p}$ with the tangent space $T_I(\mathfrak{M})$, an $SO(n)$-invariant inner product on $\mathfrak{p}$ is given by the formula $\mathrm{Tr}(AB^t)$ for $A, B \in \mathfrak{p}$.

Example 2

We have already seen that $(\mathfrak{o}(n+1), \mathfrak{o}(n))$ is an orthogonal pair corresponding to the symmetric space S^n. The same pair corresponds to $P_n(\boldsymbol{R})$.

Another example of an orthogonal pair of the compact type is provided by $(\mathfrak{o}(p+q), \mathfrak{o}(p)\oplus\mathfrak{o}(q))$. An involutive automorphism of $\mathfrak{o}(p+q)$ for which $\mathfrak{o}(p)\oplus\mathfrak{o}(q)$ is the subalgebra of all fixed points is given by the rule $A \to SAS^{-1}$ where $S = \begin{bmatrix} -I_p & 0 \\ 0 & I_q \end{bmatrix}$, I_p, I_q are the unit matrices of degree p and q, respectively. The corresponding symmetric space is the Grassman manifold described in Problem 13.

Example 3

We already know an orthogonal pair of the non-compact type, namely the pair $(\mathfrak{sl}(n; \boldsymbol{R}), \mathfrak{o}(n))$. Another interesting example is the pair $(\mathfrak{o}(1, n), \mathfrak{o}(n))$, where

$$\mathfrak{o}(1, n) = \{X \in \mathfrak{gl}(n+1; \boldsymbol{R})\colon X^tS+SX = 0\} \quad \text{for } S = \begin{bmatrix} -1 & 0 \\ 0 & I_n \end{bmatrix}$$

corresponding to the hyperboloid $-x_0^2+x_1^2+ \dots +x_n^2 = -1$. This pair and the corresponding symmetric space are described in more detail in Chapter 16.

Example 4

Let $\mathfrak{g}$ be the Lie algebra of the motion group of $\boldsymbol{R}^n$ and $\mathfrak{h}$ the subalgebra corresponding to the rotation group around a fixed point. Then the pair $(\mathfrak{g}, \mathfrak{h})$ is of the Euclidean type. A complementary subspace to $\mathfrak{h}$ is $\boldsymbol{R}^n$ because the motion group is a semi-direct product of $SO(n)$ and $\boldsymbol{R}^n$ (with a natural action of $SO(n)$ on $\boldsymbol{R}^n$), and therefore $\boldsymbol{R}^n$ is indeed an (abelian) ideal in $\mathfrak{g}$.

The type of a symmetric space is connected with the sign of the sectional curvature.

THEOREM 15.5.8 *Let $\mathfrak{M}$ be a Riemannian symmetric space. If $\mathfrak{M}$ is*

(1) *of the compact type, then $K(S) \geqslant 0$,*

(2) *of the non-compact type, then $K(S) \leqslant 0$,*

(3) *of the Euclidean type, then $K = 0$.*

Proof. We shall use the fact that the space $\mathfrak{M}$ can be represented as a homogeneous space of the group $G(\mathfrak{M})$, whose elements are isometric diffeomorphisms of $\mathfrak{M}$. In what follows let (G, K) denote a Riemann pair, associated with $\mathfrak{M}$ and $(\mathfrak{g}, \mathfrak{k})$, the corresponding pair of Lie algebras with the decomposition $\mathfrak{g} = \mathfrak{k} + \mathfrak{p}$ into eigenspaces of the involution. If $\pi: G \to \mathfrak{M}$ denotes the natural projection, then $\mathrm{d}\pi: \mathfrak{p} \to T_o(\mathfrak{M})$ is a bijection. Since $\mathrm{ad}X(\mathfrak{p}) \subset \mathfrak{p}$ for $X \in \mathfrak{k}$ and $\mathrm{Ad}_G(\exp X) = e^{\mathrm{ad}X}$, we obtain $\mathrm{Ad}_G k(\mathfrak{p}) \subset \mathfrak{p}$ and $\mathrm{d}\pi \circ \mathrm{Ad}_G k = (\mathrm{d}l_k)_o \circ \mathrm{d}\pi$.

An invariant Riemannian structure on $\mathfrak{M}$ is determined by a quadratic form on $T_o(\mathfrak{M})$, invariant under all $(\mathrm{d}l_k)_o$, $k \in K$, or, equivalently, by a form on $\mathfrak{p}$ invariant with respect to the operators $\mathrm{Ad}_G(k)|\mathfrak{p}$.

$$g\big((\mathrm{d}\pi)_e(X), (\mathrm{d}\pi)_e(Y)\big) = Q_o(X, Y), \qquad X, Y \in \mathfrak{p}.$$

The curvature tensor computed in Theorem 15.2.3 can be represented in the form:

$$R\big(\mathrm{d}\pi(X), \mathrm{d}\pi(Y)\big)_o \mathrm{d}\pi(Z) = (\mathrm{d}\pi)_o\big([[X, Y], Z]\big) \tag{15.5.2}$$

for $X, Y, Z \in \mathfrak{p}$. (see Problem 10(c)). If we choose X, Y which are orthonormal in $\mathfrak{p}$ with respect to the form Q_0, then

$$K(S) = Q_o([[X, Y], X], Y). \tag{15.5.3}$$

Let b be a symmetric operator on $\mathfrak{p}$ determined by the identity $Q_o(bX, Y) = B(X, Y)$, $X, Y \in \mathfrak{p}$ (B denotes the Killing form of the algebra $\mathfrak{g}$). If the pair $(\mathfrak{g}, \mathfrak{h})$ is of the Euclidean type, then $\mathfrak{p}$ is an abelian ideal, and therefore by formula (15.5.2) we obtain $K(S) \equiv 0$. In the compact case the operator b is non-positive, and in the non-compact case it is positive definite.

Let us perform the spectral decomposition of the operator b and let $\mathfrak{p}_i$ be the eigenspaces of this operator, corresponding to the eigenvalues λ_i; we have $\lambda_i \neq \lambda_j$ for $i \neq j$ and $\lambda_i > 0$ for the non-compact type, $\lambda_i < 0$ for the compact type.

The decomposition

$$\mathfrak{p} = \bigotimes_{i=1}^{k} \mathfrak{p}_i \tag{15.5.4}$$

is orthogonal relative to both Q and B. The operators $\operatorname{ad} X$, $X \in \mathfrak{k}$ commute with b, and so the eigenspaces satisfy the conditions

$$[\mathfrak{k}, \mathfrak{p}] \subset \mathfrak{p}_i.$$

For an arbitrary $H \in \mathfrak{k}$ the following identities hold

$$B([X_i, X_j], H) = B(X_i, [X_j, H]) = 0,$$

$$X_i \in \mathfrak{p}_i, \quad X_j \in \mathfrak{p}_j, \quad i \neq j.$$

Thus the element $[X_i, X_j] \in \mathfrak{k}$ has turned out to be orthogonal to $\mathfrak{k}$ with respect to the Killing form. Since the subspaces $\mathfrak{p}$ and $\mathfrak{k}$ of $\mathfrak{g}$ correspond to different eigenvalues of the automorphism $d\sigma$ which preserves the Killing form, we must have $\mathfrak{p} \perp \mathfrak{k}$ with respect to this form. Finally, it follows that the vector $[X_i, X_j]$ is zero because it is orthogonal to $\mathfrak{g}$ with respect to the Killing form which is non-degenerate in both the compact and the non-compact case.

Now the proof runs quickly:

Let $X = \sum_{i=1}^{k} X_i$, $Y = \sum_{i=1}^{\lambda} Y_i$ be decompositions (15.5.4). Then $[X, Y] = \sum_{i=1}^{k} [X_i, X_j]$ and by the Jacobi identity

$$[[X_i, Y_i], X] = [[X_i, Y_i], X_i],$$

$$K(S) = Q_o\big([[X, Y], X], Y\big) = \sum_i Q_o\big([[X_i, Y_i], X_i], Y_i\big)$$

$$= \sum_i \frac{1}{\lambda_i} B\big([[X_i, Y_i], X_i], Y_i\big)$$

$$= \sum_i \frac{1}{\lambda_i} B\big([X_i, Y_i], [X_i, Y_i]\big).$$

The commutators $[X_i, Y_i]$ are elements of $\mathfrak{k}$, and on this space B is negative definite. Therefore for $\lambda_i > 0$ the curvature is non-positive (the non-compact case) and for $\lambda_i < 0$ the curvature is non-negative (the compact case). □

15.6. A SYMMETRIC PAIR IS A GELFAND PAIR

In this section it will become clear why spherical functions play such an important role in the analysis on a symmetric space.

To begin with we shall prove a result of an auxiliary character.

LEMMA 15.6.1. *Let (G, σ, K) be a symmetric pair with the subgroup K compact. Then*

$$K\sigma(g)K = Kg^{-1}K \tag{15.6.1}$$

for an arbitrary $g \in G$.

Proof. Denote by P the set $\exp\mathfrak{p} \subset G$ (as before $\mathfrak{g} = \mathfrak{k}+\mathfrak{p}$ is the decomposition into eigenspaces of $\mathrm{d}\sigma$ corresponding to the eigenvalues 1 and -1). The mapping $(\mathrm{d}\pi)_e : \mathfrak{p} \to T_o(\mathfrak{M})$, $o := eK$ is a bijection. Moreover, $\mathrm{Exp}_o(\mathrm{d}\pi)_e(X) = \pi(\exp\frac{1}{2}X)$, in view of (15.4.2) and after identifying X with $X^{\#}$. Exp_o being surjective (Corollary 15.5.6(2)), it follows that $X \to \pi\,(\exp X)$ is a surjection of $\mathfrak{p}$ onto $\mathfrak{M}$. Every element $g \in G$ can be represented in the form $(\exp X)k$ where $X \in \mathfrak{p}$ and $k \in K$. Then

$$\begin{aligned}\sigma(g) &= \sigma(\exp X)\sigma(k) = \big(\exp(\mathrm{d}\sigma)_e X\big)k = \big(\exp(-X)\big)k\\ &= (\exp X)^{-1}k.\end{aligned}$$

Since $g^{-1} = k^{-1}(\exp X)^{-1}$, the assertion follows.

We pass to the main theorem.

THEOREM 15.6.2 (Gelfand). *A symmetric pair is a Gelfand pair.*

Proof. Lemma 15.7.1 ensures that for $f \in {}^1\mathscr{C}^1(G) = \mathscr{C}(K\backslash G/K)$ we have

$$f^{\sigma}(g) := f(\sigma(g)) = f(g^{-1}) = \check{f}(g).$$

In particular, for $f_i \in L^1(K\backslash G/K)$

$$(f_1 * f_2)^{\vee}(g) = (f_1 * f_2)^{\sigma}(g).$$

It follows directly from the definition of convolution that

$$(f_1 * f_2)^{\vee} = \check{f}_2 * \check{f}_1. \tag{15.6.2}$$

But on the other hand,

$$(f_1 * f_2)^{\sigma}(g) = \int_G f_1(x) f_2\big(x^{-1}\sigma(g)\big)\mathrm{d}x$$

$$= \int_G f_1(\sigma(x)) f_2(\sigma(x^{-1})\sigma(g)) dx$$

$$= \int_G f_1^\sigma(x) f_2(\sigma(x^{-1}g)) dx = f_1^\sigma * f_2^\sigma(g). \qquad (15.6.3)$$

In the above computations we have used the fact that the measure dg^σ

$$\int f dg^\sigma := \int f^\sigma(g) dg,$$

being positive and left-invarint, is positively proportional to dg, and in view of the condition $\sigma^2 = \mathrm{id}$ the coefficient of proportionality must equal 1.

By comparing (15.6.2) and (15.6.3) we conclude that the algebra $L^1(K\backslash G/K)$ is commutative.

PROBLEMS

1. Let $(\cdot|\cdot)$ be the standard inner product on $\boldsymbol{R}^n$.

(a) Prove that the mapping S_x defined by the formula

$$S_x y - 2\frac{(x|y)}{(x|x)}x - y$$

is an orthogonal reflection with respect to the line $\boldsymbol{R}x$. Prove also that the group generated by $S_x S_y$, where $x, y \in S^{n-1}$, is the group $\mathrm{SO}(n)$.

(b) Show that the canonical connection on the symmetric space (S^{n-1}, S) is the Riemannian connection defined by imbedding S^{n-1} into $\boldsymbol{R}^n$ (cf. Problem 10).

2. Let $\langle \cdot , \cdot \rangle$ be a bilinear symmetric non-degenerate form on a vector space V. Let $\mathfrak{M}$ be the complement of the null cone, i.e. $\mathfrak{M} = \{v \in V: \langle v, v\rangle \neq 0\}$. Show that the formula

$$S_x y = 2\frac{\langle x, y\rangle}{\langle y, y\rangle}x - \frac{\langle x, x\rangle}{\langle y, y\rangle}y, \qquad x, y \in \mathfrak{M},$$

defines a symmetric structure on the space $\mathfrak{M}$. Prove also that the inside $\mathfrak{M}_+ = \{v \in \mathfrak{M}: \langle v, v\rangle > 0\}$ and the outside $\mathfrak{M}_- = \{v \in \mathfrak{M}: \langle v, v\rangle < 0\}$ of the null cone are symmetric spaces under the same symmetric structure. Check that in the case of the hyperboloid $\mathfrak{M}_k = \{v \in V: \langle v, v\rangle = k \neq 0\}$ the formula $S_x y = 2k^{-1}\langle x, y\rangle x - y$ defines a symmetric structure on $\mathfrak{M}_k$.

Hint. To prove the algebraic properties of the symmetry S_x, apply the identity

$$\langle S_x y, S_x z\rangle = \frac{\langle x, x\rangle^2 \langle y, z\rangle}{\langle y, y\rangle \langle z, z\rangle}.$$

3. Show that every homomorphism φ of the symmetric space $\boldsymbol{R}^n$ into itself (the symmetries S_x are defined as $S_x y = 2x - y$) is an affine mapping of the vector space $\boldsymbol{R}^n$ into itself, i.e. has the form $\varphi(x) = Ax + b$ where $A \in M_{n\times n}(\boldsymbol{R})$, $b \in \boldsymbol{R}^n$. Also, prove that every linearly parametrized line in $\boldsymbol{R}^n$, i.e. every mapping $t \to b + at$, b, $a \in \boldsymbol{R}^n$, is a homomorphism of the symmetric space $\boldsymbol{R}$ into the symmetric space $\boldsymbol{R}^n$.

Hint. Prove first that the mapping $x \to \varphi(x) - \varphi(0)$ is homogeneous for dyadic numbers (i.e. numbers of the form $\frac{p}{2^n}$, p being an integer), and then make use of the continuity of the mapping.

4. Let G be a Lie group equipped with the symmetric structure described in Example 1 $(S_g(x) = gx^{-1}g)$. Prove that for any left-invariant vector fields X, Y on G and the canonical connection ∇ we have $\nabla_X Y = \frac{1}{2}[X, Y]$. Prove that the geodesics for this connection coincide, as in the case of a left-invariant connection (cf. Problem 10 in Chapter 13), with the translates of the one-parameter subgroups of the group G.

5. Let G be a connected Lie group with an involutive automorphism σ. Put $G_\sigma = \{g \in G\colon \sigma(g) = g\}$, $G^\sigma = \{g\sigma(g)^{-1}\colon g \in G\}$.

(a) Prove that G^σ is a symmetric subspace of the symmetric space (G, S) where $S_g(x) = gx^{-1}g$.

(By a symmetric subspace of a symmetric space $(\mathfrak{M}, S)$ we understand a pair $(\mathfrak{M}', S|\mathfrak{M}')$ where $\mathfrak{M}' \subset \mathfrak{M}$ is a subspace of $\mathfrak{M}$ such that $S|\mathfrak{M}'$ satisfies conditions (15.1.1)–(15.1.4).)

(b) Let K be a closed subgroup of G such that $(G_\sigma)_0 \subset K \subset G_\sigma$. Prove that the mapping $q\colon G/K \to G$ where $q(gk) = g\sigma(g)^{-1}$ is a homomorphism of symmetric spaces and its image is G^σ.

(c) Show that q is a covering with fibre G_σ/K and that q induces an isomorphism of G/G_σ onto G^σ.

6. Let $\mathfrak{g}$ be a Lie algebra over $\boldsymbol{R}$ and $\mathfrak{h}$, $\mathfrak{p}$ subspaces of $\mathfrak{g}$ such that $\mathfrak{g}$ is the direct sum of $\mathfrak{h}$ and $\mathfrak{p}$ (as vector spaces). Prove that there exists

an involutive automorphism σ of the algebra $\mathfrak{g}$ with the fixed point set equal to $\mathfrak{h}$ provided the following relations hold:

$$[\mathfrak{h}, \mathfrak{h}] \subset \mathfrak{h} \qquad [\mathfrak{h}, \mathfrak{p}] \subset \mathfrak{p}, \qquad [\mathfrak{p}, \mathfrak{p}] \subset \mathfrak{h}.$$

7. Let o be a fixed point in a symmetric space. Consider the operation of raising to a power with respect to o, defined in § 1. Show that if $x = \mathrm{Exp}_0 Y$ then $x^n = \mathrm{Exp}_0 nX$.

8. Let $(\mathfrak{M}, S)$ be a symmetric space. By an automorphism of $(\mathfrak{M}, S)$ we shall mean an arbitrary diffeomorphism of $\mathfrak{M}$ which is a homomorphism of the symmetric space $(\mathfrak{M}, S)$. By a differentiation of $(\mathfrak{M}, S)$ we mean every smooth vector field X on $\mathfrak{M}$ such that $X(p \cdot q) = X(p) \cdot q + p \cdot X(q)$.

(a) Prove that a smooth vector field X on $\mathfrak{M}$ is a differentiation if and only if the fields X and $1 \otimes X + X \otimes 1$ are μ-related (Here $\mu: \mathfrak{M} \times \mathfrak{M} \to \mathfrak{M}$, $\mu(p, q) = p \cdot q = S_p q$).

(b) Show that the differentiations of $\mathfrak{M}$ form a Lie subalgebra of the algebra $\mathcal{T}(\mathfrak{M})$.

(c) Show that a one-parameter group of diffeomorphisms of $\mathfrak{M}$ is a (one-parameter) group of automorphisms of $(\mathfrak{M}, S)$ if and only if the induced vector field is a differentiation.

(d) Prove that every differentiation is an affine vector field relative to the canonical connection on $\mathfrak{M}$. Prove that the converse statement is also valid under the additional assumption that $\mathfrak{M}$ is connected.

Hint. (c) For a one-parameter group of diffeomorphisms φ_t generating a differentiation X, consider the curves $t \to \varphi_t(p \cdot q)$ and $t \to \varphi_t(p) \cdot \varphi_t(q)$. Show that they are both integral curves of the field X. (d) Apply Lemma 13.6.3, the identity $(X \cdot Z)X = (YX) \cdot Z + Y \cdot (ZX)$ (prove it by methods similar to those applied in Lemma 15.2.4) and Lemma 15.2.4. To prove the converse implication we use Theorem 15.2.7 and Corollary 15.2.8.

9. We retain the terminology and notation introduced in the preceding problem. Let o be an arbitrary point in $\mathfrak{M}$.

(a) Show that the mapping $X \to dS_o(X \circ S_o)$ is an involutive automorphism of the Lie algebra of the differentiations of $\mathfrak{M}$.

(b) Let D_+, D_- be the eigenspaces corresponding to the eigenvalues $+1$, -1 of this automorphism. Prove that for an arbitrary $X \in T_o(\mathfrak{M})$

the vector field $X^{\#}$ defined as $X^{\#}(p) = X \cdot (o \cdot p)$ is an element of D_- and the rule $X \to X^{\#}$ is an isomorphism of $T_o(\mathfrak{M})$ onto D_- whose inverse assigns to a vector field half the value of the field at o.

(c) Show that D_+ is precisely the set of all differentiations vanishing at the point o.

Hint. (a) Observe that $\mathrm{d}S_o(p \cdot X_q + X_p \cdot q) = S_o(p) \cdot (\mathrm{d}S_o X_q) + (\mathrm{d}S_o X_p) \cdot S_o(q)$. (b) Verify this by direct computation based on an identity from Lemma 15.2.1 or apply the preceding problem and Proposition 15.3.1. (c) Make use of (b).

10. Let $(\mathfrak{M}, S)$ be a connected symmetric space and $o \in \mathfrak{M}$ a fixed point. Further, let G be a Lie subgroup of the group $\mathrm{Aff}\,\mathfrak{M}$ satisfying the assumptions of Theorem 15.3.3 (we retain the meaning of the symbols H, σ, $\mathfrak{g}$, $\mathfrak{h}$, $\mathfrak{p}$, etc. occurring in the theorem).

(a) Show that, after a natural identification of $\mathfrak{g}$ with a Lie subalgebra of the Lie algebra of all affine vector fields on $\mathfrak{M}$, the automorphism $\mathrm{d}\sigma$ of the Lie algebra $\mathfrak{g}$ has the form $\mathrm{d}\sigma(X) = \mathrm{d}S_o(X \circ S_o)$. Prove also that the (-1)-eigenspace of the automorphism $\mathrm{d}\sigma$ in $\mathfrak{g}$ coincides through this identification with the space consisting of all infinitesimal generators of the subgroups of the form $t \to Q(\mathrm{Exp}_o tX)$ for $X \in T_o(\mathfrak{M})$.

(b) Let $X \in \mathfrak{p}$. According to (a) we regard X as a vector field on $\mathfrak{M}$. Prove that $\mathrm{d}\pi(X) = X(o)$, where $\pi\colon G \to \mathfrak{M}$ is the natural mapping $g \to g \cdot o$. Further, prove that if $X, Y, Z \in p$ then also $[[X, Y], Z] \in \mathfrak{p}$ and $[[X, Y], Z](o) = -\frac{1}{4}\big(X(o) \cdot (Y(o) \cdot Z(o)) - Y(o) \cdot (X(o) \cdot Z(o))\big)$.

(c) Show that the curvature tensor R of the space $\mathfrak{M}$ satisfies (15.6.2), i.e.

$$R_o\big(\mathrm{d}\pi(X), \mathrm{d}\pi(Y)\big)\,\mathrm{d}\pi(Z) = -\mathrm{d}\pi\big([[X, Y], Z]\big).$$

Hint to (b). Deduce from (a) that $X = Y^{\#}$ for a certain $Y \in T_o(\mathfrak{M})$ and find $\pi(\exp tX)$ on the basis of Proposition 15.3.1. In order to compute the values of the commutator use the fact that $[X, Y](o) = 0$ for $X, Y \in \mathfrak{p}$ (cf. Problem 8) and the identity $(S^{\#} T^{\#})(o) = S^{\#}(o)T = -T^{\#}(o) \cdot S^{\#}(o)$ (here $S^{\#} T^{\#}$ denotes the composite of two differential operators of the first order).

11. Let $\mathfrak{M}$ be a symmetric space and let $p, q \in \mathfrak{M}$. For $X \in T_p(\mathfrak{M})$ and $Y \in T_q(\mathfrak{M})$ we define $X * Y \in T_{S_p q}(\mathfrak{M})$ by the formula $X * Y$

$= X \cdot q + p \cdot Y$. Show that $T(\mathfrak{M})$ is a symmetric space under the operation so defined.

12. Denote by $G_{n,p}(\boldsymbol{R})$ the set of all p-dimensional subspaces of $\boldsymbol{R}^n$. Further, let $\{e_1, \ldots, e_n\}$ denote the canonical basis in $\boldsymbol{R}^n$ and $\{e_1^*, \ldots, e_n^*\}$ its dual basis. Write $q = n - p$.

(a) Let $\alpha = \{\alpha_1, \ldots, \alpha_p\}$ where $1 \leqslant \alpha_1 < \ldots < \alpha_p \leqslant n$ and let U be the set of those subspaces S in $G_{n,p}(\boldsymbol{R})$ such that the forms $e^*_{\alpha_1|S}, \ldots, e^*_{\alpha_p|S}$ are linearly independent. We define $\varphi_\alpha: U_\alpha \to M_{p\times q}(\boldsymbol{R})$ as follows: Let $\{\alpha_{p+1}, \ldots, \alpha_{p+q}\}$ be the sequence of indices complementary to α taken in the order of growth. Then $\varphi_\alpha(S)$ is the matrix $[s_j^k]$ given by $e^*_{\alpha_{p+k}|S} = s_j^k e^*_{\alpha_j|S}$.

Show that $(U_\alpha, \varphi_\alpha)$ define a manifold structure on the space $G_{n,p}(\boldsymbol{R})$ called the *Grassman manifold.*

(b) Prove that $G_{n,p}(\boldsymbol{R}) \simeq O(n)/O(p) \times O(q)$.

(c) Show that $\sigma: A \to SAS^{-1}$ where $S = \begin{bmatrix} -I_p & 0 \\ 0 & I_q \end{bmatrix}$ is an involution in $O(n)$ and the decomposition of $\mathfrak{o}(n)$ into the eigenspaces of $d\sigma$ has the form $\mathfrak{o}(n) = \mathfrak{k} + \mathfrak{p}$ with $\mathfrak{k} = \mathfrak{o}(p) + \mathfrak{o}(q)$, $\mathfrak{p} = \left\{ \begin{bmatrix} 0 & -X^t \\ X & 0 \end{bmatrix} : X \in M_{p\times q}(\boldsymbol{R}) \right\}$. Find the form of the representation of $\operatorname{Ad} K$ on $\mathfrak{p}$ and prove that the form $\frac{1}{2} \operatorname{Tr} AB$ is positive definite on $\mathfrak{p}$.

(d) Give a geometric interpretation of the symmetry S_x of the symmetric space $G_{n,p}(\boldsymbol{R})$.

Chapter 16

General Harmonic Analysis on a Symmetric Space

In the present chapter we shall give the final explanation of the role of zonal spherical functions in harmonic analysis on a symmetric space and, even more generally, on a homogeneous space G/K where (G, K) is a Gelfand pair of Lie groups.

Zonal spherical functions were introduced in Chapter 5 as continuous solutions of the integral equation

$$\int_K \omega(gkh)\mathrm{d}k = \omega(g)\,\omega(h). \tag{16.1.1}$$

Every such solution ω defines a multiplicative functional on the convolution algebra $\mathscr{C}_0(K\backslash G/K)$ of all functions on G which are bi-invariant under K:

$$f \to \omega(f) := \int_K f(g)\,\omega(g)\mathrm{d}g.$$

Formula (16.1.1) expresses the multiplicativity of this functional. To positive definite functions ω there correspond unitary positive definite functionals:

$$\omega(f^* * f) \geqslant 0, \qquad \omega(f^*) = \overline{\omega(f)}.$$

The proper domain of the theory of spherical function are Gelfand pairs, for which the algebra $L^1(K\backslash G/K)$ is commutative.

A fundamental theorem proved in Chapter 5 states that positive definite zonal spherical functions correspond via the Gelfand–Raikov construction to the irreducible representations of the group G for which the space of K-fixed vectors is precisely 1-dimensional. The Frobenius theorem concerning compact group representations suggests that these are the representations that should play a key role in harmonic analysis on the homogeneous space G/K.

In this chapter we shall denote by U the left regular representation of the group G on the space $L^2(G/K)$. We shall decompose this representation into irreducible components by applying the von Neumann spectral theorem and then the nuclear spectral theorem due to K. Maurin. The latter theorem will allow us to employ the methods of distribution theory, which will help us to identify irreducible components of the decomposition with spherical representations.

We begin by quoting the fundamental theorem of von Neumann, proved in 1938 but published as late as 1949 (cf. Dixmier, 1964).

THEOREM 16.1.1. *Let $\mathscr{A}$ be a commutative $*$-algebra of operators on a separable Hilbert space H. Suppose $\mathscr{A}$ is closed in the operator norm and let the identity operator belong to the closure of $\mathscr{A}$ in the topology of pointwise convergence of operators. Then there exists a direct integral*

$$\mathscr{H} = \int_{\Lambda} H_\lambda \,\mathrm{d}\mu(\lambda)$$

such that Λ is a locally compact space with a countable base of neighbourhoods and μ is a finite measure on Λ and there exists a unitary operator $F: H \to \mathscr{H}$ such that

(1) *For any $A \in \mathscr{A}$*

$$F(Ax)(\lambda) = a_A(\lambda) Fx(\lambda), \tag{16.1.3}$$

where a_A is a continuous function vanishing at infinity.

(2) *The mapping $\mathscr{A} \ni A \to a_A \in \mathscr{C}(\Lambda)$ is an isomorphism of the $*$-algebra $\mathscr{A}$ onto the space $\mathscr{C}(\Lambda)$ with the norm $\|f\| := \sup_\lambda |f(\lambda)|$.*

(3) *For every operator T which commutes with $\mathscr{A}$ there exists a decomposable operator on $\mathscr{H}$ such that $T = \int T_\lambda \mathrm{d}\mu(\lambda)$ (i.e. $F(Tx)(\lambda) = T_\lambda F_x(\lambda)$).*

(4) *If U is a unitary representation on H of a locally compact group G with a countable base of neighbourhoods and if the representation operators commute with $\mathscr{A}$, then for μ almost all λ the mapping $G \ni g \to U_\lambda(g)$ is a unitary representation of the group.*

Let us now assume that (G, K) is a Gelfand pair of Lie groups. The convolution algebra $\mathscr{C}_0(K\backslash G/K)$ is commutative and we define its representation on the space $L^2(G/K)$ by setting

$$\varrho(\varphi)f := f * \varphi .$$

Owing to the identity

$$||f_1 * f_2||_{L^2} \leqslant ||f_1||_{L^2}\, ||f_2||_{L^1}$$

(cf. § 1.6) the representation ϱ is continuous as a mapping of $\mathscr{C}_0(K\backslash G/K)$ into $L(L^2(G/K))$.

The representation U, whose action on functions consists in left-translating the functions, commutes with the operators $\varrho(\varphi)$ for arbitrary $\varphi \in \mathscr{C}_0(K\backslash G/K)$.

LEMMA 16.1.2. *There exists a sequence φ_n of elements of $\mathscr{C}_0(K\backslash G/K)$ such that the sequence $\varrho(\varphi_n)$ is pointwise convergent to the identity in $L^2(G/K)$.*

Proof. Let O_n be a basis of compact neighbourhoods of the neutral element of the group, such that $O_n \searrow e$. Select a sequence ψ_n of continuous functions on G satisfying the conditions:

(1) $\psi_n \geqslant 0$,

(2) $\operatorname{supp}\psi_n \subset O_n$,

(3) $\int \psi_n(g)\,\mathrm{d}g = 1$.

We shall first verify that for every $f \in L^2(G)$

$$f * \psi_n \to f$$

in the space $L^2(G)$. To this end we compute

$$f * \psi_n(x) = \int_G f(g)\,\psi_n(g^{-1}x)\,\mathrm{d}g = \int_G f(xg^{-1})\,\psi_n(g)\,\mathrm{d}g.$$

The right regular representation of G on $L^2(G)$ being continuous, this implies that the function

$$G \ni g \to R_g^{-1} f \in L^2(G)$$

is also continuous. Further, we see that

$$||f * \psi_n - f||_{L^2}$$
$$\leqslant \int_G ||R_g^{-1} f - f||_{L^2}\,\psi_n(g)\,\mathrm{d}g \leqslant \sup_{O_n} ||R_g^{-1} f - f_{L^2}|| \xrightarrow[n\to\infty]{} 0.$$

Denote by $f \to f^1$ $(f \to {}^1f)$ the operation of right (left) averaging of a function with respect to the subgroup K:

$$f^1(g) = \int_K f(gk)\,\mathrm{d}k.$$

If $f \in L^2(G/K)$ then $f * \psi_n = f * {}^1\psi_n$. On the other hand, if $f * \psi_n = f * {}^1\psi_n \to f$ then $f * {}^1\psi_n^1 \to f^1 = f$. Finally, indicating $\varphi_n = {}^1\psi_n^1$ we obtain the assertion. □

We introduce an algebra $\mathscr{U}$ of operators on $L^2(G/K)$ by closing the set $\{\varrho(\varphi): \varphi \in C_0(K\backslash G/K)\}$ in the operator norm.

Lemma 16.1.2 ensures that all the assumptions of Theorem 16.1.1 are satisfied. The operator F will denote an isometry of $L^2(G/K)$ onto the space $\mathscr{H} = \int_\Lambda H_\lambda \, d\mu(\lambda)$ diagonalizing the algebra $\mathscr{U}$.

On account of assertion (4) of Theorem 16.1.1 we obtain a decomposition of the representation U

$$U(g) = \int_\Lambda U_\lambda(g) \, d\mu(g), \tag{16.1.1}$$

in which μ-almost all representations (H_λ, U_λ) are unitary. Moreover, for every λ we have the mapping

$$\eta_\lambda: \mathscr{C}_0(K\backslash G/K) \ni \varphi \to \varrho(\varphi) \to a_{\varrho(\varphi)}(\lambda) \in C$$

which is continuous and multiplicative. According to Theorem 5.1.7 for every $\lambda \in \Lambda$ there exists a zonal spherical function ω_λ such that

$$\eta_\lambda(\varphi) = \omega_\lambda(\varphi) = \int_G \omega_\lambda(g) f(g) \, dg. \tag{16.1.2}$$

Thus the space Λ imbeds in the set of spherical functions. One may ask whether it is (H_λ, U_λ) that constitues the spherical representation corresponding to the spherical function ω_λ. Indeed, this is the case for μ-almost all λ. The proof however will be given only for a pair (G, K) corresponding to a Lie group, in which case by applying the nuclear spectral theorem we avoid technical complications of the measure-theoretic type.

Actually we present here only a consequence of the nuclear spectral theorem in a very special case. The general theorem due to K. Maurin, together with applications, can be found in Maurin, 1968.

THEOREM 16.1.3. *Let G be a Lie group and $F: L^2(G/K) \to \int_\Lambda H_\lambda \, d\mu(\lambda)$ a decomposition into a direct integral. Then for almost all $\lambda \in \Lambda$ the mapping*

$$F_\lambda: \mathscr{D}(G/K) \ni f \to Ff(\lambda) \in H_\lambda \quad \textit{is continuous.} \tag{16.1.3}$$

We shall apply this theorem to the previously obtained decomposition of $L^2(G/K)$ into a direct integral diagonalizing the algebra $\mathcal{U}$. By neglecting a set of measure zero in Λ we restrict ourselves to those λ for which the mapping F_λ is continuous.

We define a family of $1\frac{1}{2}$-linear forms on $\mathcal{D}(G)\times\mathcal{D}(G)$ by the formula

$$\beta_\lambda(f_1, f_2) := (Ff_1^1(\lambda)|Ff_2^1(\lambda))_\lambda = (F_\lambda f_1^1|F_\lambda f_2^1), \tag{16.1.4}$$

where according to the notation of Chapter 5 we write

$$f^1 = \int_K R_k f \mathrm{d}k; \quad {}^1f = \int_K L_k f \mathrm{d}k.$$

For every λ the form β_λ is invariant under the action of the left regular representation (Theorem 16.1.1(4)) and partially continuous. Forms of this kind were studied by F. Bruhat.

THEOREM 16.1.4 (Bruhat, 1956). *A partially continuous invariant $1\frac{1}{2}$-linear form on $\mathcal{D}(G)\times\mathcal{D}(G)$ defines a distribution T such that*

$$\beta(f_1, f_2) = T(f_1^* * f_2). \tag{16.1.5}$$

This theorem is a corollary to the famous Schwartz kernel theorem.

In our case the forms β_λ are, in addition, right K-invariant which implies that the distributions defined by (16.1.5) are bi-invariant under K. Denoting by T_λ the distribution related to β_λ we can express the properties of the decomposition of $L^2(G/K)$ into a direct integral by the formulas

$$(f_1|f_2)_{L^2(G/K)} = \int_\Lambda T_\lambda(f_1^* * f_2)\mathrm{d}\mu(\lambda) \tag{16.1.6}$$

and

$$(f_1|f_2 * \varphi)_{L^2(G/K)} = \int_\Lambda \omega_\lambda(\varphi)\, T_\lambda(f_1^* * f_2)\mathrm{d}\mu(\lambda)$$

for $f_i \in \mathcal{D}(G/K)$ and $\varphi \in \mathcal{D}(K\backslash G/K)$. Assuming that also $f_2 \in \mathcal{D}(K\backslash G/K)$, we derive from the commutativity of the algebra $\mathcal{D}(K\backslash G/K)$ that

$$0 = \int_\Lambda \big(\omega_\lambda(f_2)\, T_\lambda(f_1^* * \varphi) - \omega_\lambda(\varphi)\, T_\lambda(f_1^* * f_2)\big)\mathrm{d}\mu(\lambda)$$

identically with respect to all variables.

Substituting $\varphi = \varphi_1 * \varphi_2$, $\varphi_i \in \mathscr{D}(K\backslash G/K)$, we obtain

$$0 = \int_{\Lambda} \omega_\lambda(\varphi_1)\big(T_\lambda(f_1^* * \varphi_2)\,\omega_\lambda(f_2) - \omega_\lambda(\varphi_2)\,T_\lambda(f_1^* * f_2)\big)\,\mathrm{d}\mu(\lambda).$$

When φ_1 runs through the set $\mathscr{D}(K\backslash G/K)$, the operators $\varrho(\varphi_1)$ run through a dense set in $\mathscr{U}$, which means that the functions $\Lambda \ni \lambda \to \omega_\lambda(\varphi_1)$ constitute a dense subset of $C_\infty(\Lambda)$. For every $f_1 \in \mathscr{D}(G/K)$; $f_2, \varphi \in \mathscr{D}(K\backslash G/K)$, upon removing a set of measure zero we obtain

$$T_\lambda(f_1^* * \varphi)\,\omega_\lambda(f_2) - \omega_\lambda(\varphi)\,T_\lambda(f_1^* * f_2) = 0. \tag{16.1.7}$$

Select a countable set $\mathscr{F} = \{f_i\}_1^\infty$ of elements of $\mathscr{D}(G/K)$ which is total in $\mathscr{D}(G/K)$ (i.e. the linear span of $\mathscr{F}$ is dense in $\mathscr{D}(G/K)$) and hence in $L^2(G/K)$ (G being separable!). Then the set ${}^1\mathscr{F} = \{{}^1f_i\}_1^\infty$ is total in $L^2(K\backslash G/K)$ and in $\mathscr{D}(K\backslash G/K)$. We may assume that the family $\{{}^1f_i\}_1^\infty$ contains an approximated identity φ_n whose existence is established by Lemma 16.1.2. Again by neglecting a set of measure zero we see that (16.1.7) is valid for $f_1 \in \mathscr{F}$, f_2, $\varphi \in {}^1\mathscr{F}$. This identity can be written in the form

$$\int_G \bar{f}_1(g)\big(T_\lambda(L_g^{-1}\varphi)\,\omega_\lambda(f_2) - \omega_\lambda(\varphi)\,T_\lambda(L_g^{-1}\varphi)\big)\,\mathrm{d}g = 0.$$

The function f_1 ranging over a total set in $L^2(G\backslash K)$ and the function $g \to T_\lambda(L_g^{-1}\varphi)$ being continuous, it follows that

$$T_\lambda(L_g^{-1}\varphi)\,\omega_\lambda(f_2) = T_\lambda(L_g^{-1}f_2)\,\omega_\lambda(\varphi)$$

for every $g \in G$; $\varphi, f \in {}^1\mathscr{F}$. Thus on the set of those λ for which $\omega_\lambda(f_2)$ and $\omega_\lambda(\varphi)$ are non-zero we have

$$\frac{T_\lambda(\varphi)}{\omega_\lambda(\varphi)} = \frac{T_\lambda(f_2)}{\omega_\lambda(f_2)}, \qquad \varphi, f_2 \in {}^1\mathscr{F}. \tag{16.1.8}$$

By putting

$$\alpha(\lambda) = \inf_{\varphi \in {}^1\mathscr{F}} \frac{T_\lambda(\varphi)}{\omega_\lambda(\varphi)} \tag{16.1.9}$$

we obtain a measurable function, $\mathrm{d}\mu$-almost everywhere non-negative and finite, since, for $\varphi_n \to \delta_e$, $\omega_\lambda(\varphi_n)$ is pointwise convergent to 1 on Λ. It follows from (16.1.8) that

$$T_\lambda(f_1) = \alpha(\lambda)\,\omega_\lambda(f_1) \tag{16.1.10}$$

for $f_1 \in {}^1\mathscr{F}$ and almost all $\lambda \in \Lambda$.

The distributions T_λ and $\alpha(\lambda)\omega_\lambda$, being identical on a total set in $\mathscr{D}(K\backslash G/K)$, must be equal. By (16.1.6) we now get

$$(f_1|f_2)_{L^2(G/K)} = \int_\Lambda \alpha(\lambda)\,\omega_\lambda(f_1^* * f_2)\,\mathrm{d}\mu(\lambda)$$

$$=: \int_\Lambda \omega_\lambda(f_1^* * f_2)\,\mathrm{d}\nu(\lambda), \qquad f_i \in \mathscr{D}(G/K). \quad (16.1.11)$$

Thus we obtain a densely defined isometry of $L^2(G/K)$ into the space $\int_\Lambda H_\lambda \mathrm{d}\nu(\lambda)$ where $H_\lambda := {}^{\omega_\lambda}H$ is the representation space of the unitary irreducible representation corresponding under the Gelfand–Raikov construction to the spherical function ω_λ.

The mapping $\mathscr{D}(G/K) \ni f \to [f]_{\omega_\lambda} \in H_\lambda$ will also be denoted by F_λ and we shall write Ff for the element of the direct integral of the form $\lambda \to F_\lambda f$. The mapping F uniquely extends from $\mathscr{D}(G/K)$ to a linear isometry of $L^2(G/K)$ onto $\int_\Lambda H_\lambda \mathrm{d}\nu(\lambda)$, retaining the ability of intertwining the representation

$$L_g = \int_\Lambda {}^{\omega_\lambda}U\,\mathrm{d}\nu(\lambda).$$

The decomposition contains only spherical and therefore irreducible components.

Moreover, we observe that for $\lambda_1 \neq \lambda_2$ the functionals $\mathscr{D}(K\backslash G/K) \ni f \to a_{\varrho(f)}(\lambda_i)$ are different (since the functions $a_{\varrho(f)}, f \in \mathscr{D}(K\backslash G/K)$ separate points in Λ!) and hence $\omega_{\lambda_1} \neq \omega_{\lambda_2}$ which implies that representations do not recur in the decomposition.

We have thus proved

THEOREM 16.1.5. *Suppose (G, K) is a Gelfand pair of Lie groups and let G be separable. Then the representation $\big(L^2(G/K), L\big)$ decomposes into a direct integral of non-equivalent irreducible spherical representations (H_λ, U_λ).*

$$L^2(G/K) = \int_\Lambda H_\lambda \,\mathrm{d}\nu(\lambda), \quad (16.1.12)$$

$$L = \int_\Lambda U_\lambda \,\mathrm{d}\nu(\lambda). \quad (16.1.13)$$

The Fourier transformation $f \to Ff$ is an isometry intertwining $L^2(G/K)$ with the space $L^2(\mu, H)$ (the Plancherel formula).

The measure $d\nu$ is called the *Plancherel measure* for the representation L.

We shall now interpret formula (16.1.12) as an expansion of functions on the homogeneous space G/H into spherical functions.

PROPOSITION 16.1.6. *For every function $f \in \mathscr{D}(G/K)$ the following decomposition holds*

$$f(gK) = \int_{\Lambda} f * \omega_\lambda(g) d\nu(\lambda). \tag{16.1.14}$$

The integral is absolutely convergent.

Proof. Let $f = f_1 * f_2$. We define

$$f_g(y) := \int_K f(gky) dk = \int_G \int_K f_1(x) f_2(x^{-1}gky) dk dx$$

$$= \int_G \int_K f_1(gkx) dk f_2(x^{-1}y) dx = {}^1(L_{g^{-1}} f)^* * f_2(y).$$

The function f_g is bi-invariant under K and at the same time is a convolution of two functions from $\mathscr{C}_0(K \backslash G)$.

Formula (16.1.11) can be rewritten in the form

$$\delta_e(f_1^* * f_2) = \int_{\Lambda} \omega_\lambda(f_1^* * f_2) d\nu(\lambda), \qquad f_1, f_2 \in \mathscr{C}_0(G/K),$$

where δ_e denotes the Dirac distribution. For every $\psi = f_1^* * f_2$

$$\delta_e(\psi) = \int_{\Lambda} \omega_\lambda(\psi) d\nu(\lambda).$$

Replacing ψ by the function f_g, which is of the desired form, we obtain

$$f_g(e) = \int_{\Lambda} \omega_\lambda(f_g) d\nu(\lambda).$$

But $f_g(e) = f(g)$, and owing to the Hermitian symmetry of the positive definite function ω_λ we have

$$\omega_\lambda(f_g) = \int_G \int_K \omega_\lambda(y) f(gky) dk dy = \int_G \omega_\lambda(y) f(gy) dy = f * \omega_\lambda(g).$$

Thus formula (16.1.14) is valid for $f = f_1^* * f_2$, $f_i \in \mathscr{C}_0(G/K)$. For a general f we apply

LEMMA 16.1.7 (Cartier, 1974). *Every function $f \in \mathscr{D}(G)$ is a linear combination of functions of the form $f_1 * f_2$, $f_1 \in \mathscr{D}(G)$, $f_2 \in C_0(G)$.*

Proof of the lemma. Let C be a compact neighbourhood of the neutral element in G. Every distribution with support contained in C is of the form $\sum_{n_i} \delta_e^{(n_i)} * \varphi_i$ where φ_i are continuous functions with support in C and $\delta_e^{(n_i)}$ denotes the derivative of order n of the Dirac distribution δ_e. Representing δ_e in the above form, we get

$$f = f * \delta_e = \sum_i f * \delta_e^{(n_i)} * \varphi_i .$$

Putting $f_i := f * \delta_e^{(n_i)}$ we obtain the assertion.

The proposition now follows immediately in view of Lemma 16.1.7. □

PROPOSITION 16.1.8. *For every $f \in \mathscr{D}(K\backslash G/K)$ we have the decomposition*

$$f(gK) = \int_\Lambda \hat{f}(\lambda)\, \omega_\lambda(g) \mathrm{d}\lambda , \tag{16.1.15}$$

where $\hat{f}(\lambda) := \int_G f(g) \bar{\omega}_\lambda(g) \mathrm{d}g$ is the spherical Fourier transform of the function f.

Proof. Formula (16.1.15) results from (16.1.14) by applying Theorem 5.1.7(3). □

In comparison with the general theorem on the decomposition of unitary representations into irreducible components, we have made considerable progress in the study of the representation $(L^2(G/K), U)$. Namely, we have managed to imbed the space Λ, the spectrum of the decomposition, into the space of p.d. spherical functions and to identify the components of the decomposition as spherical representations which occur only once in the decomposition.

We now face the problem of describing the space of spherical functions and of finding the Plancherel measure $\mathrm{d}\nu$.

Both problems are far from simple, but in the case of Gelfand pairs corresponding to symmetric spaces of Lie groups they have been thoroughly examined, mainly thanks to the investigations carried out by Gelfand and his collaborators, Harish-Chandra, Helgason and others.

Before passing to a more detailed harmonic analysis on symmetric spaces, however we must familiarize ourselves with their geometrical structure.

The following chapter is devoted to this task. It mainly comprises the results of E. Cartan.

Chapter 17

Semisimple Algebras. Semisimple Groups. Symmetric Spaces of the Non-Compact Type

17.1. COMPACT LIE ALGEBRAS

In Chapter 1 we introduced fundamental types of Lie algebras, namely the abelian, nilpotent, solvable, semisimple and simple algebras.

The present chapter serves the purpose of acquainting the reader with the structure of the semisimple groups and algebras and the corresponding symmetric spaces. We begin by recalling some concepts and notation introduced in Chapter 1.

A linear automorphism $A\colon \mathfrak{g} \to \mathfrak{g}$ is said to be an *automorphism of the algebra* $\mathfrak{g}$ provided the following condition holds:

$$A[X,Y] = [AX, AY], \qquad X, Y \in \mathfrak{g}. \tag{17.1.1}$$

The group of all automorphisms of an algebra $\mathfrak{g}$ is denoted by $\operatorname{Aut}\mathfrak{g}$. Every one-parameter subgroup $a(\cdot)$ of the group $\operatorname{Aut}\mathfrak{g}$ defines a differential operator $D := a'(0)$ which verifies the identity

$$D[X,Y] = [DX, Y] + [X, DY], \qquad X, Y \in \mathfrak{g}. \tag{17.1.2}$$

A linear automorphism of $\mathfrak{g}$ which satisfies (17.1.2) is called a differentiation of the algebra $\mathfrak{g}$. All differentiations constitute a Lie algebra under the natural commutator in $L(\mathfrak{g})$. The algebra is denoted by $\partial(\mathfrak{g})$. Every differentiation defines in turn a one-parameter group of automorphisms of $\mathfrak{g}$ by means of the exponential mapping

$$R \ni t \to e^{tD} \in \operatorname{Aut}\mathfrak{g}.$$

This allows us to identify the algebra of differentiations $\partial(\mathfrak{g})$ with the Lie algebra of the group $\operatorname{Aut}\mathfrak{g}$.

The Jacobi identity asserts that the operators $\operatorname{ad} X$ of the adjoint representation form a subalgebra of the algebra of differentiations. We shall denote by $\operatorname{Int}\mathfrak{g}$ the connected subgroup of the group $\operatorname{Aut}\mathfrak{g}$

whose Lie algebra is the subalgebra $\operatorname{ad}\mathfrak{g} \subset \partial(\mathfrak{g})$. The following fact plays a key role in the theory of semisimple Lie algebras:

PROPOSITION 17.1.1. *Every differentiation D of a semisimple algebra $\mathfrak{g}$ is of the form $D = \operatorname{ad} X$ for a certain $X \in \mathfrak{g}$.*

Proof. Identity (17.1.1) can be written as $\operatorname{ad}(DX) = [D, \operatorname{ad} X]$; therefore the operators of the form $\operatorname{ad} X$, $X \in \mathfrak{g}$ constitute an ideal in $\partial(\mathfrak{g})$. Let $\mathfrak{a}$ be a space orthogonal to the ideal $\operatorname{ad}\mathfrak{g}$ in $\partial(\mathfrak{g})$.

The space $\mathfrak{a}$ is an ideal in $\partial(\mathfrak{a})$ owing to property (1.8.1). The intersection of $\mathfrak{a}$ and $\operatorname{ad}\mathfrak{g}$ is an ideal in $\partial(a)$ and so its Killing form is identical with the restriction to $\mathfrak{a}$ of the Killing form of the algebra $\mathfrak{g}$ as well as that of the algebra $\partial(\mathfrak{g})$.

The first form is non-degenerate (Corollary 1.8.2) the other is zero on $\mathfrak{a}$. Hence $\mathfrak{a} \cap \operatorname{ad}\mathfrak{g} = \{0\}$. Thus for arbitrary $D \in \mathfrak{a}$ and $X \in \mathfrak{g}$ we have $\operatorname{ad}(DX) = [D, \operatorname{ad} X] \subset \operatorname{ad}\mathfrak{g} = \{0\}$. The adjoint representation of a semisimple algebra being faithful ($\operatorname{ad} X = 0 \Rightarrow X = 0$), it follows that $D = 0$. □

We have thus proved that for an arbitrary semisimple algebra $\mathfrak{g}$ the Lie algebra of the group $\operatorname{Aut}\mathfrak{g}$ is isomorphic to $\operatorname{ad}\mathfrak{g} = \partial(\mathfrak{g})$. This implies that the algebras $\mathfrak{g}$, $\operatorname{ad}\mathfrak{g}$, $\partial(\mathfrak{g})$ are all isomorphic.

We now introduce another type of algebra in a way which both justifies the terminology and explains the role of those algebras in the investigations which follow.

DEFINITION. A Lie algebra $\mathfrak{g}$ over $\boldsymbol{R}$ is called *compact* if there exists a compact Lie group whose Lie algebra is isomorphic to $\mathfrak{g}$.

One can also give a description of compact algebras, compatible with the previous classification methods.

PROPOSITION 17.1.2. *If an algebra $\mathfrak{g}$ is compact, then its Killing form is non-positive.*

Proof. Let G be a compact connected Lie group whose Lie algebra is $\mathfrak{g}$. The adjoint representation of the group G

$$G \ni g \to \operatorname{Ad} g$$

is orthogonal with respect to a suitably chosen inner product on $\mathfrak{g}$. (formula (4.5.9)). Since $\operatorname{Ad}(\exp X) = e^{\operatorname{ad} X}$ (Proposition 1.3.6), it follows that the matrices of the operators ad$X$ in an orthogonal basis of $\mathfrak{g}$ are all skew-symmetric.

Denoting by $a_{ij}(X)$ the matrix elements of the operator ad$X$, we see that

$$B(X,X) = \mathrm{Tr}(\mathrm{ad}\,X\,\mathrm{ad}X) = \sum_{ij} a_{ij}(x)a_{ji}(x) = -\sum_{ij} a_{ij}(x)^2 \leqslant 0.$$

□

The converse is not true (cf. an algebra which is nilpotent but not abelian). However, under the additional assumptions that the algebra $\mathfrak{g}$ is semisimple (the form B being non-degenerate) the compactness condition proves to be equivalent to the assertion of Proposition 17.1.1. Thus we have

THEOREM 17.1.3. *If $\mathfrak{g}$ is a semisimple Lie algebra over **R** whose Killing form is negative definite then $\mathfrak{g}$ is compact.*

Proof. We define an inner product on the vector space $\mathfrak{g}$:

$$(X|Y) := -B(X, Y), \qquad X, Y \in \mathfrak{g}.$$

Automorphisms of the algebra $\mathfrak{g}$ preserve the Killing form, and so the group Aut$\mathfrak{g}$ constitutes a closed subgroup of the orthogonal group of the space $(\mathfrak{g}, (\,|\,))$. The Lie algebra of this group is $\partial(\mathfrak{g})$ and is isomorphic (by Proposition 17.1.1) to the algebras ad$\mathfrak{g}$ and $\mathfrak{g}$ itself. □

COROLLARY 17.1.4. *If $\mathfrak{g}$ is a compact semisimple Lie algebra, then the group* Int$\mathfrak{g}$ *is compact.*

Proof. In this case the group Int $\mathfrak{g}$ is equal to the connected component of the group Aut $\mathfrak{g}$.

Example 1

Consider the Lie algebra $\mathfrak{u}(n)$ corresponding to the group $U(n)$ of unitary matrices of degree n. Recall that

$$\mathfrak{u}(n) = \{X \in M_n(C) \colon X^* = -X\},$$

where *denotes Hermitian conjugation of a matrix. Denote by $\mathfrak{d}$ the set of diagonal matrices in $\mathfrak{u}(n)$. We shall write diag$(\lambda_1, \ldots, \lambda_n)$ for the diagonal matrix with the elements $\lambda_1, \ldots, \lambda_n$ lying along the diagonal. By a diagonalization theorem we conclude that for every $X \in \mathfrak{u}(n)$

there exists a $g \in U(n)$ such that $\mathrm{Ad}\, gX = gXg^{-1} \in d$, which implies that the Killing form of the Lie algebra $\mathfrak{u}(n)$ is completely determined by its restriction to $\mathfrak{d} \times \mathfrak{d}$.

Let E_{rs} denote the element of a natural basis of $M_n(C)$ which has 1 in the (r, s)-th place (r-th row, s-th column) and zero in the remaining places. Define

$$X_{rs} = E_{rs} - E_{sr}, \qquad Y_{rs} = i(E_{rs} + E_{sr})$$

for r, s such that $r < s$. The matrices X_{rs}, Y_{rs} form a basis for a space complementary to $\mathfrak{d}$ in $\mathfrak{u}(n)$. Consider $H \in \mathfrak{d}$, $H = \mathrm{diag}(ix_1, \ldots, ix_n)$ where x_k are real numbers. Then

$$\mathrm{ad}\, H|_{\mathfrak{d}} = 0,$$

$$\mathrm{ad}\, H(X_{rs}) = (x_r - x_s)\, Y_{rs}, \qquad \mathrm{ad}\, H(Y_{rs}) = -(x_r - x_s)\, X_{rs}.$$

In particular, it follows that $\mathfrak{d}$ is a maximal abelian Lie subalgebra in $u(n)$ and the centre of the algebra $\mathfrak{u}(n)$ is the one-dimensional subspace

$$\mathfrak{z} = \{\mathrm{diag}(it, \ldots, it);\ t \in \boldsymbol{R}\}.$$

Moreover, the algebra

$$\mathfrak{h}_0 = \left\{\mathrm{diag}(ix_1, \ldots, ix_n) \in \mathfrak{d} \colon \sum_{k=1}^{n} x_k = 0\right\}$$

is a maximal abelian subalgebra of $\mathfrak{su}(n)$ and

$$\mathfrak{d} = \mathfrak{z} + \mathfrak{h}_0.$$

For $H \in \mathfrak{d}$, $\mathrm{ad}\, H$ acts on $\mathfrak{u}(n)$ as a diagonalizable operator, with eigenvalues equal to 0 and $-(x_r - x_s)^2$. In the latter case the multiplicity of the eigenvalues is 2 unless they are equal to zero. Thus

$$B(H, H) = \mathrm{Tr}((\mathrm{ad}\, H^2)) = -2 \sum_{r<s} (x_r - x_s)^2 = -\sum_{r \neq s} (x_r - x_s)^2$$

$$= -2(n-1) \sum_{r=1}^{n} x_r^2 - 4 \sum_{r \neq s} x_r x_s.$$

Since

$$-(\mathrm{Tr}\, H)^2 = \sum_{r=1}^{n} x_r^2 = \sum_{r=1}^{n} x_r^2 + 2 \sum_{r \neq s} x_r x_s,$$

we obtain

$$B(H,H) = 2n\operatorname{Tr}(H^2) - 2(\operatorname{Tr} H)^2.$$

Both sides of the above identity being Ad(g)-invariant, we get

$$B(X,X) = 2n\operatorname{Tr}(X^2) - 2(\operatorname{Tr} X)^2, \qquad X \in u(n),$$

and by the polarization formula

$$B(X,Y) = 2n\operatorname{Tr}(XY) - 2\operatorname{Tr} X \operatorname{Tr} Y, \qquad X, Y \in u(n).$$

In particular, it follows that X_{rs}, Y_{rs} are mutually orthogonal relative to the Killing form and

$$B(X_{rs}, X_{rs}) = -2 = B(Y_{rs}, Y_{rs}).$$

Moreover, for $\mathfrak{su}(n)$ we have

$$B(X,Y) = 2n\operatorname{Tr}(XY)$$

since $\mathfrak{su}(n)$ is an ideal in $\mathfrak{u}(n)$.

17.2. STRUCTURE OF SEMISIMPLE ALGEBRAS

Further investigations of the structure of a semisimple algebra will consist in considering the eigenspaces of the operators $\operatorname{ad} X$, $X \in \mathfrak{g}$. As in the case of the elementary theory of linear operators on finite-dimensional vector spaces over $\boldsymbol{R}$, it will be convenient to study a Lie algebra over $\boldsymbol{C}$ and derive hence information concerning the initial real algebra.

Denoting as before by $\mathfrak{g}$ a real Lie algebra, we introduce a space $\mathfrak{g}^C$, which is a complexification of $\mathfrak{g}$. As a vector space over $\boldsymbol{R}$, $\mathfrak{g}^C$ is isomorphic to the product $\mathfrak{g} \times \mathfrak{g}$ with multiplication by elements of $\boldsymbol{C}$ defined as follows:

$$(a+ib)(X,Y) = (aX - bY, aY + bX).$$

The pair (X, Y) is written formally as $X+iY$. The space $\mathfrak{g}^C$ possesses a natural Lie algebra structure. Namely, we define

$$[X+iY, Z+iT] := [X,Z] - [Y,T] + i([Y,Z] + [X,T]).$$

The commutator in $\mathfrak{g}^C$ is antisymmetric and C-bilinear, and satisfies the Jacobi identity. Clearly, the complex dimension of $\mathfrak{g}^C$ equals the real dimension of $\mathfrak{g}$.

Different real algebras may have isomorphic complexifications, as will be seen in Example 1. The algebra $\mathfrak{g}^C$ can be regarded as a $2n$-dimensional algebra over $\boldsymbol{R}$: we shall denote it by $\mathfrak{g}^R$.

We shall now establish some simple relations between the Killing forms of the algebras $\mathfrak{g}$, $\mathfrak{g}^C$ and $\mathfrak{g}^R$, which will be denoted by B, B^C, B^R, respectively.

Proposition 17.2.1. (1) *For* $X, Y \in \mathfrak{g}$

$$B(X, Y) = B^C(X, Y).$$

(2) $B^R(X, Y) = 2\operatorname{Re} B^C(X, Y)$ *for* $X, Y \in \mathfrak{g}^R$.

We omit the simple proof, which consists in a direct verification of these identities.

Let $\mathfrak{g}$ be a Lie algebra over $\boldsymbol{C}$. By a real form of the algebra $\mathfrak{g}$ we shall understand a real subalgebra $\mathfrak{g}_0 \subset \mathfrak{g}$ with the property that

$$\mathfrak{g} = \mathfrak{g}_0 + i\mathfrak{g}_0,$$

i.e. that every element $X \in \mathfrak{g}$ has a unique decomposition of the form $X = X_0 + iY_0$, $X_0, Y_0 \in \mathfrak{g}_0$.

A real algebra $\mathfrak{g}_0$ is a real form of its complexification. As we shall see in Example 1, there exist in general many different forms of a given complex algebra.

Example 1

(a) The identity $\mathfrak{sl}(2, \boldsymbol{C}) = \mathfrak{su}(2) + i\mathfrak{su}(2)$ noted in Chapter 8 is actually a special case of the relation

$$\mathfrak{sl}(n, \boldsymbol{C}) = \mathfrak{su}(n) + i\,\mathfrak{su}(n).$$

For a $X \in \mathfrak{sl}(n, \boldsymbol{C})$ it means that X decomposes into the anti-hermitian part $\frac{1}{2}(X - X^*)$ and the hermitian part $\frac{1}{2}(X + X^*)$. The algebra $\mathfrak{sl}(n, \boldsymbol{C})$ has also a "natural" real form, namely $\mathfrak{sl}(n, \boldsymbol{R})$. Since $\mathfrak{su}(n)$ is compact and $\mathfrak{sl}(n, \boldsymbol{R})$ is not, we see that these real forms of $\mathfrak{sl}(n, \boldsymbol{C})$ are not isomorphic. By Example 1 in Section 1 and Proposition 17.2.1 we deduce that the Killing form corresponding to each of the algebras $\mathfrak{su}(n)$, $\mathfrak{sl}(n, \boldsymbol{C})$, $\mathfrak{sl}(n, \boldsymbol{R})$ is given by the same formula

$$B(X, Y) = 2n \operatorname{Tr}(XY).$$

(b) Let $\mathfrak{o}(p, q)$ denote the Lie algebra of the group $O(p, q)$ of all matrices preserving the quadratic form

$$\sum_{k=1}^{p} x_k^2 - \sum_{k=p+1}^{p+q} x_k^2 .$$

Denoting by $I_{p,q} = \operatorname{diag}(1, \ldots, 1, -1, \ldots, -1)$ the matrix of this quadratic form and setting $n = p+q$, we have

$$\mathfrak{o}(p, q) = \{A \in M_n(R)\colon A^t I_{p,q} + I_{p,q} A = 0\}.$$

Writting A in the form the block matrix

$$\begin{bmatrix} A_{11} & A_{12} \\ A_{21} & A_{22} \end{bmatrix},$$

where $A_{11}, A_{12}, A_{21}, A_{22}$ are $p \times p$, $p \times q$, $q \times p$, $q \times q$ matrices respectively we see that $A \in \mathfrak{o}(p, q)$ if and only if

$$A_{11}^t = -A_{11}, \qquad A_{22}^t = -A_{22}, \qquad A_{12}^t = A_{21}.$$

Let $P = \operatorname{diag}(1, \ldots, 1, i, \ldots, i)$ (1—p times, i—q times); then the conjugation $A \to A' = PAP^{-1}$ maps $\mathfrak{o}(p, q)$ isomorphically onto the Lie subalgebra of the algebra $\mathfrak{o}(n, C)$ comprised of all matrices

$$A' = \begin{bmatrix} A_{11} & -iA_{12} \\ iA_{21} & A_{22} \end{bmatrix},$$

where the blocks A_{ij}, $i, j = 1, 2$ satisfy the conditions given above. In particular, $A' \in \mathfrak{o}(n, \boldsymbol{C})$ (recall that $\mathfrak{o}(n, \boldsymbol{C})$ consists of skew symmetric matrices). Since every element of $\mathfrak{o}(n, \boldsymbol{C})$ can be represented in the form $A' + iB'$, where $A', B' \in P\mathfrak{o}(p, q)P^{-1}$, it follows that $\mathfrak{o}(p, q)$ is (isomorphic to) a real form of $\mathfrak{o}(n, \boldsymbol{C})$. As in the previous example, the "natural" real form $\mathfrak{o}(n) = \mathfrak{o}(n, \boldsymbol{R})$, being compact, is not isomorphic to $\mathfrak{o}(p, q)$ for $p \neq 0 \neq q$. We note that for different (p, q) the algebras $\mathfrak{o}(p, q)$ are different except the obvious identity $\mathfrak{o}(p, q) = \mathfrak{o}(q, p)$, and in the case of odd $p+q$ they exhaust all real forms of the algebra $\mathfrak{o}(p+q, \boldsymbol{C})$. For even $p+q$ there exists yet another real form: the algebra $\mathfrak{sp}(n)$ (see Problem 7).

As before, let $\mathfrak{g}_0$ be a real form of a complex algebra $\mathfrak{g}$. The mapping $\tau\colon \mathfrak{g} \to \mathfrak{g}$ given by the formula $\tau(X_0 + iY_0) := X_0 - iY_0$ is called the *conjugation* of $\mathfrak{g}$ with respect to the real form $\mathfrak{g}_0$. The mapping τ satisfies the identity

$$\tau[X, Y] = [\tau X, \tau Y], \qquad \tau(\alpha X) = \bar{\alpha} X, \qquad \alpha \in \boldsymbol{C}. \tag{17.2.1}$$

A real form $\mathfrak{g}_0$ is said to be *compact* provided it is a compact Lie algebra over $\boldsymbol{R}$, which in the case of a semisimple algebra is equivalent, as we know, to the statement that its Killing form is negative definite.

We shall now study the implications of the assumption that a complex algebra possesses a compact real form $\mathfrak{u}$. By using the Killing form $B_{\mathfrak{u}}$ one can construct an inner product on $\mathfrak{g}$ invariant under the action of the operators $\mathrm{ad}X$, $X \in \mathfrak{u}$. Namely, for $X, X', Y, Y' \in \mathfrak{u}$ we put

$$\begin{aligned}(X+iY|X'+iY') := {} & B_{\mathfrak{g}}(-X+iY, X'+iY') \\ = {} & -B_{\mathfrak{u}}(X, X')+i\big(B_{\mathfrak{u}}(Y, X')-B_{\mathfrak{u}}(X, Y')\big) \\ & -B_{\mathfrak{u}}(Y, Y').\end{aligned} \tag{17.2.2}$$

By applying the conjugation τ we can write shortly

$$(X|Y) = -B_{\mathfrak{g}}(\tau X, Y), \qquad X, Y \in \mathfrak{g}.$$

It follows from the form (17.2.2) of the inner product that for $X \in \mathfrak{u}$

$$\big(\mathrm{ad}X(Y)|Z\big)+\big(Y|\mathrm{ad}X(Z)\big) = 0.$$

Thus the operator i ad X is Hermitian on the unitary space $\big(\mathfrak{g}, (\cdot|\cdot)\big)$, which means that the elements $\mathrm{ad}X$ for $X \in \mathfrak{u}$ and $X \in i\mathfrak{u}$ are diagonalizable operators in $\mathfrak{g}$. Choose a maximal abelian subalgebra $\mathfrak{h}_0$ of $\mathfrak{u}$. Define $\mathfrak{h} := \mathfrak{h}_0 + i\mathfrak{h}_0$. The set of operators $\{\mathrm{ad}X\colon X \in \mathfrak{h}\}$ is a commutative family of operators decomposable in $\mathfrak{g}$.

Let $\{X_i\}_{i=1}^n$ be a basis consisting of eigenvectors of the operators $\mathrm{ad}H$, $H \in \mathfrak{h}$. Thus we have

$$\mathrm{ad}H(X_i) = \alpha_i(H) X_i, \qquad \text{where} \quad \alpha_i(H) \in \boldsymbol{C}. \tag{17.2.3}$$

The rule $\mathfrak{h} \ni H \to \alpha_i(H)$ is a linear functional on $\mathfrak{h}$. It is natural to introduce the following concept.

DEFINITION. A linear functional $0 \neq \alpha \in \mathfrak{h}^*$ is called a *root* if the space

$$\mathfrak{g}^{\alpha} := \{X \in \mathfrak{g}\colon \mathrm{ad}H(X) = \alpha(H)\, X,\ H \in \mathfrak{h}\}$$

is non-zero.

We have seen above that roots indeed exist and that the algebra $\mathfrak{g}$ decomposes into

$$\mathfrak{g} = \mathfrak{h} + \sum_{\Delta \ni \alpha} \mathfrak{g}^{\alpha}, \tag{17.2.4}$$

where Δ is the set of roots, obviously finite.

By the Jacobi identity we have

$$[\mathfrak{g}^{\alpha}, \mathfrak{g}^{\beta}] \subset \mathfrak{g}^{\alpha+\beta} \quad \text{for arbitrary } \alpha, \beta \in \mathfrak{h}^*. \tag{17.2.5}$$

Decomposition (17.2.4) has the following properties:

THEOREM 17.2.2. (1) *If* α, $\beta \in \Delta$ *and* $\alpha+\beta \neq 0$, *then* $\mathfrak{g}^{\alpha} \perp \mathfrak{g}^{\beta}$ *with respect to the Killing form*;

(2) *for every* $\alpha \in \Delta$ *there exists an* $H_{\alpha} \in \mathfrak{h}$ *such that*

$$B_{\mathfrak{g}}(H, H_{\alpha}) = \alpha(H), \qquad H \in \mathfrak{h};$$

(3) $\dim \mathfrak{g}^{\alpha} = 1$, $\alpha \in \Delta$;

(4) *if* $\alpha \in \Delta$ *then* $-\alpha \in \Delta$ *and* $[\mathfrak{g}^{\alpha}, \mathfrak{g}^{-\alpha}] = CH_{\alpha}$.

Proof. Assertion (1) follows directly from the definition of the space $\mathfrak{g}^{\alpha}$. For $X \in \mathfrak{g}^{\alpha}$, $Y \in \mathfrak{g}^{\alpha}$ we have

$$\begin{aligned} B(\operatorname{ad} H(X), Y) &= \alpha(H) B(X, Y) = -B(X, \operatorname{ad} H(Y)) \\ &= -\beta(H) B(X, Y) \end{aligned}$$

for $H \in \mathfrak{h}$. Choosing H such that $(\alpha+\beta)(H) \neq 0$ we get

$$(\alpha+\beta)(H) B(X, Y) = 0$$

which means that $X \perp Y$.

Ad (2). The form $B_{\mathfrak{u}} = 2 \operatorname{Re} B_{\mathfrak{g}}|\mathfrak{u}$ is negative definite and therefore non-degenerate on the space $\mathfrak{h}_0$. As a result $B_{\mathfrak{g}}$ is non-degenerate on the complexification of $\mathfrak{h}_0$. Since the operators $i\operatorname{ad} X$ are Hermitian, they have a real spectrum. Hence it follows that roots assume purely imaginary values. This proves (2).

We shall prove (4). To this end we assume that $\mathfrak{g}^{-\alpha} = \{0\}$ while $\mathfrak{g}^{\alpha} \neq 0$. It follows from decomposition (17.2.4) that $B_{\mathfrak{g}}(X_{\alpha}, X) = 0$ for an arbitrary $X_{\alpha} \in \mathfrak{g}^{\alpha}$ and every $X \in \mathfrak{g}$. Therefore $X_{\alpha} = 0$ in contradiction with the assumption. The commutator $[X_{\alpha}, X_{-\alpha}]$ is an element of $\mathfrak{h}$. On the other hand,

$$\begin{aligned} B([X_{\alpha}, X_{-\alpha}], H) &= B(X_{-\alpha}, [H, X_{\alpha}]) = \alpha(H) B(X_{-\alpha}, X_{\alpha}) \\ &= B(H_{\alpha}, H) B(X_{-\alpha}, X_{\alpha}), \end{aligned}$$

for every $H \in \mathfrak{h}$. Since $B_{\mathfrak{g}|\mathfrak{h}\times\mathfrak{h}}$ is a non-degenerate form, it follows that

$$[X_\mu, X_{-\alpha}] = B(X_{-\alpha}, X_\alpha)H_\alpha. \tag{17.2.6}$$

Passing to the proof of (3) we assume that $\dim \mathfrak{g}^\alpha > 1$ and consequently that there exists a $0 \neq D_\alpha \in \mathfrak{g}^\alpha$ such that $B(X_{-\alpha}, D_\alpha) = 0$.

Next we introduce the sequence $D_{-1} = 0$, $D_n := (\operatorname{ad} X_\alpha)^n D_\alpha$, $n = 0, 1, \ldots$, where X_α is chosen in such a way that $B(X_\alpha, X_{-\alpha}) = 1$. We shall prove inductively that the sequence satisfies the recurrence relation

$$[X_{-\alpha}, D_n] = -\frac{n(n+1)}{2}\alpha(H_\alpha)D_{n-1} \qquad (n = 0, 1, \ldots),$$

For $n = 0$ this is precisely identity (17.2.6). Assuming that the relation is valid for $n-1$, we find

$$\begin{aligned}
[X_{-\alpha}, D_n] &= [X_{-\alpha}, [X_\alpha, D_{n-1}]] \\
&= -[D_{n-1}, [X_{-\alpha}, X_\alpha]] - [X_\alpha, [D_{n-1}, X_{-\alpha}]] \\
&= [D_{n-1}, H_\alpha] - \frac{(n-1)n}{2}\alpha(H)[X_\alpha, D_{n-2}] \\
&= -\left(\frac{(n-1)n}{2} + n\right)\alpha(H_\alpha)D_{n-1} \\
&= -\frac{n(n+1)}{2}\alpha(H_\alpha)D_{n-1},
\end{aligned}$$

where we have used the fact that $D_{n-1} \in \mathfrak{g}^{n\alpha}$. If D_n were zero for a certain n, then we would have $D_0 = 0$ in contradiction with our assumption. On the other hand the supposition that $D_n \neq 0$ for every n also leads to a contradiction, since otherwise we would have $\mathfrak{g}^{n\alpha} \neq 0$ for each n and the set of roots would be infinite. Thus $\dim \mathfrak{g}^\alpha = 1$. □

The above description of the structure of the algebra $\mathfrak{g}$ in terms of the root system was based on the assumption that $\mathfrak{g}$ possesses a compact real form. This allowed us to distinguish an abelian subalgebra $\mathfrak{h} \subset \mathfrak{g}$ for which the operators $\operatorname{ad} H$, $H \in \mathfrak{h}$ are diagonalizable.

DEFINITION. A subalgebra $\mathfrak{h}$ of a semisimple Lie algebra is called a *Cartan subalgebra* if

(1) $\mathfrak{h}$ is a maximal commutative subalgebra,

(2) every operator $\operatorname{ad} H$, $H \in \mathfrak{h}$ is decomposable.

The existence of a compact real form $\mathfrak{g}$ guarantees, as we have seen, the existence of a Cartan subalgebra. We are now in a position to appreciate the role of the following theorems, which constitute a basis for the investigation of the structure of semisimple algebras

THEOREM 17.2.3. (1) *Every semisimple algebra over* $\boldsymbol{C}$ *contains a Cartan subalgebra.*

(2) *All Cartan subalgebras of an algebra* $\mathfrak{g}$ *are conjugate under an inner automorphism.*

(3) *For every Cartan subalgebra* $\mathfrak{h} \subset \mathfrak{g}$ *there exists a compact real form.* $\mathfrak{u} \subset \mathfrak{g}$ *such that* $\mathfrak{h}_0 := \mathfrak{u} \cap \mathfrak{h}$ *is a maximal abelian subalgebra in* $\mathfrak{u}$.

The proof of this theorem, due to E. Cartan, and its further consequences can be found in numerous monographs (cf. Varadarajan, 1974, Helgason, 1968 and Serre, 1965, 1966).

In particular, it follows from the Cartan theorem that to every complex semisimple algebra there corresponds a root system, and assertion (2) suggests that the root system is actually independent of the choice of a Cartan subalgebra. It turns out that the system determines the algebra completely. Thus the problem of the classification of semisimple algebras reduces to the study of finite subsets satisfying some simple algebraic relations in finite dimensional real unitary spaces.

In order that a finite set $\Delta \subset \boldsymbol{R}^n$ be the root system of a certain semisimple complex algebra it is necessary and sufficient that

(1) $0 \in \Delta$;

(2) for every $\alpha \in \Delta$ the reflection s_α in the plane orthogonal to α maps Δ onto Δ;

(3) for arbitrary $\alpha, \beta \in \Delta$

$$s_\alpha(\beta) - \beta = n\alpha \qquad \text{for a certain } n \in \boldsymbol{Z};$$

(4) for every $\alpha \in \Delta$ the root $-\alpha$ is the only element of Δ colinear with α.

By studying the geometric structure of the root systems one arrives at a description of semisimple algebras by means of graphical diagrams—the Dynkin diagrams.

Example 3

We shall construct decomposition (17.2.4) for the algebra $\mathfrak{sl}(n, \boldsymbol{C})$ by choosing for the Cartan algebra $\mathfrak{h}$ the complexification of $\mathfrak{h}_0$ from Example 1. (We noted in it that $\mathfrak{h}_0$ is a maximal abelian subalgebra in $\mathfrak{su}(n)$.) Thus we have

$$\mathfrak{h} = \mathfrak{h}_0 + i\mathfrak{h}_0 = \left\{\operatorname{diag}(z_1, \ldots, z_n) \colon z_i \in \boldsymbol{C} \text{ and } \sum_{i=1}^{n} z_i = 0\right\}.$$

By using the form of the operators ad H for $H \in \mathfrak{h}_0$ established in Example 1 (or by direct computation) one shows that

$$\operatorname{ad} H(E_{rs}) = \tfrac{1}{2}\operatorname{ad} H(X_{rs} - iY_{rs}) = (z_r - z_s)E_{rs}, \qquad r < s,$$

$$\operatorname{ad} H(E_{rs}) = -\tfrac{1}{2}\operatorname{ad} H(X_{sr} + iY_{sr}) = (z_r - z_s)E_{rs}, \qquad s < r.$$

It follows that the roots are the forms

$$\mathfrak{h} \ni H = \operatorname{diag}(z_1, \ldots, z_n) \to \alpha_{rs}(H) = z_r - z_s \in \boldsymbol{C}$$

for $r, s = 1, \ldots, n$, $r \neq s$, with the corresponding root spaces $\mathfrak{g}^{rs} = \boldsymbol{C}E_{rs}$.

In Chapter 8 we came across a simple case of this decomposition, corresponding to $n = 2$. Namely, we proved that a matrix of trace 0 decomposes into the sum of an upper-triangular, a lower-triangular and a diagonal matrix.

Utilizing the expression for the Killing form for $\mathfrak{sl}(n, \boldsymbol{C})$ given in Example 2, we also find the root vector H_{rs} defined by the condition

$$\alpha_{rs}(H) = B(H, H_{rs}), \qquad H \in \mathfrak{h}.$$

to be

$$H_{rs} = \frac{1}{2n}(E_{rr} - E_{ss}),$$

where E_{kk} is the diagonal matric with 1 in the k-th place on the diagonal and zeros in the remaining places.

We now proceed to theorems which will allow us to transfer information concerning the structure of a complex algebra $\mathfrak{g}$ onto its real forms $\mathfrak{g}_0$. We shall first prove that for every real form $\mathfrak{g}_0$ with the conjugation σ there exists a compact form $\mathfrak{u}$ invariant under σ. More precisely, we shall prove

THEOREM 17.2.4. *Let $\mathfrak{g}_0$ be a real form of a complex algebra $\mathfrak{g}$ and let σ be the mapping $X+iY \to X-iY$, $X, Y \in \mathfrak{g}_0$. Then there exists a compact form $\mathfrak{u}$ such that the conjugation $\tau_1 : \mathfrak{u} \ni X+iY \to X-iY$ for $X, Y \in \mathfrak{u}$ is an automorphism of the algebra $\mathfrak{g}^R$ commuting with σ.*

Proof. Choose an arbitrary compact form with the conjugation τ and consider $\mathfrak{g}$ as a unitary space with the inner product

$$(X|Y) = -B_g(X, \tau Y).$$

The operators τ and σ are both antilinear mappings which leave the multiplication $[\,,]$ invariant. Therefore their composite $N = \sigma\tau$ is an automorphism of the algebra $\mathfrak{g}$ and preserves the Killing form. Hence it follows that

$$\begin{aligned}(NX|Y) &= -B_\mathfrak{g}(\sigma\tau X, \tau Y) = (\text{since } \tau^2 = 1, \sigma^2 = 1)\\ &= -B_\mathfrak{g}(X, \tau\sigma\tau Y) = (X, NY).\end{aligned}$$

Thus the automorphism N is an Hermitian mapping. The mapping $P^t := (N^2)^t$ is a positive definite symmetric automorphism of g for an arbitrary $t \in \boldsymbol{R}^1$.

We have the relation $\tau N \tau^{-1} = \tau\sigma\tau\tau^{-1} = \tau\sigma = N^{-1}$ and hence also $\tau P^1 \tau^{-1} = P^{-1}$. Thus we conclude that $\tau P^t \tau^{-1} = P^{-t}$ for an arbitrary $t \in \boldsymbol{R}^1$. We compute

$$\begin{aligned}\sigma P^t \tau P^{-t} &= \sigma\tau\tau^{-1} P^t \tau P^{-t} = NP^{-2t},\\ P^t \tau P^{-t} \sigma &= P^t \tau P^{-t} \tau^{-1} \tau\sigma = P^{2t} N^{-1}.\end{aligned}$$

Putting $t = \frac{1}{4}$ and $P^{\frac{1}{4}} \tau P^{-\frac{1}{4}} = \tau_1$, we obtain

$$\sigma\tau_1 = NP^{\frac{1}{2}} \qquad \text{and} \qquad \tau_1 \sigma = P^{\frac{1}{2}} N^{-1}.$$

It is easy to observe that $NP^{-\frac{1}{2}} = P^{\frac{1}{2}} N = |N| N^{-1}$. Hence $\tau_1 \sigma = \sigma \tau_1$.

Thus we see that the compact subalgebra $P^{\frac{1}{4}} u$ is a real form for which the conjugation is τ_1. This concludes the proof of the theorem. □

The formula $\tau_1 \sigma = \sigma \tau_1$ implies that $\tau_1 \mathfrak{g}_0 \subset \mathfrak{g}_0$ and $\sigma \mathfrak{u}_1 \subset \mathfrak{u}_1$. The operator τ_1 restricted to $\mathfrak{g}_0$ is an involutive automorphism. Thus we have the decomposition

$$\mathfrak{g}_0 = \mathfrak{k} + \mathfrak{p},$$

where $\mathfrak{k}$ is the subspace consisting of all vectors which are left fixed by τ_1 and therefore equal to $\mathfrak{u}_1 \cap \mathfrak{g}_0$, and $\mathfrak{p} = \mathfrak{g}_0 \cap i\mathfrak{u}_1$ is the subspace of all eigenvectors of τ_1 corresponding to the eigenvalue -1. This decomposition is called the Cartan decomposition of the algebra $\mathfrak{g}_0$. In precise terms:

DEFINITION. By a *Cartan decomposition of a real algebra* $\mathfrak{g}_0$ we shall understand a direct sum

$$\mathfrak{g}_0 = \mathfrak{k}+\mathfrak{p}$$

which is orthogonal with respect to the Killing form and such that the involution θ

$$\theta: X+Y \to X-Y \qquad \text{for} \qquad X \in \mathfrak{k}, \qquad Y \in \mathfrak{p}$$

is an automorphism of $\mathfrak{g}_0$ and $B \leqslant 0$ on $\mathfrak{k}$, $B \geqslant 0$ on $\mathfrak{p}$.

The decomposition defined previously satisfies the last condition thanks to the property $B = B_{\mathfrak{g}}|_{\mathfrak{g}_0 \times \mathfrak{g}_0}$ and the fact that for every compact form $\mathfrak{u}$ the decomposition

$$\mathfrak{g}^R = \mathfrak{u}+i\mathfrak{u}$$

is a Cartan decomposition of the algebra $\mathfrak{g}^R$.

It is also proper to note that, given an arbitrary Cartan decomposition of a real algebra $\mathfrak{g}$, we can find in the complexification $\mathfrak{g}^C$ a compact σ-invariant form by putting $\mathfrak{u} = \mathfrak{k}+i\mathfrak{p}$. As we know, $\sigma|_{\mathfrak{u}}$ is an automorphism of $\mathfrak{u}$. We then say that the pairs $(\mathfrak{g}_0, \mathfrak{k})$ and $(\mathfrak{u}, \mathfrak{k})$ are in *duality* with respect to each other.

Example 4

Again consider $\mathfrak{g} = \mathfrak{sl}(n, \boldsymbol{C})$. The mappings $\sigma(X) = \bar{X}$ (the matrix $\bar{X}$ conjugate to $X = [a_{ij}]$ is $\bar{X} := [\bar{a}_{ij}]$) and $\tau(X) = -X^*$ are involutive automorphisms of $\mathfrak{g}^R$ commuting with each other. Clearly, the algebra of all elements which are fixed by σ is $\mathfrak{g}_0 = \mathfrak{sl}(n, \boldsymbol{R})$ and $\mathfrak{u} = \mathfrak{su}(n)$ is the algebra of τ-fixed elements. We have

$$\mathfrak{k} = \mathfrak{g}_0 \cap \mathfrak{u} = \mathfrak{o}(n, \boldsymbol{R}),$$

$$\mathfrak{p} = \mathfrak{g}_0 \cap i\mathfrak{u} = \{X \in \mathfrak{sl}(n, \boldsymbol{R}) \colon X^t = \{X\}$$

and the Cartan decomposition $\mathfrak{g}_0 = \mathfrak{k}+\mathfrak{p}$ consists in the decomposition of a real matrix into the antisymmetric part $X^- = \frac{1}{2}(X-X^t)$ and the symmetric part $X^+ = \frac{1}{2}(X+X^t)$. The corresponding Cartan involution has the form $\theta(X) = -X^t$. On the other hand, the decomposition $\mathfrak{u} = \mathfrak{k} + i\mathfrak{p}$ is the decomposition of an anti-hermitian matrix into the real (antisymmetric) part and the purely imaginary (symmetric) part. The pairs $(\mathfrak{sl}(n, \boldsymbol{R}), \mathfrak{o}(n, \boldsymbol{R}))$ and $(\mathfrak{su}(n), \mathfrak{o}(n, \boldsymbol{R}))$ are in duality.

By studying the structure of symmetric Riemannian spaces we have arrived at the notion of an orthogonal pair of Lie algebras which is the "infinitesimal image" of a pair of groups (G, K) such that the space G/K is isomorphic to the initial symmetric space.

An orthogonal pair is determined by the Lie algebra $\mathfrak{g}$ of the group G and by its automorphism θ such that the set $\mathfrak{k}$ of θ-fixed vectors in $\mathfrak{g}$ constitutes a subalgebra isomorphic to the algebra of the group K.

We have the decomposition $\mathfrak{g} = \mathfrak{k} + \mathfrak{p}$ where $\mathfrak{p}$ is the eigenspace of the operator θ corresponding to the eigenvalues -1. The decomposition resembles the Cartan decomposition.

We shall prove that the Cartan decomposition of an arbitrary semisimple real algebra $\mathfrak{g}_0$ determines an orthogonal pair. To this end we must show that the algebra $\mathfrak{t}$ from the Cartan decomposition is compactly imbedded in $\mathfrak{g}_0$. In other words, we have to prove that the subgroup of the group $\operatorname{Int}\mathfrak{g}_0$ corresponding to the subalgebra $\mathfrak{k}$ is compact.

We first observe:

LEMMA 17.2.4. *Let* $\mathfrak{g} = \mathfrak{g}_0 + i\mathfrak{g}_0$ *where* $\mathfrak{g}_0$ *is a semisimple algebra. Then* $\operatorname{Int}\mathfrak{g}_0$ *constitutes a closed subgroup of the group* $\operatorname{Int}\mathfrak{g}^R$.

Proof. In Section 1 we proved that for an arbitrary real semi-simple algebra $\mathfrak{g}$ the group $\operatorname{Int}\mathfrak{g}$ is isomorphic to the connected component of the identity of $\operatorname{Aut}\mathfrak{g}$ and therefore closed in it.

Passing to the case in question, we observe that the group $\operatorname{Aut}\mathfrak{g}_0$ imbeds in a natural way in $\operatorname{Aut}\mathfrak{g}^R$. Namely, to an automorphism $g \in \operatorname{Aut}\mathfrak{g}_0$ we assign an element $j(g) \in \operatorname{Aut}\mathfrak{g}^R$ by putting

$$j(g)(X+iY) := gX+igY, \qquad X, Y \in \mathfrak{g}_0.$$

The restriction of the automorphism of $\mathfrak{g}^R$ to $\mathfrak{g}_0$ is a continuous mapping inverse to j since

$$j(\operatorname{Aut}\mathfrak{g}_0) = \{g \in \operatorname{Aut}\mathfrak{g}^R\colon g\mathfrak{g}_0 = \mathfrak{g}_0\}.$$

Hence the set is closed in $\operatorname{Aut}\mathfrak{g}^R$. Since the set $\operatorname{Int}\mathfrak{g}_0$ is closed in $\operatorname{Aut}\mathfrak{g}_0$ and the set $\operatorname{Int}\mathfrak{g}^R$ closed in $\operatorname{Aut}\mathfrak{g}^R$, the assertion follows. □

The constituent $\mathfrak{k}$ of the Cartan decomposition of the algebra $\mathfrak{g}_0$ has been defined as $\mathfrak{g}_0 \cap \mathfrak{u}$ where $\mathfrak{u}$ is a compact real form of the algebra $\mathfrak{g}_0^C = \mathfrak{g}_0 + i\mathfrak{g}_0$.

By imbedding $G = \operatorname{Int} \mathfrak{g}_0$ and $K = \operatorname{Int} \mathfrak{u}$ in $\operatorname{Int}(\mathfrak{g}_0 + i\mathfrak{g}_0)^R$ we obtain closed subgroups, and the second of them is even compact (by Corollary 17.1.4). The subgroup $\tilde{K} = j(\operatorname{Int} \mathfrak{g}_0 \cap \operatorname{Int} \mathfrak{u})$ is a compact subgroup of the group $\operatorname{Int}(\mathfrak{g}_0 + i\mathfrak{g}_0)^R$, whose Lie algebra is isomorphic to $\mathfrak{k}$. The connected component of K is also compact. Thus we have proved

THEOREM 17.2.5. *Let $\mathfrak{g}_0 = \mathfrak{k} + \mathfrak{p}$ be a Cartan decomposition of a semisimple real algebra $\mathfrak{g}_0$ and θ the reflection of $\mathfrak{g}_0$ in the subspace $\mathfrak{k}$. Then $(\mathfrak{g}_0, \theta)$ defines an orthogonal pair, i.e.*

$$G/K = \operatorname{Int} \mathfrak{g}_0 / \operatorname{Int}(\mathfrak{g}_0 \cap \mathfrak{u})$$

is a symmetric space.

This theorem ends an important stage of the investigations of the structure of symmetric spaces.

In Chapter 1 we obtained the description of a simply connected symmetric Riemannian space as a Cartesian product of spaces of the Euclidean, compact and non-compact types. Spaces of each of these types determine orthogonal pairs and decompositions $\mathfrak{g} = \mathfrak{k} + \mathfrak{p}$. Theorem 17.2.5 states that every Cartan decomposition of a semisimple algebra of the non-compact type leads to a symmetric space.

We shall formulate a more general result, which describes the properties of a group associated with a semisimple algebra $\mathfrak{g}_0$ of the non-compact type.

THEOREM 17.2.6. *Let $\mathfrak{g}_0$ be a non-compact semisimple algebra over the field* **R** *and let $\mathfrak{g}_0 = \mathfrak{k} + \mathfrak{p}$ be a Cartan decomposition. Suppose that (G, K) is a pair of connected groups such that $\mathfrak{g}_0$ is the Lie algebra of the group G and K is a subgroup of G corresponding to the subalgebra $\mathfrak{k}$. Then*

(1) *K is a closed subgroup of G which contains the discrete centre of the group G. K is compact if and only if the centre of G is finite.*

(2) *There exists an involutive automorphism θ of the group G such that $d\theta$ is identity on $\mathfrak{k}$ and $-\mathrm{id}$ on $\mathfrak{p}$. The pair (G, K) is a symmetric Riemannian space.*

(3) *The mapping* $\varphi\colon (X, k) \to (\exp X)k$ *is a diffeomorphism of the manifold* $\mathfrak{p} \times K$ *onto* G.

In view of what has been said above we see that assertion (2) is true of the pair $(\operatorname{Int} \mathfrak{g}_0, \operatorname{Int} \mathfrak{k})$. A complete proof of this theorem can be found in Helgason (1968) (Ch. 6).

Example 5 (*continuation of Ex.* 4).

We have already shown for $\mathfrak{g}_0 = \mathfrak{sl}(n, \boldsymbol{R})$ that the decomposition $\mathfrak{g}_0 = \mathfrak{k} + \mathfrak{p}$, where $\mathfrak{k} = \mathfrak{o}(n, \boldsymbol{R})$ and $\mathfrak{p}$ is the subspace of all symmetric traceless matrices, is a Cartan decomposition. Take $G = SL(n, \boldsymbol{R})$ and $K = SO(n)$, the group of rotations of the space $\boldsymbol{R}^n$. We know from Chapter 1 that these groups are connected (for $n > 1$). An involutive automorphism θ of the group $SL(n, \boldsymbol{R})$ for which $d\theta(X) = -X^t$ is obviously $\theta(g) = (g^{-1})^t$ and K is the subgroup of all fixed points of the involution θ.

The set $\exp(\mathfrak{p})$ is the set of all positive definite symmetric (real) matrices of degree n and the decomposition

$$g = k\exp(X), \qquad g \in G, \quad k \in K, \quad X \in \mathfrak{p}$$

is the "polar decomposition" of matrices well known from linear algebra (cf. also Example 1 in § 15.5).

17.3. IWASAWA DECOMPOSITION OF AN ALGEBRA AND OF A GROUP

A new decomposition of a semisimple non-compact algebra $\mathfrak{g}$ results from combining the Cartan decomposition of the algebra $\mathfrak{g}$ and the root space decomposition of its complexification $\mathfrak{g}^C$.

For this purpose we fix a Cartan decomposition $\mathfrak{g} = \mathfrak{k} + \mathfrak{p}$ and denote by θ the automorphism of $\mathfrak{g}$ which is a reflection in the subspace $\mathfrak{k}$.

There exists a symmetric form on $\mathfrak{g}$

$$(X|Y) = -B(X, \theta Y)$$

relative to which the operators $\operatorname{ad} X_1$, $X_1 \in \mathfrak{p}$ are symmetric:

$$\begin{aligned}(\operatorname{ad} X_1(X)|Y) &= -B([X_1, X], \theta Y) = B(X, [X_1, \theta Y]) \\ &= -B(X, \theta[X_1, Y]) = (X|\operatorname{ad} X_1(Y)).\end{aligned}$$

As in the case of a complex algebra we select a maximal commutative subspace $\mathfrak{a}$ of $\mathfrak{p}$.

The set $\{\mathrm{ad}X : X \in \mathfrak{a}\}$ constitutes a family of operators which are simultaneously decomposable in $\mathfrak{g}$.

A linear functional $\alpha \neq 0$ on $\mathfrak{a}$ is called a *reduced root* if the space

$$\mathfrak{g}_\alpha := \{X \in \mathfrak{g} : \mathrm{ad}H(X) = \alpha(H)X,\ H \in \mathfrak{a}\} \tag{17.3.1}$$

is non-empty. The set of all reduced roots will be denoted by Σ.

We have the decomposition

$$\mathfrak{g} = \sum_{\alpha \in \Sigma} \mathfrak{g}_\alpha + \mathfrak{g}^0,$$

where $\mathfrak{g}^0$ is the subspace of all vectors annihilated by the operators $\mathrm{ad}X$, $X \in \mathfrak{a}$. We define an ordering of the set of roots in the following way. Denote by $\mathfrak{a}'$ the (open) subset of $\mathfrak{a}$ on which all reduced roots assume values different from zero. (The set $\mathfrak{a}-\mathfrak{a}'$ is a union of hyperplanes of the form $\mathrm{Ker}\,\alpha$, $\alpha \in \Sigma$.) Connected components of the set $\mathfrak{a}'$ are called *Weyl chambers*. Denote by $\mathfrak{a}^+$ a fixed Weyl chamber.

We say that a root α is positive if $\alpha(H) > 0$ for $H \in \mathfrak{a}^+$. The totality of positive roots will be denoted by Σ_+. We define a subspace of $\mathfrak{g}$:

$$\mathfrak{n} = \sum_{\alpha \in \Sigma_+} \mathfrak{g}_\alpha.$$

LEMMA 17.3.1. (a) $\mathfrak{n}$ *constitutes a nilpotent subalgebra of* $\mathfrak{g}$.

(b) $\mathfrak{n} \cap \mathfrak{k} = \{0\}$.

Proof. The identity $[\mathfrak{g}_\alpha, \mathfrak{g}_\beta] \subset \mathfrak{g}_{\alpha+\beta}$ is obviously also valid for reduced roots as a consequence of the Jacobi identity. If $\alpha \in \Sigma_+$ then $[\mathfrak{g}_\alpha, \mathfrak{n}] \subset \sum_{\beta \in \alpha+\Sigma_+} \mathfrak{g}_\beta$, and hence by induction we deduce that $(\mathrm{ad}X)^n \mathfrak{n} \subset \sum_{\beta+n\alpha \in \Sigma_+} \mathfrak{g}_\beta$ for $X \in \mathfrak{g}_\alpha$. The set of roots being bounded, it follows that there exists an $n \in N$ such that $(n\alpha+\Sigma_+) \cap \Sigma = \emptyset$, i.e. $(\mathrm{ad}X)^n \mathfrak{n} = 0$, which means that the algebra $\mathfrak{n}$ is nilpotent.

In order to prove the second part of the assertion let us study the action of the automorphism θ on the root spaces. Let $X \in \mathfrak{g}_\alpha$, i.e. $[H, X] = \alpha(H)X$. Then $\alpha(H)\theta X = \theta[H, X] = [\theta H, \theta H] = -[H, \theta X]$, which means that $\theta\mathfrak{g}_\alpha \subset \mathfrak{g}_{-\alpha}$ and $\theta\mathfrak{n} = \sum_{\alpha \in \Sigma_+} \mathfrak{g}_{-\alpha} =: \bar{\mathfrak{n}}$. Clearly, $\mathfrak{n} \cap \bar{\mathfrak{n}} = 0$. But also $\mathfrak{k} \cap \mathfrak{n} = \theta(\mathfrak{k} \cap \mathfrak{n}) \subset \theta(\mathfrak{k}) \cap \theta(\mathfrak{n}) = \mathfrak{k} \cap \bar{\mathfrak{n}} \subset \mathfrak{n} \cap \bar{\mathfrak{n}} = 0$. □

We now pass to the main result of this chapter.

THEOREM 17.3.2 (Iwasawa). *A real semisimple algebra* $\mathfrak{g}$ *of the non-compact type regarded as a vector space is a direct sum of subalgebras* $\mathfrak{k}$, $\mathfrak{a}$, $\mathfrak{n}$:

$$\mathfrak{g} = \mathfrak{k}+\mathfrak{a}+\mathfrak{n}.$$

Proof. We shall first prove that the sum $\mathfrak{k}+\mathfrak{a}+\mathfrak{n}$ is direct. Let $X \in \mathfrak{k}$, $H \in \mathfrak{a}$, $N \in \mathfrak{n}$ be such that $X+H+N = 0$. Applying the automorphism θ, we obtain $X-H-\theta N = 0$ and hence $2X+N-\theta N = 0$. But $N-\theta N \in \mathfrak{p}$, and we know that $\mathfrak{k}\cap\mathfrak{p} = 0$. Therefore $X = 0$ and $N-\theta N = 0$. The latter identity means that $N \in \mathfrak{n}\cap\bar{\mathfrak{n}} = 0$, and so $N = 0$. This shows that the sum in question is indeed direct.

Next we study the possibility of obtaining the decomposition $Y = X+H+N$ for an arbitrary $Y \in \mathfrak{g}$. If such a decomposition exists, then $Y-\theta Y = 2H+N-\theta N$. We know that $\mathfrak{g} = \sum_{\alpha\in\Sigma} \mathfrak{g}_\alpha+\mathfrak{g}^0$ and $\mathfrak{g}^0\cap\mathfrak{p} = \mathfrak{a}$ since $\mathfrak{a}$ is an abelian subalgebra maximal in $\mathfrak{p}$. In particular, $\mathfrak{p} \subset \sum_{\alpha\in\Sigma} \mathfrak{g}_\alpha +\mathfrak{a}$. We shall define the constituent H of the decomposition as $\frac{1}{2}$ the component in $\mathfrak{a}$ of the vector $Y-\theta Y \in \mathfrak{p}$. The direct sum $\sum_{\alpha\in\Sigma} \mathfrak{g}_\alpha$ can be represented as $\mathfrak{n}+\bar{\mathfrak{n}}$ since an arbitrary root α belongs either to Σ_+ or $-\Sigma_+$. Thus we represent $Y-\theta Y = 2H+N+\bar{N}$, $N \in \mathfrak{n}$, $\bar{N} \in \bar{\mathfrak{n}}$. Since $Y-\theta Y-2H = N+\bar{N} \in \mathfrak{p}$, we observe that $\theta N+\theta\bar{N} = -N-\bar{N}$. However, since $\theta N \in \bar{\mathfrak{n}}$ and $\theta\bar{N} \in \mathfrak{n}$, it follows from the uniqueness of the decomposition $\mathfrak{n}+\bar{\mathfrak{n}}$ that $\theta N = -N$. Consequently $Y-\theta Y = 2H +N-\theta N$, which amounts to $Y-H-N = \theta Y+H-\theta N = \theta(Y-H-N)$. Hence $Y-H-N \in \mathfrak{k}$, which completes the proof. □

We now present the global version of the Iwasawa decomposition: the proof can be found in Helgason (1968, VI).

THEOREM 17.3.4. *Let G be a connected semisimple Lie group whose algebra* $\mathfrak{g}_0$ *has an Iwasawa decomposition*

$$\mathfrak{g}_0 = \mathfrak{k}+\mathfrak{a}+\mathfrak{n}.$$

Let K, A, N be connected subgroups of G with the Lie algebras $\mathfrak{k}$, $\mathfrak{a}$, $\mathfrak{n}$, *respectively. Then the mapping* $K\times A\times N \ni (k, a, n) \to kan \in G$ *is a diffeomorphism. The subgroups A and N are simply connected.*

Example 1

In the Cartan decomposition for $\mathfrak{sl}(n, \boldsymbol{R})$ described in Example 4 (§ 17.2) let us choose for $\mathfrak{a}$ the set of all diagonal matrices in $\mathfrak{sl}(n, \boldsymbol{R})$, i.e. $\mathfrak{a} = \mathfrak{h} \cap \mathfrak{sl}(n, \boldsymbol{R})$. The reasoning applied to establish the root decomposition for $\mathfrak{sl}(n, \boldsymbol{C})$ (cf. Example 3, § 17.2) can be repeated literally in the present case after replacing $\boldsymbol{C}$ by $\boldsymbol{R}$. In particular, the functions $\alpha_{rs|\mathfrak{a}}$ are reduced roots, i.e.

$$\alpha_{rs}(\mathrm{diag}(x_1, \ldots, x_n)) = x_r - x_s,$$

and the corresponding root spaces $\mathfrak{g}^{rs}$ are given as

$$\mathfrak{g}^{rs} = RE_{rs}.$$

Choose as $\mathfrak{a}^+$ the subset of $\mathfrak{a}$ defined as

$$\mathfrak{a}^+ = \{\mathrm{diag}(x_1, \ldots, x_n) \in \mathfrak{sl}(n, \boldsymbol{R}) \colon x_1 > x_2 > \ldots > x_n\}.$$

Then the positive roots are the functions α_{rs} with $s > r$ and $\mathfrak{n}$ is the Lie subalgebra comprised of all uppertriangular matrices (with zeros in and below the diagonal). Thus the Iwasawa decomposition for the Lie algebra $\mathfrak{sl}(n, \boldsymbol{R})$ is the decomposition of a (traceless) matrix into an antisymmetric, a diagonal and an uppertiangular matrix. The constituents of the global decomposition of the group $\mathrm{SL}(n, \boldsymbol{R})$ are the subgroups $K = \mathrm{SO}(n)$, $A = \{\mathrm{diag}(\lambda_1, \ldots, \lambda_n) \colon \lambda_1 \ldots \lambda_n = 1,\ \lambda_i > 0\}$,

$$N = \left\{ \begin{bmatrix} 1 & & * \\ & \ddots & \\ 0 & & 1 \end{bmatrix} : * \text{ denotes arbitrary entries} \right\}.$$

The subgroup $S = AN$ (semidirect product) consists of uppertriangular matrices with zeros below the diagonal and positive entries in the diagonal such that their product equals 1. Thus the Iwasawa decomposition expresses the Gram-Schmidt orthogonalization process known from linear algebra.

Example 2

Consider the Lie algebra $\mathfrak{g} = \mathfrak{o}(n, 1)$ for $n > 1$. We shall obtain a Cartan decomposition for $\mathfrak{g}$ by finding $\mathfrak{k} = \mathfrak{g} \cap \mathfrak{u}$, $\mathfrak{p} = \mathfrak{g} \cap i\mathfrak{u}$ for the compact Lie algebra (isomorphic to $\mathfrak{o}(n+1, \boldsymbol{R})$, cf. Example 2, § 17.2)

$$\mathfrak{u} = P^{-1}\mathfrak{o}(n+1, \boldsymbol{R})P,$$

where $P = \mathrm{diag}(1, \ldots, 1, i)$. Simple computation leads to the identity

$$\mathfrak{k} = \left\{\begin{bmatrix} A & 0 \\ 0 & 0 \end{bmatrix} : A \in \mathfrak{o}(n, \boldsymbol{R})\right\} \simeq \mathfrak{o}(n, \boldsymbol{R}),$$

$$\mathfrak{p} = \left\{\begin{bmatrix} 0 & x \\ x^t & 0 \end{bmatrix} : x \in \boldsymbol{R}^n\right\}$$

(here, matrices are represented in the form of $n \times n$, $1 \times n$, $n \times 1$ and 1×1 blocs) and allow us to observe that the reflection in $\mathfrak{k}$ corresponding to this decomposition is the mapping $X \to -X^t$. Also without difficulty we verify by direct computation that elements $X, Y \in \mathfrak{p}$ commute with each other if and only if their last columns are proportional—in particular, the maximal abelian subspace in $\mathfrak{p}$ has dimension 1.

Let us choose as a basis element for $\mathfrak{a}$ the matrix

$$H = \begin{bmatrix} 0 & e_n^+ \\ e_n & O_n \end{bmatrix}, \qquad e_n = \begin{pmatrix} 0 \\ \vdots \\ 1 \end{pmatrix} \in \boldsymbol{R}^n.$$

We shall say that a reduced root α is positive if $\alpha(H) > 0$. The identity

$$[H, X] = \alpha(H)X$$

for $X \in \mathfrak{o}(n, 1)$, $\alpha(H) > 0$ has as solutions only matrices of the form

$$Z(x) = \begin{bmatrix} 0 & x_1 & \ldots & x_{n-1} & 0 \\ -x_1 & 0 & \ldots & 0 & x_1 \\ \vdots & \vdots & \ldots & \vdots & \vdots \\ -x_{n-1} & 0 & \ldots & 0 & x_{n-1} \\ 0 & x_1 & \ldots & x_{n-1} & 0 \end{bmatrix},$$

$$x = \begin{bmatrix} x_1 \\ \vdots \\ x_{n-1} \end{bmatrix} \in \boldsymbol{R}^{n-1}.$$

They all correspond to the reduced root α determined by $\alpha(H) = 1$. Note that

$$[Z(x), Z(x')] = 0, \qquad x, x' \in \boldsymbol{R}^{n-1}$$

and so the nilpotent constituent of the Iwasawa decomposition is abelian in our case. The global constituents of the Iwasawa decomposition for the group $G = SO_0(n, 1)$ (the connected component of the group $SO(n, 1)$) are the following:

$$K \simeq SO(n); \quad K = \left\{ \begin{bmatrix} R & 0 \\ 0 & 1 \end{bmatrix}; \; R \in SO(n) \right\},$$

$$A = \left\{ \begin{bmatrix} \cosh t & 0 & \ldots & 0 & \sinh t \\ 0 & & & & 0 \\ \vdots & & I_{n-1} & & \vdots \\ \sinh t & 0 & \ldots & 0 & \cosh t \end{bmatrix} = \exp tH; \; t \in \boldsymbol{R} \right\} \simeq \boldsymbol{R},$$

$$N \simeq \boldsymbol{R}^{n-1}; \; N = \{\exp Z(x) \colon x \in \boldsymbol{R}^{n-1}\}$$

$$= \left\{ \begin{bmatrix} 1 - \frac{1}{2}x^2 & x_1 & \ldots & x_{n-1} & \frac{1}{2}x^2 \\ -x_1 & & & & x_1 \\ \vdots & & I_{n-1} & & \vdots \\ -x_{n-1} & & & & x_{n-1} \\ -\frac{1}{2}x^2 & x & \ldots & x_{n-1} & 1 + \frac{1}{2}x^2 \end{bmatrix} : x \in \boldsymbol{R}^{n-1} \right\},$$

where $|x|^2 = \sum x_i^2$.

17.4. THE WEYL GROUP

The construction of the Iwasawa decomposition is clearly non-unique. Beginning with the construction we choose arbitrarily the Cartan decomposition, we then fixed an abelian subspace $\mathfrak{a}$ maximal in $\mathfrak{p}$, and finally a Weyl chamber $\mathfrak{a}^+$ which determined the decomposition $\Sigma = \Sigma_+ \cup -\Sigma_+$ and consequently the nilpotent constituent of the Iwasawa decomposition.

The theorems which follow demonstrate, among other things, that the essential features of the decomposition are actually independent of the choice which we made.

THEOREM 17.4.1. *Let $\mathfrak{a}_1, \mathfrak{a}_2$ be abelian subspaces which are maximal in $\mathfrak{p}$. Let (G, K) be a symmetric pair associated with $\mathfrak{g}_0$. Then there exists a $k \in K$ such that* $\mathrm{Ad}_G(k)\mathfrak{a}_1 = a_2$

$$\mathfrak{p} = \bigcup_{k \in K} \mathrm{Ad}_G(k)\mathfrak{a}_1. \tag{17.4.1}$$

Proof. Fix $H \in \mathfrak{a}_1$ lying inside a Weyl chamber and an arbitrary vector $X \in \mathfrak{p}$ and consider the function $f: K \ni k \to B(H, \mathrm{Ad}_G(k)X)$. The group $\mathrm{Ad}_G(K)$ being compact, the function f attains its minimum at a certain point $k_0 \in K$. This being the case, it follows that for an arbitrary $Y \in \mathfrak{k}$

$$B(H, [Y, \mathrm{Ad}_G(k_0)X]) = \frac{\mathrm{d}}{\mathrm{d}t} B(H, \mathrm{e}^{t\,\mathrm{ad}\,Y}\mathrm{Ad}_G(k_0)X)|_{t=0}$$

$$= \frac{\mathrm{d}}{\mathrm{d}t} B(H, \mathrm{Ad}_G((\exp tY)k_0)X)|_{t=0} = 0.$$

Hence we derive that $B([H, \mathrm{Ad}_G(k_0)X], Y) = 0$ for an arbitrary $Y \in \mathfrak{k}$, which means that $\mathrm{ad}\,H(\mathrm{Ad}_G(k_0)X) = 0$.

We shall apply decomposition (17.3.2) to the vector $\mathrm{Ad}_G(k_0)X$. Since every root is non-zero on H, we conclude that $\mathrm{Ad}_G(k_0)X \subset \mathfrak{g}_0$. This being so, the maximality of $\mathfrak{a}_1$ means that $\mathrm{Ad}_G(k_0)X \in \mathfrak{a}_1$, and hence

$$[\mathfrak{a}_1, \mathrm{Ad}_G(k_0)X] = 0. \tag{17.4.2}$$

If we now take for X a vector $H_2 \in \mathfrak{a}_2$ which lies inside a Weyl chamber $\mathfrak{a}_2^+$, then, as is seen from (17.4.2), $\mathrm{Ad}_G(k_0^{-1})\mathfrak{a}_1$ commutes with H_2 and therefore belongs to $\mathfrak{a}_2$. This concludes the proof of the first part of the assertion. Property (17.4.1) follows directly from (17.4.2). □

The group K acts transitively on the set of all abelian subspaces of $\mathfrak{k}$. Let us examine the isotropy group of this action. Let

$$M' := \{k \in K\colon \mathrm{Ad}_G(k)\mathfrak{a} \subset \mathfrak{a}\}, \quad M = \{k \in K\colon \mathrm{Ad}_G(k)|\mathfrak{a} = \mathrm{id}\}.$$

The Lie algebras of these subgroups of K are characterized by the identities

$$\mathfrak{m}' = \{Y \in \mathfrak{k}\colon [Y, \mathfrak{a}] \subset \mathfrak{a}\}, \qquad \mathfrak{m} = \{Y \in \mathfrak{t}\colon [Y, \mathfrak{a}] = 0\}.$$

If $Y \in \mathfrak{m}'$, then for every $H \in \mathfrak{a}$

$$B([Y, H], [Y, H]) = B([[Y, H], H], Y) = 0$$

since $[[Y, H], H] = 0$. Since the form B is positive definite on $\mathfrak{p}$, it follows that $[Y, H] = 0$ and so $Y \in \mathfrak{m}$. Thus subgroups M' and M have isomorphic Lie algebras. This implies that the group $W := M'/M$ is discrete and consequently finite. We call it the *Weyl group* of the group G.

Theorem 17.4.1 shows that W is independent of the particular $\mathfrak{a}$ taken in the construction of W.

The Weyl group W acts on the vector space $\mathfrak{a}$ as a linear transformation group by the operators $\mathrm{Ad}_G(k)$, $k \in M'$. The following consideration explains the connection between the Weyl group and the structure of the group G and its algebra $\mathfrak{g}_0$.

Define an Euclidean structure on the space $\mathfrak{a}$ by restricting to $\mathfrak{a}$ the Killing form

$$(H_1|H_2) := B(H_1, H_2).$$

Let s_α denote the orthogonal reflection in the hyperplane $\{H: \alpha(H) = 0\}$ for $\alpha \in \Sigma$.

PROPOSITION 17.4.2. *$s_\alpha \in W$ for an arbitrary $\alpha \in \Sigma$.*

We invite the reader to prove this fact himself. Suggestions concerning the proof can be found in Problem 14.

The proof of the following statement is slightly more complicated.

THEOREM 17.4.3. *The reflections s_α, $\alpha \in \Sigma$ generate the Weyl group. If $w(H) = H$ for $w \neq \mathrm{id}$, then $\alpha(H) = 0$ for a certain root $\alpha \in \Sigma$. W acts transitively and freely on the set of Weyl chambers.*

This theorem not only supplies us with information concerning the structure of the Weyl group but also gives a description of the orbits of the action of W on $\mathfrak{a}$. In particular, it asserts that every Weyl chamber is mapped under the action of an arbitrary element $w \neq 0$ into another Weyl chamber. The rank of the group W equals the number of the Weyl chambers in $\mathfrak{a}$.

Example 1

We shall find the Weyl group for $G = \mathrm{SO}_0(n, 1)$ and $K = \mathrm{SO}(n)$ directly from the definition. To this end we observe that the action of K on $\mathfrak{p}$ with the help of the adjoint representation reduces to the usual action of $\mathrm{SO}(n)$ on $\boldsymbol{R}^n$, i.e. if

$$k = \begin{bmatrix} \boldsymbol{R} & 0 \\ 0 & 1 \end{bmatrix} \in K, \qquad X = \begin{bmatrix} 0 & x \\ x^t & O_n \end{bmatrix} \in \mathfrak{p},$$

then

$$\mathrm{Ad}_G(k)X = \begin{bmatrix} 0 & \boldsymbol{R}x \\ x^t\boldsymbol{R}^t & 0 \end{bmatrix}.$$

Hence we have

$$M = \left\{ \begin{bmatrix} 1 & & \\ & B & \\ & & 1 \end{bmatrix} : B \in \mathrm{SO}(n-1) \right\},$$

$$M = \left\{ \begin{bmatrix} \det B & & \\ & B & \\ & & 1 \end{bmatrix} : B \in O(n-1) \right\}$$

and consequently

$$W = M'/M = \boldsymbol{Z}_2.$$

Example 2

The Weyl group for $G = \mathrm{SL}(n, \boldsymbol{R})$ will be found on the grounds of Theorem 17.4.2 and the root space decomposition for $\mathfrak{sl}(n, \boldsymbol{R})$ described in Example 1 in § 17.3. The hyperplane in $\mathfrak{a}$ on which the root α_{rs} is zero is the set comprised of all matrices with equal diagonal entries in the r-th and s-th places. One can easily derive from the expression for the Killing form for $\mathfrak{sl}(n, \boldsymbol{R})$ that the orthogonal reflection in this hyperplane is the transposition of elements in the r-th and the s-th place. Thus the Weyl group is isomorphic to the group of permutations of n elements.

A rather unexpected but nevertheless simple consequence of Theorem 17.4.1 is the description of the symmetric space G/K in terms of the polar coordinates. Let $A^+ = \exp \mathfrak{a}^+$.

THEOREM 17.4.4 (Cartan). *The mapping* $\varphi\colon K/M \times A^+ \ni (kM, a) \to kaK \in G/K$ *is a diffeomorphism onto an open dense submanifold of* G/K.

We shall first prove

LEMMA 17.4.5. *Let* $X \in \mathfrak{a}^+$ *and let* $\mathrm{Ad}\, k(X) \in \mathfrak{a}$. *Then* $k \in M'$.

Proof of the lemma. If $\mathrm{Ad}\, k(X) \in \mathfrak{a}$ then for every $H \in \mathfrak{a}$ we have

$$[\mathrm{Ad}(k^{-1})H, X] = \mathrm{Ad}(k^{-1})[H, \mathrm{Ad}\, k(X)] = 0.$$

As we know (cf. (17.4.2)), this implies $\mathrm{Ad}(k^{-1})H \in \mathfrak{a}$. Therefore $k^{-1} \in M'$ and hence $k \in M'$. □

It follows from the lemma that the mapping φ is an injection. Namely, assuming that $\varphi(k_1 M, a_1) = \varphi(k_2 M, a_2)$ and $a_i = \exp H_i$, we obtain

$$k_1(\exp H_1)K = \exp(\mathrm{Ad}\, k_1(H_1))K = \exp(\mathrm{Ad}\, k_2(H_2))\, K,$$

and by Theorem 17.2.7. (3) we have $\mathrm{Ad}\, k_1(H_1) = \mathrm{Ad}\, k_2(H_2)$. In view of the lemma we conclude that $k_1 k_2^{-1} \in M'$. Next it follows from Theorem 17.4.3 that $k_1 k_2^{-1} \in M$, hence $k_1 M = k_2 M$ and $H_1 = H_2$.

Finally, we have to prove that the image of φ is dense in G/K. Owing to the fact that the mapping $\mathfrak{p} \ni X \to (\exp X)K$ is a diffeomorphism, it remains to show that the mapping

$$K/M \times \mathfrak{a}^+ \ni (kM, H) \to \mathrm{Ad}\, k(H)$$

has dense image in $\mathfrak{p}$.

The action of M' on $\mathfrak{a}^+$ yields a dense subset in $\mathfrak{a}$ equal to $\mathfrak{a}'$ (Theorem 17.4.3). On account of formula (17.4.1) we deduce that $\bigcup_{k \in K} \mathrm{Ad}(k)\mathfrak{a}^+ = \bigcup_{k \in K} \mathrm{Ad}(k)\mathfrak{a}'$ is a dense subset in $\mathfrak{p}$. □

On the analogy of the Euclidean space and the unit disc in C and the hyperboloid in Minkowski's space the set A^+ may be interpreted as a radius-vector and the space K/M as angular coordinates.

The results thus obtained can be presented in a weaker but very useful form. Namely we have

THEOREM 17.4.6. *The space of double cosets $K \backslash G/K$ is parametrized by the set of orbits of the Weyl group W in $\mathfrak{a}$.*

Proof. The space G/K is the bijective image of $\mathfrak{p}$ under the exponential mapping. Since $k(\exp X)K = \exp[\mathrm{Ad}k(X)]K$, it follows that $K\backslash G/K$ is parametrized by the set of orbits of $\mathrm{Ad}K$ in $\mathfrak{p}$.

By fixing a $\mathfrak{a} \subset \mathfrak{p}$ and applying (17.4.1) we reduce the problem to describing the set of orbits of the group M' in $\mathfrak{a}$, i.e. to the set of orbits of W in $\mathfrak{a}$. □

The exponential mapping assigns to vectors in $\mathfrak{a}$ in a one-to-one manner elements of the abelian group A. Transporting the action of W to A by the formula $w(\exp X) = \exp(wX)$, we can formulate Theorem 17.4.6 as follows:

$$K\backslash G/K = W\backslash A := \text{the set of orbits of } W \text{ in } A. \qquad (17.4.2)$$

A function on the group which is bi-invariant under K (in particular, a zonal spherical function) has the form

$$f(kak_1) = f^{\#}(a),$$

where $f^{\#}$ is a function on A invariant under the action of W. If the function f is continuous, then it is completely determined by its values on A^+. In this way the problems of decomposition of a function on the symmetric space G/K reduce to the problems of harmonic analysis on the commutative group A. Namely, we have seen in examples in Part II and in the considerations of the preceding chapter that the knowledge of the decomposition of functions of class $\mathscr{D}(K\backslash G/K)$ is enough to allow us to determine the Plancherel measure on the symmetric space G/K.

17.5. BOUNDARY OF A SYMMETRIC SPACE OF THE NON-COMPACT TYPE

An open convex cone C in $\mathfrak{p}$ is termed a Weyl chamber in $\mathfrak{p}$ if there exists an abelian subspace $\mathfrak{a} \subset \mathfrak{p}$ maximal in $\mathfrak{p}$ and such that $C \subset \mathfrak{a}$ and C is a Weyl chamber in $\mathfrak{a}$.

DEFINITION. The set of Weyl chambers in $\mathfrak{p}$ is called the *boundary of the symmetric space* G/K and denoted by B.

It follows from Theorems 17.4.1 and 17.4.3 that the group K acts transitively on the set of Weyl chambers in $\mathfrak{p}$ and if $\mathrm{Ad}(k)\mathfrak{a}^+ = \mathfrak{a}^+$ then $k \in M$. The set B of chambers in $\mathfrak{p}$ is therefore parametrized by points of the manifold K/M. By means of this bijection we endow B with the structure of a compact manifold of a homogeneous space of the group K. Further, by means of the isomorphism

$$G/MAN = K/M,$$

which is a consequence of the Iwasawa decomposition, we equip B with the structure of a homogeneous space of the group G.

To make sure that the set $MAN = \{g \in G\colon g = man,\ m \in M,\ a \in A,\ n \in N\}$ is indeed a subgroup of G we verify that the operators $\mathrm{Ad}\, m$, $m \in M$, operate inside the root space of $\mathfrak{g}_\alpha$, and hence the subgroup N is invariant under the inner automorphisms $\alpha(m)$, $m \in M$. The group MAN is the direct product of M and the solvable subgroup AN. The subgroup AN is called the *Borel subgroup* of G.

The easiest way to justify the term "boundary" adopted here is to consider the case of the unit disc D with the action of the group $SL(2, \boldsymbol{R})$ by homographic mappings.

In Chapter 10 we proved that the homogeneous space G/MAN is isomorphic to the boundary of the disc and that the action of the group defined above coincides with the natural extension of the action of homographic mappings from the interior of the disc to its boundary. Similarly, in Chapter 11 in the case of a symmetric space of the Lorentz group, a hyperboloid in $\boldsymbol{R}^4$, the boundary B corresponded to "the elements at infinity" of the hyperboloid, i.e. to its asymptotic directions: the lines lying on the light cone.

The boundary of a symmetric space plays a fundamental role in harmonic analysis on the space G/K.

We saw that in the cases of $SL(2, \boldsymbol{R})$ and $SL(2, \boldsymbol{C})$, the irreducible representations of these groups entering into the decomposition of $L^2(G/K)$ into a direct integral, were induced from the subgroup MAN and realized on the space $L^2(B)$.

In the general case the boundary of a symmetric space also supplies a model of a Hilbert space acted upon by those representations of the group G which are of "greatest importance" from the point of view of the analysis on $\mathfrak{M} = G/K$. Accordingly the abstract Fourier transformation on $\mathfrak{M}$ defined in Chapter 16 assumes the form of an integral operator and in the search for the Plancherel measure we gain new arguments, resulting from the compactness of the space B. These topics will be discussed in the subsequent sections. For the time being we return to the study of the properties of the boundary and of the isotropy subgroup at a point in the manifold $\mathfrak{M}$ regarded as a homogeneous space of the group G.

Definition. A group Γ is said to have the *fixed point property* if for an arbitrary continuous representation $(\mathscr{V}, E)$ of the group Γ on a linear locally convex space E and for every non-empty convex subset $Q \subset E$, invariant under the action of the operators $\mathscr{V}(\gamma)$, $\gamma \in \Gamma$, there exists an $x \in Q$ such that

$$\mathscr{V}(\gamma)x = x \quad \text{for} \quad \gamma \in \Gamma.$$

From the classical Schauder–Tichonov fixed point theorem for linear operators one can derive by induction the following result:

THEOREM 17.5.1 (see Bourbaki, 1955, p. 115). *Any solvable group has the fixed point property.*

THEOREM 17.5.2 (Furstenberg, 1965). *The group $\Gamma = MAN$ has the fixed point property.*

Proof. Let dm be the normalized Haar measure on M and $(\mathscr{V}, E)$ a representation of Γ on the space E. We define a projection P by the formula

$$Px = \int_M \mathscr{V}(m) x \, dm.$$

If Q is a convex connected Γ-invariant subspace of E, then $PQ \subset Q$. Let $x_0 \in Q$ be a fixed vector for the action of the group AN. Then $Px_0 \in Q$ is a Γ-fixed element of E, because $mANm^{-1} \subset AN$. □

It is certainly surprising that the property of the group Γ proved above constitutes the basis of the theory of harmonic functions on $\mathfrak{M}$ The formulation of the theory of harmonic functions from the point of view of representation theory is due to H. Furstenberg. His methods are outlined in § 18.1.

Example 1

We shall construct such an imbedding into the space $\boldsymbol{R}^n$ of the symmetric space associated with the orthogonal pair $(\mathfrak{o}(n, 1), \mathfrak{o}(n))$ for which the boundary in the sense here introduced is identical with the topological boundary.

Let us denote by η the quadratic form with the signature $(n, 1)$, i.e.

$$\eta(x) = x_1^2 + \dots + x_n^2 - x_{n+1}^2, \qquad x = (x_1, \dots, x_{n+1}),$$

and let $\mathscr{H}$ be the set of all half-lines in $\boldsymbol{R}^{n+1}$ of the form $R_+ \cdot v$ where $\eta(v) < 0$ and $v_{n+1} > 0$. In other words, $\mathscr{H}$ is the set of all half-lines contained inside the "upper-half of the light cone" which is defined by the equation $\eta(x) = 0$. The half-line determined by v is denoted by $r(v)$. We define the action of $G = SO_0(n, 1)$ on $\mathscr{H}$ by putting

$$g \cdot r(v) := r(gv), \qquad (*)$$

where on the right-hand side we have the "usual" action of a matrix on a vector. Under a natural differentiable structure on $\mathscr{H}$ this action is smooth—one can also prove that it is transitive (Problem 20). For $e_{n+1} = (0, \ldots, 0, 1)$ the isotropy subgroup at the half-line $r(e_{n+1})$ equals $K = \left\{\begin{bmatrix} A & 0 \\ 0 & 1 \end{bmatrix} : A \in \mathrm{SO}(n)\right\}$, and thus $\mathscr{H} = G/K$ is the symmetric space associated with the orthogonal pair $(\mathfrak{o}(n, 1), \mathfrak{o}(n))$. This space is usually called the *n-dimensional Lobatchevski space* or equally often, the *n-dimensional real hyperbolic space*.

Let $B = \{x \in \boldsymbol{R}^{n+1} : \eta(x) < 1 \text{ and } x_{n+1} = 0\}$. B is clearly an n-dimensional ball (in the Euclidean norm). We define a diffeomorphism $\varphi : B \to \mathscr{H}$

$$\varphi(x) = r(x + e_{n+1})$$

and transport the action of G from $\mathscr{H}$ to B with the help of φ. To this end it will be convenient to regard B as a subset of $\boldsymbol{R}^n$ by "forgetting" the last zero component of vectors in B. In an explicit form this action is defined as follows. If $x \in B$ and $g \in \mathrm{SO}_0(n, 1)$ and

$$g \begin{bmatrix} x_1 \\ \vdots \\ x_n \\ 1 \end{bmatrix} = \begin{bmatrix} y_1 \\ \\ \vdots \\ \\ y_{n+1} \end{bmatrix},$$

then

$$g(x) = \left(\frac{y_1}{y_{n+1}}, \ldots, \frac{y_n}{y_{n+1}}\right). \qquad (**)$$

(Indeed, $\eta(y_1, \ldots, y_{n+1}) = \eta(x_1, \ldots, x_n, 1) < 0$ owing to the invariance of the form η, and $y_{n+1} > 0$ since only the connected component of the identity in $\mathrm{SO}(n, 1)$ is considered.) The group G acts equally well on the set of half-lines lying on the "upper half" of the cone $\eta(x) = 0$, more precisely lying in the set $\mathscr{N} = \{v \in \boldsymbol{R}^{n+1} : \eta(v) = 0 \text{ and } v_{n+1} > 0\}$, and the action is still defined by formula $(*)$. The set of these half-lines is diffeomorphic to the unit sphere of dimension $n-1$. We verify this by defining for $x \in S = \{x \in \boldsymbol{R}^{n+1} : \eta(x) = 1 \text{ and } x_{n+1} = 0\}$

$$\psi(x) := r(x + e_{n+1}).$$

Arguing as above, we see that the action of G transports to an action on S, given also by formula (**). To identify S with the boundary of the symmetric space G/K it is enough to note that the group K acts transitively on S and that M is the isotropy subgroup for this action. Both facts are obvious.

17.6. PLANES AND HOROCYCLES IN A SYMMETRIC SPACE

As before, let $\pi: G \to \mathfrak{M} = G/K$ be a natural projection of a group onto a symmetric space $\mathfrak{M}$ of the non-compact type. We also assume that the centre of G is trivial, which does not affect the generality of our considerations but only simplifies the description of the manifold $\mathfrak{M}$ in terms of the group G.

Let $\mathfrak{a}$ be an abelian subspace of $\mathfrak{p}$. The submanifold $E = \pi(A) = \pi(\exp \mathfrak{a})$ has an interesting geometrical interpretation. $\mathfrak{M}$ being a symmetric space, if follows that we can apply the results of Ch. 15. Identifying the space $\mathfrak{p}$ with the tangent space to $\mathfrak{M}$ at $o = eK$, we infer by formula (17.4.2) of this chapter that

$$E = \operatorname{Exp}_o\big(d\pi_e(a)\big).$$

The curves $\gamma_X: t \to \operatorname{Exp} tX, X \in \mathfrak{a}$ are geodesics issuing from the point o. It follows from formula (15.5.3) that the sectional curvature K vanishes at o in the directions belonging to $\mathfrak{a}$.

Similarly we verify that for every point $x \in E$ the sectional curvature vanishes in the directions tangent to E since we may also represent $E = \exp \mathfrak{a} \cdot x$ and again apply formula (15.5.3). The manifold E with the (A-invariant) Riemannian structure induced from $\mathfrak{M}$ is therefore a symmetric space of the Euclidean type.

DEFINITION. A submanifold $E \subset \mathfrak{M}$ is said to be a *plane* if:

(1) Every geodesic in $\mathfrak{M}$ issuing from $o \in E$ in the direction of a vector tangent to E is wholly contained in E (we then say that E is a *geodesic submanifold*),

(2) E is a symmetric submanifold, maximal in $\mathfrak{M}$, of the Euclidean type (relative to the Riemannian structure induced from $\mathfrak{M}$).

The dimension of a plane in $\mathfrak{M}$ is called the *rank of the symmetric space* $\mathfrak{M}$ and denoted by $\operatorname{rank} \mathfrak{M}$.

The correctness of the definition will become clear if we show that all planes have the same dimension. We shall prove even more: namely, that the group G acts transitively on the set of all planes and that $E = A \cdot o$ is a plane. Accordingly $\dim \mathfrak{a} = \operatorname{rank} \mathfrak{M}$.

THEOREM 17.6.1. *Every plane passing through the point* $o = \pi(e)$ *is of the form* $\operatorname{Exp}_o(d\pi_e(\mathfrak{a}))$, where $\mathfrak{a}$ *is a malimal abelian subspace of* $\mathfrak{p}$.

Proof. Let E be a plane and let $\mathfrak{a}_0$ be a subspace of $\mathfrak{p}$ such that $d\pi(\mathfrak{a}_0)$ coincides with the space of tangent vectors to E at $o \in E$. It follows from the definition of E on the basis of formula (15.5.3) that for $X, Y \in \mathfrak{a}_0$

$$B([X, Y], [X, Y]) = B([[X, Y], X], Y) = K(S) = 0,$$

where S is the subspace of $T_0(X)$ spanned by the vectors $d\pi_e(X)$, $d\pi_e(Y)$. Since $[X, Y] \in \mathfrak{k}$, it follows that $[X, Y] = 0$, the Killing form being negative definite on $\mathfrak{k}$. Thus $\mathfrak{a}_0$ is an abelian subspace. It follows from what we said at the beginning of the section that $\mathfrak{a}_0$ is a maximal abelian subspace in $\mathfrak{p}$.

E being a geodesic submanifold, we have $\operatorname{Exp}_0(d\pi(\mathfrak{a})) \subset E$. Since E is of the Euclidean type, the exponential mapping is surjective. □

It follows from Theorems 17.4.1 and 17.6.1 that all planes passing through o are of the form kE, $k \in K$ and therefore have the same dimension, equal to $\dim \mathfrak{a}$. The action of G on $\mathfrak{M}$ being isometric and transitive, this implies that G maps planes into planes and acts transitively on the set of planes.

The results thus obtained can equivalently be formulated as follows:

COROLLARY 17.6.2. *Every plane in* $\mathfrak{M}$ *is the orbit of a subgroup of the form* gAg^{-1} *for a certain* $g \in G$ $(A = \exp \mathfrak{a})$.

The difference between Euclidean geometry and the geometry of a symmetric space of the non-compact type consists in the fact that the isotropy subgroup at a point does not act transitively on a sphere with centre at that point. This is a consequence of the Cartan theorem on the polar coordinates (Theorem 17.4.4). The counterpart of the radius vector issuing from the point o is the submanifold $\operatorname{Exp}_o d\pi_e(\mathfrak{a}^+)$ of dimension equal to the rank of $\mathfrak{M}$. Thus drawing a plane through a point in a symmetric space is an analogue of drawing a line in a Euclidean space. Accordingly, the submanifolds of $\mathfrak{M}$ which we define below, being orthogonal complements of planes, are the counterparts of hyperplanes.

DEFINITION. By a *horocycle* we shall understand the orbit in $\mathfrak{M}$ of a group of the form $g^{-1}Ng$, $g \in G$.

A horocycle is a closed submanifold of $\mathfrak{M}$. Let us observe that the orbit $N \cdot o$ is indeed orthogonal to $E = A \cdot o$ at the point o.

The tangent space to this orbit at o is comprised of all vectors $\mathrm{d}\pi(\mathfrak{n})$.

Denote by $\mathfrak{g}_1$ the subspace of $\mathfrak{p}$ such that $\mathrm{d}\pi(\mathfrak{n}) = \mathrm{d}\pi(\mathfrak{g}_1)$. If $\mathrm{d}\pi(X) = \mathrm{d}\pi(X_1)$, then $X - X_1 \in \mathfrak{k}$, and hence $B(X, H) = B(X_1, H)$ for $H \in \mathfrak{p}$. The inner product on $T_o(M)$ is the transport of the Killing form on $\mathfrak{p}$ via the mapping $\mathrm{d}\pi$.

Now, if $X \in \mathfrak{g}_1$, $H \in \mathfrak{a}$, then

$$\big(\mathrm{d}\pi_e(X)|\,\mathrm{d}\pi_e(H)\big) = B(X, H) = B(X_1, H),$$

where $X_1 \in \mathfrak{n}$. The spaces $\mathfrak{a}$ and $\mathfrak{n}$ being spanned by eigenvectors of operators $\operatorname{ad} H$, $H \in \mathfrak{a}$, corresponding to different eigenvalues, we see that

$$\big(\mathrm{d}\pi_e(X)|\mathrm{d}\pi_e(H)\big) = 0$$

for arbitrary $X \in \mathfrak{g}_1$ and $H \in \mathfrak{a}$.

In a similar manner we verify that the orbit $aNa^{-1}(a \cdot o) = Na \cdot o$ is orthogonal to E at the point $a \cdot o \in E$. It follows from the Iwasawa decomposition that every point in the manifold $\mathfrak{M}$ can be represented in a unique way in the form $x = na \cdot o$. The family of horocycles

$$\{\xi\} = \{\xi_a := Na \cdot o \colon a \in A\}$$

covers $\mathfrak{M}$ and consists of manifolds orthogonal to E and such that $\xi_a \neq \xi_{a'}$ for $a \neq a'$. This being so, we can define the horocyclic coordinates on $\mathfrak{M}$.

PROPOSITION 17.6.3. *The mapping $N \times A \ni (n, a) \to na \cdot o$ is a diffeomorphism onto $\mathfrak{M}$.*

Indeed, it is the pair (n, a) that we call the *horocyclic coordinates* of the point $na \cdot o \in \mathfrak{M}$.

LEMMA 17.6.4. *The group G acts transitively on the space of horocycles.*

Proof. The orbit of the group gNg^{-1} containing the point $g_1 \cdot o$ has the form $\xi = gNg^{-1}g_1 \cdot o$. Performing the Iwasawa decomposition of the element $g_1^{-1}g = kan$ and using the relation $aNa^{-1} = N$ we obtain $\xi = g_1 kN \cdot o$. This shows that ξ is the image of the horocycle $\xi_0 = N \cdot o$ under the translation by $g_1 k$. □

Let us examine the isotropy subgroup of the horocycle ξ_0 by studying the Iwasawa constituents of an element $g \in G$ which verifies the identity $g\xi_0 = \xi_0$ equivalent to the identity

$$gNK = NK. \tag{17.6.1}$$

Suppose that $g^{-1} = kan$. By (17.6.1) we deduce that there exist $n_1 \in N$ and $k_1 \in K$ such that

$$n^{-1}a^{-1}k^{-1} = n_1 k_1.$$

The uniqueness of the Iwasawa decomposition leads to the conclusion that $a = e$.

Clearly, the constituent n of the decomposition is arbitrary. It remains to describe the set $K_1 = \{k \subset K: k\xi_0 = \xi_0\}$. For $k \subset K_1$ we have the relation $kNk^{-1} \subset NK$. The mapping $(dl_k)_o$ leaves invariant the space of vectors tangent to ξ_0, and hence also its orthogonal complement, which, as we know, constitutes the tangent space to the plane $E = A \cdot o$. Through the "usual" identification of $T_0(\mathfrak{M})$ with $\mathfrak{p}$ (and $(dl_k)_o$ with $\mathrm{Ad}(k)$) we arrive at the condition $\mathrm{Ad}(k)\mathfrak{a} = \mathfrak{a}$, and so $k \in \mathfrak{M}'$.

Suppose that $\mathrm{Ad}(k)$ moves the Weyl chamber $\mathfrak{a}^+$ and sends it onto $\mathfrak{a}_1^+$.

There exists a root $\alpha > 0$ such that $\alpha(\mathfrak{a}_1^+) < 0$. Let $X \in \mathfrak{n}$ be a non-zero element of $\mathfrak{g}_\alpha$

$$[H, X] = \alpha(H)X.$$

Then $X' = \mathrm{Ad}(k)X$ satisfies the eigenequation

$$[H, X'] = \alpha\big(\mathrm{Ad}(k^{-1})H\big)X' = \alpha(w^{-1}H)X',$$

where $w = kM$. The root $\alpha \circ w^{-1}$ being negative, we have $X' \in \theta\mathfrak{n} = \bar{\mathfrak{n}}$. Thus the assumption that w is different from the identity of the Weyl group implies

$$(kNk^{-1}) \cap \bar{N} \neq \{e\}.$$

We know that $kNk^{-1} \subset NK$, and consequently we get $(NK) \cap \bar{N} \neq \{e\}$. This, however, is untrue. Namely, suppose that

$$\bar{n}n = k_1 \quad \text{for} \quad \bar{n} \in \bar{N},\ n \in N,\ k_1 \in K.$$

In problem 22 appended to § 17.3 we proved that the operators $\operatorname{Ad} g$ can be represented by $n \times n$ matrices; $\operatorname{Ad} k$ is then a unitary matrix, $\operatorname{Ad} n$ a triangular matrix with unities on the diagonal and zeros below and $\operatorname{Ad} \bar{n}$ a triangular matrix with unities on the diagonal and zeros above. Thus $\operatorname{Ad} \bar{n}n$ is a matrix with unities in the diagonal, and if it is unitary then $\operatorname{Ad} k_1 = \mathrm{id}$, which means that k_1 belongs to the centre of G. However, G acts on $\mathfrak{M}$ effectively and so the centre is trivial and we must have $k_1 = e$.

From the above-mentioned matrix representation of the group, we also derive $N \cap \bar{N} = \{e\}$, and conseqently $\bar{n} = n = k_1 = e$ in contradiction with our assumption. Finally, $w = \mathrm{id}$. We have thus proved

LEMMA 17.6.5. *The isotropy subgroup at the horocycle ξ_0 is MN.*

We represent the space of horocycles as G/MN and thus endow it with a manifold structure and a homogeneous space structure denoted by Ξ.

THEOREM 17.6.6 (Helgason, 1963). *The mapping*

$$B \times A \ni (kM, a) \to ka\xi_0 \in \Xi$$

is a diffeomorphism.

This is a direct consequence of the Iwasawa decomposition and Lemma 17.6.5. Theorem 17.6.6 is an analogue of the Cartan theorem on the polar coordinates in a symmetric space.

The element of the boundary of kM corresponding to the horocycle $\xi = ka\xi_0$ is said to be normal to ξ.

The orbit of the group K in Ξ of the form $K \cdot \xi_0$ corresponds to the pencil of horocycles in $\mathfrak{M}$ passing through the point $o = eK$. The orbit of the subgroup A, i.e. the set $A \cdot \xi_0$, has the properties of a pencil of parallel subspaces:

(1) The set $\xi_1 \cap \xi_2$ is either empty or equal to ξ_1.

(2) The sum of elements of the horocycles belonging to the pencil equals $\mathfrak{M}$.

(3) The geodesics $\gamma_X(t) = \exp tX \cdot o$, $X \in \mathfrak{a}$, intersect at right angles every horocycle in the pencil through which they pass.

If a horocycle ξ belongs to the pencil described above and $x \in \mathfrak{M}$, then there exists a unique element $a \in A$ such that $a \cdot x \in \xi$. That element of A is called the *complex distance* from x to ξ and denoted by $d(\xi, x)$. All points of a horocycle ξ_1 from the pencil of parallels to ξ have the same complex distance from ξ.

The pencil of parallels to a horocycle of the form $k \cdot \xi_0$ consists of the horocycles of the form $ka \cdot \xi_0 = kak^{-1}(k \cdot \xi_0)$, and hence arises either by a rotation of the pencil $A \cdot \xi_0$ or by a translation of the horocycle $k \cdot \xi_0$ by an element of the subgroup kak^{-1}. Such a pencil is denoted by $\mathscr{P}_{k\mathfrak{M}}$.

The complex distance $d(\xi, x)$ from a point $x \in \mathfrak{M}$ to $\xi = ka\xi_0 \in \mathscr{P}_{k\mathfrak{M}}$ is defined as an element $a_1 \in A$ such that

$$X \in \xi_1 = ka_1^{-1}a\xi_0,$$

Thus, if $x = ka_1 n \cdot o$ and $\xi = ka_2 \cdot \xi_0$, then $d(\xi, x) = a_1^{-1}a_2$. Without referring to the decomposition of the space Ξ described in Theorem 17.6.6 we can describe the pencil of parallel horocycles as the set of orbits in $\mathfrak{M}$ of a fixed subgroup gNg^{-1} for a certain $g \in G$.

Example 1

In order to investigate the horocycles in the Lobatchevski space $SO_0(n, 1)/SO(n)$ it will be convenient to use a slightly different model from that discussed in Example 1, § 17.5. It will now be the upper sheet of the hyperboloid $\eta(v) = -1$, i.e. the set $\mathfrak{H} = \{v = (v_1, \dots, v_{n+1}) \colon \eta(v) = -1,\ v_{n+1} > 0\}$. By assigning to a half line $r(v)$ lying inside the light cone the point of its intersection with $\mathfrak{H}$ we define a bijection between the space described in Example 1 and the space $\mathfrak{H}$. The bijection preserves the natural action of the group $G = SO_0(n, 1)$ on both sets. Denote by $\langle \cdot, \cdot \rangle$ the symmetric bilinear form associated with η, i.e. $\langle v, w \rangle = v_1 w_1 + \dots + v_n w_n - v_{n+1} w_{n+1}$, and let $\mathscr{N}$ denote the upper-half of the light cone, i.e. $\mathscr{N} = \{v \colon \eta(v) = 0 \text{ and } v_{n+1} > 0\}$. The horocycle Ne_{n+1} consists of the points $(\frac{1}{2}|x|^2, x_1, \dots, x_{n-1}, 1 + \frac{1}{2}|x|^2)$ where $x = (x_1, \dots, x_{n-1}) \in \boldsymbol{R}^{n-1}$ and is the intersection of the hyperboloid $\mathfrak{H}$ with the hyperplane given by the equation

$$\langle v, e_1 + e_{n+1} \rangle = -1.$$

It follows from the invariance of $\langle \cdot, \cdot \rangle$ that every horocycle is the intersection of $\mathfrak{H}$ with the hyperplane given by $\langle v, s \rangle = -1$, where s is a point in the cone $\mathcal{N}$, obviously a unique one satisfying that equation. Thus we have a bijection between the set of all horocycles in $\mathfrak{H}$ and the cone $\mathcal{N}$. Since $\mathcal{N} \cong G/MN$, where MN is the isotropy subgroup at the point $s_0 = e_1 + e_{n+1}$, this bijection establishes an identification of the space of horocycles with the homogeneous space G/MN.

PROBLEMS

1. (a) Let F be a bilinear form on a Lie algebra $\mathfrak{g}$ such that

$$F([X, Y], Z) = -F(Y, [X, Z]), \qquad X, Y, Z \in \mathfrak{g}$$

(sometimes such a form is referred to as an invariant form). Prove that for every ideal $\mathfrak{h} \subset \mathfrak{g}$ the set $\mathfrak{h}^{\perp} = \{X \in \mathfrak{g}: F(X, Y) = 0,\ Y \in \mathfrak{h}\}$ is an ideal.

(b) Show that a Lie algebra $\mathfrak{g}$ over $\boldsymbol{R}$ is compact if and only if it admits an invariant inner product.

(c) Show that a compact Lie algebra $\mathfrak{g}$ is the direct sum of its centre and the ideal $D\mathfrak{g} = [\mathfrak{g}, \mathfrak{g}]$, which is semisimple.

(d) Prove that $\mathfrak{g}$ is a compact Lie algebra if and only if the group $\operatorname{Int} \mathfrak{g}$ is compact.

Hint. (b) For a compact Lie group G with the Lie algebra $\mathfrak{g}$ construct an inner product invariant under the operators $\operatorname{Ad}_G(g)$, $g \in G$. Reduce the proof of the converse implication to the case of a semisimple Lie algebra by decomposing $\mathfrak{g} = \mathfrak{z} + \mathfrak{h}$ (orthogonal sum), where $\mathfrak{z}$ is the centre of $\mathfrak{g}$.

2. (a) Prove that a compact connected solvable Lie group is a torus. Prove also that a closed connected commutative subgroup of a compact Lie group is a torus.

(b) Classify all compact connected commutative Lie groups.

Hint. Observe that a compact solvable Lie algebra is commutative and use Problem 13, § 1.3.

3. Let $\mathfrak{g}$ be a semisimple Lie algebra over $\boldsymbol{C}$ and $\mathfrak{h} \subset \mathfrak{g}$ a Lie subalgebra.

(a) Show that $\mathfrak{h}$ is a Cartan subalgebra if and only if $\mathfrak{h}$ is a maximal Lie subalgebra having the property that there exists a basis for $\mathfrak{g}$ consisting of common eigenvectors for the operators ad H, $H \in \mathfrak{h}$ (i.e. ad($\mathfrak{h}$)) is a commonly diagonalizable set of operators on $\mathfrak{g}$).

(b) If $\mathfrak{h}$ is a Cartan subalgebra, then the normalizer of $\mathfrak{h}$, i.e. $\{X \in \mathfrak{g}\colon \operatorname{ad} X\mathfrak{h} \subset \mathfrak{h}\}$, equals $\mathfrak{h}$.

(c) Give an example showing that a maximal abelian subalgebra need not contain an element X such that ad X is decomposable.

4. Let $\mathfrak{g} = \mathfrak{sl}(n, \boldsymbol{C})$ and let $\mathfrak{h}$, α_{rs}, E_{rs}, $\mathfrak{g}^{rs}$, etc. have the same meaning as in Example 3.

(a) Prove that $\mathfrak{g}_{rs} := \boldsymbol{C}E_{rs} \oplus \boldsymbol{C}E_{rs} \oplus \boldsymbol{C}(E_{rr} - E_{ss})$ is a Lie subalgebra of $\mathfrak{g}$ isomorphic to $\mathfrak{sl}(2, \boldsymbol{C})$.

(b) Let $X \in \mathfrak{g}_{rs}$ and let α_{pq} be an arbitrary root. Show that if $p \neq s$ and $q \neq r$ then ad $X|_{\mathfrak{g}^{pq}} = 0$ and for $p = s$, $q \neq r$ $(q = r, p \neq s)$ the subspace $\mathfrak{g}^{pq} \oplus \mathfrak{g}^{rq}$ $(\mathfrak{g}^{pq} \oplus \mathfrak{g}^{ps})$ is a smallest space closed under the action of all ad X with $X \in \mathfrak{g}_{rs}$. On the other hand, if $p = s$ and $q = r$, then the same is true of $\mathfrak{g}_{rs} = \mathfrak{g}^{rs} + \mathfrak{g}^{sr} + \boldsymbol{C}H_{rs}$.

(c) Prove that for arbitrary r, s, p, q with $(r, s) \neq (p, q) \neq (s, r)$ the number

$$2\frac{B(H_{rs}, H_{pq})}{B(H_{pq}, H_{pq})}$$

is an integer equal either to 0 or to ± 1. (These are the so called *Cartan numbers* of the algebra $\mathfrak{sl}(n, \boldsymbol{C})$.)

(d) Denote by Γ the abelian subgroup of $\mathfrak{h}$ generated by vectors $\tau_{pq} = \dfrac{2}{B(H_{pq}, H_{pq})} H_{pq}$ and by Γ_1 the abelian subgroup of $\mathfrak{h}$ consisting of all $H \in \mathfrak{h}$ such that $\alpha(H) \in \boldsymbol{Z}$ for all roots α. Verify that $\Gamma \subset \Gamma_1$ and that $\Gamma_1/\Gamma \simeq \boldsymbol{Z}_n$ (the cyclic group of order n).

5. Let $\mathfrak{g} = \mathfrak{o}(2n, \boldsymbol{C})$.

(a) Suppose that $E = \begin{bmatrix} 0 & 1 \\ -1 & 0 \end{bmatrix}$ and let $\mathfrak{h} = \{\operatorname{diag}(x_1 E, x_2 E, \ldots, x_n E)\colon x_i \in \boldsymbol{C}\}$ where by $\operatorname{diag}(x_1 E, \ldots, x_n E)$ we denote the $2n \times 2n$ matrix formed of the blocks $x_i E$ placed along the main diagonal and having zeros in the remaining places. Show that $\mathfrak{h}$ is a Cartan subalgebra of the Lie algebra $\mathfrak{g}$ and find the root decomposition for $\mathfrak{g}$.

(b) Show that the Killing form for $o(2n, \boldsymbol{C})$ is given by the formula

$$B(X, Y) = 2(n-1)\operatorname{Tr}(XY).$$

(c) Find the root vectors H_α, the Cartan numbers $\dfrac{2B(H_\alpha, H_\beta)}{B(H_\alpha, H_\alpha)}$ and the group Γ_1/Γ where Γ is the subgroup generated by $\dfrac{2}{B(H_\alpha, H_\alpha)} H_\alpha$ and Γ_1 is the group comprised of those vectors $H \in \mathfrak{h}$ for which $\alpha(H)$ is an integer for an arbitrary root α.

6. Let $\mathfrak{g}$ be a semisimple algebra over $\boldsymbol{C}$, $\mathfrak{h} \subset \mathfrak{g}$ a Cartan subalgebra, $\Delta \subset \mathfrak{h}^*$ the set of roots of $\mathfrak{g}$ relative to $\mathfrak{h}$ and B the Killing form of $\mathfrak{g}$.

(a) Show that for $X, Y \in \mathfrak{h}$ we have the identity

$$B(X, Y) = \sum_{\alpha \in \Delta} \alpha(X)\,\alpha(Y)$$

and derive hence that the set Δ spans the space $\mathfrak{h}^*$.

(b) Let $H_\alpha \in \mathfrak{h}$ be determined by $B(H, H_\alpha) = \alpha(H)$. Define a vector $\tau_\alpha \in \mathfrak{h}$ by the formula

$$\tau_\alpha = \frac{2}{B(H_\alpha, H_\alpha)} H_\alpha$$

and choose $X_\alpha \in \mathfrak{g}^\alpha$, $X_{-\alpha} \in \mathfrak{g}^{-\alpha}$ so that $[X_\alpha, X_{-\alpha}] = \tau_\alpha$. Show that the vectors X_α, $X_{-\alpha}$, τ_α span a Lie subalgebra of $\mathfrak{g}$ isomorphic to the Lie algebra $\mathfrak{sl}(2, \boldsymbol{C})$. Utilizing the properties of representations of the algebra $\mathfrak{sl}(2, \boldsymbol{C})$ described in Chapter 8, show that for every $\beta \in \Delta$, $\beta(\tau_\alpha)$ is an integer (the numbers $\beta(\tau_\alpha) = \dfrac{2B(H_\alpha, H_\beta)}{B(H_\alpha, H_\alpha)}$ are called the *Cartan numbers of the Lie algebra* $\mathfrak{g}$).

(c) Let $\mathfrak{h}_R$ denote the real subspace of $\mathfrak{h}$ spanned by the vectors τ_α and let $\mathfrak{h}_R^*$ be the subspace of $\mathfrak{h}$ spanned by Δ. Let $H_0 \in \mathfrak{h}_R$ be such a vector that $\alpha(H_0) \neq 0$ for all roots $\alpha \in \Delta$. We shall say that a root α is positive if $\alpha(H_0) > 0$. The set of all positive roots is denoted by Δ_+. Prove that

$$\mathfrak{n} = \sum_{\alpha \in \Delta_+} \mathfrak{g}^\alpha, \qquad \mathfrak{v} = \sum_{\alpha \in \Delta_+} \mathfrak{g}^{-\alpha}$$

are nilpotent subalgebras of $\mathfrak{g}$ and the following decomposition holds

$$\mathfrak{g} = \mathfrak{v} + \mathfrak{h} + \mathfrak{n}.$$

Moreover, prove that, for an arbitrary $X \in \mathfrak{g}^\alpha$, $\operatorname{ad} X$ is a nilpotent operator.

(d) Prove that if $\mathfrak{u}$ is a compact real form of $\mathfrak{g}$ and $\mathfrak{h} = \mathfrak{h}_0 + i\mathfrak{h}_0$ where $\mathfrak{h}_0$ is a maximal abelian subalgebra in $\mathfrak{u}$ then $\mathfrak{h}_R = i\mathfrak{h}_0$.

7. Let $\mathfrak{g}$ be a semisimple Lie algebra over $\boldsymbol{C}$.

(a) Show that every involutive automorphism of the Lie algebra $\mathfrak{g}^R$, antilinear on $\mathfrak{g}$, is a conjugation with respect to one and only one real form of $\mathfrak{g}$.

(b) Let $\mathfrak{g}_1, \mathfrak{g}_2$ be two real forms of the Lie algebra $\mathfrak{g}$ and σ_1, σ_2 the corresponding conjugations of $\mathfrak{g}$. Prove that $\sigma_1\sigma_2 = \sigma_2\sigma_1$ if and only if $\sigma_1\mathfrak{g}_2 \subset \mathfrak{g}_1$ and $\sigma_2\mathfrak{g}_1 \subset \mathfrak{g}_2$.

(c) Let σ_0 be an involutive automorphism of a real form $\mathfrak{g}_1$ of the Lie algebra $\mathfrak{g}$ and let $\mathfrak{k}_0, \mathfrak{p}_0$ be its eigenspaces corresponding to the eigenvalues $+1, -1$, respectively. Show that $\mathfrak{g}_0 = \mathfrak{k}_0 + i\mathfrak{p}_0$ is a real form of $\mathfrak{g}$ such that σ_0 is the restriction to $\mathfrak{g}_1$ of the conjugation with respect to $\mathfrak{g}_0$, and, conversely, that every involutive automorphism of $\mathfrak{g}_1$ can be obtained in this way.

(d) By finding all possible conjugations for $\mathfrak{sl}(2, \boldsymbol{C})$ show that there exist only two non-isomorphic real forms of $\mathfrak{sl}(2, \boldsymbol{C})$, namely the compact form $\mathfrak{sl}(2, C)$ and the non-compact form $\mathfrak{sl}(2, \boldsymbol{C})$.

(e) Show that to the automorphism $\sigma_0\colon \mathfrak{o}(n, \boldsymbol{R}) \to \mathfrak{o}(n, \boldsymbol{R})$ defined as the conjugation by $I_{p,q} = \mathrm{diag}(1, \ldots, 1, -1, \ldots, -1)$ corresponds, via the construction in (c), an algebra isomorphic to $\mathfrak{o}(p, q)$.

(f) Find a conjugation of $\mathfrak{o}(2n, \boldsymbol{R})$ which leads via the construction in (c) to the algebra $\mathfrak{o}^*(2n) = \{X \in \mathfrak{o}(2n, \boldsymbol{C})\colon X^*I + IX = 0\}$, where I is the matrix of the symplectic form $x_1y_2 - x_2y_1 + \ldots$ $\ldots + x_{2n-1}y_{2n} - x_{2n}y_{2n-1}$.

(g) Show that all non-isomorphic real forms of $\mathfrak{o}(n, \boldsymbol{C})$ are $\mathfrak{o}(p, q)$, for $p+q = n$ and when $n = 2k$ is even—$\mathfrak{o}^*(2k)$.

8. (a) Show that for every couple of compact real forms $\mathfrak{u}_1, \mathfrak{u}_2$ of a semisimple algebra g there exists a $P \in \mathrm{Int}\,\mathfrak{g}$ such that $P\mathfrak{u}_1 = \mathfrak{u}_2$.

(b) Let $\mathfrak{g}_0 = \mathfrak{k}_1 + \mathfrak{p}_1 = \mathfrak{k}_2 + \mathfrak{p}_2$ be two Cartan decompositions of a semisimple algebra g_0. Prove that there exists a $P \in \mathrm{Int}(\mathfrak{g}_0)$ such that

$$P(\mathfrak{k}_1) = \mathfrak{k}_2, \qquad P(\mathfrak{p}_1) = \mathfrak{p}_2.$$

Hint. Combine the construction from the proof of Theorem 17.2.4 with Theorem 17.1.1 on differentiations of a semisimple Lie algebra.

9. Let G be a connected Lie group with the Lie algebra $\mathfrak{g}$. Suppose

that $\mathfrak{k} \subset \mathfrak{g}$ is a Lie subalgebra. Prove that the connected subgroup of $\operatorname{Int} \mathfrak{g}$ corresponding to the Lie subalgebra $\operatorname{ad}(\mathfrak{k})$ of the Lie algebra $\operatorname{ad}(\mathfrak{g})$ is isomorphic to $\operatorname{Ad}_G(K)$ where $K \subset G$ is the connected Lie subgroup of G corresponding to $\mathfrak{k}$.

10. Let (G, K) be a Riemann pair, $\mathfrak{k}$ the Lie algebra of the group K and $\mathfrak{z}$ the centre of the Lie algebra of the group G. Show that if $\mathfrak{k} \cap \mathfrak{z} = \{0\}$ then there exists precisely one involutive automorphism σ of the Lie group G such that $(K\sigma)_0 \subset K \subset K_\sigma$.

Hint. Owing to the connectedness of G, σ is uniquely determined by the automorphism $d\sigma$ of the Lie algebra $\mathfrak{g}$. Next observe that the assumption $\mathfrak{k} \cap \mathfrak{z} = \{0\}$ ensures that the Killing form for $\mathfrak{g}$ is negative definite on $\mathfrak{k}$ and the eigenspace for $d\sigma$ corresponding to the eigenvalue -1 is orthogonal to $\mathfrak{k}$ with respect to the Killing form.

11. Let $\mathfrak{g} = \mathfrak{sl}(2, \boldsymbol{C})$, $\mathfrak{g}_0 = \mathfrak{su}(2)$. Show that $\operatorname{Int} \mathfrak{g}^{\boldsymbol{R}} \simeq SO_0(1, 3)$ and $\operatorname{Int} \mathfrak{g}_0 \subset \operatorname{Int} \mathfrak{g}\, \boldsymbol{R}$ is the subgroup of all rotations of the three-dimensional subspace.

12. Suppose that $\mathfrak{g}$ is a semisimple Lie algebra over $\boldsymbol{C}$ with a compact real form $\mathfrak{u}$ and let σ be the conjugation with respect to $\mathfrak{u}$.

(a) Check that $\mathfrak{g}^{\boldsymbol{R}} = \mathfrak{u} + i\mathfrak{u}$ is a Cartan decomposition of the algebra $\mathfrak{g}^{\boldsymbol{R}}$. Taking $\mathfrak{a} = i\mathfrak{t}$ where $\mathfrak{t} \subset \mathfrak{u}$ is a maximal abelian Lie subalgebra of $\mathfrak{u}$, show that the set Σ of the reduced roots with respect to $\mathfrak{a}$ is identical with the set of restrictions to $\mathfrak{a}$ of the roots of $\mathfrak{g}$ with respect to the Cartan subalgebra $\mathfrak{t} + i\mathfrak{t}$.

(b) Let Σ_+ be the set of all positive reduced roots (relative to any ordering of Σ) of the Lie algebra $\mathfrak{g}^{\boldsymbol{R}}$ with respect to $\mathfrak{a} = i\mathfrak{t}$ and let Δ_+ denote the set of those roots of $\mathfrak{g}$ with respect to $\mathfrak{t} + i\mathfrak{t}$ whose restrictions to $\mathfrak{a}$ belong to Σ_+. For a root α (of the algebra $\mathfrak{g}$) we denote by α' the reduced root obtained by restricting α to $\mathfrak{a}$. Show that

$$\mathfrak{n} = \sum_{\alpha \in \Delta_+} \mathfrak{g}^\alpha = \sum_{\alpha \in \Delta_+} \mathfrak{g}^{\alpha'}$$

and that $\mathfrak{n}$ is a nilpotent subalgebra of $\mathfrak{g}$. Prove also that

$$\sigma \mathfrak{g}^\alpha = \mathfrak{g}^{-\alpha}$$

for every root $\alpha \in \Delta$ and that

$$\mathfrak{g} = \sigma\mathfrak{n} + \mathfrak{h} + \mathfrak{n}, \qquad \mathfrak{h} = \mathfrak{t} + i\mathfrak{t}.$$

Moreover, show that the Iwasawa decomposition for the Lie algebra $\mathfrak{g}^R$ is the decomposition

$$\mathfrak{g}^R = \mathfrak{u}+\mathfrak{a}+\mathfrak{n}.$$

(c) Find in this way the Iwasawa decomposition for $\mathfrak{sl}(n, \boldsymbol{C})^R$. In the case of $\mathfrak{sl}(2, \boldsymbol{C})^R$ compare the decomposition thus obtained with the decomposition of the Lie algebra $\mathfrak{o}(3, 1) \simeq \mathfrak{sl}(2, \boldsymbol{C})^R$ obtained in Chapter 7. Show that for a semisimple Lie algebra $\mathfrak{g}$ over $\boldsymbol{R}$ and a given Iwasawa decomposition

$$\mathfrak{g} = \mathfrak{k}+\mathfrak{a}+\mathfrak{n}$$

there exists a basis for $\mathfrak{g}$ in which the matrices of $\operatorname{ad}(X)$ have a triangular form for $X \in \mathfrak{n}$, a diagonal form for $X \in \mathfrak{a}$ and an antisymmetric form for $X \in \mathfrak{k}$.

13. Let $\mathfrak{g}$ be a semisimple Lie algebra over $\boldsymbol{R}$, $\mathfrak{g} = \mathfrak{k}+\mathfrak{p}$ a Cartan decomposition and θ the reflection in $\mathfrak{k}$ corresponding to this decomposition (the Cartan involution).

(a) Show that every subalgebra $\mathfrak{a}$ of the algebra $\mathfrak{g}$ contained in $\mathfrak{p}$ is abelian.

(b) Let $\mathfrak{a}$ be a maximal subalgebra of the algebra $\mathfrak{g}$ contained in $\mathfrak{p}$ and let $\mathfrak{h} \subset \mathfrak{g}$ be a maximal abelian subalgebra of $\mathfrak{g}$ containing $\mathfrak{a}$. Prove that $\theta\mathfrak{n} = \mathfrak{n}$ and that $\mathfrak{n}_C \subset \mathfrak{g}_C$ is a Cartan subalgebra of the complexification $\mathfrak{g}_C$ of the Lie algebra $\mathfrak{g}$.

(c) Let Σ be the set of reduced roots with respect to a maximal subalgebra $\mathfrak{a} \subset \mathfrak{p}$. For $\lambda \in \Sigma$ let $\mathfrak{g}_\lambda$ be the root space, $\mathfrak{h}_C$ the Cartan subalgebra constructed in (b), and Δ the set of roots corresponding to that subalgebra. Prove that

$$\mathfrak{g}_\lambda = \Big(\sum_{\substack{\alpha\in\Delta \\ \alpha|\mathfrak{a}=\lambda}} \mathfrak{g}^\alpha\Big)\cap\mathfrak{g}.$$

Prove also that if $\mathfrak{m}_C$ is the complexification of $\mathfrak{m}$ then

$$\mathfrak{m}_C = (\mathfrak{n}\cap\mathfrak{k})+\sum_{\substack{\alpha\in\Delta \\ \alpha|\mathfrak{a}=0}} \mathfrak{g}^\alpha.$$

(d) If $\mathfrak{g} = \mathfrak{g}^0+\sum_{\lambda\in\Sigma} \mathfrak{g}_\lambda$ is the decomposition into common eigenspaces for $\operatorname{ad}H$, $H \in \mathfrak{a}$, then

$$\mathfrak{g}_0 = \mathfrak{a}+\mathfrak{m}.$$

14. Let $\mathfrak{g}$, $\mathfrak{g}_\alpha$, M, M', etc. have the same meaning as in § 17.3.

(a) Show that $\mathrm{Ad}(k)\mathfrak{g}_\alpha = \mathfrak{g}_\alpha$ for $k \in M$.

(b) Let W act on $\mathfrak{a}'$ contragrediently to the action on $\mathfrak{a}$. Prove that Σ is an invariant subset of $\mathfrak{a}'$.

(c) Show that the reflection S_α is given by the formula

$$S_\alpha(H) = H - 2\frac{\alpha(H)}{\alpha(H_\alpha)}H_\alpha, \qquad H \in \mathfrak{a},$$

where. $H_\alpha \in \mathfrak{a}$ is determined by $B(H_\alpha, H) \equiv \alpha(H)$, $H \in \mathfrak{a}$.

(d) Let Z_α be a non-zero vector in $\mathfrak{g}_\alpha$(i.e. $[H, Z_\alpha] = \alpha(H)Z$ for $H \in a$) and let $Z_\alpha = K_\alpha + P_\alpha$ with $K_\alpha \in \mathfrak{k}$, $P_\alpha \in \mathfrak{p}$. Show that $[H, K_\alpha] = \alpha(H)P_\alpha$, $[H, P_\alpha] = \alpha(H)K_\alpha$ and $[K_\alpha, P_\alpha] = cH_\alpha$ with a $c > 0$.

(e) By choosing Z_α so that $[K_\alpha, P_\alpha] = H_\alpha$ show that $\mathrm{Ad}(\exp tK_\alpha)H = H$ for H such that $\alpha(H) = 0$.

(f) Prove the existence of a $t_0 \in R$ such that

$$\mathrm{Ad}(\exp t_0 K_\alpha)H_\alpha = -H_\alpha.$$

Hint. (c) Use the commutation relation for k and p, the maximality of $\mathfrak{a}$ in $\mathfrak{p}$ and the non-degeneracy of the Killing form on $\mathfrak{a}$. (e) and (f). Apply the formula $\mathrm{Ad}(\exp X) = e^{\mathrm{ad} X}$ and the power series expansion of the exponential function.

15. Prove that the complex distance from the point $o = eK$ to the horocycle $\xi = ka \cdot \xi_0$ equals a. Show also that the point $ka \cdot o \in \xi$ is a unique point of ξ lying at the minimum distance from o. For $a = \exp H$, where $H \in \mathfrak{a}$, the distance equals $\sqrt{B(H, H)}$ (here the distance is determined by the Riemannian structure on $\mathfrak{M}$).

16. Show that for every $x \in G/K$ and $kM \in B$ there exists only one horocycle in the pencil F_{kM} passing through x.

17. Let $\mathfrak{g}$ be a Lie algebra which admits a decomposition into the direct sum $\mathfrak{g} = \mathfrak{g}_1 + \mathfrak{g}_2$ where $\mathfrak{g}_1$, $\mathfrak{g}_2$ are subalgebras (not necessarily ideals) of the Lie algebra $\mathfrak{g}$. Let G be the connected Lie subgroup corresponding to $\mathfrak{g}$ and G_1, G_2 the connected Lie subgroups corresponding to the subalgebras $\mathfrak{g}_1$, $\mathfrak{g}_2$. Show that the mapping

$$\psi\colon G_1 \times G_2 \ni (x_1, x_2) \to x_1 x_2 \in G$$

is an immersion (i.e. the differential $d\psi(x_1, x_2)$ is an isomorphism of the respective tangent spaces for every point $(x_1, x_2) \in G_1 \times G_2$)).

Hint. It suffices to prove the desired property of the differential of the mapping ψ for the point (e, e). Make use of the isomorphism

$$T_{(e,e)}(G_1 \times G_2) \simeq \mathfrak{g}_1 \oplus \mathfrak{g}_2 .$$

18. Let K, A, N be constituents of a global Iwasawa decomposition for a group G. Let $M \subset K$ be the centralizer of A in K.

(a) Prove that $AN = NA = S$ is a solvable subgroup of G and the mapping $A \times N \ni (a, n) \to na \in S$ is an isomorphism of the semi-direct product $A \times_\tau N$ onto S where $\tau: A \to \operatorname{Aut} N$ is defined as $\tau(a)n = ana^{-1}$. Prove also that MAN is a subgroup isomorphic to the semi-direct product of M and S.

(b) By using a problem from § 1.6 find the invariant measures and the modular functions for the groups $S = AN$ and $\Gamma = MAN$.

Hint. For a nilpotent Lie group N with the Lie algebra $\mathfrak{n}$ the exponential mapping $\exp: \mathfrak{n} \to N$ is a diffeomorphism and the Hurwitz measure on N is the transport of the Lebesgue measure on the vector space $\mathfrak{n}$

$$\int_N f(n)\,\mathrm{d}n = \int_{\mathfrak{n}} f(\exp X)\,\mathrm{d}X .$$

19. (a) Prove Theorem 17.3.4 (global Iwasawa decomposition) for the group $G = \mathrm{SL}(n, \boldsymbol{R})$ and the decomposition of its Lie algbra

$$\mathfrak{sl}(n, R) = \mathfrak{k} + \mathfrak{a} + \mathfrak{n}$$

given in Example 1, § 17.3.

Hint. For the subgroups K, A, N one should take $K = \mathrm{SO}(n)$, $N = \{[a_{ij}]: a_{ij} = 0 \text{ for } i > j,\ a_{ii} = 1 \text{ for } i = 1, \ldots, n\}$, $A = \{\operatorname{diag}(x_1, \ldots, x_n): x_i > 0,\ x_1 \ldots x_n = 1\}$. Note that AN is a closed subgroup and KAN a closed subset and use Problem 17 and the connectedness of G to prove the equality $G = KAN$.

(b) Prove this theorem for the group $\operatorname{Int}(\mathfrak{g})$ where $\mathfrak{g}$ is a semisimple Lie group with a given Iwasawa decomposition $\mathfrak{g} = \mathfrak{k} + \mathfrak{a} + \mathfrak{n}$.

20. Let $\mathscr{H}$ be the symmetric space defined in Example 10.

(a) Let $\boldsymbol{R}^{n+1} \supset \Omega := \{v \in \boldsymbol{R}^{n+1}: \eta(v) < 0 \text{ and } v^{n+1} > 0\}$. Show that $\mathscr{H}$ has a unique differentiable manifold structure such that the natural surjection $\pi: \Omega \ni v \to r(v) \in \mathscr{H}$, where $r(v) = \boldsymbol{R}_+ v$, is a smooth mapping, and under this structure the pair $(G, \mathscr{H})$ is a transitive Lie transformation group.

(b) Give a detailed proof of all those statements in Example 1, § 17.5. which, in your opinion, require such a proof.

(c) Prove that the G-invariant measure on $\mathscr{H}$ has the form

$$dm(x) = \frac{dv^1 \dots dv^n}{v^{n+1}}, \qquad x = r(v), \quad v = (v^1, \dots, v^{n+1}).$$

21. (a) Let ξ be a horocycle in G/K. Prove that for an arbitrary $g \in G$ the set $g\xi = \{gx: x \in \xi\}$ is a horocycle. If $E \subset G/K$ is a plane, is $gE = \{gx: x \in E\}$ also a plane?

(b) Let $\mathfrak{H} = G/K$ be the symmetric space from Example 1, § 17.6 and let $\xi \subset \mathfrak{H}$ be a horocycle. Prove in detail that there exists precisely one vector s_ξ in the cone $\mathcal{N}$ such that $\xi = \{v \in \mathfrak{H}: \langle v, s_\xi \rangle = -1\}$ and the transport via the mapping $\xi \to s_\xi$ of the action of the group G on the space of horocycles Σ given by $\xi \to g\xi$ is the natural action $s \to gs$ of the group G on the cone $\mathcal{N}$.

(c) Show that every plane in $\mathfrak{H}$ passing through e_{n+1} is of the form $R\gamma_0$ where $R \in K$ and $\gamma_0: R \to \mathfrak{H}$ is a curve given by $\gamma_0(t) = \exp(tH)e_{n+1} = (\sinh t, 0, \dots, 0, \cosh t)$. Deduce hence that the mapping $K \times \boldsymbol{R}_+ \ni (R, t) \to R\gamma_0(t) \in \mathfrak{H}$ is a diffeomorphism of $S^{n-1} \times \boldsymbol{R}_+$ onto $\mathfrak{H} - \{e_{n+1}\}$ and find the explicit form of the coordinate system on $\mathfrak{H} - \{e_{n+1}\}$ corresponding to this diffeomorphism (the spherical coordinates).

Hint. (b) Note that $N(e_1 + e_{n+1}) = e_1 + e_{n+1}$; for $\xi = gNg^{-1}he_{n+1}$ express s_ξ in terms of the Iwasawa decomposition.

22. Let $\mathfrak{g}$ be a real semisimple Lie algebra. Prove that in $\mathfrak{g}$ there exists a basis such that the matrices of the operators ad X are triangular for $X \in \mathfrak{n}$, diagonal for $X \in \mathfrak{a}$ and antisymmetric for $X \in \mathfrak{k}$.

Chapter 18

Harmonic Analysis on Symmetric Spaces of the Non-Compact Type

18.1. PLANE WAVES AND SPHERICAL FUNCTIONS

Let $G = KAN$ denote, as before, the Iwasawa decomposition of a semi-simple group G of the non-compact type. The constituent A of this decomposition is a vector group and thus the mapping

$$\exp\colon \mathfrak{a} \to A$$

is a diffeomorphism.

It follows that for every $g \in G$ there exists a unique element $H(g)$ such that

$$g = k \exp H(g) n, \qquad k \in K, \qquad n \in N.$$

The restriction of the mapping H to the subgroup A will be denoted by $\log\colon A \to \mathfrak{a}$ as the function inverse to the exponential mapping.

Further, we introduce the mapping $A\colon \mathfrak{M} \times B \to \mathfrak{a}$ defined as

$$A(gK, kM) := -H(g^{-1}k). \tag{18.1.1}$$

Naturally, we should check that the function is defined correctly, i.e. that the value $H(g^{-1}k)$ is independent of the choice of the representations of the classes gK and kM. There is no difficulty in proving this, but we can easily dispel any doubts by explaining the geometrical sense of the element $A(x, b)$.

LEMMA 18.1.1. *If $x \in \mathfrak{M}$ and $b \in B$, then $\exp A(x, b)$ is the complex distance from the point $o = eK \in \mathfrak{M}$ to the horocycle passing through x and belonging to the pencil with the normal $b \in B$.*

Proof. For a fixed pencil of parallel horocycles with the normal $b \in kM$ and an arbitrary $x \in \mathfrak{M}$ there exist unique elements $a \in A$ and $n \in N$ such that $x = kan \cdot o$. The element a is just the complex distance from the horocycle $\xi = ka \cdot \xi_0$ to $o \in M$. We compute

$$\begin{aligned} A(x, b) &= -H(n^{-1}a^{-1}k^{-1}k) = -H(n^{-1}a^{-1}) \\ &= -H(a^{-1}\alpha_a(n^{-1})) = H(a) = \log a. \end{aligned}$$

□

DEFINITION. Let λ be an $\boldsymbol{R}$-linear function on $\mathfrak{a}$ with values in $\boldsymbol{C}$. By a *plane wave on M with frequency λ and the normal $b \in B$* we mean the function

$$e_{\lambda,b}(x) = e^{\lambda(A(x,b))}. \tag{18.1.2}$$

The following property of the plane wave $e_{\lambda,b}$ results directly from Lemma 18.1.1:

COROLLARY 18.1.2. *The plane wave $e_{\lambda,b}$ is constant on every horocycle in the pencil $\mathscr{P}_b$.*

We shall now regard the boundary $B = K/M = G/MAN$ as a homogeneous space with respect to the action of the group G. Also the group K will be viewed as the space G/AN and the action of the group G on K will be denoted by

$$G \times K \ni (g, k) \to g \cdot k \in K.$$

PROPOSITION 18.1.3. *A plane wave satisfies the identity*

$$e_{\lambda,b}(g_1 g_2 K) = e_{\lambda,g_1^{-1}b}(g_2 K)\, e_{\lambda,b}(g_1 K). \tag{18.1.3}$$

Proof. We shall prove that the function H has the property

$$H(g_1 g_2 k) = H\big(g_1(g_2 \cdot k)\big) + H(g_2 k). \tag{18.1.4}$$

To this end we shall make use of the uniqueness of the Iwasawa decomposition. Let $g_2 k = k_2 a_2 n_2$, $g_1 k_2 = k_1 a_1 n_1$. Then

$$g_1 g_2 k = g_1 k_2 a_1 n_2 = k_1 a_1 n_1 a_2 n_2 = k_1 a_1 a_2 n_1' n_2$$

where

$$n_1' = a_2 n_2 a_2^{-1} \in N.$$

Observe that $k_2 = g_2 \cdot k$ and thus

$$a_1 = \exp H\big(g_1(g_2 \cdot k)\big) \quad \text{and} \quad a_2 = \exp H(g_2 k).$$

Hence follows identity (18.1.3). Finally, substituting $A(gK, kM) = -H(g^{-1}k)$, we obtain the assertion of the proposition. □

REMARK. It follows from Proposition 18.1.3 or directly from equation (18.1.3) that the function $G \times B \ni (g, b) \to \sigma_\lambda(g, b) := e_{\lambda,b}(g^{-1}K) = e^{-\lambda(H(gk))}$ is a multiplier. Note also that $\sigma_\lambda(k_1, b) = 1$ identically on $K \times B$ and hence σ_λ is a K-multiplier. Property (18.1.3) of a plane wave e_λ allows us to relate to this function a representation of the group

G on the space of functions on the boundary $B = G/MAN$. Namely we put for a function f on B:

$$\mathcal{U}^\lambda(g)f(b) := e_{\lambda,b}(gK)f(g^{-1} \cdot b).$$

Identity (18.1.3) ensures that

$$\mathcal{U}^\lambda(g_1 g_2) = \mathcal{U}^\lambda(g_1)\mathcal{U}^\lambda(g_2).$$

Now it is easy to pass to the description of $\mathcal{U}^\lambda$ as a representation induced from the subgroup MAN.

Owing to the Iwasawa decomposition we can assign to every function f on B a function on the group G by setting

$$\tilde{f}(kan) = f(kM)e^{-\langle\lambda, \log a\rangle}.$$

The function $\tilde{f}$ satisfies the condition

$$\tilde{f}(gman) = e^{-\langle\lambda,\log a\rangle}f(g). \tag{18.1.5}$$

Every function on G satisfying condition (18.1.5) is uniquely determined by its values on the subgroup K and moreover drops onto the homogeneous space $B = K/M$. Therefore the rule $f \to \tilde{f}$ is a bijection.

We shall compute the action of the left regular representation of the group G on the function $\tilde{f}|_K$:

$$L_g\tilde{f}(k) = \tilde{f}(g^{-1}k).$$

The Iwasawa decomposition of the element $g^{-1}k$ can be written as

$$g^{-1}k = (g^{-1} \cdot k)\exp H(g^{-1}k)n_1, \qquad n_1 \in N,$$

and then

$$L_g\tilde{f}(k) = e^{-\langle\lambda, H(g^{-1}k)\rangle}\tilde{f}(g^{-1}k) = e_{\lambda,kM}(gK)f(g^{-1} \cdot (kM)).$$

Thus the representation $\mathcal{U}^\lambda$ has proved to be the multiplier description of the induced representation. Knowing the properties of induced representations, we can determine those values of the parameter λ for which the representation $\mathcal{U}^\lambda$ is unitary in $L^2(B)$. For this purpose it is enough to find the form of the invariant measure on the subgroup MAN and of its modular function. We avail ourselves of this opportunity to give further information about measures on those groups and homogeneous spaces which interest us.

The group G is diffeomorphic to $K \times S$ ($S = AN$), and so the Haar measure on G is equivalent to the product measure $dk \otimes ds$ where dk, ds are the right Haar measures on K and S, respectively. Thus there exists a function J on $K \times S$ such that

$$\int_G f(g)\mathrm{d}g = \int_K \int_S f(ks)\, J(k,s)\mathrm{d}k\mathrm{d}s$$

for every $f \in \mathscr{C}_0(G)$.

The compact group K being unimodular, it follows that dk is also the left Haar measure. Hence

$$\int_G f(g)\mathrm{d}g = \int_G f(k_1 g)\mathrm{d}g = \int_K \int_S f(k_1 ks)\, J(k,s)\mathrm{d}k\mathrm{d}s$$

$$= \int_K \int_S f(ks) J(k_1^{-1}k, s)\mathrm{d}k\mathrm{d}s = \int_K \int_S f(ks) J(k,s)\mathrm{d}k\mathrm{d}s$$

for arbitrary $k_1 \in K$ and $f \in \mathscr{C}_0(G)$. Thus the identity $J(k,s) = J(e,s)$ is valid for dk-almost all k. The same argument, when applied to the right translate of the function by an element of S, shows that $J(k,s) = J(k,e)$. It follows that the function J is constant (up to a set of measure zero). Therefore the measure dk ds on G is bi-invariant.

The subgroup $S = AN$ is a semi-direct product of the unimodular groups A and N. The left invariant measure for such a subgroup has the form

$$\int_S f(s)\mathrm{d}_l s = \int_A \int_N f(an)\mathrm{d}a\mathrm{d}n, \tag{18.1.6}$$

da, dn denoting the invariant measures on the subgroups A and N, respectively.

We know that if δ_S is the modular function of the group S then the measure $\delta_S^{-1} d_l s$ is right-invariant. We shall find the function δ_S by using a formula of Chapter 1, which shows that

$$\delta_S(s) = |\det \mathrm{Ad}_S s|^{-1}. \tag{18.1.7}$$

Clearly, $\det \mathrm{Ad}_S(an) = \det \mathrm{Ad}_S(a) \det \mathrm{Ad}_S(n) = \det \mathrm{Ad}_S(a)$ since N is a unimodular group and the inner automorphism α_n acts trivially on A. We have the following

LEMMA 18.1.4. $\det \mathrm{Ad}_S(\exp H) = \mathrm{e}^{2\varrho(H)}$, *where* $\varrho = \frac{1}{2} \sum_{\beta \in \Sigma_+} (\dim g_\beta)$.

Proof. We compute

$$\det \mathrm{Ad}(\exp H) = \det \mathrm{e}^{\mathrm{ad}\, H} = \mathrm{e}^{\mathrm{tr}\, \mathrm{ad}\, H}.$$

Every element $Y \in \mathfrak{n}$ is of the form $Y = \sum_{\beta \in \Sigma_+} Y_\beta$, $Y_\beta \in g_\beta$. Hence

$$\operatorname{ad} H(Y) = \sum_{\beta \in \Sigma_+} \beta(H)(Y_\beta) \qquad \text{for} \qquad H \in \mathfrak{a}.$$

Finally, $\operatorname{trace} H = 2\varrho(H)$ and the assertion follows. □

The modular function of the group S has the form

$$\delta_S(an) = e^{-2\varrho(\log a)}.$$

The left Haar measure on the group G can be represented in the form

$$\int_G f(g)\mathrm{d}g = \int_K \int_A \int_N f(kan) e^{2\varrho(\log a)} \mathrm{d}k \, \mathrm{d}a \, \mathrm{d}n. \tag{18.1.8}$$

We shall also give formulas which represent the invariant measures on M and Ξ:

$$\int_M f(x)\mathrm{d}mx = \int_A \int_N f(an)\mathrm{d}a\,\mathrm{d}n, \tag{18.1.9}$$

$$\int_\Xi f(\xi)\mathrm{d}\xi = \int_K \int_A f(ka\xi_0) e^{2\varrho(\log a)} \mathrm{d}k\,\mathrm{d}a. \tag{18.1.10}$$

Proposition 18.1.5.

$$P(gK, kM) = \frac{\mathrm{d}gb}{\mathrm{d}b} := e^{-2\varrho(H(g^{-1}k))} = e_{2\varrho,kM}(g). \tag{18.1.11}$$

Proof. We write formula (18.1.8) in the form

$$\int_G f(g) e^{-2\varrho(H(g))} \mathrm{d}g = \int_K \int_A \int_N f(kan)\mathrm{d}k\,\mathrm{d}a\,\mathrm{d}n = \int_K \int_S f(ks)\mathrm{d}k\,\mathrm{d}_l s$$

$$= \int_B \mathrm{d}b \int_\Gamma f(k\gamma)\mathrm{d}\gamma,$$

where $\mathrm{d}_l s$ denotes the left Haar measure on $S = AN$ and $\mathrm{d}\gamma$ the left measure on $\Gamma = MAN$. Let us compare this formula with the decomposition of the Haar measure on G with respect to the measure $\mathrm{d}k$, which is a G-quasi-invariant measure on G/S. The function $G \ni g \to e^{-2\varrho(H(g))}$ is the ϱ-function (cf. Theorem 1.7.3) for the measure $\mathrm{d}b$ on B. On account of formula (1.7.6) we obtain

$$\frac{\mathrm{d}g \cdot b}{\mathrm{d}b} = e^{-2\varrho(H(g^{-1}k))} = e_{2\varrho,kM}(gK).$$

PROPOSITION 18.1.6. *The representation $\mathscr{U}^{i\lambda+\varrho}$ on $L^2(B)$ is unitary for real functionals on* $\mathfrak{a}$.

Proof. The representation $\mathscr{U}^{i\lambda+\varrho}$ is equivalent to the representation induced by the unitary character of the subgroup MAN given by the formula

$$\chi_\lambda(man) = e^{i\langle\lambda, \log a\rangle}.$$

The modular function of the group MAN is of the form $man \to e^{-2\varrho(\log a)}$, and hence the assertion follows. □

Let us observe that the restriction of the representation $\mathscr{U}^\lambda$ to the subgroup K coincides for an arbitrary λ with the quasi-regular representation of K on $L^2(K/M)$, which is unitary. The subspace of K-fixed vectors for this representation is 1-dimensional. Let us compute the zonal spherical function of the representation U^λ, which we shall denote by ω_λ.

$$\omega_\lambda(g) = (f_0|\mathscr{U}_g^\lambda f_0) = \int_B e_{\lambda,b}(gK)\mathrm{d}b = \int_B e_{\lambda,b}(gK)\mathrm{d}b$$

$$= \int_K e^{-\lambda(H(g^{-1}k))}\mathrm{d}k. \tag{18.1.12}$$

In particular the zonal spherical function of the unitary representation $U^{i\lambda+\varrho}$ has the form

$$\omega_{i\lambda+\varrho}(g) = \int_B e_{i\lambda+\varrho,b}(gK)\mathrm{d}b = \int_K e^{(-i\lambda-\varrho)(H(g^{-1}k))}\mathrm{d}k.$$

We introduce the notation

$$\phi_\lambda(g) = \int_B e_{i\lambda+\varrho,b}(gK)\mathrm{d}b = \int_K e^{-(i\lambda+\varrho)(H(g^{-1}k))}\mathrm{d}k. \tag{18.1.13}$$

Harish-Chandra (1958) proved

LEMMA 18.1.7.

$$\phi_\lambda(g^{-1}) = \phi_{-\lambda}(g).$$

Proof. Utilizing the algebraic identities (18.1.3), we obtain

$$1 = e_{\lambda,b}(gg^{-1}K) = e_{\lambda,g^{-1}b}(g^{-1}K)e_{\lambda,b}(gK).$$

Hence

$$e_{\lambda,b}(g^{-1}K) = e_{-\lambda,gb}(gK). \tag{18.1.14}$$

Further, we compute

$$\phi_\lambda(g^{-1}) = \int_B e_{i\lambda+\varrho,b}(g^{-1}K)\mathrm{d}b = \int_B e_{-i\lambda-\varrho,gb}(gK)\mathrm{d}b$$

$$= \int_B e_{-i\lambda-\varrho,b}(gK)\frac{\mathrm{d}g\cdot b}{\mathrm{d}b}\mathrm{d}b = \int_B e_{-i\lambda+\varrho,b}(gK)\mathrm{d}b$$

$$= \phi_{-\lambda}(g). \qquad \square$$

It follows from the lemma that the function ϕ_λ can also be represented as

$$\phi_\lambda(g) = \int_B \mathrm{e}^{(i\lambda-\varrho)H(gk)}\mathrm{d}k = \int_B \sigma_{i\lambda+\varrho}(g, b)\mathrm{d}b.$$

The zonal spherical function of the unitary representation $\mathcal{U}^{i\lambda+\varrho}$ can now be written in the form

$$\omega_{i\lambda+\varrho}(g) = \int_B e_{i\lambda+\varrho,b}(gK)\mathrm{d}b = \phi_\lambda(g) = \int_B \mathrm{e}^{(i\lambda-\varrho)H(gk)}\mathrm{d}k$$

$$= \int_B \sigma_{-i\lambda+\varrho}(g, b)\mathrm{d}b. \tag{18.1.15}$$

Example 1

Plane waves and spherical functions for the n-dimensional Lobatchevski space.

Denote by $\langle\cdot, \cdot\rangle$ the bilinear symmetric form on $\boldsymbol{R}^{n+1}$

$$\langle x, y\rangle := x^1y^1 + \ldots + x^ny^n - x^{n+1}y^{n+1},$$

by $(\cdot\,|\,\cdot)$ the natural inner product on $\boldsymbol{R}^n$ and by $\|\cdot\|$ the corresponding norm on $\boldsymbol{R}^n$.

The connected group of all automorphisms of $\boldsymbol{R}^{n+1}$ preserving the form $\langle\cdot, \cdot\rangle$ is denoted by G, $G = SO_0(n, 1)$. Let K, A, N be selected as in Example 2 in § 17.3. Let us define a function t: $G \to \boldsymbol{R}$ by the formula

$$t(g) := \alpha(H(g)),$$

where $H(g) \in \mathfrak{a}$ is given by the Iwasawa decomposition for G according to the formula $g = k(g)\exp H(g)n$, and $\alpha \in \mathfrak{a}'$ is the unique positive root. We identify the space $\mathfrak{a}'_C$ with C by means of $C \ni \lambda \to \lambda\alpha \in \mathfrak{a}'_C$. Then we have

$$\exp H(g)(e_1+e_{n+1}) = e^{t(g)}(e_1+e_{n+1}),$$

and since $Ke_{n+1} = e_{n+1}$, $N(e_1+e_{n+1})$, also

$$|\langle e_{n+1}, g(e_1+e_{n+1})\rangle| = e^{t(g)}.$$

For the plane wave $e_{\lambda,b}$

$$e_{\lambda,kM}(gK) = e^{\lambda\alpha(H(g^{-1}k))} = e^{\lambda t(g^{-1}k)}$$

we obtain

$$\begin{aligned} e_{\lambda,kM}(gK) &= |\langle e_{n+1}, g^{-1}k(e_1+e_{n+1})\rangle|^{\lambda} \\ &= |\langle ge_{n+1}, k(e_1+e_{n+1})\rangle|^{\lambda}. \end{aligned}$$

Writing $x = ge_{n+1} \in \mathscr{H}$, $\tilde{s} = ke_1+e_{n+1} = s+e_{n+1}$ with $s \in S$ (cf. Ex. 1, § 17.4), we get

$$e_{\lambda,s}(x) = |\langle x, \tilde{s}\rangle|^{\lambda}, \qquad x \in \mathscr{H}, \qquad s \in S.$$

Let φ_λ be the zonal spherical function given by formula (18.1.13) and let $x_t = \exp(tH)e_{n+1}$. Then

$$x_t = \exp(tH)e_{n+1} = (\sinh t, 0, \ldots, 0, \cosh t)$$

and for $\tilde{s} = (s^1, \ldots, s^n, 1)$ we obtain

$$e_{\lambda,s}(x_t) = (\cosh t - s^1 \sinh t)^{\lambda}.$$

Next using the identification of K/M with S, which transports the normalized K-invariant measure on K/M into the normalized $SO(n)$-invariant measure ds on S, we obtain

$$\varphi_\lambda(x_t) = \int_S (\cosh t - s^1 \sinh t)^{-i\lambda - \frac{n-1}{2}} ds.$$

In terms of spherical coordinates on the sphere S (cf. Chapter 9)

$$\begin{aligned} s^1 &= \cos\theta_1, \\ s^2 &= \sin\theta_1 \cos\theta_2, \\ &\cdots\cdots\cdots\cdots \\ s^n &= \sin\theta_1 \ldots \sin\theta_{n-2} \sin\theta_{n-1} \end{aligned}$$

the measure ds takes the form

$$ds = \frac{\Gamma\left(\frac{n}{2}\right)}{2\pi^{\frac{n}{2}}} \sin^{n-2}\theta_1 \ldots \sin\theta_{n-2}\, d\theta_1 \ldots d\theta_{n-1},$$

and hence

$$\varphi_\lambda(x_t) = \frac{\Gamma\left(\frac{n}{2}\right)}{\pi^{\frac{1}{2}}\Gamma\left(\frac{n-1}{2}\right)} \int_0^\pi (\cosh t - \cos\varphi \sinh t)^{-i\lambda-\frac{n-1}{2}} \sin^{n-2}\varphi \, \mathrm{d}\varphi.$$

On account of the integral representation of the Legendre function P^μ_ν (see Problem 13, Chapter 10) we get

$$\varphi(x_t) = 2^{\frac{n}{2}-1} \Gamma\left(\frac{n}{2}\right) \sinh^{1-\frac{n}{2}}(t) \, P^{1-\frac{n}{2}}_{-i\lambda-\frac{1}{2}}(\cosh t).$$

From Problem 13, Chapter 10, we also obtain the following formulas:

(a) $n = 2m+1$,

$$\varphi_\lambda(x_t) = \frac{1\cdot 3 \ldots (2m-1)}{\lambda^2(\lambda^2+1^2)\ldots(\lambda^2+(m-1)^2)} \left(\frac{-1}{\sinh t}\frac{\mathrm{d}}{\mathrm{d}t}\right)^m \cos\lambda t,$$

(b) $n = 2m$,

$$\varphi_\lambda(x_t) = \frac{2^{m-1}(m-1)!}{\left(\lambda^2+\left(\frac{1}{2}\right)^2\right)\left(\lambda^2+\left(\frac{3}{2}\right)^2\right)\ldots\left(\lambda^2+\left(m-\frac{3}{2}\right)^2\right)} \times \left(\frac{-1}{\sinh t}\frac{\mathrm{d}}{\mathrm{d}t}\right)^{m-1} P_{\frac{1}{2}+i\lambda} \cosh t$$

found by Takahashi (1963).

We complete this section with a digression concerning the decomposition of a harmonic function on $\mathfrak{M}$ into plane waves.

The notion of a harmonic function on a homogeneous space was introduced by R. Godement, who also proved the fundamental mean values theorem (Godement, 1952). Next, Furstenberg formulated and proved the Poisson theorem for harmonic functions on a symmetric space of the non-compact type (Furstenberg, 1963).

We adopt the following definition of a harmonic function:

DEFINITION. A smooth function f on the symmetric space $\mathfrak{M} = G/K$ is said to be *harmonic* if it satisfies the integral equation

$$\int_K f(gkh\cdot o)\mathrm{d}k = f(g\cdot o). \tag{18.1.16}$$

From the point of view of the theory of spherical functions a harmonic function is a spherical function associated with the zonal spherical function identically equal to one.

It follows from Theorem 5.4.2 that any harmonic function satisfies the system of differential equations

$$Df = 0 \tag{18.1.17}$$

with $D \in \mathrm{D}(\mathfrak{M})$. The result of Godement mentioned above states that conditions (18.1.16) and (18.1.17) are equivalent.

The Poisson formula for a harmonic function corresponds to the fixed point property of the Borel subgroup of the group G.

For a fixed bounded harmonic function f on $\mathfrak{M}$ we consider the set $Q \subset L^\infty(G)$ consisting of those functions ψ which satisfy

(1) $f(g \cdot o) = \int_K \psi(gkh \cdot o)\mathrm{d}k, \qquad g, h \in G,$

(2) $\|\psi\|_\infty \leqslant \|f\|_\infty$.

The set Q is non-empty since the function

$$G \ni g \to f(g \cdot o)$$

belongs to it.

Moreover, it is a convex closed bounded set in $L^\infty(G)$—the dual space to the Banach space $L^1(G)$, and therefore is compact in the weak topology on $L^\infty(G)$.

Conditions (1) and (2) are both invariant under right translations from G. Since right translations act continuously on $L^\infty(G)$ in the weak topology, we can apply Theorem 17.5.2 from the previous chapter to obtain

LEMMA 18.1.8. *For every bounded harmonic function f on $\mathfrak{M}$ there exists a measurable function $f^b \in L^\infty(B)$ such that*

$$f(g \cdot o) = \int_K f^b(fgkMAN)\mathrm{d}k \tag{18.1.18}$$

and

$$\|f^b\|_\infty = \|f\|.$$

In formula (18.1.18) we identified the fixed point h of the right regular representation of the group MAN on Q with a function $f^b \in L^\infty(B)$ in the following way

$$f^b(gMAN) = h(g).$$

The functional on $L^\infty(B)$ which to a function φ assigns the number $\int_K \varphi(kMAN)\mathrm{d}k$ is a unique K-invariant probability measure on B, denoted by $\mathrm{d}b$. Formula (18.1.18) now assumes the form

$$f(g\cdot o) = \int_B f^b(g\cdot b)\mathrm{d}b = \int_B f^b(b)\frac{\mathrm{d}gb}{\mathrm{d}b}\,\mathrm{d}b.$$

The Radon–Nikodym derivative of the measure $\mathrm{d}gb$ with respect to $\mathrm{d}b$, regarded as a function on $G\times B$, is right K-invariant in the first argument:

$$\mathrm{d}(gk)b = \delta_{gk} * \mathrm{d}b = \delta_g * \delta_k * db = \delta_g * \mathrm{d}b = \mathrm{d}gb.$$

As before, we write

$$P(gk, b) := \frac{\mathrm{d}gb}{\mathrm{d}b}(b).$$

The function P is called the *Poisson kernel* and the formula

$$f(x) - \int_B f^b(b)P(x, b)\mathrm{d}b = \int_B e_{2\varrho, b}(x)\mathrm{d}b \tag{18.1.19}$$

is called the *Poisson formula*. This is just the expansion of a harmonic function into plane waves mentioned above.

18.2. THE FOURIER TRANSFORMATION ON A SYMMETRIC SPACE

To begin with we define the Fourier transformation on the space $\mathscr{C}_0(\mathfrak{M})$. Namely, by the *Fourier transform* of a function $f\in\mathscr{C}_0(\mathfrak{M})$ we understand the function defined on $a^*_{\mathbf{C}}\in B$ by the formula

$$a^*_{\mathbf{C}}\times B\ni(\lambda, b)\to\hat{f}(\lambda, b) = \int_M f(x)e_{-i\lambda+\varrho, b}(x)\mathrm{d}m(x)$$

$$= \int_G f(gK)\mathrm{e}^{(i\lambda-\varrho)\,(H(g^{-1}k))}\,\mathrm{d}g, \tag{18.2.1}$$

for $b = kM$.

Defining the transform in this way, we are in agreement with the terminology adopted in harmonic analysis on the space $\boldsymbol{R}^n$. The Fourier transform of a function is the integral of the product of the function and a plane wave. However, it should be pointed out that the analogy is deeper if we regard the space $\boldsymbol{R}^n$ as a homogeneous manifold of the motion group and not only of the translation group.

Plane waves on $\boldsymbol{R}^n$ can be parametrized by elements of $\boldsymbol{R}^n$ if we assign to a point $p \in \boldsymbol{R}^n$ the wave

$$\boldsymbol{R}^n \ni x \to e^{i\langle p, x\rangle}.$$

We can also write $p = rs$ where $r \in S^{n-1}$ and then $f(p) = f(r, s)$ is a function on $\boldsymbol{R}_+ \times S^{n-1}$.

Having represented the motion group $M(n)$ as the semi-direct product of the rotations $O(n)$ and the translations of $\boldsymbol{R}^n$, we can regard S^{n-1} as $M(n)/M_0 \times_\tau \boldsymbol{R}^n$ where M_0 is the isotropy subgroup of a non-zero vector in $\boldsymbol{R}^n$. Under the above identifications the sphere is the counterpart of the boundary $B = G/MAN$.

We now pass to the description of the Fourier transformation regarded as an operator intertwining the regular representation of the group on $\mathscr{C}_0(\mathfrak{M})$ with the induced representation. To this end we fix a $\lambda \in a_c^*$ and compute

$$\begin{aligned}(L_g f)^\wedge(\lambda, b) &= \int_M f(g^{-1}x)e_{-i\lambda+\varrho,b}(x)\mathrm{d}m(x)\\ &= \int_M f(x)e_{-i\lambda+\varrho,b}(g\cdot x)\mathrm{d}m(x)\\ &= e_{-i\lambda+\varrho,b}(gK)\int_M f(x)e_{-i\lambda+\varrho\cdot g^{-1},g^{-1}\cdot b}(x)\mathrm{d}m(x)\\ &= (\mathscr{U}^{-i\lambda+\varrho}(g)\hat{f})(\lambda, b). \end{aligned} \tag{18.2.2}$$

The operator $\mathscr{F}_\lambda: f \to \hat{f}(\lambda, \cdot)$ intertwines the left regular representation with the induced representation $\mathscr{U}^{-i\lambda+\varrho}$ which for real values of the parameter λ is unitary, and as will be subsequently verified, also irreducible.

The zonal spherical function for this representation equals

$$\omega_{-i\lambda+\varrho}(g) = \int_B e_{-i\lambda+\varrho,b}(gK)\mathrm{d}b = \Phi_{-\lambda}(g).$$

To every positive definite zonal spherical function ω we can assign by means of the Gelfand–Raikov construction (§ 5.2) an irreducible unitary representation $({}^{\omega}H, {}^{\omega}U)$. The construction consists in assigning to the function $f \in \mathscr{C}_0(G)$ the class $f+N$ where $N = \{f: \omega(f^* * f) = 0\}$. The representation operators act on the classes by left translations. Thus for a fixed $\lambda \in a^*$ we have two operators intertwining the left regular representation on $\mathscr{C}_0(G)$ with the representation $\mathscr{U}^{-i\lambda+\varrho}$ and ${}^{\Phi_{-\lambda}}U$

$$\begin{array}{ccc} & (\mathscr{C}_0(\mathfrak{M}), L) & \\ \swarrow & & \searrow \\ (L^2(B), \mathscr{U}^{-i\lambda+\varrho}) & & ({}^{\Phi_{-\lambda}}H, {}^{\Phi_{-\lambda}}U) \end{array}$$

The representation ${}^{\Phi_\lambda}U$ is irreducible and spherical. As regards the representation $\mathscr{U}^{-i\lambda+\varrho}$, we only know that the subspace of K-fixed vectors of this representation is 1-dimensional. It would be irreducible if this subspace were cyclic for $\mathscr{U}^{-i\lambda+\varrho}$.

Denote by $f_0 \in L^2(B)$ the function equal to 1 identically. Further, let $f \in \mathscr{C}_0(G)$. We compute the value

$$(\mathscr{U}^{-i\lambda+\varrho}(f)f_0)(b) = \int_G f(g)e_{-i\lambda+\varrho,b}(gK)\,dg.$$

If $f \in \mathscr{C}_0(G/K)$ then

$$(\mathscr{U}^{-i\lambda+\varrho}(f)f_0)(b) = \hat{f}(\lambda, b).$$

The vector $f_0 \in L^2(B)$ is cyclic if and only if the Fourier transformation $\mathscr{F}_\lambda: \mathscr{G}_0(\mathfrak{M}) \to L^2(B)$ has the image dense in $L^2(B)$. Thus it must be the case that the representation $\mathscr{U}^{-i\lambda+\varrho}$ is spherical and therefore irreducible and equivalent to the representation ${}^{\Phi_{-\lambda}}U$.

The representations $\mathscr{U}^{i\lambda+\varrho}$ for $\lambda \in a^*$ enter into the so-called principal series of representations for the group G. The irreducibility of these representations was invastigated by F. Bruhat, who in 1956 showed that if $\lambda \neq \lambda \circ w$ for an arbitrary element $w \neq e$ of the Weyl group then the representation $\mathscr{U}^{i\lambda+\varrho}$ is irreducible.

In 1969 B. Kostant strengthened this result by proving the irreducibility of almost all representations in the principal series (cf. Th. 18.5.4).

Owing to this result we can identify for $\lambda \in a^*$ the representations $\mathscr{U}^{i\lambda+\varrho}$ and ${}^{\Phi_\lambda}U$, and the corresponding intertwining operators

$$\begin{array}{ccc} & (\mathscr{C}_0(\mathfrak{M}), L) & \\ {\scriptstyle F_\lambda}\swarrow & & \searrow{\scriptstyle T} \\ (L^2(B), \mathscr{U}^{-i\lambda+\varrho}) & \simeq & ({}^{\Phi_{-\lambda}}H, {}^{\Phi_{-\lambda}}U). \end{array}$$

We shall now find the form of the Fourier transformation for those functions in $\mathscr{C}_0(\mathfrak{M})$ which are left K-invariant and thus can be identified with elements of the space $\mathscr{C}_0(K\backslash G/K)$. The function $f(\lambda, \cdot)$ on B is then also K-invariant and therefore constant. The restriction of the Fourier transformation to $\mathscr{C}_0(K\backslash G/K)$ is called the *spherical Fourier transformation* on M. It is the operator which to a function $f \in \mathscr{C}_0(K\backslash G/K)$ assigns the scalar function on a_C^*:

$$\begin{aligned}\hat{f}(\lambda, b) &:= \int_{\mathfrak{M}} e_{-i\lambda+\varrho,b}(x)f(x)\mathrm{d}m(x)\\ &= \int_{\mathfrak{M}} e_{-i\lambda+\varrho,b}(x) \int_K f(kx)\mathrm{d}k\,\mathrm{d}m(x)\\ &= \int_{\mathfrak{M}}\int_K e_{-i\lambda+\varrho,b}(kx)\mathrm{d}k\, f(x)\mathrm{d}m(x)\\ &= \int_{\mathfrak{M}}\int_K e_{-i\lambda+\varrho,kb}(x)\mathrm{d}k\, f(x)\mathrm{d}m(x) = \int_{\mathfrak{M}} \Phi_{-\lambda}(x)f(x)\mathrm{d}m(x).\end{aligned}$$

We introduce the notation

$$\tilde{f}(x) := \int_{\mathfrak{M}} \Phi_{-\lambda}(x)f(x)\mathrm{d}m(x). \tag{18.2.3}$$

We know from the general theory of spherical functions that the functional

$$C_0(K\backslash G/K) \ni f \to \tilde{f} \in \boldsymbol{C}$$

is multiplicative and, for real values of λ, also unitary.

The examples of the groups $\mathrm{SL}(2, \boldsymbol{R})$ and $\mathrm{SL}(2, \boldsymbol{C})$ presented earlier, show that the knowledge of the inversion formula for the spherical Fourier transformation, is sufficient to obtain the Plancherel formula and the inversion formula, for the general Fourier transformation. These problems will be dealt with in § 18.6. But first we shall acquaint ourselves with the properties of spherical functions of a semisimple group.

18.3. PROPERTIES OF SPHERICAL FUNCTIONS

A plane wave satisfies equation (18.1.3), which can be written in the form

$$e_{\lambda,b}(gk \cdot x) = e_{\lambda,k^{-1}g^{-1}\cdot b}(x)\, e_{\lambda,b}(gK).$$

Integrating both sides over K, we obtain

$$\int_K e_{\lambda,b}(gk\cdot x)dk = e_{\lambda,b}(gK)\omega_\lambda(x), \tag{18.3.1}$$

where ω_λ is the zonal spherical function of $\mathscr{U}^\lambda$(cf.(18.1.12)). We have thus proved that the plane wave $e_{\lambda,b}$ is a spherical function associated with the zonal spherical function ω^λ. According to Theorem 5.4.2 the plane wave is an eigenfunction of every G-invariant differential operator on the space $\mathfrak{M}$.

To compute the corresponding eigenvalue we shall view the plane wave as a function on the group

$$G \ni g \to e_{\lambda,g}(gK) \in C.$$

Let X be an element of the envelopping algebra $\mathscr{G}$ of the Lie algebra $\mathfrak{g}$. We choose a basis for $\mathfrak{g} = \mathfrak{k}+\mathfrak{a}+\mathfrak{n}$ is such a way that the vectors $X_1, \ldots, X_k$ span the subalgebra $\mathfrak{k}$, the vectors $H_1, \ldots, H_r$ span the subalgebra $\mathfrak{a}$ and the vectors $N_1, \ldots, N_s$ span the subalgebra $\mathfrak{n}$ from an Iwasawa decomposition for $\mathfrak{g}$.

In virtue of the Poincaré–Birkhoff–Witt theorem there exist unique constants such that

$$X = \sum_{\alpha,\beta,\gamma} b_{\alpha,\beta,\gamma} N_1^{\alpha_1} \ldots N_s^{\alpha_s} H_1^{\beta_1} \ldots H_r^{\beta_r} X_1^{\gamma_1} \ldots H_k^{\gamma_k}, \tag{18.3.3}$$

α, β, γ being multi-indices.

The element X can uniquely be represented in the form $X = X_{\mathfrak{k}} + X_{\mathfrak{a}} + X_{\mathfrak{n}}$, where $X_{\mathfrak{k}}$ is the sum of all those constituents of the decomposition (18.3.3) for which the multi-index γ is non-zero:

$$X_{\mathfrak{a}} = \sum_\beta b_\beta H_1^{\beta_1} \ldots H_r^{\beta_r}. \tag{18.3.4}$$

The constituent $X_{\mathfrak{n}}$ is composed of all those elements of decomposition (18.3.3) which do not enter into $X_{\mathfrak{k}}$ and moreover contain a non-zero multi-index α.

Let $\tilde{X}$ denote the left-invariant differential operator on G determined by $X \in \mathscr{G}$.

The function $g \to e_{\lambda,b}(gK)$ being right K-invariant, this implies $\tilde{X}_{\mathfrak{k}} e_{\lambda,b} = 0$. Further, since $e_{\lambda,b}(nK) \underset{\mathfrak{n}}{=} 1$, we get

$$\tilde{X}_{\mathfrak{n}} e_{\lambda,b}(eK) = 0.$$

Finally $\tilde{X}e_{\lambda,b}(eK) = \tilde{X}_{\mathfrak{a}}e_{\lambda,b}(eK)$. Keeping in mind that $e_{\lambda,eM}(aK) = e^{\lambda}(\log a)$, we conclude that

$$\tilde{X}e_{\lambda,eM}(eK) = \sum_{\beta} b_{\beta}\langle\lambda, H_1\rangle^{\beta_1} \ldots \langle\lambda, H_r\rangle^{\beta_r}.$$

We have thus constructed a rule which to an element $X \in \mathscr{G}$ assigns a polynomial function w_X on $\mathfrak{a}_{\mathbf{C}}^*$ or equivalently, an element of the complexification of the symmetric tensor algebra $S(\mathfrak{a})$.

We shall apply the mapping $\mathscr{G} \ni X \to \omega_X \in S(\mathfrak{a})$ to get a description of the algebra $\mathrm{D}(\mathfrak{M})$ of all G-invariant differential operators on $\mathfrak{M}$. Denote by $\mathscr{G}_0$ the subalgebra comprised of those elements of $\mathscr{G}$ for which

$$\operatorname{Ad} k(X) = X \qquad \text{for } k \in K. \tag{18.3.5}$$

LEMMA 18.3.1. *The algebra* $\mathrm{D}(\mathfrak{M})$ *is isomorphic to the restriction to* $\mathscr{E}(G/K)$ *of the algebra of left-invariant differential operators on* G. *In precise terms*:

for every $D \in \mathrm{D}(\mathfrak{M})$ *there exists a* $\tilde{D} \in \mathscr{G}_0$ *such that for* $f \in \mathscr{E}(G/K)$

$$(Df) \circ \pi = \tilde{D}(f \circ \pi) \tag{18.3.6}$$

(as before, $\pi: G \to G/K$ *denotes the natural projection).*

Proof. Let $\mathfrak{g} = \mathfrak{k} + \mathfrak{p}$ be a Cartan decomposition of the Lie algebra of the group G_0. The mapping $(\mathrm{d}\pi)_e$ maps $\mathfrak{p}$ onto $T_0(G/K)$ injectively, which allows us to identify these spaces. We have the representation

$$Df(o) = P(Y_1, \ldots, Y_p) f \circ \pi(e),$$

where P is a real polynomial and $\{Y_i\}_{i=1}^p$ is a basis for $\mathfrak{p}$. Then

$$Df(g \circ o) = P(\tilde{Y}_1, \ldots, \tilde{Y}_p) f \circ \pi(g),$$

where $\tilde{Y}_1, \ldots, \tilde{Y}_p$ denote the left-invariant vector fields on G determined by $Y_1, \ldots, Y_p$.

We define an operator $\tilde{D}_0$ on G by setting

$$\tilde{D}_0 = P(\tilde{Y}_1, \ldots, \tilde{Y}_p).$$

It remains to show that $\tilde{D}$ can be chosen from $\mathscr{G}_0$.

For $f \in \mathscr{E}(G/K)$ we have the formula

$$\begin{aligned}(r_k \circ \tilde{D}_0 \circ r_k^{-1})(f \circ \pi)(e) &= r_k(\tilde{D}(f \circ \pi))(e) = \tilde{D}(f \circ \pi)(k) \\ &= (Df)(k \circ o) = Df(o).\end{aligned}$$

The operator

$$\tilde{D} := \int_K r_k \tilde{D}_0 r_k^{-1} \mathrm{d}k$$

belongs to G_0 and satisfies (18.3.6).

The simple verification that the mapping $D \to \tilde{D}$ is bijective is left to the reader. □

Now we are in a position to define the mapping

$$\mathrm{D}(\mathfrak{M}) \ni D \to \Gamma_D \in S(\mathfrak{a})$$

where

$$\Gamma_D(\lambda) := \omega_{\tilde{D}}(\lambda + \varrho). \tag{18.3.7}$$

Identity (18.3.2) now takes the form

$$De_{i\lambda+\varrho,b} = \Gamma_D(i\lambda)e_{i\lambda+\varrho,b}. \tag{18.3.8}$$

Example

For the Laplace–Beltrami operator we have

$$\Gamma_D(i\lambda) = -(\langle\lambda|\lambda\rangle - \langle\varrho|\varrho\rangle),$$

where $\langle\cdot|\cdot\rangle$ is the bilinear form on $\mathfrak{a}_C^*$ defined by means of the Killing form in the following way:

for $\lambda \in \mathfrak{a}_C^*$ we denote by H_λ the element of $\mathfrak{a}_C$ given by $\lambda(H) = B(H_\lambda, H)$. Next we define

$$\langle\lambda|\mu\rangle := B(H_\lambda, H_\mu).$$

The form $\langle\cdot|\cdot\rangle$ so defined is W-invariant.

Denote by $I(\mathfrak{a})$ the subalgebra of $S(\mathfrak{a})$ consisting of all tensors invariant under the action of the Weyl group on $\mathfrak{a}$.

The structure of the algebra $\mathrm{D}(\mathfrak{M})$ is described by the following theorem, due to Harish-Chandra.

THEOREM 18.3.2. *The mapping* $\mathrm{D}(\mathfrak{M}) \ni D \to \Gamma_D$ *is an isomorphism of the algebra* $\mathrm{D}(\mathfrak{M})$ *onto the algebra* $I(\mathfrak{a})$.

The proof of the above theorem and a detailed exposition of the remaining subjects presented in this section can be found in Helgason, 1968. We shall study the consequences of this purely algebraic theorem which provides a solution of the basic problems of harmonic analysis on the space $\mathfrak{M}$.

COROLLARY 18.3.3. *The algebra* $\mathrm{D}(\mathfrak{M})$ *is a commutative algebra generated by* r (= rank M) *linearly independent generators.*

The corollary can be deduced from Theorem 18.3.2 by applying Chevalley's theorem (see Chevalley, 1965), which states that if a finite group W generated by reflections acts on an r-dimensional space V then the algebra of polynomials on V invariant under the action of the group W is generated by r linearly independent polynomials.

Observe that in the case of a symmetric space of rank 1 the algebra $\mathrm{D}(\mathfrak{M})$ is generated by the Laplace–Beltrami operator.

COROLLARY 18.3.4. $\Phi_{s\lambda} = \Phi_\lambda$ *for an arbitrary* $\lambda \in \mathfrak{a}_C^*$, $s \in W$.

Proof. We define a projection of the algebra $\mathscr{G}$ onto $\mathscr{G}_0 \cong \mathrm{D}(\mathfrak{M})$ by putting

$$\mathscr{G} \ni X \to X_0 := \int_K \mathrm{Ad}\, k(X) \mathrm{d}k.$$

If $f \in \mathscr{E}(K \backslash G/K)$ then $Xf(e) = X_0 f(e)$. Therefore

$$X\Phi_{s\lambda}(e) = X_0 \Phi_{s\lambda}(e) = \Gamma_{X_0}(s\lambda) = \Gamma_{X_0}(\lambda) = X_0 \Phi_\lambda(e) = X\Phi_\lambda(e)$$

for an arbitrary $X \in \mathscr{G}$. The definition of the function Φ_λ shows directly that it is analytic. All derivatives of the functions $\Phi_{s\lambda}$ and Φ_λ at the point e being equal, we conclude that the functions are identical, as desired. □

Finally let us outline in what manner one can derive from Theorem 18.3.2 the fundamental theorem of Harish-Chandra, which gives a complete description of the space of spherical functions of the group G.

THEOREM 18.3.5 (Harish-Chandra, 1958). *For every spherical function* ω *on* G *there exists a functional* $\lambda \in \mathfrak{a}_C^*$ *such that*

$$\omega(g) = \int_B \mathrm{e}^{(i\lambda - \varrho)H(gk)} \mathrm{d}k.$$

Proof. A zonal spherical function defines a homomorphism $\varkappa$ of the algebra $\mathrm{D}(\mathfrak{M}) \cong I(\mathfrak{a})$ into $\boldsymbol{C}$. Namely, we put

$$\varkappa(\Gamma_D) := D\omega(o) \in \boldsymbol{C}.$$

It turns out that every homomorphism of the algebra $I(\mathfrak{a})$ extends to a homomorphism of the algebra $S(\mathfrak{a})$ (see Helgason, 1968).

Denote by j the natural inclusion mapping $\mathfrak{a}^C \to S(\mathfrak{a})$. We can define an $\boldsymbol{R}$-linear functional on $\mathfrak{a}^C$ by requiring $i\lambda(H) := \tilde{\varkappa}(j(H))$, where $\tilde{\varkappa}$ denotes the extension of $\varkappa$ to $S(\mathfrak{a})$.

Viewing $\Gamma_D \in I(\mathfrak{a})$ as a polynomial function on $\mathfrak{a}_C^*$, we can write

$$\varkappa(\Gamma_D) = \Gamma_D(\lambda).$$

It follows that the functions ω and $\Phi_{-\lambda}$ satisfy the same system of equations

$$D\varphi = \Gamma_D(i\lambda)\varphi,$$

and this implies, as demonstrated in the proof of Corollary 18.3.4, the identity

$$\omega = \Phi_\lambda = \int_B e^{(i\lambda-\varrho)H(gk)}dk. \qquad \square$$

The theorem of Harish-Chandra parametrizes the set of all spherical functions by the elements of the set $W\backslash\mathfrak{a}_C^*$. The positive definite functions of the principal series correspond through this parametrization to the subset $W\backslash\mathfrak{a}^* \subset W\backslash\mathfrak{a}_C^*$.

18.4. ASYMPTOTIC BEHAVIOUR OF A SPHERICAL FUNCTION. THE HARISH-CHANDRA $c(\cdot)$-FUNCTION

In the present section we outline Harish-Chandra's method of investigating the asymptotic behaviour of a spherical function. It consists in expanding the function Φ_λ into a series of exponential functions by using a method resembling the classical technique of finding solutions of ordinary differential equations in the form of power series (this technique was presented in Chapter 13).

We obtain the expansion by studying the equation

$$\Omega\Phi_\lambda = \Gamma_\Omega(i\lambda)\Phi_\lambda. \qquad (18.4.1)$$

The dominating term which appears in the expansion contains the factor $c(\lambda)$ known as the Harish-Chandra $c(\cdot)$-function.

The $c(\cdot)$-function describes the Plancherel measure for the space $L^2(G/K)$. Namely, the following formula holds

$$d\mu = |c(\lambda)|^{-2}d\lambda,$$

where $d\lambda$ denotes the Lebesgue measure on $\mathfrak{a}^*$.

A spherical function, like any function of class $\mathscr{C}(K\backslash G/K)$ is fully determined by its values on the plane $E = A \cdot o$ or even on the subset $A^+ = (\operatorname{Exp}\mathfrak{a}^+) \cdot o$. This fact will allow us to reduce the problems which interest us to an analysis on a Euclidean space.

Let us introduce an operation $f \to f|$ which assigns to a K-invariant function f on $\mathfrak{M}$ the following function on A^+:

$$f|(a) := f(a \cdot o), \qquad a \in A^+.$$

By applying the Cartan theorem on the polar decomposition of a symmetric space we observe that a differential operator $D \in \mathrm{D}(\mathfrak{M})$ determines a unique operator $R(D)$ on A^+ satisfying

$$R(D)(f|) = (Df)| \qquad \text{for } f \in \mathscr{E}(K\backslash G/K). \tag{18.4.2}$$

To this end we argue as follows:

Given a function $\varphi \in \mathscr{E}(A^+)$, we define an element $f \in \mathscr{E}(KA^+ \cdot o)$ by putting

$$f(k \cdot a \cdot o) = \varphi(a).$$

The Cartan theorem tells us that $O = KA^+ \cdot c$ is an open subset of $\mathfrak{M}$ diffeomorphic to $K \times A^+$. The operator D is well defined on the subset $O \subset M$; therefore we can compute the value $Df \in \mathscr{E}(O)$. The invariance of D (K-invariance would be sufficient) implies that Df is a K-invariant function.

We define

$$R(D)\varphi := Df|.$$

The theorem of Peetre characterizes differential operators on a manifold V as those linear mappings of $\mathscr{C}^\infty(V)$ into itself which do not increase the support of a function. Clearly the operator $R(D)$ satisfies this condition and thus is a differential operator. The operator $R(D)$ is called the *radial part* of the operator D. The following expression for $R(D)$ is given in a paper by Harish-Chandra, 1958.

$$R(\Omega) = \Omega_A + \sum_{\alpha \in \Sigma_+} m_\alpha (\coth \alpha) H_\alpha, \tag{18.4.3}$$

where Ω_A is the Laplacian on A, $m_\alpha := \dim \mathfrak{g}_\alpha$ and $H_\alpha = \langle \alpha, \cdot \rangle$ is an element of $\mathfrak{a}$ acting as an invariant differential operator on A. The system of positive roots Σ^+ is not, in general, linearly independent; however, it always spans the space $\mathfrak{a}^*$. Therefore we can choose a subset

$\Sigma_s \subset \Sigma_+$ which constitutes a basis for the space $\mathfrak{a}^*$ and every root has the decomposition

$$\beta = \sum_{\alpha \in \Sigma_s} n_\alpha \alpha, \qquad n_\alpha \in Z.$$

The set Σ_s is called the set of simple roots. The spherical function Φ_λ is a solution of the equation

$$R(\Omega)\Phi_\lambda = -(\langle \lambda, \lambda \rangle + \langle \varrho, \varrho \rangle)\Phi_\lambda, \tag{18.4.4}$$

We will seek solutions of this equation in the form of a power series with respect to the variables, on A^+, determined by the identities:

$$\xi_i(\exp H) = e^{\alpha_i(H)}, \qquad \alpha_i \in \Sigma_s.$$

We assume that the solutions have the following form

$$\psi_\lambda(\exp H) = e^{(-i\lambda-\varrho)(H)} \sum_{\mu \in L} \gamma_\mu(\lambda) e^{-\mu(H)} \tag{18.4.5}$$

for $H \in \mathfrak{a}^+$, where L denotes the lattice in $\mathfrak{a}^*$ formed by linear combinations of simple roots with non-negative integer coefficients. Let us substitute into (18.4.3) the expansion

$$\coth \alpha = 1 + 2 \sum_{k \geqslant 1} e^{-2k\alpha}.$$

Equation (18.4.4) leads to the following recurrence relations for the coefficients $\gamma_\mu(\lambda)$:

$$\{\langle \mu, \mu \rangle - 2i\mu\}\gamma_\mu = 2 \sum_{\alpha \in \Sigma_+} m_\alpha \sum_k \gamma_{\mu-2k}\{\langle \mu + \varrho - 2k\alpha | \alpha \rangle - i\alpha\},$$

where k runs through the set of integers $\geqslant 1$ such that $\mu - 2k\alpha \in L$. Assuming $\gamma_0 = 1$, we obtain an expression for the coefficients γ_λ in the form of rational functions on $\mathfrak{a}_C^*$.

The conditions of convergence of series (18.4.5) were investigated by Harish-Chandra, 1958. S. Helgason proved its convergence on a larger domain. His result is quoted below.

THEOREM 18.4.1. *Let Λ be the complementary domain in $\mathfrak{a}_C^*$ of the set of hyperplanes*

$$S_\mu = \{\lambda : \langle \mu, \mu \rangle = 2i\langle \mu, \lambda \rangle\}, \qquad \mu \in L.$$

For $\lambda \in \Lambda$, $H \in \mathfrak{a}^+$ the series in (18.4.5) *is absolutely convergent.*

It follows directly from the construction of the function ψ_λ that it is a solution of the equation

$$R(\Omega)\psi = -(\langle\lambda, \lambda\rangle + \langle\varrho, \varrho\rangle)\psi$$

while the spherical function is determined by the system of equations

$$R(D)\Phi_\lambda = \Gamma_D(i\lambda)\Phi_\lambda, \qquad D \in D(\mathfrak{M}). \tag{18.4.6}$$

It turns out, however, (Harish-Chandra, 1958) that the function ψ_λ also satisfies the system of equations (18.4.6). This does not mean that ψ_λ must be identical with the restriction to $\mathfrak{a}^+$ of the spherical function Φ_λ. In fact, for every $s \in W$ the function $\psi_{s\lambda}$ also satisfies the system (18.4.6), because the polynomial $\Gamma_D(i\lambda)$ is invariant under the Weyl group.

If we confine ourselves to functionals λ such that $i(w\lambda - s\lambda) \notin L$ for $w \neq s$, then we obtain a linearly independent system of functions $\psi_{s\lambda}$, $s \in W$ which constitutes a basis for the space of all solutions of the system (18.4.6). In this way we arrive at the following

THEOREM 18.4.2 (Harish-Chandra, 1958). *Let Ω' be the set of those $\lambda \in \mathfrak{a}_c^*$ which possess the following properties*:

(1) $i(s\lambda - w\lambda) \notin L$ *for* $s \neq w$, s, $w \in W$,

(2) $\langle\mu, \mu\rangle \neq 2i\langle\mu, w\lambda\rangle$ *for* $\mu \in L - \{0\}$ *and* $w \in W$.

Then there exists a function $c(\cdot)$ *on* Ω' *such that*

$$\begin{aligned}\Phi_\lambda(\exp H) &= \sum_{s\in W} c(s\lambda)\psi_{s\lambda}(\exp H)\\ &= \sum_{s\in W} c(s\lambda)e^{(is\lambda-\varrho)(H)} \sum_{\mu\in L} \gamma_\mu(s\lambda)e^{-\mu(H)}\end{aligned} \tag{18.4.8}$$

for $H \in \mathfrak{a}^+$.

The Harish-Chandra $c(\cdot)$-function determines the asymptotic behaviour of the spherical function ϕ_λ. The dominating term of this sum for large values of $H \in \mathfrak{a}^+$ is the function

$$e^{-\varrho(H)} \sum_{s\in W} c(s\lambda)e^{is\lambda(H)}. \tag{18.4.9}$$

18.5. PROPERTIES OF THE HARISH-CHANDRA $c(\cdot)$-FUNCTION

The Harish-Chandra function, which appeared in the study of asymptotic properties of spherical functions, is found in solutions of central problems of harmonic analysis on the group G and on a symmetric space.

It allows us to describe the Plancherel measure on a symmetric space; in the proof of the Paley–Wiener theorem for a symmetric space a decisive role will be played by the asymptotic estimates for the $c(\cdot)$-function. Moreover, the irreducibility of the representation $\mathscr{U}^{-i\lambda+\varrho}$ depends, as we shall see, upon the regularity of the function c at the points λ and $-\lambda$.

In the present section we list formulas describing the form of the $c(\cdot)$-function and present their simple consequences. The most important are Harish-Chandra's integral formula (Theorem 18.5.2) and the formula in Theorem 18.5.1 which was proved in the case of a symmetric space of rank 1 by Harish-Chandra and in the general case by Gindikin and Karpelewicz.

We begin by establishing an explicit expression for the $c(\cdot)$-function in the case of rank 1.

Example (The case of a symmetric space of rank 1 (for details we refer the reader to Flensted-Jensen (1970, 1972))).

Let α and, if necessary, 2α be positive roots on $\mathfrak{g}$. Write $p = \dim \mathfrak{g}_\alpha$, $q = \dim \mathfrak{g}_{2\alpha}$. By choosing a $H \in \mathfrak{a}$ such that $\alpha(H) = 1$ we identify $\mathfrak{a}^+$ with $]0, \infty[$, $\mathfrak{a}_C^*$ with C and A^+ with $]0, \infty[$ via exp and the previous identification of $\mathfrak{a}^+$ with the half-axis. As a result the radial part of the Laplace–Beltrami operator assumes the form of an ordinary differential operator

$$R(\Delta) = \frac{1}{2(p+4q)}\left[\frac{d^2}{dt^2} + g(t)\frac{d}{dt}\right],$$

where $g(t) = p\coth t + 2q\coth 2t$. Thus the restriction to A^+ of the spherical function φ_λ satisfies the equation

$$\left(\frac{d^2}{dt^2} + g(t)\frac{d}{dt}\right)f + (\lambda^2+\varrho^2)f = 0 \tag{18.5.1}$$

with singularities at the points 0 and ∞. Substituting $F(t) = e^{(-i\lambda+\varrho)t}f(t)$, we transform equation (18.5.1) into

$$\frac{d^2F}{dt^2} + (-2i\lambda+2\varrho+g)\frac{dF}{dt} + 2\varrho(\varrho+i\lambda)(1-g)F = 0.$$

Since for $t > 0$ the function g has the expansion

$$g(t) = p\left(1+2\sum_{n=1}^{\infty} e^{-2nt}\right)+2q\left(1+2\sum_{n=1}^{\infty} e^{-4nt}\right),$$

it is reasonable to seek the solution F_λ in the form of the series $F_\lambda = \sum_{m=0}^{\infty} \Gamma_m e^{-mt}$. Then, for $\lambda \neq -in$, $n \in N$, the recurrence formulas take the form

$$\Gamma_0 = 1, \qquad \Gamma_{2n-1} = 0,$$

and, for $m = 2n$ and $n_0 \in N$ such that $\frac{1}{2}n-1 < n_0 \leqslant \frac{1}{2}n$, the form

$$\Gamma_m = \frac{1}{m(m-2i)}\left[2p\sum_{k=1}^{n} \Gamma_{m-2k}(m-2k-i\lambda+\varrho)+\right.$$

$$\left.+2q\sum_{k=1}^{n_0} \Gamma_{m-4k}(m-4k-i\lambda+\varrho\right].$$

Here all Γ's are functions of the parameter λ. The series $\sum_{m=0}^{\infty} \Gamma_m(\lambda)e^{-mt}$ is convergent on sets of the form $[\varepsilon, \infty[\times K$ where $\varepsilon > 0$ and $K \subset C - \{-iN\}$ is compact. Hence it follows that for $t \to \infty$ the function $f_\lambda(t) = e^{(i\lambda-\varrho)t}F_\lambda(t)$, which is a solution of (18.5.1), behaves as $e^{(i\lambda-\varrho)t}$. Thus, for $\lambda \notin iZ$, $f_\lambda(t)$ and $f_{-\lambda}(t)$ are linearly independent solutions of (18.5.1) and hence the spherical function φ_λ is a combination of these functions

$$\varphi_\lambda(t) = c(\lambda)f_\lambda(t)+c(-\lambda)f_{-\lambda}(t),$$

for $\lambda \notin iZ$.

The function $c(\lambda)$ can be represented in this case in the form

$$c(\lambda) = \frac{2^{\left(\frac{1}{2}p+q-i\lambda\right)}\Gamma(i\lambda)\Gamma\left(\frac{p+q+1}{2}\right)}{\Gamma\left(\frac{p+2+2i\lambda}{4}\right)\Gamma\left(\frac{p+2q+2i\lambda}{4}\right)}. \tag{18.5.2}$$

The method by which Gindikin and Karpelewič (1962) generalized formula (18.5.2) to the case of an arbitrary semisimple group is based on the following observation:

In the set of positive roots Σ^+ we choose the subset

$$\Sigma_0^+ = \{\alpha \in \Sigma^+ : \tfrac{1}{2}\alpha \notin \Sigma^+\}.$$

The elements of the set Σ_0^+ are called *indivisible roots*. The root spaces $\mathfrak{g}_\alpha$ and $\mathfrak{g}_{-\alpha}$ generate a subalgebra $\mathfrak{g}^{(\alpha)}$ in $\mathfrak{g}$ which is semisimple and admits the Cartan decomposition

$$\mathfrak{g}^{(\alpha)} = \mathfrak{k}^{(\alpha)} + \mathfrak{p}^{(\alpha)}$$

where $\mathfrak{k}^{(\alpha)} = \mathfrak{g}^{(\alpha)} \cap \mathfrak{k}$, $\mathfrak{p}^{(\alpha)} = \mathfrak{g}^{(\alpha)} \cap \mathfrak{p}$. A maximal commutative subalgebra of $\mathfrak{p}^{(\alpha)}$ is one-dimensional and equals $\mathfrak{a}^{(\alpha)} = [\mathfrak{g}_\alpha, \mathfrak{g}_{-\alpha}] = RH_\alpha$.

Denote by $G^{(\alpha)}$, $K^{(\alpha)}$, $A^{(\alpha)}$ the connected subgroups of G corresponding to the algebras $\mathfrak{g}^{(\alpha)}$, $\mathfrak{h}^{(\alpha)}$, $\mathfrak{a}^{(\alpha)}$, respectively. The main result of Gindikin and Karpelewič (1962) is the multiplicativity of the $c(\cdot)$-function in the following sense. Let $c^{(\alpha)}(\cdot)$ denote the Harish-Chandra function for the pair $G^{(\alpha)}$, $K^{(\alpha)}$. Then there exists a constant a such that

$$c(\lambda) = a \prod_{\alpha \in \Sigma_0^+} c^{(\alpha)}(\lambda^\alpha). \qquad (18.5.3)$$

where $\lambda^{(\alpha)} = \lambda_{\mathfrak{a}(\alpha)}$. The expression for the $c(\cdot)$-function presented below can be derived most easily from formulas (18.5.2) and (18.5.3).

THEOREM 18.5.1. *The $c(\cdot)$-function has the form*

$$c(\lambda) = I(i\lambda)/I(\varrho),$$

where

$$I(\mu) = \prod_{\alpha \in \Sigma^+} B\left(\tfrac{1}{2} m_\alpha, \tfrac{1}{4} m_{\frac{1}{2}\alpha} + \frac{\langle \mu, \alpha \rangle}{\langle \alpha, \alpha \rangle}\right)$$

and B is the Euler function.

In Helgason (1970) the above formula is transformed into

$$c(\lambda) = \prod_{\alpha \in \Sigma_0^+} \Gamma\left(\tfrac{1}{2}(m_\alpha + m_{2\alpha} + 1)\right) 2^{\frac{1}{2}m_\alpha + m_{2\alpha}} e(\lambda) d(\lambda), \qquad (18.5.4)$$

where

$$d(\lambda) = \prod_{\alpha \in \Sigma_0^+} \Gamma\left(\frac{\langle i\lambda, \lambda \rangle}{\langle \alpha, \alpha \rangle}\right) 2^{-\frac{\langle i\lambda, \alpha \rangle}{\langle \alpha, \alpha \rangle}} \qquad (18.5.5)$$

and

$$e(\lambda)^{-1} = \sum_{\alpha\in\Sigma_0^+} \Gamma\left(\tfrac{1}{2}\left(\tfrac{1}{2}m_\alpha+1+\frac{\langle i\lambda, \alpha\rangle}{\langle \alpha, \alpha\rangle}\right)\right)$$

$$\times \Gamma\left(\tfrac{1}{2}\left(\tfrac{1}{2}m_\alpha+m_{2\alpha}+\frac{\langle i\lambda, \alpha\rangle}{\langle \alpha, \alpha\rangle}\right)\right). \qquad (18.5.6)$$

There is also an integral expression for the $c(\cdot)$-function, given by Harish-Chandra.

THEOREM 18.5.2. *Let* $d\bar{n}$ *be the invariant measure on* $\bar{N}$ *verifying the normalization condition*

$$\int_{\bar{N}} e^{-2\varrho(H(\bar{n}))} d\bar{n} = 1.$$

Then $c(\lambda) = \int_{\bar{N}} e^{-(i\lambda+\varrho)H(\bar{n})} d\bar{n}$.

The following properties of the $c(\cdot)$-function can be read directly from Theorem 18.5.1:

(1) The $c(\cdot)$-function is meromorphic on a dense and open subset of $\mathfrak{a}_C^*$.

(2) The function $c(\cdot)^{-1}$ is holomorphic in a certain cylinder containing $\mathfrak{a}^*$. The only zero of the function c^{-1} in that cylinder is $\lambda = 0$.

Further, in view of the asymptotic properties of the Γ-function, one can prove (see Warner, 1972, II, p. 327) that

(3) There exist constants a, $b > 0$ such that

$$\frac{1}{|c(\lambda)|} \leqslant a(1+|\lambda|)^b \qquad \text{for } \lambda \in \mathfrak{a}^*.$$

This property implies that the function $\frac{1}{c}$ is a tempered distribution on $\mathfrak{a}^*$.

As we remember, the function $z \to \Gamma(z+n)/\Gamma(z)$ for $z \neq 0, -1, -2, \ldots, n \geqslant 1$, is a polynomial. Thus, if all roots are indivisible and the dimensions of the root spaces are even, then the function $c^{-1}(\cdot)$ is a polynomial. We have the following

THEOREM 18.5.3. *If all Cartan subgroups of G are conjugate under inner automorphisms, then the roots of the algebra are indivisible* ($\Sigma_0^+ = \Sigma^+$) *and* m_α *is an even number for all* α.

We note the following property

(4) If all Cartan subgroups of the group G are conjugate, then the function c^{-1} is a polynomial.

S. Helgason proved that the function $e(\cdot)$, i.e. the denominator in expression (18.5.4) for the $c(\cdot)$-function, describes the totality of those $\lambda \in \mathfrak{a}_C^*$ for which the representations $\mathscr{U}^{i\lambda+\varrho}$ are irreducible.

THEOREM 18.5.4. *The representation* $\mathscr{U}^{i\lambda+\varrho}$ *on* $L^2(B)$ *is irreducible if and only if* $e(\lambda)e(-\lambda) \neq 0$.

A functional $\lambda \in \mathfrak{a}_C^*$ is said to be *simple* if the image of the space $\mathscr{C}_0(\mathfrak{M})$ under the Fourier transformation at the point λ is a dense subset of $L^2(B)$.

We have the following characterization of simple functionals:

THEOREM 18.5.5 (Helgason, 1970). *The following statements are equivalent*:

(1) *a functional* λ *is simple*;

(2) *the vector* $\psi_B = 1$ *is a cyclic element of the representation* $\mathscr{U}^{-i\lambda+\varrho}$;

(3) $e(\lambda) \neq 0$.

The function $e(\cdot)$ is different from zero in a certain cylinder containing the set $\mathfrak{a}^*$. Therefore the representations $\mathscr{U}^{i\lambda+\varrho}$ are all irreducible for real values of the parameter λ.

18.6. THE PLANCHEREL FORMULA FOR THE FOURIER TRANSFORMATION ON A SYMMETRIC SPACE

The form of the Plancherel measure for the spherical Fourier transformation on a symmetric space is described by the following

THEOREM 18.6.1 (Harish-Chandra, 1958). *For any* $f \in \mathscr{D}(K\backslash G/K)$ *the function* $\tilde{f}$ *is integrable on* $\mathfrak{a}^*$ *with respect to the measure* $d\mu(\lambda) = |c(\lambda)|^{-2}d\lambda$ *and the following formulas hold*:

$$\int_G |f|^2(x)\,dm(x) = \frac{1}{|W|}\int_{\mathfrak{a}^*} |\tilde{f}(\lambda)|^2|c(\lambda)|^{-2}d\lambda$$

$$= \int_{W\backslash\mathfrak{a}^*} |\tilde{f}(\lambda)|^2|c(\lambda)|^{-2}d\lambda, \tag{18.6.1}$$

where $|W|$ *denotes the rank of the Weyl group;*

$$f(x) = \int_{W\backslash\mathfrak{a}^*} \tilde{f}(\lambda)\Phi_\lambda(x)|c(\lambda)|^{-2}d\lambda. \tag{18.6.2}$$

Observe that formula (18.6.1) follows from (18.6.2) applied to the function $\varphi = f^* * f$.

Theorem 18.6.1 leads in a standard way to the Plancherel formula and the inversion formula for the general Fourier transformation. To this end we proceed as follows:

Let $f \in \mathscr{D}(G/K)$. Denote by F_g the function on the space G/K given by the formula

$$F_g(x) = \int_K f(gk \cdot x)dk = \int_K L_k L_{g^{-1}} f(x)dk.$$

The function F_g belongs to the class $\mathscr{D}(K\backslash G/K)$. Obviously we have the relation

$$F_g(o) = f(g \cdot o).$$

By applying formula (18.6.2) we obtain

$$f(g \cdot o) = F_g(o) = \int_{W\backslash\mathfrak{a}^*} \tilde{F}_g(\lambda)|c(\lambda)|^2 d\lambda. \tag{18.6.3}$$

We shall express the value $\tilde{F}_g(\lambda)$ by means of the Fourier transform of the function f

$$\begin{aligned}
\tilde{F}_g(\lambda) &= \int_G\int_K f(gkx)\Phi_{-\lambda}(x)dm(x) = \int_G f(gx)\Phi_{-\lambda}(x)dm(x) \\
&= \int_G f(x) \int_B e_{-i\lambda+\varrho,b}(g^{-1} \cdot x)db\,dm(x) \\
&= \int_{G/K}\int_B e_{-i\lambda+\varrho,g\cdot b}(x)e_{-i\lambda+\varrho,b}(g \cdot o)db f(x)dm(x) \\
&= \int_{G/K}\int_B e_{-i\lambda+\varrho,b}(x)e_{-i\lambda+\varrho\cdot g^{-1}\cdot b}(g \cdot o)e_{2\varrho,b}(g \cdot o)db f(x)dm(x) \\
&= \int_{G/K}\int_B e_{-i\lambda+\varrho,b}(x)e_{-i\lambda+\varrho,b}(g)db f(x)dm(x) \\
&= \int_B e_{i\lambda+\varrho,b}(g) \int_{G/K} f(x)e_{-i\lambda+\varrho,b}(x)dm(x) \\
&= \int_B \hat{f}(\lambda, b)e_{i\lambda+\varrho,b}(g)db.
\end{aligned}$$

In the above computations we have applied twice formulas (18.1.3) and (18.1.10) and also Fubini's theorem.

Finally, by substituting the above result into formula (18.6.3), we obtain

THEOREM 18.6.2. *For $f \in \mathscr{D}(G/K)$ we have the formulas*:

$$f(y) = \int_{W\backslash\mathfrak{a}^*} \int_B \hat{f}(\lambda, b) e_{i\lambda+\varrho,b}(y)|c(\lambda)|^{-2} d\lambda db, \tag{18.6.4}$$

$$\int_{G/K} |f(y)|^2 dm(y) = \int_{W\backslash\mathfrak{a}^*} \int_B |\hat{f}(\lambda, b)|^2 |c(x)|^{-2} d\lambda db. \tag{18.6.5}$$

Formula (18.6.5) asserts the isommetricity of the operator F which to a function $f \in \mathscr{D}(G/K)$ assigns the $L^2(B)$-valued function on $\mathfrak{a}^*$:

$$\mathfrak{a}^* \ni \lambda \to \hat{f}(\lambda, \cdot) := Ff(\lambda)(\cdot).$$

As we know from § 18.1, the Fourier transformation intertwines for every λ the left regular representation with the action of the representation $\mathscr{U}^{-i\lambda+\varrho}$ on $L^2(B)$.

The operator F is an isometry intertwining the regular representation on $L^2(G/K)$ with the representation

$$\int_{W\backslash\mathfrak{a}^*} \mathscr{U}^{-i\lambda+\varrho}|c(\lambda)|^{-2} d\lambda.$$

The Plancherel measure, introduced in Chapter 15 as a measure on the space of p.d. spherical functions of the group G, has turned out to be supported by the set of those spherical functions which correspond to real values of the parameter λ.

18.7. THE RADON TRANSFORMATION

As hypersurfaces in a symmetric space, horocycles appear in the theory of an integral transformation, which in view of its connection with the Fourier transformation, is a counterpart of the Radon transformation from the analysis on a Euclidean space.

The concept of an integral transformation of this type is known from the papers of Gelfand and Graev and has been applied to obtain the Plancherel formulas on the symmetric spaces of the classical groups and on other homogeneous spaces of those groups.

The theory of the Radon transformation on arbitrary symmetric spaces has been worked out in detail by Helgason (see Helgason, 1980 b). Our exposition of this theory, presented below, is based on his papers. We begin by recalling certain facts from the classical Radon transformation on a Euclidean space.

The Classical Case of a Euclidean Space

In this subsection we shall denote by M the Euclidean space $\boldsymbol{R}^n$. Ξ will be the set of all hyperplanes in M.

Let G denote the group of proper motions of the space M and K its subgroup isomorphic to $\mathrm{SO}(n)$.

M, as a homogeneous space, is isomorphic to G/K. Ξ is also a homogeneous space of the motion group. If $M \ni x \to g \cdot x \in M$ denotes the action of an element $g \in G$ on M, then for every $\xi \in \Xi$ the set $g \cdot \xi$ is also a hyperplane. This action of G is transitive and the isotropy subgroup at a point ξ is isomorphic to the group of proper motions of the $(n-1)$-dimensional Euclidean space.

We shall introduce polar coordinates in the space Ξ. Namely, to a parameter $(n, t) \in S^n \times \boldsymbol{R}$ we assign the hyperplane $\xi_{n,t}$ normal to the vector n and passing through the point tn, i.e. satisfying the equation

$$(n|x) = t.$$

This parametrization is obviously not one-to-one, since the points (n, t) and $(-n, -t)$ represent the same hyperplane. Denote by ξ_0 the hyperplane given by the equation $x^1 = 0$. Let $\mathrm{d}x$ be the Lebesgue measure on M and $\mathrm{d}\omega_{\xi_0}$ the Lebesgue measure on ξ_0. Since for every $\xi \in \Xi$ there exists a $g \in G$ such that $\xi = g \cdot \xi_0$, we can transport the measure $\mathrm{d}\omega_{\xi_0}$ to ξ and the measure thus obtained will not depend on the particular $g \in G$ chosen. We shall denote this measure on ξ by $\mathrm{d}\omega_\xi$. The same symbol will also be applied to denote the measure on M which assigns to a function $f \in \mathscr{C}_0(M)$ the number $\int_\xi f|\xi \,\mathrm{d}\omega_\xi$.

By the *Radon transform* of a function $f \in \mathscr{C}_0(M)$ we understand the function defined on Ξ by the formula

$$\hat{f}(\xi) = \int_\xi f \,\mathrm{d}\omega_\xi. \tag{18.7.1}$$

The value $\hat{f}(\xi_{n,t})$ will be denoted by $\hat{f}(n, t)$. It follows directly from the definition that

$$\widehat{L_g f}(\xi) = \hat{f}(g^{-1}\xi) = L_g\hat{f}(\xi). \qquad (18.7.2)$$

This means that the Radon transformation intertwines the actions of the group G on the spaces of functions on M and Ξ.

Example 1

Let $n \in S^{n-1}$ be a fixed vector and, for an arbitrary $t \in \boldsymbol{R}$, let $d\omega_t$ denote the measure $d\omega_\xi$, defined above, on the hyperplane $\xi = \xi_{n,t}$. It is easy to check that we then have

$$\int_M f(x)dx = \int_R \int_{\xi_{n,t}} f(x)d\omega_t(x)dt, \qquad f \in \mathscr{C}_0(M).$$

(This formula obviously holds true for $n = (1, 0, \ldots, 0)$ since then $d\omega_t = dx^2 \ldots dx^n$. Next, it follows from the invariance of the Lebesgue measure with respect to proper motions that it is also valid for an arbitrary $n \in S^{n-1}$.)

In certain cases the above formula allows us to determine the Radon transform without computations by means of purely geometric considerations. For instance, if f is the characteristic function of the ball of radius r, then $\hat{f}(n, t)$ is equal to the measure of the intersection of the ball with the hyperplane $(x|n) = t$ and thus it is

$$\hat{f}(n, t) = \begin{cases} \dfrac{\pi^{\frac{1}{2}(n-1)}}{\Gamma(\frac{1}{2}(n+1))}(r^2 - t^2)^{\frac{1}{2}(n-1)}, & |t| \leqslant r, \\ 0, & |t| > r. \end{cases}$$

Example 2

Let g be the motion given by the formula $g(x) = Ax + b$ where $A \in SO(n)$, $b \in \boldsymbol{R}^n$ and let ξ denote the hyperplane $(x|n) = t$, with $n \in S^{n-1}$. Then $g^{-1}\xi$ is the hyperplane given by the formula $(x|A^{-1}n) = t - (b|n)$, and thus formula (18.7.2) assumes the form $\widehat{L_g f}(n, t) = \hat{f}(A^{-1}n, t - (b|n))$, $f \in \mathscr{C}_0(M)$. Hence we obtain the following identity which is important in applications:

$$\widehat{\left(\frac{\partial f}{\partial x_i}\right)}(n, t) = n^i \frac{\partial \hat{f}}{\partial t}(n, t).$$

Applying a formula from Example 1 and formula (18.7.2) in the case of the translation $x \to x+b$ one can easily obtain an expression for the Radon transform of convolution. Namely we have

$$\widehat{f_1 * f_2}(n, t) = \int_R \hat{f}_1(n, s)\hat{f}_2(n, t-s)\mathrm{d}s.$$

The question arises whether it is possible to recover the function f provided we know its transform $\hat{f}$, i.e. the mean values of the function f on arbitrary hyperplanes. The answer is yes.

We shall obtain this result by applying the Fourier transformation. First we shall examine the relation between the Fourier and the Radon transforms of a function f.

LEMMA 18.7.1. *For* $f \in \mathscr{C}_0(M)$ *we have the relation*

$$\hat{f}(p) = \int_R \hat{f}\left(\frac{p}{\|p\|}, t\right)\mathrm{e}^{i\|p\|t}\mathrm{d}t.$$

Proof. Fix a point $p \in M$. The measure $\mathrm{d}x$ on M can be represented as follows:

$$\int_M \varphi\mathrm{d}x = \int_R \int \varphi\left(t\frac{p}{\|p\|}+y\right)\mathrm{d}\omega_{\xi_{\frac{p}{\|p\|}, 0}}\mathrm{d}t = \int_R \hat{\varphi}\left(\frac{p}{\|p\|}, t\right)\mathrm{d}t.$$

Taking for φ the function $x \to f(x)\mathrm{e}^{i(p|x)}$, we obtain the required result. □

According to the inversion formula for the Fourier transformation we have

$$f(0) = \frac{1}{(2\pi)^n}\int_{R^n} \hat{f}(p)\mathrm{d}p = \frac{1}{(2\pi)^n}\int_{R^n}\int_R \hat{f}\left(\frac{p}{\|p\|}, t\right)\mathrm{e}^{i\|p\|t}\mathrm{d}t\mathrm{d}p. \tag{18.7.3}$$

Next we apply the polar coordinates in $\boldsymbol{R}^n$ to compute the integral with respect to p. Writing $p = rn$, $n \in S^n$, we get

$$f(0) = \frac{1}{(2\pi)^n}\int_{S^n}\int_{R_+} r^{n-1}\int_R \hat{f}(n, t)\mathrm{e}^{irt}\mathrm{d}t\mathrm{d}r\mathrm{d}n. \tag{18.7.4}$$

Here by $\mathrm{d}n$ we denote the surface element of the unit sphere.

Let us now consider separately the case where $n = 2k+1$ is an odd integer. Again, by successive changes of variables we obtain

$$\int_{-\infty}^{0} r^{n-1} \int_R \hat{f}(n, t) e^{irt} dt dr = \int_{0}^{\infty} r^{n-1} \int_R \hat{f}(n, t) e^{-irt} dt dr$$

$$= \int_{0}^{\infty} r^{n-1} \int_R \hat{f}(n, t) e^{irt} dt dr.$$

We recall, however, that the parameters $(-n, t)$ and $(n, -t)$ describe the same point of the space Ξ and thus $f(n, -t) = f(-n, t)$. Finally, owing to the invariance of the measure dn with respect to the central symmetry, we obtain

$$f(0) = \frac{1}{2(2\pi)^n} \int_{S^n} \int_R r^{n-1} \int_R \hat{f}(n, t) e^{irt} dt dr dn. \tag{18.7.5}$$

It follows from the basic properties of the Fourier transformation that

$$\int_R r^m \int_R \varphi(t) e^{irt} dt = 2\pi i^{n-1} \frac{d^m}{dt^m} \varphi(0).$$

Thus formula (18.7.5) assumes the form

$$f(0) = \frac{(-1)^{\frac{n-1}{2}}}{2(2\pi)^{n-1}} \int_{Sn} \frac{\partial^{n-1}}{\partial t^{n-1}} \hat{f}(n, 0) dn. \tag{18.7.6}$$

The value $f(x)$ for an arbitrary $x \in M$ is obtained by applying formula (18.7.6) to the function $L_{-x} f$ because

$$f(x) = L_{-x}f(0) \qquad \text{and} \qquad (L_{-x}f)^\frown = L_{-x}\hat{f}.$$

Thus we have the relation

$$\widehat{L_x f}(n, t) = \hat{f}(n, t + (x|n)),$$

which leads to the final formula

$$f(x) = \frac{(-1)^{\frac{n-1}{2}}}{2(2\pi)^{n-1}} \int_{Sn} \frac{\partial^{n-1}}{\partial t^{n-1}} f(n, (n|x)) dn. \tag{18.7.7}$$

We see that the operator which maps $\hat{f}$ at f is a composition of a differential operator acting on $\mathscr{C}^\infty(\Xi)$ and an integral operator from $\mathscr{C}^\infty(\Xi)$ into $\mathscr{C}^\infty(\mathfrak{M})$, which has a clear geometric interpretation.

For a fixed $x \in M$ the set of hyperplanes with parameters $(n, (n|x))$, $n \in S^n$, is the pencil of hyperplanes passing through the point x.

For $\varphi \in \mathscr{C}^\infty(\Xi)$ we denote

$$\breve{\varphi}(x) := \int_{S^n} \varphi(n, (n|x))\, dn.$$

We call "$\breve{\ }$" the *operator dual to the Radon transformation.*

$$f = \frac{(-1)^{\frac{n-1}{2}}}{2(2\pi)^{n-1}} \left(\frac{\partial^{n-1}}{\partial t^{n-1}} \hat{f} \right)^{\breve{}}. \tag{18.7.8}$$

In the case of even dimension n we can only write

$$f = \frac{1}{(2\pi)^n} (L\hat{f})^{\breve{}}, \tag{18.7.9}$$

where L is the operator on $\mathscr{C}_0(\Xi)$ given by the formula

$$L\varphi(n, s) = \int_0^\infty r^{n-1} e^{-irt} \int_{-\infty}^{+\infty} \varphi(n, t) e^{irt} dt\, dr.$$

Thus L is a pseudo-differential operator whose symbol is the function $r^{n-1} \cdot 1_{R_+}$.

The formulas which describe the inverse operation to the Radon transformation are given here the form which corresponds best with the theory of the Radon transformation on symmetric spaces.

For a different approach to these problems the reader is referred to John (1958) and Gelfand Graev and Vilenkin (1962).

The Radon Transformation on a Symmetric Space of the Non-Compact Type

We return to the notation applied in the preceding chapters. Manifold $\mathfrak{M}$ is a symmetric space of the non-compact type and is represented as a homogeneous space of the form $\mathfrak{M} = G/K$ where G is a semisimple group with a maximal compact subgroup K.

Denote by Ξ the space of horocycles in $\mathfrak{M}$. We have $\Xi = G/MN$ as homogeneous spaces. The horocycle ξ_0 whose isotropy subgroup equals MN is the orbit of the point $o = eK \in G/K$. Every horocycle can be represented as $\xi = g \cdot \xi_0 = gNg^{-1}(gK)$ for a certain $g \in G$.

The group N is unimodular. Let us fix a Haar measure dn for this group. The mapping $N \ni n \to gnK \in \xi \subset G/K$ allows us to transport the measure dn to the symmetric space $\mathfrak{M}$ and the support of the measure thus obtained is concentrated on the horocycle $\xi = g \cdot \xi_0$. We shall denote this measure by $d\omega_\xi$.

We define the Radon transformation on the space M by putting for $f \in \mathscr{C}_0(\mathfrak{M})$ and $\xi = g \cdot \xi_0$:

$$\check{f}(\xi) = \int f d\omega_\xi = \int_N f(gnK) dn. \tag{18.7.10}$$

This is the mean value of the function f over the horocycle ξ. We shall also define a transformation dual to "$\smile$". The pencil of horocycles through the point $x = gK$ is obtained by rotating the horocycle $g \cdot \xi_0$ about the point x with the help of the action of the subgroup $gKg^{-1} \subset G$. Thus every element of this pencil can be represented (non-uniquely) as

$$\xi = gk\xi_0 \quad \text{for} \quad k \in K.$$

For $x = g\xi_0$ we write

$$\check{x} = \{\xi \in \Xi : \xi = gk\xi_0, k \in K\}.$$

Let dk be the normalized Haar measure on K. The mapping $K \ni k \to gk \cdot \xi_0 \in \Xi$ transports this measure to Ξ and the support of the measure $dv_{\check{x}}$ thus obtained is concentrated on the set $\check{x}$.

We define for $\varphi \in \mathscr{C}(\Xi)$, $x = gK$

$$\check{\varphi}(x) = \int_{\check{x}} \varphi(\xi) dv_{\check{x}}(\xi) = \int_K \varphi(gk \cdot \xi_0) dk. \tag{18.7.11}$$

The value $\check{\varphi}(x)$ equals the mean value of the function φ on the pencil $\check{x}$.

As in the classical case, we shall focus our attention on the problem of finding the inverse transformation to the Radon transformation. As before, the key formula will be the relation between the Fourier and the Radon transforms.

To this end we shall apply formula (18.1.9). According to that formula and to definition (18.2.1) of the Fourier transformation we get

$$\hat{f}(\lambda, kM) = \int_G f(gK) e^{(i\lambda-\varrho)H(g^{-1}kK)} dg = \int_G f(kgK) e^{(i\lambda-\varrho)H(g^{-1})} dg$$

$$= \int_A \int_N f(kanK) e^{(i\lambda-\varrho)H(a^{-1})} da\, dn$$

$$= \int_A e^{(-i\lambda+\varrho)\log a} \widehat{f}(ka\xi_0) da. \tag{18.7.12}$$

Thus the value $\hat{f}(\lambda, kM)$ is obtained by computing the classical Fourier transformation of the function $a \to \widehat{f}(ka\xi_0)$.

Now we shall apply the inversion formula for the Fourier transformation on a symmetric space

$$f(o) = \frac{1}{|W|} \int_K \int_{\mathfrak{a}^*} \hat{f}(\lambda, kM) |c(\lambda)|^{-2} d\lambda\, dk$$

$$= \frac{1}{|W|} \int_K \int_{\mathfrak{a}^*} \int_A e^{(-i\lambda+\varrho)\log a} \widehat{f}(ka\xi_0) da |c(\lambda)|^{-2} d\lambda\, dk. \tag{18.7.13}$$

We intend to find an operator on the space $\mathscr{E}(\Xi)$ such that the last formula will assume the form

$$f(o) = (L\widehat{f})^{\smile}(o).$$

Let $\mathscr{S}(\mathfrak{a})$, $\mathscr{S}(\mathfrak{a}^*)$ denote the spaces of Schwartz functions on the spaces $\mathfrak{a}$ and $\mathfrak{a}^*$, respectively. By $\mathscr{S}(A)$ we shall denote the set of functions f on A such that $\mathfrak{a} \ni X \to f(\exp X)$ is of class $\mathscr{S}(\mathfrak{a})$. Further we shall write $f \in \mathscr{S}_\varrho(A)$ if $a \to f(a) e^{\varrho(\log a)}$ is a function of class $\mathscr{S}(A)$. We now recall that the function c^{-1} of Harish-Chandra has polynomial growth on the space $\mathfrak{a}^*$, and therefore the operator of multiplication by c^{-1} is a continuous endomorphism of the space $\mathscr{S}(\mathfrak{a}^*)$.

Thus we can define on the space $\mathscr{S}(A)$ a continuous operator A_+ by the formula

$$F(A_+ f) := c^{-1} F f,$$

where F denotes the Fourier transformation on the abelian group A.

We also introduce an operator A_-:

$$F(A_- f) := c^{-1} F f. \tag{18.7.14}$$

The next step consists in introducing operators $S_\pm$ on the space $\mathscr{S}_\varrho(A)$:

$$S_\pm f(a) = e^{-\varrho(\log a)} A_\pm (e^{\varrho(\log a)} f)(a).$$

The reader will note that the operators $S_\pm$ are continuous endomorphisms of the space $\mathscr{S}(A)$ commuting with the left regular representation of A on this space.

Finally, we define a space of functions on Ξ which is the proper domain of the required operator.

DEFINITION. We say that a function φ on Ξ is of *class* $\mathscr{S}_\varrho(\Xi)$ if for every $k \in K$ the function

$$f_k\colon A \ni a \to f(ka\xi_0) \quad \text{is of class } \mathscr{S}_\varrho(A).$$

PROPOSITION 18.7.2. *The operators defined as*

$$A_\pm f(ka\xi_0) := S_\pm f_k(a)$$

are endomorphisms of the space $\mathscr{S}_\varrho(\Xi)$ *commuting with the action of the group* G *on* Ξ.

Proof. It is not immediately obvious that the function

$$G \ni g = kan \to S_\pm f_k(a)$$

is right MN-invariant, i.e. drops onto the quotient space $\Xi = G/MN$. To show this we shall prove the following

LEMMA 18.7.3. *The mapping* $\tilde{r}_a\colon \Xi \ni g \cdot \xi_0 \to (ga) \cdot \xi_0$ *is a diffeomorphism of* Ξ *commuting with the action of the group* G.

Proof. We shall verify that the following diagram is commutative.

$$\begin{array}{ccc} G & \xrightarrow{r_a} & G \\ \pi \downarrow & & \downarrow \pi \\ G/MN & \xrightarrow{\tilde{r}_a} & G/MN \end{array}$$

The operation r_a is the right multiplication by a in the group G.

If $\pi(g) = \pi(g')$, i.e. if $gmn = g'$ for certain $m \in N$, $n \in N$, then $gmna = gmaa^{-1}na = gmam^{-1}ma^{-1}na = gamn$ since the actions by inner automorphisms of A on N and of M on A are trivial.

Hence it follows that $\pi(ga) = \pi(g'a)$, which means that $\tilde{r}_a$ is a well defined mapping on Ξ which commutes with the action of the group.

The inverse mapping to $\tilde{r}_a$ is $\tilde{r}_{a^{-1}}$. This proves the lemma. □

Comparing formulas (18.7.11), (18.7.13) and (18.7.14) we observe that for $f \in \mathscr{C}_0(\mathfrak{M})$ we have

$$f(o) = (\Lambda_- \Lambda_+ f)\,(o). \tag{18.7.15}$$

The operations $\check{}$, $\hat{}$, Λ_+, Λ_- being all commutative, we obtain

$$\begin{aligned} f(g^{-1}o) &= L_g f(o) = (\Lambda_- \Lambda_+ \widehat{L_g f})\check{}\,(o) \\ &= \big(l_g(\Lambda_- \Lambda_+ \hat{f})\big)(o) = (\Lambda_- \Lambda_+ \hat{f})\check{}\,(g^{-1} \cdot o). \end{aligned} \tag{18.7.16}$$

Below we present the fundamental theorem of the theory of the Radon transformation on a symmetric space.

THEOREM 18.7.4 (Helgason, 1970). *For every function $f \in \mathscr{D}(\mathfrak{M})$ we have the inversion formula*

$$f = (\Lambda_- \Lambda_+ \hat{f})\check{} \tag{18.7.17}$$

and the Plancherel formula

$$\int_{\mathfrak{M}} |f(x)|^2 \mathrm{d}m(x) = \int_{\Xi} |(\Lambda_+ \hat{f})|^2 \mathrm{d}\xi. \tag{18.7.18}$$

The latter formula arises from the Plancherel formula for $\hat{f}$ by applying the relation between the transforms $\hat{f}$ and $\hat{f}$ and the description of the invariant measure on the space of horocycles.

For certain symmetric spaces the inversion formula assumes a simpler form. Namely, if the function c^{-1} is a polynomial, then the operators $A_\pm$, and hence also $\Lambda_\pm$ are differentiations.

Formula (18.7.17) is a close counterpart of the well-known formula from the classical theory corresponding to an odd-dimensional Euclidean space. Helgason's paper contains the following version of the inversion formula for the Radon transformation.

THEOREM 18.7.5. *Let the group G satisfy the assumptions of Theorem 18.5.3. Then there exists a differential operator $\square$ on the symmetric space $\mathfrak{M} = G/K$ such that $f = \square(\hat{f})\check{}$.*

18.8. THE PALEY–WIENER THEOREM

Besides the inversion formulas and the Plancherel formulas, the Fourier transforms of various classes of functions also play an important role in the theory of the Fourier transformation.

A description of the image of the space $\mathscr{D}(\boldsymbol{R}^n)$ is provided by the classical Paley–Wiener theorem. The role of the space $\mathscr{S}(\boldsymbol{R}^n)$ of the Schwartz functions results from its invariance with respect to the Fourier transformation.

In the present section we present counterparts of these results in the case of analysis on a symmetric space.

The Paley–Wiener theorem for the group SL(2, $\boldsymbol{R}$) has been proved by Ehrenpreis and Mautner; the case of the group SL(2, $\boldsymbol{C}$) is due to Gelfand and Graev. S. Helgason first proved the theorem for the class of group comprising complex groups and the case of rank $M = 1$, and then, applying the results of Gangolli, extended his proof to the general case.

The image of the functions of class $\mathscr{D}(\mathfrak{M})$ under the Fourier transformation on M has certain symmetry properties with respect to the action of the Weyl group on $\mathfrak{a}_C^*$. These properties result from the invariance of spherical functions with respect to the action of the Weyl group on the parameter $\lambda \in \mathfrak{a}_C^*$:

$$\Phi_{w\lambda} = \Phi_\lambda .$$

We shall write explicitly the identity

$$f * \Phi_\lambda = f * \Phi_{w\lambda} .$$

Owing to property (18.1.3) of plane waves and formula (18.1.11), we find

$$\begin{aligned} f * \Phi_\lambda(x) &= \int_G f(g\cdot o)\int_B e_{i\lambda+\varrho,b}(g^{-1}\cdot x)\,db\,dg \\ &= \int_G f(g\cdot o)\int_B e_{i\lambda+\varrho,b}(x)e_{-i\lambda+\varrho,b}(g\cdot o)\,db\,dg \\ &= \int_B e_{i\lambda+\varrho,b}(x)\int_{\mathfrak{M}} f(y)e_{-i\lambda+\varrho,b}(y)\,dm(y)\,db \\ &= \int_B e_{i\lambda+\varrho,b}(x)\tilde{f}(\lambda, b)\,db . \end{aligned}$$

Thus the symmetry of the spherical function Φ_λ implies that

$$\int_B e_{i\lambda+\varrho,b}(x)\tilde{f}(\lambda, b)\,db = \int_B e_{iw\lambda+\varrho,b}(x)\tilde{f}(\lambda, b)\,db \tag{18.8.1}$$

identically with respect to $x \in M$, $w \in W$.

If the function f is, in addition, K-invariant, then

$$\tilde{f}(\lambda, b) = \tilde{f}(w\lambda, b), \qquad b \in B, \quad w \in W.$$

Thus the Fourier transforms of functions of class $\mathscr{D}(K\backslash M)$ are W-invariant functions on $\mathfrak{a}_C^*$.

We shall say that a function ψ on $\mathfrak{a}_C^* \times B$ has *exponential growth* if there exists a constant $R > 0$ such that

$$\sup_{\lambda \in \mathfrak{a}_C^*, \, b \in B} e^{-R|\mathrm{Im}\lambda|} |P(\lambda)\psi(\lambda, b)| < \infty \tag{18.8.2}$$

for an arbitrary polynomial P on $\mathfrak{a}_C^*$.

We shall prove that the Fourier transforms of functions of class $\mathscr{D}(\mathfrak{M})$ are functions with exponential growth.

Let supp$f$ be contained in a compact set $C \subset \mathfrak{M}$. For an arbitrary differential operator $D \in \mathrm{D}(\mathfrak{M})$ we have

$$\widetilde{Df}(\lambda, b) = \int_{\mathfrak{M}} Df(x) e_{-i\lambda+\varrho, b}(x) \mathrm{d}m(x)$$

$$= \int_{\mathfrak{M}} f(x) D^+ e_{-i\lambda+\varrho, b}(x) \mathrm{d}m(x) = \Gamma_{D^+}(-i\lambda)\tilde{f}(\lambda, b),$$

where D^+ denotes the formal adjoint of D.

Further, we have the estimates

$$|f(\lambda, b)| \leqslant \sup f \int_C e_{-i\lambda+\varrho, b}(x) \mathrm{d}m(x)$$

and

$$|e_{-i\lambda+\varrho, b}(g \cdot o)| = |e^{(i\lambda-\varrho)H(g^{-1}k)}|$$

$$\leqslant e^{(-|\mathrm{Im}\lambda|-\varrho)H(g^{-1}k)}, \qquad \text{where} \quad kM = b.$$

Taking $R > \sup_{K \times C} \| H(g^{-1}k) \|$, we see that

$$\sup e^{-R|\mathrm{Im}\lambda|} | \tilde{f}(\lambda, b)| < \infty, \tag{18.8.3}$$

Applying this estimate to the function Df_1, we obtain

$$\sup_{\lambda, b} e^{-R\|\mathrm{Im}\lambda\|} |\tilde{\Gamma}_{D^+}(-i\lambda) f_1(\lambda, b)| < \infty.$$

We know that every W-invariant polynomial on $\mathfrak{a}_C^*$ can be obtained as Γ_D for a certain operator $D \in \mathrm{D}(\mathfrak{M})$. Next, an arbitrary polynomial can be dominated by a W-invariant polynomial. Thus we obtain estimate (18.8.3) for an arbitraty function $f \in \mathscr{D}(\mathfrak{M})$ and a polynomial P.

The necessary conditions obtained above prove also to be sufficient.

THEOREM 18.8.1 (Helgason, 1966). (1) *The Fourier transformation maps the space* $\mathscr{D}(\mathfrak{M})$ *onto the space of holomorphic functions on* $\mathfrak{a}_C^* \times B$ *having exponential growth and satisfying condition* (18.8.1).

(2) *The space of functions* $\mathscr{D}(K \backslash G/K)$ *is mapped under the spherical Fourier transformation onto the space of* W*-invariant functions on* $\mathfrak{a}_C^*$ *of exponential growth.*

Table of Formulas

The Euler Γ- and B-Functions

Definition:

$$\Gamma(z) := \int_0^\infty e^{-t} t^{z-1} dt = \int_0^1 \left(\log \frac{1}{t}\right)^{z-1} dt \qquad \text{for } \operatorname{Re} z > 0.$$

Basic properties:

$$\Gamma(n) = (n-1)!, \qquad n = 1, 2, \ldots,$$

$$\Gamma\left(n - \frac{1}{2}\right) = \frac{(2n)!}{2^{2n} n!} \sqrt{\pi},$$

$$\Gamma(z) = \frac{\Gamma(z-n)}{(z-n-1)(z-n-2)\ldots z}, \qquad n \in N, \operatorname{Re} z > 0.$$

By means of this formula the function Γ is extended to $C - \{0, -1, -2, \ldots\}$.

$$\Gamma(z)\Gamma(1-z) = \frac{\pi}{\sin \pi z}.$$

For real $x \neq 0$

$$|\Gamma(ix)|^2 = \frac{\pi}{x \sin \pi x},$$

$$\Gamma(\tfrac{1}{2} + ix)^2 = \frac{\pi}{\cosh \pi x}.$$

$$\Gamma(2x) = \frac{2^{2x-1}\Gamma(x+\frac{1}{2})\Gamma(z)}{\sqrt{\pi}}.$$

$$\Gamma(z) = \lim_{n\to\infty} \int_0^n \left(1 - \frac{t}{n}\right)^n t^{z-1} dt, \qquad \operatorname{Re} z > 0$$

$$= \lim_{n\to\infty} \frac{n!\, z^n}{z(z+1)(z+2)\ldots(z+n)},$$

$z = 0, -1, -2, \ldots$.

$$B(x, y) := \int_0^1 t^{x-1}(1-t)^{y-1} dt = \int_0^\infty t^{x-1}(1-t)^{-x-y} dt$$

For $\operatorname{Re}x > 0$, $\operatorname{Re}y > 0$.

$$B(x, y) = \frac{\Gamma(x)\Gamma(y)}{\Gamma(x+y)}, \qquad B(x, y) = B(y, x),$$

$$B(x, y+1) = \frac{y}{x} B(x+1, y) = \frac{y}{x+y} B(x, y),$$

$$B(x, y) = 2\int_0^{2\pi} \cos^{2x-1}\varphi \sin^{2y-1}\varphi \, d\varphi,$$

$$B(x, y) = \int_0^{\infty} \frac{t^{x-1}}{(1+t)^{x+y}} \, dt,$$

$$B\left(\frac{x+1}{2}, y-\frac{x+1}{2}\right) = 2\int_0^{\infty} \frac{t^x}{1+t^2} \, dt.$$

The Riemann ζ-Function

Definition:

$$\zeta(z) := \sum_{n=1}^{\infty} \frac{1}{n^z} \qquad \text{for} \quad \operatorname{Re}z > 1.$$

Basic properties:

$$\zeta(z) = \frac{1}{\Gamma(z)} \int_0^{\infty} \frac{x^{z-1}}{e^x - 1} \, dx = \frac{\mathscr{M}((e^x-1)^{-1})}{\Gamma(z)}$$

$$\zeta(z) \prod_{p \in \Omega} (1 - p^{-z}) = 1, \qquad \operatorname{Re}z > 0,$$

Ω being the set of prime numbers.

The Hypergeometric Function

The hypergeometric equation:

$$z(1-z)\frac{d^2u}{dz^2} + \{c-(a+b+1)z\}\frac{du}{dz} - abu = 0.$$

A solution regular at 0 with free term 1 is given by the hypergeometric series

$$F(a, b; c; z) = 1+\frac{ab}{c}z+\frac{a(a+1)b(b+1)}{c(c+1)2!}z^2$$

$$+\frac{a(a+1)(a+2)b(b+1)(b+2)}{c(c+1)(c+2)3!}z^3+\ldots$$

$$=\frac{\Gamma(c)}{\Gamma(a)\Gamma(b)}\sum_{s=0}^{\infty}\frac{\Gamma(a+s)\Gamma(b+s)}{\Gamma(c+s)s!}z^s,$$

for those c which are not negative integers. The radius of convergence of the series is 1. If c is a negative integer then a solution regular at zero is given by

$$z^{1-c}F(1+a-c, 1+b-c; 2-c; z).$$

I. For $\operatorname{Re} c > \operatorname{Re} a > 0$ the following function is an analytic continuation of the hypergeometric series to $\Omega = \boldsymbol{C}-[1, \infty[$:

$$F(a, b; c; z) = B(c-a, a)^{-1}\int_0^1 s^{a-1}(1-s)^{c-a-1}(1-zs)^{-b}ds.$$

II. Let D be a fixed double loop in $\boldsymbol{C}$ around 0, 1 and let $\Omega = \left\{z \in \boldsymbol{C}: \frac{1}{z} \text{ is not inside } D\right\}$. Then the function

$$\boldsymbol{C}^1 \ni z \to \frac{\int_D s^{a-1}(1-s)^{c-a-1}(1-zs)^{-b}ds}{\int_D s^{a-1}(1-s)^{c-a-1}ds}$$

is an analytic continuation of the function F to Ω. (When computing both integrals one should use the same branch of the multivalued integrand.)

$$F(a, b; c; 1) = B(b, c-b-a)/B(c-a, a)$$

for $\operatorname{Re}(c-a-b) > 0$.

Elementary properties:

$$F(a, b; c; z) = F(b, a; c; z),$$

$$F(a, b; c; z) = (1-z)^{-a} F\left(a, c-b; c; \frac{z}{z-1}\right)$$

$$= (1-z)^{-b} F\left(b, c-a; c; \frac{z}{z-1}\right),$$

$$F(a, b; c; z) = (1-z)^{c-a-b} F(c-a, c-b; c; z),$$

$$F(a, b+1; c; z) - F(a, b; c; z) = \frac{az}{c} F(a+1, b+1; c+1; z),$$

$$F(a+1, b+1; c; z) - F(a, b; c; z)$$

$$= \frac{abz}{c} F(a+1, b+1; c+1; z),$$

$$(1-t)^{b-c}(1-t+zt)^{-b}$$

$$= \sum_{n-0}^{\infty} \frac{c(c+1) \dots (c+n-1)}{n!} F(-n, b; c; z) t^n,$$

$$\frac{\mathrm{d}^n}{\mathrm{d}z^n} F(a, b; c; z)$$

$$= \frac{a(a+1) \dots (a+n-1) b(b+1) \dots (b+n-1)}{c(c+1) \dots (c+n-1)}$$

$$\times F(a+n, b+n; c+n; z),$$

$$F(2a, 2b; a+b+\tfrac{1}{2}; z) = F\left(a, b; a+b+\tfrac{1}{2}; \frac{4z}{z-1}\right).$$

The Confluent Hypergeometric Function

The confluent hypergeometric equation:

$$x \frac{\mathrm{d}^2 f}{\mathrm{d}x^2} + (c-x) \frac{\mathrm{d}f}{\mathrm{d}x} - af = 0.$$

$${}_1F_1(a; c; x) = 1 + \frac{a}{c} \frac{x}{1!} + \frac{a(a+1)x^2}{c(c+1)2!} + \dots,$$

$$ {}_1F_1(a;c;x) = \frac{\Gamma(c)}{\Gamma(a)\Gamma(c-a)} \int_0^1 e^{xu} u^{a-1} (1-u)^{c-a-1} du, $$

$$ \operatorname{Re} c > \operatorname{Re} a > 0, $$

$$ \frac{d^n}{dx^n}\, {}_1F_1(a;c;x) = \frac{a(a+1)\ldots(a+n-1)}{c(c+1)\ldots(c+n-1)} \times {}_1F_1(a+n;c+n;x), $$

$$ (c-a)\,{}_1F_1(a-1;c;x) + (2a-c+x)\,{}_1F_1(a;c;x) - a_1F_1(a+1,c;x) = 0, $$

$$ (a-c+1)\,{}_1F_1(a;c;x) - a\,{}_1F_1(a+1;c;x) + (c-1)\,{}_1F_1(a,c-1;x) = 0. $$

Bessel Functions

The Bessel functions of order ν:

$$ J_\nu(z) = \sum_{m=0}^{\infty} \frac{(-1)^m (\frac{1}{2}z)^{2m+\nu}}{m!\,\Gamma(m+\nu+1)} $$

$$ = \frac{(\frac{1}{2}z)^\nu}{\Gamma(\nu+1)} \exp(-iz)\,{}_1F_1(\nu+\tfrac{1}{2}; 2\nu+1; 2iz). $$

The Bessel equation:

$$ z^2 J_\nu''(z) + z J_\nu'(z) + (z^2 - \nu^2) J_\nu(z) = 0. $$

The Bessel functions of integer order ($n = 0, 1, 2, \ldots$):

$$ J_n(z) = (\tfrac{1}{2}z)^n \sum_{l=0}^{\infty} \frac{(-1)^l}{(n+1)!\,l!} (\tfrac{1}{2}z)^{2l} $$

$$ = \frac{1}{2\pi} \int_0^{2\pi} \exp(iz\sin\varphi - in\varphi)\, d\varphi. $$

The generating functions:

$$\exp\left(\tfrac{1}{2}r(z-z^{-1})\right) = \sum_{m=-\infty}^{\infty} J_m(r)z^m,$$

$$\exp(ir\sin\varphi) = \sum_{n=-\infty}^{\infty} J_n(r)\mathrm{e}^{in\varphi},$$

$$(\tfrac{1}{2}z)^n = \sum_{m=0}^{\infty} \frac{(n+2m)\Gamma(n+m)}{m!} J_{n+2m}(z),$$

$$\tfrac{1}{2}z\cos z = \sum_{n=0}^{\infty} (-1)^n(2n+1)^2 J_{2n+1}(z).$$

Integral representations:
the Poisson formula:

$$J_n(z) = \frac{1}{\Gamma(n+\frac{1}{2})\,\Gamma(\frac{1}{2})}(\tfrac{1}{2}z)^n \int_{-1}^{1} \exp(izt)\,(1-t^2)^{n-\frac{1}{2}}\mathrm{d}t,$$

the Sonine integral ($n > 0$):

$$J_n(z) = (\tfrac{1}{2}z)^n \frac{2}{(n-1)!}\int_0^{\frac{1}{2}\pi} \sin\varphi(\cos\varphi)^{2n-1}J_0(z\sin\varphi)\,\mathrm{d}\varphi.$$

For $n = 0, 1, 2, \ldots$

$$\pi J_n(z) = \int_0^{\pi} \cos(z\sin\varphi - n\varphi)\,\mathrm{d}\varphi = i^{-n}\int_0^{\pi} \mathrm{e}^{iz\cos\varphi}\cos n\varphi\,\mathrm{d}\varphi,$$

$$\pi J_{2n}(z) = 2\int_0^{\frac{1}{2}\pi} \cos(z\sin\varphi)\cos(2n\varphi)\,\mathrm{d}\varphi,$$

$$\pi J_{2n+1}(z) = 2\int_0^{\frac{1}{2}\pi} \sin(z\sin\varphi)\sin\left((2n+1)\varphi\right)\mathrm{d}\varphi,$$

$$\tfrac{1}{2}\pi(J_n(z))^2 = \int_0^{\frac{1}{2}\pi} J_{2n}(2z\cos\varphi)\,\mathrm{d}\varphi.$$

Symmetries with respect to indices:

$$J_n = (-1)^n J_{-n}, \qquad J_n(-z) = (-1)^n J_n(z).$$

Differential relations:

$$J_n(z) = \frac{2^n n!}{(2n)!} z^n \left(1+\frac{\mathrm{d}^2}{\mathrm{d}z^2}\right)^n J_0(z),$$

$$zJ_{n+1} = nJ_n - zJ_n', \qquad zJ_{n-1} = nJ_n + zJ_n',$$

$$\left(\frac{1}{z}\frac{\mathrm{d}}{\mathrm{d}z}\right)^n (z^\nu J_\nu(z)) = z^{\nu-n}J_{\nu-n}(z),$$

$$\left(\frac{1}{z}\frac{\mathrm{d}}{\mathrm{d}z}\right)^n (z^{-\nu} J_\nu(z)) = (-1)^n z^{-\nu-n}J_{\nu+n}(z).$$

Addition formulas:

$$J_m(r)\mathrm{e}^{im\theta} = \sum_{k=-\infty}^{\infty} J_{m-k}(r_1)J_k(r_2)\mathrm{e}^{ik\psi} \qquad \text{for } r\mathrm{e}^{i\theta} = r_1 + r_2\mathrm{e}^{i\psi},$$

$$J_n(r_1+r_2) = \sum_{k=-\infty}^{\infty} J_{n-k}(r_1)J_k(r_2),$$

$$J_n(r_1-r_2) = \sum_{k=-\infty}^{\infty} (-1)^k J_{n-k}(r_1)J_k(r_2),$$

$$J_0(r) = \sum_{k=-\infty}^{\infty} (-1)^k \mathrm{e}^{ik\psi} J_k(r_1)J_k(r_2),$$

$$J_0(\lambda z) = \sum_{m=0}^{\infty} \frac{((1-\lambda^2)z)^m}{2^m m!} J_m(z),$$

$$J_n(\lambda z) = \lambda^n \sum_{m=0}^{\infty} \frac{((1-\lambda^2)z)^m}{2^m m!} J_{m+n}(z) \qquad \text{for } n > 0.$$

Integral equations:

$$\frac{1}{2\pi}\int_0^{2\pi} \exp i(m\psi - n\theta) J_n(r)\,\mathrm{d}\psi = J_{n-m}(r_1)J_m(r_2),$$

where $re^{i\theta} = r_1 + r_2 e^{i\varphi}$;

$$\frac{1}{\pi}\int_0^{\pi} \exp\big(i(m-2n)\theta\big) J_m(2r\cos\theta)\,d\theta = J_{m-n}(r)J_n(r).$$

Jacobi and Legendre Polynomials

$$P_n^{(\alpha,\beta)}(x) := \frac{(-1)^n}{2^n n!}(1-x)^{-\alpha}(1+x)^{-\beta}\frac{d^n}{dx^n}\big((1-x)^{\alpha+n}(1+x)^{\beta+n}\big)$$

$$= \binom{n+\alpha}{n} F\big(-n, n+\alpha+\beta+1; \alpha+1; \tfrac{1}{2}(1-x)\big)$$

$$= (-1)^n \binom{n+\beta}{n} F\big(-n, n+\alpha+\beta+1; \beta+1; \tfrac{1}{2}(1+x)\big),$$

$$P_n^{(\alpha,\beta)}(1) = \binom{n+\alpha}{n}.$$

The functions P^l_{kj}

$l = 0, \frac{1}{2}, 1, \frac{3}{2}, 2, \ldots, k, j$ — assume integer values when l is an integer and half-integer values when l is a half-integer number, $|k|, |j| < l$:

$$P^l_{kj}(x) := \sqrt{\frac{(l-j)!}{(l+k)!(l-k)!(l+j)!}}\,(1+x)^{\frac{1}{2}(k+j)}(1-x)^{\frac{1}{2}(j-k)}$$

$$\times 2^{j-l} i^{(j-k)} \frac{d^{l+j}}{dx^{l+j}}\big((x-1)^{l+k}(x+1)^{l-k}\big)$$

$$= P_{l+j}^{(k-j,-k-j)}(x)(1-x)^{\frac{1}{2}(k-j)}(1+x)^{-\frac{1}{2}(k+j)} i^{k-j}$$

$$\times \sqrt{\frac{(l-k)!(l+k)!}{(l-j)!(l+j)!}}$$

$$= \big((l+k)!(l-k)!(l+j)!(l-j)!\big) i^{k-j} \left(\frac{1+x}{2}\right)^l \left(\frac{1-x}{1+x}\right)^{\frac{1}{2}(k-j)}$$

$$\times \sum_{\substack{0\le t\le l-k\\ j-k\le t\le l-j}} \frac{(-1)^t}{(l+j-t)!(k-j+t)!(l-k-t)!t!}\left(\frac{1-x}{1+x}\right)^t$$

$$= \sqrt{\frac{(l+j)!(l+k)!}{(l-j)!(l-k)!}}\, i^{k+j}\left(\tfrac{1}{2}(1-x)\right)^{\frac{1}{2}(k+j)}$$

$$\times \left(\tfrac{1}{2}(1+x)\right)^{\frac{1}{2}(k-j)}$$

$$\times \sum_{p=\max(k,j)}^{l} \frac{(l+p)!}{(l-p)!(j+p)!(k+p)!}\left(\frac{1-x}{1+x}\right)^{p},$$

$$P^l_{mj}(\cos\theta) = \frac{1}{2\pi}\frac{(l+m)!(l-m)!}{(l+j)!(l-j)!}\oint_{\Gamma}(\zeta\cos\tfrac{1}{2}\theta - i\sin\tfrac{1}{2}\theta)^{l-j}$$

$$\times (i\zeta\sin\tfrac{1}{2}\theta + \cos\tfrac{1}{2}\theta)^{l+j}\zeta^{-l-m+1}\mathrm{d}\zeta,$$

Γ denoting a contour which encircles zero once in the positive direction.

Symmetries with respect to indices:

$$P^l_{k,-j} = P^l_{-k,j}, \qquad P^l_{kj} = P^l_{jk}, \qquad \bar{P}^l_{kj} = (-1)^{k-j}P^l_{kj}.$$

Orthogonality relations:

$$\int_{-1}^{1} \bar{P}^l_{mn}(x)P^{l'}_{mn}(x)\mathrm{d}x = \frac{2}{2l+1}\,\delta_{ll'}.$$

Differential equations:

$$(1-z^2)^{\frac{1}{2}}\frac{\mathrm{d}}{\mathrm{d}z}P^l_{mk} - \frac{m-kz}{(1-z^2)^{\frac{1}{2}}}P^l_{mk}$$

$$= -i\left((l-k)(l+k+1)\right)^{\frac{1}{2}}P^l_{m,k+1},$$

$$(1-z^2)^{\frac{1}{2}}\frac{\mathrm{d}}{\mathrm{d}z}P^l_{mk} + \frac{m-kz}{(1-z^2)^{\frac{1}{2}}}P^l_{mk}$$

$$= -i\left((l+k)(l-k+1)\right)^{\frac{1}{2}}P^l_{m,k-1},$$

$$(1-z^2)\frac{\mathrm{d}^2}{\mathrm{d}z^2}P^l_{mk} - 2z\frac{\mathrm{d}}{\mathrm{d}z}P^l_{mk} - \frac{m^2+k^2-2mkz}{1-z^2}P^l_{mk}$$

$$= -l(l+1)P^l_{mk},$$

$$i\,\frac{m-kz}{(1-z^2)^{\frac{1}{2}}}P^l_{mk} = \tfrac{1}{2}\Big[\left((l+k)(l-k+1)\right)^{\frac{1}{2}}P^l_{m,k-1}$$

$$-\left((l-k)(l+k+1)\right)^{\frac{1}{2}}P^l_{m,k+1}\Big],$$

Integral equations:
For arbitrary angles θ_1, θ_2 and $-l \leqslant j \leqslant l$ we have the formula

$$\frac{1}{2\pi}\int_{-\pi}^{\pi} P^l_{mn}(\cos\theta)\exp i(m\varphi+n\psi-j\gamma)\,\mathrm{d}\gamma = P^l_{mj}(\cos\theta_1)P^l_{jn}(\cos\theta_2),$$

where φ, ψ, θ are determined by θ_1, θ_2, γ by the equations

$$\cos\theta = \cos\theta_1\cos\theta_2 - \sin\theta_1\sin\theta_2\cos\gamma,$$

$$\mathrm{e}^{i\varphi} = \frac{\sin\theta_1\cos\theta_2+\cos\theta_1\sin\theta_2\cos\gamma+i\sin\theta_2\sin\gamma}{\sin\theta},$$

$$\mathrm{e}^{\frac{1}{2}i(\varphi+\psi)} = \frac{\cos\frac{1}{2}\theta_1\cos\frac{1}{2}\theta_2\mathrm{e}^{i\frac{1}{2}\gamma}-\sin\frac{1}{2}\theta_1\sin\frac{1}{2}\theta_2\mathrm{e}^{-i\frac{1}{2}\gamma}}{\cos\frac{1}{2}\theta}.$$

Addition formulas:
For the angles related as above we have:

$$P^l_{mn}(\cos\theta)\mathrm{e}^{i(m\varphi+n\psi)} = \sum_{k=-l}^{l} P^l_{mk}(\cos\theta_1)P^l_{kn}(\cos\theta_2)\mathrm{e}^{ik\gamma}.$$

Special cases:

$$P^l_{mn}(\cos(\theta_1+\theta_2)) = \sum_{k=-l}^{l} P^l_{mk}(\cos\theta_1)P^l_{kn}(\cos\theta_2),$$

$$P^l_{mn}(\cos(\theta_1-\theta_2)) = \sum_{k=-l}^{l} (-1)^{n-k}P^l_{mk}(\cos\theta_1)P^l_{kn}(\cos\theta_2),$$

$$\sum_{k=-l}^{l} \overline{P}^l_{mk}P^l_{nk} = \delta_{mn},$$

$$P^l_{ll}(z) = \left(\tfrac{1}{2}(1+z)\right)^l, \qquad P^l_{l,-l}(z) = (-1)^l\left(\tfrac{1}{2}(1-z)\right)^l,$$

$$P^l_{l0}(z) = (-1)^{\frac{1}{2}l}\frac{((2l)!)^{\frac{1}{2}}}{2^l l!}(1-z^2)^{\frac{1}{2}l},$$

$$\sum_{m=-l}^{l} P^l_{mm}(\cos\theta) = \frac{\sin(l+\frac{1}{2})\theta}{\sin\frac{1}{2}\theta}.$$

Associated Legendre Functions

$$P_l^n(x) = \frac{\Gamma(l+n+1)}{\Gamma(l-n+1)n!}\left(\frac{x-1}{x+1}\right)^{\frac{1}{2}n} F(l+1,\ -l; n+l; \tfrac{1}{2}(1-x)$$

$$= \frac{(-1)^{l+n}}{2^l l!}(1-x^2)^{\frac{1}{2}n}\frac{\mathrm{d}^{l+n}}{\mathrm{d}x^{l+n}}(1-x^2)^l$$

$$= (-1)^n(1-x^2)^{\frac{1}{2}n}\frac{\mathrm{d}^n}{\mathrm{d}x^n}P_l(x)$$

$$= \frac{(-1)^{l+n}}{2^l}(1-x^2)^{\frac{1}{2}n}\sum_k \frac{(-1)^k(2k)!\,x^{2k-l-n}}{k!(l-k)!(2k-l-n)!},$$

where $0 \leqslant k \leqslant l$ and $2k \geqslant l+n$,

$$P_l^n(x) = \left(\frac{1+z}{1-z}\right)^{\frac{1}{2}n}\sum_{\max(n,0)}^{l}\frac{(-1)^j(l+j)!}{(l-j)!(j-n)!j!}\left(\frac{1-z}{2}\right)^j,$$

$$P_l^n(\cos\theta) = \frac{i^n(l+n)!}{2\pi l!}\int_0^{2\pi}(\cos\theta+i\sin\theta\cos\varphi)^l \mathrm{e}^{in\varphi}\mathrm{d}\varphi.$$

For the angles φ, ψ, θ and θ_1, θ_2, γ related in the same manner as before we have:

$$\mathrm{e}^{-nl\varphi}P_l^n(\cos\theta)$$

$$= i^n\sqrt{\frac{(l+n)!}{(l-n)!}}\sum_{k=-l}^{l} i^{-k}\sqrt{\frac{(l-k)!}{(l+k)!}}\ \mathrm{e}^{-ik\gamma}P_{nk}^l(\cos\theta_1)P_l^k(\cos\theta_2),$$

$$P_l(\cos\theta_1\cos\theta_2-\sin\theta_1\sin\theta_2\cos\gamma)$$

$$= \sum_{k=-l}^{l}\mathrm{e}^{-ik\gamma}P_l^k(\cos\theta_1)P_l^{-k}(\cos\theta_2),$$

$$\frac{1}{2\pi}\int_{-\pi}^{\pi}\mathrm{e}^{-ik\gamma}P_l(\cos\theta_1\cos\theta_2-\sin\theta_1\sin\theta_2\cos\gamma)\mathrm{d}\gamma$$

$$= P_l^k(\cos\theta_1)P_l^{-k}(\cos\theta_2),$$

$$P_l^m(z) = \frac{(-1)^m(1-z^2)^{\frac{1}{2}m}}{2^l l!}\ \frac{(l-m)!}{2\pi i}\int_\Gamma \frac{(t^2-1)^l\mathrm{d}t}{(t-z)^{l+m-1}},$$

Γ is the unit circle encircled in the positive direction

$$\frac{\mathrm{d}}{\mathrm{d}z} P_l^m(z) = -(1-z^2)^{-\frac{1}{2}} P_l^{m+1}(z) - \frac{mz}{1-z^2} P_l^m(z),$$

$$(2l+1)(1-z^2)^{\frac{1}{2}} P_l^m(z) = P_{l+1}^{m+1}(z) - P_{l-1}^{m+1}(z),$$

$$P_l^l(\cos\theta) = (-1)^l \frac{(2l)!}{2^l l!} \sin^l\theta,$$

$$\frac{\mathrm{d}}{\mathrm{d}z}\left[(1-z^2)\frac{\mathrm{d}}{\mathrm{d}z} P_l^m(z)\right] + \left[l(l+1) - \frac{m^2}{1-z^2}\right] P_l^m(z) = 0.$$

Legendre Polynomials

$$P^l = P_{00}^l = P_l^{(0,0)} = P_l^0,$$

$$P_l(x) = \frac{(-1)^l}{2^l l!} \frac{\mathrm{d}^l}{\mathrm{d}x^l} (1-z^2)^l = F(-l, l+1; 1; \tfrac{1}{2}(1-z))$$

$$= \sum_{k=0}^{l} \frac{(-1)^k (l+k)!}{(l-k)!(k!)^2} \left(\frac{1-z}{2}\right)^k$$

$$= (l!)^2 \left(\frac{1+z}{2}\right)^l \sum_{k=0}^{l} \frac{(-1)^k}{(k!)^2((l-k)!)^2} \left(\frac{1-z}{1+z}\right)^k,$$

$$P_l(\cos\theta) = \frac{1}{2\pi} \int_0^{2\pi} (\cos\theta + i\sin\theta\cos\varphi)^l \mathrm{d}\varphi$$

$$= \frac{1}{2\pi i} \int_{\Gamma'} \frac{z^l \mathrm{d}z}{(z^2 - 2zx + 1)^{\frac{1}{2}}} = \frac{1}{2^l} \frac{1}{2\pi i} \int_{\Gamma} \frac{(t^2-1)^l}{(t-x)^{l+1}} \mathrm{d}t$$

$$= \frac{2}{\pi} \int_\theta^\pi \frac{\sin t(l+\frac{1}{2})\mathrm{d}t}{\sqrt{\cos^2\frac{1}{2}\theta - \cos^2\frac{1}{2}t}}$$

$$= \frac{2}{\pi} \int_\theta^\pi \frac{\sin t(l+\frac{1}{2})\mathrm{d}t}{\sqrt{2(\cos\theta - \cos t)}},$$

Γ—the unit circle, Γ' a circle of radius > 1 around $t = 0$, $x := \cos\theta$.

$$zP_l' = lP_l + P_{l-1}', \qquad (2l+1)P_l = P_{l+1}' - P_{l-1}',$$

$$(1-z^2)P_l' + lzP_l = lP_{l-1},$$

$$(1-z^2)P_l' - (l+1)zP_l = -(l+1)P_{l+1},$$

$$(2l+1)zP_l = (l+1)P_{l-1}' + lP_{l+1}'.$$

The Legendre equation:

$$(1-z^2)P_l'' - 2zP_l' + l(l+1)P_l = 0.$$

Gegenbauer Polynomials

$$C_l^p(x) := \frac{\Gamma(l+2p)}{\Gamma(2p)l!} F(-l, l+2p; p+\tfrac{1}{2}, \tfrac{1}{2}(1-x))$$

$$= \frac{\Gamma(2p+l)}{\Gamma(2p)} \frac{\Gamma(p+\frac{1}{2})}{\Gamma(p+\frac{1}{2}+l)} P_l^{(p-\frac{1}{2}, p-\frac{1}{2})}(x)$$

$$= \sum_{m=0}^{E(\frac{1}{2}l)} \frac{(-1)^m p(p+1)\ldots(p+l-m-1)}{m!(l-2m)!} (2x)^{l-2m},$$

for $\quad p > -\frac{1}{2}, \quad l = 0, 1, 2, \ldots,$

$E(t)$ is the integer part of the number t.

$$C_{2m+1}^p(t) = (-1)^m \frac{\Gamma(p+m+1)}{m!\Gamma(p)} 2tF(-m, m+p+1; \tfrac{3}{2}; t^2),$$

$$C_{2m}^p(t) = (-1)^m \frac{\Gamma(p+m)}{m!\Gamma(p)} F(-m, m+p; \tfrac{1}{2}, t^2).$$

The generating function:

$$(1-2tx+t^2)^{-p} = \sum_{l=0}^{\infty} C_l^p(x)t^l.$$

The Rodrigues formula:

$$2^l l! \frac{\Gamma(p+l+\frac{1}{2})}{\Gamma(p+\frac{1}{2})} (1-x^2)^{p-\frac{1}{2}} C_l^p(x)$$

$$= (-1)^l \frac{\Gamma(2p+l)}{\Gamma(2p)} \frac{d^l}{dx^l} (1-x^2)^{l+p-\frac{1}{2}}.$$

Special cases:

$$C_l^{\frac{1}{2}} = P^l, \qquad C_{l-p}^{p+\frac{1}{2}}(t) = \frac{(-1)^p \sqrt{\pi}\,(1-t^2)^{-\frac{1}{2}p}}{2^p \Gamma(p+\frac{1}{2})} P_l^p(t),$$

$$C_l^p(1) = \frac{\Gamma(2p+l)}{l!\,\Gamma(2p)},$$

$$C_0^p(x) = 1,$$

$$C_1^p(x) = 2px,$$

$$C_2^p(x) = 2p(p+1)\left(x^2 - \frac{1}{2p+2}\right),$$

$$C_3^p(x) = \frac{4}{3}p(p+1)(p+2)\left(x^3 - \frac{3}{2p+4}x\right).$$

Symmetry:

$$C_l^p(-x) = (-1)^l C_l^p(x).$$

Recurrences and differential equations:

$$\frac{\mathrm{d}}{\mathrm{d}x} C_l^p = 2pC_{l-1}^{p+1},$$

$$\left[(1-x^2)\frac{\mathrm{d}^2}{\mathrm{d}x^2} - (2p+1)x\frac{\mathrm{d}}{\mathrm{d}x} + l(2p+l)\right] C_l^p = 0,$$

$$(l+1)C_{l+1}^p(x) = (2p+l)xC_l^p(x) - 2p(1-x^2)C_{l-1}^{p+1}(x),$$

$$(2p+l)C_l^p(x) = 2p\left(C_l^{p+1}(x) - xC_{l-1}^{p+1}(x)\right).$$

Integral formulas:

$$C_l^p(\cos\varphi) = \frac{\Gamma(2p+l)}{2^{2p-1}(\Gamma(p))^2 l!}\int_0^\pi (\cos\varphi - i\sin\varphi\cos\theta)^l \sin^{2p-1}\mathrm{d}\theta$$

$$= \frac{l+p}{\pi}\int_0^\pi\int_0^\pi C_l^p(\cos\theta\cos\varphi + \sin\theta\sin\varphi\cos\theta_1)$$

$$\times\, C_l^p(\cos\theta)\sin^{2p}\theta\sin^{2p-1}\theta_1\,\mathrm{d}\theta\,\mathrm{d}\theta_1,$$

$$\int_0^\pi C_l^p(\cos\theta)C_{l'}^p(\cos\theta)\sin^{2p}\theta\, d\theta = \delta_{ll'}\frac{\pi\Gamma(2p+l)}{2^{2p-1}l!(l+p)\Gamma^2(p)},$$

$$\int_0^\pi C_l^p(\cos\theta\cos\varphi+\sin\theta\sin\varphi\cos\theta_1)\sin^{2p-1}\theta_1\, d\theta_1$$

$$= \frac{2^{2p-1}l!\Gamma^2(p)}{\Gamma(2p+l)} C_l^p(\cos\varphi)C_l^p(\cos\theta).$$

Jacobi and Legendre Functions

l—a complex number, m, n—real numbers both integers or both dyadic numbers, $n \geqslant m$.

$$\mathfrak{P}_{mn}^l(z) = \frac{\Gamma(l+n+1)}{2^l\Gamma(l+m+1)(n-m)!}(z-1)^{\frac{n-m}{2}}(z+1)^{l-\frac{n-m}{2}}$$

$$\times F\left(-l-m, n-l; n-m+1; \frac{z-1}{z+1}\right)$$

$$= \frac{\Gamma(l+n+1)}{2^n\Gamma(l+m+1)(n-m)!}(z-1)^{\frac{n-m}{2}}(z+1)^{\frac{m+n}{2}}$$

$$\times F(n+l+1, n-l; n-m+1; \tfrac{1}{2}(1-z)),$$

$$\mathfrak{P}_{mn}^l(\cosh\tau) = \frac{1}{2\pi i}\int_{|z|=1}(\cosh\tfrac{1}{2}\tau+z\sinh\tfrac{1}{2}\tau)^{l+n}$$

$$\times(z\cosh\tfrac{1}{2}\tau+\sinh\tfrac{1}{2}\tau)^{l-n}z^{m-n-1}dz.$$

Symmetries:

$$\mathfrak{P}_{mn}^l = \mathfrak{P}_{-m,-n}^l, \qquad \mathfrak{P}_{mn}^l = (-1)^{m-n}\mathfrak{P}_{mn}^{-l-1},$$

$$\overline{\mathfrak{P}_{mn}^l} = \mathfrak{P}_{mn}^{\bar{l}}, \qquad \mathfrak{P}_{nn}^{i\nu-\frac{1}{2}} = \mathfrak{P}_{nn}^{-i\nu-\frac{1}{2}}.$$

Addition formulas:

With τ_1, τ_2, γ and τ, φ, ψ related by the formulas

$$\cosh\tau = \cosh\tau_1\cosh\tau_2+\sinh\tau_1\sinh\tau_2\cos\gamma,$$

$$e^{i\varphi} = \frac{\sinh\tau_1\cos\tau_2+\cosh\tau_1\sinh\tau_2\cos\gamma+i\sinh\tau_2\sin\gamma}{\sinh\frac{1}{2}\tau},$$

$$e^{\frac{1}{2}i(\varphi+\psi)} = \frac{\cosh\frac{1}{2}\tau_1\cosh\frac{1}{2}\tau_2 e^{\frac{1}{2}i\gamma}+\sinh\frac{1}{2}\tau_1\sinh\frac{1}{2}\tau_2 e^{-\frac{1}{2}i\gamma}}{\cosh\frac{1}{2}\tau}$$

we have

$$e^{-i(\varphi m+n\psi)}\mathfrak{P}^l_{mn}(\cosh\tau) = \sum_{k=-\infty}^{\infty} e^{-ik\gamma}\mathfrak{P}^l_{mk}(\cosh\tau_1)\mathfrak{P}^l_{kn}(\cosh\tau_2).$$

Summation over k runs through the set of integers if m and n are integers and through the set of dyadic numbers if both m and n are dyadic. In particular,

$$\mathfrak{P}^l_{mn}(\cosh(\tau_1+\tau_2)) = \sum_k \mathfrak{P}^l_{mk}(\cosh\tau_1)\mathfrak{P}^l_{kn}(\cosh\tau_2),$$

$$\mathfrak{P}^l_{mn}(\cosh(\tau_1-\tau_2)) = \sum_k (-1)^{n-k}\mathfrak{P}^l_{mk}(\cosh\tau_1)\mathfrak{P}^l_{kn}(\cosh\tau_2),$$

$$\sum_k \mathfrak{P}^{-i\nu-\frac{1}{2}}_{mk}(\cosh\tau)\mathfrak{P}^{i\nu-\frac{1}{2}}_{kn}(\cosh\tau) = \delta_{mn}, \qquad \nu\in \boldsymbol{R},$$

$$(z\cosh\tfrac{1}{2}\tau+\sinh\tfrac{1}{2}\tau)^{l+n}(z\sinh\tfrac{1}{2}\tau+\cosh\tfrac{1}{2}\tau)^{l-n}$$

$$= \sum_{m=-\infty}^{\infty} \mathfrak{P}^l_{mn}(\cosh\tau)z^{l+m}.$$

With the angles satisfying the same relations as above we have:

$$\frac{1}{2\pi}\int_0^{2\pi} e^{i(k\gamma-m\varphi-n\psi)}\mathfrak{P}^l_{mn}(\cosh\tau)\,d\gamma = \mathfrak{P}^l_{mk}(\cosh\tau_1)\mathfrak{P}^l_{kn}(\cosh\tau_2).$$

Recurrence formulas:

$$\sqrt{z^2-1}\,\frac{d}{dz}\mathfrak{P}^l_{mn}+\frac{m-nz}{\sqrt{z^2-1}}\mathfrak{P}^l_{mn} = (l-n)\mathfrak{P}^l_{m,n+1},$$

$$\sqrt{z^2-1}\,\frac{d}{dz}\mathfrak{P}^l_{mn}+\frac{nz-m}{\sqrt{z^2-1}}\mathfrak{P}^l_{mn} = (l+n)\mathfrak{P}^l_{m,n-1},$$

$$\sqrt{z^2-1}\,\frac{d}{dz}\mathfrak{P}^l_{mn} = \frac{l+n}{2}\mathfrak{P}^l_{m,n-1}+\frac{l-n}{2}\mathfrak{P}^l_{m,n+1},$$

$$\frac{m-nz}{\sqrt{z^2-1}}\mathfrak{P}^l_{mn} = -\frac{l+n}{2}\mathfrak{P}^l_{m,n-1}+\frac{l-n}{2}\mathfrak{P}^l_{m,n+1},$$

$$\mathfrak{P}^{l+\frac{1}{2}}_{m+\frac{1}{2},n-\frac{1}{2}}(\cosh\tau) = \sinh\tfrac{1}{2}\tau\,\mathfrak{P}^l_{mn}(\cosh\tau)$$

$$+\cosh\tfrac{1}{2}\tau\,\mathfrak{P}^l_{m+1,n}(\cosh\tau),$$

$$(l+n)\cosh\tfrac{1}{2}\tau\mathfrak{P}^{l-\frac{1}{2}}_{m-\frac{1}{2},n-\frac{1}{2}}(\cosh\tau)+(l-n)\sinh\tfrac{1}{2}\tau$$

$$\times\mathfrak{P}^{l-\frac{1}{2}}_{m-\frac{1}{2},n+\frac{1}{2}}(\cosh\tau) = (l+m)\mathfrak{P}^{l}_{mn}(\cosh\tau).$$

Differential equation:

$$(z^2-1)\frac{d^2}{dz^2}\mathfrak{P}^l_{mn}+2z\frac{d}{dz}\mathfrak{P}^l_{mn}-\frac{m^2+n^2-2mnz}{z^2-1}\mathfrak{P}^l_{mn}$$

$$= l(l+1)\mathfrak{P}^l_{mn}.$$

Associated Legendre functions

$$\mathfrak{P}^m_l = \mathfrak{P}^l_{m0}\frac{\Gamma(l+m+1)}{\Gamma(l+1)}, \qquad \mathfrak{P}^m_l = P^m_l \qquad \text{for } l\in N,\ |m|\leqslant l,$$

$$\mathfrak{P}^n_l(z) = \frac{\Gamma(l+n+1)}{\Gamma(l-n+1)n!}\left(\frac{z-1}{z+1}\right)^{\frac{1}{2}n} F(l+1,-l;n+1;\tfrac{1}{2}(1-z))$$

$$\mathfrak{P}^n_l(\cosh\tau) = \frac{1}{2\pi i}\frac{\Gamma(l+n+1)}{\Gamma(l+1)}$$

$$\times\int_\Gamma\left(\cosh\tau+\frac{z^2+1}{2}z\sinh\tau\right)^l z^{n-1}dz,$$

Γ being the unit circle in $\boldsymbol{C}$.

$$\mathfrak{P}^m_l(z) = \frac{\Gamma(l+m+1)}{\Gamma(l+1)}(z^2-1)^{\frac{1}{2}m}\frac{1}{2\pi i}\int_\Gamma\frac{(t^2-1)^l dt}{2^l(t-z)^{l+m+1}},$$

Γ an arbitrary simple contour containing z and 1 and leaving -1 outside.

$$\mathfrak{P}^m_l(z) = \frac{\Gamma(l+m+1)}{2\pi\Gamma(l+1)}\int_0^{2\pi}(z+\sqrt{z^2-1}\cos\theta)^l e^{im\theta}d\theta,$$

$$\mathfrak{P}^n_l(\cosh\tau) = \frac{\Gamma(l+n+1)}{2\pi\Gamma(l+1)}\int_{-\tau}^{\tau}\frac{\exp(l+\frac{1}{2})t\,T_n\left(\dfrac{e^t-\cosh t}{\sinh t}\right)}{(\cosh^2\frac{1}{2}\tau-\cosh^2\frac{1}{2}t)^{\frac{1}{2}}}dt,$$

$T_n(x) := \cos(n \arccos x)$—Tchebyshev polynomial,

$$\frac{\cosh^l \tau}{\Gamma(l+1)} = \sum_{m=-\infty}^{\infty} \frac{i^m}{\Gamma(l+m+1)} \mathfrak{P}_l^m(\cosh \tau),$$

$$(z^2-1)\frac{d^2}{dz^2}\mathfrak{P}_l^m + 2z\frac{d}{dz}\mathfrak{P}_l^m - \frac{m^2}{z^2-1}\mathfrak{P}_l^m = l(l+1)\mathfrak{P}_l^m,$$

$$\mathfrak{P}_l^{-m} = \frac{\Gamma(l-m+1)}{\Gamma(l+m+1)}\mathfrak{P}_l^m, \qquad \mathfrak{P}_l^m - \mathfrak{P}_{-l-1}^m \qquad \text{for } m \in \boldsymbol{Z}.$$

Legendre Functions

$$\mathfrak{P}_l := \mathfrak{P}_{00}^l = \mathfrak{P}_l^0, \qquad \mathfrak{P}_l(z) = F(l+1, -l; 1; \tfrac{1}{2}(1-z)),$$

$\mathfrak{P}_l = P_l$ for l a positive integer,

$$\mathfrak{P}_l(z) = \frac{1}{2\pi i}\int_\Gamma \frac{(t^2-1)^l dt}{2^l(t-z)^{l+1}},$$

Γ—a simple contour encircling z and 1 and not containing -1.

$$\mathfrak{P}_l(\cosh \tau) = \frac{1}{2\pi}\int_0^{2\pi}(\cosh \tau + \sinh \tau \cos \theta)^l d\theta$$

$$= \frac{\sqrt{2}\sin \pi l}{\pi}\int_0^\infty \frac{\cosh(l+\frac{1}{2})t}{\sqrt{\cosh t + \cosh \tau}} dt,$$

$$\mathfrak{P}_l(\cosh \tau_1 \cosh \tau_2 + \sinh \tau_1 \sinh \tau_2 \cos \gamma)$$

$$= \sum_{k=-\infty}^{\infty} e^{-ik\gamma}\mathfrak{P}_l^k(\cosh \tau_1)\mathfrak{P}_l^{-k}(\cosh \tau_2),$$

$$\frac{1}{2\pi}\int_0^{2\pi} e^{ik\gamma}\mathfrak{P}_l(\cosh \tau_1 \cosh \tau_2 + \sinh \tau_1 \sinh \tau_2 \cos \gamma) d\gamma$$

$$= \mathfrak{P}_l^k(\cosh \tau_1)\mathfrak{P}_l^{-k}(\cosh \tau_2),$$

$$(z^2-1)\frac{d^2}{dz^2}\mathfrak{P}_l + 2z\frac{d}{dz}\mathfrak{P}_l = l(l+1)\mathfrak{P}_l.$$

$$\mathfrak{P}_{i\nu-\frac{1}{2}} = \mathfrak{P}_{-i\nu-\frac{1}{2}}, \qquad \nu \in \boldsymbol{R}.$$

Laguerre Polynomials

$$L_n^\alpha(z) := e^z \frac{z^{-\alpha}}{n!} \frac{d^n}{dz^n} (e^{-z} z^{n+\alpha}), \qquad \alpha > -1,$$

$$n! e^{-n} z^{m-n} L_n^{m-n}(z) = \frac{d^n}{dz^n} (e^{-z} z^m),$$

$$L_k^{m-k}(x) = (-1)^{k-m} \frac{\Gamma(m+1)}{\Gamma(k+1)} x^{k-m} L_m^{k-m}(x).$$

The generating function:

$$\exp(-\lambda z)(z+1)^m = \sum_{n=0}^{\infty} L_n^{m-n}(z) z^n,$$

$$(1-z)^{-\alpha-1} \exp \frac{-xz}{1-z} = \sum_{n=0}^{\infty} L_n^\alpha(x) z^n.$$

Addition formula:

For r_1, r_2, φ and r, ψ related by the formula $r_1 + r_2 e^{i\varphi} = re^{i\psi}$ we have:

$$\exp(r_1 r_2 e^{-i\varphi}) \sum_{k=0}^{\infty} r_1^{k-n} L_n^{k-n}(r_1^2)(r_2 e^{i\varphi})^{m-k} L_k^{m-k}(r_2^2)$$

$$= (re^{i\psi})^{m-k} L_n^{m-n}(r^2),$$

$$\sum_l \left(\frac{x}{y}\right)^l L_l^{m-l}(x^2) L_n^{l-n}(y^2)$$

$$= \left(1 + \frac{y}{x}\right)^m \left(1 + \frac{x}{y}\right)^n e^{-xy} L_n^{m-n}((x-y)^2),$$

$$\sum_l \left(-\frac{iy}{x}\right)^l L_l^{m-l}(x^2) L_n^{l-n}(y^2)$$

$$= \left(1 - \frac{iy}{x}\right)^m \left(1 + \frac{ix}{y}\right)^{-n} e^{ixy} L_n^{m-n}(x^2+y^2).$$

Integral formula:

$$L_k^l(w) = \frac{w^{k+1}}{\pi k!} \int_C \exp(-w(z+|z|))^2 (z+1)^{k+1} \bar{z}^k dx\, dy,$$

$$z = x + iy.$$

Recurrences and differential equations:

$$\frac{d}{dx} L_n^k = -L_{n-1}^{k+1},$$

$$x\frac{d}{dx} L_n^k + (k-n) L_n^k = (n+1) L_{n+1}^{k-1},$$

$$L_n^k = L_n^{k+1} - L_{n-1}^{k+1},$$

$$x\frac{d^2}{dx^2} L_n^k + (k+1-x)\frac{d}{dx} L_n^k + nL_n^k = 0.$$

Orthogonality relations:

$$\int_0^\infty e^{-r} r^k L_m^k(r) L_n^k(r)\,dr = \frac{n!}{(n+k)!}\delta_{nm},$$

$$\int_0^\infty x^p e^{-x} L_m^p(xy) L_{m-s}^p(x)\,dx = \frac{(p+m)!}{s!(m-s)!}(1-y)^s y^{m-s},$$

$$L_n^l(x) = \sum_{k=0}^{n} \frac{(n+l)!}{(k+l)!}\frac{(-x)^k}{k!(n-k)!} = \frac{\Gamma(n+l+1)}{n!\Gamma(l+1)}\, {}_1F_1(-n; l+1; z).$$

Relations with the Jacobi polynomials:
For each $r > 0$ and θ, any integers l_1, l_2, j, m

$$\sum_{k=0}^{\min(2l_1, 2l_2)} (-1)^{k-j} \sqrt{\frac{(2l_1-k)!}{(2l_2-k)!}}\, P_{l_1-j, l_1-k}^{l_1}(\cos\theta)\, P_{l_2-m, l_2-k}^{l_2}(\cos\theta)$$

$$\times L_{2l_1-k}^{2(l_2-l_1)}(r) = \sqrt{\frac{j!(2l_1-j)!}{m!(2l_2-m)!}}(i\sin\tfrac{1}{2}\theta)^{m-j}(\cos\tfrac{1}{2}\theta)^{2l_2-2l_1-m+j}$$

$$\times L_j^{m-j}(r\sin^2\tfrac{1}{2}\theta) L_{2l_1-j}^{2l_2-2l_1-m+j}(r\cos^2\tfrac{1}{2}\theta),$$

$$\sum_{k=0}^{2l} (P_{l-j, l-k}^{l}(\cos\theta))^2 L_{2l-k}^0(r) = L_j^0(r\sin^2\tfrac{1}{2}\theta)\, L_{2l-j}^0(r\cos^2\tfrac{1}{2}\theta),$$

$$\sum_{k=0}^{m} \frac{(p+m)!}{k!(p+m-k)!}(1-y)^k y^{m-k} L_{m-k}^p(x) = L_m^p(xy),$$

$$x > 0, \qquad 0 \leqslant y \leqslant 1,$$

$$\int_0^\pi L_j^0(r\sin^2\tfrac{1}{2}\theta)\,L_{2l-j}^0(r\cos^2\tfrac{1}{2}\theta)\sin\theta\,\mathrm{d}\theta = \frac{2}{2l+1}\sum_{k=0}^{2l} L_{2l-k}^0(r),$$

$$\sqrt{\frac{(2l_1-k)!}{(2l_2-k)!}}\,P_{l_1-j,l_1-k}^{l_1}(\cos\theta)\,L_{2l_1-k}^{2l_2-2l_1}(r)$$

$$= \sum_{m=0}^{2l_2}\sqrt{\frac{j!(2l_1-j)!}{m!(2l_2-m)!}}(i\sin\tfrac{1}{2}\theta)^{m-j}(\cos\tfrac{1}{2}\theta)^{2l_2-2l_1-m+j}$$

$$\times\, L_j^{m-j}(r\sin\tfrac{1}{2}\theta)\,L_{2l_1-j}^{2l_2-2l_1-m+j}(r\cos^2\tfrac{1}{2}\theta)\,P_{l_2-m,l_2-k}^{l_2}(\cos\theta).$$

Hermite Polynomials

$$H_n := (-1)^n e^{x^2}\frac{\mathrm{d}^n}{\mathrm{d}x^n}(e^{-x^2}),$$

$$H_n(x) := n!\sum_{m=0}^{E\left(\frac{1}{2}n\right)}\frac{(-1)^n(2x)^{n-2m}}{m!(n-2m)!},$$

$E(\cdot)$—the function entier,

$$H_{2m}(x) = (-1)^m 2^{2m} m!\,L_m^{-\frac{1}{2}}(x^2),$$

$$H_{2m+1}(x) = (-1)^m 2^{2m+1} m!\,x L_m^{-\frac{1}{2}}(x^2),$$

$$H_{n+1} - 2xH_n + 2nH_{n-1} = 0,$$

$$\exp(2xz - z^2) = \sum_{n=0}^{\infty} H_n(x)\frac{z^n}{n!},$$

$$H_n(-x) = (-1)^n H_n(x), \qquad H_n' = 2nH_{n-1},$$

$$H_n'' - 2xH_n' + 2nH_n = 0,$$

$$\int_{-\infty}^{\infty} H_n(x)H_n(x)\exp(-x^2)\,\mathrm{d}x = \delta_{mn}2^n n!\,\pi^{\frac{1}{2}},$$

The Fourier Transformation

For $f \in L^1(\boldsymbol{R}^n)$

$$\hat{f}(p) := \int_{\boldsymbol{R}^n} f(x) \exp i(p|x) \mathrm{d}x.$$

For $f \in \mathscr{S}(\boldsymbol{R}^n)$

$$f(x) = \frac{1}{(2\pi)^n} \int_{\boldsymbol{R}^n} \hat{f}(p) \exp(-i(p|x)) \mathrm{d}p$$

and

$$\int_{\boldsymbol{R}^n} |f|^2(x) \mathrm{d}x = \frac{1}{(2\pi)^n} \int_{\boldsymbol{R}^n} |\hat{f}|^2(p) \mathrm{d}p.$$

The Mellin Transformation

For $f \in \mathscr{C}_0(\boldsymbol{R}_+)$

$$\mathscr{M}f(z) := \int_{\boldsymbol{R}_+} f(v) v^{z-1} \mathrm{d}v$$

the following formulas hold:

$$f(t) = \frac{1}{2\pi i} \int_{-i\infty} \mathscr{M}f(x) t^{-x} \mathrm{d}x;$$

and

$$\int_0^\infty |f(t)|^2 \frac{\mathrm{d}t}{t} = \frac{1}{2\pi} \int_{-\infty}^{\infty} \mathscr{M}f(iy)^2 \mathrm{d}y.$$

For $f \in \mathscr{C}_0(\boldsymbol{R})$ with support contained in $\boldsymbol{R}_+$ we have:

$$\mathscr{M}f(z) = \frac{\Gamma(z)}{2\pi} \mathrm{e}^{-i\frac{\pi}{2}z} \int_{-\infty}^{\infty} f(x) x^{-z} \mathrm{d}x$$

for $\operatorname{Re} z > 0$.

The Fourier Transformation on the Circle Group

For $f \in \mathscr{C}(\boldsymbol{T})$

$$\hat{f}(n) := \int_0^1 f(\mathrm{e}^{2\pi i \varphi}) \mathrm{e}^{2\pi i n \varphi} \mathrm{d}\varphi,$$

and we have the formulas:

$$f(e^{i\varphi}) = \sum_{n\in Z} \hat{f}(n)e^{-in\varphi}$$

and

$$\int_0^1 |f|^2(e^{2\pi i\varphi})d\varphi = \sum_{n\in Z} |\hat{f}|^2(n).$$

The Poisson Summation Formula

For $f\in\mathscr{D}(\boldsymbol{R})$ we have

$$\sum_{n=-\infty}^{\infty} f(n) = \sum_{m=-\infty}^{\infty} \mathscr{F}f(2\pi m),$$

$\mathscr{F}f$ being the classical Fourier transform of a function on $\boldsymbol{R}$.

Integral Transformations Associated with Bessel Functions

I. For every function $f\in\mathscr{C}_0(\boldsymbol{C})$

$$f(\xi) = \frac{1}{2\pi}\int_0^\infty f^r(\xi)r\,dr,$$

where

$$f^r(\xi) = \frac{i}{2}\int_C f(z)J_0(r|z-\xi|)dz\wedge d\bar{z}.$$

II. For every function $f\in\mathscr{C}_0(\boldsymbol{C})$

$$f(\xi) = \int_0^\infty r\,dr\sum_{k=-\infty}^{\infty}\left(\frac{\xi}{|\xi|}\right)^k J_k(r|\xi|)\boldsymbol{F}f(k,r),$$

where

$$\boldsymbol{F}f(k,r) := \frac{i}{4\pi}\int_C f(z)\left(\frac{z}{|z|}\right)^{-k} J_k(r|z|)dz\wedge d\bar{z}.$$

III. The Fourier–Bessel transformation.

$$f(r) = \int_0^\infty \varrho\,d\varrho J_k(\varrho r)Ff(k,\varrho),$$

where

$$Ff(k,\varrho) := \int_0^\infty r\,dr\,J_k(\varrho r)f(r).$$

Expansions of Functions on the Sphere

The classical harmonics are defined as

$$Y_{lk}(\varphi, \theta) := e^{-ik\varphi} P^l_{k0}(\cos\theta), \qquad -l \leqslant k \leqslant l.$$

For $f \in L^2(S^2)$ we have the decomposition

$$f(\varphi, \theta) = \sum_{l=0}^{\infty} (2l+1) \sum_{k=-l}^{l} c_{lk} Y_{lk}(\varphi, \theta),$$

where

$$c_{lk} = \frac{1}{4\pi} \int_0^{2\pi}\int_0^{\pi} f(\varphi, \theta) \overline{Y_{lk}(\varphi, \theta)} \sin\theta \, d\theta \, d\varphi,$$

and the following Plancherel formula is valid:

$$||f||_{L^2(S^2)} = \sum_{l=0}^{\infty} (2l+1) \sum_{k=-l}^{l} |c_{kl}|^2.$$

Moreover,

$$f(\varphi, \theta) = \frac{1}{4\pi} \sum_{l=0}^{\infty} (2l+1) \int_0^{2\pi}\int_0^{\pi} f(\varphi_1, \theta_1) P_l(\cos\theta\cos\theta_1 - \sin\theta\sin\theta_1\cos(\varphi_1 - \varphi)) \, d\theta_1 \, d\varphi_1.$$

The functions Y_{lk} satisfy the differential equation:

$$\Delta Y_{lk} = -l(l+1) Y_{lk}, \qquad -l \leqslant k \leqslant l,$$

$$\Delta := \frac{\partial^2}{\partial\theta^2} + \cot\theta \frac{\partial}{\partial\theta} + \frac{1}{\sin^2\theta} \frac{\partial^2}{\partial\varphi^2}.$$

Expansions of Functions on the Group SU(2):

For any $f \in L^2(\mathrm{SU}(2))$

$$f(\varphi, \theta, \psi) := \sum_{l} \sum_{m=-l}^{l} \sum_{n=-l}^{l} c^l_{mn} e^{-i(m\varphi + n\psi)} P^l_{mn} \cos\theta,$$

where l runs over the set of dyadic numbers and

$$c^l_{mn} := \frac{(-1)^{m-n}(2l+1)}{16\pi^2} \int_{-2\pi}^{2\pi}\int_0^{2\pi}\int_0^{\pi} f(\varphi, \theta, \psi) e^{i(m\varphi + n\psi)} \times P^l_{mn}(\cos\theta) \sin\theta \, d\theta \, d\varphi \, d\psi.$$

The series is convergent in $L^2(\mathrm{SU}(2))$ and the following Plancherel formula holds:

$$\|f\|_{L^2(\mathrm{SU}(2))} = \sum_l \sum_{m=-l}^{l} \sum_{n=-l}^{l} |c_{mn}^l|^2.$$

The paremeters φ, θ, ψ are the Euler angles of an element of the group.

Expansions of Functions Corresponding to Orthogonal Polynomials

I. For $\alpha, \beta > -1$ the polynomials $P_n^{(\alpha,\beta)}$, $n = 0, 1, 2 \ldots$, form a complete orthonormal system in $L^2([-1, 1], (1-x)^\alpha(1+x)^\beta \mathrm{d}x)$.

II. For an arbitrary $l \geqslant 0$ the sequence of functions

$$\frac{(2l+k+1)k!^{\frac{1}{2}}}{2(2l+k)} P_{l+k}^l, \qquad k = 0, 1, 2, \ldots$$

is an orthonormal complete system in $L^2([-1, 1], \mathrm{d}x)$.

III. The system of functions $(l+\frac{1}{2})^{\frac{1}{2}}\ P_l$ constitutes a complete orthonormal system in $L^2([-1, 1], \mathrm{d}x)$.

IV. For an arbitrary p the system

$$\psi_l(t) := 2^{p-1}\Gamma(p)\left[\frac{2(l+p)l!}{\pi\Gamma(2p+l)}\right]^{\frac{1}{2}} C_l^p(t), \qquad l = 0, 1, 2, \ldots$$

is a complete orthonormal system in $L^2([-1, 1], \mathrm{d}\mu)$, where

$$\mathrm{d}\mu(t) = (1-t^2)^{p-\frac{1}{2}}\mathrm{d}t.$$

V. The functions

$$\varphi_n(z) = \pi^{-\frac{1}{4}}(n!)^{-\frac{1}{2}}(-1)^n 2^{-\frac{1}{2}n}\exp(-\tfrac{1}{2}z^2)H_n(z)$$

form an orthonormal basis for $L^2(\boldsymbol{R})$.

The Mehler–Fok Transformation

Let $f \in \mathscr{C}_0(\boldsymbol{R}_+)$. Define

$$\tilde{f}(\nu) = \int_0^\infty f(t)\mathfrak{P}_{i\nu-\frac{1}{2}}(\cosh t)\mathrm{d}t.$$

Then

$$f(\tau) = \frac{1}{2\pi} \int_0^\infty \tilde{f}(\nu) \mathfrak{P}_{-i\nu-\frac{1}{2}}(\cosh\tau)\nu \tanh(\pi\nu)\,d\nu$$

and

$$\int_0^\infty |f|^2(\tau) \sinh\tau\,d\tau = \frac{1}{2\pi} \int_0^\infty |\tilde{f}|^2(\nu)\nu \tanh(\pi\nu)\,d\nu.$$

References

Adams, J. F.: (1969), *Lectures on Lie groups*, W. A. Benjamin, New York.

Araki, S.: (1962), 'On root systems and an infinitesimal classification of irreducible symmetric spaces', *J. of Math. Osaka City Univ.* **13**, 1–34.

Askey, R. A.: (1975), *Orthogonal polynomials and special functions*, SIAM, CBMS Regional Conf., Philadelphia.

Askey, R. A., (ed.): (1975), *Theory and application of special functions*, Academic Press, New York.

Atiyah, M. F. (ed.): (1979), *Representation theory of Lie groups*, University Press, Cambridge.

Auslander, L. and Kostant, B.: (1967), 'Quantization and representations of solvable Lie groups', *Bull. Amer. Math. Soc.* **73**, 692–695.

Bargmann, V., (1947), 'Irreducible representations of the Lorentz group', *Ann. of Math.* **48**, 568–640.

Bargmann, V., (ed.): (1970), *Group representations in mathematics and physics*, Battelle Seattle 1969 Rencontres, Springer Verlag, Berlin.

Barut, A. O. and Rączka, R.: (1977), *Theory of group representations and applications*, PWN–Polish Scientific Publishers, Warszawa.

Belifante, J. G. F. and Kolman, B.: (1972), *A survey of Lie groups and Lie algebras*, SIAM, Philadelphia.

Bhanu-Murthy, T. S.: (1966), 'Mera Plansherelya dlya faktorprostranstva SL($n, \boldsymbol{R}$), SO($n, \boldsymbol{R}$)', *DANSSSR* **133**, 3, 503–506.

Bhanu-Murthy, T. S.: (1960), 'Asimptoticheskie provedenie zonalnykh sfericheskikh funktsii na verkhnei poluploskosti Zigelya', *DANSSSR* **135**, 5, 1027–1030.

Bishop, R. L. and Crittenden, R. L.: (1964), *Geometry of manifolds*, Academic Press, New York.

Blattner, R. J.: (1961), 'On induced representations', *Amer. J. of Math.* **83**, 79–98.

Blattner, R. J.: (1961b), 'On induced representations, II, Infinitesimal induction', *Amer. J. of Math.* **83**, 499–512.

Blattner, R. J.: (1963), 'Positive definite measures', *Proc. Amer. Math. Soc.* **14**, 423–428.

Bochner, S.; (1933), 'Monotone Funktionen, Stjeltjes Integralen, harmonische Analyse', *Math. Ann.* 96.

Boseck, H.; (1973), *Grundlagen der Darstellungstheorie*, VEB Deutscher Verlag der Wissenschaft, Berlin.

Bourbaki, N.: (1952), *Éléments de mathématique, Intégration*, Livre VI, Chapitre I–V, Hermann, Paris.

Bourbaki, N.: (1955), *Éléments de mathématique, Espaces vectoriels topologiques*, Hermann, Paris.

Bourbaki, N.: (1959), *Éléments de mathematique, Intégration*, Livre VI, Chapitre VI, Hermann, Paris.

Bourbaki, N.: (1960), *Éléments de mathématique, Groupes et algèbres de Lie*, Chapitre I, Hermann, Paris.

Bourbaki, N.: (1963), *Éléments de mathématique, Intégration*, Livre VI, Chapitre VII–VIII, Hermann, Paris.

Bourbaki, N.: (1968), *Éléments de mathématique, Groupes et algèbres de Lie*, Chapitre IV–VI, Hermann, Paris.

Bredon, G.: (1972), *Introduction to compact transformation groups*, Academic Press, New York.

Bruhat, F.: (1956), 'Sur les representations induites des groupes de Lie', *Bull. Soc. Math. France* **84**, 97–205.

Cartan, E.: (1914), 'Les groupes réeles simples finis et continus', *Ann. Sci. École Norm. Sup.* **31**, 263–335.

Cartan, E.: (1926), 'Sur une classe remarquable d'espaces de Riemann', *Bull. Soc. Math. France* **54**, 214–262.

Cartan, E.: (1927), 'Sur une classe remarquable d'espaces de Riemann', *Bull. Soc. Math. France* **55**, 114–132.

Cartan, E.: (1929), 'Groupes simples clos et ouverts et géométrie riemannienne', *J. Math. Pures Appl.* **8**, 1–13.

Cartan, E.: (1929b), 'Sur la détermination d'une système orthogonal complet dans un espace de Riemann symétrique clos', *Rend. Circ. Mat. Palermo* **53**, 217–252.

Cartier, P.: (1966), 'Quantum-mechanical commutation relations and θ-function', *Proc. Symp. Pure Math. IX, Am. Math. Soc.*, 361–383.

Cartier, P.: (1974), 'Vecteurs différentiables dans les représentations unitaires des groupes de Lie', *Seminaire Bourbaki* **454**, 1–14.

Chevalley, C.: (1946), *Theory of Lie groups*, Princeton University Press, Princeton.

Chevalley, C.: (1955), 'Invariants of finite groups generated by reflection', *Amer. J. Math.* **77**, 778–782.

Coifman, R. R. and Weis, G.: (1969), 'Representations of compact groups and spherical harmonics', *L' Enseignement Mathematique* **14**, 121–175.

Dieudonné, J.: (1980), *Special functions and linear representations of Lie groups*, AMS CBMS, Providence.

Dixmier, J.: (1964), *Les C*-algèbres et leurs représentations*, Gauthier-Villars, Paris.

Dixmier, J.: (1969), *Les algèbres d'operateurs dans l'espace Hilbertien*, Gauthier-Villars, Paris.

Dym, H. and McKean, H. D.: (1972), *Fourier series and integrals*, Academic Press, New York.

Dyson, F. (ed.), (1966), *Symmetry groups in nuclear and particle physics*, W. A. Benjamin, New York.

Ehrenpreis, L. and Mautner, F. I.: (1957), 'Some properties of the Fourier transform on semisimple Lie groups, II', *Trans. Amer. Math. Soc.* **84**, 1–55.

Ehrenpreis, L. and Mautner, F. I.: (1959), 'Some properties of the Fourier transform on semisimple Lie groups, III', *Trans. Amer. Math. Soc.* **90**, 431–484.

Flensted-Jensen, M.: (1970), *On the Fourier transform on a symmetric space of rank one*, Institut Mittag-Leffler, Djursholm.

Flensted-Jensen, M.: (1972), 'Paley–Wiener theorems for differential operator associated with symmetric spaces', Arkiv för Matematik **10**.

Freudenthal, H. and De Vries, H.: (1969), *Linear Lie groups*, Academic Press, New York.

Frobenius, G.: (1898), 'Über Relationen zwischen den Charakteren einer Gruppe und denen ihrer Untergruppe, '*Sitz. Preus. Akad. Wiss.*, 501–515.

Furstenberg, H.: (1963), 'A Poisson formula for semisimple Lie groups', *Ann. of Math.* **77**, 335–386.

Furstenberg, H: (1965), 'Translation-invariant cones of functions on semisimple Lie groups', *Bull. Amer. Math. Soc.* **71**, 271–326.

Gelfand, I. M. and Raikov, D. A.: (1943), 'Neprivodimye unitarnye predstavleniya lokalno bikompaktnykh grupp', *Mat. Sb.* **13**, 301–346.

Gelfand, I. M. and Naimark, M. A.: (1947), 'Unitarnye predstavleniya gruppy Lorentsa', *Izvestiya ANSSSR* **11**, 5, 411–504.

Gelfand, I. M.: (1950), 'Sfericheskie funktsii na simmetricheskikh rimanovykh prostranstvakh', *DANSSSR* **70**, 5–8.

Gelfand, I. M. and Naimark, M. A.: (1950), 'Unitarnye predstavleniya klassicheskikh grupp', *Tr. Matem. instituta ANSSSR im. V. A. Steklova* **34**.

Gelfand, I. M. and Graev, M. I.: (1959), 'Geometriya odnorodnykh prostranstv, predstavleniya grupp v odnorodnykh prostranstvakh i svyazannye s nimi voprosy integralnoi geometrii', *Tr. Mosk. Matem. obshchestva* **8**, 321–390.

Gelfand, I. M.: (1960), 'Integralnaya geometriya i ee svyaz s teoriei predstavlenii' *Usp. Matem. Nauk* **15**, 2, 155–164.

Gelfand, I. M.: (1962), 'Primenenie metoda orisfer k spektralnomu analizu funktsii v veshchestvennom i mnimom prostranstvakh Lobachevskogo', *Tr. Mosk. Matem. obshchestva* **11**, 243–308.

Gelfand, I. M., Graev, M. I. and Vilenkin, N. Ya.: (1962), *Obobshchennye funktsii*, V, *Integralnaya geometriya i teoriya predstavlenii grupp*, Gos. Izd. Fiz. i Matem. Literatury, Moskva.

Gelfand, I. M., Graev, M. I. and Pyatetskii-Shapiro, I. I.: (1966), *Obobshchennye funktsii*, VI, *Teoriya predstavlenii i avtomorfnye funktsii*, Gos. Izd. Fiz. i Matem. Literatury, Moskva.

Godement, R.: (1947), 'Sur les relations d'orthogonalité de V. Bargman, I, II, *C. R. Acad. Sc. Paris* **225**, 521–523, 657–659.

Godement, R.: (1948), 'Les fonctions de type positif et la théorie des groupes', *Trans. Amer. Math. Soc.* **63**, 1–84.

Godement, R.: (1952), 'A theory of spherical functions', *Trans. Amer. Math. Soc.* **73**, 496–556.

Godement, R.: (1952b), 'Une généralisation du théorème de la moyenne pour les fonctions harmoniques', *C. R. Acad. Sci. Paris* **234**, 2137–2139.

Greub, W., Halperin, S. and Vanstone, R.: (1972), *Connections, curvature and cohomology of manifolds and vector bundles*, Academic Press, New York.

Gunning, R. C.: (1976), *Riemann surfaces and generalized theta functions*, Springer Verlag, Berlin.

Hamermesh, M: (1962), *Group theory and its applications to physical problems*, Addison-Wesley, Reading.

Harish-Chandra: (1952), 'Plancherel formula for 2×2 real unimodular group', *Proc. Nat. Acad. Sci. USA* **38**, 337–342.

Harish-Chandra: (1953), 'Representations of semi-simple Lie groups, I', *Trans. Amer. Math. Soc.* **75**, 185–242.

Harish-Chandra: (1954), 'Representations of semi-simple Lie groups, II', *Trans. Amer. Math. Soc.* **76**, 26–65.

Harish-Chandra: (1954b), 'Representations of semi-simple Lie groups, III', *Trans. Amer. Math. Soc.* **76**, 234–253.

Harish-Chandra: (1954c), 'On the Plancherel formula for the right invariant functions on semi-simple Lie group', *Proc. Nat. Acad. Sci. USA* **40**, 200–204.

Harish-Chandra: (1954d), 'The Plancherel formula for complex semi-simple Lie groups', *Trans. Amer. Math. Soc.* **76**, 485–528.

Harish-Chandra: (1955), 'Representations of semi-simple Lie groups, IV', *Amer. J. of Math.* **77**, 743–777.

Harish-Chandra: (1956), 'Representations of semi-simple Lie groups, V', *Amer. J. of Math.* **78**, 1–41.

Harish-Chandra: (1956b), 'Representations of semi-simple Lie groups, VI', *Amer. J. of Math.* **78**, 564–628.

Harish-Chandra: (1958), 'Spherical functions on semi-simple Lie group, I, II', *Amer. J. of Math.*, **80**, 241–310; 553–613.

Harish-Chandra: (1968), *Automorphic forms on semi-simple Lie groups*, Springer Verlag, Berlin.

Harish-Chandra: (1970), 'Harmonic analysis on semi-simple Lie groups', *Bull. Amer. Math. Soc.* **76**, 529–551.

Harish-Chandra; (1975), 'Harmonic analysis on real reductive groups, I. The theory of constant term', *Funct. Anal.* **19**, 2, 104–204.

Heine, V.: (1960), *Group Theory in Quantum Mechanics*, Pergamon Press, London.

Helgason, S.: (1962), *Differential geometry and symmetric spaces*, Academic Press, New York.

Helgason, S.: (1963), 'Duality and Radon transform for symmetric spaces', *Amer. J. of Math.* **85**, 667–692.

Helgason, S.: (1966), 'An analog of the Paley–Wiener theorem for the Fourier transform on certain symmetric spaces', *Math. Ann.* **165**, 297–308.

Helgason, S.: (1968), *Lie groups and symmetric spaces*, Battele Rencontres, 1967,

Lectures in Math. and Phys., ed. C. M. DeWitt and J. A. Wheeler, W. A. Benjamin, New York.

Helgason, S. and Johnson, K.: (1969), 'The bounded spherical functions on symmetric spaces', *Adv. in Math.* **3**, 586–593.

Helgason, S.: (1970), 'A duality for symmetric spaces with applications to group representations', *Adv. in Math.* **5**, 1–154.

Helgason, S.: (1972), *Functions on symmetric spaces*, AMS Summer Inst., preprint.

Helgason, S.: (1973), 'Paley–Wiener theorem and surjectivity of invariant differential operators on symmetric spaces and Lie groups', *Bull. Am. Math. Soc.* **79**, 1, 129–132.

Helgason, S.: (1973b), 'The surjectivity of invariant differential operators on symmetric spaces I', *Ann. of Math.* **98**, 3, 451–479.

Helgason, S.: (1976), 'A duality for symmetric spaces with applications to group representations II' **22**, 2, 187–218.

Helgason, S.: (1980), *The X-ray transform on symmetric spaces*, Proc. Conf. Diff. Geom. and Global Analysis Berlin 1979, Springer Verlag, Berlin.

Helgason, S.: (1980b), *The Radon transform*, Birkhäuser, Boston.

Helgason, S.: (1981) *Topics in harmonic analysis on homogeneous spaces*, Birkhäuser, Boston.

Hewitt, E. and Ross, K. A.: (1963) *Abstract harmonic analysis I*, Springer Verlag, Berlin.

Hewitt, E. and Ross, K. A.: (1970), *Abstract harmonic analysis II*, Springer Verlag, Berlin.

Hirai, T.: (1970), 'The Plancherel formula for $\mathrm{SU}(p, q)$', *J. Math. Soc. Japan* **22**, 134–179.

Hochschild, G.: (1965), *The structure of Lie groups*, Holden-Day, San Francisco.

John, F.: (1955), *Plane waves and spherical means applied to partial differential equations*, Interscience Publishers, New York.

Karpelevič, F. I.: (1965), 'The geometry of geodesics and the eigenfunctions of the Beltrami–Laplace operator on symmetric spaces', *Trans. Amer. Math. Soc.* **14**, 48–185.

Kirillov, A. A.: (1976), *Elements of the theory of representations*, Springer Verlag, Berlin.

Kobayashi, S.: (1972), *Transformation groups in differential geometry*, Springer Verlag, Berlin.

Kobayashi, S. and Nomizu, K.: (1963), *Foundations of differential geometry I*, Interscience Publishers, New York.

Kobayashi, S. and Nomizu, K.: (1969), *Foundations of differential geometry II*, Interscience Publishers, New York.

Kostant, B.: (1969), 'On the existence and irreducibility of certain series of representations', *Bull. Amer. Math. Soc.* **75**, 627–642.

Kratzer, A. and Franz, W.: (1960), *Transzendente Funktionen*, Akadem. Verlagsgeselschaft, Leipzig.

Lang, S.: (1972), *Differential manifolds*, Addison-Wesley, Reading.

Loos, O.: (1969), *Symmetric spaces, I, II*, W. A. Benjamin, New York.

Mackey, G. W.: (1952), 'Induced representations of locally compact groups, I', *Ann. of Math.* **55**, 101–139.

Mackey, G. W.: (1953), 'Induced representations of locally compact groups, II', *Ann. of Math.* **58**, 193–221.

Mackey, G. W.: (1956), *Mimeographed notes on group representations*, Chicago.

Mackey, G. W.: (1963), 'Infinite dimensional group representations', *Bull. Amer. Math. Soc.* **69**, 628–686.

Matsushima, Y.: (1972), *Differentiable manifolds*, Marcel Dekker, New York.

Maurin, K.: (1968), *General eigenfunction expansions and unitary representations of topological groups*, PWN—Polish Scientific Publishers, Warsaw.

Maurin, K.: (1976), *Analysis I*, PWN—Polish Scientific Publishers, D. Reidel Publishing Company, Warsaw, Dordrecht.

Maurin, K.: (1980), *Analysis II*, PWN—Polish Scientific Publishers, D. Reidel Publishing Company, Warsaw, Dotdrecht.

Mautner, F. I.: (1951), 'On the decomposition of unitary representations of Lie. groups', *Proc. Amer. Math. Soc.* **2**, 490–496.

McBride, E. B.: (1971), *Obtaining generating functions*, Springer Verlag, Berlin.

Miller, W., Jr: (1968), *Lie theory and special functions*, Academic Press, New York,

Miller, W., Jr: (1972), *Symmetry groups and their applications*, Academic Press, New York.

Miller, W., Jr: (1977), *Symmetry and separation of variables*, Addison-Wesley Reading.

Montgomery, D. and Zippin, L.: (1955), *Topological transformation groups*, Interscience Publishers, New York.

Naimark, M. A.: (1964), *Linear representations of the Lorentz group*, The Macmillan Co., New York.

Naimark, M. A.: (1959), *Normed rings*, Noordhoff N. V., Groningen.

Nomizu, K.: (1954), 'Invariant affine connections on homogeneous spaces', *Amer. J. of Math.* **76**, 33–65.

Nomizu, K.: (1963), *Lie groups and differential geometry I*, New York.

Nomizu, K.: (1969), *Lie groups and differential geometry II*, New York.

Olver, F. W. J.: (1974), *Asymptotics and special functions*, Academic Press, New York.

Palais, R. S.: (1957), 'A global formulation of the Lie theory of transformation groups', *Memoirs Amer. Math. Soc.* **22**, Providence.

Paley, R. A. C. and Wiener, N.: (1934), *Fourier transforms in complex domain*, AMS Coll. Publ., New York.

Peter, F. and Weyl, H.: (1927), 'Die Vollständigkeit der primitiven Darstellungen einer geschlossenen kontinuierlichen Gruppe', *Math. Ann.* **97**, 737–755.

Pontriagin, L.: (1958), *Topological groups*, Princeton University Press, Princeton.

Poulsen, N. S.: (1971), 'On C^∞-vectors and intertwining bilinear forms for representations of Lie groups', *J. of Funct. Anal.* **9**, 87–120.

Pukanszky, L.: (1964), 'The Plancherel formula for the universal covering group of SL(2, ***R***)', *Math. Ann.* **156**, 96–143.

Pukanszky, L.: (1967), *Leçons sur les representations des groupes*, Dunod, Paris.

Rainville, E. D.: (1960), *Special functions*, Macmillan Comp., New York.

Rieffel, M. A.: (1972), 'Unitary representations induced from compact subgroups', *Studia Math.* **42**, 145–175.

Rudin, W.: (1962), *Fourier analysis on groups*, Interscience Publ., New York.

Rudin, W.: (1966), *Real and complex analysis*, McGraw-Hill.

Rühl, W.: (1970), *The Lorentz group and harmonic analysis*, W. A. Benjamin, New York.

Schwartz, L.: (1957), *Théorie des distributions I, II*, Hermann, Paris.

Serre, J.-P.: (1965), *Lie algèbras and Lie groups*, W. A. Benjamin, New York.

Serre, J.-P.: (1966), *Algèbres de Lie semisimple complexes*, W. A. Benjamin, New York.

Serre, J.-P.: (1971), *Représentations lineaires des groupes finis*, Hermann, Paris.

Shin'ýa, H.: (1976), *Spherical functions and spherical matrix functions on locally compact groups*, Lectures in Math., Dept. of Math., Tokyo.

Slater, L. J.: (1966), *Generalized hypergeometric functions*, University Press, Cambridge.

Sneddon, I. H.: (1961), *Special functions of mathematical physics*, Reinhold, Edinburgh.

Sneddon, I. H.: (1972), *The use of integral transforms*, McGraw-Hill Book Co., New York.

Spain, B., Smith, M. G.: (1970), *Functions of mathematical physics*, Reinhold, Edinburgh.

Stein, E. M. and Weiss, G.: (1971), *Introduction to Fourier analysis on Euclidean spaces*, Princeton Univ. Press.

Sternberg, S.: (1964), *Lectures on differential geometry*, Prentice Hall, Englewood Cliffs.

Strasburger, A.: (1975), 'Inducing spherical representations of semisimple Lie groups', *Diss. Math.* **122**, 1–52.

Takahashi, L.: (1963), 'Sur les représentations unitaires des groupes de Lorentz généralisés', *Bull. Soc. Math. France* **91**, 289–443.

Takahashi, T.: (1952), 'Generalized spherical harmonics as representations', *J. Phys. Soc. Japan* **7**, 307–312.

Talman, J. D.: (1968), *Special functions, group representation approach*, W. A. Benjamin, New York.

Tranter, C. J.: (1968), *Bessel functions with some physical applications*, Univ. Press, London.

Varadarajan, V. S.: (1974), *Lie groups, Lie algebras and their representations*, Prentice Hall, Englewood Cliffs.

Vilenkin, N. Ya.: (1963), 'Spetsiyalnye funktsii svyazannye s predstavleniyami klassa

I grupy dvizhenii prostranstv postoyannoi krivizny', *Tr. Mosk. Matem. obshchestva* **12**, 185–257.

Vilenkin, N. Ya.: (1965), *Spetsiyalnye funktsii i teoriya predstavlenii grupp*, Izdatelstvo "Nauka", Moskva.

Vilenkin, N. Ya.: (1968), Integral transforms of functions on hyperboloids, *Mat. Sb.*, **74**, 109–119.

Vilenkin, N. Ya.: (1968b), 'Laguerre poynomials, Whittaker functions and the representations of groups of bordered matrices', *Mat. Sb.* **75**, 399–409.

Wallach, N.: (1973), *Harmonic analysis on homogeneous spaces*, Marcel Dekker, New York.

Warner, G.: (1972), *Harmonic analysis on semi-simple Lie groups I, II*, Springer Verlag, Berlin.

Watson, G. N.: (1938), *A treatise on the theory of Bessel functions*, Cambridge Univ. Press, Cambridge.

Wawrzyńczyk, A.: (1968), 'On tempered distributions and Bochner–Schwartz theorem on arbitrary locally compact Abelian groups', *Coll. Math.* **19**, 305–318.

Wawrzyńczyk, A.: (1978), *Współczesna teoria funkcji specjalnych*, PWN—Polish Scientific Publishers, Warszawa.

Wawrzyńczyk, A.: *Spectral analysis and synthesis on symmetric spaces*, in preparation.

Weil, A.: (1951), *L'intégration dans les groupes topologiques et ses applications* Hermann, Paris.

Weyl, H.: (1928), *Gruppentheorie und Quantenmechanik*, Verl. S. Hirzel, Leipzig.

Weyl, H.: (1934), 'Harmonics on homogeneous manifolds', *Ann. of Math.* **35**, 486–499.

Weyl, H.: (1939), *The classical groups*, Princeton Univ. Press, Princeton.

Weyl, H.: (1952), *Symmetry*, Princeton Univ. Press, Princeton.

Whittaker, E. T. and Watson, G. N.: (1963), *A course of modern analysis*, Cambridge Univ. Press, Cambridge.

Wiener, N.: (1933), *The Fourier integral and certain of its applications*, Dover Publications, New York.

Wigner, E. P.: (1939), 'On the unitary representations of inhomogeneous Lorentz group', *Ann. of Math.* **40**, 149–204.

Wigner, E. P.: (1955), *Application of group theory to the special functions of mathematical physics*, unpublished lecture notes, Princeton.

Wigner, E. P.: (1959), *Group theory and its applications to the quantum mechanics of atomic spectra*, Academic Press, New York.

Wolf, J. A.: (1967), *Spaces of constant curvature*, McGraw-Hill Book Co., New York.

Zhelobenko, D. P.: (1970), *Kompaktnye gruppy Li i ikh predstavleniya*, Izdatelstvo "Nauka", Moskva.

Zhelobenko, D. P.: (1974), *Garmonicheskii analiz po poluprostykh kompleksnykh gruppakh Li*, Izdatelstvo "Nauka", Moskva.

List of Symbols

Author Index

Subject Index

Mathematics and Its Applications

1. Willem Kuyk, *Complementarity in Mathematics, A First Introduction to the Foundations of Mathematics and Its History*. 1977.
2. Peter H. Sellers, *Combinatorial Complexes, A Mathematical Theory of Algorithms*. 1979.
3. Jacques Chaillou, *Hyperbolic Differential Polynomials and their Singular Perturbations*. 1979.
4. Svtopluk Fučik, *Solvability of Nonlinear Equations and Boundary Value Problems* 1980.
5. Williard L. Miranker, *Numerical Methods for Stiff Equations and Singular Perturbation Problems*. 1980.
6. P. M. Cohn, *Universal Algebra*. 1981.
7. Vasile I. Istrăţescu, *Fixed Point Theory, An Introduction*. 1981.